PDE Control of String-Actuated Motion

PRINCETON SERIES IN APPLIED MATHEMATICS

Ingrid Daubechies (Duke University); Weinan E (Princeton University); Jan Karel Lenstra (Centrum Wiskunde & Informatica, Amsterdam); Endre Süli (University of Oxford), Series Editors

The Princeton Series in Applied Mathematics features high-quality advanced texts and monographs in all areas of applied mathematics. The series includes books of a theoretical and general nature as well as those that deal with the mathematics of specific applications and real-world scenarios.

For a full list of titles in the series, go to https://press.princeton.edu/series /princeton-series-in-applied-mathematics

PDE Control of String-Actuated Motion, *Ji Wang and Miroslav Krstic*

Delay-Adaptive Linear Control, *Yang Zhu and Miroslav Krstic*

Statistical Inference via Convex Optimization, *Anatoli Juditsky and Arkadi Nemirovski*

A Dynamical Systems Theory of Thermodynamics, *Wassim M. Haddad*

Formal Verification of Control System Software, *Pierre-Loïc Garoche*

Rays, Waves, and Scattering: Topics in Classical Mathematical Physics, *John A. Adam*

Mathematical Methods in Elasticity Imaging, *Habib Ammari, Elie Bretin, Josselin Garnier, Hyeonbae Kang, Hyundae Lee, and Abdul Wahab*

Hidden Markov Processes: Theory and Applications to Biology, *M. Vidyasagar*

Topics in Quaternion Linear Algebra, *Leiba Rodman*

Mathematical Analysis of Deterministic and Stochastic Problems in Complex Media Electromagnetics, *G. F. Roach, I. G. Stratis, and A. N. Yannacopoulos*

Stability and Control of Large-Scale Dynamical Systems: A Vector Dissipative Systems Approach, *Wassim M. Haddad and Sergey G. Nersesov*

Matrix Completions, Moments, and Sums of Hermitian Squares, *Mihály Bakonyi and Hugo J. Woerdeman*

Modern Anti-windup Synthesis: Control Augmentation for Actuator Saturation, *Luca Zaccarian and Andrew R. Teel*

Totally Nonnegative Matrices, *Shaun M. Fallat and Charles R. Johnson*

Graph Theoretic Methods in Multiagent Networks, *Mehran Mesbahi and Magnus Egerstedt*

Matrices, Moments and Quadrature with Applications, *Gene H. Golub and Gérard Meuran*

PDE Control of String-Actuated Motion

Ji Wang
Miroslav Krstic

PRINCETON UNIVERSITY PRESS
PRINCETON AND OXFORD

Published by Princeton University Press
41 William Street, Princeton, New Jersey 08540
99 Banbury Road, Oxford OX2 6JX

press.princeton.edu

All Rights Reserved
ISBN 9780691233482
ISBN (pbk.) 9780691233499
ISBN (e-book) 9780691233505

British Library Cataloging-in-Publication Data is available

Editorial: Diana Gillooly, Kristen Hop, Kiran Pandey
Jacket/Cover Design: Heather Hansen
Production: Lauren Reese
Publicity: Matthew Taylor, Charlotte Coyne
Copyeditor: Wendy Lawrence

Jacket/Cover Credit: Marko Čelebonović (1902–1986), *Coastal Motif*, 1956, oil on canvas, exhibited at the 1957 Biennial in São Paulo. After graduating with a law degree from the Sorbonne, Čelebonović studied sculpture with Antoine Bourdelle and then devoted the rest of his life to painting. He was a member of the Serbian Academy of Sciences and Arts.

This book has been composed by using LaTeX

10 9 8 7 6 5 4 3 2 1

Contents

List of Figures

List of Tables

Preface

This book deals with the boundary control of wave partial differential equations (PDEs) in one dimension, with a moving actuated boundary, or with finite-dimensional dynamics (modeled as ordinary differential equations; ODEs) at either the unactuated boundary or in the actuation path. These structures are inspired by applications involving cables and strings that move mechanical loads.

Cables moving mechanical loads are most conspicuously employed in elevators (both in buildings, where they may be hundreds of meters long, and in deep mining, where they may be kilometers long), but also in undersea construction, such as laying telecommunication cables along uneven seafloor or building artificial reefs that have an environmentally beneficial purpose for the sea life, and in other applications, including many yet to be revealed in other domains.

In deep-sea drilling, the so-called drill string is not a cable but a kilometers-long thin cylinder, with a drill bit on its unactuated end, and its dynamics of rotation (torsional dynamics) are governed by the wave equation; that is, they are mathematically equivalent to string or cable vibration.

Hence, our book's overarching title for this multitude of physical configurations, load types, and operating purposes — *PDE Control of String-Actuated Motion*.

The actuation of mechanical loads by means of strings (cables) has its advantages over rigid connections. It allows for significant improvements in energy efficiency, weight, size of the operation workspace, operation speed, and maximum payload, compared with rigid-body mechanisms, due to the string's properties of lower weight, resisting relatively large axial loads, and low bending and torsional stiffness.

However, the distributed parameter nature of a string or cable makes the control design of the cable-actuated mechanisms more challenging than the traditional ODE-based control designs for lumped parameter rigid-body mechanisms, giving rise to many new problems in boundary control of PDEs. The theoretical challenges and practical significance have led us to carry out research on many topics involving cables and strings of time-varying length, with moving loads, and with actuator dynamics. This book presents this collection of methodologies, whose meaning is predominantly mathematical, but whose inspiration comes entirely from applications and technology.

Cable elevators and other lifting and depositing tasks, as well as drilling at a high penetration rate, introduce a heretofore unstudied problem of boundary control of wave PDEs with moving boundaries—that is, on time-varying domains. This is the central issue of our book—wave PDE control on one-dimensional domains of time-varying lengths—that is, vibration suppression in strings of time-varying lengths. One should note that, as the cable length varies with time, possibly fast, even if one could be talking about *eigenvalues* and *eigenfunctions* (that rapidly change), the spectral approaches to control design certainly would not be applicable. We approach this challenge using time-domain and Lyapunov-based approaches.

The second key challenge is that, at the moving end of a time-varying string, a load is present, and the motion of this load needs to be controlled with an actuator on the opposite end of the string. The objective is to suppress the vibration in both the string of varying length and in the load at the distal end from the actuator. Using feedback control to add damping artificially, where physical viscous damping might be absent or insufficient, would be easy at the location of (i.e., proximal to) the actuator. However, the actuator being at a boundary makes emulating viscous damping along the entire string, and at the load on the string's distal end, challenging.

This challenge is met using the method of PDE backstepping. Backstepping employs two tools, a Volterra transformation of the infinite-dimensional state, and feedback, to add damping at locations other than the actuator. But, prior to employing the backstepping transformation, we usually employ first a transformation of the state into the Riemann variables, which is a canonical representation for coupled hyperbolic PDEs, to which backstepping is readily applicable.

However, the actuator in most boundary-controlled cable and string systems does not act directly and instantly. The actuator, be it hydraulic or electrical, has its own considerable inertia—namely, its own lumped-parameter dynamics modeled by an ODE. These dynamics themselves have to be overcome using finite-dimensional backstepping (the classical *integrator backstepping*).

Hence, the overall system that arises in string-actuated motion is often a sandwiched ODE-PDE-ODE configuration, with an input acting on only one of the two ODEs at the end of the PDE, and not directly on the PDE.

The reality of applications gives rise to additional effects: nonlinearities, disturbances, unknown parameters, input delays, sampled (or event-triggered) sensing, as well as many more which we deem beyond the page and time limits of this book.

A vast literature exists on control of overhead cranes and gantry cables. A good entry point into this literature, in terms of both the theory and applications, is the tutorial article [23].

A large portion of the research on this topic justifiably focuses on transverse motion and employs the techniques of differential flatness, finite-time motion planning, and finite-horizon optimal control. Many, if not most of these approaches, are concerned with designing open-loop control signals. While some of our work in this book is applicable to the transversal motion of overhead cranes and gantry cables, these problems are not our focus here. Instead, we focus on vibration suppression, in axial and other directions, cables of varying length, Lyapunov stabilization techniques, and handling of uncertainties.

What Does the Book Cover?

The book comprises three parts. The first part is devoted to various control applications, as drivers for control design and theoretical study. Control problems for mining cable elevators are introduced in chapters 2–5, focusing on single-cable elevators in chapter 2, dual-cable elevators in chapter 3, and airflow disturbances and the influence of flexible guideways in chapter 4 and chapter 5, respectively. In addition to the mining cable elevator, the deep-sea construction vessel for undersea moving is also a cable-actuated manipulator to move mechanical loads. Its basic control design is introduced in chapter 6, and additional real-world effects—that is,

Table 1. The motion types

Chapter	Axial	Transversal	Torsional
2	√		
3	√		
4	√		
5		√	
6	√	√	
7		√	
8			√
10		√	
11	√		
13		√	
15	√		

sensor signal delays occurring in large distance signal transmission through acoustic devices and the requirement of reducing changes in the actuator signal considering the massive ship-mounted crane—are dealt with in chapter 7. Apart from the cable-actuated mining elevator and the deep-sea construction vessel, another distinct but kindred application, deep-sea drilling, is tackled in chapter 8.

Inspired by but going beyond the applications in part I, generalized control problems are dealt with in part II—that is, boundary control of sandwich hyperbolic PDEs in particular. Control of the sandwich systems is covered in chapters 9–12, with the basic control design presented in chapter 9 and then extended to a variety of more challenging problems, including control of the sandwich systems with sensor delay in chapter 10, with event-triggered design in chapter 11, and with nonlinearities in chapter 12. The general results in part II are justified by applications in part I.

The last of the three parts presents triggered-type adaptive control of hyperbolic PDEs. Three triggered adaptive control schemes (event-triggered control, regulation-triggered parameter estimation, and a combination of both) for hyperbolic PDE-ODE systems are developed in chapters 13–15, respectively. The triggered-type adaptive control results in part III are also verified in the applications in part I.

The book deals with all of the three possible motions of strings or cables: longitudinal/axial/stretching, lateral/transversal/bending, and angular/rotational/torsional. However, it is only in one chapter that we deal with more than one of these three motions. In chapter 6 we deal with coupled longitudinal-lateral vibrations. Table 1 gives an overview of the motion types that each chapter covers. The suppression of axial vibrations dominates our exposition, with transversal vibrations a close second.

Table 2. Configurations

Chapter	ODE at distal end of PDE	ODE at proximal end of PDE
2	√	
3	√	
4	√	
5	√	
6	√	
7	√	
8	√	
9	√	√
10	√	√
11	√	√
12	√	√
13	√	
14	√	
15	√	

All of the problems considered in the book incorporate at least one PDE and one ODE. Some consider a second ODE as well. The configurations considered are given in table 2. Those configurations that include ODEs at both distal and proximal ends in chapters 9–12 (namely, about a third of the book) are sandwich systems. In addition, chapter 10 contains a sensor delay at the distal end.

We consider both PDE-ODE systems that are fully known and those that contain unknown or unmeasured quantities—such as unmeasured states, unmeasured disturbances, and unknown parameters. In table 3 we overview the contents of the book based on the unmeasured and unknown effects. Virtually all of our exposition is for cables and (drill) strings that are not instrumented with distributed sensing, as is consistent with reality. Disturbances and unknown parameters occupy a large share of the book and create some of the most significant challenges for design and analysis.

When it comes to adaptive control, a topic dealt with comprehensively for coupled hyperbolic PDEs in [9] (and for parabolic PDEs a decade earlier in [166]) is tackled in chapters 5, 8, 13, 14, and 15 of this book, as indicated in table 3. Chapters 5 and 8 employ a classical continuous-in-time Lyapunov-based approach. On the other hand, chapters 13–15 employ novel event-triggered approaches. Chapters 13 and 14 are very different in what event triggering is employed for. An adaptive controller consists of two components: the control law and the parameter estimator (update law). Both of these components can employ piecewise-constant values—the control input and the parameter estimate. And the changes in the piecewise-constant values in both of these components can be triggered in various ways. In

Table 3. Unmeasured and unknown effects

Chapter	Unmeasured PDE state	Unmeasured ODE state	Unmeasured disturbance	Unknown parameters
2	✓			
3	✓			
4	✓		✓	
5	✓		✓	✓
6	✓	✓		
7	✓	✓		
8	✓		✓	✓
10	✓	✓		
11	✓	✓		
12	✓	✓		
13	✓			✓
14				✓
15				✓

chapter 13 we employ continuous parameter updates and event-triggered control inputs. Conversely, in chapter 14 we employ continuous control inputs (except for finite time instants) and event-triggered parameter updates. Because the parameter estimator is the more delicate of the two components of an adaptive control since it is generally not endowed with convergence guarantees, it is chapter 14 that is considerably more challenging of the two chapters. In chapter 15, simultaneous triggering is employed for the parameter update law and the control law, the result of which is that both the parameter estimates and the control input employ piecewise-constant values.

For the researcher in coupled hyperbolic PDE systems who is interested in going beyond the basic 2×2 case, which is superbly covered in [16] and [9], there are interesting designs for 4×4 cases in this book—for example, in chapter 3 and chapter 6. Chapter 3 deals with axial oscillations in a pair of cables connected by a payload at the distal boundary. So the 4×4 system in chapter 3 is a set of two 2×2 pairs that are coupled not along the domain but at the boundary. In contrast, chapter 6 deals with a single cable but with axial and lateral oscillations that bring domain-wide coupling into the plant and, therefore, a fully coupled 4×4 hyperbolic system. While in multiphase flows, in both oil drilling and production, as well as in congested multiclass traffic flow, a larger number of first-order hyperbolic PDEs arise in the direction toward the actuated end than away from it, this interesting occurrence of underactuated heterodirectional hyperbolic PDEs does not arise with cables.

What Niche in the Literature Does This Book Fill?

The main inspiration for this book comes from the cascade PDE-ODE configurations in [116]. The book [111] develops the cascade ideas from [116] in the parabolic PDE realm for applications in additive manufacturing.

The present book expands the reach of [116] into cable-operated systems and on time-varying domains. But this book's closest cousin may be [77], a major volume spawned from the classic [132]. While [77, 132] employ collocated static feedbacks for wave PDEs on static domains, and in the absence of ODEs, our focus is broader in terms of plant structure (varying domain, ODE included), methodology (backstepping controller and observer designs), and the emphasis on applications. We do not, however, deal with beam systems.

For Whom Is This Book

This book should be valuable to researchers working on control and dynamic systems—engineers, graduate students, and PDE system specialists in academia.

Mathematicians with interest in control of distributed parameter systems will find the book stimulating, because it tackles and opens a door for control of sandwich PDE systems, which present many stimulating challenges and opportunities for further research on the stabilization of ever-expanding classes of unstable sandwich PDE systems.

Engineers in mechanical, aerospace, and civil/structural engineering, focusing on vibration or motion control, especially for flexible structures or manipulators, will learn some new and useful methodologies for designing controller/observer algorithms, and addressing some problems they have no doubt faced in practice: time delay, disturbances, uncertainties, and so on.

The background required to read this book includes little beyond the basics of function spaces and Lyapunov theory for ODEs. We hope that the reader will regard the book not as a collection of problems that have been solved but as a collection of tools and techniques that are applicable to open problems, particularly in interconnected systems of ODEs and PDEs, and to physical problems modeled by PDE-ODE coupled systems.

ACKNOWLEDGMENTS

We would like to thank Shumon Koga for his contribution to chapter 2, Yangjun Pi for his contributions to chapters 2, 3, and 4, Shuxia Tang for her contributions to chapters 4, 5, and 8, and Iasson Karafyllis for his contribution to chapter 14.

We are grateful to Rafael Vazquez, Jean-Michel Coron, Ole Morten Aamo, Florent Di Meglio, Federico Bribiesca Argomedo, Long Hu, Joachim Deutscher, Henrik Anfinsen, Delphine Bresch-Pietri, and their coauthors for their inspiring early work on boundary control of the 2×2 linear hyperbolic system. The pioneering work of Wei Guo, Hongyinping Feng, and Bao-Zhu Guo on extending the active disturbance rejection control (ADRC), introduced by Jingqing Han for finite dimension systems, to infinite dimension systems inspired chapters 4 and 8. We also would like to thank Nicolas Espitia, Christophe Prieur, and their coauthors for the original work on event-triggered designs in hyperbolic PDEs, from which we benefited in developing chapters 7, 11, and 13. We also thank CITIC Heavy Industries Co. Ltd.

for providing practical motivations and expertise on the mining cable elevator in chapters 2–5.

Ji Wang would like to give a special thanks to his family for their unwavering support and love. Miroslav Krstic thanks the National Science Foundation, the Air Force Office of Scientific Research, the Los Alamos National Laboratory, and Dr. Dan Alspach for research support, as well as his colleagues at General Atomics for enriching discussions on cable-actuated control systems.

La Jolla, California
July 2021

Ji Wang
Miroslav Krstic

Chapter One

Introduction

This chapter presents the motives, as well as a limited literature summary, for both the applied and the theoretical control design problems that we pursue in this book.

1.1 STRING-ACTUATED MECHANISMS

In this section, we introduce four string-actuated mechanisms: mining cable elevators, deep-sea construction vessels (DCVs), unmanned aerial vehicles (UAVs), and oil-drilling systems.

Mining Cable Elevators

In mining exploitation, a cable elevator, which is used to transport the cargo and miners between the ground and the working platform underground, is an indispensable piece of equipment [99]. The mining cable elevator is a cable-actuated moving load system. A common arrangement is a single-drum system [178]: a single-cable mining elevator comprising a driving winding drum, a steel wire cable, a head sheave, and a cage. The cable plays a vital role in mining elevators because its advantages of low bending and torsional stiffness, resisting relatively large axial loads, are helpful to heavy load and large-depth transportation. However, the compliance property, or stretching and contracting abilities of cables, tends to cause mechanical vibrations, especially when the elevator is running at high speed, which leads to tension oscillations and premature fatigue fracture [87, 88, 91]. Therefore, the importance of suppressing the vibrations and tension oscillations cannot be overestimated, for the safety of both personnel and profitability. The economical and convenient way to suppress the vibrations is by designing an appropriate control input without modifying the original structure of the mining cable elevator. In chapter 2, the control design for axial vibration suppression of a high-speed, single-cable mining cable elevator is presented.

For operation at a greater depth, such as over 2000 m, and carrying a heavier load, the single-cable elevator is not suitable. Because a very thick cable is required to bear the heavy load, such a thick cable, at high bending, suffers from problems in the winding on the winder drum. A dual-cable mining elevator [37], shown in figure 3.1 in chapter 3, is proposed to solve this problem, removing the requirement of a very thick cable because two cables tow the cage. However, an imbalance problem, such as cage roll, frequently appears in the dual-cable elevator [184], as shown in figure 3.1, for which, taut cables are used as flexible guide rails because traditional steel rails come with a high cost of manufacture and installation in deep mines. Cage roll will increase the differences in oscillation tension between two cables and enlarge the oscillation amplitude of the tension, which accelerates premature fatigue and requires inspections and costly repairs. One feasible arrangement to balance the cage roll and suppress the vibrations and tension oscillations in cables of the dual-cable mining elevator is to design and apply additional control forces through actuators at floating head sheaves, as shown in figure 3.1. The control design for suppressing of the axial vibrations and tension oscillations and balancing the cage roll in a dual-cable mining elevator is presented in chapter 3.

In an actual operating environment, the cage is usually subject to uncertain airflow disturbances. Additionally, the flexible guides, with their uncertain properties, may affect the smooth and steady running of the mining cable elevator. These factors inspired the control designs in chapters 4 and 5.

Deep-Sea Construction Vessels

In deep-sea oil exploration, some equipment, such as a subsea manifold, a subsea pump station, and a subsea distribution unit along with associated foundations, flow lines and umbilicals, is installed at designated locations [167, 168] around the drill center on the seafloor. The installation of the equipment is completed by a DCV [168, 182] because the installation sites are located outside a radius of 45 m from the floating drilling platform (see figure 2 in [168]) and cannot be accessed by the huge floating drilling platform, which has limited access and mobility [168], and because some of the equipment, such as flow lines and umbilicals, is installed in advance to prepare to hook up the floating drilling platform when it arrives. A DCV is shown in figure 6.1 in chapter 6, where the top of the cable is attached to a crane on a vessel at the ocean surface, and the cable's bottom is attached to equipment to be installed at the sea floor. The traditional method in underwater installation by DCVs is to regulate the vessel dynamics position and manipulate the crane to obtain the desired heading for the payload [94]. Such a method is not suitable for the deeper water construction in offshore oil drilling (more than 1000 m) because the cable is very long when the payload is near the seabed, which would introduce large cable oscillations [94, 182], causing a large offset between the payload and the desired heading position of the crane—namely, the designated installation location. In chapter 6, the control forces at the onboard crane are designed to reduce the cable oscillations and then place the equipment in the target area on the sea floor. In chapter 7, we employ a piecewise-constant control input that is more suitable for the massive ship-mounted crane and compensate for delays in the transmission of sensing signals from the seabed to the vessel on the ocean surface through acoustic devices.

Unmanned Aerial Vehicles

In addition to DCVs, the control design in chapter 6 can also be applied to unmanned aerial vehicles (UAVs) used to aid delivery to dangerous and inaccessible areas, such as to flood, earthquake, fire, and industrial disaster victims [70, 140]. Food and first-aid kits are tied to the bottom of a cable, whose other end is attached to the UAV. The swing/oscillation of the cable-payload may appear during the transportation motion due to the properties of the cable and external disturbances, such as wind, which may cause damage to the suspended object, the environment, and the people nearby [70]. At the end of the transport motion, when the UAV arrives at the location directly over the rescue site and is ready to land the aid supplies, the suspended object naturally continues to swing [140], which makes precisely placing the aid supplies at the target position difficult. Therefore, rapid suppression of the oscillations of the cable and suspended object through a control force provided by the rotor wings of the UAV is required. In addition to aid delivery in disaster relief, UAV delivery is also used in some commercial cases to reduce labor cost. For example, some companies use UAVs to transport cargo in storehouses or lift and position building elements in architectural construction [191]. Some logistics companies have also begun to use UAVs to deliver packages in small areas [70].

Oil Drilling

Oil-drilling systems used to drill deep boreholes for hydrocarbon exploration and production often suffer severe vibrations, which can cause the premature failure of drilling string components, damage to the borehole wall, and problems with precise control [98]. The vibrations also cause significant wastage of drilling energy [53]. The suppression of vibrations in the oil-drilling system is thus required for the economic interest and improvement of drilling performance [156].

There are three main types of vibration in oil-drilling systems [154]: vertical vibration, also called the bit-bouncing phenomenon, lateral vibration due to an out-of-balance drill string, which is called whirl motion, and torsional vibration, which appears due to friction between the bit and the rock. This nonlinear torsional interaction between the drill bit and the rock will cause the bit to slow down and even stall while the rotary table is still in motion. Once enough energy is accumulated, the bit will suddenly be released and start rotating at very high speed before slowing down again [24], settling into a limit cycling motion. This is called the stick-slip phenomenon. Several physical laws of bit-rock friction [156] are used to roughly describe stick-slip behavior in the oil-drilling system, such as the velocity-weakening law [31], the stiction plus Coulomb friction model [160], a class of Karnopp model [38, 107, 136], and so on. The stick-slip oscillations lead to instability from the lower end to travel up the drill string to the rotary table, which results in distributed instabilities and is the primary cause of fatigue to the drill collar connections as well as damage to the drill bit [154]. Therefore, suppressing torsional vibrations (stick-slip oscillations) in the oil-drilling system is important. In addition waves at the sea surface causing a heaving motion of the drilling rig [1] in an offshore rotary oil-drilling system [186] introduce an external disturbance at the bit, which is another instability source.

As will be seen in chapter 8, the designed control input at the rotary table goes down from the rotating table, through the drill string, to the drill bit, to eliminate the stick-slip instability and, as a result, reduce the oscillations of the angular displacement and velocity of the drill bit.

1.2 HYPERBOLIC PDE-ODE SYSTEMS

The dynamics of the aforementioned string-actuated mechanisms are governed by hyperbolic partial differential equation-ordinary differential equation (PDE-ODE) systems. The design of controllers for such hyperbolic PDE-ODE systems requires boundary control approaches because the control input can only be applied at one end of the string in such mechanisms. In this section, we review boundary control of elementary wave PDE-ODE systems, as well as of a class of coupled first-order hyperbolic PDE models, with the possible inclusion of an ODE in cascade with the hyperbolic PDEs.

Wave PDE-ODE Systems

A wave PDE-ODE system serves as a basic model of a string-actuated mechanism in which the wave PDE describes a vibrating string (without in-domain damping), and the attached payload is modeled as an ODE. Classical results on backstepping boundary control for wave PDEs with anti-damping terms in domain or on the uncontrolled boundary are found in [121, 162, 165]. In the past decade, many results on boundary control of wave PDE-ODE systems have been reported. The very first result on boundary control of a wave PDE-ODE plant was presented in [114], where the interconnection is of the Dirichlet type. Boundary control design for a wave PDE-ODE cascade system with a Neumann-type interconnection was also addressed in [170]. The boundary control problem was also tackled in [18, 28, 29] for a wave PDE-nonlinear ODE system.

Coupled First-Order Hyperbolic PDEs

For the sake of greater clarity in control design and analysis, wave PDEs can be converted to a class of heterodirectional coupled first-order hyperbolic PDE systems via the Riemann transformations [26]. Especially when considering the in-domain viscous damping terms describing the string material damping, there would exist in-domain coupling in the resulting coupled first-order hyperbolic PDE systems [147], which makes the control design more challenging. Some theoretical results on boundary control of coupled first-order hyperbolic PDEs have emerged over the last decade. The basic boundary stabilization problem of 2×2 coupled linear transport PDEs was addressed in [32, 177] by backstepping, based on which the extended results on boundary control of these 2×2 systems were presented in [4, 39]. The sliding mode approach and the proportional integral (PI) controller design applied to the control of such a 2×2 system was also considered in [127] and [173], respectively. Boundary control of the 2×2 transport PDE system was further extended to that of an $n + 1$ system in [50]. For a more general coupled transport PDE system where the number of PDEs in either direction is arbitrary, a boundary stabilization law was first designed by backstepping in [96], which is a systematic framework for control design for this class of systems, and other extended results were proposed in [6, 14, 40, 41, 97]. In addition to the applications to string/cable models, the boundary control design for coupled first-order hyperbolic PDEs has also been applied to water-level dynamics [45, 47, 51, 143, 142] and traffic flow [101, 194, 195, 197].

In the past five years, some results on the control of linear coupled hyperbolic PDEs cascaded with ODEs have also been reported. The first is [48], which addressed the state-feedback stabilization of a general linear hyperbolic PDE-ODE coupled system. The state-feedback boundary control design of a 2×2 linear hyperbolic PDE-ODE coupled system with nonlocal terms was also dealt with in [169]. Based on the observer design, in [44] an output-feedback controller with anti-collocated measurements was proposed to stabilize general linear heterodirectional hyperbolic PDE-ODE systems with spatially varying coefficients.

1.3 "SANDWICH" PDEs

In the results discussed in section 1.2, the control input enters the PDE boundary directly, neglecting the inertia and dynamics of the actuator. However, in some applications, actuator dynamics cannot be neglected, especially when the actuator has its own considerable inertia. Incorporating the actuator dynamics into the control design of the string-actuated mechanisms modeled by hyperbolic PDE-ODE systems gives rise to a more challenging problem: control of what we call *sandwich* ODE-PDE-ODE systems.

The first backstepping state-feedback control design for *sandwich* hyperbolic systems was proposed in [113], which considered a transport PDE-ODE system with an integration (first-order ODE) at the input of the transport PDE. Also, the control problem of an ODE with input delay and unmodeled bandwidth-limiting actuator dynamics, which is represented by an ODE-transport PDE-ODE system where the input ODE is first order, was addressed in [118]. The boundary control design of a transport PDE sandwiched between two ODEs describing actuator and sensor dynamics was also proposed in [8]. Regarding coupled transport PDEs, state-feedback control of heterodirectional coupled transport PDEs sandwiched between two ODEs was proposed in [151, 152, 183]. Adding observer designs, output-feedback control of this type of sandwich systems was developed in [43, 49, 180, 181]. Parameter identification of a drill string, modeled as a wave PDE sandwich system from experimental boundary data, was studied in [150]. Regarding other types of PDEs, boundary control of viscous Burgers PDE, heat PDE, and n coupled parabolic PDE sandwich systems was also addressed in [130, 179] and [42], respectively. A fairly complete theory for boundary control of sandwich hyperbolic PDEs is derived in chapters 9–12, including basic design, delay compensation, event-triggered design, and design with nonlinearities.

1.4 ADVANCED BOUNDARY CONTROL OF HYPERBOLIC PDEs

Apart from the basic boundary control designs mentioned in sections 1.3 and 1.2, in this section we review some extended results on disturbance attenuation, adaptive control, delay compensation, and event-triggered control for hyperbolic PDEs.

Disturbance Rejection/Cancellation

Most research on disturbance rejection and adaptive cancellation for PDE systems focuses on disturbances collocated with control. Sliding mode control (SMC) schemes designed for heat, Euler-Bernoulli beam, and Schrödinger equations with boundary input disturbances were presented in [73, 76, 189]. The internal model principle [63] on the basis of the estimation/cancellation strategy was utilized in the beam equation [145]. For wave PDEs, adaptive disturbance cancellation was used in the output-feedback asymptotic stabilization of one-dimensional wave equations subject to harmonic disturbances at the controlled end and at the measured output in [82, 83] and [81], respectively. By applying the active disturbance rejection control method introduced by Han [86] for ODEs, state-feedback or output-feedback control designs for wave PDEs with matched disturbances were presented in [60, 74, 75, 80, 172, 198].

The task of rejection or adaptive cancellation of unmatched disturbances—that is, the disturbances anti-collocated with the control input—is more difficult. While several results for this task have been developed for ODE systems, such as those found in [78, 129, 192] and so on, the literature is less ample in PDE systems, where the disturbance is on the far (distal) end from the control input. A state-feedback controller that practically stabilizes a Schrödinger equation-ODE cascade system in the presence of an unmatched disturbance assumed to be small and measurable was presented in [100]. An output-feedback controller was designed for output reference tracking in a wave equation with an anti-collocated harmonic disturbance at a stable damping boundary in [84]. The output regulation problem for a wave equation with a harmonic anti-collocated disturbance at a free boundary was further dealt with in [85]. In chapters 4, 5, and 8, the asymptotic rejection and adaptive cancellation of unmatched disturbances in hyperbolic PDEs are proposed, respectively, along with applications in the control of disturbed mining cable elevators and oil-drilling systems.

Adaptive Control

Three traditional adaptive schemes for PDEs with uncertain parameters are the Lyapunov design, the passivity-based design, and the swapping design [112, 124]. Using the three design methods initially developed for ODEs in [122], the same three adaptive control approaches were proposed for parabolic PDEs in [123, 163, 164]. For adaptive control of hyperbolic PDEs, many results based on the three conventional methods have also been achieved, as follows. In [24, 25, 26, 117], adaptive control laws were presented for a one-dimensional wave PDE that had an actuator on one boundary and an anti-damping instability with an unknown coefficient on the other boundary. The first result on adaptive control of general first-order hyperbolic partial integro-differential equations was proposed in [20]. An adaptive boundary control design of coupled first-order hyperbolic PDEs with uncertain boundary and spatially varying in-domain coefficients was developed in [5]. In [7], two adaptive boundary controllers of coupled hyperbolic PDEs with unknown in-domain and boundary parameters were proposed using identifier and swapping designs, respectively. More adaptive control results of coupled first-order hyperbolic PDEs have been collected in [9]. Adaptive control design methods for hyperbolic PDEs are employed in chapters 5 and 8 to deal with parameter uncertainties in mining cable elevator and oil-drilling systems. Also, an event-triggered adaptive controller is proposed in chapter 13.

Recently, a new adaptive scheme using a regulation-triggered batch least-squares identifier was introduced in [102, 104], which has some advantages over the traditional adaptive approaches, such as guaranteeing exponential regulation of the states to zero as well as finite-time convergence of the estimates to the true values. This method has been successfully applied in the adaptive control of a parabolic PDE where the unknown parameters are the reaction coefficient and the high-frequency gain [106]. Regarding hyperbolic PDEs, using a scalar least-squares identifier updated at a sequence of times with fixed intervals, backstepping adaptive boundary control of a first-order hyperbolic PDE with an unknown transport speed was proposed in [11]. In chapters 14 and 15, adaptive controllers based on batch least-squares identifiers are designed for 2×2 hyperbolic PDE-ODE systems where the transport speeds of both transport PDEs and the coefficients of the in-domain couplings are unknown, respectively.

Delay Compensation

Time delays often exist in practical control systems and may destroy system stability [79]. For example, in a subsea installation by a DCV, sensor delays [94] exist due to the fact that the sensor signal is transmitted over a large distance from the seafloor to the vessel on the ocean surface, through a set of acoustic devices. Such sensor signal transmission may result in information distortion or even make the control system lose stability. Therefore, the time delay is a vital issue that should be considered in the control design.

Recently, boundary control designs for hyperbolic PDEs have been proposed that take time delays into consideration. For example, delay-robust stabilizing feedback control designs for coupled first-order hyperbolic PDEs were introduced in [12, 13], achieving robustness to small delays in actuation. In order to compensate arbitrarily long delays, a delay compensation technique was developed in [116, 125], where the delay was captured as a transport PDE, and the original ODE plant with a sensor delay was treated as an ODE-transport PDE cascaded system in the controller and observer designs. The observer was built as a "full-order" type, which estimates both the plant states and the sensor states, compared with some classical results on delay-compensated observer designs [3, 27, 67], which only estimate plant states — namely, observers of the "reduced-order" type. While compensation for arbitrarily long delays by this technique are commonly available for finite-dimensional systems, very few examples exist where such compensation has been achieved for PDEs, including parabolic (reaction-diffusion) PDEs [108, 115]. Delay compensation for the wave PDEs with arbitrarily long delays is more complex than that for reaction-diffusion PDEs. The primary reason is the second-order-in-time character of the wave equation. In [119], by treating the delay as a transport PDE and applying a backstepping design, a boundary controller was developed for a wave PDE with compensation for an arbitrarily long input delay and with a guarantee of exponential stability for the closed-loop system. In chapter 10, we design a delay-compensated control scheme for a sandwich hyperbolic PDE in the presence of a sensor delay of arbitrary length by treating the delay as a transport PDE.

Event-Triggered Control

When implementing the designed PDE control laws in an actual mining cable elevator, two challenges caused by high-frequency elements in the control law appear: 1) the massive actuator, comprising a hydraulic cylinder and head sheave, shown in figure 11.1 in chapter 11, is incapable of supporting the high-frequency control signal, and 2) the high-frequency components in the control input may in turn become vibration sources for the cable. It is thus necessary to reduce the actuation frequency and ensure the effective suppression of the vibrations in the mining cable elevator. Designing sampling schemes to apply to the control input is a potential solution. Designs of sampled-data control laws of parabolic and hyperbolic PDEs were presented in [64, 105] and [35, 103], respectively. Compared with the periodic sampled-data control, event-triggered control, where the input to the massive actuator is changed only at the necessary times determined by an event-triggering mechanism that acts by evaluating the operation of the elevator, is more feasible for the mining cable elevator from the point of view of energy saving. This motivates us to design event-triggered PDE backstepping control laws.

Most of the current designs on event-triggering mechanisms are for ODE systems, such as those in [69, 93, 133, 159, 171]. Designs of event-based controls for PDE systems are still rare in the existing literature. There exist results on the distributed (in-domain) control of PDEs, such as [158, 193]. For boundary control, an event-triggered feedback law was proposed for a reaction-diffusion PDE in [58]. For first-order linear hyperbolic PDEs with dissipativity boundary conditions, an event-triggered boundary control law was originally proposed in [55, 56]. Furthermore, a state-feedback event-based boundary controller for a class of 2×2 coupled linear hyperbolic PDEs was designed in [57]. Based on the observer design, the output-feedback event-triggered boundary control of 2×2 coupled linear hyperbolic PDEs was developed in [54]. In chapter 11, an event-triggered backstepping boundary control law is derived for a sandwich hyperbolic PDE, and an adaptive event-triggered boundary controller is further developed in chapter 13.

1.5   NOTES

Frequently used notations in this book are given next, with more specialized notational conventions introduced in the coming chapters.

The partial derivatives and total derivatives are denoted as

$$f_x(x,t) = \frac{\partial f}{\partial x}(x,t), \quad f_t(x,t) = \frac{\partial f}{\partial t}(x,t),$$

$$f'(x) = \frac{df(x)}{dx}, \quad \dot{f}(t) = \frac{df(t)}{dt}.$$

By $C^k(A)$, where $k \geq 1$, we denote the class of continuous functions on A. The single bars $|\cdot|$ denote the Euclidean norm for a finite-dimensional vector $X(t)$. In contrast, norms of functions (of x) are denoted by double bars. For $u(x,t), x \in [0, D]$, by default, $\|\cdot\|$ denotes the L^2-norm of a function of x, namely,

$$\|u(\cdot,t)\| = \sqrt{\int_0^D u(x,t)^2 dx},$$

and Sobolev norms such as $H^1[0, D]$ or even $H^2[0, D]$ are defined by

$$\|u(\cdot,t)\|_{H^1} = \sqrt{\int_0^D u(x,t)^2 dx + \int_0^D u_x(x,t)^2 dx},$$

$$\|u(\cdot,t)\|_{H^2} = \sqrt{\int_0^D u(x,t)^2 dx + \int_0^D u_x(x,t)^2 dx + \int_0^D u_{xx}(x,t)^2 dx},$$

and the ∞-norm is denoted by

$$\|u(\cdot,t)\|_\infty = \sup_{x \in [0,D]} \{|u(x,t)|\}.$$

Part I

Applications

Chapter Two

Single-Cable Mining Elevators

In a high-speed mining cable elevator, the cable's property of *compliance*, or its ability to stretch and contract, results in mechanical vibrations, leading to imprecise positioning and premature fatigue fracture [87]. For safe operation, vibration suppression is essential. Hence, an effective and feasible control design for suppressing the vibration of a mining cable elevator is needed.

In section 2.1, the axial vibration dynamics of a single-cable mining elevator—that is, a varying-length string lifting a cage, shown in figure 2.1—is modeled as a coupled wave partial differential equation-ordinary differential equation (PDE-ODE) system with a Neumann interconnection, on a time-varying spatial domain. The control force can only be applied at the head sheave instead of at the cage—that is, a possibly unstable ODE on the far end from the control input of the wave PDE is not directly accessible for control. This is a more challenging task than the setting with the classical collocated *boundary damper* feedback control. Using the backstepping method, a state-feedback controller with explicit gains is designed to stabilize the coupled wave PDE-ODE system in section 2.2. To build an observer-based output-feedback controller, in section 2.3 an observer with explicit gains is designed to estimate the full distributed states of the varying-length string using only measurable boundary states in an anti-collocated setup. The exponential stability of the observer-based output-feedback control system is proved via Lyapunov analysis in section 2.4. The result is verified in the axial vibration suppression of a single-cable mining elevator via numerical simulations in section 2.5.

2.1 MODELING

Moving Boundary Wave PDE Model

A schematic of a mining cable elevator is given in figure 2.1. Because the catenary cable, the part between the drum and the floating head sheave, in figure 2.1(*a*) is much shorter than the vertical cable (70 m compared to 2000 m), we suppose that the vibrations on the catenary part are negligible, which gives the simplified model of the varying-length cable with a cage shown in figure 2.1(*b*). Due to the help of the steel guides, the transverse vibrations in the vertical cable can be neglected since they are much smaller than the axial vibrations.

Two external forces act on the system. One is the motion control force $U_a(t)$ driven by a motor at the drum, and the other is the vibration control force $U_v(t)$ manipulated by a hydraulic actuator at the floating head sheave. The axial transport motion $z^*(t)$ is the rigid-body motion neglecting the compliant property of the cable in the fixed coordinate system O', whereas $\dot{z}^*(t)$ and $\ddot{z}^*(t)$ are the resulting

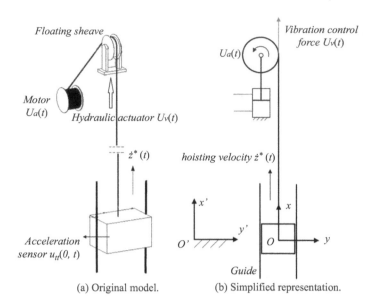

Figure 2.1. The mining cable elevator.

velocity and acceleration. The dynamics of axial elastic deformations (vibration displacements) $u(x,t)$, with $u_t(x,t)$ being the vibration velocity, are in reference to the moving coordinate system O associated with the motion $z^*(t)$. Here we assume that the motion $z^*(t)$ is controlled perfectly by the control force $U_a(t)$ and plays the role of the known target hoisting trajectory. By using the extended Hamilton's principle [134, 178], the axial vibration dynamics $u(x,t)$ on a prescribed time-varying domain $[0, l(t)]$, where

$$l(t) = L - z^*(t),\tag{2.1}$$

is derived as

$$-\rho(u_{tt}(x,t) + \ddot{z}^*(t)) + E_A u_{xx}(x,t) = 0,\tag{2.2}$$

$$M(u_{tt}(0,t) + \ddot{z}^*(t)) + E_A u_x(0,t) + c_1 u_t(0,t) = 0,\tag{2.3}$$

$$U_v(t) - E_A u_x(l(t),t) - (M + \rho l(t))g - \dot{l}(t)\rho(u_t(l(t),t) + \dot{z}^*(t))$$

$$-\frac{J_D}{R_D^2}(\ddot{u}(l(t),t) + \ddot{z}^*(t)) - c_2(\dot{u}(l(t),t) + \dot{z}^*(t)) = 0,\tag{2.4}$$

where the physical parameters are given as

M mass of the load,
L maximal length of the cable,
$l(t)$ time-varying length of the cable,
E Young's modulus of the cable,
A_a cross-sectional area of the cable,
J_D moment of inertia of the drum,
R_D radius of the drum,
ρ linear density of the cable,

g gravitational acceleration,
c_1 cage-guide damping coefficient,
c_2 cable-head sheave damping coefficient.

We neglect the target reference acceleration $\ddot{z}^*(t)$ in the model (2.2)–(2.4) for the sake of control design simplicity. The simplified model is obtained as

$$-\rho u_{tt}(x,t) + E_A u_{xx}(x,t) = 0, \qquad (2.5)$$

$$M u_{tt}(0,t) + E_A u_x(0,t) + c_1 u_t(0,t) = 0, \qquad (2.6)$$

$$U_v(t) - E_A u_x(l(t),t) - (M + \rho l(t))g - \dot{l}(t)\rho(u_t(l(t),t) + \dot{z}^*(t))$$

$$- \frac{J_D}{R_D^2} \ddot{u}(l(t),t) - c_2(\dot{u}(l(t),t) + \dot{z}^*(t)) = 0. \qquad (2.7)$$

Remark 2.1. The simplification $\ddot{z}^*(t) = 0$ is reasonable because the velocity $\dot{z}^*(t)$ of the reference motion $z^*(t)$ can be set to be constant except for starting and stopping time intervals in the practical operation of the elevator. In the simulation, the controller designed based on the simplified model (2.5)–(2.7) will be tested in the accurate model (2.2)–(2.4) to show that our control design is effective in vibration suppression of the mining cable elevator. It should be noted that $\ddot{z}^*(t) = 0$ is only for simplified presentation and is not required in our design and theory results.

Toward bringing the model into the more readily recognizable form for control of wave equations, let us define the vibration control force to consist of two components,

$$U_v(t) = U_{v1}(t) + U_{v2}(t), \qquad (2.8)$$

and let us choose the component $U_{v2}(t)$ as

$$U_{v2}(t) = (M + \rho l(t))g + \dot{l}(t)\rho(u_t(l(t),t) + \dot{z}^*(t))$$

$$+ \frac{J_D}{R_D^2} \ddot{u}(l(t),t) + c_2(\dot{u}(l(t),t) + \dot{z}^*(t)), \qquad (2.9)$$

which simplifies (2.5)–(2.7) to

$$\rho u_{tt}(x,t) = E_A u_{xx}(x,t), \ \forall (x,t) \in [0, l(t)] \times [0, \infty), \qquad (2.10)$$

$$M u_{tt}(0,t) + E_A u_x(0,t) + c_1 u_t(0,t) = 0, \qquad (2.11)$$

$$U_{v1}(t) = E_A u_x(l(t),t), \qquad (2.12)$$

where the control input $U_{v1}(t)$ is yet to be designed. Before proceeding, we point out that feedback (2.9) employs the measurement of the floating head sheave acceleration $\ddot{u}(l(t),t)$. Standard accelerometers make such a measurement readily available.

Coupled PDE-ODE Model on Time-Varying Domain

The axial vibration dynamic system (2.10)–(2.12) is a wave PDE with the boundary conditions (2.12) and (2.11) which is a second-order ODE in time. To convert this boundary condition into the standard state-space form, we introduce new variables

$x_1(t)$ and $x_2(t)$ defined by

$$x_1(t) = u(0, t), \tag{2.13}$$
$$x_2(t) = u_t(0, t) \tag{2.14}$$

as the vibration displacement and the vibration velocity of the payload. Then the following relation is obtained:

$$\dot{x}_1(t) = x_2(t), \tag{2.15}$$
$$\dot{x}_2(t) = -\frac{c_1}{M} x_2(t) - \frac{E_A}{M} u_x(0, t). \tag{2.16}$$

Let $X(t) \in \mathbb{R}^{2 \times 1}$ be a state vector defined by

$$X(t) = [x_1(t), x_2(t)]^T. \tag{2.17}$$

Through the definition (2.17), we rewrite (2.10)–(2.12) as the interconnected (at $x = 0$) PDE-ODE system

$$\dot{X}(t) = AX(t) + Bu_x(0, t), \tag{2.18}$$
$$u(0, t) = CX(t), \tag{2.19}$$
$$u_{tt}(x, t) = \frac{E_A}{\rho} u_{xx}(x, t), \tag{2.20}$$
$$E_A u_x(l(t), t) = U_{v1}(t), \tag{2.21}$$

where

$$A = \begin{bmatrix} 0 & 1 \\ 0 & \frac{-c_1}{M} \end{bmatrix}, \quad B = \frac{E_A}{M} \begin{bmatrix} 0 \\ -1 \end{bmatrix}, \quad C = [1, 0], \tag{2.22}$$

which is indeed the case for a rigid payload model. The Neumann interconnection in ODE (2.18) physically amounts to the force acting on the cage.

The cage-guide boundary is damped when the damping coefficients $c_1 > 0$ in (2.16). We develop our control design based on a more general model where c_1 is arbitrary. In other words, we allow the uncontrolled boundary in the wave equation to be damped ($c_1 > 0$), undamped ($c_1 = 0$), or even anti-damped ($c_1 < 0$).

The general wave PDE-ODE model for which we pursue our control design is

$$\dot{X}(t) = AX(t) + Bu_x(0, t), \tag{2.23}$$
$$u_{tt}(x, t) = qu_{xx}(x, t), \tag{2.24}$$
$$u(0, t) = CX(t), \tag{2.25}$$
$$u_x(l(t), t) = U(t), \tag{2.26}$$

$\forall (x, t) \in [0, l(t)] \times [0, \infty)$, where q is an arbitrary positive constant. Matrices $A \in \mathbb{R}^{2 \times 2}$, $B \in \mathbb{R}^{2 \times 1}$, $C \in \mathbb{R}^{1 \times 2}$ are assumed to be such that the pair (A, B) is controllable, the pair (A, C) is observable, and $CB = 0$. The function $X(t) \in \mathbb{R}^2$ is the ODE state, and $u(x, t) \in \mathbb{R}$ is the state of the wave PDE. The control input $U(t) = \frac{1}{E_A} U_{v1}(t)$ is to be designed.

In the control design in this chapter, the time-varying spatial domain $[0, l(t)]$ is assumed to have the following properties, which are reasonable for the string length of the ascending mining cable elevator.

Assumption 2.1. *Only the ascent of the cable elevator is considered, and the varying length $l(t)$ is decreasing and bounded—that is, $\dot{l}(t) \leq 0$, and $0 < l(t) \leq l(0) = L$, $\forall t \geq 0$.*

We only consider the ascending process—that is, $l(t)$ being decreasing—in this chapter, because the vibratory energy of the cable is increasing as the cable length is being shortened [199], making ascent a harder control problem than descent.

Assumption 2.2. *The hoisting speed $\left| \dot{l}(t) \right| < \sqrt{q}$.*

Assumption 2.2—that is, the speed of the moving boundary is smaller than the wave speed $\sqrt{q}$—is only for ensuring the well-posedness of the initial boundary value problem (2.23)–(2.26) according to [71, 72], and is not needed in the control design and stability proof for the ascending elevator in this chapter. Assumption 2.2 indeed holds in the mining cable elevator because the value of $\sqrt{q}$ in the cable is much larger than the speed of the cage.

In remark 2.1 we indicate that we let $\ddot{z}^*(t) = 0$ and explain why. While this would formally imply, from (2.1), that $\ddot{l}(t) = 0$, we do not impose $\ddot{l}(t) = 0$; that is, we do not restrict $l(t)$ to be (piecewise) linear in time. We pursue control design, and provide theory, for general $l(t)$, subject only to assumptions 2.1 and 2.2.

2.2   STATE-FEEDBACK FOR VIBRATION SUPPRESSION

In this section, we design the state-feedback controller that stabilizes the system (2.23)–(2.26), suppressing the axial vibration amplitudes of the cable and cage with the full-state measurements $u(x, t)$ for $\forall x \in [0, l(t)]$ and $X(t)$. We seek an invertible transformation that converts the (X, u)-system into the following stable target system, described as

$$\dot{X}(t) = (A + BK)X(t) + Bw_x(0, t), \qquad (2.27)$$

$$w_{tt}(x, t) = qw_{xx}(x, t), \qquad (2.28)$$

$$w(0, t) = 0, \qquad (2.29)$$

$$w_x(l(t), t) = -dw_t(l(t), t), \qquad (2.30)$$

where $d > 0$ is a positive arbitrary damping gain. The control parameter K is chosen to make $A + BK$ Hurwitz. Based on [114] and [116], the backstepping transformation is formulated as

$$w(x, t) = u(x, t) - \int_0^x \gamma(x, y)u(y, t)dy$$
$$- \int_0^x h(x, y)u_t(y, t)dy - \beta(x)X(t), \qquad (2.31)$$

where the kernel functions $\gamma(x, y) \in R$, $h(x, y) \in R$, and $\beta(x) \in \mathbb{R}^{1 \times 2}$ are to be determined. Taking the second derivatives of (2.31) with respect to x and t, respectively, along the solution of (2.23)–(2.26), we get

$$w_{tt}(x,t) - qw_{xx}(x,t)$$

$$= 2q\left(\frac{d}{dx}\gamma(x,x)\right)u(x,t) + q\int_0^x (h_{xx}(x,y) - h_{yy}(x,y))u_t(y,t)dy$$

$$+ q\int_0^x (\gamma_{xx}(x,y) - \gamma_{yy}(x,y))u(y,t)dy + 2q\left(\frac{d}{dx}h(x,x)\right)u_t(x,t)$$

$$- (\beta(x)AB - q\gamma(x,0) + qh_y(x,0)CB)u_x(0,t)$$

$$+ (qh(x,0) - \beta(x)B)u_{xt}(0,t) + (q\beta''(x) - \beta(x)A^2 - q\gamma_y(x,0)C$$

$$- qh_y(x,0)CA)X(t) = 0. \tag{2.32}$$

For (2.32) to hold, the following conditions must be satisfied:

$$\frac{d}{dx}\gamma(x,x) = 0, \tag{2.33}$$

$$\gamma_{xx}(x,y) = \gamma_{yy}(x,y), \tag{2.34}$$

$$\frac{d}{dx}h(x,x) = 0, \tag{2.35}$$

$$h_{xx}(x,y) = h_{yy}(x,y), \tag{2.36}$$

$$qh(x,0) = \beta(x)B, \tag{2.37}$$

$$\beta(x)AB = q\gamma(x,0) - qh_y(x,0)CB, \tag{2.38}$$

$$q\beta''(x) = \beta(x)A^2 + q\gamma_y(x,0)C + qh_y(x,0)CA. \tag{2.39}$$

Substituting the transformation (2.31) into (2.27) and (2.29) and comparing them with (2.23) and (2.25), we can choose $\beta(x)$ to satisfy

$$\beta'(0) = K - \gamma(0,0)C - h(0,0)CA, \tag{2.40}$$

$$\beta(0) = C. \tag{2.41}$$

By conditions (2.33)–(2.36), $\gamma(x,y)$ and $h(x,y)$ can be written as

$$\gamma(x,y) = m(x-y), \tag{2.42}$$

$$h(x,y) = n(x-y). \tag{2.43}$$

Let $D \in \mathbb{R}^{4\times4}$ and $\Lambda \in \mathbb{R}^{1\times2}$ be defined as

$$D = \begin{bmatrix} 0 & \frac{1}{q}A^2 \\ I & -\frac{1}{q}(BCA + ABC) \end{bmatrix}, \tag{2.44}$$

$$\Lambda = \frac{1}{q}CABC. \tag{2.45}$$

Solving (2.37)–(2.41) with the help of (2.42) and (2.43), the explicit solutions of $\beta(x)$, $\gamma(x,y)$, and $h(x,y)$ are obtained as

$$\beta(x) = \begin{bmatrix} C, & K - \Lambda \end{bmatrix} e^{Dx} \begin{bmatrix} I \\ 0 \end{bmatrix}, \tag{2.46}$$

$$\gamma(x,y) = \frac{1}{q}\beta(x-y)AB, \tag{2.47}$$

$$h(x,y) = \frac{1}{q}\beta(x-y)B, \tag{2.48}$$

where $I \in \mathbb{R}^{2\times 2}$ is an identity matrix. For the mining elevator modeled in section 2, the solutions of gain kernels (2.46)–(2.48) are written as

$$\beta(x) = -\frac{M}{\rho}\left[k_1 - \left(k_1 + \frac{\rho}{M}\right)e^{\frac{\rho}{M}x}, k_2 - k_2 e^{\frac{\rho}{M}x}\right], \tag{2.49}$$

$$\gamma(x,y) = k_1 - \left(k_1 + \frac{\rho}{M}\right)e^{\frac{\rho}{M}(x-y)}, \tag{2.50}$$

$$h(x,y) = k_2 - k_2 e^{\frac{\rho}{M}(x-y)}, \tag{2.51}$$

where $k_1 > 0$, $k_2 > 0$ are controller gains such that

$$K = [k_1, k_2]$$

makes $(A + BK)$ Hurwitz. For the boundary equation (2.30) to hold, the state-feedback controller is given by

$$\begin{aligned}
U(t) = \frac{1}{N_1}\Bigg(&N_2 u_t(l(t),t) + N_3 u(l(t),t) \\
&+ N_4 u_x(0,l) + N_5 u(0,l) + N_6 X(l) \\
&+ \int_0^{l(t)} N_7 u(x,t)dx + \int_0^{l(t)} N_8 u_t(x,t)dx\Bigg), \tag{2.52}
\end{aligned}$$

where

$$N_1 = 1 - dKB, \tag{2.53}$$

$$N_2 = -d, \tag{2.54}$$

$$N_3(l(t)) = \gamma(l(t),l(t)) - qh_{xy}(l(t),l(t)), \tag{2.55}$$

$$N_4(l(t)) = dqh_x(l(t),0) - d\beta(l(t))B, \tag{2.56}$$

$$N_5(l(t)) = qdh_{xy}(l(t),0), \tag{2.57}$$

$$N_6(l(t)) = \beta_x(l(t)) + d\beta(l(t))A, \tag{2.58}$$

$$N_7(l(t),x) = \gamma_x(l(t),x) + qh_{xyy}(l(t),x), \tag{2.59}$$

$$N_8(l(t),x) = h_x(l(t),x) + d\gamma(l(t),x). \tag{2.60}$$

In the same manner to obtain the direct transformation, we also obtain the inverse transformation

$$\begin{aligned}
u(x,t) = w(x,t) &- \int_0^x \varphi(x,y)w(y,t)dy \\
&- \int_0^x \lambda(x,y)w_t(y,t)dy - \alpha(x)X(t), \tag{2.61}
\end{aligned}$$

with

$$\alpha(x) = \begin{bmatrix} -C & -K \end{bmatrix} e^{Zx}\begin{bmatrix} I \\ 0 \end{bmatrix}, \tag{2.62}$$

$$\varphi(x,y) = \frac{1}{q}\alpha(x-y)(A+BK)B, \tag{2.63}$$

$$\lambda(x,y) = \frac{1}{q}\alpha(x-y)B, \tag{2.64}$$

where

$$Z = \begin{bmatrix} 0 & \frac{1}{q}(A+BK)^2 \\ I & 0 \end{bmatrix}. \tag{2.65}$$

The detailed procedure to derive the inverse transformation (2.61) is shown in appendix 2.6.

2.3 OBSERVER AND OUTPUT-FEEDBACK CONTROLLER USING CAGE SENSING

In section 2.2, a state-feedback controller was designed to stabilize the system. However, the designed state-feedback control law requires an infinite number of sensors to obtain the distributed states in the whole domain, which is not feasible. In this section, we propose an observer-based output-feedback control law that requires only a few boundary values as measurements. An exponentially convergent observer is designed to reconstruct the distributed states using a finite number of available boundary measurements, and the output-feedback control law based on the observer is proposed.

We assume that the full ODE state X is available for measurement, which does not make the problem a lot easier, as X is not collocated with the actuator.

In the mining cable elevator, we usually place the acceleration sensor at the cage and use integration to obtain $X(t) = [u(0,t), u_t(0,t)]$, where the initial value of the vibration displacement at the cage can be obtained by the static equilibrium equation, and the initial cage velocity is zero. This paragraph only explains an acquisition method for X in practical mining cable elevators. No restrictions are imposed on any initial conditions in the design and theory.

Observer Design

The observer structure consists of a copy of the plant (2.23)–(2.26) plus the boundary state error injection, described as

$$\dot{\hat{X}}(t) = A\hat{X}(t) + B\hat{u}_x(0,t) + \bar{L}C(X(t) - \hat{X}(t)), \tag{2.66}$$

$$\hat{u}_{tt}(x,t) = q\hat{u}_{xx}(x,t) - D_1(X(t) - \hat{X}(t)), \tag{2.67}$$

$$\hat{u}(0,t) = CX(t) - D_2(X(t) - \hat{X}(t)), \tag{2.68}$$

$$\hat{u}_x(l(t),t) = U(t). \tag{2.69}$$

We feed the full ODE state X into the observer component (2.66). This is not to estimate the measured X by $\hat{X}$ but to estimate the unmeasured PDE state $u(x,t)$ by $\hat{u}(x,t)$. In other words, while one could pursue the design of a reduced-order observer, as for ODEs, we pursue a full-order observer here.

The observer gains D_1, D_2 and

$$\bar{L} = [\bar{l}_1, \bar{l}_2]^T$$

are to be determined. Define the observer errors as

$$\tilde{u}(x,t) = u(x,t) - \hat{u}(x,t), \tag{2.70}$$

$$\tilde{X}(t) = X(t) - \hat{X}(t). \tag{2.71}$$

Then, subtracting (2.66)–(2.69) from (2.23)–(2.26) provides the observer error system, written as

$$\dot{\tilde{X}}(t) = (A - \bar{L}C)\tilde{X}(t) + B\tilde{u}_x(0,t), \tag{2.72}$$

$$\tilde{u}_{tt}(x,t) = q\tilde{u}_{xx}(x,t) + D_1\tilde{X}(t), \tag{2.73}$$

$$\tilde{u}(0,t) = D_2\tilde{X}(t), \tag{2.74}$$

$$\tilde{u}_x(l(t),t) = 0. \tag{2.75}$$

To convert the system (2.72)–(2.75) into the following exponentially stable target system described as

$$\dot{\tilde{X}}(t) = (A - \bar{L}C)\tilde{X}(t) + B\tilde{w}_x(0,t), \tag{2.76}$$

$$\tilde{w}_{tt}(x,t) = q\tilde{w}_{xx}(x,t), \tag{2.77}$$

$$\tilde{w}(0,t) = 0, \tag{2.78}$$

$$\tilde{w}_x(l(t),t) = -\bar{d}\tilde{w}_t(l(t),t), \tag{2.79}$$

where $\bar{L}$ is chosen to make $A - \bar{L}C$ Hurwitz and $\bar{d}$ is an arbitrary positive design parameter, the following direct and inverse transformations are formulated:

$$\tilde{u}(x,t) = \tilde{w}(x,t) - \int_0^x d_0(x,y)\tilde{w}(y,t)dy$$
$$- \int_0^x d_1(x,y)\tilde{w}_t(y,t)dy - \Gamma(x)\tilde{X}(t), \tag{2.80}$$

$$\tilde{w}(x,t) - \tilde{u}(x,t) - \int_0^x d_2(x,y)\tilde{u}(y,t)dy$$
$$- \int_0^x d_3(x,y)\tilde{u}_t(y,t)dy - \psi(x)\tilde{X}(t). \tag{2.81}$$

By matching (2.72)–(2.75) and (2.76) (2.79), the following conditions are obtained:

$$\Gamma(x)AB = qd_0(x,0), \tag{2.82}$$

$$qd_1(x,0) = \Gamma(x)B, \tag{2.83}$$

$$q\Gamma''(x) = \Gamma(x)(A - \bar{L}C)^2 + D_1, \tag{2.84}$$

$$D_2 = -\Gamma(0), \tag{2.85}$$

$$\Gamma'(0) = 0, \tag{2.86}$$

$$d_1(l(t),l(t)) = -\bar{d}, \tag{2.87}$$

$$d_0(l(t),l(t)) = 0, \tag{2.88}$$

$$d_{0x}(l(t),y) = 0, \tag{2.89}$$

$$d_{1x}(l(t),y) = 0. \tag{2.90}$$

The solutions of the kernel needed in (2.80) are obtained as

$$\Gamma(x) = -[0, q\bar{d}][AB, B]^{-1}, \tag{2.91}$$

$$d_0(x, y) = 0, \tag{2.92}$$

$$d_1(x, y) = -\bar{d}. \tag{2.93}$$

The observer gains are obtained as

$$D_1 = [0, q\bar{d}][AB, B]^{-1}(A - \bar{L}C)^2, \tag{2.94}$$

$$D_2 = [0, q\bar{d}][AB, B]^{-1}. \tag{2.95}$$

Here the matrix $[AB, B]$ is invertible since the pair (A, B) is controllable.

For the mining elevator modeled in section 2, the solutions of the gain kernels (2.91)–(2.93) are written as

$$\Gamma = \frac{\bar{d}}{\rho}[c_1, M], \tag{2.96}$$

$$d_0 = 0, \tag{2.97}$$

$$d_1 = -\bar{d}, \tag{2.98}$$

and the observer gains (2.94) and (2.95) are obtained as

$$D_1 = -\frac{\bar{d}}{\rho}\left[c_1(l_1^2 - l_2) + M\left(l_1 l_2 + \frac{c_1 l_2}{M}\right), \right.$$
$$\left. - c_1\left(\frac{c_1}{M} + l_1\right) + M\left(\frac{c_1^2}{M^2} - l_2\right)\right], \tag{2.99}$$

$$D_2 = -\frac{\bar{d}}{\rho}[c_1, M]. \tag{2.100}$$

The matrix $A - \bar{L}C$ can be Hurwitz by choosing the positive parameters $\bar{l}_1 > 0$ and $\bar{l}_2 > 0$.

Output-Feedback Control Design

To design the output-feedback controller, we consider the target $(\hat{X}, \hat{w})$-subsystem, which is constructed by the direct and inverse transformations with the same gain kernels as the state-feedback (2.31) and (2.61). Hence, we introduce the following transformations from $(\hat{X}, \hat{u})$ to $(\hat{X}, \hat{w})$, described as

$$\hat{w}(x, t) = \hat{u}(x, t) - \int_0^x \gamma(x, y)\hat{u}(y, t)dy$$
$$- \int_0^x h(x, y)\hat{u}_t(y, t)dy - \beta(x)\hat{X}(t), \tag{2.101}$$

$$\hat{u}(x, t) = \hat{w}(x, t) - \int_0^x \varphi(x, y)\hat{w}(y, t)dy$$
$$- \int_0^x \lambda(x, y)\hat{w}_t(y, t)dy - \alpha(x)\hat{X}(t). \tag{2.102}$$

Taking the time and spatial derivatives of (2.101) with the help of gain kernels (2.46)–(2.48) and $(\hat{X}, \hat{u})$-system (2.66)–(2.69), we derive the following coupled PDE-ODE $(\hat{X}, \hat{w})$-system:

$$\dot{\hat{X}}(t) = (A + BK)\hat{X}(t) + B\hat{w}_x(0, t)$$

$$+ \left(\bar{L}C + B\gamma(0, 0)(C - D_2) \right)\tilde{X}(t), \qquad (2.103)$$

$$\hat{w}_{tt}(x, t) = q\hat{w}_{xx}(x, t) - f_1(x)\tilde{X}(t) - f_2(x)\tilde{w}_x(0, t), \qquad (2.104)$$

$$\hat{w}(0, t) = (C - D_2)\tilde{X}(t), \qquad (2.105)$$

$$\hat{w}_x(l(t), t) = -d\hat{w}_t(l(t), t), \qquad (2.106)$$

where

$$f_1(x) = \beta(x)A\bar{L}C + \beta(x)\bar{L}C(A - \bar{L}C)$$

$$- \int_0^x h(x, y)D_1(A - \bar{L}C)dy - \int_0^x \gamma(x, y)D_1 dy, \qquad (2.107)$$

$$f_2(x) = -\int_0^x h(x, y)D_1 B dy + \beta(x)\bar{L}CB. \qquad (2.108)$$

By (2.106), the output-feedback controller is designed as

$$U(t) = \frac{1}{N_1} \Big(N_2\hat{u}_t(l(t), t) + N_3\hat{u}(l(t), t)$$

$$+ N_4(l(t))\hat{u}_x(0, t) + N_5(l(t))\hat{u}(0, t)$$

$$+ N_6(l(t))\hat{X}(t) + \int_0^{l(t)} N_7(l(t), x)\hat{u}(x, t)dx$$

$$+ \int_0^{l(t)} N_8(l(t), x)\hat{u}_t(x, t)dx \Big), \qquad (2.109)$$

where $N_1, \ldots, N_8$ are defined in (2.53)–(2.60).

2.4 STABILITY ANALYSIS

In this section, we establish the stability proof of the target system via Lyapunov analysis for PDEs. The equivalent stability property between the target system and the original system is ensured due to the invertibility of the backstepping transformation. The main theorem of this chapter is stated next.

Theorem 2.1. *For all initial estimates* $(\hat{u}(x, 0), \hat{u}_t(x, 0), \hat{X}(0))$ *compatible with the control law (2.109) and the initial values* $(u(x, 0), u_t(x, 0), X(0))$ *which belong to* $H^1(0, L) \times L^2(0, L) \times \mathbb{R}^2$, *the closed-loop system consisting of the plant (2.23)–(2.26) and the observer design (2.66)–(2.69) with the output-feedback control law (2.109) is exponentially stable in the sense of the norm*

$$\left(\int_0^{l(t)} {u_t}^2(x,t)dx + \int_0^{l(t)} {u_x}^2(x,t)dx + \int_0^{l(t)} \hat{u}_t^2(x,t)dx \right.$$

$$\left. + \int_0^{l(t)} \hat{u}_x^2(x,t)dx + |X(t)|^2 + \left| \hat{X}(t) \right|^2 \right)^{1/2}. \tag{2.110}$$

Proof. First, we show the stability of the $(\tilde{X}, \tilde{w})$-subsystem. Define

$$\Omega_1(t) = \|\tilde{u}_t(t)\|^2 + \|\tilde{u}_x(t)\|^2 + \left| \tilde{X}(t) \right|^2, \tag{2.111}$$

$$\Xi_1(t) = \|\tilde{w}_t(t)\|^2 + \|\tilde{w}_x(t)\|^2 + \left| \tilde{X}(t) \right|^2, \tag{2.112}$$

where $\|\tilde{u}(t)\|^2$ is a compact notation for $\int_0^{l(t)} \tilde{u}(x,t)^2 dx$. In addition, we employ the Lyapunov function

$$V_1 = \tilde{X}^T(t)P_1\tilde{X}(t) + \phi_1 E_1(t), \tag{2.113}$$

where the matrix $P_1 = P_1^T > 0$ is the solution to the Lyapunov equation

$$P_1(A - \bar{L}C) + (A - \bar{L}C)^T P_1 = -Q_1, \tag{2.114}$$

for some $Q_1 = Q_1^T > 0$. The positive parameter ϕ_1 is to be chosen later. The function $E_1(t)$ is defined as

$$E_1(t) = \frac{1}{2}\|\tilde{w}_t(t)\|^2 + \frac{q}{2}\|\tilde{w}_x(t)\|^2 + \delta_1 \int_0^{l(t)} (1+x)\tilde{w}_x(x,t)\tilde{w}_t(x,t)dx, \tag{2.115}$$

where the parameter δ_1 should satisfy

$$0 < \delta_1 < \frac{1}{1+L} \min\{1, q\}. \tag{2.116}$$

Then we get

$$\theta_{11}\Xi_1(t) \leq V_1(t) \leq \theta_{12}\Xi_1(t), \tag{2.117}$$

where

$$\theta_{11} = \min \left\{ \lambda_{\min}(P_1), \frac{\phi_1}{2}(1 - \delta_1(1+L)), \frac{\phi_1}{2}(q - \delta_1(1+L)) \right\} > 0, \tag{2.118}$$

$$\theta_{12} = \max \left\{ \lambda_{\max}(P_1), \frac{\phi_1}{2}(1 + \delta_1(1+L)), \frac{\phi_1}{2}(q + \delta_1(1+L)) \right\} > 0. \tag{2.119}$$

The time derivative of V_1 along (2.76)–(2.79) is obtained as

$$\dot{V}_1 = -\bar{d}q\phi_1\tilde{w}_t^2(l(t),t) - \frac{1}{2}\left|\dot{l}(t)\right|\phi_1\tilde{w}_t^2(l(t),t)$$

$$- \frac{q}{2}\left|\dot{l}(t)\right|\phi_1\tilde{w}_x^2(l(t),t) - \tilde{X}^T(t)Q_1\tilde{X}(t)$$

$$+ 2B\Gamma_1\tilde{w}_x(0,t)\tilde{X}(t) + \frac{\delta_2}{2}(1 + l(t))\phi_1\tilde{w}_t^2(l(t),t)$$

$$+ q\bar{d}^2\frac{\delta_1}{2}(1 + l(t))\phi_1\tilde{w}_t^2(l(t),t)$$

$$-q\frac{\delta_1}{2}\phi_1\tilde{w}_x^2(0,t)-\frac{\delta_1}{2}\phi_1\|\tilde{w}_t\|^2-\frac{\delta_1}{2}q\phi_1\|\tilde{w}_x\|^2$$

$$+\left|\dot{l}(t)\right|\bar{d}\delta_1(1+l(t))\phi_1\tilde{w}_t^2(l(t),t),\tag{2.120}$$

where assumption 2.1 yielding $\dot{l}(t)=-\left|\dot{l}(t)\right|$ has been used.

Applying Young's inequality to (2.120), we obtain the following inequality:

$$\dot{V}_1\leq-\frac{1}{2}\lambda_{\min}(Q_1)\left|\tilde{X}(t)\right|^2-\frac{\delta_1}{2}\phi_1\|\tilde{w}_t\|^2-\frac{\delta_1}{2}q\phi_1\|\tilde{w}_x\|^2$$

$$-\left(\bar{d}q-\frac{\delta_1(1+L)}{2}(1+q\bar{d}^2)\right)\phi_1\tilde{w}_t^2(l(t),t)$$

$$-\left|\dot{l}(t)\right|\left(\frac{1}{2}+\frac{\bar{d}^2q}{2}-\bar{d}\delta_1(1+L)\right)\phi_1\tilde{w}_t^2(l(t),t)$$

$$-\left(q\frac{\delta_1}{2}\phi_1-\frac{2|P_1B|^2}{\lambda_{\min}(Q_1)}\right)\tilde{w}_x^2(0,t).\tag{2.121}$$

Therefore, together with (2.116), the parameters δ_1 and ϕ_1 are chosen to satisfy the following:

$$0<\delta_1<\frac{1}{1+L}\min\left\{1,q,\frac{2\bar{d}q}{1+q\bar{d}^2},\frac{1+q\bar{d}^2}{2\bar{d}}\right\},\tag{2.122}$$

$$\phi_1>\frac{4|P_1B|^2}{q\delta_1\lambda_{\min}(Q_1)}+\varpi,\tag{2.123}$$

with a positive parameter ϖ. Then we arrive at

$$\dot{V}_1\leq-\sigma_1V_1-\varpi\tilde{w}_x^2(0,t)<-\sigma_1V_1,\tag{2.124}$$

where

$$\sigma_1=\frac{1}{\theta_{12}}\min\left\{\frac{\delta_1}{2}\phi_1,\frac{\delta_1}{2}q\phi_1,\frac{1}{2}\lambda_{\min}(Q_1)\right\}.\tag{2.125}$$

Next, we show the stability of the $(\hat{X},\hat{w})$-subsystem. Define

$$\Omega_2(t)=\|\hat{u}_t(t)\|^2+\|\hat{u}_x(t)\|^2+\left|\hat{X}(t)\right|^2,\tag{2.126}$$

$$\Xi_2(t)=\|\hat{w}_t(t)\|^2+\|\hat{w}_x(t)\|^2+\left|\hat{X}(t)\right|^2.\tag{2.127}$$

Let V_2 be a Lyapunov function written as

$$V_2=\hat{X}^T(t)P_2\hat{X}(t)+\phi_2E_2(t),\tag{2.128}$$

where the matrix $P_2=P_2^T>0$ is the solution to the Lyapunov equation

$$P_2(A+BK)+(A+BK)^TP_2=-Q_2,\tag{2.129}$$

for some $Q_2 = Q_2{}^T > 0$. The positive parameter ϕ_2 is to be chosen later. Define $E_2(t)$ as

$$E_2(t) = \frac{1}{2}\|\hat{w}_t(t)\|^2 + \frac{q}{2}\|\hat{w}_x(t)\|^2 + \delta_2 \int_0^{l(t)} (1+x)\hat{w}_x(x,t)\hat{w}_t(x,t)dx, \qquad (2.130)$$

where the parameter δ_2 must be chosen to satisfy

$$0 < \delta_2 < \frac{1}{1+L}\min\{1, q\}. \qquad (2.131)$$

Similarly to (2.117)–(2.119), we get

$$\theta_{21}\Xi_2(t) \le V_2(t) \le \theta_{22}\Xi_2(t), \qquad (2.132)$$

where

$$\theta_{21} = \min\left\{\lambda_{\min}(P_2), \frac{\phi_2}{2}(1 - \delta_2(1+L)), \frac{\phi_2}{2}(q - \delta_2(1+L))\right\} > 0, \qquad (2.133)$$

$$\theta_{22} = \max\left\{\lambda_{\max}(P_2), \frac{\phi_2}{2}(1 + \delta_2(1+L)), \frac{\phi_2}{2}(q + \delta_2(1+L))\right\} > 0. \qquad (2.134)$$

Taking the time derivative of V_2 along (2.103)–(2.106), we get

$$\begin{aligned}
\dot{V}_2 =\ & \phi_2 q \int_0^{l(t)} \hat{w}_t(x,t)\hat{w}_{xx}(x,t)dx + \phi_2 q \int_0^{l(t)} \hat{w}_x(x,t)\hat{w}_{xt}(x,t)dx \\
& - \phi_2 \int_0^{l(t)} \hat{w}_t(x,t)\left[f_1(x)\tilde{X}(t) + f_2(x)\tilde{w}_x(0,t)\right]dx \\
& - \phi_2\left|\dot{l}(t)\right|\frac{1}{2}\hat{w}_t^2(l(t),t) - q\phi_2\left|\dot{l}(t)\right|\frac{1}{2}\hat{w}_x^2(l(t),t) \\
& + \frac{1}{2}\phi_2 q\delta_2(1+l(t))\hat{w}_x^2(l(t),t) - \frac{1}{2}\phi_2 q\delta_2\hat{w}_x^2(0,t) \\
& - \frac{1}{2}\phi_2 q\delta_2\|\hat{w}_x\|^2 + \frac{1}{2}\phi_2\delta_2(1+l(t))\hat{w}_t^2(l(t),t) \\
& - \frac{1}{2}\phi_2\delta_2\hat{w}_t^2(0,t) - \frac{1}{2}\phi_2\delta_2\|\hat{w}_t\|^2 + \phi_2\dot{l}(t)\delta_2(1+l(t))\hat{w}_t(l(t),t)\hat{w}_x(l(t),t) \\
& + \dot{\hat{X}}^T(t)P_2\hat{X}(t) + \hat{X}^T(t)P_2\dot{\hat{X}}(t). \qquad (2.135)
\end{aligned}$$

Applying Young's inequality to (2.135), as in (2.121), we obtain the following inequality,

$$\begin{aligned}
\dot{V}_2 \le\ & -\left(\frac{1}{2}\phi_2\delta_2 - (B_1 + B_2)L\right)\|\hat{w}_t\|^2 - \frac{1}{2}q\phi_2\delta_2\|\hat{w}_x\|^2 - \frac{1}{2}\lambda_{\min}(Q_2)\left|\hat{X}(t)\right|^2 \\
& - \left(\frac{1}{2}q\phi_2\delta_2 - \frac{1}{2}q^2 - \frac{4|P_2 B|^2}{\lambda_{\min}(Q_2)}\right)\hat{w}_x^2(0,t) \\
& - \phi_2\left(qd - \frac{\delta_2}{2}(1+L)(1 + qd^2)\right)\hat{w}_t^2(l(t),t) \\
& - \phi_2\left|\dot{l}(t)\right|\left(\frac{1}{2} + \frac{qd^2}{2} - d\delta_2(1+L)\right)\hat{w}_t^2(l(t),t)
\end{aligned}$$

$$+ \left(\frac{1}{4}\phi_2{}^2 + \frac{1}{2}\phi_2{}^2 (C - D_2)^2 (A - \bar{L}C)^2 \right.$$

$$\left. + \frac{4 \left| P_2(\bar{L}C + B\gamma(0,0)(C - D_2)) \right|^2}{\lambda_{\min}(Q_2)} \right) \left| \tilde{X}(t) \right|^2 + \frac{1}{4}\phi_2{}^2 \tilde{w}_x^2(0,t), \qquad (2.136)$$

where B_i for $i = 1, 2$ is defined as

$$B_i = \max_{x \in [0,L]} \left\{ |f_i(x)|^2 \right\}.$$

Therefore, by choosing the parameters δ_2 and ϕ_2 as

$$0 < \delta_2 < \frac{1}{1 + L} \min \left\{ 1, q, \frac{2dq}{1 + qd^2}, \frac{1 + qd^2}{2d} \right\}, \qquad (2.137)$$

$$\phi_2 = \frac{2}{\delta_2} \max \left\{ 2(B_1 + B_2)L, \frac{q}{2} + \frac{4 |P_2 B|^2}{q \lambda_{\min}(Q_2)} \right\}, \qquad (2.138)$$

we arrive at

$$\dot{V}_2 \leq -\sigma_2 V_2 + \xi_1 \left| \tilde{X}(t) \right|^2 + \xi_2 \tilde{w}_x^2(0, t), \qquad (2.139)$$

where

$$\sigma_2 = \mu_2 / \theta_{22} > 0,$$

and

$$\mu_2 = \min \left\{ \frac{1}{4}\phi_2 \delta_2, \frac{1}{2} q \phi_2 \delta_2, \frac{1}{2} \lambda_{\min}(Q_2) \right\}, \qquad (2.140)$$

$$\xi_1 = \frac{1}{4}\phi_2{}^2 + \frac{1}{2}\phi_2{}^2 (C - D_2)^2 (A - \bar{L}C)^2$$

$$+ \frac{4 |P_2(\bar{L}C + B\gamma(0,0)(C - D_2))|^2}{\lambda_{\min}(Q_2)}, \qquad (2.141)$$

$$\xi_2 = \frac{1}{4}\phi_2{}^2. \qquad (2.142)$$

Let V be the Lyapunov function of the overall $(\tilde{X}, \tilde{w}, \hat{X}, \hat{w})$-system defined as

$$V = RV_1 + V_2. \qquad (2.143)$$

Taking the time derivative of (2.143) and using (2.117), (2.124), and (2.139), we get

$$\dot{V} \leq -\frac{R\sigma_1}{2}V_1 - \sigma_2 V_2 - \left(\frac{R\sigma_1 \theta_{12}}{2} - \xi_1 \right) |\tilde{X}(t)|^2$$

$$- (R\varpi - \xi_2) \, \tilde{w}_x^2(0, t). \qquad (2.144)$$

Therefore, choosing R sufficiently large, finally we arrive at

$$\dot{V} \leq -\sigma V \qquad (2.145)$$

for some positive σ. The differential inequality (2.145) implies that there exists a positive parameter $\eta_1 > 0$ such that

$$\|\tilde{w}_t\|^2 + \|\tilde{w}_x\|^2 + \left|\tilde{X}(t)\right|^2 + \|\hat{w}_t\|^2 + \|\hat{w}_x\|^2 + \left|\hat{X}(t)\right|^2$$

$$\leq \eta_1 \left(\|\tilde{w}_t(0)\|^2 + \|\tilde{w}_x(0)\|^2 + \left|\tilde{X}(0)\right|^2 \right.$$

$$\left. + \|\hat{w}_t(0)\|^2 + \|\hat{w}_x(0)\|^2 + \left|\hat{X}(0)\right|^2 \right) e^{-\sigma t}. \tag{2.146}$$

Therefore, the overall target $(\tilde{w}, \tilde{X}, \hat{w}, \hat{X})$-system is exponentially stable. Due to the invertibility of the transformations (2.81) and (2.101) as explicitly written in (2.80) and (2.102), applying Poincare's, Young's, and the Cauchy-Schwarz inequalities to (2.146) in a similar manner as theorem 16.1 in [116] yields

$$\|\tilde{u}_t\|^2 + \|\tilde{u}_x\|^2 + \left|\tilde{X}(t)\right|^2 + \|\hat{u}_t\|^2 + \|\hat{u}_x\|^2 + \left|\hat{X}(t)\right|^2$$

$$\leq \eta_2 \left(\|\tilde{u}_t(0)\|^2 + \|\tilde{u}_x(0)\|^2 + \left|\tilde{X}(0)\right|^2 \right.$$

$$\left. + \|\hat{u}_t(0)\|^2 + \|\hat{u}_x(0)\|^2 + \left|\hat{X}(0)\right|^2 \right) e^{-\sigma t} \tag{2.147}$$

for some positive η_2. Therefore, the exponential stability of the overall original $(\tilde{u}, \tilde{u}_t, \tilde{X}, \hat{u}, \hat{u}_t, \hat{X})$-system in the sense of (2.111) and (2.126) is proved, which concludes theorem 2.1 with the help of (2.70) and (2.71). $\qquad\square$

2.5   SIMULATION TEST IN A SINGLE-CABLE MINING ELEVATOR

The simulation is performed based on the simplified model and the accurate model. In the first case, the simulation based on the simplified model (2.18)–(2.21) under the designed state-feedback control law (2.52) and the output-feedback control law (2.109) is conducted to verify the theoretical result in theorem 2.1. In the second case, the simulation based on the accurate model (2.2)–(2.4) with (2.9), under the designed output-feedback control law (2.109), is conducted to test the controller performance on vibration suppression. The control input U_{v1} applied in both cases is $U_{v1} = E_A U(t)$, where $U(t)$ is the designed control law, and the constant $E_A = E \times A_a$.

The physical parameters of the mining cable elevator used in the simulation are shown in table 2.1. To highlight the controller performance on vibration suppression, we make the damping coefficient c_1 in the elevator zero. The designed reference of the hoisting velocity $\dot{z}^*(t) = \dot{l}(t)$ is plotted in figure 2.2. The initial profile of the vibration displacement is $u(x,0) = -(\rho x g + M g)/E_A$, which is obtained by the force balance equation at the static state, and the initial velocity is defined as $u_t(x,0) - 0$ because the initial velocity of each point in the cable is zero. The initial conditions $u(x,0)$ and $u_t(x,0)$ defined in the simulations obviously impose no restrictions on the initial conditions in theorem 2.1. The closed-loop responses with the proposed control law (2.109) and the proportional-derivative (PD) control law,

Table 2.1. Physical parameters of the mining cable elevator.

Parameters (units)	Values
Initial length L (m)	2000
Final length (m)	200
Cable effective steel area A_a (m^2)	0.47×10^{-3}
Cable effective Young's modulus E (N/m^2)	1.03×10^{10}
Cable linear density ρ (kg/m)	8.1
Total hoisted mass M (kg)	15000
Gravitational acceleration g (m/s^2)	9.8
Maximum hoisting velocities $V_{\max}$ (m/s)	15
Total hoisting time t_f(s)	150

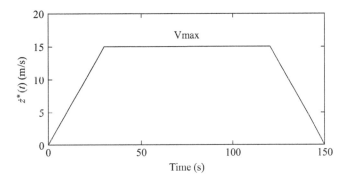

Figure 2.2. The hoisting velocity $\dot{z}^*(t)$.

which is traditionally used in industry, are examined to compare their performance in suppressing the axial vibrations of the mining cable. The PD control law is

$$U_{pd}(t) = k_p u(l(t), t) + k_d \dot{u}(l(t), t), \qquad (2.148)$$

where k_p, k_d are gain parameters. The values of k_p and k_d are tuned to attain efficient control performance. We have tested different values of k_p and k_d, and the best regulating performance is achieved with $k_p = 2000, k_d = 7000$. The gains of the proposed controller $d, \bar{d}$, and $K = [k_1, k_2]$ are chosen as $d = \bar{d} = 1$ and $[k_1, k_2] = [0.0035, 0.03]$ in the simulation. The numerical simulation is performed by the finite-difference method for the discretization in time and space after converting the time-varying domain PDE to the PDE on a fixed domain $[0, 1]$ but with time-varying coefficients by introducing $\hat{\eta} = \frac{x}{l(t)}$ [184]. The time step and space step are chosen as 0.001 and 0.01, respectively.

Comparison with the PD Control

Figure 2.3 shows the open-loop responses of the plant (2.18)–(2.21). It illustrates that the large vibration occurs, at both the cage and the midpoint of the string during the total hoisting time. To suppress the vibration, the closed-loop responses

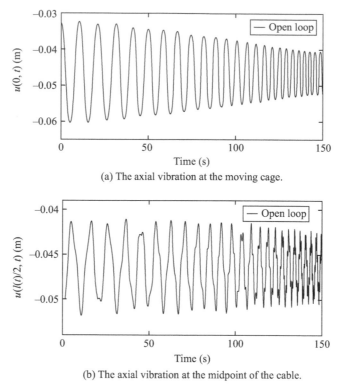

(a) The axial vibration at the moving cage.

(b) The axial vibration at the midpoint of the cable.

Figure 2.3. The open-loop responses of the plant (2.18)–(2.21). The large vibration is caused both at the moving cage and at the midpoint of the cable.

with the PD control law (2.148) and the proposed control law are investigated and shown in figure 2.4. It shows that the vibration is suppressed and converges to zero on both the proposed control law and the PD control. Moreover, it can be observed that the responses with the proposed control law have faster convergence and less overshoot than the responses with the PD control law. Thus, the proposed control law shows better performance than the classical PD control.

The Observer-Based Output-Feedback Responses

With the available boundary measurements of the displacement and the velocity of the axial vibration at the cage $u(0,t)$ and $u_t(0,t)$, the estimated variables of the distributed states required in the control law are obtained by the proposed observer (2.66)–(2.69). The closed-loop responses with an observer-based output-feedback controller are simulated with the initial observer error $\tilde{u}(x,0) = 0.002$ (m) uniformly. Then the initial conditions of the observer used in the simulation are $\hat{u}(x,0) = u(x,0) + 0.002$ and $\hat{X}(0) = X(0) + [0.002, 0]^T$, which satisfy the conditions in theorem 2.1. The dynamics of the observer error and the vibration displacement at the midpoint of the string are shown in figure 2.5(a) and (b), respectively. Because the locations of the actuator and the sensor are at opposite boundaries, the stabilization and the estimation of the vibration at the midpoint $x = l(t)/2$ is most challenging due to its accessibility. Figure 2.5(a) shows that the observer

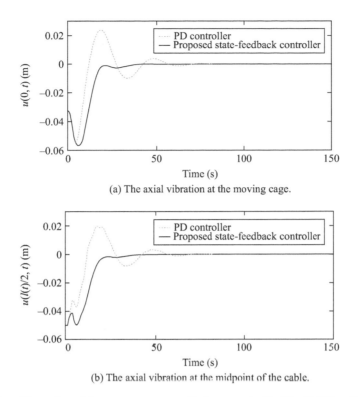

(a) The axial vibration at the moving cage.

(b) The axial vibration at the midpoint of the cable.

Figure 2.4. The closed-loop responses of the plant (2.18)–(2.21) with the PD controller (2.148) (*dashed line*) and the proposed state-feedback controller (2.52) (*solid line*).

error converges to zero quickly, which implies that the estimates of the vibration displacements reconstruct their actual distributed states. Figure 2.5(b) shows that the convergence to zero of the vibration state at the midpoint of the cable is achieved with the output-feedback control law as well, although the initial observer error affects the controller performance in the initial stage compared with the state-feedback response in figure 2.4.

Tests on the Accurate Model

We process the controller design and stability analysis based on the simplified model. In this subsection, we test the performance of our controller based on the accurate model (2.2)–(2.4) with (2.9), which includes the boundary and distributed force disturbances from the motion acceleration $\ddot{z}^*(t)$ shown in figure 2.6. After applying the proposed output-feedback controller used in section 2.5, and the PD controller with the coefficients $k_p = 3500$, $k_d = 9000$, which are adjusted to obtain efficient performance, the results under the two controllers are compared in figure 2.7. We can see that our controller also performs better at vibration suppression than the PD controller, even though the boundary and distributed force disturbances from the motion acceleration $\ddot{z}^*(t)$ are incorporated, which introduce the possibility of saltation at 30 s and 120 s.

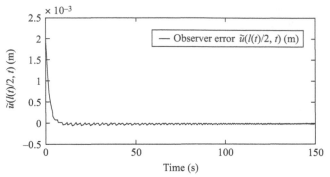

(a) The observer error of the axial vibration at the midpoint of the cable.

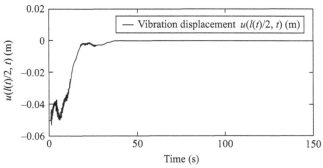

(b) The axial vibration at the midpoint of the cable.

Figure 2.5. The responses of the closed-loop system (2.18)–(2.21) under the observer (2.66)–(2.69) and the output-feedback control law (2.109). The observer achieves convergence to the actual distributed state, and the associated output-feedback controller performs similarly to the state-feedback.

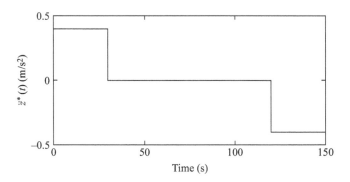

Figure 2.6. The hoisting acceleration.

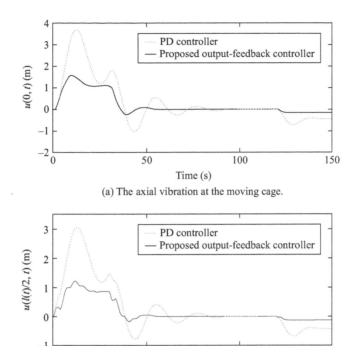

(a) The axial vibration at the moving cage.

(b) The axial vibration at the midpoint of the cable.

Figure 2.7. The closed-loop responses of the accurate plant (2.2)–(2.4) with (2.9) under the PD controller (2.148) (*dashed line*) and the proposed output-feedback controller (2.109) (*solid line*).

2.6 APPENDIX

The inverse transformation is defined as

$$u(x,t) = w(x,t) - \int_0^x \varphi(x,y)w(y,t)dy,$$

$$- \int_0^x \lambda(x,y)w_t(y,t)dy - \alpha(x)X(t), \qquad (2.149)$$

where kernel functions $\varphi(x,y) \in \mathbb{R}$, $\lambda(x,y) \in \mathbb{R}$, and $\alpha(x) \in \mathbb{R}^{1\times 2}$ are to be determined. Taking the second derivatives of (2.149) with respect to x and t, respectively, and substituting them into (2.24), recalling (2.27)–(2.30), we get

$$u_{tt}(x,t) - qu_{xx}(x,t)$$

$$= 2q\big(\varphi_y(x,x) + \varphi_x(x,x)\big)w(x,t) + q\int_0^x \big(\lambda_{xx}(x,y) - \lambda_{yy}(x,y)\big)w_t(y,t)dy$$

$$+ q\int_0^x \big(\varphi_{xx}(x,y) - \varphi_{yy}(x,y)\big)w(y,t)dy + 2q(\lambda_x(x,x) + \lambda_y(x,x))w_t(x,t)$$

$$-(\alpha(x)\tilde{A}B - q\varphi(x,0))w_x(0,t) + (q\lambda(x,0) - \alpha(x)B)w_{xt}(0,t)$$
$$+\left(q\alpha''(x) - \alpha(x)\tilde{A}^2\right)X(t) = 0, \tag{2.150}$$

where $\tilde{A} = A + BK$.

Recalling the ODE (2.23) in the system (2.23)–(2.26) and the ODE (2.27) in the target system, we have

$$\begin{aligned}
\dot{X}(t) &= AX(t) + Bu_x(0,t) \\
&= AX(t) + Bw_x(0,t) - B\alpha'(0)X(t) \\
&= (A + BK)X(t) + Bw_x(0,t).
\end{aligned} \tag{2.151}$$

By virtue of (2.25) and (2.29), we get

$$u(0,t) = w(0,t) - \alpha(0)X(t) = CX(t). \tag{2.152}$$

According to (2.150)–(2.152), we get the following conditions on the kernel functions in the inverse transformation:

$$\varphi_y(x,x) + \varphi_x(x,x) = 0, \tag{2.153}$$

$$\lambda_y(x,x) + \lambda_x(x,x) = 0, \tag{2.154}$$

$$\varphi_{yy}(x,y) - \varphi_{xx}(x,y) = 0, \tag{2.155}$$

$$\lambda_{yy}(x,y) - \lambda_{xx}(x,y) = 0, \tag{2.156}$$

$$\alpha''(x) - \frac{1}{q}\alpha(x)\tilde{A}^2 = 0, \tag{2.157}$$

$$q\lambda(x,0) - \alpha(x)B = 0, \tag{2.158}$$

$$q\varphi(x,0) - \alpha(x)\tilde{A}B = 0, \tag{2.159}$$

$$-\alpha'(0) = K, \tag{2.160}$$

$$-\alpha(0) = C. \tag{2.161}$$

According to (2.157), (2.160), and (2.161), the solution of $\alpha(x)$ can be obtained as

$$\alpha(x) = \begin{bmatrix} -C & -K \end{bmatrix} e^{Zx} \begin{bmatrix} I \\ 0 \end{bmatrix}, \tag{2.162}$$

where

$$Z = \begin{bmatrix} 0 & \frac{1}{q}\tilde{A}^2 \\ I & 0 \end{bmatrix}.$$

According to (2.158), (2.159), and (2.162), we get

$$\varphi(x,y) = \frac{1}{q}\begin{bmatrix} -C, & -K \end{bmatrix} e^{Z(x-y)} \begin{bmatrix} I \\ 0 \end{bmatrix} \tilde{A}B, \tag{2.163}$$

$$\lambda(x,y) = \frac{1}{q}\begin{bmatrix} -C, & -K \end{bmatrix} e^{Z(x-y)} \begin{bmatrix} I \\ 0 \end{bmatrix} B. \tag{2.164}$$

2.7 NOTES

Most of the existing studies on the vibration control of compliant strings, modeled by a wave PDE, focus on the fixed length [65, 88, 89, 137, 138, 139]. The time-varying length has a significant effect on the vibration dynamic characteristics of compliant string systems [157, 199, 201] and makes the design of the controller more challenging. Relatively few studies deal with the boundary vibration control problems of varying-length cables. One control design is presented in [90], where a boundary control law was developed to stabilize the transverse vibrations of a moving string system with varying length. In the above literature, however, either the actuators must be placed at both boundaries of the cable or the uncontrolled boundaries are assumed to be the fixed/damped types. For the control design in this chapter, we apply a control input in only one boundary of the cable with time-varying length and allow the uncontrolled boundary to be anti-stable.

We deal with a wave PDE with a time-dependent moving boundary in this chapter, and the results regarding the boundary control of wave PDEs with a state-dependent moving boundary can be seen in [28, 29]. Boundary control of other types of PDEs with moving boundaries can be found in [19, 46, 109, 110, 195].

While in this chapter we neglect the cable material damping and derive the controller based on an undamped wave PDE, control designs for a more realistic cable model that includes in-domain viscous damping can be found in chapters 3, 5, and 6.

Chapter Three

Dual-Cable Elevators

Unlike chapter 2 in which we solved the problem of axial vibration suppression in a single-cable ascending mining elevator, in this chapter we advance to not only addressing the suppression of the axial vibrations and tension oscillations but also balancing the cage roll in an ascending/descending dual-cable mining elevator.

In this chapter, we also employ a more realistic cable model that includes internal material damping—that is, a wave partial differential equation (PDE) with an in-domain viscous damping term. From a mathematical point of view, this introduction of damping leads to a more challenging control problem because of the PDE in-domain couplings that result from the in-domain viscous damping term [148]. The additional mathematical challenge is the requirement of stronger exponential stability estimates in the sense of the H^2 norm of the wave PDE regarding the suppression of tension oscillations.

In this chapter, by representing the wave PDE-modeled system in the Riemann coordinates in section 3.1, the vibration dynamics of the dual-cable mining elevator are modeled by two pairs of 2×2 heterodirectional coupled hyperbolic PDEs on a time-varying domain, and all four PDE uncontrolled boundaries are coupled at one ordinary differential equation (ODE). The control task of suppressing axial vibrations and tension oscillations, as well as of balancing the cage roll in the dual-cable mining elevator, can be mathematically described as output-feedback boundary exponential stabilization of the aforementioned coupled hyperbolic PDE-ODE system in the sense of the H^1 norm. To this end, a state observer is designed and proved exponentially convergent to the plant in section 3.2, and then an observer-based output-feedback boundary controller is designed via backstepping in section 3.3. The required exponential stability of the closed-loop system and the exponential convergence of the control inputs are proved in section 3.4. A simulation test on a dual-cable mining elevator is provided in section 3.5.

Throughout this chapter the cable numbers $(1, 2)$ in the dual-cable mining elevator are denoted as the subscripts $i = 1, 2$ or $j = 2, 1, j \neq i$.

3.1 DUAL-CABLE MINING ELEVATOR DYNAMICS AND REFORMULATION

Dynamics of Dual-Cable Mining Elevators

Following the modeling process in chapter 2, and with the inclusion of the material damping of the cables, the vibration dynamics of a dual-cable mining elevator are modeled as

$$\dot{X}(t) = AX(t) + B[u_x(0, t), v_x(0, t)]^T, \qquad (3.1)$$

$$u_{tt}(x,t) = qu_{xx}(x,t) - cu_t(x,t), \tag{3.2}$$

$$v_{tt}(x,t) = qv_{xx}(x,t) - cv_t(x,t), \tag{3.3}$$

$$C_3 X(t) + C_4 X(t) l_1 = u_t(0,t), \tag{3.4}$$

$$C_3 X(t) - C_4 X(t) l_1 = v_t(0,t), \tag{3.5}$$

$$u_x(l(t),t) = U_1(t), \tag{3.6}$$

$$v_x(l(t),t) = U_2(t), \tag{3.7}$$

$x \in [0, l(t)], t \in [0, \infty)$, where $u(x,t), v(x,t)$ are the axial vibrations in two cables in a moving coordinate system associated with the elevator axial motion $l(t)$ where the origin is located at the cage. Matrices A and $B = [\bar{B}_1, \bar{B}_2]$ are

$$A = \begin{pmatrix} 0 & 0 & 1 & 0 \\ 0 & 0 & 0 & 1 \\ 0 & 0 & \frac{-c_d}{M} & 0 \\ 0 & 0 & 0 & \frac{-c_a}{J_c} \end{pmatrix}, B = E_A \begin{pmatrix} 0 & 0 \\ 0 & 0 \\ \frac{-1}{M} & \frac{-1}{M} \\ \frac{-l_1}{J_c} & \frac{l_1}{J_c} \end{pmatrix}, \tag{3.8}$$

where M denotes the mass of the load. The matrices C_3, C_4 are

$$C_3 = [0,0,1,0], \quad C_4 = [0,0,0,1].$$

The function $l(t)$ describes the time-varying length of the cables. The parameter J_c is the moment of inertia of the cage, and $q = \frac{E \times A_a}{\rho}$, where E, A_a, and ρ are Young's modulus, the cross-sectional area, and the linear density of the cables, respectively. The parameters c_d and c_a denote the damping coefficients of cage axial and roll motion, respectively, and $c = \frac{\bar{c}}{\rho}$, where $\bar{c}$ is the material damping coefficient of the steel cables. The parameter l_1 is the cage dimension shown in figure 3.1.

The PDE states $u(x,t), v(x,t)$ in (3.2), (3.3) describe the axial vibration displacements of the distributed points in cable 1 and cable 2, respectively. The functions $u_x(x,t)$ and $v_x(x,t)$ denote the distributed strain in the cables, and the tension oscillations are represented as $E_A u_x(x,t)$ and $E_A v_x(x,t)$, where the constant $E_A = E \times A_a$. The functions $E_A u_x(0,t)$ and $E_A v_x(0,t)$ that denote forces acting on the cage drive the ODE dynamics (3.1), where the ODE state $X(t) = [y(t), \theta(t), \dot{y}(t), \dot{\theta}(t)]^T$ describes the cage dynamics. The functions $y(t), \dot{y}(t)$ are the axial vibration displacement and velocity of the centroid of the cage, and $\theta(t), \dot{\theta}(t)$ are the cage roll angle and roll rate around the axis which is vertical to the door and through the centroid of the cage. Equations (3.4), (3.5) describe the velocity relationship between the cage and the bottom boundaries of the two cables. Equations (3.6), (3.7) come from $E_A u_x(l(t),t) = U_{1v}(t)$ and $E_A v_x(l(t),t) = U_{2v}(t)$ with the definition of $U_{1v}(t) = E_A U_1(t)$ and $U_{2v}(t) = E_A U_2(t)$, from which the two actual control forces at the two floating sheaves $U_{1v}(t), U_{2v}(t)$ can be obtained by $U_1(t)$, $U_2(t)$ to be designed in this chapter.

We pursue the control design for a general model where the damping coefficients c_a, c_d, c in (3.1)–(3.7) are arbitrary. In other words, these damping coefficients can be damped (> 0), undamped ($= 0$), or even anti-damped (< 0).

Axial motion dynamics $l(t)$ are regulated by a separate controller $U_a(t)$ at the drum. We neglect the effect of the vibration dynamics on the motion dynamics because the vibration displacements $u(x,t), v(x,t)$ are much smaller than the hoisting motion $l(t)$ between 2000 m underground and the surface platform. We can then

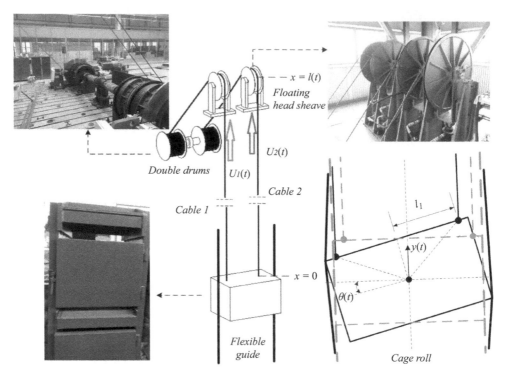

Figure 3.1. Diagram and prototype of a dual-cable mining elevator with flexible guide rails. (The prototype was built by CITIC Heavy Industries Co. Ltd.)

consider that the $l(t)$ reference governed by an independent ODE (motion dynamics) driven by $U_a(t)$ is the known hoisting trajectory. Hence, we focus on the control design $U_i(t)$ at the floating head sheave for the PDE vibration dynamics (3.1)–(3.7), where $[0, l(t)]$ acts as a known time-varying domain.

Assumption 3.1. *The varying length $l(t)$ is bounded—that is, $0 < l(t) \leq L$, $\forall t \geq 0$, where the positive constant L is the maximal length.*

Assumption 3.2. *The velocity $\dot{l}(t)$ of the moving boundary is bounded by*

$$\left| \dot{l}(t) \right| \leq \bar{v}_{\max} < \sqrt{q}, \tag{3.9}$$

where $\bar{v}_{\max}$ is the maximum velocity of the mining cable elevator.

As mentioned in chapter 2, in the mining cable elevator the wave speed, which may assume a value like $\sqrt{q} = r/\rho = 7.5 \times 10^3$, is much larger than the value of the maximum hoisting velocity, which may be $\bar{v}_{\max} = 16.25$ m/s, that is, $\bar{v}_{\max} \ll \sqrt{q}$. According to [71, 72], the fact that the speed of the moving boundary $|\dot{l}(t)|$ is smaller than the wave speed $\sqrt{q}$ ensures the well-posedness of the initial boundary value problem (3.1)–(3.7). Since the value of $\sqrt{q}$ in the string is much larger than the speed of the moving payload, the assumption that leads to well-posedness indeed usually holds in almost all string-actuated motion mechanisms.

In addition to the suppression of oscillations of tension in each cable, the suppression of tension oscillation discrepancy between two cables is required. Therefore, the error between tension oscillations in two cables should be a part of the model used to design a controller. Toward that end we introduce the new error and mean variables

$$e(x,t) = v(x,t) - u(x,t), \tag{3.10}$$

$$s(x,t) = v(x,t) + u(x,t), \tag{3.11}$$

where

$$E_A e_x(x,t) = E_A v_x(x,t) - E_A u_x(x,t)$$

is the tension oscillation discrepancy between two cables, and $E_A s_x(x,t)$ is the total of the tension oscillations in the two cables.

The control objectives are formulated in terms of the following physical behavior and theoretical results:

- Physically, suppress the oscillations of tension in each cable as fast as possible. Theoretically, make the (u,v)-system exponentially stable in the sense of terms including $\|u_{xx}(\cdot,t)\| + \|v_{xx}(\cdot,t)\|$, where the exponential decay rate can be chosen.
- Physically, reduce the error of oscillations of tension between two cables as fast as possible. Theoretically, make the (e,s)-system exponentially stable in the sense of terms including $\|e_{xx}(\cdot,t)\| + \|s_{xx}(\cdot,t)\|$, where the exponential decay rate can be chosen.
- Physically, suppress the axial vibration displacement $y(t)$ and the roll angle $\theta(t)$ of the cage as fast as possible. Theoretically, make $|X(t)|$ exponentially convergent to zero when $t \to \infty$, where the exponential decay rate can be chosen.

This may be a convenient place to remind the reader of the preceding paragraphs that $\|u(\cdot,t)\|$ is a compact notation for $\sqrt{\int_0^{l(t)} u(x,t)^2 dx}$.

Reformulation in Riemann Coordinates

In order to convert the wave models to first-order systems, for the sake of enabling the applicability of backstepping designs for first-order hyperbolic PDE-ODE systems, we introduce the following Riemann coordinates:

$$z_1(x,t) = s_t(x,t) - \sqrt{q}s_x(x,t), \tag{3.12}$$

$$w_1(x,t) = s_t(x,t) + \sqrt{q}s_x(x,t), \tag{3.13}$$

$$z_2(x,t) = e_t(x,t) - \sqrt{q}c_x(x,t), \tag{3.14}$$

$$w_2(x,t) = e_t(x,t) + \sqrt{q}e_x(x,t). \tag{3.15}$$

With (3.10), (3.11), the system (3.1)–(3.7) is rewritten as

$$\dot{X}(t) = \bar{A}X(t) + \sum_{i=1}^{2} \frac{B_i}{\sqrt{q}} w_i(0,t), \tag{3.16}$$

$$z_{it}(x,t) = -\sqrt{q}z_{ix}(x,t) - \frac{c}{2}(z_i(x,t) + w_i(x,t)), \tag{3.17}$$

$$w_{it}(x,t) = \sqrt{q}w_{ix}(x,t) - \frac{c}{2}(z_i(x,t) + w_i(x,t)), \tag{3.18}$$

$$z_i(0,t) = D_i X(t) - w_i(0,t), \tag{3.19}$$

$$w_i(l(t),t) = z_i(l(t),t) + 2\sqrt{q}U_{ei}(t), \tag{3.20}$$

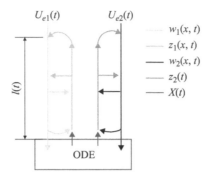

Figure 3.2. Diagram of the plant dynamics (3.16)–(3.20).

for $x \in [0, l(t)], t \in [0, \infty)$, where $i = 1, 2$, and

$$B_1 = (\bar{B}_1 + \bar{B}_2)/2 = \left[0, 0, \frac{E_A}{M}, 0 \right]^T, \tag{3.21}$$

$$B_2 = (\bar{B}_2 - \bar{B}_1)/2 = \left[0, 0, 0, \frac{E_A l_1}{J_c} \right]^T, \tag{3.22}$$

$$\bar{A} = A - \frac{2B_1}{\sqrt{q}} C_3 + \frac{2B_2}{\sqrt{q}} C_4 l_1 = \begin{bmatrix} 0 & 0 & 1 & 0 \\ 0 & 0 & 0 & 1 \\ 0 & 0 & \frac{-c_d}{M} - \frac{2E_A}{M\sqrt{q}} & 0 \\ 0 & 0 & 0 & \frac{-c_a}{J_c} + \frac{2E_A l_1^2}{J_c \sqrt{q}} \end{bmatrix}, \tag{3.23}$$

$$D_1 = 4C_3 = [0, 0, 4, 0]^T, \quad D_2 = -4l_1 C_4 = [0, 0, 0, -4l_1]^T, \tag{3.24}$$

and

$$U_{e1}(t) = U_2(t) + U_1(t), \tag{3.25}$$

$$U_{e2}(t) = U_2(t) - U_1(t). \tag{3.26}$$

The PDEs (3.17)–(3.20) are two pairs of 2×2 hyperbolic systems with states $w_i(x, t), z_i(x, t), i = 1, 2$ and with source terms that result from in-domain damping. The PDEs governing $w_i(x, t), z_i(x, t)$ are coupled with an ODE (3.16) at the distal boundary. The first transport PDE pair w_i is actuated at one boundary (3.20), and is driving the ODE (3.16) through its other boundary. The second transport PDE pair z_i convects backward and is driven by the state of the ODE $X(t)$ and the boundary states $w_i(0, t)$ in (3.19). The diagram describing these plant dynamics is shown in figure 3.2.

Unlike [147, 148, 149], which achieve results robust to small positive in-domain viscous damping in wave PDEs, in our case the in-domain damping coefficient c can be an arbitrary constant.

3.2   OBSERVER FOR CABLE TENSION

In this section we pursue observer design to estimate the distributed states $z_i(x, t)$, $w_i(x, t)$ associated with tension oscillations $u_x(x, t), v_x(x, t)$ in the two cables according to (3.10), (3.11), (3.12)–(3.15). Given that distributed tension usually cannot be measured in practice, though it is needed in the controller, we design an observer of distributed tension using only the available boundary measurements.

Observer Structure

The available measurements in the mining cable elevator are

- axial vibration acceleration $\ddot{y}(t)$ and roll angular acceleration $\ddot{\theta}(t)$ in the cage where accelerometers are placed. Measuring acceleration is the prevalent method in the control of vibrating mechanical systems because accelerations are easier to measure with accelerometers than displacements or velocities [17];
- force $E_A u_x(l(t), t)$, $E_A v_x(l(t), t)$ and velocity $u_t(l(t), t)$, $v_t(l(t), t)$ feedback signals of the actuators at two floating sheaves.

The observer is constructed as a copy of the plant (3.16)–(3.20) with output error injections:

$$\dot{\hat{X}}(t) = \bar{A}\hat{X}(t) + \sum_{i=1}^{2} \frac{B_i}{\sqrt{q}} \hat{w}_i(0, t)$$

$$+ \bar{L}\left((y(t) + \theta(t)) - (C_1 + C_2)\hat{X}(t) \right), \tag{3.27}$$

$$\hat{z}_{it}(x, t) = -\sqrt{q}\hat{z}_{ix}(x, t) - \frac{c}{2}(\hat{z}_i(x, t) + \hat{w}_i(x, t))$$

$$+ \bar{\Gamma}_i(x, t)(z_i(l(t), t) - \hat{z}_i(l(t), t)), \tag{3.28}$$

$$\hat{w}_{it}(x, t) = \sqrt{q}\hat{w}_{ix}(x, t) - \frac{c}{2}(\hat{z}_i(x, t) + \hat{w}_i(x, t))$$

$$+ \Gamma_i(x, t)(z_i(l(t), t) - \hat{z}_i(l(t), t)), \tag{3.29}$$

$$\hat{z}_1(0, t) = D_1 X(t) - \hat{w}_1(0, t) = 4\dot{y}(t) - \hat{w}_1(0, t), \tag{3.30}$$

$$\hat{z}_2(0, t) = D_2 X(t) - \hat{w}_2(0, t) = -4l_1\dot{\theta}(t) - \hat{w}_2(0, t), \tag{3.31}$$

$$\hat{w}_i(l(t), t) = z_i(l(t), t) + 2\sqrt{q}U_{ei}(t), \tag{3.32}$$

where $z_i(l(t), t)$ can be obtained by using $s_x(l(t), t)$, $s_t(l(t), t)$, $e_x(l(t), t)$, $e_t(l(t), t)$, which are computed by the measurements $u_x(l(t), t)$, $v_x(l(t), t)$, $u_t(l(t), t)$, $v_t(l(t), t)$. Because it is the acceleration (rather than displacement) sensor that is more convenient to place at the cage, $y(t)$, $\theta(t)$, $\dot{y}(t)$, $\dot{\theta}(t)$ in $X(t)$ are calculated by integrating the measured accelerations $\ddot{y}(t)$, $\ddot{\theta}(t)$ with the known initial conditions $y(0)$, $\theta(0)$, $\dot{y}(0)$, $\dot{\theta}(0)$ (this is only the explanation of an implementable signal acquisition method in practice, and no restrictions are imposed on any initial conditions in the design and theory). The ODE measurements $y(t), \theta(t)$ and the matrix $\bar{A}$ in (3.27) form an observable matrix pair $(\bar{A}, C_1 + C_2)$, where $C_1 = [1, 0, 0, 0]$, $C_2 = [0, 1, 0, 0]$.

Observer Error Dynamics and Backstepping

Defining the observer errors

$$(\tilde{z}_i(x, t), \tilde{w}_i(x, t), \tilde{X}(t)) = (z_i(x, t), w_i(x, t), X(t))$$

$$- (\hat{z}_i(x, t), \hat{w}_i(x, t), \hat{X}(t)), \tag{3.33}$$

we have the observer error dynamics between the plant and the observer as

$$\dot{\tilde{X}}(t) = \hat{A}\tilde{X}(t) + \sum_{i=1}^{2} \frac{B_i}{\sqrt{q}} \tilde{w}_i(0,t), \tag{3.34}$$

$$\tilde{z}_{it}(x,t) = -\sqrt{q}\tilde{z}_{ix}(x,t) - \frac{c}{2}(\tilde{z}_i(x,t) + \tilde{w}_i(x,t)) - \bar{\Gamma}_i(x,t)\tilde{z}_i(l(t),t), \tag{3.35}$$

$$\tilde{w}_{it}(x,t) = \sqrt{q}\tilde{w}_{ix}(x,t) - \frac{c}{2}(\tilde{z}_i(x,t) + \tilde{w}_i(x,t)) - \Gamma_i(x,t)\tilde{z}_i(l(t),t), \tag{3.36}$$

$$\tilde{z}_i(0,t) = -\tilde{w}_i(0,t), \tag{3.37}$$

$$\tilde{w}_i(l(t),t) = 0, \tag{3.38}$$

where the system matrix

$$\hat{A} = \bar{A} - \bar{L}(C_1 + C_2)$$

in (3.34) is made Hurwitz by choosing $\bar{L}$. We would like to design the observer gains $\bar{\Gamma}_1(x,t)$, $\bar{\Gamma}_2(x,t)$, $\Gamma_1(x,t)$, $\Gamma_2(x,t)$ to make sure the observer error dynamics (3.34)–(3.38) are exponentially stable.

Using the backstepping transformation

$$\tilde{z}_i(x,t) = \tilde{\alpha}_i(x,t) - \int_x^{l(t)} \bar{\phi}_i(x,y)\tilde{\alpha}_i(y,t)dy, \tag{3.39}$$

$$\tilde{w}_i(x,t) = \tilde{\beta}_i(x,t) - \int_x^{l(t)} \bar{\psi}_i(x,y)\tilde{\alpha}_i(y,t)dy, \tag{3.40}$$

we would like to convert the observer error dynamics (3.34)–(3.38) into the target observer error system as

$$\dot{\tilde{X}}(t) = \hat{A}\tilde{X}(t) + \sum_{i=1}^{2} \frac{B_i}{\sqrt{q}} \left(\tilde{\beta}_i(0,t) + \int_0^{l(t)} \bar{\psi}_i(0,y)\tilde{\alpha}_i(y,t)dy \right), \tag{3.41}$$

$$\tilde{\alpha}_{it}(x,t) = -\sqrt{q}\tilde{\alpha}_{ix}(x,t) + \int_x^{l(t)} \bar{M}_i(x,y)\tilde{\beta}_i(y,t)dy$$
$$- \frac{c}{2}\tilde{\alpha}_i(x,t) - \frac{c}{2}\tilde{\beta}_i(x,t), \tag{3.42}$$

$$\tilde{\beta}_{it}(x,t) = \sqrt{q}\tilde{\beta}_{ix}(x,t) - \frac{c}{2}\tilde{\beta}_i(x,t) + \int_x^{l(t)} \bar{N}_i(x,y)\tilde{\beta}_i(y,t)dy, \tag{3.43}$$

$$\tilde{\alpha}_i(0,t) = -\tilde{\beta}_i(0,t), \tag{3.44}$$

$$\tilde{\beta}_i(l(t),t) = 0. \tag{3.45}$$

The target system (3.41)–(3.45) is a PDE-ODE cascaded system where the PDE (3.42)–(3.45) has the same structure as the target system (17)–(20) in [21], and the PDE states flow into the ODE (3.41) with a Hurwitz system matrix $\hat{A}$.

Remark that the heterodirectional coupled hyperbolic PDEs (3.35), (3.36), which include in-domain unstable sources, especially the coupling terms $\tilde{z}_i(x,t)$, $\tilde{z}_i(l(t),t)$ in (3.36) whose propagate direction is from the right boundary to the left boundary and the ODE—that is, the "forward" direction, are intentionally converted into the target system where there is no coupling term (the states form the transport PDE in the reverse direction) in the "forward" transport PDE (3.43). This motivation is also used to construct the target system for (3.27)–(3.32), which will be shown in section 3.3.

The kernels $\bar{\phi}_i(x,y)$, $\bar{\psi}_i(x,y)$ in the transformation (3.39), (3.40), $\bar{M}_i(x,y)$, $\bar{N}_i(x,y)$ in (3.42), (3.43), and the observer gains $\bar{\Gamma}_i(x,t)$, $\Gamma_i(x,t)$ are determined next.

Calculation of the Kernels and Observer Gains

Substituting the transformations (3.39), (3.40) into (3.35), and inserting (3.42), through a lengthy calculation, we get

$$
\tilde{z}_{it}(x,t) + \sqrt{q}\tilde{z}_{ix}(x,t) + \bar{\Gamma}_i(x,t)\tilde{z}_i(l(t),t) + \frac{c}{2}(\tilde{z}_i(x,t) + \tilde{w}_i(x,t))
$$
$$
= \int_x^{l(t)} \left(-\sqrt{q}\bar{\phi}_{ix}(x,y) - \sqrt{q}\bar{\phi}_{iy}(x,y) - \frac{c}{2}\bar{\psi}_i(x,y) \right) \tilde{\alpha}_i(y,t)dy
$$
$$
- \int_x^{l(t)} \left(\int_x^y \bar{\phi}_i(x,z)\bar{M}_i(z,y)dz - \bar{M}_i(x,y) - \bar{\phi}_i(x,y)\frac{c}{2} \right) \tilde{\beta}_i(y,t)dy
$$
$$
+ \left(\bar{\Gamma}_i(x,t) - \dot{l}(t)\bar{\phi}_i(x,l(t)) + \sqrt{q}\bar{\phi}_i(x,l(t)) \right) \tilde{\alpha}_i(l(t),t). \tag{3.46}
$$

Substituting the transformation (3.39), (3.40) into (3.36), and inserting (3.42), (3.43) with further lengthy calculation, we arrive at

$$
\tilde{w}_{it}(x,t) - \sqrt{q}\tilde{w}_{ix}(x,t) + \Gamma_i(x,t)\tilde{z}_i(l(t),t) + \frac{c}{2}(\tilde{w}_i(x,t) + \tilde{z}_i(x,t))
$$
$$
= \left(\frac{c}{2} - 2\sqrt{q}\bar{\psi}_i(x,x) \right) \tilde{\alpha}_i(x,l)
$$
$$
+ \int_x^{l(t)} \left(-\sqrt{q}\bar{\psi}_{iy}(x,y) + \sqrt{q}\bar{\psi}_{ix}(x,y) - \frac{c}{2}\bar{\phi}_i(x,y) \right) \tilde{\alpha}_i(y,t)dy
$$
$$
+ \int_x^{l(t)} \left(\frac{c}{2}\bar{\psi}_i(x,y) + \bar{N}_i(x,y) - \int_x^y \bar{\psi}_i(x,z)\bar{M}_i(z,y)dz \right) \tilde{\beta}_i(y,t)dy
$$
$$
+ \left(\Gamma_i(x,t) - \dot{l}(t)\bar{\psi}_i(x,l(t)) + \sqrt{q}\bar{\psi}_i(x,l(t)) \right) \tilde{\alpha}_i(l(t),t). \tag{3.47}
$$

To make sure the right-hand sides of the equal signs in (3.46), (3.47) are equal to zero, and matching (3.44) and (3.37), the kernels $\bar{\phi}_i(x,y), \bar{\psi}_i(x,y)$ in the backstepping transformation (3.39), (3.40) should satisfy

$$
-\sqrt{q}\bar{\phi}_{ix}(x,y) - \sqrt{q}\bar{\phi}_{iy}(x,y) = \frac{c}{2}\bar{\psi}_i(x,y), \tag{3.48}
$$

$$
-\sqrt{q}\bar{\psi}_{iy}(x,y) + \sqrt{q}\bar{\psi}_{ix}(x,y) = \frac{c}{2}\bar{\phi}_i(x,y), \tag{3.49}
$$

$$
\bar{\phi}_i(0,y) = -\bar{\psi}_i(0,y), \tag{3.50}
$$

$$
\bar{\psi}_i(x,x) = \frac{c}{4\sqrt{q}}, \tag{3.51}
$$

and $\bar{M}_i(x,y), \bar{N}_i(x,y)$ in (3.42) and (3.43) should satisfy

$$
\bar{M}_i(x,y) = -\frac{c}{2}\bar{\phi}_i(x,y) + \int_x^y \bar{\phi}_i(x,z)\bar{M}_i(z,y)dz, \tag{3.52}
$$

$$\bar{N}_i(x,y) = -\frac{c}{2}\bar{\psi}_i(x,y) + \int_x^y \bar{\psi}_i(x,z)\bar{M}_i(z,y)dz. \tag{3.53}$$

The observer gains $\Gamma_i(x,t), \bar{\Gamma}_i(x,t)$ are thus obtained as

$$\Gamma_i(x,t) = \dot{l}(t)\bar{\psi}_i(x,l(t)) - \sqrt{q}\bar{\psi}_i(x,l(t)), \tag{3.54}$$

$$\bar{\Gamma}_i(x,t) = \dot{l}(t)\bar{\phi}_i(x,l(t)) - \sqrt{q}\bar{\phi}_i(x,l(t)). \tag{3.55}$$

Lemma 3.1. *The kernel equations (3.48)–(3.51) have a unique continuous solution* $(\bar{\psi}_i, \bar{\phi}_i)$ *on* $\mathcal{D}_1 = \{(x,y)|0 \le x \le y \le l(t)\}$.

The proof of lemma 3.1 is shown in appendix 3.6.

Exponential Convergence of Observer Errors

After obtaining the observer gains $\bar{\Gamma}_i(x,t)$, $\Gamma_i(x,t)$, we prove the exponential stability of the observer error dynamics (3.34)–(3.38) with the designed gains $\bar{\Gamma}_i(x,t)$, $\Gamma_i(x,t)$ in the following lemma, whose proof is shown in appendix 3.6.

Lemma 3.2. *For all initial values* $(\tilde{z}_i(x,t_0), \tilde{w}_i(x,t_0)) \in L^2(0,L_0)$, *the observer error system (3.34)–(3.38) is uniformly exponentially stable in the sense of the norm*

$$\left(\sum_{i=1}^2 \left(\|\tilde{z}_i(\cdot,t)\|^2 + \|\tilde{w}_i(\cdot,t)\|^2 \right) + \left|\tilde{X}(t)\right|^2 \right)^{\frac{1}{2}}, \tag{3.56}$$

where $L^2(0,L_0)$ *is the usual Hilbert space, with* $L_0 = l(t_0)$ *the initial length of the cable that is being retracted.*

The word "uniformly" in lemma 3.2 refers to uniformity in the initial time t_0. Using lemma 3.2 and (3.33), it is straightforward to prove the following theorem.

Theorem 3.1. *For all initial values* $(z_i(x,t_0), w_i(x,t_0)) \in L^2(0,L_0)$ *and* $(\hat{z}_i(x,t_0), \hat{w}_i(x,t_0)) \in L^2(0,L_0)$, *the observer (3.27)–(3.32) can track the system (3.16)–(3.20) with uniformly exponentially convergent errors in the sense of*

$$\sum_{i=1}^2 \left(\|z_i(\cdot,t) - \hat{z}_i(\cdot,t)\|^2 + \|w_i(\cdot,t) - \hat{w}_i(\cdot,t)\|^2 \right) + \left|X(t) - \hat{X}(t)\right|^2. \tag{3.57}$$

Theorem 3.1 shows that the proposed observer recovers the distributed states of the plant (3.16)–(3.20) using only the available boundary measurements. Moreover, the following lemma, whose proof is shown in appendix 3.6 holds as well.

Lemma 3.3. *For all initial data* $(\tilde{z}_i(x,t_0), \tilde{w}_i(x,t_0)) \in H^1(0,L_0)$, *the uniform exponential stability estimate of the observer error system (3.34)–(3.38) is obtained in the sense of*

$$\sum_{i=1}^2 \left(\|\tilde{z}_{ix}(\cdot,t)\|^2 + \|\tilde{w}_{ix}(\cdot,t)\|^2 \right)^{\frac{1}{2}}, \tag{3.58}$$

where $H^1(0,L_0) = \{u|u(\cdot,t) \in L^2(0,L_0), u_x(\cdot,t) \in L^2(0,L_0)\}$.

3.3   CONTROLLER FOR CABLE TENSION OSCILLATION SUPPRESSION AND CAGE BALANCE

In section 3.2, we obtained an observer that can exponentially track the distributed states of the system (3.16)–(3.20). In this section, we design output-feedback control laws $U_1(t)$, $U_2(t)$ by using the states recovered from the observer via the backstepping method [116],[126].

Backstepping Transformation and Target System

The design of the observer-based output-feedback controller is based on the observer (3.27)–(3.32). Using the backstepping transformation

$$\alpha_i(x,t) \equiv \hat{z}_i(x,t), \tag{3.59}$$

$$\beta_i(x,t) = \hat{w}_i(x,t) - \int_0^x \psi_i(x,y)\hat{z}_i(y,t)dy$$
$$- \int_0^x \phi_i(x,y)\hat{w}_i(y,t)dy - \gamma_i(x)\hat{X}(t), \tag{3.60}$$

we would like to convert the observer system (3.27)–(3.32) to the following target system:

$$\dot{\hat{X}}(t) = \left(\bar{A} + \sum_{i=1}^2 B_i\kappa_i\right)\hat{X}(t) + \sum_{i=1}^2 \frac{B_i}{\sqrt{q}}\beta_i(0,t)$$
$$+ \bar{L}(C_1 + C_2)\tilde{X}(t), \tag{3.61}$$

$$\alpha_{it}(x,t) = -\sqrt{q}\alpha_{ix}(x,t) - \frac{c}{2}\beta_i(x,t) - \frac{c}{2}\alpha_i(x,t)$$
$$- \frac{c}{2}\int_0^x \hat{M}_i(x,y)\alpha_i(y,t)dy - \frac{c}{2}\int_0^x \hat{N}_i(x,y)\beta_i(y,t)dy$$
$$- \frac{c}{2}\hat{\vartheta}_i(x)\hat{X}(t) + \bar{\Gamma}_i(x,t)\hat{z}_i(l(t),t), \tag{3.62}$$

$$\beta_{it}(x,t) = \sqrt{q}\beta_{ix}(x,t) - \frac{c}{2}\beta_i(x,t)$$
$$- \mathcal{N}_i(x,t)\tilde{z}_i(l(t),t) - \mathcal{N}_{1i}(x)\tilde{X}(t), \tag{3.63}$$

$$\alpha_i(0,t) = \bar{D}_i\hat{X}(t) - \beta_i(0,t) + D_i\tilde{X}(t), \tag{3.64}$$

$$\beta_i(l(t),t) = 0, \tag{3.65}$$

where

$$\mathcal{N}_i(x,t) = \int_0^x (\phi_i(x,y)\Gamma_i(y,t) + \psi_i(x,y)\bar{\Gamma}_i(y,t))dy - \Gamma_i(x,t), \tag{3.66}$$

$$\mathcal{N}_{1i}(x) = \gamma_i(x)\bar{L}(C_1 + C_2) + \sqrt{q}\psi_i(x,0)D_i, \tag{3.67}$$

and where

$$\hat{N}_i(x,y) = \int_y^x \hat{N}_i(x,\delta)\phi_i(\delta,y)d\delta + \phi_i(x,y), \tag{3.68}$$

$$\hat{M}_i(x,y) = \int_y^x \hat{N}_i(x,\delta)\psi_i(\delta,y)d\delta + \psi_i(x,y), \tag{3.69}$$

$$\hat{\vartheta}_i(x) = \int_0^x \hat{N}_i(x,y)\gamma_i(y)dy + \gamma_i(x). \tag{3.70}$$

Recalling (3.21), (3.22), (3.23), the matrix

$$\acute{A} = \bar{A} + \sum_{i=1}^2 B_i\kappa_i$$

$$= \begin{bmatrix} 0 & 0 & 1 & 0 \\ 0 & 0 & 0 & 1 \\ \frac{E_A\hat{\kappa}_{11}}{M} & 0 & -\frac{2E_A}{M\sqrt{q}} + \frac{E_A\hat{\kappa}_{13}-c_d}{M} & 0 \\ 0 & \frac{E_Al_1\hat{\kappa}_{22}}{J_c} & 0 & \frac{2E_Al_1{}^2}{J_c\sqrt{q}} + \frac{E_Al_1\hat{\kappa}_{24}-c_a}{J_c} \end{bmatrix} \tag{3.71}$$

is made Hurwitz by choosing the row vectors

$$\kappa_1 = [\hat{k}_{11}, 0, \hat{k}_{13}, 0], \tag{3.72}$$

$$\kappa_2 = [0, \hat{k}_{22}, 0, \hat{k}_{24}]. \tag{3.73}$$

The matrices $\bar{D}_i$ are $\bar{D}_i = D_i - \gamma_i(0)$. In the following section, the kernels $\psi_i(x,y)$, $\phi_i(x,y)$, $\gamma_i(x)$ are determined by mapping the observer system (3.27)–(3.32) and the target system (3.61)–(3.65) via the transformations (3.59), (3.60).

Calculation of Kernels

Substituting (3.59), (3.60) into (3.63), we get

$$\beta_{it}(x,t) - \sqrt{q}\beta_{ix}(x,t) + \frac{c}{2}\beta_i(x,t) + \mathcal{N}_i(x,t)\tilde{z}_i(l(t),t) + \mathcal{N}_{1i}(x)\tilde{X}(t)$$
$$= \left(-\frac{c}{2} + 2\sqrt{q}\psi_i(x,x)\right)\hat{z}_i(x,t)$$
$$+ \int_0^x \left(\frac{c}{2}\psi_i(x,y) + \sqrt{q}\phi_{ix}(x,y) + \sqrt{q}\phi_{iy}(x,y)\right)\hat{w}_i(y,t)dy$$
$$+ \int_0^x \left(\frac{c}{2}\phi_i(x,y) + \sqrt{q}\psi_{ix}(x,y) - \sqrt{q}\psi_{iy}(x,y)\right)\hat{z}_i(y,t)dy$$
$$+ \left(\sqrt{q}\gamma_i'(x) - \gamma_i(x)\left(\bar{A} + \frac{c}{2}I_4\right) - \psi_i(x,0)D_i\right)\hat{X}(t)$$
$$+ \left(\sqrt{q}\phi_i(x,0) - \frac{1}{\sqrt{q}}\gamma_i(x)B_i + \sqrt{q}\psi_i(x,0)\right)\hat{w}_i(0,t)$$
$$- \frac{1}{\sqrt{q}}\gamma_i(x)B_j\hat{w}_j(0,t), \tag{3.74}$$

where I_4 is an identity matrix with dimension 4. To guarantee that the right-hand side of the equal sign in (3.74) is equal to zero, which ensures (3.63), and that the ODE (3.27) is mapped into (3.61), we get the following kernel conditions:

$$\sqrt{q}\phi_{ix}(x,y) + \sqrt{q}\phi_{iy}(x,y) = -\frac{c}{2}\psi_i(x,y), \tag{3.75}$$

$$\sqrt{q}\psi_{ix}(x,y) - \sqrt{q}\psi_{iy}(x,y) = -\frac{c}{2}\phi_i(x,y), \qquad (3.76)$$

$$\phi_i(x,0) = \frac{1}{q}\gamma_i(x)B_i - \psi_i(x,0), \qquad (3.77)$$

$$\psi_i(x,x) = \frac{c}{4\sqrt{q}}, \qquad (3.78)$$

$$\gamma_i{}'(x) - \frac{1}{\sqrt{q}}\gamma_i(x)\left(\bar{A} + \frac{c}{2}I_4\right) = \frac{1}{\sqrt{q}}\psi_i(x,0)D_i, \qquad (3.79)$$

$$\gamma_i(0) = \sqrt{q}\kappa_i. \qquad (3.80)$$

According to (3.23), (3.24), (3.72), (3.73), it follows from (3.79), (3.80) that $\gamma_1(x)$ and $\gamma_2(x)$ are in the forms

$$\gamma_1(x) = [\hat{\gamma}_{11}(x), 0, \hat{\gamma}_{13}(x), 0], \qquad (3.81)$$
$$\gamma_2(x) = [0, \hat{\gamma}_{22}(x), 0, \hat{\gamma}_{24}(x)]. \qquad (3.82)$$

Recalling (3.21), (3.22), we have

$$\gamma_i(x)B_j = 0, \quad i \neq j \qquad (3.83)$$

in (3.74).

The following lemma shows that there exists a unique continuous solution $(\psi_i, \phi_i, \gamma_i)$ of (3.75)–(3.80). The proof is shown in appendix 3.6.

Lemma 3.4. *For a given κ_1, κ_2, the kernel equations (3.75)–(3.80) have a unique continuous solution $(\psi_i, \phi_i, \gamma_i)$ on $\mathcal{D} = \{(x,y)|0 \leq y \leq x \leq l(t)\}$.*

Equations (3.62), (3.64) are obtained straightforwardly, recalling (3.28), (3.30), (3.31) and the transformations (3.59), (3.60). The boundary condition (3.65) will be achieved by choosing the control inputs in the next subsection.

The inverse transformations are given by

$$\hat{z}_i(x,t) \equiv \alpha_i(x,t), \qquad (3.84)$$

$$\hat{w}_i(x,t) = \beta_i(x,t) - \int_0^x \psi_i^I(x,y)\alpha_i(y,t)dy$$

$$- \int_0^x \phi_i^I(x,y)\beta_i(y,t)dy - \gamma_i^I(x)\hat{X}(t), \qquad (3.85)$$

and the existence of kernels $\psi_i^I(x,y)$, $\phi_i^I(x,y)$, $\gamma_i^I(x)$ refers to section 2.4 in [183].

Control Law

Taking into account the boundary condition (3.65) in the target system, the boundary condition (3.32) in the observer, and the transformation (3.60), we derive the controller as

$$U_{e1}(t) = \frac{-1}{2\sqrt{q}}\left(z_1(l(t),t) - \int_0^{l(t)} \psi_1(l(t),y)\hat{z}_1(y,t)dy\right.$$

$$\left. - \int_0^{l(t)} \phi_1(l(t),y)\hat{w}_1(y,t)dy - \gamma_1(l(t))\hat{X}(t)\right), \qquad (3.86)$$

$$U_{e2}(t) = \frac{-1}{2\sqrt{q}} \left(z_2(l(t),t) - \int_0^{l(t)} \psi_2(l(t),y)\hat{z}_2(y,t)dy \right.$$
$$\left. - \int_0^{l(t)} \phi_2(l(t),y)\hat{w}_2(y,t)dy - \gamma_2(l(t))\hat{X}(t) \right). \qquad (3.87)$$

The signals $\hat{z}_i(x,t)$, $\hat{w}_i(x,t)$, $\hat{X}(t)$ are obtained from the observer (3.27)–(3.32) constructed by the measurements $\ddot{y}(t)$, $\ddot{\theta}(t)$ and $u_x(l(t),t)$, $v_x(l(t),t)$, $u_t(l(t),t)$, $v_t(l(t),t)$ which are used to calculate $z_i(l(t),t)$. The gains $(\psi_i(x,y), \phi_i(x,y), \gamma_i(x))$ are the solution of (3.75)–(3.80).

Using (3.86), (3.87), the two control inputs $U_1(t)$ and $U_2(t)$ of (3.1)–(3.7) are derived as

$$U_1(t) = U_{e1}(t) - U_{e2}(t), \quad U_2(t) = U_{e1}(t) + U_{e2}(t). \qquad (3.88)$$

The proposed controller requires only readily available measurements of the mining cable elevator. In particular, the highest time derivative signals used in the controller are the first-order derivatives $u_t(l(t),t)$, $v_t(l(t),t)$, $\dot{y}(t)$, $\dot{\theta}(t)$. Physically, they are velocities and are measurable or easily deducible from acceleration measurements.

3.4   STABILITY ANALYSIS

The following lemma establishes the exponential stability of the system (3.27)–(3.32) under the control (3.86), (3.87). The proof is shown in appendix 3.6.

Lemma 3.5. *For all initial values* $(\hat{z}_i(x,t_0), \hat{w}_i(x,t_0)) \in L^2(0,L_0)$, *the system* (3.27)–(3.32) *under the control law* (3.86), (3.87) *is uniformly exponentially stable in the sense of the norm*

$$\left(\sum_{i=1}^2 \left(\|\hat{z}_i(\cdot,t)\|^2 + \|\hat{w}_i(\cdot,t)\|^2 \right) + \left| \hat{X}(t) \right|^2 \right)^{1/2}. \qquad (3.89)$$

Based on the exponential stability result of the $(\hat{z}_i, \hat{w}_i, \hat{X})$-system in the sense of $\|\hat{z}_i(\cdot,t)\|^2 + \|\hat{w}_i(\cdot,t)\|^2 + |\hat{X}(t)|^2$, we can obtain the exponential stability estimate in the sense of $\|\hat{z}_{ix}(\cdot,t)\|^2 + \|\hat{w}_{ix}(\cdot,t)\|^2$ in the following lemma, whose proof is shown in appendix 3.6.

Lemma 3.6. *For all initial values* $(\hat{z}_i(x,t_0), \hat{w}_i(x,t_0)) \in H^1(0,L_0)$, *the uniform exponential stability estimate of the system* (3.27)–(3.32) *under the control law* (3.86), (3.87) *is obtained in the sense of*

$$\left(\sum_{i=1}^2 \left(\|\hat{z}_{ix}(\cdot,t)\|^2 + \|\hat{w}_{ix}(\cdot,t)\|^2 \right) + \left| \hat{X}(t) \right|^2 \right)^{\frac{1}{2}}. \qquad (3.90)$$

The following theorems are used to show the achievement of the control objectives formulated in section 3.1 under the proposed output-feedback controller, which is bounded and exponentially convergent to zero.

Theorem 3.2. *For all initial values* $(\hat{z}_i(x,t_0),\ \hat{w}_i(x,t_0),\ z_i(x,t_0),\ w_i(x,t_0)) \in$ $H^1(0,L_0)$, *the closed-loop system including the plant* (3.16)–(3.20), *the controller* (3.86), (3.87), *and the observer* (3.27)–(3.32) *has the following properties.*

1) The closed-loop system is uniformly exponentially stable in the sense of the norm

$$\left(\sum_{i=1}^{2} \left(\|\hat{z}_i(\cdot,t)\|^2 + \|\hat{w}_i(\cdot,t)\|^2 + \|z_i(\cdot,t)\|^2 + \|w_i(\cdot,t)\|^2 + \|\hat{z}_{ix}(\cdot,t)\|^2 \right. \right.$$
$$\left. \left. + \|\hat{w}_{ix}(\cdot,t)\|^2 + \|z_{ix}(\cdot,t)\|^2 + \|w_{ix}(\cdot,t)\|^2 \right) + \left|\hat{X}(t)\right|^2 + |X(t)|^2 \right)^{1/2} \tag{3.91}$$

with a decay rate σ_{all} *that can be adjusted by the choices of the control parameters* $\kappa_i, \bar{L}$.

2) In the closed-loop system, there exist the positive constants σ_{Ui} *and* Υ_{0i}, *making* $U_{e1}(t)$, $U_{e2}(t)$ *bounded and exponentially convergent to zero in the sense of*

$$|U_{e1}(t)| \leq \Upsilon_{01} e^{-\sigma_{U1}t}, \quad |U_{e2}(t)| \leq \Upsilon_{02} e^{-\sigma_{U2}t}. \tag{3.92}$$

Proof. 1) We now prove the first of the two portions of the theorem. Recalling the exponential stability result in the sense of $\|\hat{z}_1(\cdot,t)\|^2 + \|\hat{w}_1(\cdot,t)\|^2 + \|\hat{z}_2(\cdot,t)\|^2 +$ $\|\hat{w}_2(\cdot,t)\|^2 + |\hat{X}(t)|^2$ proved in lemma 3.5 with the decay rate σ, and the exponential stability result in the sense of $\|\tilde{z}_1(\cdot,t)\|^2 + \|\tilde{w}_1(\cdot,t)\|^2 + \|\tilde{z}_2(\cdot,t)\|^2 + \|\tilde{w}_2(\cdot,t)\|^2 +$ $|\tilde{X}(t)|^2$ proved in lemma 3.2 with the decay rate σ_e, we get the exponential stability result in the sense of $\|z_1(\cdot,t)\|^2 + \|w_1(\cdot,t)\|^2 + \|z_2(\cdot,t)\|^2 + \|w_2(\cdot,t)\|^2 + |X(t)|^2$ via (3.33) with a decay rate that can be adjusted by the control parameters κ_i and $\bar{L}$.

Similarly, recalling the exponential stability estimate in the sense of $\|\hat{z}_{1x}(\cdot,t)\|^2$ $+\|\hat{w}_{1x}(\cdot,t)\|^2+\|\hat{z}_{2x}(\cdot,t)\|^2+\|\hat{w}_{2x}(\cdot,t)\|^2$ proved in lemma 3.6 with the decay rate σ_H and the exponential stability estimate in the sense of $\|\tilde{z}_{1x}(\cdot,t)\|^2 + \|\tilde{w}_{1x}(\cdot,t)\|^2 +$ $\|\tilde{z}_{2x}(\cdot,t)\|^2 + \|\tilde{w}_{2x}(\cdot,t)\|^2$ proved in lemma 3.3 with the decay rate σ_{eH}, we obtain the exponential stability estimate in the sense of $\|z_{1x}(\cdot,t)\|^2 + \|w_{1x}(\cdot,t)\|^2 +$ $\|z_{2x}(\cdot,t)\|^2 + \|w_{2x}(\cdot,t)\|^2$ with a decay rate that can be adjusted by the control parameters κ_i and $\bar{L}$ as well.

The proof of property (1) in theorem 3.2 is complete.

2) We now prove the second portion of the theorem. Applying the Cauchy-Schwarz inequality to (3.86), (3.87), we get

$$|U_{e1}(t)|^2 + |U_{e2}(t)|^2 \leq \sum_{i=1}^{2} \left(\frac{1}{q} z_i(l(t),t)^2 - \frac{1}{q} M_{10i} L \|\hat{z}_i(\cdot,t)\|^2 \right.$$
$$\left. - \frac{1}{q} M_{11i} L \|\hat{w}_i(\cdot,t)\|^2 - \frac{1}{q} M_{12i} \left|\hat{X}(t)\right|^2 \right), \tag{3.93}$$

where

$$M_{10i} = \max\left\{ \psi_i(l(t),y)^2 \right\},$$
$$M_{11i} = \max\left\{ \phi_i(l(t),y)^2 \right\},$$
$$M_{12i} = \max\left\{ \gamma_i(l(t))^2 \right\}$$

for $0 \leq y \leq l(t) \leq L$, with L being the total length of the cables.

Using the Cauchy-Schwarz inequality and (3.65), we get

$$|\beta_i(0,t)|^2 \le |\beta_i(l(t),t)|^2 + \left|\int_0^{l(t)} \beta_{ix}(x,t)dx\right|^2 \le L\|\beta_{ix}(\cdot,t)\|^2. \qquad (3.94)$$

According to the exponential stability estimate in the sense of $\|\beta_{ix}(\cdot,t)\|^2$ in the proof of lemma 3.6, we establish that $\beta_i(0,t)$ is exponentially convergent to zero. Recalling (3.60) at $x=0$ and the exponential convergence of $|\hat{X}(t)|$ proved in lemma 3.5, we find that $\hat{w}_i(0,t)$ is exponentially convergent to zero. Recalling the relationships (3.30), (3.31) and the exponential convergence of $|X(t)|$ proved in property (1), we obtain the exponential convergence of $\hat{z}_i(0,t)$.

Similarly, using the Cauchy-Schwarz inequality and (3.45), recalling the exponential stability estimate in the sense of $\|\tilde{\beta}_{ix}(\cdot,t)\|^2$ in the proof of lemma 3.3, we find that $\tilde{\beta}_i(0,t)$ is exponentially convergent to zero and then obtain the exponential convergence of $\tilde{w}_i(0,t)$ via (3.40) at $x=0$. Together with the exponential convergence of $\hat{w}_i(0,t)$ proved above, we have that $w_i(0,t)$ is exponentially convergent to zero. Through (3.19), together with the exponential convergence of $|X(t)|$, we find that $z_i(0,t)$ is exponentially convergent to zero as well. Using the Cauchy-Schwarz inequality, we also have

$$|z_i(l(t),t)|^2 \le |z_i(0,t)|^2 + \left|\int_0^{l(t)} z_{ix}(x,t)dx\right|^2$$
$$\le |z_i(0,t)|^2 + L\|z_{ix}(\cdot,t)\|^2. \qquad (3.95)$$

We then find that $z_i(l(t),t)$ is exponentially convergent to zero when $t \to \infty$ by recalling the exponential stability estimate in the sense of $\|z_{ix}(\cdot,t)\|^2$ proved in property (1) and the exponential convergence of $z_i(0,t)$ proved above. Together with the exponential stability in the sense of $\|\hat{z}_1(\cdot,t)\|^2 + \|\hat{w}_1(\cdot,t)\|^2 + \|\hat{z}_2(\cdot,t)\|^2 + \|\hat{w}_2(\cdot,t)\|^2 + |\hat{X}(t)|^2$ proved in lemma 3.5, we ascertain that the control inputs $U_{e1}(t), U_{e2}(t)$ are bounded and exponentially convergent to zero according to (3.93). The proof of property (2) in theorem 3.2 is complete. $\qquad \square$

According to theorem 3.2, we can obtain the stability properties of the (u,v) and (e,s) systems in the following theorem.

Theorem 3.3. *For all initial values* $(u(x,t_0),v(x,t_0)) \in H^2(0,L_0)$, $(e(x,t_0),s(x,t_0)) \in H^2(0,L_0)$, *the original closed-loop* (u,v)*-system including the plant* (3.1)– (3.7) *with the controllers* (3.88) *is uniformly exponentially stable in the sense of*

$$\|u_t(\cdot,t)\| + \|u_x(\cdot,t)\| + \|v_t(\cdot,t)\| + \|v_x(\cdot,t)\| + \|u_{xt}(\cdot,t)\|$$
$$+ \|u_{xx}(\cdot,t)\| + \|v_{xt}(\cdot,t)\| + \|v_{xx}(\cdot,t)\|, \qquad (3.96)$$

where $H^2(0,L_0) = \{u|u(\cdot,t) \in L^2(0,L_0), u_x(\cdot,t) \in L^2(0,L_0), u_{xx}(\cdot,t) \in L^2(0,L_0)\}$. *The* (e,s)*-system obtained from* (3.10), (3.11) *is also uniformly exponentially stable in the sense of*

$$\|e_t(\cdot,t)\| + \|e_x(\cdot,t)\| + \|s_t(\cdot,t)\| + \|s_x(\cdot,t)\| + \|e_{xt}(\cdot,t)\|$$
$$+ \|e_{xx}(\cdot,t)\| + \|s_{xt}(\cdot,t)\| + \|s_{xx}(\cdot,t)\|. \qquad (3.97)$$

The exponential decay rates can be adjusted by the control parameters $\kappa_i, \bar{L}$. Moreover, the controllers $U_1(t), U_2(t)$ realized by hydraulic actuators at the floating sheaves are bounded and exponentially convergent to zero.

Proof. Recalling (3.12)–(3.15) and applying the Cauchy-Schwarz inequality, we get

$$\|e_t(\cdot,t)\|^2 \leq \frac{1}{2}\|z_i(\cdot,t)\|^2 + \frac{1}{2}\|w_i(\cdot,t)\|^2, \tag{3.98}$$

$$\|e_x(\cdot,t)\|^2 \leq \frac{1}{2q}\|w_i(\cdot,t)\|^2 + \frac{1}{2q}\|z_i(\cdot,t)\|^2, \tag{3.99}$$

$$\|e_{xt}(\cdot,t)\|^2 \leq \frac{1}{2}\|z_{ix}(\cdot,t)\|^2 + \frac{1}{2}\|w_{ix}(\cdot,t)\|^2, \tag{3.100}$$

$$\|e_{xx}(\cdot,t)\|^2 \leq \frac{1}{2q}\|w_{ix}(\cdot,t)\|^2 + \frac{1}{2q}\|z_{ix}(\cdot,t)\|^2 \tag{3.101}$$

for $i=2$, and $\|s_t(\cdot,t)\|^2, \|s_x(\cdot,t)\|^2, \|s_{xt}(\cdot,t)\|^2, \|s_{xx}(\cdot,t)\|^2$ have the same inequality relationships with (3.98)–(3.101) for $i=1$. Therefore, recalling the exponential stability with the decay rate σ_{all} in the sense of $\|z_i(\cdot,t)\|^2 + \|w_i(\cdot,t)\|^2 + \|z_{ix}(\cdot,t)\|^2 + \|w_{ix}(\cdot,t)\|^2$ proved in property (1) in theorem 3.2, we see that the (e,s)-system obtained from (3.10), (3.11) is exponentially stable in the sense of (3.97) with a decay rate of at least σ_{all}.

According to the definition (3.10), (3.11), it is then straightforward to ascertain that the system (3.1)–(3.7) is exponentially stable in the sense of (3.96), with a decay rate of at least σ_{all} that can be adjusted by the control parameters $\kappa_i, \bar{L}$.

Recalling (3.88) and property (2) in theorem 3.2, we show that the controllers $U_1(t), U_2(t)$ at the floating sheaves are bounded and exponentially convergent to zero. The proof of theorem 3.3 is complete. □

According to the exponential convergence of $|X(t)|$ proved in property (1) of theorem 3.2 and the exponential stability estimate in the sense of the norms including $\|u_{xx}(\cdot,t)\| + \|v_{xx}(\cdot,t)\|$ and $\|c_{xx}(\cdot,t)\| + \|s_{xx}(\cdot,t)\|$ proved in theorem 3.3, with adjustable exponential decay rates, we can state that the control objectives proposed in section 3.1 are achieved.

3.5   SIMULATION TEST FOR A DUAL-CABLE MINING ELEVATOR

The physical parameters of the mining cable elevator (3.1)–(3.7) used in the simulation are shown in table 3.1. The design reference of the hoisting velocity $\dot{l}(t)$ is plotted in figure 3.3. The control parameters chosen here are $\bar{L} = [1,1,1,1]^T$, $\kappa_1 = [0.0016, 0, 0.03, 0]^T$, and $\kappa_2 = [0, -0.0018, 0, -0.03]^T$.

We also apply the boundary damper, which is traditionally utilized in industry at the head sheaves, to compare with the proposed output-feedback controller. The boundary damper feedback laws are given as

$$U_{d1}(t) = k_{d1}\dot{u}(l(t),t)$$
$$= k_{d1}(u_t(l(t),t) + \dot{l}(t)u_x(l(t),t))$$

$$= k_{d1}\left(\frac{1}{4}(z_1(l(t),t) + w_1(l(t),t) - z_2(l(t),t) - w_2(l(t),t))\right.$$

$$\left. + \frac{\dot{l}(t)}{4\sqrt{q}}((w_1(l(t),t) - z_1(l(t),t)) - (w_2(l(t),t) - z_2(l(t),t)))\right), \quad (3.102)$$

$$U_{d2}(t) = k_{d2}\dot{v}(l(t),t)$$

$$= k_{d2}(v_t(l(t),t) + \dot{l}(t)v_x(l(t),t))$$

$$= k_{d2}\left(\frac{1}{4}(z_1(l(t),t) + w_1(l(t),t) + z_2(l(t),t) + w_2(l(t),t))\right.$$

$$\left. + \frac{\dot{l}(t)}{4\sqrt{q}}((w_1(l(t),t) - z_1(l(t),t)) + (w_2(l(t),t) - z_2(l(t),t)))\right), \quad (3.103)$$

where k_{d1}, k_{d2} are tuned to attain efficient control performance. We have tested different values of k_{d1}, k_{d2}, and the best regulating performance is achieved with $k_{d1} = -0.33, k_{d2} = -0.31$.

Computational Method

The simulation is performed for the model (3.16)–(3.20) with the control law (3.86), (3.87) and the observer (3.27)–(3.32). The responses of tension oscillations $E_A u_x(x,t), E_A v_x(x,t)$ in the cables can be calculated by those of z_1, w_1, z_2, w_2 through (3.12)–(3.15) and (3.10), (3.11). The actual control forces $E_A U_1(t), E_A U_2(t)$ applied at the two floating head sheaves are obtained via (3.88).

The simulation is performed by the finite-difference method for the discretization in time and space after converting the time-varying domain PDE to a fixed-domain PDE via introducing $\check{\xi} = \frac{x}{l(t)}$ [184], and the time step and space step are chosen as

Table 3.1. Physical parameters of the dual-cable mining elevator

Parameters (units)	Values
Initial length L (m)	2000
Final length (m)	50
Cable effective steel area A_a (m^2)	0.47×10^{-3}
Cable effective Young's modulus E (N/m^2)	2.1×10^{10}
Cable linear density ρ (kg/m)	8.1
Total hoisted mass M (kg)	15000
Moment of inertia of the cage J_c (kg·m^2)	17500
Gravitational acceleration g (m/s^2)	9.8
Maximum hoisting velocities $\bar{v}_{\max}$ (m/s)	16.25
Total hoisting time t_f (s)	150
Cable material damping coefficient $\bar{c}$ (N·s/m)	0.6
Cage axial damping coefficient c_d (N·s/m)	0.4
Cage roll damping coefficient c_a (N·m·s/rad)	0.4
Cage dimension l_1 (m)	2.5

Note: cage dimensions referred to figure 3.1.

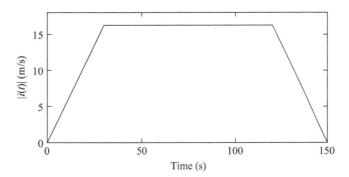

Figure 3.3. The hoisting velocity $\dot{l}(t)$.

0.001 and 0.05, respectively. The kernel equations (3.48)–(3.51) and (3.75)–(3.80), which are coupled linear hyperbolic PDEs, are also solved by the finite-difference method.

Initial Conditions

The initial conditions of the plant (3.1)–(3.7) are obtained from the physical conditions of the dual-cable mining elevator with an initial unbalance. In detail, the initial profiles of the axial strain $u_x(x,0)$, $v_x(x,0)$ are obtained by the force balance equations at the static state, which are written as

$$u_x(x,0) = \frac{\rho x g + M_{e1}g}{E_A}, \; v_x(x,0) = \frac{\rho x g + M_{e2}g}{E_A}, \tag{3.104}$$

where M_{e1} and M_{e2} are loads at the bottoms of the two cables. Define the loads M_{e1}, M_{e2} as 9250 kg and 5750 kg. The difference between the loads supported by the two cables might come from the imprecise manufacturing and installation of the two cables, or eccentricity of the cage, which always happens in loading, resulting in the initial strain error of the two cables according to (3.104). This initial strain error of the plant is unknown in the control system design. The initial vibration velocities of the two cables are defined as

$$u_t(x,0) = 0, \; v_t(x,0) = 0$$

because the initial vibration velocity of each point in the cable is zero. The initial condition of $X(t)$ is defined as

$$X(0) = [y(0), \theta(0), \dot{y}(0), \; \dot{\theta}(0)]^T = [0,0,0,0]^T.$$

From the initial conditions of $u_t(x,0)$, $v_t(x,0)$, $u_x(x,0)$, $v_x(x,0)$ and (3.10), (3.11), (3.12)–(3.15), the initial conditions $(z_i(x,0), w_i(x,0))$ of (3.16)–(3.20) tested in simulation can be determined. Similarly, the initial conditions $\hat{z}_i(x,0)$, $\hat{w}_i(x,0)$ of the observer (3.27)–(3.32) are defined. We use $M_{e1} = M_{e2} = M/2$ in (3.104) to define the observer initial conditions because the initial error between loads M_{e1}, M_{e2} is unknown in the observer design. Thus, parts of the observer initial conditions are equal to those of the plant initial conditions as $\hat{z}_1(x,0) = z_1(x,0)$, $\hat{w}_1(x,0) = w_1(x,0)$, and the others are different from the plant initial conditions as $\hat{z}_2(x,0) \neq z_2(x,0)$, $\hat{w}_2(x,0) \neq w_2(x,0)$.

Simulation Results

Tension oscillations in cables: Tension is a crucial physical variable when investigating the strength of a cable. The suppression of tension oscillations is beneficial for easing fatigue damage and prolonging the service life of the hoisting cables in the mining elevator. The responses of tension oscillations at the midpoint of cable 1 and cable 2 are shown in figures 3.4 and 3.5, respectively, with the open loop, the boundary damper, and the proposed control law. The dot-and-dash line, depicting the open-loop response in figures 3.4 and 3.5, shows that large tension oscillations would be caused in the accelerated ascending process for the hoisting velocity curve $\dot{l}(t)$ in figure 3.3. To suppress the oscillations of tension, the proposed output-feedback controller and the boundary damper are applied at the head sheaves, respectively, and the responses are shown in solid and dashed lines in figures 3.4 and 3.5. It can be seen that tension oscillations are suppressed by both the proposed control law and the boundary damper. However, we observe that the responses with the proposed control law have faster convergence and less overshoot than the responses with the boundary damper.

Error of tension oscillations between cables: The error of tension oscillations between cables is also an important physical index to investigate fatigue damage and to prolong the service life of the cables in dual-cable mining elevators. Because the constant mass of the cage is carried by two cables, a larger tension oscillation

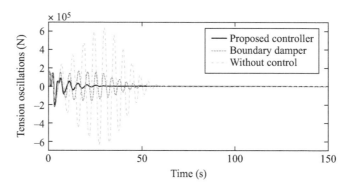

Figure 3.4. Tension oscillations $E_A \times u_x(l(t)/2, t)$ at the midpoint of cable 1.

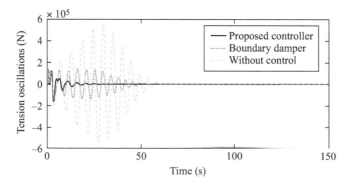

Figure 3.5. Tension oscillations $E_A \times v_x(l(t)/2, t)$ at the midpoint of cable 2.

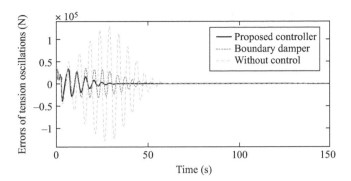

Figure 3.6. Errors $E_A \times (u_x(l(t)/2, t) - v_x(l(t)/2, t))$ of tension oscillations between cable 1 and cable 2.

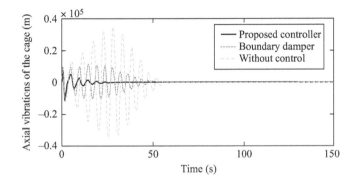

Figure 3.7. Axial vibration displacements $y(t)$ of the cage.

error between cables means that a larger maximum load would be supported by one of the cables, whose service life would be shortened due to more serious fatigue damage. The errors between cables in the open loop, under the boundary damper, and under the proposed control law are shown in figure 3.6. It shows that the tension oscillation error increases in 0–30 s, which is the accelerated ascending process, in the open-loop case, while it is reduced and convergent to zero under the proposed output-feedback controller and, moreover, has a faster convergence than the boundary damper.

Axial and roll vibrations of the cage: Axial and roll vibrations of the cage not only bring discomfort to passengers but also increase the tension error between two cables, which would cause overburdening to one of the cables and result in fatigue fracture. The responses of the axial and roll vibrations of the cage are shown in figures 3.7 and 3.8 in the cases without control, with the boundary damper, and with the proposed output-feedback controller. It can be observed that the large axial vibrations and roll of the cage in the open loop are suppressed to zero more efficiently under the proposed output-feedback controller than the traditional boundary damper, with faster convergence and less overshoot.

Observer errors and output-feedback control forces: The states used in the output-feedback control forces (3.86), (3.87) are the states recovered from the observer (3.27)–(3.32). Figure 3.9 shows the convergent observer errors between $\hat{z}_2$,

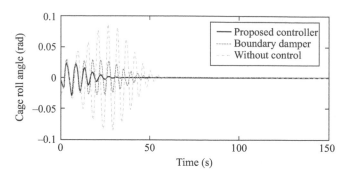

Figure 3.8. Cage roll angles $\theta(t)$.

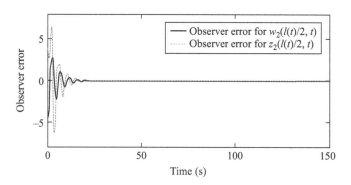

Figure 3.9. Observer errors $z_2(l(t)/2,t) - \hat{z}_2(l(t)/2,t)$ and $w_2(l(t)/2,t) - \hat{w}_2(l(t)/2,t)$ between the plant (3.16)–(3.20) and the observer (3.27)–(3.32).

$\hat{w}_2$ and the plant states z_2, w_2 at the midpoint of the domain $[0, l(t)]$, which indicates that the observer (3.27)–(3.32) can reconstruct the actual distributed states in (3.16)–(3.20). Because the locations of the actuator and the sensor are at the boundaries, the estimation of the states at the midpoint $x = l(t)/2$ is most challenging due to its accessibility. The initial conditions of the observer $\hat{z}_1(x,0), \hat{w}_1(x,0)$ are the same as the plant $z_1(x,0), w_1(x,0)$, and the observer errors of z_1, w_1 are thus at a very small magnitude 10^{-14} and convergent to zero as well. We only show the observer errors of z_2, w_2 here to avoid repetition and excessive use of space. The control forces $E_A U_1(t), E_A U_2(t)$ obtained from (3.86), (3.87) with (3.88) at the two head sheaves in the closed-loop system are given in figure 3.10, which shows that the control forces $E_A U_1(t), E_A U_2(t)$ are bounded and convergent to zero.

3.6   APPENDIX

A. Proof of lemma 3.1

Extending the domain $\mathcal{D}_1 = \{(x,y)|0 \leq x \leq y \leq l(t)\}$ to a fixed triangular domain $\{(x,y)|0 \leq x \leq y \leq L\}$ (the boundary conditions (3.50), (3.51) are given along the lines $y = x$ and $x = 0$ rather than on $y = l(t)$), the kernel equations (3.48)–(3.51) have the same form as the kernel equations (29)–(33) in [96]. In detail, (3.48) and (3.49) are two coupled transport PDEs with boundary conditions (3.50) and (3.51), which

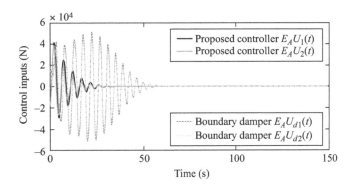

Figure 3.10. The proposed observer-based output-feedback control forces $E_A U_1(t) = E_A(U_{e2} + U_{e1})$, $E_A U_2(t) = E_A(U_{e2} - U_{e1})$ and the boundary dampers (3.102), (3.103) at the two head sheaves.

are the same form as the two coupled transport PDEs (33), (32) with boundary conditions (31), (29) in [96] through setting $m = n = 1$ and replacing K, L with $\bar{\phi}_i$, $\bar{\psi}_i$. The kernel equations of (29)–(33) have been proved well-posed in [96]. Lemma 3.1 is then proved.

B. Proof of lemma 3.2

The Lyapunov function V_e for the $(\tilde{\alpha}_1, \tilde{\beta}_1, \tilde{\alpha}_2, \tilde{\beta}_2, \tilde{X})$-system is defined as

$$V_e(t) = \sum_{i=1}^{2} V_{ei}(t), \tag{3.105}$$

where

$$V_{ei}(t) = \tilde{X}^T(t) P_2 \tilde{X}(t) + \frac{a_i}{2} \int_0^{l(t)} e^{\bar{\delta}_{2i} x} \tilde{\beta}_i(x,t)^2 dx$$

$$+ \frac{\bar{b}_i}{2} \int_0^{l(t)} e^{-\bar{\delta}_{1i} x} \tilde{\alpha}_i(x,t)^2 dx, \tag{3.106}$$

where the matrix $P_2 = P_2^T > 0$ is the solution to the Lyapunov equation

$$P_2 \hat{A} + \hat{A}^T P_2 = -Q_2,$$

for some $Q_2 = Q_2^T > 0$. The positive parameters $\bar{a}_i, \bar{b}_i, \bar{\delta}_{1i}, \bar{\delta}_{2i}$ are to be chosen later. Defining

$$\Omega_e(t) = \sum_{i=1}^{2} (\|\tilde{\alpha}_i\|^2 + \|\tilde{\beta}_i\|^2) + \left|\tilde{X}(t)\right|^2$$

yields the two positive constants θ_{e1}, θ_{e2} holding

$$\theta_{e1} \Omega_e(t) \leq V_e(t) \leq \theta_{e2} \Omega_e(t).$$

Taking the derivative of V_e along (3.41)–(3.45), using Young's inequality and the Cauchy-Schwarz inequality, through a lengthy but straightforward calculation, recalling assumption 3.2, and choosing

$$\bar{\delta}_{1i} > \frac{2}{\sqrt{q}} \max\left\{ \frac{\sqrt{q}}{2}, \xi_{0i} + \frac{c(1 + \xi_{0i})}{2} \right\},$$

$$\bar{\delta}_{2i} > \frac{2}{\sqrt{q}}\left(-\frac{c}{2\bar{a}_i} + \xi_{0i} + \frac{\bar{b}_i \xi_{0i}}{\bar{a}_i} + \frac{c\xi_{0i}\bar{b}_i}{2\bar{a}_i} \right),$$

$$\bar{a}_i > 4r_0\bar{b}_i + \frac{64|P_2B_i|^2}{q\sqrt{q}\lambda_{\min}(Q_2)},$$

where ξ_{0i} is a positive constant only depending on the plant parameters, and large enough $\bar{b}_i$ and r_0, we get

$$\dot{V}_e(t) \leq -\sigma_e V_e(t) - \frac{\sqrt{q}}{4} \sum_{i=1}^{2} \bar{a}_i \tilde{\beta}_i(0, t)^2$$

$$- \sum_{i=1}^{2} \left(\sqrt{q} - \dot{l}(t) \right) \frac{\bar{b}_i}{2} e^{-\bar{\delta}_{1i}L} \tilde{\alpha}_i(l(t), t)^2, \tag{3.107}$$

where the decay rate σ_e can be adjusted by the control parameter $\bar{L}$ which affects $\lambda_{\min}(Q_2)$.

Therefore, we obtain the exponential stability result of the target observer error system (3.41)–(3.45) in the sense of $\Omega_e(t)$. Due to the invertibility of the transformations (3.39), (3.40), the exponential stability with a decay rate at least σ_e of the $(\tilde{z}_i, \tilde{w}_i, \hat{X})$-system in the sense of the norm (3.56) is proved.

C. Proof of lemma 3.3

Differentiating (3.42) and (3.43) with respect to x, and differentiating (3.44) and (3.45) with respect to t, we get

$$\tilde{\alpha}_{ixt}(x, t) = -\sqrt{q}\tilde{\alpha}_{ixx}(x, t) - \frac{c}{2}\tilde{\alpha}_{ix}(x, t) - \frac{c}{2}\tilde{\beta}_{ix}(x, t) + \bar{\eta}_{1i}(x, t), \tag{3.108}$$

$$\tilde{\beta}_{ixt}(x, t) = \sqrt{q}\tilde{\beta}_{ixx}(x, t) - \frac{c}{2}\tilde{\beta}_{ix}(x, t) + \bar{\eta}_{2i}(x, t), \tag{3.109}$$

$$\tilde{\alpha}_{ix}(0, t) = \tilde{\beta}_{ix}(0, t) - \frac{c}{\sqrt{q}}\tilde{\beta}_i(0, t) - \frac{1}{\sqrt{q}}\bar{\eta}_{3i}(t) - \frac{c}{2\sqrt{q}}\tilde{\alpha}_i(0, t)$$

$$\tilde{\beta}_{ix}(l(t), t) = 0, \tag{3.110}$$

where

$$\bar{\eta}_{1i}(x, t) = \int_x^{l(t)} \bar{M}_{ix}(x, y)\tilde{\beta}_i(y, t)dy - \bar{M}_i(x, x)\tilde{\beta}_i(x, t), \tag{3.111}$$

$$\bar{\eta}_{2i}(x, t) = \int_x^{l(t)} \bar{N}_{ix}(x, y)\tilde{\beta}_i(y, t)dy - \bar{N}_i(x, x)\tilde{\beta}_i(x, t), \tag{3.112}$$

$$\bar{\eta}_{3i}(t) = -\int_0^{l(t)} (\bar{M}_i(0, x) + \bar{N}_i(0, x))\tilde{\beta}_i(x, t)dx. \tag{3.113}$$

Taking the derivative of (3.45), we have

$$\dot{l}(t)\tilde{\beta}_{ix}(l(t), t) + \tilde{\beta}_{it}(l(t), t) = 0.$$

Inserting (3.43), we get

$$(\dot{l}(t) + \sqrt{q})\tilde{\beta}_{ix}(l(t), t) = 0,$$

which gives (3.110), recalling assumption 3.2.

Applying the Cauchy-Schwarz inequality into (3.111)–(3.113) yields positive constants M_{7i}, M_{8i}, M_{9i} such that

$$\|\bar{\eta}_{1i}(\cdot, t)\|^2 \le M_{8i} \|\tilde{\beta}_i(\cdot, t)\|^2, \tag{3.114}$$

$$\|\bar{\eta}_{2i}(\cdot, t)\|^2 \le M_{9i} \|\tilde{\beta}_i(\cdot, t)\|^2, \tag{3.115}$$

$$|\bar{\eta}_{3i}(t)|^2 \le M_{7i} \|\tilde{\beta}_i(\cdot, t)\|^2, \tag{3.116}$$

for $i = 1, 2$.

Define a Lyapunov function

$$V_{ieH}(t) = \frac{\breve{a}_i}{2} \int_0^{l(t)} e^{\breve{\delta}_{2i}x} \tilde{\beta}_{ix}(x, t)^2 dx + R_4 V_e(t)$$

$$+ \frac{\breve{b}_i}{2} \int_0^{l(t)} e^{-\breve{\delta}_{1i}x} \tilde{\alpha}_{ix}(x, t)^2 dx, \tag{3.117}$$

where the positive constants $\breve{a}_i, \breve{b}_i, \breve{\delta}_{1i}, \breve{\delta}_{2i}, R_4$ are to be chosen later. Taking the derivative of (3.117) along (3.108) (3.110), substituting the results of $\dot{V}_e(t)$ in (3.107), using Young's inequality and the Cauchy-Schwarz inequality, and substituting (3.114)–(3.116), through a lengthy calculation we obtain

$$\dot{V}_{ieH}(t) \le -\frac{R_4 \sigma_e}{2} V_e(t) - \left(\frac{1}{2}\sqrt{q}\breve{a}_i - 3\sqrt{q}\breve{b}_i\right) \tilde{\beta}_{ix}(0, t)^2$$

$$- \left(\left(\frac{1}{2}\sqrt{q}\breve{\delta}_{2i} - \frac{|c|}{2} - \frac{L}{r_{10i}}\right)\breve{a}_i - \frac{L\breve{b}_i|c|}{4r_{11i}}\right) \int_0^{l(t)} e^{\breve{\delta}_{2i}x}\tilde{\beta}_{ix}(x, t)^2 dx$$

$$- \left(\frac{1}{2}\sqrt{q}\breve{\delta}_{1i} - \frac{|c|}{2} - |c|Lr_{11i} - \frac{L}{r_{12i}}\right)\breve{b}_i \int_0^{l(t)} e^{-\breve{\delta}_{1i}x}\tilde{\alpha}_{ix}(x, t)^2 dx$$

$$- \frac{(\sqrt{q} - \dot{l}(t))\breve{b}_i e^{-\breve{\delta}_{1i}L}}{2}\tilde{\alpha}_{ix}(l(t), t)^2$$

$$- \left(\frac{R_4}{2}\sigma_e\theta_{e1} - \frac{3\breve{b}_i M_{7i}}{\sqrt{q}} - \frac{M_{8i}r_{12i}\breve{b}_i L}{4} - \frac{M_{9i}\breve{a}_i r_{10i}e^{\breve{\delta}_{2i}L}L}{4}\right)\|\tilde{\beta}_i(\cdot, t)\|^2$$

$$- \left(\frac{R_4}{4}\sqrt{q}\bar{a}_i - \frac{6\breve{b}_i c^2}{\sqrt{q}}\right)\tilde{\beta}_i(0, t)^2$$

$$- \left(R_4(\sqrt{q} - \dot{l}(t))\frac{\bar{b}_i e^{-\breve{\delta}_{1i}L}}{2}\right)\tilde{\alpha}_i(l(t), t)^2, \tag{3.118}$$

where (3.44) has been used. The positive constants r_{10i}, r_{11i}, r_{12i} are from using Young's inequality.

Recalling assumption 3.2 and choosing $\breve{\delta}_{1i}, \breve{\delta}_{2i}, \breve{a}_i$ to satisfy

$$\breve{\delta}_{1i} > \frac{2}{\sqrt{q}}\left(cLr_{11i} + \frac{L}{r_{12i}} + \frac{|c|}{2}\right), \breve{\delta}_{2i} > \frac{2}{\sqrt{q}}\left(\frac{L}{r_{10i}} + \frac{|c|}{2}\right), \breve{a}_i > 6\breve{b}_i,$$

with large enough r_{11i}, R_4 and arbitrary $\breve{b}_i, r_{10i}, r_{12i}$, we thus arrive at

$$\dot{V}_{ieH}(t) \leq -\sigma_{ieH} V_{ieH}(t),$$

where $\sigma_{ieH} > 0$.

Defining a Lyapunov function

$$V_{eH}(t) = V_{1eH}(t) + V_{2eH}(t)$$

and taking the derivative of $V_{eH}(t)$, we get

$$\dot{V}_{eH}(t) \leq -\sigma_{eH} V_{eH}(t), \tag{3.119}$$

where the decay rate $\sigma_{eH} > 0$ can be adjusted by the control parameter $\bar{L}$ which affects σ_e.

We thus obtain the exponential stability estimate in the sense of

$$\|\tilde{\alpha}_{1x}(\cdot,t)\|^2 + \|\tilde{\beta}_{1x}(\cdot,t)\|^2 + \|\tilde{\alpha}_{2x}(\cdot,t)\|^2 + \|\tilde{\beta}_{2x}(\cdot,t)\|^2.$$

Due to the invertibility of the transformation (3.39), (3.40), we can obtain the exponential stability estimate in the sense of (3.58) with a decay rate σ_{eH}. The proof of lemma 3.3 is complete.

D. Proof of lemma 3.4

Extending the domain $\mathcal{D} = \{(x,y)|0 \leq y \leq x \leq l(t)\}$ to a fixed triangular domain $\{(x,y)|0 \leq y \leq x \leq L\}$ (the boundary conditions (3.77), (3.78) are given along the lines $y = x$ and $y = 0$ rather than on $x = l(t)$), the kernel equations (3.75)–(3.80) have the same form as the kernel equations (17)–(23) in [48]. In detail, (3.75), (3.76) are two coupled linear hyperbolic PDEs that correspond to (17), (18) in [48] by setting $m = n = 1$ and replacing K, L with ψ_i, ϕ_i. The boundary conditions (3.78), (3.77) correspond to (19), (21) in [48]. The ODE (3.79) with the initial condition (3.80) corresponds to (22) and (23) in [48]. The well-posedness of (17)–(23) in [48] has been proved. Lemma 3.4 can then be proved.

E. Proof of lemma 3.5

We establish the stability proof of the target $(\alpha_1, \beta_1, \alpha_2, \beta_2, \hat{X})$-system via Lyapunov analysis. The equivalent stability properties between the target system and the original $(\hat{z}_1, \hat{w}_1, \hat{z}_2, \hat{w}_2, \hat{X})$-system are ensured due to the invertibility of the backstepping transformation (3.59), (3.60).

Step 1. Consider a Lyapunov function for the $(\alpha_1, \beta_1, \alpha_2, \beta_2, \hat{X})$-system, as follows:

$$V_i(t) = \hat{X}^T(t) P_1 \hat{X}(t) + \frac{a_i}{2} \int_0^{l(t)} e^{\delta_{i2}x} \beta_i(x,t)^2 dx$$

$$+ \frac{b_i}{2} \int_0^{l(t)} e^{-\delta_{i1}x} \alpha_i(x,t)^2 dx, \tag{3.120}$$

where the matrix $P_1 = P_1^T > 0$ is the solution to the Lyapunov equation

$$P_1 \acute{A} + \acute{A}^T P_1 = -Q_1$$

for some

$$Q_1 = Q_1^T > 0.$$

The positive parameters $a_i, b_i, \delta_{i1}, \delta_{i2}$ are to be chosen later. Defining

$$\Omega_{1i}(t) = \|\beta_i(\cdot, t)\|^2 + \|\alpha_i(\cdot, t)\|^2 + |\hat{X}(t)|^2,$$

we get

$$\theta_{11i}\Omega_{1i}(t) \leq V_i(t) \leq \theta_{12i}\Omega_{1i}(t),$$

where

$$\theta_{11i} = \min\left\{\lambda_{\min}(P_1), \frac{a_i}{2}, \frac{b_i e^{-\delta_{i1}L}}{2}\right\} > 0,$$

$$\theta_{12i} = \max\left\{\lambda_{\max}(P_1), \frac{a_i e^{\delta_{i2}L}}{2}, \frac{b_i}{2}\right\} > 0.$$

Taking the derivative of $V_i(t)$ along (3.61)–(3.65), applying Young's inequality, and choosing the parameters $b_i, \delta_{i1}, \delta_{i2}$ to satisfy

$$0 < b_i < \frac{\lambda_{\min}(Q_1)}{2\sqrt{q}\,|\bar{D}_i|^2}, \quad \delta_{i2} > \frac{8\xi_i b_i}{a_i\sqrt{q}} + \frac{2|c|}{\sqrt{q}}, \tag{3.121}$$

$$\delta_{i1} > \frac{1}{\sqrt{q}}\max\left\{\sqrt{q}, 12\xi_i + \frac{4\xi_i^2 b_i}{\lambda_{\min}(Q_1)} + 2|c|\right\}, \tag{3.122}$$

where ξ_i are positive constants only depending on plant parameters and the design parameters κ_i, together with small enough positive constants r_{4i}, r_{3i} obtained from applying Young's inequality, we get

$$\dot{V}_i(t) \leq -\sigma_i V_i(t) - \left(\frac{\sqrt{q}}{2}a_i - \frac{3\sqrt{q}}{2}b_i - \frac{4|P_1 B_i|}{q\lambda_{\min}(Q_1)}\right)\beta_i(0, t)^2$$

$$- \left(\frac{\sqrt{q}}{2}b_i e^{-\delta_{i1}l(t)} - \dot{l}(t)\frac{b_i}{2}e^{-\delta_{i1}l(t)}\right)\alpha_i(l(t), t)^2$$

$$+ \frac{4|P_1 B_j|}{q\lambda_{\min}(Q_1)}\beta_j(0, t)^2 + \left(\frac{a_i^2 e^{\delta_{i2}L}L}{r_{4i}}H_i + \frac{b_i^2 L}{2r_{3i}}G_i\right)\tilde{z}_i(l(t), t)^2$$

$$+ \left(\frac{3\sqrt{q}b_i}{2}|D_i|^2 + \frac{a_i^2 e^{\delta_{i2}L}L}{r_{4i}}Y_i\right)\left|\tilde{X}(t)\right|^2, \tag{3.123}$$

for some positive σ_i that can be adjusted by the control parameters κ_i, where a_i are chosen later. The constants H_i, Y_i, G_i are defined as

$$H_i = \max\left\{|\mathcal{N}_i(x, t)|\right\}, Y_i = \max\left\{|\mathcal{N}_{1i}(x)|\right\}, \tag{3.124}$$

$$G_i = \max\left\{|\bar{\Gamma}_i(x, t)|\right\} \tag{3.125}$$

for $x \in [0, L], t \in [0, \infty)$.

Step 2. Recalling the exponential stability result in the sense of $\|\tilde{z}_i(\cdot, t)\|$, $\|\tilde{w}_i(\cdot, t)\|$, $|\tilde{X}(t)|$ and $\|\tilde{z}_{ix}(\cdot, t)\|$, $\|\tilde{w}_{ix}(\cdot, t)\|$ proved in lemma 3.2 and lemma 3.3, by virtue of $\tilde{w}_i(l(t), t) = 0$ in (3.38) and using the Cauchy-Schwarz inequality, similarly to (3.94), we can arrive at the fact that $\tilde{w}_i(0, t)$ is exponentially convergent to zero. Recalling $\tilde{w}_i(0, t) + \tilde{z}_i(0, t) = 0$ in (3.38) and using the Cauchy-Schwarz inequality again, similarly to (3.95), we get that $\tilde{z}_i(l(t), t)$ is exponentially convergent. Then we get that signals $\tilde{X}(t)$ and $\tilde{z}_i(l(t), t)$ are exponentially convergent to zero in the sense of

$$\max\left\{\left|\tilde{z}_i(l(t),t)\right|,\left|\tilde{X}(t)\right|\right\}\leq\lambda_0 e^{-\eta t}:=\check{\eta}_m(t) \tag{3.126}$$

for $i=1,2$, where the decay rate $\eta>0$, which can be adjusted by the control parameters $\bar{L}$, and λ_0 is a positive constant which depends on initial conditions only.

Now we consider a Lyapunov function candidate

$$V(t)=\sum_{i=1}^{2}V_i(t)+R\check{\eta}_m(t)^2, \tag{3.127}$$

where $R>0$, to be determined later. Defining

$$\Omega_a(t)=\|\beta_1(\cdot,t)\|^2+\|\alpha_1(\cdot,t)\|^2+\|\beta_2(\cdot,t)\|^2+\|\alpha_2(\cdot,t)\|^2+|\hat{X}(t)|^2+\check{\eta}_m(t)^2,$$

we obtain

$$\theta_{a1}\Omega_a(t)\leq V(t)\leq\theta_{a2}\Omega_a(t)$$

with two positive constants θ_{a1},θ_{a2}.

Taking the derivative of (3.127) and recalling (3.123), we get

$$\begin{aligned}
\dot{V}(t)\leq-\sum_{i=1}^{2}&\left[\left(\sqrt{q}-\dot{l}(t)\right)\frac{b_i}{2}e^{-\delta_{i1}l(t)}\alpha_i(l(t),t)^2\right.\\
&+\left(\frac{\sqrt{q}}{2}a_i-\frac{3\sqrt{q}}{2}b_i-\frac{8\left|P_1B_i\right|}{q\lambda_{\min}(Q_1)}\right)\beta_i(0,t)^2\\
&+\left(\frac{R}{2}\eta-\frac{a_i^2 e^{\delta_{i2}L}L}{r_{4i}}H_i-\frac{b_i^2 L}{2r_{3i}}G_i\right)\check{\eta}_m(t)^2+\sigma_i V_i(t)\\
&+\left.\left(\frac{R}{2}\eta-\frac{3\sqrt{q}b_i}{2}\left|D_i\right|^2-\frac{a_i^2 e^{\delta_{i2}L}L}{r_{4i}}Y_i\right)\check{\eta}_m(t)^2\right], \tag{3.128}
\end{aligned}$$

where (3.126) is used to replace $\tilde{z}_i(l(t),t)^2$ and $\tilde{X}(t)^2$ in the last two lines of (3.123) as $\check{\eta}_m(t)^2$, respectively. We choose

$$a_i>3b_i+\frac{16\left|P_1B_i\right|}{q\sqrt{q}\lambda_{\min}(Q_1)} \tag{3.129}$$

and choose large enough R to make sure the coefficients before $\check{\eta}_m(t)^2$ are positive. Recalling assumption 3.2—that is, $\dot{l}(t)<\sqrt{q}$—the coefficients of $\alpha_i(l(t),t)^2$ in (3.128) are positive in all ascending ($\dot{l}(t)<0$), descending ($\dot{l}(t)>0$), and stop ($\dot{l}(t)=0$) cases. We thus arrive at

$$\dot{V}(t)\leq-\sigma V(t)-\sum_{i=1}^{2}\bar{s}_i\beta_i(0,t)^2, \tag{3.130}$$

where

$$\begin{aligned}
\sigma=\min\left\{\sigma_1,\sigma_2,\frac{1}{R}\sum_{i=1}^{2}\left(R\eta-\frac{3\sqrt{q}b_i}{2}\left|D_i\right|^2-\frac{a_i^2 e^{\delta_{i2}L}L}{r_{4i}}Y_i\right.\right.\\
\left.\left.-\frac{a_i^2 e^{\delta_{i2}L}L}{r_{4i}}H_i-\frac{b_i^2 L}{2r_{3i}}G_i\right)\right\}>0, \tag{3.131}
\end{aligned}$$

which can be adjusted by the control parameters κ_i and $\bar{L}$ and where

$$\bar{s}_i = \frac{\sqrt{q}}{2}a_i - \frac{3\sqrt{q}}{2}b_i - \frac{8\,|P_1 B_i|}{q\lambda_{\min}(Q_1)} > 0.$$

Therefore, we obtain the exponential stability result in the sense of $\|\alpha_i(\cdot,t)\|^2 + \|\beta_i(\cdot,t)\|^2 + |\hat{X}(t)|^2$. Due to the invertibility of the transformations (3.59), (3.60), the exponential stability with a decay rate at least σ of the $(\hat{z}_i, \hat{w}_i, \hat{X})$-system in the sense of the norm (3.89) in lemma 3.5 is proved.

F. Proof of lemma 3.6

Differentiating (3.62) and (3.63) with respect to x, and differentiating (3.64) and (3.65) with respect to t, we get

$$\alpha_{ixt}(x,t) = -\sqrt{q}\,\alpha_{ixx}(x,t) - \frac{c}{2}\beta_{ix}(x,t) - \frac{c}{2}\alpha_{ix}(x,t) + \eta_{i1}(x,t), \qquad (3.132)$$

$$\beta_{ixt}(x,t) = \sqrt{q}\,\beta_{ixx}(x,t) - \frac{c}{2}\beta_{ix}(x,t) + \eta_{i2}(x,t), \qquad (3.133)$$

$$\alpha_{ix}(0,t) = \beta_{ix}(0,t) - \frac{c}{2\sqrt{q}}\beta_i(0,t)$$

$$- D_i \sum_{i=1}^{2} \frac{B_i}{q}(\tilde{\beta}_i(0,t) + \beta_i(0,t)) - \frac{1}{\sqrt{q}}\eta_{i3}(t), \qquad (3.134)$$

$$\beta_{ix}(l(t),t) = \frac{\mathcal{N}_i(l(t),t)}{l(t) + \sqrt{q}}\tilde{z}_i(l(t),t) + \frac{\mathcal{N}_{1i}(l(t))}{l(t) + \sqrt{q}}\tilde{X}(t), \qquad (3.135)$$

where

$$\eta_{i1}(x,t) = -\frac{c}{2}\int_0^x \hat{M}_{ix}(x,y)\alpha_i(y,t)dy - \frac{c}{2}\int_0^x \hat{N}_{ix}(x,y)\beta_i(y,t)dy$$

$$- \frac{c}{2}\hat{\vartheta}'(x)\hat{X}(t) + \bar{\Gamma}_i'(x,t)\tilde{z}_i(l(t),t) \quad \frac{c}{2}\hat{M}_i(x,x)\alpha_i(x,t)$$

$$- \frac{c}{2}\hat{N}_i(x,x)\beta_i(x,t), \qquad (3.136)$$

$$\eta_{i2}(x,t) = -\mathcal{N}_{1i}'(x)\tilde{X}(t) - \mathcal{N}_i'(x,t)\tilde{z}_i(l(t),t), \qquad (3.137)$$

$$\eta_{i3}(t) = \left[\bar{D}_i\bar{L}(C_1 + C_2) + D_i\hat{A} + \frac{c}{2}D_i + \mathcal{N}_{1i}(0)\right]\tilde{X}(t) \qquad (3.138)$$

$$+ \left[\bar{D}_i\left(\bar{A} + \sum_{i=1}^{2} B_i\kappa_i\right) + \frac{c}{2}\vartheta(0) + \frac{c}{2}\bar{D}_i\right]\hat{X}(t) \qquad (3.139)$$

$$- (\bar{\Gamma}_i(0,t) + \Gamma_i(0,t))\tilde{z}_i(l(t),t) \qquad (3.140)$$

$$+ D_i \sum_{i=1}^{2} \frac{B_i}{\sqrt{q}}\int_0^{l(t)} \bar{\psi}_i(0,y)\tilde{\alpha}_i(y,t)dy. \qquad (3.141)$$

Applying the Cauchy-Schwarz inequality into (3.136), (3.137) yields the positive constants $N_{1i}, N_{2i}, N_{3i}, N_{4i}, N_{5i}, N_{6i}$ such that

$$\|\eta_{i1}(\cdot,t)\|^2 \leq N_{1i}\|\alpha_i(\cdot,t)\|^2 + N_{2i}\left|\hat{X}(t)\right|^2$$

$$+ N_{3i}\|\beta_i(\cdot,t)\|^2 + N_{4i}\tilde{z}_i(l(t),t)^2, \qquad (3.142)$$

$$\|\eta_{i2}(\cdot,t)\|^2 \leq N_{5i}\tilde{z}_i(l(t),t)^2 + N_{6i}\left|\tilde{X}(t)\right|^2. \tag{3.143}$$

According to (3.126) and the exponential stability results proved in lemma 3.2 and lemma 3.5, we see that the signals $\eta_{i3}(t)$ are exponentially convergent to zero in the sense of

$$|\eta_{i3}(t)| \leq \lambda_{03i}e^{-\eta_{0i}t} := \eta_{ima}(t), \tag{3.144}$$

where $\lambda_{03i} > 0$ only depends on the initial values, and η_{0i} are positive constants.

Define a Lyapunov function

$$V_H(t) = \sum_{i=1}^{2}\left[\frac{\hat{a}_i}{2}\int_0^{l(t)} e^{\hat{\delta}_{2i}x}\beta_{ix}(x,t)^2 dx + \frac{\hat{b}_i}{2}\int_0^{l(t)} e^{-\hat{\delta}_{1i}x}\alpha_{ix}(x,t)^2 dx\right.$$
$$\left. + \frac{1}{2}R_3\eta_{ima}(t)^2\right] + R_2\breve{\eta}_m(t)^2 + R_1 V(t) + R_0 V_e(t), \tag{3.145}$$

where the positive constants $\hat{a}_i, \hat{b}_i, \hat{\delta}_{1i}, \hat{\delta}_{2i}, R_0, R_1, R_2, R_3$ are determined later, and $\eta_m(t)$ is defined by (3.126). Taking the derivative of (3.145) along (3.132)–(3.135); using the Young and Cauchy-Schwarz inequalities; substituting the results of $\dot{V}(t)$ (3.130), $\dot{V}_e(t)$ (3.107) and (3.142), (3.143); and recalling assumption 3.2, through a lengthy calculation we obtain

$$\dot{V}_H(t) \leq -\frac{R_1}{2}\sigma V(t) + \sum_{i=1}^{2}\left[-\left(\frac{\sqrt{q}\hat{a}_i}{2} - 2\sqrt{q}\hat{b}_i\right)\beta_{ix}(0,t)^2\right.$$

$$-\left(\left(\frac{1}{2}\sqrt{q}\hat{\delta}_{2i} - \frac{|c|}{2} - \frac{1}{r_{6i}}\right)\hat{a}_i - \frac{r_{7i}\hat{b}_i|c|}{4}\right)\int_0^{l(t)} e^{\hat{\delta}_{2i}x}\beta_{ix}(x,t)^2 dx$$

$$-\left(\frac{1}{2}\sqrt{q}\hat{\delta}_{1i} - \frac{|c|}{2} - \frac{|c|}{r_{7i}} - \frac{1}{r_{8i}}\right)\hat{b}_i\int_0^{l(t)} e^{-\hat{\delta}_{1i}x}\alpha_{ix}(x,t)^2 dx$$

$$-\left(\frac{R_1\sigma\theta_{a1}}{4} - \frac{r_{8i}\hat{b}_i N_{1i}}{4}\right)\|\alpha_i(\cdot,t)\|^2 - \left(\frac{R_1\sigma\theta_{a1}}{4} - \frac{r_{8i}\hat{b}_i N_{3i}}{4}\right)\|\beta_i(\cdot,t)\|^2$$

$$-\left(\frac{R_1\sigma\theta_{a1}}{4} - \frac{r_{8i}\hat{b}_i N_{2i}}{4}\right)\left|\hat{X}(t)\right|^2 - \left(R_3\eta_{0i} - \frac{2\hat{b}_i}{\sqrt{q}}\right)\eta_{ima}(t)^2$$

$$-\left(R_2\eta - \frac{a_i r_{6i}e^{\hat{\delta}_{2i}L}N_{5i}}{4} - \frac{r_{8i}\hat{b}_i N_{4i}}{4} - \frac{(\sqrt{q}+\bar{v}_{\max})\hat{a}_i e^{\hat{\delta}_{2i}L}H_i^2}{(\sqrt{q}-\bar{v}_{\max})^2}\right)\breve{\eta}_m(t)^2$$

$$-\left(\frac{R_1}{4}\bar{s}_i - \frac{8\max\{r_{9i}\hat{b}_i\bar{D}_i^2\}B_i^2}{q\sqrt{q}} - \frac{c^2\hat{b}_i}{2\sqrt{q}}\right)\beta_i(0,t)^2$$

$$-\left(\frac{R_1\sigma\theta_{a1}}{4} - \frac{a_i r_{6i}e^{\hat{\delta}_{2i}L}N_{6i}}{4} - \frac{(\sqrt{q}+\bar{v}_{\max})\hat{a}_i e^{\hat{\delta}_{2i}L}Y_i^2}{(\sqrt{q}-\bar{v}_{\max})^2}\right)\left|\tilde{X}(t)\right|^2$$

$$-\left(R_0\frac{\sqrt{q}}{4}\bar{a}_i - \frac{8\max\{r_{9i}\hat{b}_i D_i^2\}B_i^2}{q\sqrt{q}}\right)\tilde{\beta}_i(0,t)^2\right] - R_0\sigma_c V_c(t) \tag{3.146}$$

for $i = 1, 2$, and the expressions H_i, Y_i are shown in (3.124). We have used (3.126) and (3.144) to replace $\tilde{z}_i(l(t),t)^2$ and $\eta_{i3}(t)^2$ with positive signs by $\breve{\eta}_m(t)^2$ and

$\eta_{ima}(t)^2$, respectively. From using Young's inequality, r_{6i}, r_{7i}, r_{8i} are arbitrary positive constants. Choose the positive constants $\hat{a}_i, \hat{\delta}_{1i}, \hat{\delta}_{2i}$ satisfying

$$\hat{a}_i > 4\hat{b}_i, \quad \hat{\delta}_{1i} > \frac{2}{\sqrt{q}}\left(\frac{|c|}{2} + \frac{|c|}{r_{7i}} + \frac{1}{r_{8i}}\right), \tag{3.147}$$

$$\hat{\delta}_{2i} > \frac{r_{7i}\hat{b}_i|c|}{2\hat{a}_i\sqrt{q}} + \frac{|c|}{\sqrt{q}} + \frac{2}{\sqrt{q}r_{6i}}, \tag{3.148}$$

with large enough R_0, R_1, R_2, R_3 and arbitrary $\hat{b}_i$, and then we arrive at

$$\dot{V}_H(t) \leq -\sigma_H V_H(t), \tag{3.149}$$

where $\sigma_H > 0$ can be adjusted by the design parameters κ_i and $\bar{L}$.

We thus obtain the exponential stability estimate in the sense of $\|\alpha_{ix}(\cdot, t)\|^2 + \|\beta_{ix}(\cdot, t)\|^2$. Differentiating (3.84), (3.85) with respect to x, using the Young and Cauchy-Schwarz inequalities, we get the exponential stability estimate in the sense of $\|\hat{z}_{ix}(\cdot, t)\|^2 + \|\hat{w}_{ix}(\cdot, t)\|^2$. Together with the exponential convergence of $|\hat{X}(t)|$ proved in lemma 3.5, we can obtain the exponential stability estimate in the sense of (3.90) with the decay rate σ_H.

3.7   NOTES

The in-domain viscous damping of cables in the plant has been considered here. It results in in-domain couplings between transport PDEs after applying Riemann transformations into the original wave PDEs, and the control design comes down to the boundary stabilization of heterodirectional coupled hyperbolic PDEs on a time-varying domain. The basic results and systematic framework of the boundary control of this kind of system on a fixed domain are provided in [96], [177].

In this chapter, we only considered an ideal mining cable elevator model without uncertainties and disturbances. These are further dealt with in chapters 4 and 5.

Chapter Four

Elevators with Disturbances

In chapters 2 and 3 we conducted control designs based on the nominal models of mining cable elevators—that is, in the absence of disturbances or uncertainties. In an actual operating environment, the cage is usually subject to uncertain airflow disturbances, which would affect the smooth and steady running of the mining cable elevator. For this reason, in this chapter we pursue disturbance rejection control design for the mining cable elevator. When the disturbance is as far from the control input as being at the opposite boundary, and the control action has to be propagated through the wave partial differential equation (PDE) dynamics to achieve disturbance rejection, a new technical challenge arises: how to achieve the rejection of the uncertain external disturbance anti-collocated with the control input in a wave PDE.

We start by presenting the wave PDE-modeled vibration dynamics of a cable elevator with an airflow disturbance at the cage in section 4.1, for which a disturbance estimator and a state observer are designed in section 4.2 and section 4.3, respectively, which allow exponential tracking of the unknown disturbance and unaccessible actual states. In section 4.4, based on the disturbance estimator and state observer, we present a two-step control design, where the first step is to move the anti-collocated disturbance terms to the controlled boundary, and the second step is to cancel the collocated disturbance terms and stabilize the overall system using the backstepping control design introduced in chapter 2. This is followed by proof of the exponential convergence of the state at the uncontrolled boundary and the uniform boundedness of all the states in the closed-loop system. The simulation results in section 4.5 show the achievement of rejection of the disturbance and of suppressing the vibrations at the moving cage via the control force at the head sheave.

4.1   PROBLEM FORMULATION

A mining cable elevator with an axial disturbance at the cage is depicted in figure 4.1. The drum drives the cable through the floating sheave to lift the cage which is subject to the disturbance. Input $U_v(t)$ generated by the hydraulic actuator at the floating sheave is a control force for disturbance rejection and vibration suppression. Input $U_a(t)$ at the drum is a separate control force to regulate motion dynamics. Following the modeling process in chapter 2, with the inclusion of an external disturbance force at the cage, the axial vibration dynamics of the mining cable elevator are obtained as

$$u_{tt}(x,t) = q u_{xx}(x,t), \tag{4.1}$$

$$u_x(0,t) = -\frac{m}{r}u_{tt}(0,t) - \frac{1}{r}d(t), \tag{4.2}$$

$$u_x(l(t),t) = U(t), \tag{4.3}$$

where $x \in [0, l(t)]$ denotes the position coordinate along the cable in a moving coordinate system associated with the elevator axial motion $l(t)$ and whose origin is located at the cage at the initial moment, and $t \in [0, \infty)$ presents time. The physical parameters are shown in table 4.1 and $r = E \times A_a$, $q = E \times A_a/\rho$. The function $u(x,t)$ denotes the axial vibration displacement distributed along the cable in the moving coordinate system. Equation (4.2) describes the cage dynamics. Equation (4.3) comes from $ru_x(l(t),t) = U_v(t)$ with the definition of $U_v(t) = rU(t)$, where $U(t)$ is the control input to be designed in this chapter. Input $U_v(t)$ is obtained by multiplying $U(t)$ designed here by the constant gain r. The disturbance force $d(t)$ caused by airflow at the cage $x = 0$ is anti-collocated with the control input $U(t)$. We only consider the airflow disturbance acting on the cage and neglect that acting at the cable. Since the axial vibration displacement is parallel to the direction of

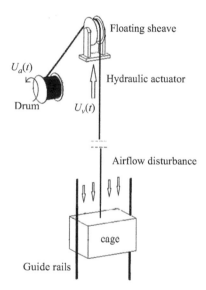

Figure 4.1. The mining cable elevator with airflow disturbances.

Table 4.1. Physical parameters of the mining elevator

Parameters (units)	Values
Time-varying length of the cable (m)	$l(t)$
Maximal length (m)	L
Cable effective steel area (m^2)	A_a
Cable effective Young's modulus (N/m^2)	E
Cable linear density (kg/m)	ρ
Total hoisted mass (kg)	m
Maximum hoisting velocity (m/s)	$\bar{v}$

the airflow disturbance and the cross section of the cable is much smaller than that of the cage, the airflow disturbance essentially affects only the cage.

In practice, the measurements available in this system are $\dot{u}(l(t), t)$, $u_{tt}(0, t)$, and $u(0, t)$. The measurement $\dot{u}(l(t), t)$ is at the controlled boundary, which can be directly obtained from the velocity feedback signals of the actuator acting at the floating sheave. The cage vibration acceleration $u_{tt}(0, t)$ is measured directly by an accelerometer placed at the cage, and then the vibration displacement $u(0, t)$ is obtained by integrating twice the measured acceleration with the known initial conditions $u_t(0, 0)$, $u(0, 0)$. This paragraph is only to explain the signal acquisition for the implementation of the feedback law, imposing no restrictions on any initial conditions in the design and theory.

Assumption 4.1. *Only the ascent of the cable elevator is considered, and the varying length $l(t)$ is decreasing and bounded—that is, $\dot{l}(t) \leq 0$ and $0 < l(t) \leq l(0) = L$, $\forall t \geq 0$.*

Assumption 4.2. *The hoisting velocity $\dot{l}(t)$ is bounded:*

$$-\bar{v} \leq \dot{l}(t) \leq 0,$$

where $\bar{v}$, which is the maximum hoisting velocity, satisfies $\bar{v} < \sqrt{q}$.

Assumption 4.2 ensures well-posedness of the initial boundary value problem (4.1)–(4.3) and holds in the mining cable elevator according to the comments below assumption 3.2 in chapter 3.

Considering that the hoisting is usually at a constant speed except for the starting and stopping time intervals in the practical operation of the elevator, we apply the following assumption.

Assumption 4.3. *The hoisting acceleration $\ddot{l}(t) = 0$.*

Assumption 4.3 is not required in the other chapters but only here, for unmatched disturbance rejection in the wave PDE.

Assumption 4.4. *The disturbance $d(t)$ is of the harmonic form as*

$$d(t) = \sum_{j=1}^{N} [\bar{a}_j \cos(\alpha_j t) + \bar{b}_j \sin(\alpha_j t)], \tag{4.4}$$

where N is an arbitrary positive integer. The amplitudes $\bar{a}_j, \bar{b}_j$ are arbitrary and unknown constants, whereas the frequencies α_j are known. That means, in particular, that the disturbance $d(t)$ is bounded by constant lower and upper bounds as $|d(t)| \leq \overline{D}$, where $\overline{D}$ is an unknown constant.

Because the periodic disturbance may cause resonance and thus result in serious vibrations in the flexible/compliant structure, it needs to be taken into account. The harmonic form (4.4) approximates all periodic signals to arbitrary accuracy. Therefore, assuming the disturbance in the harmonic form is reasonable. The disturbance estimator in section 4.2 is designed for a general disturbance (without knowing the amplitude and frequency information). The assumption that the disturbance frequencies α_j are known is required in the control design in section 4.4, which is also reasonable because the frequencies can be obtained by using the real-time spectrum analyzer in practice, based on the disturbance estimator designed in section 4.2.

The control objective in this chapter is to ensure the exponential convergence of $u(0,t)$ and the uniform boundedness of all states in the closed-loop system with the control input realized by the hydraulic actuator.

4.2 DISTURBANCE ESTIMATOR

The disturbance estimation is a key step in active disturbance rejection control. For the disturbed wave PDE (4.1)–(4.3) without the second-order term in (4.2), [187] poses an estimator for the disturbance, based on which we build a disturbance estimator for (4.1)–(4.3) by using the available measurements $\dot{u}(l(t),t)$, $u(0,t)$, and $u_{tt}(0,t)$.

By virtue of (4.2), the disturbance $d(t)$ can be tracked by estimating the boundary state $u_x(0,t)$ due to $u_{tt}(0,t)$ being measurable. Thus, we would like to build the disturbance estimator by copying the form of the plant and obtaining an error system, whose derivative is exponentially stable (because $d(t)$ is connected with $u_x(0,t)$).

We obtain the error system by building the estimator, including two subsystems $\bar{u}(x,t)$ and $\bar{d}(x,t)$. The $\bar{u}$ subsystem is built as

$$\bar{u}_{tt}(x,t) = q\bar{u}_{xx}(x,t), \tag{4.5}$$

$$\bar{u}_x(0,t) = -\frac{m}{r}u_{tt}(0,t), \tag{4.6}$$

$$\bar{u}_x(l(t),t) = (1 - a_1\dot{l}(t))U(t) + a_1\dot{u}(l(t),t) - a_1\bar{u}_t(l(t),t), \tag{4.7}$$

where the constant a_1 is to be chosen within the interval

$$\bar{v}/q < a_1 < 1/\bar{v}, \tag{4.8}$$

and this choice will be justified in the proof of lemma 4.3, in appendix 4.6.

By substituting the relation

$$\dot{u}(l(t),t) = u_t(l(t),t) + \dot{l}(t)u_x(l(t),t)$$

into (4.7) and recalling (4.3), (4.7) is shown equal to

$$\bar{u}_x(l(t),t) = U(t) + a_1(u_t(l(t),t) - \bar{u}_t(l(t),t)). \tag{4.9}$$

From the original plant (4.1)–(4.3), we subtract the $\bar{u}$-subsystem, obtaining the following intermediate error system $\acute{u}(x,t) = u(x,t) - \bar{u}(x,t)$,

$$\acute{u}_{tt}(x,t) = q\acute{u}_{xx}(x,t), \tag{4.10}$$

$$\acute{u}_x(0,t) = -\frac{1}{r}d(t), \tag{4.11}$$

$$\acute{u}_x(l(t),t) = -a_1\acute{u}_t(l(t),t), \tag{4.12}$$

where the disturbance $d(t)$ shows up in (4.11), and the other boundary is constructed to be of the damping type.

The following lemma shows the uniform boundedness of the $\acute{u}(x,t)$-system, which is crucial to ensure all internal subsystems of the disturbance estimator are uniformly bounded. The proof is shown in appendix 4.6.

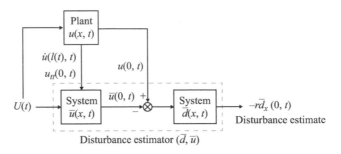

Figure 4.2. Block diagram of the disturbance estimator.

Lemma 4.1. *For all initial values* $(\acute{u}(x,0), \acute{u}_t(x,0)) \in H^1(0, L) \times L^2(0, L)$, *defining the system's spatial norm*

$$(\|\acute{u}_t(\cdot, t)\|^2 + \|\acute{u}_x(\cdot, t)\|^2)^{1/2}, \tag{4.13}$$

then the system (4.10)–(4.12) *is uniformly bounded in the sense of*

$$\sup_{t \geq 0}(\|\acute{u}_t(\cdot, t)\|^2 + \|\acute{u}_x(\cdot, t)\|^2) \leq \frac{V_{\acute{u}}(0)}{\theta_{\acute{u}1}},$$

where $V_{\acute{u}}(0)$ *and* $\theta_{\acute{u}1}$ *are positive constants.*

By virtue of (4.11), we build a $\bar{d}$-subsystem to exponentially recover the intermediate error system $\acute{u}$ and its derivative $\acute{u}_t$. The $\bar{d}$-subsystem is designed based on a copy of the $\acute{u}$-system (4.10)–(4.12), as follows:

$$\bar{d}_{tt}(x, t) = q\bar{d}_{xx}(x, t), \tag{4.14}$$

$$\bar{d}(0, t) = u(0, t) - \bar{u}(0, t), \tag{4.15}$$

$$\bar{d}_x(l(t), t) = -a_1\bar{d}_t(l(t), t), \tag{4.16}$$

where (4.15) ensures $\bar{d}(0, t) = \acute{u}(0, t)$, which is an important condition to obtain that the final error system $u(x, t) - \bar{u}(x, t) - \bar{d}(x, t)$ and its derivative are exponentially stable.

Based on the $\bar{d}$-subsystem (4.14)–(4.16), we define the disturbance estimate $\hat{d}(t)$ as

$$\hat{d}(t) = -r\bar{d}_x(0, t). \tag{4.17}$$

The process of estimating the disturbance is shown in figure 4.2.

The cascaded PDE systems (4.5)–(4.7) and (4.14)–(4.16) are completely known, since they are determined by the input and outputs of the original system (4.1)–(4.3)—that is, $U(t)$, $\dot{u}(l(t), t)$, $u(0, t)$, and $u_{tt}(0, t)$.

The following theorem tells us that the disturbance estimator defined in (4.17) can track the actual disturbance $d(t)$ exponentially.

Theorem 4.1. *The error* $\tilde{d}(t)$ *between the disturbance estimate* $\hat{d}(t)$ *defined by* (4.17) *with* (4.5)–(4.7), (4.14)–(4.16) *and the actual disturbance* $d(t)$ *is exponentially convergent in the sense of the following inequality:*

$$\left|\tilde{d}(t)\right| = \left|d(t) - \hat{d}(t)\right| \leq \mu_{\tilde{d}}e^{-\sigma_{\tilde{d}}t}, \forall t \geq 0,$$

where $\sigma_{\tilde{d}} > 0$ depends on a_1, and $\mu_{\tilde{d}}$ is a positive constant that depends on the initial values only.

Then we present two lemmas that are useful to complete the proof of theorem 4.1. The first lemma tells us that $\bar{d}(x,t)$ is exponentially convergent to $\acute{u}(x,t)$. The proof is shown in appendix 4.6.

Lemma 4.2. *For any initial values $(\tilde{v}(x,0), \tilde{v}_t(x,0))$ that belong to $H^1(0,L) \times L^2(0,L)$, the error system defined by $\tilde{v}(x,t) = \acute{u}(x,t) - \bar{d}(x,t)$ is exponentially stable in the sense of the norm*

$$\left(\|\tilde{v}_t(\cdot,t)\|^2 + \|\tilde{v}_x(\cdot,t)\|^2 \right)^{1/2}, \tag{4.18}$$

where $\|\cdot\|$ denotes the $L^2(0,l(t))$ norm. The decay rate depends on a_1.

Based on lemma 4.2, the subsequent lemma states that $\bar{d}_t(x,t)$ is exponentially convergent to $\acute{u}_t(x,t) = u_t(x,t) - \bar{u}_t(x,t)$, from which $\bar{d}_x(0,t)$ governed by (4.14)–(4.16) can be proved exponentially convergent to $\acute{u}_x(0,t)$ by using the Cauchy-Schwarz inequality. The proof is shown in appendix 4.6.

Lemma 4.3. *For any a_1 that satisfies (4.8), the system defined by $e(x,t) = \tilde{v}_t(x,t)$ with initial values $(e(x,0), e_t(x,0))$ which belong to $H^1(0,L) \times L^2(0,L)$, is exponentially stable in the sense of the norm*

$$\left(\|e_t(\cdot,t)\|^2 + \|e_x(\cdot,t)\|^2 \right)^{1/2}. \tag{4.19}$$

Furthermore, $|\tilde{v}_x(0,t)| \le \mu_{\tilde{v}} c^{-\sigma_{\tilde{d}} t}, \forall t \ge 0$, where $\sigma_{\tilde{d}} > 0$, and $\mu_{\tilde{v}}$ is a positive constant that only depends on the initial values.

With lemma 4.3, we prove theorem 4.1 as follows.

Proof. According to (4.17) and (4.11), the estimation error of the proposed disturbance estimator is obtained as

$$\tilde{d}(t) = d(t) - \hat{d}(t) = -r\acute{u}_x(0,t) + r\bar{d}_x(0,t) = -r\tilde{v}_x(0,t). \tag{4.20}$$

With lemma 4.3, we conclude theorem 4.1. $\qquad\qquad\square$

Theorem 4.1 can be regarded as an independent contribution about exponentially tracking a general disturbance $d(t)$ in time-varying interval wave PDEs.

In addition to the exponential convergence result of $\tilde{d}(t)$ proved in theorem 4.1, we obtain the exponential convergence estimates of $\dot{d}(t), \ddot{d}(t)$ in the following lemma. The proof is shown in appendix 4.6.

Lemma 4.4. *For all initial values $(\tilde{v}_t(x,0), \tilde{v}_{xx}(x,0)) \in H^3(0,L) \times H^2(0,L)$, the derivatives $\dot{\tilde{d}}(t), \ddot{\tilde{d}}(t)$ of the estimation error $\tilde{d}(t)$ are exponentially convergent to zero.*

Furthermore, we can estimate each sinusoidal component in the harmonic disturbance (4.4) by using the disturbance estimate (4.17) in the following steps. Define

$$Z(t)_{2N \times 1} = \left[\bar{a}_1 \cos(\alpha_1 t), \bar{b}_1 \sin(\alpha_1 t), \dots, \bar{a}_N \cos(\alpha_N t), \bar{b}_N \sin(\alpha_N t) \right]^T. \tag{4.21}$$

The disturbance (4.4) can be written as

$$\ddot{Z}(t) = -A_z Z(t), \ d(t) = C_z Z(t), \tag{4.22}$$

where

$$A_z = \text{diag}\left[\begin{pmatrix} \alpha_1^2 & 0 \\ 0 & \alpha_1^2 \end{pmatrix}, \dots, \begin{pmatrix} \alpha_N^2 & 0 \\ 0 & \alpha_N^2 \end{pmatrix}\right], C_z = [1, \dots, 1]_{1 \times 2N}. \tag{4.23}$$

According to the exponentially convergent disturbance estimate $\hat{d}(t) = -r\bar{d}_x(0, t)$ in theorem 4.1, the matrix A_z consisting of the disturbance frequencies α_j can be regarded as known because the frequencies of the disturbance estimation $\hat{d}(t)$ are considered to be equal to those of the actual periodic disturbance $d(t)$ after eliminating the high-frequency noise with an appropriate cutoff frequency, which can be seen in figure 4.5 in section 4.5. In practice, we can use the real-time spectrum analyzer to obtain the frequencies $\alpha_j, j = 1, 2, \dots, N$ by analyzing the signal $-r\bar{d}_x(0, t)$.

Denoting $Y(t) = [\dot{Z}(t)^T, Z(t)^T]^T$, system (4.22) can be written as

$$\dot{Y}(t) = \hat{A}_z Y(t), \ d(t) = \hat{C}_z Y(t), \tag{4.24}$$

with

$$\hat{A}_z = \begin{pmatrix} 0 & -A_z \\ I & 0 \end{pmatrix}_{4N \times 4N}, \ \hat{C}_z = [0, C_z]_{1 \times 4N}, \tag{4.25}$$

where I is an identity matrix with the appropriate dimension, and $(\hat{A}_z, \hat{C}_z)$ is observable.

Using the disturbance estimate $\hat{d}(t)$ defined in (4.17), we can design an observer to estimate the state $Y(t)$ that includes sinusoidal components of the harmonic disturbance. The observer is proposed in the form

$$\dot{\hat{Y}}(t) = \hat{A}_z \hat{Y}(t) + L_z(\hat{d}(t) - \hat{C}_z \hat{Y}(t)), \tag{4.26}$$

where L_z is designed to make $\hat{A}_z - L_z \hat{C}_z$ Hurwitz.

Subtracting (4.26) from (4.24), we obtain the error system $\tilde{Y}(t) = Y(t) - \hat{Y}(t)$ as

$$\dot{\tilde{Y}}(t) = (\hat{A}_z - L_z \hat{C}_z)\tilde{Y}(t) + L_z \tilde{d}(t). \tag{4.27}$$

The following lemma holds.

Lemma 4.5. *The state $\tilde{Y}(t)$ of the system (4.27) is exponentially convergent to zero.*

Proof. According to the exponential convergence of $\tilde{d}(t)$ proved in theorem 4.1, recalling that $\hat{A}_z - L_z \hat{C}_z$ is Hurwitz, it is straightforward to obtain lemma 4.5. $\square$

Defining

$$\hat{Z}(t) = \hat{C}\hat{Y}(t), \tag{4.28}$$

with $\hat{C} = [0, \text{diag}(C_z)]_{2N \times 4N}$, we obtain

$$\tilde{Z}(t) = Z(t) - \hat{Z}(t) = \hat{C}Y(t) - \hat{C}\hat{Y}(t) = \hat{C}\tilde{Y}(t).$$

According to lemma 4.5, we obtain that $|\hat{Z}(t)|$ is exponentially convergent to zero, which yields

$$\left|\tilde{Z}(t)\right| \leq \eta_{\tilde{Z}}(t) := \Upsilon_{\tilde{Z}} e^{-\sigma_{\tilde{Z}} t} \tag{4.29}$$

for some positive $\Upsilon_{\tilde{Z}}$, where $\sigma_{\tilde{Z}} > 0$ depends on the exponential decay rate of the $\tilde{Y}$-system.

4.3   OBSERVER OF CABLE-AND-CAGE STATE

With the disturbance estimator in section 4.2, we can design a state observer to reconstruct the states in the system (4.1)–(4.3)—that is, axial vibration displacements in the cable and cage. Recalling that the control objective is to ensure the exponential convergence of $u(0,t)$, in addition to the challenges from the anti-collocated disturbance, the second-order boundary condition (4.2), which is also anti-collocated with the control input in the system (4.1)–(4.3), poses difficulties to the control problem as well. A new variable $X(t) = [u(0,t), u_t(0,t)]^T$ is introduced to rewrite the system (4.1)–(4.3) as a PDE-ordinary differential equation (ODE) coupled system, as follows:

$$\dot{X}(t) = AX(t) + Bu_x(0,t) + \frac{1}{r}Bd(t), \tag{4.30}$$

$$u_{tt}(x,t) = qu_{xx}(x,t), \tag{4.31}$$

$$u(0,t) = CX(t), \tag{4.32}$$

$$u_x(l(t),t) = U(t), \tag{4.33}$$

with

$$A = \begin{bmatrix} 0 & 1 \\ 0 & 0 \end{bmatrix}, B = \frac{r}{m}\begin{bmatrix} 0 \\ -1 \end{bmatrix}, C = [1,0], \tag{4.34}$$

where the boundary order is reduced and $CB = 0$. By virtue of (4.34), the pair (A, B) is stabilizable, and the pair (A, C) is observable.

In order to facilitate the design of observer-based output-feedback control, which depends on the construction of the observer, a copy of the plant (4.30)–(4.33) is used to build the observer by using the available measurements $u(0,t), \dot{u}(l(t),t)$. Consider the observer to be

$$\dot{\hat{X}}(t) = A\hat{X}(t) + B\hat{u}_x(0,t) + \bar{L}(u(0,t) - C\hat{X}(t)) - B\bar{d}_x(0,t), \tag{4.35}$$

$$\hat{u}_{tt}(x,t) = q\hat{u}_{xx}(x,t), \tag{4.36}$$

$$\hat{u}(0,t) = u(0,t), \tag{4.37}$$

$$\hat{u}_x(l(t),t) = (1 - a_2\dot{l}(t))U(t) + a_2\dot{u}(l(t),t) - a_2\hat{u}_t(l(t),t), \tag{4.38}$$

where $\bar{d}_x(0,t)$ is governed by (4.14)–(4.16). The parameter a_2 is a positive damping gain, and $\bar{L}$ is chosen to make $A - \bar{L}C$ Hurwitz.

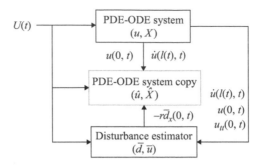

Figure 4.3. Block diagram of the state observer.

Substituting the relation

$$\dot{u}(l(t), t) = u_t(l(t), t) + \dot{l}(t) u_x(l(t), t) \tag{4.39}$$

into (4.38) and recalling (4.3), the equation (4.38) is equal to

$$\hat{u}_x(l(t), t) = U(t) + a_2(u_t(l(t), t) - \hat{u}_t(l(t), t)). \tag{4.40}$$

A block diagram of the state observer is shown in figure 4.3.

Define the observer errors as

$$\tilde{u}(x, t) = u(x, t) - \hat{u}(x, t), \quad \tilde{X}(t) = X(t) - \hat{X}(t). \tag{4.41}$$

Then the observer error system is

$$\dot{\tilde{X}}(t) = (A - \bar{L}C)\tilde{X}(t) + B\tilde{u}_x(0, t) + \frac{1}{r}B\tilde{d}(t), \tag{4.42}$$

$$\tilde{u}_{tt}(x, t) = q\tilde{u}_{xx}(x, t), \tag{4.43}$$

$$\tilde{u}(0, t) = 0, \tag{4.44}$$

$$\tilde{u}_x(l(t), t) = -a_2\tilde{u}_t(l(t), t). \tag{4.45}$$

Define

$$\mathcal{H} = H^2(0, L) \times H^1(0, L).$$

The following lemma holds, indicating that the errors of the observer (4.35)–(4.38) exponentially converge to zero. The proof is shown in appendix 4.6.

Lemma 4.6. *For all initial values* $(\tilde{u}(x, 0), \tilde{u}_t(x, 0)) \in \mathcal{H}$, $\tilde{X}(0) \in \mathbb{R}^2$, *there exists a unique solution* $(\tilde{u}, \tilde{u}_t)^T \in C([0, \infty); \mathcal{H})$, $\tilde{X} \in \mathbb{R}^2$ *to the system* (4.43)–(4.45) *and its states are exponentially convergent to zero in the sense of*

$$\left(\|\tilde{u}_t(\cdot, t)\|^2 + \|\tilde{u}_x(\cdot, t)\|^2 + \left|\tilde{X}(t)\right|^2 \right)^{\frac{1}{2}}. \tag{4.46}$$

According to lemma 4.6, we can state that the observer (4.35)–(4.38) exponentially converges to the original system (4.1)–(4.3). We then straightforwardly obtain the following theorem.

Theorem 4.2. *For all initial values* $(\hat{u}(x, 0), \hat{u}_t(x, 0)) \in \mathcal{H}$, $(u(x, 0), u_t(x, 0)) \in \mathcal{H}$, *the observer* (4.35)–(4.38) *exponentially converges to the original system* (4.1)–(4.3) *in the sense of*

$$\left(\|u_t(\cdot,t) - \hat{u}_t(\cdot,t)\|^2 + \|u_x(\cdot,t) - \hat{u}_x(\cdot,t)\|^2 \right.$$

$$\left. + |u(0,t) - \hat{x}_1(t)|^2 + |u_t(0,t) - \hat{x}_2(t)|^2 \right)^{\frac{1}{2}}, \qquad (4.47)$$

where $[\hat{x}_1(t), \hat{x}_2(t)]^T = \hat{X}(t)$.

4.4 CONTROL DESIGN FOR REJECTION OF DISTURBANCES AT THE CAGE

We obtained the disturbance estimator in section 4.2 and the state observer in section 4.3 with the measurements $u(0,t)$, $u_{tt}(0,t)$, and $\dot{u}(l(t),t)$. In this section, we design an output-feedback control law $U(t)$ using the state and disturbance information recovered from the state observer and the disturbance estimator. The objective is to achieve rejection of the disturbance at the end $x = 0$ (cage) and regulate exponentially $u(0,t)$ (cage axial vibrations) by using the control signal $U(t)$ at the other end, $x = l(t)$ (head sheave). Because of (4.37), the objective can be achieved by guaranteeing the exponential convergence of $\hat{u}(0,t)$ in the state observer via designing a control input at the actuated boundary, $x = l(t)$.

In the process of the controller design, we convert the system (4.35)–(4.38) to an exponentially stable target PDE-ODE system without the disturbance terms. This conversion is achieved in two stages. The anti-collocated disturbance terms are first moved to the actuated boundary, and $\hat{u}(0,t) = \hat{z}(0,t)$ is ensured via the first invertible transformation. Then the collocated disturbance terms are canceled, and the target PDE-ODE coupled system is made exponentially stable via the second backstepping transformation.

Conversion to an Intermediate System

We transform the system (4.35)–(4.38) into the following intermediate system:

$$\dot{\hat{X}}(t) = A\hat{X}(t) + B\hat{z}_x(0,t) + \bar{L}(u(0,t) - C\hat{X}(t))$$

$$- \frac{1}{r}B\tilde{d}(t) + \frac{1}{r}BC_z\tilde{Z}(t), \qquad (4.48)$$

$$\hat{z}_{tt}(x,t) = q\hat{z}_{xx}(x,t) + \eta(x,t), \qquad (4.49)$$

$$\hat{z}(0,t) = \hat{u}(0,t), \qquad (4.50)$$

$$\hat{z}_x(l(t),t) = U(t) + a_2\tilde{u}_t(l(t),t) + \vartheta'(l(t))\hat{Z}(t), \qquad (4.51)$$

where $\eta(x,t)$ and $\vartheta(x)$ will be given later.

Our goal is to make the control and the anti-collocated disturbance appear as if they are collocated, with $\hat{z}(0,t)$ set equal to $\hat{u}(0,t)$ in the intermediate system (4.48)–(4.51). In the later derivation, the collocated disturbance can be easily canceled through control input design. Moreover, we can get the exponential stability of the intermediate system (4.48)–(4.51) via designing a control law through backstepping in the next stage so that $\hat{u}(0,t)$ is exponentially convergent to zero by virtue of (4.50).

The transformation is defined as

$$\hat{z}(x,t) = \hat{u}(x,t) + \vartheta(x)\hat{Z}(t). \tag{4.52}$$

Also, (4.52) can be written as

$$\hat{z}(x,t) = \hat{u}(x,t) + \vartheta(x)Z(t) - \vartheta(x)\tilde{Z}(t). \tag{4.53}$$

Taking the second partial derivative of (4.53) with respect to x and t, respectively, we get

$$\hat{z}_{tt}(x,t) - q\hat{z}_{xx}(x,t) = -\left(\vartheta(x)A_z + q\vartheta''(x)\right)Z(t) + q\vartheta''(x)\tilde{Z}(t) - \vartheta(x)\ddot{\tilde{Z}}(t). \tag{4.54}$$

Set $\eta(x,t)$ in (4.49) as

$$\begin{aligned}
\eta(x,t) &= q\vartheta''(x)\tilde{Z}(t) - \vartheta(x)\ddot{\tilde{Z}}(t) \\
&= \left[q\vartheta''(x)\hat{C} - \vartheta(x)\hat{C}(\hat{A}_z - L_z\hat{C}_z)^2\right]\tilde{Y}(t) \\
&\quad - \vartheta(x)\hat{C}(\hat{A}_z - L_z\hat{C}_z)L_z\tilde{d}(t) - \vartheta(x)\hat{C}L_z\dot{\tilde{d}}(t),
\end{aligned} \tag{4.55}$$

where $\tilde{Z}(t) = \hat{C}\tilde{Y}(t)$ and (4.27) have been used.

Then we obtain that $\vartheta(x)$ satisfies the ODE

$$q\vartheta''(x) + \vartheta(x)A_z = 0, \tag{4.56}$$

$$\vartheta'(0) = \frac{1}{r}C_z, \tag{4.57}$$

$$\vartheta(0) = 0, \tag{4.58}$$

where (4.56) is obtained by comparing (4.54) with (4.49), and (4.57) is obtained by comparing (4.35) with the result of the substitution of (4.52) into (4.48) with (4.17), (4.22). For the boundary condition (4.50) to hold, the condition (4.58) is obtained.

From (4.56)–(4.58), the solution of $\vartheta(x)$ is

$$\vartheta(x) = \frac{C_z}{r}\sqrt{\frac{q}{A_z}}\sin\left(\sqrt{\frac{A_z}{q}}x\right). \tag{4.59}$$

Conversion from the Intermediate System to the Target System

Having completed the conversion from the state observer (4.35)–(4.38) to the intermediate system (4.48)–(4.51) where the collocated disturbance can be easily canceled and $\hat{z}(0,t) = \hat{u}(0,t)$, we next convert the intermediate system (4.48)–(4.51) into a target system via the PDE backstepping approach [114].

The backstepping transformation is formulated as

$$\hat{w}(x,t) - \hat{z}(x,t) + \int_0^x \gamma(x,y)\hat{z}(y,t)dy$$

$$+ \int_0^x h(x,y)\hat{z}_t(y,t)dy + \beta(x)\hat{X}(t), \tag{4.60}$$

where the kernel functions $\gamma(x,y), h(x,y)$, and $\beta(x)$ in (4.60) are to be determined.

The target system is

$$\dot{\hat{X}}(t) = (A + BK)\hat{X}(t) + B\hat{w}_x(0,t) - \frac{1}{r}B\tilde{d}(t)$$

$$+ (\bar{L}C - B\gamma(0,0)C)\tilde{X}(t) + \frac{1}{r}BC_z\tilde{Z}(t), \tag{4.61}$$

$$\hat{w}_{tt}(x,t) = q\hat{w}_{xx}(x,t) - \bar{f}_1(x)\tilde{X}(t) + \bar{\eta}(x,t), \tag{4.62}$$

$$\hat{w}(0,t) = C\tilde{X}(t), \tag{4.63}$$

$$\hat{w}_x(l(t),t) = -a_3\hat{w}_t(l(t),t), \tag{4.64}$$

where a_3 in (4.64) is a positive damping gain, and K in (4.61) is chosen to make $A + BK$ Hurwitz. The function $\bar{f}_1(x)$ in (4.62) is

$$\bar{f}_1(x) = \beta(x)\bar{L}C(A - \bar{L}C) + \beta(x)A\bar{L}C + q\gamma_y(x,0)C$$
$$+ qh_y(x,0)C\bar{L}C + qh_y(x,0)CA,$$

and $\bar{\eta}(x,t)$ in (4.62) is

$$\bar{\eta}(x,t) = \eta(x,t) + \int_0^x \gamma(x,y)\eta(y,t)dy + \int_0^x h(x,y)\eta_t(y,t)dy. \tag{4.65}$$

Applying the Cauchy-Schwarz inequality to (4.65), recalling (4.55), with theorem 4.1, lemma 4.4, and lemma 4.5, we get

$$\max_{0 \le x \le L} |\bar{\eta}(x,t)| \le \max_{0 \le x \le L} \{|C_2(x)|, |C_3(x)|, |C_4(x)|, |C_5(x)|\}$$

$$\times \left(|\tilde{Y}(t)| + |\tilde{d}(t)| + |\dot{\tilde{d}}(t)| + |\ddot{\tilde{d}}(t)| \right)$$

$$\le C_{\max}\Upsilon_{\bar{\eta}}e^{-\sigma_{\bar{\eta}}t}, \tag{4.66}$$

where $C_2(x), C_3(x), C_4(x), C_5(x)$ are some bounded gains, and $C_{\max}$, $\Upsilon_{\bar{\eta}}$, $\sigma_{\bar{\eta}}$ are positive constants. Defining

$$\bar{\eta}_m(t) = \Upsilon_{\bar{\eta}}e^{-\sigma_{\bar{\eta}}t}, \tag{4.67}$$

it follows that

$$\frac{d\bar{\eta}_m(t)^2}{dt} = -2\sigma_\eta\bar{\eta}_m(t)^2. \tag{4.68}$$

The following lemma shows the exponential regulation of the target system (4.61)–(4.64). The proof is shown in appendix 4.6.

Lemma 4.7. *For all initial values $(\hat{w}(x,0), \hat{w}_t(x,0)) \in \mathcal{H}$, $\hat{X}(0) \in \mathbb{R}^2$, there exists a unique solution $(\hat{w}, \hat{w}_t)^T \in C([0,\infty); \mathcal{H})$, $\hat{X} \in \mathbb{R}^2$ to the target system (4.61)–(4.64), and its states are exponentially convergent to zero in the sense of*

$$\left(\|\hat{w}_t(\cdot,t)\|^2 + \|\hat{w}_x(\cdot,t)\|^2 + |\hat{X}(t)|^2 \right)^{\frac{1}{2}}. \tag{4.69}$$

By matching the system (4.48)–(4.51) with the system (4.61)–(4.64) through (4.60), we obtain the conditions for the kernels to be determined in the transformation (4.60) as

$$\gamma_{xx}(x,y) = \gamma_{yy}(x,y), \tag{4.70}$$

$$\frac{d}{dx}\gamma(x,x) = 0, \tag{4.71}$$

$$\gamma(x,0) = \frac{1}{q}\beta(x)AB + h_y(x,0)CB, \tag{4.72}$$

$$h_{xx}(x,y) = h_{yy}(x,y), \tag{4.73}$$

$$h(x,0) = \frac{1}{q}\beta(x)B, \tag{4.74}$$

$$\frac{d}{dx}h(x,x) = 0, \tag{4.75}$$

$$\beta''(x) = \frac{1}{q}\beta(x)A^2 + \gamma_y(x,0)C + h_y(x,0)CA, \tag{4.76}$$

$$\beta'(0) = K - \gamma(0,0)C - h(0,0)CA, \tag{4.77}$$

$$\beta(0) = -C, \tag{4.78}$$

which is a coupled system of an ODE and two PDEs. According to (4.70)–(4.78), the kernel functions $\gamma(x,y)$, $h(x,y)$, and $\beta(x)$ in (4.60) are calculated as

$$\gamma(x,y) = \frac{1}{q}\begin{bmatrix} -C & \Lambda - K \end{bmatrix} e^{D(x-y)} \begin{bmatrix} I \\ 0 \end{bmatrix} AB, \tag{4.79}$$

$$h(x,y) = \frac{1}{q}\begin{bmatrix} -C & \Lambda - K \end{bmatrix} e^{D(x-y)} \begin{bmatrix} I \\ 0 \end{bmatrix} B, \tag{4.80}$$

$$\beta(x) = \begin{bmatrix} -C & \Lambda - K \end{bmatrix} e^{Dx} \begin{bmatrix} I \\ 0 \end{bmatrix}, \tag{4.81}$$

where I denotes the identity matrix of the appropriate dimension, and D, Λ are defined as

$$D = \begin{bmatrix} 0 & \frac{1}{q}A^2 \\ I & -\frac{1}{q}(BCA + ABC) \end{bmatrix}, \quad \Lambda = \frac{1}{q}CABC.$$

Similarly, the inverse transformation of (4.60) can be obtained.

For the boundary condition (4.64) to hold, the controller $U(t)$ can be obtained as

$$
\begin{aligned}
U(t) = \frac{1}{c_1}\Big(& c_2\hat{z}_t(l(t),t) + f_3(l(t))\hat{z}(l(t),t) \\
& + f_4(l(t))\hat{z}_x(0,t) + f_5(l(t))\hat{z}(0,t) + f_6(l(t))\hat{X}(t) \\
& + \int_0^{l(t)} f_7(l(t),x)\hat{z}(x,t)dx + \int_0^{l(t)} f_8(l(t),x)\hat{z}_t(x,t)dx \Big) \\
& - a_2\tilde{u}_t(l(t),t) - \vartheta'(l(t))\hat{Z}(t),
\end{aligned}
\tag{4.82}
$$

where

$$c_1 = 1 - a_3KB, \tag{4.83}$$

$$c_2 = -a_3, \tag{4.84}$$

$$f_3(l(t)) = \gamma(l(t),l(t)) - qh_{xy}(l(t),l(t)), \tag{4.85}$$

$$f_4(l(t)) = a_3 q h_x(l(t), 0) - a_3 \beta(l(t)) B, \tag{4.86}$$

$$f_5(l(t)) = q a_3 h_{xy}(l(t), 0), \tag{4.87}$$

$$f_6(l(t)) = \beta_x(l(t)) + a_3 \beta(l(t)) A, \tag{4.88}$$

$$f_7(l(t), x) = \gamma_x(l(t), x) + q h_{xyy}(l(t), x), \tag{4.89}$$

$$f_8(l(t), x) = h_x(l(t), x) + a_3 \gamma(l(t), x). \tag{4.90}$$

The following lemma shows the exponential regulation of the closed-loop $(\hat{z}, \hat{X})$-system, which will be used in the proof of the exponential convergence of $u(0, t)$ in the original system because the $(\hat{z}, \hat{X})$-system and the original system are connected at $x = 0$.

Lemma 4.8. *For all initial values $(\hat{z}(x, 0), \hat{z}_t(x, 0)) \in \mathcal{H}$, the states of the system consisting of the plant (4.48)–(4.51) and the control law (4.82) are exponentially convergent to zero in the sense of*

$$\left(\|\hat{z}_t(\cdot, t)\|^2 + \|\hat{z}_x(\cdot, t)\|^2 + \left| \hat{X}(t) \right|^2 \right)^{1/2}.$$

Proof. Based on lemma 4.7 and the invertibility and continuity of the transformation (4.60), the proof is straightforward. $\qquad\square$

Stability of the Closed-Loop System

Controller (4.82) can be written by $\hat{u}$ and the available measurements as

$$
\begin{aligned}
U(t) = {} & \frac{1}{(1 + a_2 |\dot{l}(t)|)} \left[\frac{1}{c_1} \left(c_2 \hat{u}_t(l(t), t) + f_3(l(t)) \hat{u}(l(t), t) \right. \right. \\
& + f_4(l(t)) \hat{u}_x(0, t) + f_5(l(t)) \hat{u}(0, t) \\
& + f_6(l(t)) \hat{X}(t) + \int_0^{l(t)} f_7(l(t), x) \hat{u}(x, t) dx \\
& \left. + \int_0^{l(t)} f_8(l(t), x) \hat{u}_t(x, t) dx \right) - a_2 \dot{u}(l(t), t) \\
& + a_2 \hat{u}_t(l(t), t) + r \mathcal{L}(l(t)) \hat{C} L_z \bar{d}_x(0, t) \\
& \left. - [\mathcal{P}(l(t)) + \mathcal{L}(l(t))(\hat{A}_z - L_z \hat{C}_z) \hat{C}] \hat{Z}(t) \right],
\end{aligned}
\tag{4.91}
$$

where

$$
\begin{aligned}
\mathcal{P}(l(t)) = {} & \left[\frac{f_3(l(t))}{c_1} \vartheta(l(t)) + \frac{f_4(l(t))}{c_1} \right. \\
& \left. + \int_0^{l(t)} \frac{f_7(l(t), x)}{c_1} \vartheta(x) dx + \vartheta'(l(t)) \right],
\end{aligned}
\tag{4.92}
$$

$$
\mathcal{L}(l(t)) = \left[\frac{c_2}{c_1} \vartheta(l(t)) + \int_0^{l(t)} \frac{f_8(l(t), x)}{c_1} \vartheta(x) dx \right],
\tag{4.93}
$$

and $\vartheta(x)$ is defined in (4.59). All signals required in the control law (4.91) are obtained from the measurable boundary quantities $\dot{u}(l(t), t)$, $u(0, t)$, and $u_{tt}(0, t)$.

The measurements $\dot{u}(l(t), t)$, $u(0, t)$ are used to construct the observer (4.35)–(4.38) to estimate the distributed states $u(x, t)$ and $X(t)$—that is, to obtain $\hat{u}(x, t)$, $\hat{X}(t)$. The measurements $\dot{u}(l(t), t)$, $u(0, t)$, and, $u_{tt}(0, t)$ are also used to implement the disturbance estimator (4.5)–(4.7) and (4.14)–(4.16) to obtain the disturbance estimate $\hat{d}(t) = -r\bar{d}_x(0, t)$, which is used to get $\hat{Z}(t)$ based on (4.26)–(4.28). In practice, the estimator and the observer can be calculated by using the finite-difference method, where different spatial steps are chosen by considering the trade-off between the model accuracy and the computational speed in different cases, whereas the time step depends on the sample period of the data acquisition.

With the controller (4.91), which uses the information from the disturbance estimator in section 4.2 and the state observer in section 4.3, the complete closed-loop system is

$$u_{tt}(x, t) = qu_{xx}(x, t), \tag{4.94}$$

$$u_x(0, t) = -\frac{m}{r}u_{tt}(0, t) - \frac{1}{r}d(t), \tag{4.95}$$

$$u_x(l(t), t) = U(t), \tag{4.96}$$

$$\dot{\hat{X}}(t) = A\hat{X}(t) + B\hat{u}_x(0, t) + \bar{L}(u(0, t) - C\hat{X}(t)) - B\bar{d}_x(0, t), \tag{4.97}$$

$$\hat{u}_{tt}(x, t) = q\hat{u}_{xx}(x, t), \tag{4.98}$$

$$\hat{u}(0, t) = u(0, t), \tag{4.99}$$

$$\hat{u}_x(l(t), t) = (1 - a_2\dot{l}(t))U(t) + a_2\dot{u}(l(t), t) - a_2\hat{u}_t(l(t), t), \tag{4.100}$$

$$\bar{d}_{tt}(x, t) = q\bar{d}_{xx}(x, t), \tag{4.101}$$

$$\bar{d}(0, t) = u(0, t) - \bar{u}(0, t), \tag{4.102}$$

$$\bar{d}_x(l(t), t) = -a_1\bar{d}_t(l(t), t), \tag{4.103}$$

$$\bar{u}_{tt}(x, t) = q\bar{u}_{xx}(x, t), \tag{4.104}$$

$$\bar{u}_x(0, t) = -\frac{m}{r}u_{tt}(0, t), \tag{4.105}$$

$$\bar{u}_x(l(t), t) = (1 - a_1\dot{l}(t))U(t) + a_1\dot{u}(l(t), t) - a_1\bar{u}_t(l(t), t), \tag{4.106}$$

$$\dot{\hat{Y}}(t) = (\hat{A}_z - L_z\hat{C}_z)\hat{Y}(t) - L_z r\bar{d}_x(0, t), \tag{4.107}$$

$$\hat{Z}(t) = \hat{C}\hat{Y}(t), \tag{4.108}$$

where $U(t)$ is shown in (4.91)–(4.93).

We next present the main theorem of this chapter.

Theorem 4.3. *The closed-loop system consisting of the plant (4.94)–(4.96) with the unmatched disturbance $d(t)$; the disturbance estimator (4.101)–(4.106) and (4.107), (4.108); the state observer (4.97)–(4.100); and the controller (4.91) has the following properties.*

1. *There exist positive constants μ_1 and σ such that the output state $u(0, t)$ of the closed-loop system is exponentially convergent to zero in the sense of*

$$|u(0, t)| \le \mu_1 e^{-\sigma t}, \forall t \ge 0.$$

2. *All states in the closed-loop system are uniformly bounded in the sense of*

$$\sup_{t \geq 0} \left[\int_0^{l(t)} \left(u_t^2(x,t) + u_x^2(x,t) + \hat{u}_t^2(x,t) + \hat{u}_x^2(x,t) + \bar{d}_t^2(x,t) + \bar{d}_x^2(x,t) \right. \right.$$

$$\left. \left. + \bar{u}_t^2(x,t) + \bar{u}_x^2(x,t) \right) dx + \left| \hat{X}(t) \right|^2 + \left| \hat{Y}(t) \right|^2 + \left| \hat{Z}(t) \right|^2 \right] < \infty. \quad (4.109)$$

3. *The control input $U(t)$ in (4.91) is bounded—that is,*

$$\sup_{t \geq 0} |U(t)| < \infty.$$

Proof. 1) We now prove the first of the three portions of the theorem. According to (4.99) and (4.50), we get $u(0,t) = \hat{u}(0,t) = \hat{z}(0,t)$. This, together with lemma 4.8, from which we can infer that $\hat{z}(0,t)$ is exponentially convergent to zero with the decay rate σ, gives property 1.

2) We now prove the second of the three portions of the theorem. According to assumption 4.4 and theorem 4.1, we know that $\hat{d} = -r\bar{d}_x(0,t)$ is bounded. Then $\hat{Z}(t)$ obtained from (4.108) is bounded because $\hat{A}_z - L_z\hat{C}_z$ in (4.107) is Hurwitz, which shows the boundedness of $\hat{Y}(t)$ thanks to the boundedness of $\hat{d} = -r\bar{d}_x(0,t)$. Together with lemma 4.8 and the invertible transformation (4.52), we obtain that $\|\hat{u}_t(x,t)\|^2, \|\hat{u}_x(x,t)\|^2, |\hat{X}(t)|^2$ are uniformly bounded. For the sake of brevity, when we mention boundedness, we refer to the corresponding state norms in (4.109). Then, from (4.41), with lemma 4.6, we conclude that $u(x,t)$ is uniformly bounded. Based on lemma 4.1, which proves the uniform boundedness of the system $\acute{u}(x,t)$, and lemma 4.2, which means the exponential stability of the $\tilde{v}(x,t)$ system, we get $\bar{d}(x,t)$ uniformly bounded thanks to $\tilde{v}(x,t) = \acute{u}(x,t) - \bar{d}(x,t)$. Then we find that $\bar{u}(x,t)$ is also uniformly bounded because of $\tilde{v}(x,t) = u(x,t) - \bar{u}(x,t) - \bar{d}(x,t)$. Therefore, all subsystems in the closed-loop system (4.94)–(4.108) are uniformly bounded as (4.109). Thus, we get property 2.

3) We now prove the third and last of the three portions of the theorem. In the proof of property 2, we have proved the boundedness of all states in the closed-loop system in the sense of (4.109). Now we prove the boundedness of the control input $U(t)$ in (4.91). Due to (4.91) and property 2, we know that the boundedness analysis of the four signals $\hat{u}_t(l(t),t), \hat{u}_x(0,t), u_x(l(t),t), u_t(l(t),t)$ in (4.91) needs to be conducted, for which the L_2 estimates need to be produced for $u_{xx}(x,t), u_{xt}(x,t), \hat{u}_{xx}(x,t), \hat{u}_{xt}(x,t)$. That is, estimates of $\|\tilde{u}_{xx}(\cdot,t)\|, \|\tilde{u}_{xt}(\cdot,t)\|$ and $\|\hat{u}_{xx}(\cdot,t)\|, \|\hat{u}_{xt}(\cdot,t)\|$ need to be found.

Toward that end, we now present two lemmas. The first one shows the bounded estimates in terms of $\|\hat{u}_{xx}(\cdot,t)\|^2 + \|\hat{u}_{xt}(\cdot,t)\|^2$. The proof is shown in appendix 4.6. The second one gives the bounded estimates in terms of $\|\tilde{u}_{xx}(\cdot,t)\|^2 + \|\tilde{u}_{xt}(\cdot,t)\|^2$.

Lemma 4.9. *For all initial values $(\hat{u}(x,0), \hat{u}_t(x,0)) \in \mathcal{H}$, the states of the $\hat{u}(x,t)$-system are bounded in the sense of $\|\hat{u}_{xx}(\cdot,t)\|^2 + \|\hat{u}_{xt}(\cdot,t)\|^2$.*

Through a similar process as in the proof of lemma 4.9, it is straightforward to prove the following lemma.

Lemma 4.10. *For all initial values $(\tilde{u}(x,0), \tilde{u}_t(x,0)) \in \mathcal{H}$, the states of the $\tilde{u}(x,t)$-system are bounded in the sense of $\|\tilde{u}_{xx}(\cdot,t)\|^2 + \|\tilde{u}_{xt}(\cdot,t)\|^2$.*

Recalling the bounded estimate for the norm $\|\hat{u}_{xx}(\cdot,t)\| + \|\hat{u}_{xt}(\cdot,t)\|$ proved in lemma 4.9 and using the Sobolev inequality, we show that $\hat{u}_x(l(t),t), \hat{u}_x(0,t)$, and $\hat{u}_t(l(t),t)$ are bounded.

Similarly, using lemma 4.10, we obtain the boundedness of $\tilde{u}_x(l(t),t), \tilde{u}_t(l(t),t)$.

According to the boundedness of $\hat{u}_x(0,t)$, $\hat{u}_x(l(t),t)$, $\hat{u}_t(l(t),t)$, $\tilde{u}_x(l(t),t)$, $\tilde{u}_t(l(t),t)$, we can obtain the boundedness of the four signals $\hat{u}_t(l(t),t)$, $\hat{u}_x(0,t)$, $u_x(l(t),t)$, $u_t(l(t),t)$ required for proving that $U(t)$ is bounded. Thus, we get property 3.

This completes the proof of theorem 4.3.   □

4.5   SIMULATION FOR A DISTURBED ELEVATOR

The system parameters of the mining cable elevator used in the simulation are shown in table 4.2. Consider the cage subject to an airflow disturbance of the form

$$d(t) = 150\sin(0.3t) + 100\sin(0.4t) + 200\cos(0.2t) + 140\cos(0.25t). \quad (4.110)$$

The simulation is performed based on a priori-known $l(t)$ in figure 4.4, which is considered to be a monotonically decreasing curve from 2000 m to 200 m with a maximum velocity $\bar{v} = 15$ s during the total hoisting time of 150 s.

The control force is (4.91) multiplied by the constant $r = E \times A_a$ mentioned in section 4.1, with the gain functions (4.83)–(4.90) and (4.92), (4.93), where kernels $\gamma(x,y)$, $h(x,y)$, $\beta(x)$, $\vartheta(x)$ are defined in (4.79)–(4.81) and (4.59), respectively. In the controller, the states $\hat{u}(x,t), \hat{X}(t)$ are defined by the state observer (4.97)–(4.100). The function $\bar{d}_x(0,t)$ is defined by the disturbance estimator (4.101)–(4.106), and $\hat{Z}(t)$ is defined by (4.107), (4.108). The constant control parameters required in the controller are shown next. The parameters $K = [k_1, k_2]$ are chosen as $[0.0012, 0.011]$. The parameters $\bar{L} = [l_1, l_2]$ are chosen as $[1.5, 1]$, and $L_z = [1, \ldots, 1]_{1 \times 16}$. Other design parameters are $a_1 = 0.022$, $a_2 = 0.07$, $a_3 = 0.01$. The PDE on the time-varying domain $[0, l(t)]$ is converted to a PDE on the fixed domain $[0, 1]$ with time-varying coefficients by introducing $\check{\xi} = \frac{x}{l(t)}$, and then the

Table 4.2. Simulation parameters of the mining elevator

Parameters	L	r	ρ	q	m
Values	2000	0.48×10^7	8.1	5.9×10^5	15000

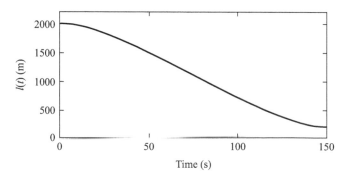

Figure 4.4. The target hoisting trajectory $l(t)$.

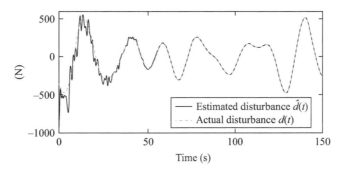

Figure 4.5. The disturbance estimate $\hat{d}(t) = -r\bar{d}_x(0, t)$ and the actual disturbance $d(t)$ (4.110) (*dashed line*).

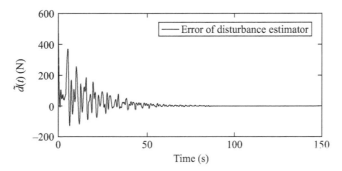

Figure 4.6. The estimation error $\tilde{d}(t) = d(t) - \hat{d}(t)$ between the actual disturbance $d(t)$ (4.110) and the disturbance estimate $\hat{d}(t) = -r\bar{d}_x(0, t)$.

simulation is conducted based on the finite-difference method with a time step of 0.001 and a space step of 0.05.

By using the available boundary measurements $u(0, t)$, $u_{tt}(0, t)$, and $\dot{u}(l(t), t)$, the disturbance estimator (4.101)–(4.106) is implemented with the initial conditions $\bar{d}(x, 0) = 0$ and $\bar{u}(x, 0) = 0$. Figure 4.5 shows that the estimate from the disturbance estimator (4.101)–(4.106) can quickly track the actual unknown disturbance (4.110). In practice, the high-frequency noise at the beginning of the estimation process can be eliminated with a low-pass filter by setting an appropriate cutoff frequency. The error between the estimated and actual values of the disturbance (4.110) is shown in figure 4.6. The estimated variables of the distributed states are obtained by the proposed observer (4.35) (4.38) with available boundary measurements $u(0, t)$ and $\dot{u}(l(t), t)$. The initial errors of the observer are defined as 0.005 m. Because the locations of the actuator and the sensor are at opposite boundaries, the estimation of the state at the midpoint $x = l(t)/2$ is most challenging due to its accessibility. Figure 4.7 shows that the observer error at the midpoint of the cable converges to zero quickly, which implies that the estimates from the state observer (4.35)–(4.38) can reconstruct the distributed states.

The closed-loop responses under the proposed control law (4.91) and the proportional-derivative (PD) control law, which is traditional in industry, are examined, to compare their performance at suppressing the axial vibrations at the cage. Consider the PD control law

$$U_{pd}(t) = k_p u(l(t), t) + k_d \dot{u}(l(t), t), \tag{4.111}$$

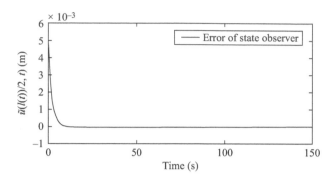

Figure 4.7. The observer error $\tilde{u}(l(t)/2, t)$ at the midpoint of the cable.

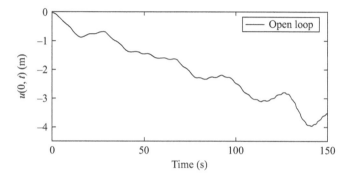

Figure 4.8. The open-loop response $u(0,t)$ of the plant (4.1)–(4.3) under the disturbance (4.110) at $x = 0$.

where k_p, k_d are gain parameters. The values of k_p, k_d are tuned to attain efficient control performance. Here we choose $k_p = 700, k_d = 14000$. From figure 4.8, we observe that the oscillation appearing at the cage is becoming larger and larger because of the disturbance. Figure 4.9 shows that both the proposed output-feedback control law and the PD control law suppress the enlargement of the vibration displacement. Moreover, the proposed control law can regulate the vibration displacement $u(0,t)$ of the cage to zero with faster convergence and less overshoot despite the disturbance at the cage. In addition, according to figure 4.10, we can see that the proposed control law also has better control performance for the internal state, such as the midpoint $u(l(t)/2, t)$ of the cable. This illustrates that the states in the domain's interior are uniformly bounded. Figure 4.11 shows that the output-feedback control input in the closed-loop system is uniformly bounded. The control input is not zero at the final moment $t = 150$ s because the disturbance in figure 4.5 is not zero at that time, so the action of disturbance rejection is continuing in the controller to ensure the convergence of the controlled states.

The model parameter error between the actual plant and the nominal plant often appears in practice. In order to test the robustness of the proposed controller to the model parameter error, we change some plant parameters with respect to their nominal values in table 4.2, such as $r = 0.56 \times 10^7$, $\rho = 8.5$, $q = r/\rho = 6.6 \times 10^5$, and $M = 15500$. These plant parameters are considered as actual plant parameters and those in table 4.2 are nominal plant parameters, and the difference between them is the model error. For the actual plant, a comparison of the proposed controller

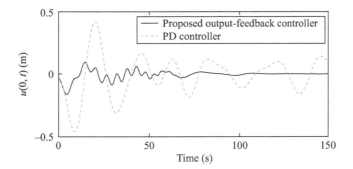

Figure 4.9. The output responses $u(0,t)$ of the closed-loop system (4.94)–(4.108) under the disturbance (4.110) at $x = 0$ with the proposed output-feedback controller (4.91) (*solid line*) and PD controller (4.111) (*dashed line*).

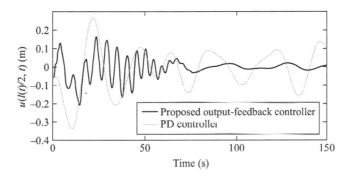

Figure 4.10. The responses $u(l(t)/2,t)$ of the closed-loop system (4.94)–(4.108) under the disturbance (4.110) at $x = 0$ with the proposed output-feedback controller (4.91) (*solid line*) and the PD controller (4.111) (*dashed line*).

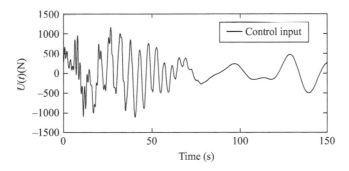

Figure 4.11. The output-feedback controller.

based on the nominal plant parameters (under the model error), the proposed controller based on the actual plant parameters (without the model error), and the PD controller with new control gains $k_{\mathrm{p}} = 630$, $k_{\mathrm{d}} = 15000$, which are tuned to attain efficient control performance, is presented in figure 4.12 and figure 4.13, which show the vibration responses of the cage and the midpoint of the cable, respectively. We observe that, although the vibration amplitudes under the proposed controller with

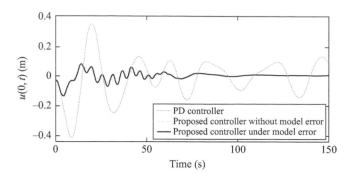

Figure 4.12. The output responses $u(0,t)$ of the closed-loop system (4.94)–(4.108) under the disturbance (4.110) at $x=0$ with the proposed output-feedback controller (4.91) under the model error (*solid line*) and without the model error (*dot-and-dash line*), as well as the PD controller (4.111) (*dashed line*).

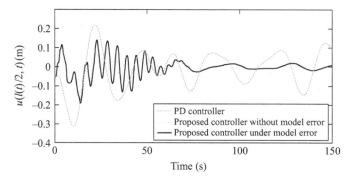

Figure 4.13. The output responses $u(l(t)/2, t)$ of the closed-loop system (4.94)–(4.108) under the disturbance (4.110) at $x=0$ with the proposed output-feedback controller (4.91) under the model error (*solid line*) and without the model error (*dot-and-dash line*), as well as the PD controller (4.111) (*dashed line*).

the model error are slightly larger than the vibration amplitudes under the proposed controller without the model error before 65 s, both exhibit similar good results as time goes on and perform better than the standard PD controller.

4.6   APPENDIX

A. Proof of lemma 4.1

As in [121], we employ a Lyapunov function

$$V_{\acute{u}}(t) = \frac{1}{2}\|\acute{u}_t(\cdot,t)\|^2 + \frac{q}{2}\|\acute{u}_x(\cdot,t)\|^2$$
$$+ \delta_{\acute{u}} \int_0^{l(t)} (1+x)\acute{u}_x(x,t)\acute{u}_t(x,t)dx, \qquad (4.112)$$

where the parameter $\delta_{\acute{u}}$ is to be determined and needs to at least satisfy

$$0 < \delta_{\acute{u}} < 1/(1+L)\min\{1,q\}$$

to guarantee the positive definiteness of $V_{\acute{u}}(t)$. Defining $\Omega_{\acute{u}}(t)$ as the square of the norm (4.13), we get the inequality

$$\theta_{\acute{u}1}\Omega_{\acute{u}}(t) \leq V_{\acute{u}}(t) \leq \theta_{\acute{u}2}\Omega_{\acute{u}}(t), \tag{4.113}$$

where

$$\theta_{\acute{u}1} = \frac{1}{2}[\min\{1,q\} - \delta_{\acute{u}}(1+L)] > 0,$$

$$\theta_{\acute{u}2} = \frac{1}{2}[\max\{1,q\} + \delta_{\acute{u}}(1+L)] > 0.$$

Taking the derivative of $V_{\acute{u}}(t)$ along the system (4.10)–(4.12), applying Young's inequality and $0 < l(t) \leq L$, and choosing

$$0 < \delta_{\acute{u}} < \frac{1}{1+L}\min\left\{1, q, \frac{2a_1 q}{1+qa_1{}^2}, \frac{1+qa_1^2}{2a_1}\right\}, \tag{4.114}$$

we obtain

$$\dot{V}_{\acute{u}} \leq -\lambda_{\acute{u}}V_{\acute{u}} + \overline{M}, \tag{4.115}$$

where

$$\lambda_{\acute{u}} = \delta_{\acute{u}}/(2\theta_{\acute{u}2}),$$

$$\overline{M} = 1/r^2[1/(2r_1) - \delta_{\acute{u}}/2]\overline{D}^2$$

with $0 < r_1 < \delta_{\acute{u}}/q^2$.

Multiplying both sides of (4.115) by $e^{\lambda_{\acute{u}}t}$, we obtain

$$\frac{d(V_{\acute{u}}e^{\lambda_{\acute{u}}t})}{dl} \leq \overline{M}e^{\lambda_{\acute{u}}t}. \tag{4.116}$$

Integration of (4.116) yields

$$\Omega_{\acute{u}}(t) \leq \frac{1}{\theta_{\acute{u}1}}V_{\acute{u}}(t) \leq \frac{1}{\theta_{\acute{u}1}}\left(V_{\acute{u}}(0) - \frac{\overline{M}}{\lambda_{\acute{u}}}\right)e^{-\lambda_{\acute{u}}t} + \frac{\overline{M}}{\theta_{\acute{u}1}\lambda_{\acute{u}}},$$

which implies that $\Omega_{\acute{u}}(t)$ is uniformly bounded by $V_{\acute{u}}(0)/\theta_{\acute{u}1}$. Moreover, it is uniformly ultimately bounded with the ultimate bound $\overline{M}/(\theta_{\acute{u}1}\lambda_{\acute{u}})$. The proof is complete.

B. Proof of lemma 4.2

According to (4.10)–(4.12) and (4.14)–(4.16), the system $\tilde{v}$ is governed by

$$\tilde{v}_{tt}(x,t) = q\tilde{v}_{xx}(x,t), \tag{4.117}$$

$$\tilde{v}(0,t) = 0, \tag{4.118}$$

$$\tilde{v}_x(l(t),t) = -a_1\tilde{v}_t(l(t),t). \tag{4.119}$$

We employ a Lyapunov function

$$V_{\tilde{v}}(t) = \frac{1}{2}\|\tilde{v}_t(\cdot,t)\|^2 + \frac{q}{2}\|\tilde{v}_x(\cdot,t)\|^2$$
$$+ \delta_{\tilde{v}} \int_0^{l(t)} (1+x)\tilde{v}_x(x,t)\tilde{v}_t(x,t)dx, \qquad (4.120)$$

where the parameter $\delta_{\tilde{v}}$ is to be determined and needs to at least satisfy

$$0 < \delta_{\tilde{v}} < 1/(1+L)\min\{1,q\}$$

to guarantee the positive definiteness of $V_{\tilde{v}}(t)$. Defining $\Omega_{\tilde{v}}(t)$ as the square of the norm (4.18), we can then get the inequality

$$\theta_{\tilde{v}1}\Omega_{\tilde{v}}(t) \le V_{\tilde{v}}(t) \le \theta_{\tilde{v}2}\Omega_{\tilde{v}}(t), \qquad (4.121)$$

where

$$\theta_{\tilde{v}1} = \frac{1}{2}[\min\{1,q\} - \delta_{\tilde{v}}(1+L)] > 0,$$

$$\theta_{\tilde{v}2} = \frac{1}{2}[\max\{1,q\} + \delta_{\tilde{v}}(1+L)] > 0.$$

Taking the derivative of $V_{\tilde{v}}(t)$ along the trajectory of the system (4.117)–(4.119), we obtain

$$\dot{V}_{\tilde{v}} \le -\frac{\delta_{\tilde{v}}}{2}\|\tilde{v}_t\|^2 - \frac{\delta_{\tilde{v}}}{2}q\|\tilde{v}_x\|^2 - \frac{\delta_{\tilde{v}}}{2}\tilde{v}_x^2(0,t)$$
$$- \left(a_1 q - \frac{\delta_{\tilde{v}}(1+L)}{2}(1+qa_1{}^2)\right)\tilde{v}_t^2(l(t),t)$$
$$- |\dot{l}(t)|\left(\frac{1}{2} + \frac{a_1{}^2 q}{2} - a_1\delta_{\tilde{v}}(1+L)\right)\tilde{v}_t^2(l(t),t), \qquad (4.122)$$

where $0 < l(t) \le L$ and $\dot{l}(t) \le 0$ in assumption 4.1 have been used.

Choosing $\delta_{\tilde{v}}$ to satisfy

$$0 < \delta_{\tilde{v}} < \frac{1}{1+L}\min\left\{1, q, \frac{2a_1 q}{1+qa_1^2}, \frac{1+qa_1^2}{2a_1}\right\},$$

we obtain

$$\dot{V}_{\tilde{v}} \le -\lambda_{\tilde{v}} V_{\tilde{v}},$$

where

$$\lambda_{\tilde{v}} = \delta_{\tilde{v}}/(2\theta_{\tilde{v}2}) > 0.$$

With (4.121), we get

$$\Omega_{\tilde{v}}(t) \le \frac{\theta_{\tilde{v}2}}{\theta_{\tilde{v}1}}\Omega_{\tilde{v}}(0)e^{-\lambda_{\tilde{v}}t}.$$

The proof is complete.

C. Proof of lemma 4.3

According to the system (4.117)–(4.119), the e-system can be written as

$$e_{tt}(x,t) = qe_{xx}(x,t), \tag{4.123}$$
$$e(0,t) = 0, \tag{4.124}$$
$$e_x(l(t),t) = -b_1 e_t(l(t),t), \tag{4.125}$$

where

$$b_1 = \frac{qa_1 - |\dot{l}(t)|}{q - qa_1|\dot{l}(t)|}. \tag{4.126}$$

With the choice

$$\bar{v}/q < a_1 < 1/\bar{v}, \tag{4.127}$$

we ensure $b_1 > 0$, given that $\bar{v}/q < 1/\bar{v}$ in assumption 4.2.

Consider a Lyapunov function for the system (4.123)–(4.125):

$$V_e(t) = \frac{1}{2}\|e_t(\cdot,t)\|^2 + \frac{q}{2}\|e_x(\cdot,t)\|^2$$
$$+ \delta_e \int_0^{l(t)} (1+x)e_x(x,t)e_t(x,t)dx, \tag{4.128}$$

where the parameter δ_e is to be determined and needs to at least satisfy

$$0 < \delta_e < 1/(1+L)\min\{1,q\}$$

to guarantee the positive definiteness of $V_e(t)$. Defining $\Omega_e(t)$ as the square of the norm (4.19), we get the inequality

$$\theta_{e1}\Omega_e(t) \leq V_e(t) \leq \theta_{e2}\Omega_e(t), \tag{4.129}$$

where

$$\theta_{e1} = \frac{1}{2}[\min\{1,q\} - \delta_e(1+L)] > 0,$$
$$\theta_{e2} = \frac{1}{2}[\max\{1,q\} + \delta_e(1+L)] > 0.$$

Taking the derivative of $V_e(t)$ along the trajectory of the system (4.123)–(4.125), through a similar computation as (4.122), and using (4.129), we get the exponential stability of the system $e(x,t)$:

$$\dot{V}_e \leq -\sigma_{\tilde{d}}V_e, \tag{4.130}$$

where

$$\sigma_{\tilde{d}} = \delta_e/(2\theta_{e2})$$

and δ_e satisfy

$$0 < \delta_e < \frac{1}{1+L}\min\left\{1, q, \frac{2b_1 q}{1+qb_1^2}, \frac{1+qb_1^2}{2b_1}\right\}. \tag{4.131}$$

We can then get

$$\sigma_{\tilde{d}} = \frac{\delta_e}{\max\{1, q\} + \delta_e(1 + L)}. \tag{4.132}$$

From (4.131) and (4.132), it can be seen that the decay rate $\sigma_{\tilde{d}}$ depends on b_1. By combining with (4.126), we observe that $\sigma_{\tilde{d}}$ depends on a_1.

From (4.129) and (4.130), we obtain

$$\Omega_e(t) \le \frac{\theta_{e2}}{\theta_{e1}} \Omega_e(0) e^{-\sigma_{\tilde{d}} t}. \tag{4.133}$$

According to the $\tilde{v}$-system (4.117)–(4.119) and using (4.133), from the Cauchy-Schwarz inequality, we obtain

$$\begin{aligned}
|\tilde{v}_x(0, t)| &\le |\tilde{v}_x(l(t), t)| + \left| \int_0^{l(t)} \tilde{v}_{xx}(x, t) dx \right| \\
&\le |a_1 \tilde{v}_t(l(t), t)| + \frac{\sqrt{L}}{q} \left(\int_0^{l(t)} |\tilde{v}_{tt}(x, t)|^2 dx \right)^{\frac{1}{2}} \\
&= |a_1 e(l(t), t)| + \frac{\sqrt{L}}{q} \|e_t(\cdot, t)\| \le \mu_{\tilde{v}} e^{-\sigma_{\tilde{d}} t},
\end{aligned} \tag{4.134}$$

where the positive constant $\mu_{\tilde{v}}$ depends only on the initial data. The proof is complete.

D. Proof of lemma 4.4

Defining $\bar{y} = e_t$ and taking the derivative of (4.123)–(4.125), recalling assumption 4.3, we obtain

$$\bar{y}_{tt}(x, t) = q \bar{y}_{xx}(x, t), \tag{4.135}$$

$$\bar{y}(0, t) = 0, \tag{4.136}$$

$$\bar{y}_x(l(t), t) = -b_2 \bar{y}_t(l(t), t), \tag{4.137}$$

where

$$b_2 = \frac{q b_1 - |\dot{l}(t)|}{q - q b_1 |\dot{l}(t)|}. \tag{4.138}$$

Substituting (4.126) into (4.138), we get

$$b_2 = \frac{q(q + |\dot{l}(t)|)a_1 - 2q|\dot{l}(t)|}{q^2 + q|\dot{l}(t)|^2 - 2q^2 a_1 |\dot{l}(t)|}.$$

There exists a positive constant a_1 to make $b_2 > 0$. This is because there is a mapping between the dependent variable interval $b_1 \in (0, \infty)$ and the independent variable interval $a_1 \in (\bar{v}/q, 1/\bar{v})$ according to the function (4.126). Therefore, we can choose some a_1 in the range

$$(\bar{v}/q, 1/\bar{v})$$

to make b_1 stay in the range $0 < \bar{v}/q < b_1 < 1/\bar{v} < \infty$, which yields $b_2 > 0$ according to (4.138).

A calculation similar to (4.128)–(4.134), leads to

$$\left|\dot{\tilde{d}}(t)\right| = r|\tilde{v}_{xt}(0,t)| \le \mu_{\dot{\tilde{d}}_t} e^{-\sigma_{\dot{\tilde{d}}_t} t}$$

for some positive constants $\sigma_{\dot{\tilde{d}}_t}$ and $\mu_{\dot{\tilde{d}}_t}$ which depends on the system initial values only.

Similarly, we can prove the exponential convergence to zero of $\left|\ddot{\tilde{d}}(t)\right| = r|\tilde{v}_{xtt}(0,t)|$. The proof is complete.

E. Proof of lemma 4.6

First, we illustrate the well-posedness of the observer error system (4.42)–(4.45).
Define an operator $\mathcal{A}\colon D(\mathcal{A}) \to \mathcal{H}$ by

$$\mathcal{A}(z,v)^T = (v,qz'')^T, \forall (z,v) \in D(\mathcal{A}),$$
$$D(\mathcal{A}) = \{(z,v) \in \mathcal{H} | z(0) = v(0) = 0, z'(l(t)) = -a_2 v(l(t))\}.$$

The system (4.43)–(4.45) can be written as

$$\frac{d}{dt}\left(\begin{array}{c} \tilde{u}(\cdot,t) \\ \tilde{u}_t(\cdot,t) \end{array} \right) = \mathcal{A}\left(\begin{array}{c} \tilde{u}(\cdot,t) \\ \tilde{u}_t(\cdot,t) \end{array} \right).$$

Under the boundedness and regularity assumptions (4.1–4.3) on $l(t)$, according to [121], $\mathcal{A}$ generates an exponential stable C_0 semigroup, which also can be obtained through the Lyapunov analysis in the sequel of this proof. Then there exist $\mathcal{K}, \mu_2 > 0$ such that $\|e^{\mathcal{A}t}\| \le \mathcal{K}e^{-\mu_2 t}$. By [190], one concludes that for all initial values $(\tilde{u}(x,0), \tilde{u}_t(x,0))^T \in \mathcal{H}$, and there exists a unique solution $(\tilde{u}, \tilde{u}_t)^T \in C([0,\infty);\mathcal{H})$ to the system (4.43)–(4.45) as

$$\left(\begin{array}{c} \tilde{u}(\cdot,t) \\ \tilde{u}_t(\cdot,t) \end{array} \right) = e^{\mathcal{A}t}\left(\begin{array}{c} \tilde{u}(\cdot,0) \\ \tilde{u}_t(\cdot,0) \end{array} \right).$$

With the initial values $X(0) \in \mathbb{R}^2$, it is straightforward to demonstrate that there exists a unique solution $X \in \mathbb{R}^2$ to the ODE (4.42), which is cascaded with the $\tilde{u}$-PDE proved to be well-posed above. The signal

$$\tilde{d}(t) = d(t) - \hat{d}(t) = d(t) - (-r\bar{d}_x(0,t)) \tag{4.139}$$

in the ODE (4.42) is well-defined, because $\bar{d}_x(0,t)$ is defined by three well-posed systems where the well-posedness of the $\bar{d}$-system in (4.14)–(4.16) and the $\bar{u}$-system in (4.5)–(4.7) are proved in [121], whereas the well-posedness of the u-system in (4.1)–(4.3) is proved in section 5 of [190]. The well-posedness of (4.42)–(4.45) can then be obtained.

Next, we prove the exponential stability of the observer error system (4.42)–(4.45) via Lyapunov analysis.

Recalling theorem 4.1, define

$$\eta_{\tilde{d}}(t) = \mu_{\tilde{d}}e^{-\sigma_{\tilde{d}}t}. \tag{4.140}$$

We employ a Lyapunov function

$$V_{\tilde{u}}(t) = \tilde{X}^T(t)P_1\tilde{X}(t) + \phi_{\tilde{u}}E_{\tilde{u}}(t) + \eta_1\eta_{\tilde{d}}(t)^2, \tag{4.141}$$

where the positive parameters $\phi_{\tilde{u}}$ and η_1 are to be chosen later. The matrix $P_1 = P_1^T > 0$ is the unique solution to the Lyapunov equation

$$P_1(A - \bar{L}C) + (A - \bar{L}C)^T P_1 = -Q_1$$

for some $Q_1 = Q_1^T > 0$, and $E_{\tilde{u}}(t)$ is defined as

$$E_{\tilde{u}}(t) = \frac{1}{2}\|\tilde{u}_t(\cdot, t)\|^2 + \frac{q}{2}\|\tilde{u}_x(\cdot, t)\|^2$$
$$+ \delta_{\tilde{u}} \int_0^{l(t)} (1+x)\tilde{u}_x(x, t)\tilde{u}_t(x, t)dx, \qquad (4.142)$$

where $\delta_{\tilde{u}}$ should at least satisfy

$$0 < \delta_{\tilde{u}} < 1/(1+L)\min\{1, q\}$$

to guarantee the positive definiteness of $E_{\tilde{u}}(t)$.

Taking the derivative of $V_{\tilde{u}}$ along (4.42)–(4.45), (4.140); applying Young's inequality and $0 < l(t) \leq L$; and choosing the parameters $\delta_{\tilde{u}}$, η_1, and $\phi_{\tilde{u}}$ to satisfy the inequalities

$$0 < \delta_{\tilde{u}} < \frac{1}{1+L}\min\left\{1, q, \frac{2a_2 q}{1+qa_2^2}, \frac{1+qa_2^2}{2a_2}\right\},$$

$$\eta_1 > \frac{8\left|\frac{1}{r}P_1 B\right|^2}{\sigma_{\tilde{d}}\lambda_{\min}(Q_1)},$$

$$\phi_{\tilde{u}} > \frac{4|P_1 B|^2}{q\delta_{\tilde{u}}\lambda_{\min}(Q_1)},$$

we arrive at

$$\dot{V}_{\tilde{u}} \leq -\sigma_{\tilde{u}}V_{\tilde{u}}, \qquad (4.143)$$

where

$$\sigma_{\tilde{u}} = \frac{1}{\theta_{\tilde{u}2}}\min\left\{\frac{\delta_{\tilde{u}}}{2}\phi_{\tilde{u}}, \frac{\delta_{\tilde{u}}}{2}q\phi_{\tilde{u}}, \frac{1}{2}\lambda_{\min}(Q_1), \eta_1\sigma_{\tilde{d}} - \frac{4\left|\frac{1}{r}P_1 B\right|^2}{\lambda_{\min}(Q_1)}\right\} > 0. \qquad (4.144)$$

The proof of lemma 4.6 is complete.

F. Proof of lemma 4.7

First, we establish well-posedness of the target $(\hat{w}, \hat{X})$-system in (4.61)–(4.64).

Define an operator $\mathcal{A}_1: D(\mathcal{A}_1) \to \mathcal{H}$ by

$$\mathcal{A}_1(z, v)^T = (v, qz'')^T, \forall (z, v) \in D(\mathcal{A}_1),$$
$$D(\mathcal{A}_1) = \{(z, v) \in \mathcal{H} | z(0) = v(0) = 0, z'(l(t)) = -a_3 v(l(t))\}.$$

The system (4.62)–(4.64) can be written as

$$\frac{d}{dt}\begin{pmatrix} \hat{w}(\cdot, t) \\ \hat{w}_t(\cdot, t) \end{pmatrix} = \mathcal{A}_1\begin{pmatrix} \hat{w}(\cdot, t) \\ \hat{w}_t(\cdot, t) \end{pmatrix} + \begin{pmatrix} 0 \\ f(\cdot, t) \end{pmatrix} + \mathcal{B}C\tilde{X}(t),$$

where

$$f(x,t) = -\bar{f}_1(x)\tilde{X}(t) + \bar{\eta}(x,t)$$

and

$$\mathcal{B} = (0, \delta(x))^T.$$

Similar to $\mathcal{A}$ in lemma 4.6, $\mathcal{A}_1$ generates an exponentially stable C_0-semigroup, which also can be obtained through the following Lyapunov analysis. Then there exist $\mathcal{K}_1, \mu_3 > 0$ such that $\|e^{\mathcal{A}_1 t}\| \leq \mathcal{K}_1 e^{-\mu_3 t}$.

It is a routine exercise that $\mathcal{B}$ is admissible for $\mathcal{A}_1$. By [190], recalling lemma 4.6 and (4.65), one concludes that for all initial values $(\hat{w}(x,0), \hat{w}_t(x,0))^T \in \mathcal{H}$ there exists a unique solution $(\hat{w}, \hat{w}_t)^T \in C([0, \infty); \mathcal{H})$ to the system (4.62)–(4.64) as

$$\begin{pmatrix} \hat{w}(\cdot, t) \\ \hat{w}_t(\cdot, t) \end{pmatrix} = e^{\mathcal{A}_1 t} \begin{pmatrix} \hat{w}(\cdot, 0) \\ \hat{w}_t(\cdot, 0) \end{pmatrix} + \int_0^t e^{\mathcal{A}_1 (t-s)} \begin{pmatrix} 0 \\ f(\cdot, s) \end{pmatrix} ds$$
$$+ \int_0^t e^{\mathcal{A}_1 (t-s)} \mathcal{B}C\tilde{X}(s) ds.$$

With the initial value $\hat{X}(0) \in \mathbb{R}^2$, it is straightforward to show that there exists a unique solution $X \in \mathbb{R}^2$ to the ODE (4.61) cascaded with the $\hat{w}$-subsystem proved as well-posed above. It should be noted that the signal $\tilde{Z}(t)$ in the ODE (4.61) depends on the ODE $\hat{Y}$ in (4.26), which is obviously a well-posed system. The well-posedness of (4.61)–(4.64) is thus obtained.

Next, we prove the exponential regulation of the target system (4.61)–(4.64) in the sense of (4.69) via Lyapunov analysis.

Let $V_{\hat{w}}(t)$ be a Lyapunov function defined as

$$V_{\hat{w}}(t) = \hat{X}^T(t) P_2 \hat{X}(t) + \phi_{\hat{w}} E_{\hat{w}}(t) + \xi_3 \bar{\eta}_m(t)^2 + \xi_4 \eta_{\tilde{d}}(t)^2 + \xi_5 \eta_{\tilde{Z}}(t)^2, \qquad (4.145)$$

where $\bar{\eta}_m(t)^2$, $\eta_{\tilde{Z}}(t)$, $\eta_{\tilde{d}}(t)$ are defined in (4.67), (4.29), (4.140), and the matrix $P_2 = P_2^T > 0$ is the unique solution to the Lyapunov equation

$$P_2(A + BK) + (A + BK)^T P_2 = -Q_2$$

for some matrix $Q_2 = Q_2^T > 0$. The function $E_{\hat{w}}(t)$ in (4.145) is defined as

$$E_{\hat{w}}(t) = \frac{1}{2}\|\hat{w}_t(\cdot, t)\|^2 + \frac{q}{2}\|\hat{w}_x(\cdot, t)\|^2$$
$$+ \delta_{\hat{w}} \int_0^{l(t)} (1 + x)\hat{w}_x(x,t)\hat{w}_t(x,t)dx, \qquad (4.146)$$

where the parameter $\delta_{\hat{w}}$ is to be determined and needs to at least satisfy

$$0 < \delta_{\hat{w}} < 1/(1 + L) \min\{1, q\}$$

to guarantee the positive definiteness of $E_{\hat{w}}(t)$. The positive parameters $\phi_{\hat{w}}$ and ξ_3, ξ_4, ξ_5 are to be chosen later.

Taking the derivative of $V_{\hat{w}}$ along (4.61)–(4.64), recalling (4.29), (4.140) and (4.66)–(4.68), and applying Young's inequality through the similar computation of (4.141)–(4.144), we arrive at

$$\dot{V}_{\hat{w}} \leq -\sigma_{\hat{w}} V_{\hat{w}} + \xi_2 \left|\tilde{X}(t)\right|^2 \qquad (4.147)$$

for some positive $\sigma_{\hat{w}}$ and

$$\xi_2 = \frac{1}{4}\phi_{\hat{w}}^2 + \frac{1}{2}\phi_{\hat{w}}^2 C^2 (A - \bar{L}C)^2 + \frac{4\left|P_2(\bar{L}C + B\gamma(0,0)C)\right|^2}{\lambda_{\min}(Q_2)}.$$

The parameters $\xi_3, \xi_4, \xi_5, \delta_{\hat{w}}, \phi_{\hat{w}}$ should satisfy

$$\xi_3 > \frac{C_{max}^2 L}{4r_0 \sigma_{\bar{\eta}}},$$

$$\xi_4 > \frac{8|P_2 B|}{\sigma_{\tilde{d}} r^2 \lambda_{\min}(Q_2)},$$

$$\xi_5 > \frac{8|P_2 B C_z|}{\sigma_{\tilde{Z}} r^2 \lambda_{\min}(Q_2)},$$

$$0 < \delta_{\hat{w}} < \frac{1}{1+L} \min\left\{1, q, \frac{2a_3 q}{1 + qa_3^2}, \frac{1 + qa_3^2}{2a_3}\right\},$$

$$\phi_{\hat{w}} > \frac{2}{\delta_{\hat{w}}} \max\left\{2C_1 L, \frac{q}{2} + \frac{4|P_2 B|^2}{q\lambda_{\min}(Q_2)}\right\},$$

$$0 < r_0 < \frac{1}{4}\phi_{\hat{w}}\delta_{\hat{w}} - C_1 L,$$

where r_0 comes from using Young's inequality, and $C_1 = \max\limits_{x \in [0,L]} \left\{\left|\bar{f}_1(x)\right|^2\right\}$.

Consider the observer error system (4.42)–(4.45). The Lyapunov function for the overall $(\tilde{X}, \tilde{u}, \hat{X}, \hat{w})$-state is chosen as

$$V(t) = \lambda V_{\tilde{u}}(t) + V_{\hat{w}}(t), \tag{4.148}$$

where a positive constant λ is to be determined. Taking the derivative of (4.148) and using (4.141), (4.143), (4.147), we get

$$\dot{V} \leq -\frac{\lambda\sigma_{\tilde{u}}}{2}V_{\tilde{u}} - \sigma_{\hat{w}}V_{\hat{w}} - \left(\frac{\lambda\sigma_{\tilde{u}}\lambda_{\min}(P_1)}{2} - \xi_2\right)\left|\tilde{X}(t)\right|^2. \tag{4.149}$$

Choosing λ sufficiently large, we arrive at

$$\dot{V} \leq -\frac{\lambda\sigma_{\tilde{u}}}{2}V_{\tilde{u}} - \sigma_{\hat{w}}V_{\hat{w}} \leq -\sigma V, \tag{4.150}$$

where

$$\sigma = \min\left\{\frac{\sigma_{\tilde{u}}}{2}, \sigma_{\hat{w}}\right\} > 0. \tag{4.151}$$

The proof of lemma 4.7 is complete.

G. Proof of lemma 4.9

Differentiating (4.62) with respect to x, and differentiating (4.63), (4.64) with respect to t, we obtain

$$\hat{w}_{ttx}(x,t) = q\hat{w}_{xxx}(x,t) - \bar{f}_1'(x)\tilde{X}(t) + \bar{\eta}_x(x,t), \tag{4.152}$$

$$\hat{w}_t(0,t) = C(A - \bar{L}C)\tilde{X}(t), \tag{4.153}$$

$$\hat{w}_{tt}(l(t),t) = -b_3 \hat{w}_{xt}(l(t),t) - \frac{\dot{l}(t)\bar{f}_1(l(t))}{a_3q + \dot{l}(t)}\tilde{X}(t)$$

$$+ \frac{\dot{l}(t)}{a_3q + \dot{l}(t)}\bar{\eta}(l(t),t), \tag{4.154}$$

where (4.42) and $CB = 0$ are recalled, and where $b_3 > 0$ by choosing

$$\bar{v}/q < a_3 < 1/\bar{v}.$$

We know that $\bar{\eta}(l(t),t)$ is exponentially convergent to zero according to (4.66). Similarly, we find that $\bar{\eta}_x(x,t)$ is also exponentially convergent to zero by using (4.55), (4.65), (4.66) and theorem 4.1, lemma 4.4, and lemma 4.5. According to lemma 4.6, we know that $\tilde{X}(t)$ is exponentially convergent to zero. Through a similar calculation with the proof of lemma 4.7, we obtain the exponential stability of the system (4.152)–(4.154) in the sense of $\|\hat{w}_{xt}(\cdot,t)\|^2 + \|\hat{w}_{xx}(\cdot,t)\|^2$.

Through the invertible transformations (4.60), we get an exponential stability estimate for the norm $(\|\hat{z}_{xt}(\cdot,t)\|^2 + \|\hat{z}_{xx}(\cdot,t)\|^2)^{1/2}$. Recalling the transformation (4.52), we get

$$\|\hat{u}_{xx}(\cdot,t)\|^2 \le 2\|\hat{z}_{xx}(\cdot,t)\|^2 + 2L|\vartheta_m''\hat{Z}(t)|^2,$$

$$\|\hat{u}_{xt}(\cdot,t)\|^2 \le 2\|\hat{z}_{xt}(\cdot,t)\|^2 + 2L|\vartheta_m'\dot{\hat{Z}}(t)|^2,$$

where

$$\vartheta_m'' = \max_{x \in [0,L]}\{|\vartheta''(x)|\},$$

$$\vartheta_m' = \max_{x \in [0,L]}\{|\vartheta'(x)|\}.$$

The terms $|\vartheta_m''\hat{Z}(t)|$ and

$$\left|\vartheta_m'\dot{\hat{Z}}(t)\right| = |\vartheta_m'|\left|\hat{C}[(\hat{A}_z - L_z\hat{C}_z)\hat{Y}(t) - L_z r\bar{d}_x(0,t)]\right|$$

are bounded. We thus obtain the bounded estimates for the norm

$$(\|\hat{u}_{xt}(\cdot,t)\|^2 + \|\hat{u}_{xx}(\cdot,t)\|^2)^{1/2}.$$

The proof of lemma 4.9 is complete.

4.7 NOTES

The first results on dealing with anti-collocated disturbances in wave PDEs is given in [85] using an adaptive cancellation scheme, where the PDE is on a fixed domain and the asymptotic convergence of the output state is achieved. In this chapter, we proposed a disturbance rejection scheme for a wave PDE-ODE system on a time-varying domain, where a disturbance estimator that generates estimates exponentially convergent to external uncertain disturbances was designed, and the exponential convergence of the state at the uncontrolled and disturbed boundary was achieved.

Differing from this chapter's topic of disturbance rejection in wave PDEs, chapter 5 will present adaptive cancellation of unmatched disturbances in more complex coupled hyperbolic PDEs with spatially varying coefficients.

Chapter Five

Elevators with Flexible Guides

In chapters 2–4 we developed control designs for suppression of the axial vibrations in mining cable elevators with steel guideways. In this chapter, we address suppression of the lateral vibrations in mining cable elevators moving along flexible guideways. The elastic support of flexible guides is usually approximated as a spring-damper system [174, 200], where the stiffness and damping coefficients are not known exactly. This uncertainty leads to unknown parameters existing in the system matrix of the ordinary differential equation (ODE) (cage dynamics) at the uncontrolled boundary of the partial differential equation (PDE) with a time-varying domain. Moreover, as in chapter 4, the cage is always subject to an airflow disturbance, which increases the unmatched uncertainties in the PDE system and makes the control design more challenging.

The objective in this chapter is to design a control law at the top of a vibrating cable with a time-varying length to regulate the cage at the cable's bottom, where information about the viscoelastic guides (parameters in the system matrix of the ODE) are unknown, and the cage is subject to uncertain external disturbances. The content of this chapter is organized as follows. In section 5.1, the lateral vibration dynamics of the mining cable elevator with flexible guides are shown, and the general mathematical problem is introduced. With the ODE state fully measured, an observer is designed to estimate the PDE states in section 5.2. The design of the output-feedback controller via the backstepping method is proposed in section 5.3. Adaptive update laws for the unknown parameters are given in section 5.4. In section 5.5, the adaptive output-feedback control law is presented, and the stability result of the closed-loop control system is proved, followed by simulation tests in a mining cable elevator in section 5.6.

5.1 DESCRIPTION OF FLEXIBLE GUIDES AND GENERALIZED MODEL

Model of the Mining Cable Elevator with Flexible Guides

For lateral vibrations, an important factor of influence is the interaction between the cage and the flexible guides. The elastic support of flexible guides is approximated as a spring-damper system—that is, as a viscoelastic guide [174, 200] where the stiffness and damping coefficients k_c, c_d are not exactly known (see figure 5.1). The wave PDE-modeled lateral vibration dynamics of the mining cable elevator are given by

$$\rho u_{tt} = T(x)u_{xx}(x,t) + T'(x)u_x(x,t) - \bar{c}u_t(x,t), \qquad (5.1)$$

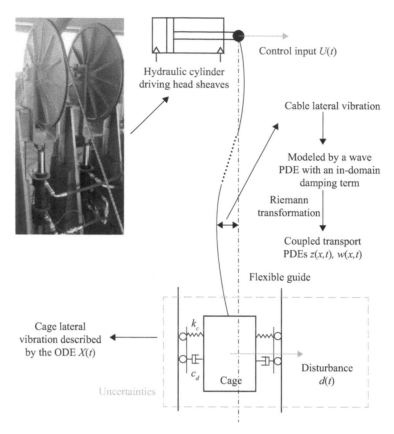

Figure 5.1. Lateral vibration control of a mining cable elevator with viscoelastic guides.

$$M_c u_{tt}(0,t) = -k_c u(0,t) - c_d u_t(0,t) + T(0)u_x(0,t) + d(t), \qquad (5.2)$$

$$-T(l(t))u_x(l(t),t) = U(t), \qquad\qquad\qquad\qquad\qquad\qquad\qquad (5.3)$$

where $u(x,t)$ denotes the lateral vibration displacements along the cable shown in figure 5.1, and $x \in [0, l(t)]$ are the positions along the cable in a moving coordinate system associated with the motion $l(t)$, with the origin located at the cage. The function $T(x) = M_c g + x\rho g$ is the static tension along the cable, and ρ is the linear density of the cable. The coefficient $\bar{c}$ is the material damping of the cable. The signal $d(t)$ is the uncertain airflow disturbance [188] acting at the cage. The constants k_c, c_d are the unknown equivalent stiffness and damping coefficients of the viscoelastic guide. The modeling process of the lateral vibration dynamics of the mining cable elevator (5.1)–(5.3) is based on [30].

By applying the Riemann transformations

$$z(x,t) = u_t(x,t) - \sqrt{\frac{T(x)}{\rho}} u_x(x,t), \qquad\qquad\qquad (5.4)$$

$$w(x,t) = u_t(x,t) + \sqrt{\frac{T(x)}{\rho}} u_x(x,t) \qquad\qquad\qquad (5.5)$$

and defining

$$X(t) = [x_1(t), x_2(t)]^T = [u(0,t), u_t(0,t)]^T,$$

which physically means the lateral displacement and velocity of the cage, we convert (5.1)–(5.3) into a 2×2 hyperbolic system coupled with an ODE, given by

$$\dot{X}(t) = AX(t) + Bw(0,t) + B_1 d(t), \tag{5.6}$$

$$z(0,t) = CX(t) - p_1 w(0,t), \tag{5.7}$$

$$z_t(x,t) = -q_1(x)z_x(x,t) + c_1(x)z(x,t) + c_2(x)w(x,t), \tag{5.8}$$

$$w_t(x,t) = q_2(x)w_x(x,t) + c_3(x)z(x,t) + c_4(x)w(x,t), \tag{5.9}$$

$$w(l(t),t) = U(t) + p_2 z(l(t),t), \tag{5.10}$$

with $x \in [0, l(t)], t \in [0, \infty)$, and the coefficients defined as

$$q_1(x) = q_2(x) = \sqrt{\frac{T(x)}{\rho}}, c_1(x) = c_3(x) = \frac{-\bar{c}}{2\rho} - \frac{T'(x)}{4\sqrt{\rho T(x)}}, \tag{5.11}$$

$$c_2(x) = c_4(x) = \frac{-\bar{c}}{2\rho} + \frac{T'(x)}{4\sqrt{\rho T(x)}}, \quad p_1 = p_2 = 1, \tag{5.12}$$

$$A = \frac{1}{M_c} \begin{bmatrix} 0 & M_c \\ -k_c & -c_d - \sqrt{M_c \rho g} \end{bmatrix}, B = \begin{bmatrix} 0 \\ \sqrt{\frac{\rho g}{M_c}} \end{bmatrix}, \tag{5.13}$$

$$B_1 = \begin{bmatrix} 0 \\ \frac{1}{M_c} \end{bmatrix}, C = [0, 2]. \tag{5.14}$$

It should be noted that the control input designed based on (5.6)–(5.10) with the above coefficients should be multiplied by $-\frac{\sqrt{\rho T(l(t))}}{2}$ in order to convert the input $U(t)$ in (5.10) into a control force in the practical mining cable elevator—that is, into the control input $U(t)$ in the boundary condition (5.3) in the wave PDE model (5.1)–(5.3). In the practical mining cable elevator, $l(t)$ is obtained by the product of the radius and the angular displacement of the rotating drum driving the cable, where the angular displacement is measured by the angular displacement sensor at the drum.

Generalization

In this chapter, we conduct the control design based on (5.6)–(5.10) in a general form with the following conditions. The vector $X(t) \in \mathbb{R}^n$ is an ODE state, whereas $z(x,t), w(x,t)$ are PDE states. The spatially varying transport speeds q_1, q_2 are positive-valued $C^1([0,L])$ functions, and c_1, c_2, c_3, c_4 are $C^0([0,L])$ functions where the positive constant L is the upper bound of $l(t)$, as will be seen in assumption 5.3. The constant p_1 is nonzero, and the constant p_2 is arbitrary. The matrix $C \in \mathbb{R}^{1 \times n}$ is arbitrary. The matrix $A \in \mathbb{R}^{n \times n}$ is the system matrix, and $B \in \mathbb{R}^{n \times 1}$ is the input matrix and $B_1 = Bb_d$, where b_d is an arbitrary constant. The matrices A, B and the signal $d(t)$ are expected to satisfy the following assumptions:

Assumption 5.1. *The matrices A, B are in the form of*

$$A = \begin{pmatrix} 0 & 1 & 0 & 0 & \cdots & 0 \\ 0 & 0 & 1 & 0 & \cdots & 0 \\ & & \vdots & & & \\ 0 & 0 & 0 & 0 & \cdots & 1 \\ g_1 & g_2 & g_3 & \cdots & g_{n-1} & g_n \end{pmatrix}, B = \begin{pmatrix} 0 \\ 0 \\ 0 \\ 0 \\ h_n \end{pmatrix}, \tag{5.15}$$

where the constants $g_1, g_2, g_3, \ldots, g_{n-1}, g_n$ are unknown and arbitrary, and their lower and upper bounds are known and arbitrary. The constant h_n is nonzero and known.

Assumption 5.1 indicates that the ODE is in the controllable form, which covers many practical models, including the cage dynamics modeled in (5.6), (5.13).

Choose a target Hurwitz matrix

$$A_{\mathrm{m}} = \begin{pmatrix} 0 & 1 & 0 & 0 & \cdots & 0 \\ 0 & 0 & 1 & 0 & \cdots & 0 \\ & & \vdots & & & \\ 0 & 0 & 0 & 0 & \cdots & 1 \\ \bar{g}_1 & \bar{g}_2 & \bar{g}_3 & \cdots & \bar{g}_{n-1} & \bar{g}_n \end{pmatrix}, \tag{5.16}$$

where $\bar{g}_1, \bar{g}_2, \bar{g}_3, \ldots, \bar{g}_{n-1}, \bar{g}_n$ are determined by the user according to the desired performance for the specific application, such as the required stiffness coefficient and damping coefficient of the cage in figure 5.1.

According to assumption 5.1 and (5.16), we know that there exists a unique, though unknown, row vector

$$K_{1 \times n} = [k_1, \ldots, k_n] \tag{5.17}$$

such that

$$A_{\mathrm{m}} = A + BK, \tag{5.18}$$

and

$$\bar{g}_i = g_i + h_n k_i, \quad i = 1, 2, \ldots, n. \tag{5.19}$$

By virtue of (5.18), while the k_i's are unknown, the lower and upper bounds on the k_i's—that is, $[\underline{k}_i, \bar{k}_i]$, $i = 1, 2, \ldots, n$, are thus known, given that the lower and upper bounds of the g_i's are known in assumption 5.1, and the $\bar{g}_i$'s are chosen by the user.

Assumption 5.2. *The disturbance $d(t)$ is of the general harmonic form as*

$$d(t) = \sum_{j=1}^{N} [a_j \cos(\theta_j t) + b_j \sin(\theta_j t)], \tag{5.20}$$

where the integer N is arbitrary. The frequencies θ_j, $j \in \{1, 2, \ldots, N\}$ are known and arbitrary constants. The amplitudes a_j, b_j are unknown constants bounded by the known and arbitrary positive constants $\bar{a}_j, \bar{b}_j$, that is, $a_j \in [0, \bar{a}_j], b_j \in [0, \bar{b}_j]$.

Assumption 5.2 can model all periodic disturbance signals to an arbitrarily high degree of accuracy by choosing N sufficiently large.

The time-varying domain $[0, l(t)]$ associated with the moving boundary $l(t)$, which is a known time-varying function, is under the following two assumptions.

Assumption 5.3. *The function $l(t)$ is bounded—that is, $0 < l(t) \leq L$, $\forall t \geq 0$, where L is a positive constant.*

The constant L is the maximal length of the cable in the application of vibration suppression of mining cable elevators.

Assumption 5.4. *The function $\dot{l}(t)$ is bounded as*

$$\left|\dot{l}(t)\right| < \min_{0 \leq x \leq L} \{q_1(x), q_2(x)\}, \forall t \geq 0. \tag{5.21}$$

As in chapters 2–4, the limit of the speed of the moving boundary in assumption 5.4 ensures the well-posedness of the initial boundary value problem (5.6)–(5.10) according to [71, 72]. This assumption holds in the applications of the mining cable elevator, as we shall see in the simulation section.

5.2   OBSERVER FOR DISTRIBUTED STATES OF THE CABLE

Observer Structure

To estimate the PDE states $(z(x,t),\ w(x,t))^T$—that is, the distributed state in the cable—which usually cannot be fully measured in practice but are required in the controller, an observer using the measurements $X(t), z(l(t), t)$ is introduced as

$$\dot{X}(t) = AX(t) + B\hat{w}(0,t) + B\tilde{w}(0,t) + B_1 d(t), \tag{5.22}$$

$$\hat{z}(0,t) = CX(t) - p_1 \hat{w}(0,t), \tag{5.23}$$

$$\hat{z}_t(x,t) = -q_1(x)\hat{z}_x(x,t) + c_1(x)\hat{z}(x,t) + c_2(x)\hat{w}(x,t)$$
$$+ \Phi_2(x,t)(z(l(t),t) - \hat{z}(l(t),t)), \tag{5.24}$$

$$\hat{w}_t(x,t) = q_2(x)\hat{w}_x(x,t) + c_3(x)\hat{z}(x,t) + c_4(x)\hat{w}(x,t)$$
$$+ \Phi_3(x,t)(z(l(t),t) - \hat{z}(l(t),t)), \tag{5.25}$$

$$\hat{w}(l(t),t) = U(t) + p_2 z(l(t),t), \tag{5.26}$$

where (5.22) is exactly the ODE (5.6) with $w(0,t) = \hat{w}(0,t) + \tilde{w}(0,t)$. Equation (5.22) is a part of the plant dynamics instead of representing a part of the observer. The functions $\Phi_2(x,t)$, $\Phi_3(x,t)$ are observer gains to be determined. The ODE state $X(t)$ physically means the vibration displacement and velocity of the cage in the mining cable elevator, which can be obtained by an acceleration sensor placed at the cage plus an integral algorithm [178]. Because the ODE state is available, the task for the observer (5.23)–(5.26) is only to estimate the PDE states. Define the observer error state as

$$(\tilde{z}(x,t), \tilde{w}(x,t)) = (z(x,t), w(x,t)) - (\hat{z}(x,t), \hat{w}(x,t)). \tag{5.27}$$

According to (5.7)–(5.10) and (5.23)–(5.26), the observer error system is

$$\tilde{z}(0,t) = -p_1 \tilde{w}(0,t), \tag{5.28}$$

$$\tilde{z}_t(x,t) = -q_1(x)\tilde{z}_x(x,t) + c_1(x)\tilde{z}(x,t) + c_2(x)\tilde{w}(x,t)$$
$$- \Phi_2(x,t)\tilde{z}(l(t),t), \tag{5.29}$$

$$\tilde{w}_t(x,t) = q_2(x)\tilde{w}_x(x,t) + c_3(x)\tilde{z}(x,t) + c_4(x)\tilde{w}(x,t)$$
$$- \Phi_3(x,t)\tilde{z}(l(t),t), \tag{5.30}$$

$$\tilde{w}(l(t),t) = 0. \tag{5.31}$$

Observer gains $\Phi_2(x,t)$, $\Phi_3(x,t)$ are to be designed to ensure convergence to zero of the observer errors.

Determining Observer Gains via Backstepping

Postulate the invertible backstepping transformation

$$\tilde{z}(x,t) = \tilde{\alpha}(x,t) - \int_x^{l(t)} \bar{\phi}(x,y)\tilde{\alpha}(y,t)dy$$

$$- \int_x^{l(t)} \check{\phi}(x,y)\tilde{\beta}(y,t)dy, \tag{5.32}$$

$$\tilde{w}(x,t) = \tilde{\beta}(x,t) - \int_x^{l(t)} \bar{\psi}(x,y)\tilde{\alpha}(y,t)dy$$

$$- \int_x^{l(t)} \check{\psi}(x,y)\tilde{\beta}(y,t)dy \tag{5.33}$$

to convert the original observer error system (5.28)–(5.31) to the following target observer error system:

$$\tilde{\alpha}(0,t) = -p_1\tilde{\beta}(0,t), \tag{5.34}$$

$$\tilde{\alpha}_t(x,t) = -q_1(x)\tilde{\alpha}_x(x,t) + c_1(x)\tilde{\alpha}(x,t), \tag{5.35}$$

$$\tilde{\beta}_t(x,t) = q_2(x)\tilde{\beta}_x(x,t) + c_4(x)\tilde{\beta}(x,t), \tag{5.36}$$

$$\tilde{\beta}(l(t),t) = 0. \tag{5.37}$$

The form of the backstepping transformation (5.32), (5.33) for coupled hyperbolic PDEs is taken from [96]. The integration interval chosen in the transformation (5.32), (5.33) is $[x, l(t)]$ because the PDE boundary measurement used in the observer is at the boundary $x = l(t)$. Even though the integration interval is time varying, the kernels in (5.32), (5.33) need not include the time argument because the extra terms introduced by the time-varying integration interval during the calculation of the kernel conditions will be "absorbed" by the time-dependent observer gains $\Phi_2(x,t)$, $\Phi_3(x,t)$, which will be seen clearly later. By matching (5.28)–(5.31) and (5.34)–(5.37) using (5.32), (5.33) (the details are shown in appendix 5.7A), the conditions on the kernels $\bar{\phi}(x,y), \bar{\psi}(x,y), \check{\phi}(x,y), \check{\psi}(x,y)$ in (5.32), (5.33) are obtained as the following two well-posed hyperbolic systems:

$$\bar{\psi}(x,x) = \frac{-c_3(x)}{q_1(x) + q_2(x)}, \tag{5.38}$$

$$\bar{\phi}(0,y) = -p_1\bar{\psi}(0,y), \tag{5.39}$$

$$q_2(x)\bar{\psi}_x(x,y) - q_1(y)\bar{\psi}_y(x,y)$$
$$+(c_4(x) - c_1(y) - q_1'(y))\bar{\psi}(x,y) + c_3(x)\bar{\phi}(x,y) = 0, \tag{5.40}$$
$$-q_1(x)\bar{\phi}_x(x,y) - q_1(y)\bar{\phi}_y(x,y)$$
$$+(c_1(x) - c_1(y) - q_1'(y))\bar{\phi}(x,y) + c_2(x)\bar{\psi}(x,y) = 0, \tag{5.41}$$

and

$$\check{\phi}(x,x) = \frac{c_2(x)}{q_1(x) + q_2(x)}, \tag{5.42}$$

$$\check{\psi}(0,y) = -\frac{1}{p_1}\check{\phi}(0,y), \tag{5.43}$$

$$q_2(x)\check{\psi}_x(x,y) + q_2(y)\check{\psi}_y(x,y)$$
$$+(c_4(x) - c_4(y) + q_2'(y))\check{\psi}(x,y) + c_3(x)\check{\phi}(x,y) = 0, \tag{5.44}$$
$$-q_1(x)\check{\phi}_x(x,y) + q_2(y)\check{\phi}_y(x,y)$$
$$+(c_1(x) - c_4(y) + q_2'(y))\check{\phi}(x,y) + c_2(x)\check{\psi}(x,y) = 0, \tag{5.45}$$

on $\mathcal{D} = \{0 \leq x \leq y \leq l(t)\}$. The observer gains are thus determined as

$$\Phi_2(x,t) = \dot{l}(t)\bar{\phi}(x,l(t)) - q_1(l(t))\bar{\phi}(x,l(t)), \tag{5.46}$$

$$\Phi_3(x,t) = \dot{l}(t)\bar{\psi}(x,l(t)) - q_1(l(t))\psi(x,l(t)). \tag{5.47}$$

Remark 5.1. The equations (5.38)–(5.41) and (5.42)–(5.45) are in the same form as the kernel equations (24)–(31) in [177] if we extend the domain $\mathcal{D}$ to a fixed triangular domain. The boundary conditions in (5.38)–(5.41) and (5.42)–(5.45) on the triangular domain $\mathcal{D} = \{0 \leq x \leq y \leq l(t)\}$ are given along the lines $y = x$ and $x = 0$ rather than on $y = l(t)$. Therefore, it is feasible to extend the domain $\mathcal{D}$ to a fixed triangular domain $\mathcal{D}_1 = \{0 \leq x \leq y \leq L\}$ (L is defined in assumption 5.3) and obtain the solutions of (5.38)–(5.41) and (5.42)–(5.45) on $\mathcal{D}$ by solving (5.38)–(5.41) and (5.42)–(5.45) on $\mathcal{D}_1$, whose well-posedness is proved in [177]. This ensures the existence of the observer gains $\Phi_2(x,t)$ and $\Phi_3(x,t)$ in (5.46), (5.47) consisting of $\bar{\phi}(x,l(t))$, $\bar{\psi}(x,l(t))$ obtained by extracting the results along $y = l(t)$ in the solution of (5.38)–(5.41) on $\mathcal{D}_1$.

Stability of the Observer Error System

Lemma 5.1. *For all initial data $(\tilde{z}(\cdot,0), \tilde{w}(\cdot,0)) \in H^1(0,L)$, the states $\tilde{z}(\cdot,t), \tilde{w}(\cdot,t)$ of the observer error system (5.28)–(5.31) with the observer gains (5.46), (5.47) become and remain zero no later than the time $t = t_a$, where*

$$t_a = \frac{L}{\min_{0 \leq x \leq L}\{q_1(x)\}} + \frac{L}{\min_{0 \leq x \leq L}\{q_2(x)\}}.$$

Proof. According to the target observer error system (5.34)–(5.37) and the result in [96], we know that $\tilde{\alpha}(x,t), \tilde{\beta}(x,t)$ reach zero by the time t_a, at the latest. Applying the Cauchy-Schwarz inequality into (5.32), (5.33), the proof of this lemma is complete. $\square$

Lemma 5.1 physically means that the designed observer (5.23)–(5.26), which uses only boundary measurements, can effectively recover the actual distributed states of the vibrating string.

5.3   ADAPTIVE DISTURBANCE CANCELLATION AND STABILIZATION

In this section, we design an observer-based output-feedback controller. We conduct the state-feedback control design based on the observer using the backstepping method, which makes the resulting control law employ only the observer states. Three transformations are used to convert the observer (5.22)–(5.26) to a target system, with the intention of adaptively canceling the unmatched disturbance (the disturbance at the cage), removing the coupling in the PDE domain, and making the system matrix of the ODE Hurwitz (stabilizing the vibrating cable and cage).

The output injection signals $\tilde{z}(l(t), t), \tilde{w}(0, t)$ in the observer are regarded as zero in the control design, following which the separation principle, which is verified by the fact that the stability of the observer error system is independent of the control design according to lemma 5.1, which shows that the observer errors vanish in finite time only depending on the plant parameters, is applied in the stability analysis of the resulting closed-loop system.

The First Transformation for Adaptively Canceling the Unmatched Disturbance

We introduce the transformation $(\hat{w}, \hat{z}) \to (\hat{v}, \hat{s})$:

$$\hat{v}(x, t) = \hat{w}(x, t) + \Gamma(x, t) Z(t), \tag{5.48}$$

$$\hat{s}(x, t) = \hat{z}(x, t) + \Gamma_1(x, t) Z(t), \tag{5.49}$$

where $\Gamma(x, t), \Gamma_1(x, t)$ are to be determined, and

$$Z(t) = [\cos(\theta_1 t), \sin(\theta_1 t), \dots, \cos(\theta_N t), \sin(\theta_N t)]^T. \tag{5.50}$$

We then have

$$\dot{Z}(t) = A_z Z(t), \tag{5.51}$$

where

$$A_z = \mathrm{diag}\left[\begin{pmatrix} 0 & -\theta_1 \\ \theta_1 & 0 \end{pmatrix}, \dots, \begin{pmatrix} 0 & -\theta_N \\ \theta_N & 0 \end{pmatrix} \right]. \tag{5.52}$$

According to assumption 5.2, the disturbance can be written as

$$d(t) = [a_1, b_1, \dots, a_N, b_N] Z(t),$$

and we define the disturbance estimate $\hat{d}(t)$ as

$$\hat{d}(t) = [\hat{a}_1(t), \hat{b}_1(t), \dots, \hat{a}_N(t), \hat{b}_N(t)] Z(t), \tag{5.53}$$

where $\hat{a}_1(t)$, $\hat{b}_1(t), \ldots, \hat{a}_N(t)$, $\hat{b}_N(t)$ are estimates of $a_1, b_1, \ldots, a_N, b_N$, which will be shown in section 5.4.

Through (5.48), (5.49), we convert the system (5.22)–(5.26) into the following system:

$$\dot{X}(t) = AX(t) + B\hat{v}(0,t) + B_1\tilde{d}(t), \tag{5.54}$$

$$\hat{s}(0,t) + p_1\hat{v}(0,t) = CX(t), \tag{5.55}$$

$$\hat{s}_t(x,t) = -q_1(x)\hat{s}_x(x,t) + c_1(x)\hat{s}(x,t) + c_2(x)\hat{v}(x,t)$$
$$+ \Gamma_{1t}(x,t)Z(t), \tag{5.56}$$

$$\hat{v}_t(x,t) = q_2(x)\hat{v}_x(x,t) + c_3(x)\hat{s}(x,t) + c_4(x)\hat{v}(x,t)$$
$$+ \Gamma_t(x,t)Z(t), \tag{5.57}$$

$$\hat{v}(l(t),t) = U(t) + p_2\hat{s}(l(t),t)$$
$$+ (\Gamma(l(t),t) - p_2\Gamma_1(l(t),t))Z(t), \tag{5.58}$$

where $\tilde{d}(t)$ is given as

$$\tilde{d}(t) = d(t) - \hat{d}(t)$$
$$= \sum_{j=1}^{N}[(a_j - \hat{a}_j(t))\cos(\theta_j t) + (b_j - \hat{b}_j(t))\sin(\theta_j t)]$$
$$= \sum_{j=1}^{N}[\tilde{a}_j(t)\cos(\theta_j t) + \tilde{b}_j(t)\sin(\theta_j t)]. \tag{5.59}$$

The functions $\Gamma_1(x,t), \Gamma(x,t)$ in (5.48), (5.49) are determined as follows. Taking the time and spatial derivatives of (5.48), (5.49), substituting the result into (5.56), (5.57), and recalling (5.24), (5.25), and (5.51), we get

$$\hat{s}_t(x,t) + q_1(x)\hat{s}_x(x,t) - c_1(x)\hat{s}(x,t)$$
$$- c_2(x)\hat{v}(x,t) - \Gamma_{1t}(x,t)Z(t)$$
$$= \hat{z}_t(x,t) + q_1(x)\hat{z}_x(x,t) + \Gamma_{1t}(x,t)Z(t) + \Gamma_1(x,t)A_zZ(t)$$
$$- c_2(x)\hat{w}(x,t) - c_1(x)\hat{z}(x,t) + q_1(x)\Gamma_{1x}(x,t)Z(t)$$
$$- c_2(x)\Gamma(x,t)Z(t) - c_1(x)\Gamma_1(x,t)Z(t) - \Gamma_{1t}(x,t)Z(t)$$
$$= (\Gamma_1(x,t)A_z + q_1(x)\Gamma_{1x}(x,t)$$
$$- c_2(x)\Gamma(x,t) - c_1(x)\Gamma_1(x,t))Z(t) = 0, \tag{5.60}$$

and

$$\hat{v}_t(x,t) - q_2(x)\hat{v}_x(x,t) - c_3(x)\hat{s}(x,t)$$
$$- c_4(x)\hat{v}(x,t) - \Gamma_t(x,t)Z(t)$$
$$= \hat{w}_t(x,t) - q_2(x)\hat{w}_x(x,t) + \Gamma_t(x,t)Z(t) + \Gamma(x,t)A_zZ(t)$$
$$- c_4(x)\hat{w}(x,t) - c_3(x)\hat{z}(x,t) - q_2(x)\Gamma_x(x,t)Z(t)$$
$$- c_4(x)\Gamma(x,t)Z(t) - c_3(x)\Gamma_1(x,t)Z(t) - \Gamma_t(x,t)Z(t)$$
$$= (\Gamma(x,t)A_z - q_2(x)\Gamma_x(x,t)$$
$$- c_4(x)\Gamma(x,t) - c_3(x)\Gamma_1(x,t))Z(t) = 0. \tag{5.61}$$

For (5.60), (5.61) to hold, we obtain the conditions

$$-q_2(x)\Gamma_x(x,t) + \Gamma(x,t)(A_z - c_4(x)I_{2N}) - c_3(x)\Gamma_1(x,t) = 0, \qquad (5.62)$$

$$q_1(x)\Gamma_{1x}(x,t) + \Gamma_1(x,t)(A_z - c_1(x)I_{2N}) - c_2(x)\Gamma(x,t) = 0, \qquad (5.63)$$

where I_{2N} is an identity matrix with dimension $2N$.

Defining $\zeta(x,t) = [\Gamma(x,t), \Gamma_1(x,t)]$, we rewrite (5.62), (5.63) as

$$\zeta_x(x,t) = -\zeta(x,t)\bar{A}(x), \qquad (5.64)$$

where

$$\bar{A}(x) = \begin{pmatrix} A_z - c_4(x)I_{2N} & -c_2(x)I_{2N} \\ -c_3(x)I_{2N} & A_z - c_1(x)I_{2N} \end{pmatrix}$$
$$\times \begin{pmatrix} -q_2(x)I_{2N} & 0_{2N} \\ 0_{2N} & q_1(x)I_{2N} \end{pmatrix}^{-1}.$$

By mapping (5.22), (5.23) and (5.54), (5.55) through the transformation (5.48), (5.49), recalling (5.53) and $B_1 = Bb_d$, we obtain the condition

$$\begin{aligned} \zeta(0,t) &= [\Gamma(0,t), \Gamma_1(0,t)] \\ &= b_d[\hat{a}_1(t), \hat{b}_1(t), \ldots, \hat{a}_N(t), \hat{b}_N(t), \\ &\quad -p_1\hat{a}_1(t), -p_1\hat{b}_1(t), \ldots, -p_1\hat{a}_N(t), -p_1\hat{b}_N(t)]. \end{aligned} \qquad (5.65)$$

The solution to (5.64), (5.65) is

$$\zeta(x,t) = \zeta(0,t)\bar{H}(x), \qquad (5.66)$$

where $\bar{H}(x)$ is the unique solution of the following initial value problem:

$$\bar{H}_x(x) = -\bar{H}(x)\bar{A}(x), \quad \bar{H}(0) = I_{4N} \qquad (5.67)$$

for $x \in [0, L]$.

The Second Transformation for Decoupling PDEs

We postulate the backstepping transformation

$$\begin{aligned} \hat{\alpha}(x,t) &= \hat{s}(x,t) - \int_0^x \bar{\lambda}(x,y)\hat{s}(y,t)dy \\ &\quad - \int_0^x \check{\lambda}(x,y)\hat{v}(y,t)dy, \end{aligned} \qquad (5.68)$$

$$\begin{aligned} \hat{\beta}(x,t) &= \hat{v}(x,t) - \int_0^x \bar{\Upsilon}(x,y)\hat{s}(y,t)dy \\ &\quad - \int_0^x \check{\Upsilon}(x,y)\hat{v}(y,t)dy \end{aligned} \qquad (5.69)$$

to convert $\hat{s}, \hat{v}, X$ (5.54)–(5.58) into the following system:

$$\dot{X}(t) = AX(t) + B\hat{\beta}(0,t) + B_1\tilde{d}(t), \qquad (5.70)$$

$$\hat{\alpha}(0,t) = CX(t) - p_1\hat{\beta}(0,t), \tag{5.71}$$

$$\begin{aligned}
\hat{\alpha}_t(x,t) = &-q_1(x)\hat{\alpha}_x(x,t) + c_1(x)\hat{\alpha}(x,t) \\
&- \bar{\lambda}(x,0)q_1(0)CX(t) \\
&+ \left(\Gamma_{1t}(x,t) - \int_0^x \bar{\lambda}(x,y)\Gamma_{1t}(y,t)dy \right. \\
&\left. - \int_0^x \check{\lambda}(x,y)\Gamma_t(y,t)dy \right) Z(t),
\end{aligned} \tag{5.72}$$

$$\begin{aligned}
\hat{\beta}_t(x,t) = &\, q_2(x)\hat{\beta}_x(x,t) + c_4(x)\hat{\beta}(x,t) \\
&- \check{\Upsilon}(x,0)q_1(0)CX(t) \\
&- \left(\int_0^x \check{\Upsilon}(x,y)\Gamma_t(y,t)dy \right. \\
&\left. + \int_0^x \bar{\Upsilon}(x,y)\Gamma_{1t}(y,t)dy - \Gamma_t(x,t) \right) Z(t),
\end{aligned} \tag{5.73}$$

$$\begin{aligned}
\hat{\beta}(l(t),t) = &\, U(t) + p_2\hat{s}(l(t),t) \\
&+ (\Gamma(l(t),t) - p_2\Gamma_1(l(t),t))Z(t) \\
&- \int_0^{l(t)} \bar{\Upsilon}(l(t),y)\hat{s}(y,t)dy \\
&- \int_0^{l(t)} \check{\Upsilon}(l(t),y)\hat{v}(y,t)dy.
\end{aligned} \tag{5.74}$$

By matching (5.70)–(5.74) and (5.54)–(5.58) via (5.68), (5.69) (the details are shown in appendix 5.7B), the conditions of kernels $\bar{\lambda}(x,y)$, $\check{\lambda}(x,y)$, $\bar{\Upsilon}(x,y)$, $\check{\Upsilon}(x,y)$ are obtained as the following two well-posed hyperbolic systems:

$$\check{\lambda}(x,x) = \frac{c_2(x)}{q_1(x) + q_2(x)}, \tag{5.75}$$

$$\bar{\lambda}(x,0) = -\frac{q_2(0)}{q_1(0)p_1}\check{\lambda}(x,0), \tag{5.76}$$

$$\begin{aligned}
-q_1(x)\check{\lambda}_x(x,y) + q_2(y)\check{\lambda}_y(x,y) \\
+(q_2{}'(y) + c_1(x) - c_4(y))\check{\lambda}(x,y) - c_2(y)\bar{\lambda}(x,y) = 0,
\end{aligned} \tag{5.77}$$

$$\begin{aligned}
q_1(x)\bar{\lambda}_x(x,y) + q_1(y)\bar{\lambda}_y(x,y) \\
+(q_1{}'(y) + c_1(y) - c_1(x))\lambda(x,y) + c_3(y)\check{\lambda}(x,y) = 0,
\end{aligned} \tag{5.78}$$

and

$$\bar{\Upsilon}(x,x) = -\frac{c_3(x)}{q_1(x) + q_2(x)}, \tag{5.79}$$

$$\check{\Upsilon}(x,0) = -\frac{q_1(0)p_1}{q_2(0)}\bar{\Upsilon}(x,0), \tag{5.80}$$

$$\begin{aligned}
q_2(x)\check{\Upsilon}_x(x,y) + q_2(y)\check{\Upsilon}_y(x,y) \\
+(q_2{}'(y) + c_4(x) - c_4(y))\check{\Upsilon}(x,y) - c_2(y)\bar{\Upsilon}(x,y) = 0,
\end{aligned} \tag{5.81}$$

$$\begin{aligned}
-q_2(x)\bar{\Upsilon}_x(x,y) + q_1(y)\bar{\Upsilon}_y(x,y) \\
+(q_1{}'(y) + c_1(y) - c_4(x))\bar{\Upsilon}(x,y) + c_3(y)\check{\Upsilon}(x,y) = 0
\end{aligned} \tag{5.82}$$

on $\{0 \leq y \leq x \leq l(t)\}$. The equation sets (5.75)–(5.78) and (5.79)–(5.82) are in the same form as (5.38)–(5.41) and (5.42)–(5.45). Please refer to remark 5.1 and appendix B of [6] for the well-posedness of (5.75)–(5.78) and (5.79)–(5.82).

The Third Transformation for a Stable ODE

We postulate the backstepping transformation

$$\hat{\eta}(x,t) = \hat{\beta}(x,t) - \int_0^x \hat{N}(x,y;\hat{K}(t))\hat{\beta}(y,t)dy - D(x;\hat{K}(t))X(t), \qquad (5.83)$$

where the update law for the control gain $\hat{K}(t) \in R^{1 \times n}$ is developed in the next section. The conditions for the kernels $\hat{N}(x,y;\hat{K}(t)), D(x;\hat{K}(t))$ are to be determined later. The inverse transformation is defined as

$$\hat{\beta}(x,t) = \hat{\eta}(x,t) - \int_0^x \hat{N}_I(x,y;\hat{K}(t))\hat{\eta}(y,t)dy - D_I(x;\hat{K}(t))X(t), \qquad (5.84)$$

where $\hat{N}_I, D_I$ are the kernels whose existence and continuity will be shown later.

Through the transformation (5.83), we convert (5.70)–(5.74) into the following target system:

$$\dot{X}(t) = A_{\mathrm{m}}X(t) + B\hat{\eta}(0,t) + B_1\tilde{d}(t) - B\tilde{K}(t)X(t), \qquad (5.85)$$

$$\hat{\alpha}(0,t) = (C - p_1 D(0;\hat{K}(t)))X(t) - p_1\hat{\eta}(0,t), \qquad (5.86)$$

$$\hat{\alpha}_t(x,t) = -q_1(x)\hat{\alpha}_x(x,t) + c_1(x)\hat{\alpha}(x,t) - \bar{\lambda}(x,0)q_1(0)CX(t)$$
$$+ \left(\Gamma_{1t}(x,t) - \int_0^x \bar{\lambda}(x,y)\Gamma_{1t}(y,t)dy\right.$$
$$\left. - \int_0^x \check{\lambda}(x,y)\Gamma_t(y,t)dy\right)Z(t), \qquad (5.87)$$

$$\hat{\eta}_t(x,t) = q_2(x)\hat{\eta}_x(x,t) + c_4(x)\hat{\eta}(x,t)$$
$$+ \left[\Gamma_t(x,t) - \int_0^x \check{\Upsilon}(x,y)\Gamma_t(y,t)dy\right.$$
$$- \int_0^x \bar{\Upsilon}(x,y)\Gamma_{1t}(y,t)dy$$
$$- \int_0^x \hat{N}(x,y;\hat{K}(t))\left(-\int_0^y \check{\Upsilon}(y,z)\Gamma_t(z,t)dz\right.$$
$$\left. - \int_0^y \bar{\Upsilon}(y,z)\Gamma_{1t}(z,t)dz + \Gamma_t(y,t)\right)dy\right]Z(t)$$
$$+ \left(D(x;\hat{K}(t))B\tilde{K}(t) - \dot{K}(t)D_{\hat{K}(t)}(x;\hat{K}(t))\right)X(t)$$
$$- \dot{\hat{K}}(t)R(x,t) - D(x;\hat{K}(t))B_1\tilde{d}(t), \qquad (5.88)$$

$$\hat{\eta}(l(t),t) = 0, \qquad (5.89)$$

where

$$\tilde{K}(t) = K - \hat{K}(t), \qquad (5.90)$$

and where

$$R(x,t) = \int_0^x \hat{N}_{\hat{K}(t)}(x,y;\hat{K}(t))\hat{\beta}(y,t)dy$$

$$= \int_0^x \hat{N}_{\hat{K}(t)}(x,y;\hat{K}(t))\Big[\hat{\eta}(y,t)$$

$$- \int_0^y \hat{N}_I(y,\sigma;\hat{K}(t))\hat{\eta}(\sigma,t)d\sigma - D_I(y;\hat{K}(t))X(t)\Big]dy \tag{5.91}$$

and

$$D_{\hat{K}(t)}(x;\hat{K}(t)) = \frac{\partial D(x;\hat{K}(t))}{\partial \hat{K}(t)}, \tag{5.92}$$

$$\hat{N}_{\hat{K}(t)}(x,y;\hat{K}(t)) = \frac{\partial \hat{N}(x,y;\hat{K}(t))}{\partial \hat{K}(t)}. \tag{5.93}$$

By matching (5.70)–(5.74) and (5.85)–(5.89) with the aid of (5.83) (the details are shown in appendix 5.7C), the conditions on the kernels $N(x,y;\hat{K}(t))$, $D(x;\hat{K}(t))$ in (5.83) are determined as follows:

$$D(0;\hat{K}(t)) = \hat{K}(t), \tag{5.94}$$

$$-q_2(x)D'(x;\hat{K}(t)) + D(x;\hat{K}(t))(A_m - c_4(x)I_n - B\hat{K}(t))$$

$$+\bar{\Upsilon}(x,0)q_1(0)C - \int_0^x \hat{N}(x,y;\hat{K}(t))\bar{\Upsilon}(y,0)q_1(0)Cdy = 0, \tag{5.95}$$

$$q_2(y)\hat{N}_y(x,y;\hat{K}(t)) + q_2(x)\hat{N}_x(x,y;\hat{K}(t))$$

$$+ q_2'(y)\hat{N}(x,y;\hat{K}(t)) = 0, \tag{5.96}$$

$$q_2(0)\hat{N}(x,0;\hat{K}(t)) - D(x;\hat{K}(t))B = 0. \tag{5.97}$$

The equation set (5.94)–(5.97) is a transport PDE-ODE coupled system consisting of the transport PDE (5.96) with the boundary condition (5.97) on $\{(x,y)|0 \le y \le x \le l(t)\}$ and the ODE (5.95) with the initial value (5.94) on $\{0 \le x \le l(t)\}$. It should be noted that $\hat{K}(t)$ is a parameter rather than a variable in the transport PDE (5.96), (5.97) with respect to the independent variables x,y and in the ODE (5.94), (5.95) with respect to the independent variable x when solving (5.94)–(5.97).

To establish the well-posedness of (5.94)–(5.97), the transport PDE state $\hat{N}(x,y;\hat{K}(t))$ can be represented by its boundary value $D(x;\hat{K}(t))B$. By substituting the result into ODE (5.95) to replace $\hat{N}(x,y;\hat{K}(t))$, the solution of the ODE $\hat{D}(x;\hat{K}(t))$ can be obtained. The well-posedness of the transport PDE $\hat{N}(x,y;\hat{K}(t))$ (5.96), (5.97) is thus obtained because of the well-defined boundary condition (5.97).

The existence and continuity of the kernels $\hat{N}_I, D_I$ in the inverse transformation (5.84) are shown as follows. Rewrite (5.83) as

$$\hat{\eta}(x,t) + D(x;\hat{K}(t))X(t) = \hat{\beta}(x,t) - \int_0^x \hat{N}(x,y;\hat{K}(t))\hat{\beta}(y,t)dy. \tag{5.98}$$

Because $\hat{N}(x, y; \hat{K}(t))$ is continuous, according to [169], there exists a unique continuous $\varrho(x, y; \hat{K}(t))$ on $\{(x, y) | 0 \leq x \leq y \leq l(t)\}$ such that

$$
\begin{aligned}
\hat{\beta}(x, t) &= \hat{\eta}(x, t) + D(x; \hat{K}(t))X(t) \\
&\quad + \int_0^x \varrho(x, y; \hat{K}(t))(\hat{\eta}(y, t) + D(y; \hat{K}(t))X(t))dy \\
&= \hat{\eta}(x, t) + \int_0^x \varrho(x, y; \hat{K}(t))\hat{\eta}(y, t)dy \\
&\quad + \left(\int_0^x \varrho(x, y; \hat{K}(t))D(y; \hat{K}(t))dy + D(x; \hat{K}(t)) \right)X(t). \quad (5.99)
\end{aligned}
$$

Comparing with (5.99) and the inverse transformation (5.84), we obtain the existence and continuity of the kernels $\hat{N}_I(x, y; \hat{K}(t))$, $D_I(x; \hat{K}(t))$.

Finally, for (5.89) to hold, recalling (5.83) and (5.74), we derive the boundary control input $U(t)$, the expression for which is shown in section 5.5.

5.4 ADAPTIVE UPDATE LAWS

Using normalization and projection operators to guarantee boundedness, as is typical in adaptive control designs, the adaptive update laws for the self-tuned control gains

$$
\hat{K}(t) = [\hat{k}_1(t), \dots, \hat{k}_n(t)] \quad (5.100)
$$

and for the unknown parameters $\hat{a}_j(t), \hat{b}_j(t)$, $j \in \{1, \dots, N\}$ are built as

$$
\dot{\hat{k}}_i(t) = \text{Proj}_{[\underline{k}_i, \bar{k}_i]}\left(\tau_i(t), \hat{k}_i(t) \right), \quad (5.101)
$$

$$
\dot{\hat{a}}_j(t) = \text{Proj}_{[0, \bar{a}_j]}\left(\tau_{1j}(t), \hat{a}_j(t) \right), \quad (5.102)
$$

$$
\dot{\hat{b}}_j(t) = \text{Proj}_{[0, \bar{b}_j]}\left(\tau_{2j}(t), \hat{b}_j(t) \right), \quad (5.103)
$$

$i \in \{1, \dots, n\}$, $j \in \{1, \dots, N\}$. For any $m \leq M$ and any r, p, $\text{Proj}_{[m, M]}$ is the standard projection operator given by

$$
\text{Proj}_{[m, M]}(r, p) = \begin{cases} 0, & \text{if } p = m \text{ and } r < 0, \\ 0, & \text{if } p = M \text{ and } r > 0, \\ r, & \text{else.} \end{cases}
$$

The role of the projection operator is to keep the parameter estimates bounded. The bounds $\underline{k}_i, \bar{k}_i$ and $\bar{a}_j, \bar{b}_j$ are defined in section 5.1. The functions $\tau_i, \tau_{1j}, \tau_{2j}$ in (5.101)–(5.103) are defined as

$$
\begin{aligned}
& [\tau_1(t), \dots, \tau_n(t)]^T \\
& = \gamma_c \frac{-2XB^T PX + r_a \int_0^{l(t)} e^{\delta x}\hat{\eta}(x, t)XB^T D(x; \hat{K}(t))^T dx}{1 + \Omega(t)}, \quad (5.104)
\end{aligned}
$$

$$\tau_{1j}(t)$$

$$= \gamma_{aj} \frac{\left(2X^T P B_1 - r_a \int_0^{l(t)} e^{\delta x} \hat{\eta}(x,t) D(x; \hat{K}(t)) B_1 dx\right) \cos(\theta_j t)}{1 + \Omega(t)}, \tag{5.105}$$

$$\tau_{2j}(t)$$

$$= \gamma_{bj} \frac{\left(2X^T P B_1 - r_a \int_0^{l(t)} e^{\delta x} \hat{\eta}(x,t) D(x; \hat{K}(t)) B_1 dx\right) \sin(\theta_j t)}{1 + \Omega(t)}, \tag{5.106}$$

where

$$\gamma_c = \mathrm{diag}\{\gamma_{c1}, \ldots, \gamma_{cn}\}. \tag{5.107}$$

The positive update gains $\gamma_{ci}, \gamma_{aj}, \gamma_{bj}$ are to be chosen by the user, and

$$\delta > \max\left\{ \frac{2\overline{c_4} + \overline{q'_2}}{\underline{q_2}}, \frac{2\overline{c_1} + 1 + \overline{q'_1}}{\underline{q_1}} \right\} \tag{5.108}$$

with

$$\underline{q_1} = \min_{0 \le x \le L} \{q_1(x)\}, \quad \overline{q'_1} = \max_{0 \le x \le L} \{|q'_1(x)|\}, \tag{5.109}$$

$$\underline{q_2} = \min_{0 \le x \le L} \{q_2(x)\}, \quad \overline{q'_2} = \max_{0 \le x \le L} \{|q'_2(x)|\}, \tag{5.110}$$

$$\overline{c_1} = \max_{0 \le x \le L} \{|c_1(x)|\}, \quad \overline{c_4} = \max_{0 \le x \le L} \{|c_4(x)|\}. \tag{5.111}$$

The scalar $\Omega(t)$ is defined as

$$\Omega(t) = X^T P X(t) + \frac{1}{2} r_a \int_0^{l(t)} e^{\delta x} \hat{\eta}(x, t)^2 dx$$

$$+ \frac{1}{2} r_b \int_0^{l(t)} e^{-\delta x} \hat{\alpha}(x, t)^2 dx. \tag{5.112}$$

The determination of the positive constants r_a, r_b will be shown in the next section. The matrix $P = P^T > 0$ is the unique solution to the following Lyapunov equation:

$$P A_{\mathrm{m}} + A_{\mathrm{m}}^T P = -Q \tag{5.113}$$

for some $Q = Q^T > 0$. Because A_{m} in (5.16) is known (chosen by the user, in spite of A and K being unknown), the matrix P is known.

The normalization $\Omega(t) + 1$ is introduced in the denominator in (5.104)–(5.106) to limit the rates of change of the parameter estimates—that is, $\dot{\hat{k}}_i(t)$ and $\dot{\hat{a}}_j(t), \dot{\hat{b}}_j(t)$. The functions $\hat{\eta}(\cdot, t)$ and $\hat{\alpha}(\cdot, t)$ in (5.104)–(5.106), (5.112) can be represented by the observer states through (5.48), (5.49), (5.68), (5.69), (5.83). The idea of constructing the adaptive update laws $\hat{k}_i(t), \hat{a}_j(t), \hat{b}_j(t)$ $i \in \{1, \ldots, n\}, j \in \{1, \ldots, N\}$ in (5.101)–(5.106) will be clear from the Lyapunov analysis in the next section.

5.5   CONTROL LAW AND STABILITY ANALYSIS

Control Law

According to (5.89), (5.83), and (5.74), $U(t)$ is obtained as

$$U(t) = -p_2 \hat{s}(l(t), t) - (\Gamma(l(t), t) - p_2 \Gamma_1(l(t), t)) Z(t)$$
$$+ \int_0^{l(t)} \bar{\Upsilon}(l(t), y) \hat{s}(y, t) dy + \int_0^{l(t)} \check{\Upsilon}(l(t), y) \hat{v}(y, t) dy$$
$$+ \int_0^{l(t)} \hat{N}(l(t), y; \hat{K}(t)) \hat{\beta}(y, t) dy + D(l(t); \hat{K}(t)) X(t). \qquad (5.114)$$

Recalling (5.83), (5.69), (5.48), (5.49), (5.26), the control law (5.114) is rewritten as

$$U(t) = - p_2 z(l(t), t) - \Gamma(l(t), t) Z(t)$$
$$+ \int_0^{l(t)} \bar{\Upsilon}(l(t), y) (\hat{z}(y, t) + \Gamma_1(y, t) Z(t)) dy$$
$$+ \int_0^{l(t)} \check{\Upsilon}(l(t), y) (\hat{w}(y, t) + \Gamma(y, t) Z(t)) dy$$
$$+ \int_0^{l(t)} \hat{N}(l(t), y; \hat{K}(t)) \Big[\hat{w}(y, t) + \Gamma(y, t) Z(t)$$
$$- \int_0^y \bar{\Upsilon}(y, \sigma) (\hat{z}(\sigma, t) + \Gamma_1(\sigma, t) Z(t)) d\sigma$$
$$- \int_0^y \check{\Upsilon}(y, \sigma) (\hat{w}(\sigma, t) + \Gamma(\sigma, t) Z(t)) d\sigma \Big] dy + D(l(t); \hat{K}(t)) X(t), \qquad (5.115)$$

which is the output-feedback adaptive controller sought in this chapter. The signals $z(l(t), t)$, $X(t)$ are measurements, and $Z(t)$ is defined in (5.50). The states $\hat{z}(x, t)$, $\hat{w}(x, t)$ are obtained from the observer (5.23)–(5.26). The functions $\Gamma_1(y, t)$ and $\Gamma(y, t)$ are solutions of (5.66), (5.67) where the adaptive estimates $\hat{a}_j(t), \hat{b}_j(t)$ are defined in (5.102), (5.103) and (5.105), (5.106). The functions $\bar{\Upsilon}(y, \sigma)$, $\check{\Upsilon}(y, \sigma)$ are solutions of (5.79)–(5.82). The functions $\Gamma(l(t), t), \bar{\Upsilon}(l(t), y), \check{\Upsilon}(l(t), y), \hat{N}(l(t), y;$ $\hat{K}(t))$, $D(l(t); \hat{K}(t))$ are the solutions of (5.66), (5.67), (5.79)–(5.82), and (5.94)–(5.97) on $x = l(t)$, respectively. The adaptive estimate $\hat{K}(t)$ is defined in (5.101), (5.104).

The block diagram of the closed-loop system is shown in figure 5.2, whose stability result is given in the next subsection.

Stability Analysis

Lemma 5.2. *For all initial data* $(\hat{\alpha}(\cdot, 0), \hat{\eta}(\cdot, 0)) \in H^1(0, L)$, $X(0) \in \mathbb{R}^n$, $\hat{K}(0) \in \mathbb{R}^n$, $\hat{a}_j(0) \in \mathbb{R}$, $\hat{b}_j(0) \in \mathbb{R}$, $j = 1, \ldots, N$, *the target system* (5.85)–(5.89) *is asymptotically regulated in the sense of*

$$\lim_{l \to \infty} (\|\hat{\alpha}(\cdot, t)\| + \|\hat{\eta}(\cdot, t)\| + |X(t)|) = 0. \qquad (5.116)$$

Proof. Define

$$\Theta(t) = \|\hat{\eta}(\cdot, t)\|^2 + \|\hat{\alpha}(\cdot, t)\|^2 + |X(t)|^2, \qquad (5.117)$$

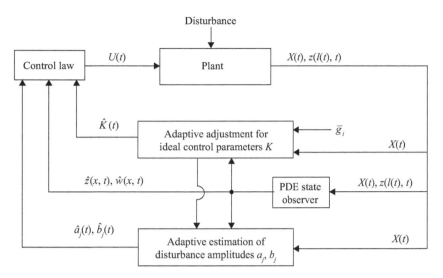

Figure 5.2. Block of the closed-loop system.

where $\|\cdot\|$ denotes the L_2 norm. Recalling (5.112), we get

$$\mu_1 \Theta(t) \le \Omega(t) \le \mu_2 \Theta(t), \tag{5.118}$$

with positive μ_1, μ_2 determined as

$$\mu_1 = \frac{1}{2}\min\{r_a, r_b e^{-\delta L}, \lambda_{\min}(P)\}, \tag{5.119}$$

$$\mu_2 = \frac{1}{2}\min\{r_a e^{\delta L}, r_b, \lambda_{\max}(P)\}, \tag{5.120}$$

where $\lambda_{\min}$ and $\lambda_{\max}$ denote the minimum and maximum eigenvalues of the corresponding matrix.

Let us choose a Lyapunov function as

$$V(t) = \ln\left(1 + \Omega(t)\right) + \sum_{j=1}^{N} \frac{1}{2\gamma_{aj}}\tilde{a}_j(t)^2 + \sum_{j=1}^{N} \frac{1}{2\gamma_{bj}}\tilde{b}_j(t)^2$$

$$+ \frac{1}{2}\tilde{K}(t)\gamma_c^{-1}\tilde{K}(t)^T, \tag{5.121}$$

where $\tilde{a}_j, \tilde{b}_j, \tilde{K}$ are given in (5.59), (5.90). Recalling (5.112), we rewrite the Lyapunov function as

$$V(t) = \ln\left(X^T P X(t) + \frac{1}{2}r_a \int_0^{l(t)} e^{\delta x}\hat{\eta}(x,t)^2 dx\right.$$

$$\left. + \frac{1}{2}r_b \int_0^{l(t)} e^{-\delta x}\hat{\alpha}(x,t)^2 dx + 1\right) + \sum_{j=1}^{N} \frac{1}{2\gamma_{aj}}\tilde{a}_j(t)^2$$

$$+ \sum_{j=1}^{N} \frac{1}{2\gamma_{bj}}\tilde{b}_j(t)^2 + \frac{1}{2}\tilde{K}(t)\gamma_c^{-1}\tilde{K}(t)^T. \tag{5.122}$$

Taking the derivative of $V(t)$, through a lengthy calculation in appendix 5.7D we obtain

$$
\begin{aligned}
\dot{V}(t) \le \frac{1}{1+\Omega(t)} &\left[-\left(\frac{3}{4}\lambda_{\min}(Q) - q_1(0)r_b\bar{D}^2 - \frac{q_1(0)^2}{2}Lr_b\bar{J}^2\,|C|^2 \right)|X(t)|^2 \right. \\
&- \left(\frac{1}{2}q_2(0)r_a - q_1(0)r_bp_1^2 - \frac{8}{\lambda_{\min}(Q)}|PB|^2 \right)\hat{\eta}(0,t)^2 \\
&- \left(\frac{1}{2}\delta\underline{q_2}r_a - r_a\overline{c_4} - \frac{1}{2}r_a\overline{q_2'} \right) \int_0^{l(t)} e^{\delta x}\hat{\eta}(x,t)^2 dx \\
&- \frac{1}{2}\left(q_1(l(t)) - \dot{l}(t) \right) r_b e^{-\delta L}\hat{\alpha}(l(t),t)^2 \\
&\left. - \left(\frac{1}{2}\underline{q_1}\delta r_b - r_b\overline{c_1} - \frac{1}{2}r_b - \frac{1}{2}r_b\overline{q_1'} \right) \int_0^{l(t)} e^{-\delta x}\hat{\alpha}(x,t)^2 dx \right] \\
&- \tilde{K}(t)\left[\gamma_c^{-1}\dot{\hat{K}}(t)^T \right. \\
&\left. - \frac{-2X(t)B^TPX(t) + r_a\int_0^{l(t)} e^{\delta x}\hat{\eta}(x,t)X(t)B^TD(x;\hat{K}(t))^T dx}{1+\Omega(t)} \right] \\
&- \sum_{j=1}^N \tilde{a}_j(t)\left[\frac{1}{\gamma_{aj}}\dot{\hat{a}}_j(t) - \frac{(2X^TPB_1 - r_a\int_0^{l(t)} e^{\delta x}\hat{\eta}(x,t)D(x;\hat{K}(t))B_1 dx)\cos(\theta_j t)}{1+\Omega(t)} \right] \\
&- \sum_{j=1}^N \tilde{b}_j(t)\left[\frac{1}{\gamma_{bj}}\dot{\hat{b}}_j(t) - \frac{(2X^TPB_1 - r_a\int_0^{l(t)} e^{\delta x}\hat{\eta}(x,t)D(x;\hat{K}(t))B_1 dx)\sin(\theta_j t)}{1+\Omega(t)} \right] \\
&+ \left[r_a\int_0^{l(t)} e^{\delta x}\hat{\eta}(x,t)\left(H_bZ(t) + \dot{\hat{K}}D_{\hat{K}(t)}X(t) - \dot{\hat{K}}R(x,t) \right) dx \right. \\
&\left. + r_b\int_0^{l(t)} e^{-\delta x}\hat{\alpha}(x,t)H_aZ(t)dx \right]\frac{1}{1+\Omega(t)}.
\end{aligned}
\tag{5.123}
$$

Regarding (5.101), (5.102), (5.103), we know that there exist positive constants m_2, m_3 such that

$$
\max_{j\in\{1,\ldots,N\}}\left\{ \left|\dot{\hat{a}}_j(t)\right|^2, \left|\dot{\hat{b}}_j(t)\right|^2 \right\} \le m_2 \max_{j\in\{1,\ldots,N\}}\{\gamma_{aj}^2, \gamma_{bj}^2\}(|X(t)|^2 + \|\hat{\eta}\|^2), \tag{5.124}
$$

$$
\left|\dot{\hat{K}}(t)\right|^2 \le m_3 \max_{i\in\{1,\ldots,n\}}\{\gamma_{ci}^2\}, \tag{5.125}
$$

where m_2, m_3 only depend on the parameters of the plant and the coefficients introduced in the Lyapunov function.

With (5.66), we obtain

$$
\begin{aligned}
\max_{x\in[0,L], t\in[0,\infty)} &\left\{ |\Gamma_t(x,t)|^2, |\Gamma_{1t}(x,t)|^2 \right\} \\
&\le 2N \max_{j\in\{1,\ldots,N\}}\left\{ \left|\dot{\hat{a}}_j(t)\right|^2, \left|\dot{\hat{b}}_j(t)\right|^2 \right\}\bar{h}_m^2 \\
&\le 2Nm_2 \max_{j\in\{1,\ldots,N\}}\{\gamma_{aj}^2, \gamma_{bj}^2\}(|X(t)|^2 + \|\hat{\eta}\|^2)\bar{h}_m^2,
\end{aligned}
\tag{5.126}
$$

where $\bar{h}_m$ is defined as

$$\bar{h}_m = \max_{0 \leq x \leq L} \{\bar{\sigma}(\bar{h}(x))\}, \tag{5.127}$$

and $\bar{H}(x)$ is the solution of (5.67), and $\bar{\sigma}$ stands for the largest singular value at x.

According to the definitions of H_a, H_b in (5.154), (5.155), recalling (5.125), (5.126), there exist positive constants ξ_1, ξ_2 such that

$$\max\{|H_a Z(t)|^2, |H_b Z(t)|^2\} \leq \xi_1 \max_{j \in \{1, \ldots, N\}} \{\gamma_{aj}^2, \gamma_{bj}^2\}(|X(t)|^2 + \|\hat{\eta}\|^2), \tag{5.128}$$

$$\left| \dot{\hat{K}} D_{\hat{K}(t)} X(t) \right|^2 \leq \xi_2 \max_{i \in \{1, \ldots, n\}} \{\gamma_{ci}^2\}|X(t)|^2, \tag{5.129}$$

with ξ_1, ξ_2 depending only on the kernels, the parameters of the plant, and the coefficients used in the Lyapunov analysis but not on $\gamma_{ci}, \gamma_{aj}, \gamma_{bj}$.

Applying the Young and Cauchy-Schwarz inequalities, we obtain the inequality

$$\begin{aligned}
r_a \int_0^{l(t)} e^{\delta x} \hat{\eta}(x,t) H_b Z(t) dx &\leq \frac{1}{2} r_a \int_0^{l(t)} e^{\delta x} \hat{\eta}(x,t)^2 dx \\
&\quad + \frac{1}{2} r_a e^{\delta L} L \xi_1 \max_{j \in \{1, \ldots, N\}} \{\gamma_{aj}^2, \gamma_{bj}^2\}(|X(t)|^2 + \|\hat{\eta}\|^2),
\end{aligned} \tag{5.130}$$

where we have used (5.128); the inequality

$$\begin{aligned}
-r_a \int_0^{l(t)} e^{\delta x} \hat{\eta}(x,t) \dot{\hat{K}} D_{\hat{K}(t)} X(t) dx \\
\leq \frac{1}{2} r_a \int_0^{l(t)} e^{\delta x} \hat{\eta}(x,t)^2 dx + \frac{1}{2} r_a e^{\delta L} L \max_{i \in \{1, \ldots, n\}} \{\gamma_{ci}^2\} \xi_2 |X(t)|^2,
\end{aligned} \tag{5.131}$$

where we have used (5.129); and the inequality

$$\begin{aligned}
-r_a \int_0^{l(t)} e^{\delta x} \hat{\eta}(x,t) \dot{\hat{K}} R(x,t) dx \\
\leq \frac{1}{2} r_a \int_0^{l(t)} e^{\delta x} \hat{\eta}(x,t)^2 dx \\
+ \frac{1}{2} r_a e^{\delta L} m_3 \max_{i \in \{1, \ldots, n\}} \{\gamma_{ci}^2\} \xi_3 (|X(t)|^2 + \|\hat{\eta}\|^2)
\end{aligned} \tag{5.132}$$

for which we have employed (5.125), (5.91). The positive constant ξ_3 in (5.132) only depends on kernels $\hat{N}_I, \hat{D}_I, \hat{N}_{\hat{K}(t)}$. Finally, we also obtain the inequality

$$\begin{aligned}
r_b \int_0^{l(t)} e^{-\delta x} \hat{\alpha}(x,t) H_a Z(t) dx \\
\leq \frac{r_b}{2} \int_0^{l(t)} e^{-\delta x} \hat{\alpha}(x,t)^2 dx \\
+ \frac{1}{2} r_b L \xi_1 \max_{j \in \{1, \ldots, N\}} \{\gamma_{aj}^2, \gamma_{bj}^2\}(|X(t)|^2 + \|\hat{\eta}\|^2)
\end{aligned} \tag{5.133}$$

by applying (5.128). Applying (5.130)–(5.133), we obtain

$$r_a \int_0^{l(t)} e^{\delta x} \hat{\eta}(x,t) \left(H_b Z(t) - \dot{\hat{K}} D_{\hat{K}(t)} X(t) - \dot{\hat{K}} R(x,t) \right) dx$$

$$+ r_b \int_0^{l(t)} e^{-\delta x} \hat{\alpha}(x,t) H_a Z(t) dx$$

$$\leq \left[\frac{1}{2} r_a e^{\delta L} L \max_{i \in \{1,\dots,n\}} \{\gamma_{ci}^2\} \xi_2 + \frac{1}{2} r_a e^{\delta L} L \xi_1 \max_{j \in \{1,\dots,N\}} \{\gamma_{aj}^2, \gamma_{bj}^2\} \right.$$

$$+ \frac{1}{2} r_a e^{\delta L} m_3 \max_{i \in \{1,\dots,n\}} \{\gamma_{ci}^2\} \xi_3 L$$

$$\left. + \frac{1}{2} r_b L \xi_1 \max_{j \in \{1,\dots,N\}} \{\gamma_{aj}^2, \gamma_{bj}^2\} \right] |X(t)|^2$$

$$+ \left[\frac{3}{2} r_a e^{\delta L} + \frac{1}{2} r_a e^{\delta L} L \xi_1 \max_{j \in \{1,\dots,N\}} \{\gamma_{aj}^2, \gamma_{bj}^2\} \right.$$

$$+ \frac{1}{2} r_a e^{\delta L} m_3 \max_{i \in \{1,\dots,n\}} \{\gamma_{ci}^2\} \xi_3 L$$

$$\left. + \frac{1}{2} r_b L \xi_1 \max_{j \in \{1,\dots,N\}} \{\gamma_{aj}^2, \gamma_{bj}^2\} \right] \|\hat{\eta}\|^2 + \frac{r_b}{2} \|\hat{\alpha}\|^2$$

$$\leq \max_{i \in \{1,\dots,n\}, j \in \{1,\dots,N\}} \{\gamma_{ci}^2, \gamma_{aj}^2, \gamma_{bj}^2\} \lambda_b \left(|X(t)|^2 + \|\hat{\eta}\|^2 + \|\hat{\alpha}\|^2 \right), \qquad (5.134)$$

where $\lambda_b > 0$ depends only on the kernels, the parameters of the plant, and the coefficients used in the Lyapunov analysis.

Choosing

$$r_b < \frac{\frac{3}{4} \lambda_{\min}(Q)}{q_1(0) |\bar{D}|^2 + \frac{q_1(0)^2}{2} L \bar{J}^2 |C|^2}, \qquad (5.135)$$

$$r_a > \frac{2}{q_2(0)} \left(q_1(0) r_b p_1^2 - \frac{8}{\lambda_{\min}(Q)} |PB|^2 \right), \qquad (5.136)$$

inserting (5.134), recalling (5.108), and applying the adaptive laws (5.101)–(5.106), the inequality (5.123) becomes

$$\dot{V}(t) \leq \frac{1}{1+\Omega} \left[-\lambda_a \left(|X(t)|^2 + \hat{\eta}(0,t)^2 + \|\hat{\eta}(\cdot,t)\|^2 \right. \right.$$

$$+ \hat{\alpha}(l(t),t)^2 + \|\hat{\alpha}(\cdot,t)\|^2 \right)$$

$$+ \max_{i \in \{1,\dots,n\}, j \in \{1,\dots,N\}} \{\gamma_{ci}^2, \gamma_{aj}^2, \gamma_{bj}^2\} \lambda_b (|X(t)|^2$$

$$\left. + \|\hat{\eta}(\cdot,t)\|^2 + \|\hat{\alpha}(\cdot,t)\|^2 \right].$$

The coefficients $\gamma_{aj}, \gamma_{bj}, \gamma_{ci}$ in the adaptive law are independent of λ_a, λ_b, which only depend on the kernels, the plant parameters, and the coefficients used in the Lyapunov analysis. Choosing $\gamma_{aj}, \gamma_{bj}, \gamma_{ci}$ to satisfy

$$\max_{i\in\{1,\dots,n\},j\in\{1,\dots,N\}}\{\gamma_{ci}^2,\gamma_{aj}^2,\gamma_{bj}^2\}:=\gamma_0^2<\frac{\lambda_a}{\lambda_b}, \tag{5.137}$$

we arrive at

$$\dot{V}(t)\leq\frac{-\bar{\lambda}_a}{1+\Omega}\left(|X(t)|^2+\hat{\eta}(0,t)^2+\|\hat{\eta}(\cdot,t)\|^2\right.$$
$$\left.+\hat{\alpha}(l(t),t)^2+\|\hat{\alpha}(\cdot,t)\|^2\right)\leq 0, \tag{5.138}$$

where

$$\bar{\lambda}_a=\lambda_a-\gamma_0^2\lambda_b>0.$$

Hence, we obtain

$$V(t)\leq V(0),\ \forall t\geq 0.$$

One can easily see that $|\tilde{K}(t)|$, $\tilde{d}(t)$, and $\Omega(t)$ are uniformly bounded. Therefore, we ascertain that $\|\hat{\eta}(\cdot,t)\|$, $\|\hat{\alpha}(\cdot,t)\|$, $|X(t)|$ are uniformly bounded. Taking the time derivative of $|X(t)|^2$, $\|\hat{\alpha}(\cdot,t)\|^2$, and $\|\hat{\eta}(\cdot,t)\|^2$ along (5.85)–(5.89), we obtain

$$\frac{d}{dt}|X(t)|^2=2X^T(t)(A_mX(t)+B\hat{\eta}(0,t)+B_1\tilde{d}(t)-B\tilde{K}(t)X(t)), \tag{5.139}$$

$$\frac{d}{dt}\|\hat{\eta}(\cdot,t)\|^2=(q_2(l(t))+\dot{l}(t))\hat{\eta}(l(t),t)^2-q_2(0)\hat{\eta}(0,t)^2$$
$$-\int_0^{l(t)}(q_2'(x)-2c_4(x))\hat{\eta}(x,t)^2dx+2\int_0^{l(t)}\hat{\eta}(\cdot,t)\left[\left(\Gamma_t(x,t)\right.\right.$$
$$-\int_0^x\check{\Upsilon}(x,y)\Gamma_t(y,t)dy-\int_0^x\bar{\Upsilon}(x,y)\Gamma_{1t}(y,t)dy$$
$$-\int_0^x\hat{N}(x,y;\hat{K}(t))\left(-\int_0^y\check{\Upsilon}(y,z)\Gamma_t(z,t)dz\right.$$
$$\left.-\int_0^y\bar{\Upsilon}(y,z)\Gamma_{1t}(z,t)dz+\Gamma_t(y,t)\right)dy\bigg)Z(t)$$
$$+\left(D(0;\hat{K}(t))B\tilde{K}(t)-\dot{\hat{K}}(t)D_{\hat{K}(t)}(x;\hat{K}(t))\right)X(t)$$
$$\left.-\dot{\hat{K}}(t)R(x,t)-D(0;\hat{K}(t))B_1\tilde{d}(t)\right]dx, \tag{5.140}$$

$$\frac{d}{dt}\|\hat{\alpha}(\cdot,t)\|^2=-(q_1(l(t))-\dot{l}(t))\hat{\alpha}(l(t),t)^2$$
$$+q_1(0)\hat{\alpha}(0,t)^2+\int_0^{l(t)}(q_1'(x)+2c_1(x))\hat{\alpha}(x,t)^2dx$$
$$+2\int_0^{l(t)}\hat{\alpha}(x,t)\left[-\bar{\lambda}(x,0)q_1(0)CX(t)+\left(\Gamma_{1t}(x,t)\right.\right.$$
$$\left.-\int_0^x\bar{\lambda}(x,y)\Gamma_{1t}(y,t)dy-\int_0^x\check{\lambda}(x,y)\Gamma_t(y,t)dy\right)Z(t)\bigg]dx. \tag{5.141}$$

Recalling the boundedness results of $\dot{\hat{K}},\Gamma_t,\Gamma_{1t}$ in (5.125), (5.126) and the boundedness of $\dot{l}(t)$ in assumption 5.4, according to (5.88), (5.89), we find that $\hat{\eta}(0,t)$ is bounded. The signal $\hat{\alpha}(0,t)$ is also bounded via (5.86), and then $\hat{\alpha}(l(t),t)$ is bounded through transport PDE (5.88). Therefore, we see that $\frac{d}{dt}|X(t)|^2$, $\frac{d}{dt}\|\hat{\eta}(\cdot,t)\|^2$ and $\frac{d}{dt}\|\hat{\alpha}(\cdot,t)\|^2$ are uniformly bounded according to (5.139)–(5.141).

Finally, integrating (5.138) from 0 to ∞, it follows that $|X(t)|$, $\|\hat{\alpha}(\cdot,t)\|$, $\|\hat{\eta}(\cdot,t)\|$ are square integrable. By Barbalat's lemma, $|X(t)|$, $\|\hat{\alpha}(\cdot,t)\|$, $\|\hat{\eta}(\cdot,t)\|$ tend to zero as $t \to \infty$. □

The achievement of the control objective of adaptively canceling the unmatched disturbance and stabilizing the vibrating string is shown as the following theorem.

Theorem 5.1. *For all initial data* $(z(\cdot,0), w(\cdot,0)) \in H^1(0,L)$, $X(0) \in \mathbb{R}^n$, $\hat{K}(0) \in \mathbb{R}^n$, $\hat{a}_j(0) \in \mathbb{R}$, $\hat{b}_j(0) \in \mathbb{R}$, $j=1,\ldots,N$, *the closed-loop system including the plant* (5.6)–(5.10), *the observer* (5.23)–(5.26), *the adaptive update laws* (5.101)–(5.106), *and the control law* (5.115) *has the following properties.*

1. The ODE state $X(t)$ *is asymptotically convergent to zero in the sense of*

$$\lim_{t\to\infty} |X(t)| = 0. \tag{5.142}$$

2. The PDE states are uniformly ultimately bounded in the sense of the norm

$$\|z(\cdot,t)\| + \|w(\cdot,t)\|.$$

Proof. The separation principle can be verified by the fact that the stability of the observer error system is independent of the control design. Recalling the asymptotic stability result proved in lemma 5.2, property 1 is immediate.

Due to the invertibility and continuity of the backstepping transformations (5.68), (5.69), (5.83), we obtain the asymptotic convergence to zero of $\|\hat{v}(\cdot,t)\| + \|\hat{s}(\cdot,t)\|$. According to (5.48), (5.49), we obtain

$$\|\hat{w}(\cdot,t)\|^2 + \|\hat{z}(\cdot,t)\|^2$$
$$= 2\|\hat{v}(\cdot,t)\|^2 + 2N\|\Gamma(\cdot,t)\|^2 + 2\|\hat{s}(\cdot,t)\|^2 + 2N\|\Gamma_1(\cdot,t)\|^2,$$
$$\leq 2\|\hat{v}(\cdot,t)\|^2 + 2\|\hat{s}(\cdot,t)\|^2$$
$$+ \bar{\gamma}_1 \max_{j\in\{1,\ldots,N\}}\{\bar{a}_j^2, \bar{b}_j^2\} + \bar{\gamma}_2 \max_{j\in\{1,\ldots,N\}}\{\bar{a}_j^2, \bar{b}_j^2\}, \tag{5.143}$$

with the positive constants

$$\bar{\gamma}_1 = 4N^2 b_d^2 \bar{h}_m^2 L, \quad \bar{\gamma}_2 = 4N^2 b_d^2 p_1^2 \bar{h}_m^2 L.$$

Recalling the asymptotic convergence to zero of $\|\hat{v}(\cdot,t)\| + \|\hat{s}(\cdot,t)\|$, and with

$$\bar{\gamma}_1 \max_{j\in\{1,\ldots,N\}}\{\bar{a}_j, \bar{b}_j\}, \quad \bar{\gamma}_2 \max_{j\in\{1,\ldots,N\}}\{\bar{a}_j, \bar{b}_j\}$$

as positive constants, we obtain the uniform ultimate boundedness of $\|\hat{w}(\cdot,t)\| + \|\hat{z}(\cdot,t)\|$. Applying the separation principle, recalling lemma 5.1 and (5.27), we obtain property 2.

The proof of theorem 5.1 is complete. □

5.6   SIMULATION FOR A FLEXIBLE-GUIDE ELEVATOR

Simulation Model

The simulation is conducted based on (5.6)–(5.10) with the coefficients (5.11)–(5.13). The values of the physical parameters of the mining cable elevator tested in the simulation are shown in table 5.1.

The unknown damping coefficient and stiffness coefficient shown in figure 5.1 of the flexible guide are $c_d = 0.4$ and $k_c = 1000$ in (5.13). The unknown disturbance $d(t)$ in (5.6) is modeled as

$$d(t) = a_1 \cos\left(\frac{\pi}{8}t\right) + b_1 \sin\left(\frac{\pi}{8}t\right), \tag{5.144}$$

where $a_1 = 5, b_1 = 2$ are unknown amplitudes. We assume that we only know their upper bounds $\bar{a}_1, \bar{b}_1$ of $10, 5$, respectively.

The target system matrix of the ODE is set as

$$A_{\mathrm{m}} = \begin{pmatrix} 0 & 1 \\ -0.73 & -2.25 \end{pmatrix}, \tag{5.145}$$

whose eigenvalues are $-1.8, -0.4$. The unknown target control parameters k_1, k_2 are turned online by the adaptive mechanism (5.101), (5.104) to achieve the target system matrix A_{m}. The ranges of the unknown control parameters k_1, k_2 in the adaptive estimates are $[-10, 0], [-40, 0]$. The parameters $\gamma_{c1}, \gamma_{c2}, \gamma_{a1}, \gamma_{b1}$ in the adaptive law are chosen as $2, 6.8, 1.2, 0.6$.

The initial conditions are given as

$$w(x, 0) = 0.2 \sin(2\pi x / L), \quad z(x, 0) = 0.3 \sin(3\pi x / L),$$

$$x_2(0) - 0.5w(0, 0) + 0.5z(0, 0) = 0, \quad x_1(0) = 0.$$

The hoisting velocity curve is shown in figure 2.2. The maximum velocity of the moving boundary—that is, the maximum hoisting velocity $\bar{v}_{\mathrm{max}}$ shown in table 5.1—satisfies the limit of the changing rate of the time-varying domain proposed in assumption 5.4.

Table 5.1. Physical parameters of the mining cable elevator

Parameters (units)	Values
Initial length L (m)	2000
Final length (m)	200
Cable linear density ρ (kg/m)	8.1
Total hoisted mass M_c (kg)	15000
Gravitational acceleration g (m/s^2)	9.8
Maximum hoisting velocities $\bar{v}_{\mathrm{max}}$ (m/s)	15
Total hoisting time t_f (s)	150
Cable material damping coefficient $\bar{c}$ (N·s/m)	0.4

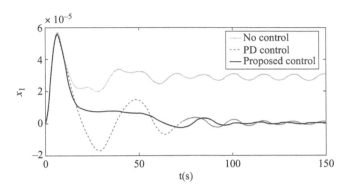

Figure 5.3. Responses of $x_1(t)$.

Numerical Methods

In the simulation program, the closed-loop system on the time-varying domain $[0, l(t)]$ is converted to the one on the fixed domain $[0, 1]$ with time-varying coefficients via introducing a new variable $\ell_1 = x/l(t)$, where the hoisting trajectory $l(t)$ is a predetermined time-varying function and then the computation for the plant and observer is conducted using the finite-difference method with the step sizes of t and ℓ_1 as 0.001, 0.05, respectively. The obtained responses are then represented back in the domain $[0, l(t)]$, as shown in this section. The approximate solutions of the kernel PDEs (5.75)–(5.82) are also solved by the finite-difference method, where, in addition to $\ell_1 = x/l(t)$ mentioned above, the variable $\ell_2 = y/l(t)$ is introduced, and the lower triangular domain $0 \leq \ell_2 \leq \ell_1 \leq 1$ is discretized as a grid with the uniform interval of 0.05. Notice that the derivatives on the left side of (5.78), (5.81) on $x = y$ are represented by a finite-difference scheme in the direction of the information flow—that is, along the line $x = y$ for the sake of avoiding using the points outside the domain when performing the finite-difference method. The approximate solutions of the kernel PDEs (5.94)–(5.97) are obtained with a similar process. For the approximate solutions of the kernel PDEs (5.38)–(5.45), a similar process is conducted on the upper triangular domain $0 \leq \ell_1 \leq \ell_2 \leq 1$.

Simulation Results

We compare the performance of the proposed controller with a traditional proportional-derivative (PD) controller $U_{pd}(t)$:

$$U_{pd}(t) = -800x_1(t) - 3400x_2(t) - p_2 z(l(t), t), \qquad (5.146)$$

where the design parameters are chosen by trial and error, and the term $-p_2 z(l(t), t)$ is to compensate the proximal reflection term for a fair comparison with the proposed controller. Figures 5.3 and 5.4 show that the proposed control guarantees a fast reduction of the lateral vibration amplitude and velocity of the cage. Moreover, the proposed controller suppresses the lateral vibration amplitude and velocity of the cage faster than the PD controller. Similar results are observed in the PDE domain—that is, the closed-loop responses at the midpoint of the spatial domain, shown in figure 5.5.

According to the uncertain system matrix A, the input matrix B in (5.13), and the target system matrix A_{m}, we know that the ideal control parameters k_1, k_2

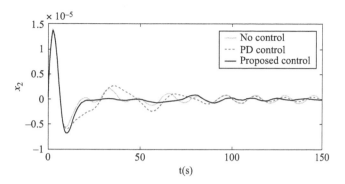

Figure 5.4. Responses of $x_2(t)$.

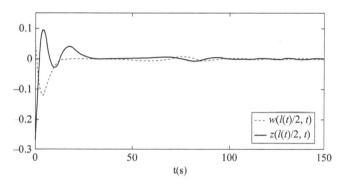

Figure 5.5. Responses of the PDE states $w(l(t)/2, t), z(l(t)/2, t)$ under the proposed control.

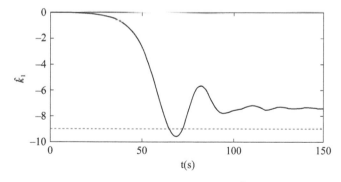

Figure 5.6. Self-adjustment of the control parameter $\hat{k}_1$ to approach the ideal value $k_1 = -9$.

are $-9, -30$ according to (5.18). Figures 5.6–5.9 show that our adaptive design can adjust online the control parameters $\hat{k}_1(t), \hat{k}_2(t)$ to approach the ideal values k_1, k_2 and estimate the unknown disturbance amplitudes a_1, b_1. Figure 5.10 shows that the observer errors of the PDE state at the midpoint of the time-varying spatial domain are convergent to zero. The PD and the proposed control inputs applied in (5.10) are given in figure 5.11. The proposed control input is not completely

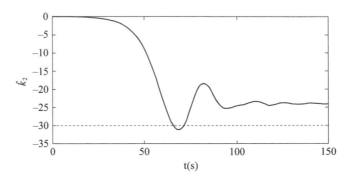

Figure 5.7. Self-adjustment of the control parameter $\hat{k}_2$ to approach the ideal value $k_2 = -30$.

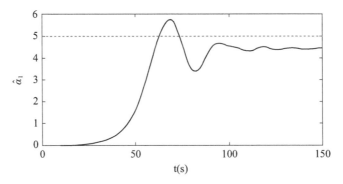

Figure 5.8. Estimation $\hat{a}_1$ of the disturbance amplitude $a_1 = 5$.

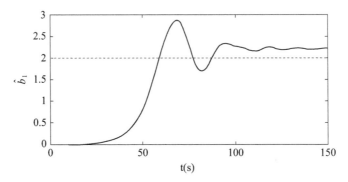

Figure 5.9. Estimation $\hat{b}_1$ of the disturbance amplitude $b_1 = 2$.

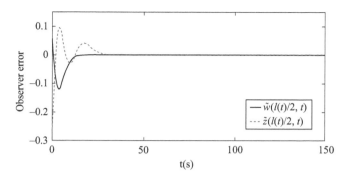

Figure 5.10. Responses of observer errors.

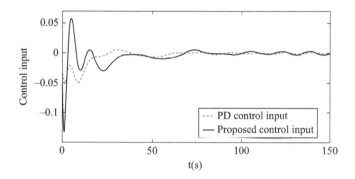

Figure 5.11. Control input.

convergent to zero because the estimates for the harmonic disturbance are not zero in the control law.

5.7   APPENDIX

A. Calculating the conditions of kernels $\bar{\phi}(x,y)$, $\check{\phi}(x,y)$, $\bar{\psi}(x,y)$, $\check{\psi}(x,y)$

Substituting (5.32), (5.33) into (5.29) along (5.34)–(5.37), we get

$$
\ddot{\tilde{z}}_t(x,t) + q_1(x)\tilde{z}_x(x,l) - c_1(x)\tilde{z}(x,t) - c_2(x)\tilde{w}(x,t) + \Phi_2(x,t)\tilde{z}(l(t),t)
$$

$$
= c_1(x)\tilde{\alpha}(x,t) + \int_x^{l(t)} \bar{\phi}(x,y)q_1(y)\tilde{\alpha}_x(y,t)dy
$$

$$
+ \int_x^{l(t)} \Big(-q_1(x)\bar{\phi}_x(x,y) - c_1(y)\bar{\phi}(x,y)
$$

$$
+ c_1(x)\bar{\phi}(x,y) + c_2(x)\bar{\psi}(x,y) \Big)\tilde{\alpha}(y,t)dy
$$

$$
+ q_1(x)\bar{\phi}(x,x)\tilde{\alpha}(x,t) - \dot{l}(t)\bar{\phi}(x,l(t))\tilde{\alpha}(l(t),t)
$$

$$
+ \Phi_2(x,t)\tilde{\alpha}(l(t),t) - c_1(x)\tilde{\alpha}(x,t) - c_2(x)\tilde{\beta}(x,t)
$$

$$-\int_x^{l(t)} \check{\phi}(x,y) \left(q_2(y)\tilde{\beta}_x(y,t) + c_4(y)\tilde{\beta}(y,t) \right) dy$$

$$-q_1(x) \int_x^{l(t)} \check{\phi}_x(x,y)\tilde{\beta}(y,t)dy$$

$$+q_1(x)\check{\phi}(x,x)\tilde{\beta}(x,t) + c_1(x) \int_x^{l(t)} \check{\phi}(x,y)\tilde{\beta}(y,t)dy$$

$$+c_2(x) \int_x^{l(t)} \check{\psi}(x,y)\tilde{\beta}(y,t)dy$$

$$= \int_x^{l(t)} \left(-q_1(x)\bar{\phi}_x(x,y) - c_1(y)\bar{\phi}(x,y) + c_1(x)\bar{\phi}(x,y) \right.$$

$$\left. +c_2(x)\bar{\psi}(x,y) - q_1(y)\bar{\phi}_y(x,y) - q_1{}'(y)\bar{\phi}(x,y) \right) \tilde{\alpha}(y,t)dy$$

$$+ \int_x^{l(t)} \left(\check{\phi}_y(x,y)q_2(y) + \check{\phi}(x,y)q_2{}'(y) - q_1(x)\check{\phi}_x(x,y) \right.$$

$$\left. +(c_1(x) - c_4(y))\check{\phi}(x,y) + c_2(x)\check{\psi}(x,y) \right) \tilde{\beta}(y,t)dy$$

$$- \left(\dot{l}(t)\bar{\phi}(x,l(t)) - \Phi_2(x,t) - q_1(l(t))\bar{\phi}(x,l(t)) \right) \tilde{\alpha}(l(t),t)$$

$$+ \left(\check{\phi}(x,x)(q_2(x) + q_1(x)) - c_2(x) \right) \tilde{\beta}(x,t) = 0. \tag{5.147}$$

Substituting (5.32), (5.33) into (5.30) along (5.34)–(5.37), we obtain

$$\tilde{w}_t(x,t) - q_2(x)\tilde{w}_x(x,t) - c_3(x)\tilde{z}(x,t)$$

$$-c_4(x)\tilde{w}(x,t) + \Phi_3(x,t)\tilde{z}(l(t),t)$$

$$= \int_x^{l(t)} \bar{\psi}(x,y)q_1(y)\tilde{\alpha}_x(y,t)dy - \int_x^{l(t)} c_1(y)\bar{\psi}(x,y)\tilde{\alpha}(y,t)dy$$

$$+q_2(x) \int_x^{l(t)} \bar{\psi}_x(x,y)\tilde{\alpha}(y,t)dy - q_2(x)\bar{\psi}(x,x)\tilde{\alpha}(x,t)$$

$$-\dot{l}(t)\bar{\psi}(x,l(t))\tilde{\alpha}(l(t),t)$$

$$-c_3(x)\tilde{\alpha}(x,t) + c_3(x) \int_x^{l(t)} \bar{\phi}(x,y)\tilde{\alpha}(y,t)dy$$

$$+c_4(x) \int_x^{l(t)} \bar{\psi}(x,y)\tilde{\alpha}(y,t)dy + \Phi_3(x,t)\tilde{\alpha}(l(t),t)$$

$$+ \int_x^{l(t)} \left(q_2(x)\check{\psi}_x(x,y) + c_3(x)\check{\phi}(x,y) + (c_4(x) - c_4(y))\check{\psi}(x,y) \right.$$

$$\left. +\check{\psi}_y(x,y)q_2(y) + \check{\psi}(x,y)q_2{}'(y) \right) \tilde{\beta}(y,t)dy$$

$$= \left(-c_3(x) - (q_1(x) + q_2(x))\bar{\psi}(x,x) \right) \tilde{\alpha}(x,t)$$

$$+ \int_x^{l(t)} \left(q_2(x)\check{\psi}_x(x,y) + \check{\psi}_y(x,y)q_2(y) + c_3(x)\check{\phi}(x,y) \right.$$

$$+ (c_4(x) - c_4(y))\check{\psi}(x,y) + \check{\psi}(x,y)q_2'(y)\Big)\bar{\beta}(y,t)dy$$

$$+ \int_x^{l(t)} \Big(-q_1(y)\bar{\psi}_y(x,y) + q_2(x)\bar{\psi}_x(x,y) - q_1'(y)\bar{\psi}(x,y)$$

$$+ c_3(x)\bar{\phi}(x,y) + (c_4(x) - c_1(y))\bar{\psi}(x,y)\Big)\tilde{\alpha}(y,t)dy$$

$$+ \Big(\Phi_3(x,t) - \dot{l}(t)\bar{\psi}(x,l(t)) + q_1(l(t))\bar{\psi}(x,l(t))\Big)\tilde{\alpha}(l(t),t)$$

$$= 0. \tag{5.148}$$

Inserting (5.32), (5.33) into (5.28) and recalling (5.34), we get

$$\tilde{z}(0,t) + p_1\tilde{w}(0,t) = -\int_0^{l(t)} (p_1\bar{\psi}(0,y) + \bar{\phi}(0,y))\tilde{\alpha}(y,t)dy$$

$$-\int_0^{l(t)} (p_1\check{\psi}(0,y) + \check{\phi}(0,y))\tilde{\beta}(y,t)dy = 0. \tag{5.149}$$

For (5.147)–(5.149) to hold, the conditions of the kernels are obtained as (5.38)–(5.45).

B. Calculating the conditions of kernels $\bar{\lambda}(x,y)$, $\check{\lambda}(x,y)$, $\bar{\Upsilon}(x,y)$, $\check{\Upsilon}(x,y)$

Substituting (5.68), (5.69) into (5.72) along (5.54)–(5.58), we obtain

$$\hat{\alpha}_t(x,t) + q_1(x)\hat{\alpha}_x(x,t) - c_1(x)\hat{\alpha}(x,t)$$

$$+ \bar{\lambda}(x,0)q_1(0)CX(t) - \Big(\Gamma_{1t}(x,t) - \int_0^x \bar{\lambda}(x,y)\Gamma_{1t}(y,t)dy$$

$$- \int_0^x \check{\lambda}(x,y)\Gamma_l(y,t)dy \Big) Z(t)$$

$$= \hat{s}_t(x,t) - \int_0^x \bar{\lambda}(x,y)\hat{s}_t(y,t)dy - \int_0^x \check{\lambda}(x,y)\hat{v}_t(y,t)dy$$

$$+ q_1(x)\hat{s}_x(x,t) - q_1(x)\int_0^x \bar{\lambda}_x(x,y)\hat{s}(y,t)dy$$

$$- q_1(x)\int_0^x \check{\lambda}_x(x,y)\hat{v}(y,t)dy$$

$$- q_1(x)\bar{\lambda}(x,x)\hat{s}(x,t) - q_1(x)\check{\lambda}(x,x)\hat{v}(x,t)$$

$$- c_1(x)\hat{s}(x,t) + c_1(x)\int_0^x \bar{\lambda}(x,y)\hat{s}(y,t)dy + c_1(x)\int_0^x \check{\lambda}(x,y)\hat{v}(y,t)dy$$

$$+ \bar{\lambda}(x,0)q_1(0)CX(t) - \Big(\Gamma_{1t}(x,t) - \int_0^x \bar{\lambda}(x,y)\Gamma_{1t}(y,t)dy$$

$$- \int_0^x \check{\lambda}(x,y)\Gamma_t(y,t)dy \Big) Z(t)$$

$$= \Big(\check{\lambda}(x,0)q_2(0) + \bar{\lambda}(x,0)q_1(0)p_1\Big) \hat{v}(0,t)$$

$$+ \big(c_2(x) - (q_1(x) + q_2(x))\check{\lambda}(x,x)\big)\,\hat{v}(x,t)$$

$$+ \int_0^x \bigg(\check{\lambda}_y(x,y)q_2(y) + \check{\lambda}(x,y)q_2{}'(y) - \bar{\lambda}(x,y)c_2(y)$$

$$+ (c_1(x) - c_4(y))\check{\lambda}(x,y) - q_1(x)\check{\lambda}_x(x,y)\bigg)\hat{v}(y,t)dy$$

$$- \int_0^x \bigg(\check{\lambda}(x,y)c_3(y) + \bar{\lambda}_y(x,y)q_1(y) + \bar{\lambda}(x,y)q_1{}'(y)$$

$$+ \bar{\lambda}(x,y)(c_1(y) - c_1(x)) + q_1(x)\bar{\lambda}_x(x,y)\bigg)\hat{s}(y,t)dy = 0. \tag{5.150}$$

Substituting (5.68), (5.69) into (5.73) along (5.54)–(5.58), we obtain

$$\hat{\beta}_t(x,t) - q_2(x)\hat{\beta}_x(x,t) - c_4(x)\hat{\beta}(x,t)$$

$$+ \bar{\Upsilon}(x,0)q_1(0)CX(t) + \bigg(\int_0^x \bar{\Upsilon}(x,y)\Gamma_t(y,t)dy$$

$$+ \int_0^x \bar{\Upsilon}(x,y)\Gamma_{1t}(y,t)dy - \Gamma_t(x,t)\bigg)Z(t)$$

$$= \hat{v}_t(x,t) - \int_0^x \bar{\Upsilon}(x,y)\hat{s}_t(y,t)dy - \int_0^x \check{\Upsilon}(x,y)\hat{v}_t(y,t)dy$$

$$- q_2(x)\hat{v}_x(x,t) + q_2(x)\int_0^x \bar{\Upsilon}_x(x,y)\hat{s}(y,t)dy + q_2(x)\int_0^x \check{\Upsilon}_x(x,y)\hat{v}(y,t)dy$$

$$+ q_2(x)\bar{\Upsilon}(x,x)\hat{s}(x,t) + q_2(x)N(x,x)\hat{v}(x,t)$$

$$- c_4(x)\hat{v}(x,t) + c_4(x)\int_0^x \bar{\Upsilon}(x,y)\hat{s}(y,t)dy + c_4(x)\int_0^x \check{\Upsilon}(x,y)\hat{v}(y,t)dy$$

$$+ \bar{\Upsilon}(x,0)q_1(0)CX(t) + \bigg(\int_0^x \check{\Upsilon}(x,y)\Gamma_t(y,t)dy$$

$$+ \int_0^x \bar{\Upsilon}(x,y)\Gamma_{1t}(y,t)dy - \Gamma_t(x,t)\bigg)Z(t)$$

$$= \big(c_3(x) + (q_1(x) + q_2(x))\bar{\Upsilon}(x,x)\big)\,\hat{s}(x,t)$$

$$+ \big(\check{\Upsilon}(x,0)q_2(0) + \bar{\Upsilon}(x,0)q_1(0)p_1\big)\,\hat{v}(0,t)$$

$$+ \int_0^x \bigg(\check{\Upsilon}_y(x,y)q_2(y) + \check{\Upsilon}(x,y)q_2{}'(y) - \bar{\Upsilon}(x,y)c_2(y)$$

$$+ q_2(x)\check{\Upsilon}_x(x,y) + (c_4(x) - c_4(y))\check{\Upsilon}(x,y)\bigg)\hat{v}(y,t)dy$$

$$- \int_0^x \bigg(\check{\Upsilon}(x,y)c_3(y) + \bar{\Upsilon}_y(x,y)q_1(y) + \bar{\Upsilon}(x,y)q_1{}'(y)$$

$$- q_2(x)\bar{\Upsilon}_x(x,y) + \bar{\Upsilon}(x,y)(c_1(y) - c_4(x))\bigg)\hat{s}(y,t)dy = 0. \tag{5.151}$$

Equations (5.70), (5.71) hold straightforwardly because of $\hat{\alpha}(0,t) = \hat{s}(0,t)$, $\hat{\beta}(0,t) = \hat{v}(0,t)$ from (5.68), (5.69). Equation (5.74) is obtained by using (5.69), (5.58). According to (5.150), (5.151), the kernels $\bar{\lambda}(x,y)$, $\check{\lambda}(x,y)$, $\bar{\Upsilon}(x,y)$, $\check{\Upsilon}(x,y)$ should satisfy (5.75)–(5.82).

C. Calculating conditions of kernels $\hat{N}(x, y; \hat{K}(t))$, $D(x; \hat{K}(t))$

Inserting (5.83) at $x = 0$ into (5.70) and matching with (5.85),

$$
\begin{aligned}
\dot{X}(t) &= A_\mathrm{m} X(t) - (A_\mathrm{m} - A - BD(0; \hat{K}(t)))X(t) \\
&\quad + B\hat{\eta}(0, t) + B_1 \tilde{d}(t) \\
&= A_\mathrm{m} X(t) - B(K - D(0; \hat{K}(t)))X(t) + B\hat{\eta}(0, t) + B_1 \tilde{d}(t) \\
&= A_\mathrm{m} X(t) + B\hat{\eta}(0, t) - B\tilde{K}(t)X(t) + B_1 \tilde{d}(t),
\end{aligned}
\tag{5.152}
$$

and we obtain the condition for $D(0; \hat{K}(t))$. Equation (5.86) holds directly by inserting (5.83) at $x = 0$ into (5.71).

Inserting (5.83) into (5.88), recalling (5.85) and (5.73), we get

$$
\begin{aligned}
&\hat{\eta}_t(x, t) - q_2(x)\hat{\eta}_x(x, t) - c_4(x)\hat{\eta}(x, t) \\
&\quad - \left[\Gamma_t(x, t) - \int_0^x \check{\Upsilon}(x, y)\Gamma_t(y, t)dy - \int_0^x \bar{\Upsilon}(x, y)\Gamma_{1t}(y, t)dy \right.\\
&\qquad - \int_0^x \hat{N}(x, y; \hat{K}(t))\left(-\int_0^y \check{\Upsilon}(y, z)\Gamma_t(z, t)dz \right. \\
&\qquad \left. \left. - \int_0^y \bar{\Upsilon}(y, z)\Gamma_{1t}(z, t)dz + \Gamma_t(y, t) \right) dy \right] Z(t) \\
&\quad - \left(D(x; \hat{K}(t))B\tilde{K}(t) - \dot{\hat{K}}(t)D_{\hat{K}(t)}(x; \hat{K}(t)) \right) X(t) \\
&\quad + \dot{\hat{K}}(t)R(x, t) + D(x; \hat{K}(t))B_1 \tilde{d}(t) \\
&= \hat{\beta}_t(x, t) - \int_0^x \hat{N}(x, y; \hat{K}(t))\hat{\beta}_t(y, t)dy - \dot{\hat{K}}(t)R(x, t) \\
&\quad - D(x; \hat{K}(t))\dot{X}(t) - \dot{\hat{K}}(t)D_{\hat{K}(t)}(x; \hat{K}(t))X(t) \\
&\quad - q_2(x)\hat{\beta}_x(x, t) + q_2(x)\int_0^x \hat{N}_x(x, y; \hat{K}(t))\hat{\beta}(y, t)dy \\
&\quad + q_2(x)\hat{N}(x, x, \hat{K}(t))\hat{\beta}(x, t) + q_2(x)D'(x; \hat{K}(t))X(t) \\
&\quad - c_4(x)\hat{\beta}(x, t) + c_4(x)\int_0^x \hat{N}(x, y; \hat{K}(t))\hat{\beta}(y, t)dy + c_4(x)D(x; \hat{K}(t))X(t) \\
&\quad - \left[\Gamma_t(x, t) - \int_0^x \check{\Upsilon}(x, y)\Gamma_t(y, t)dy - \int_0^x \bar{\Upsilon}(x, y)\Gamma_{1t}(y, t)dy \right. \\
&\qquad - \int_0^x \hat{N}(x, y; \hat{K}(t))\left(-\int_0^y \check{\Upsilon}(y, z)\Gamma_t(z, t)dz \right. \\
&\qquad \left. \left. - \int_0^y \bar{\Upsilon}(y, z)\Gamma_{1t}(z, t)dz + \Gamma_t(y, t) \right) dy \right] Z(t) \\
&\quad - \left(D(x; \hat{K}(t))B\tilde{K}(t) - \dot{\hat{K}}(t)D_{\hat{K}(t)}(x; \hat{K}(t)) \right) X(t) \\
&\quad + \dot{\hat{K}}(t)R(x, t) + D(x; \hat{K}(t))B_1 \tilde{d}(t)
\end{aligned}
$$

$$= \Big(q_2(0)\hat{N}(x,0,\hat{K}(t)) - D(x;\hat{K}(t))B \Big)\hat{\eta}(0,t)$$

$$- \Big(-q_2(x)D'(x;\hat{K}(t)) + D(x;\hat{K}(t))(A_{\mathrm{m}} - c_4(x)I_n)$$

$$+ \bar{\Upsilon}(x,0)q_1(0)C - \int_0^x \hat{N}(x,y;\hat{K}(t))\bar{\Upsilon}(y,0)q_1(0)C dy$$

$$- q_2(0)\hat{N}(x,0,\hat{K}(t))D(0;\hat{K}(t)) \Big)X(t)$$

$$+ \int_0^x \Big(q_2(y)\hat{N}_y(x,y;\hat{K}(t)) + q_2{}'(y)\hat{N}(x,y;\hat{K}(t))$$

$$+ q_2(x)\hat{N}_x(x,y;\hat{K}(t)) \Big)\hat{\beta}(y,t)dy = 0. \tag{5.153}$$

For (5.153) to hold and recalling the condition derived from (5.152), the conditions of kernels $\hat{N}, D$ are obtained as (5.94)–(5.97).

D. Calculating $\dot{V}(t)$ (5.123)

Taking the derivative of (5.122) along (5.85)–(5.89), we obtain

$$\dot{V}(t) \leq \frac{1}{1+\Omega(t)} \bigg[-\lambda_{\min}(Q)X(t)^2 + 2X^T PB\hat{\eta}(0,t)$$

$$+ 2X^T PB_1\tilde{d}(t) - 2\tilde{K}(t)X(t)B^T PX(t)$$

$$+ r_a \int_0^{l(t)} e^{\delta x}\hat{\eta}(x,t)\hat{\eta}_t(x,t)dx$$

$$+ r_b \int_0^{l(t)} e^{-\delta x}\hat{\alpha}(x,t)\hat{\alpha}_t(x,t)dx + \frac{\dot{l}(t)}{2}r_b e^{-\delta l(t)}\hat{\alpha}(l(t),t)^2 \bigg]$$

$$+ \tilde{K}(t)\gamma_c^{-1}\dot{\tilde{K}}(t)^T + \sum_{j=1}^N \frac{1}{\gamma_{aj}}\dot{\tilde{a}}_j(t)\tilde{a}_j(t) + \sum_{j=1}^N \frac{1}{\gamma_{bj}}\dot{\tilde{b}}_j(t)\tilde{b}_j(t)$$

$$\leq \frac{1}{1+\Omega(t)} \bigg[-\lambda_{\min}(Q)|X(t)|^2 + 2X^T PB\hat{\eta}(0,t)$$

$$+ 2X^T PB_1\tilde{d}(t) - 2\tilde{K}(t)X(t)B^T PX(t)$$

$$+ r_a \int_0^{l(t)} e^{\delta x}\hat{\eta}(x,t)\left(q_2(x)\hat{\eta}_x(x,t) + c_4(x)\hat{\eta}(x,t) \right)dx$$

$$+ r_b \int_0^{l(t)} e^{-\delta x}\hat{\alpha}(x,t)\left(-q_1(x)\hat{\alpha}_x(x,t) + c_1(x)\hat{\alpha}(x,t) \right)dx$$

$$+ \frac{\dot{l}(t)}{2}r_b e^{-\delta l(t)}\hat{\alpha}(l(t),t)^2 - r_a \int_0^{l(t)} e^{\delta x}\hat{\eta}(x,t)$$

$$\times \left(D(x;\hat{K}(t))B_1\tilde{d}(t) - D(x;\hat{K}(t))B\tilde{K}(t)X(t) \right)dx$$

$$+ r_a \int_0^{l(t)} e^{\delta x}\hat{\eta}(x,t)\left(H_b Z(t) - \dot{\hat{K}}D_{\hat{K}(t)}X(t) - \dot{\hat{K}}R(x,t) \right)dx$$

$$+ r_b \int_0^{l(t)} e^{-\delta x} \hat{\alpha}(x,t) \left(H_a Z(t) - \bar{\lambda}(x,0) q_1(0) C X(t) \right) dx \Bigg]$$

$$+ \tilde{K}(t) \gamma_c^{-1} \dot{\tilde{K}}(t)^T + \sum_{j=1}^N \frac{1}{\gamma_{aj}} \dot{\tilde{a}}_j(t) \tilde{a}_j(t) + \sum_{j=1}^N \frac{1}{\gamma_{bj}} \dot{\tilde{b}}_j(t) \tilde{b}_j(t),$$

where

$$\begin{aligned}
H_a &= \Gamma_{1t}(x,t) - \int_0^x \bar{\lambda}(x,y) \Gamma_{1t}(y,t) dy \\
&\quad - \int_0^x \check{\lambda}(x,y) \Gamma_t(y,t) dy, \quad\quad\quad\quad (5.154) \\
H_b &= -\int_0^x \tilde{\Upsilon}(x,y) \Gamma_t(y,t) dy - \int_0^x \bar{\Upsilon}(x,y) \Gamma_{1t}(y,t) dy \\
&\quad + \Gamma_t(x,t) - \int_0^x \hat{N}(x,y; \hat{K}(t)) \left[-\int_0^y \check{\Upsilon}(y,z) \Gamma_t(z,t) dz \right. \\
&\quad \left. - \int_0^y \bar{\Upsilon}(y,z) \Gamma_{1t}(z,t) dz + \Gamma_t(y,t) \right] dy. \quad (5.155)
\end{aligned}$$

Using integration by parts and

$$\dot{\tilde{K}}(t) = -\dot{\hat{K}}(t), \;\; \dot{\tilde{a}}_j(t) = -\dot{\hat{a}}_j(t), \;\; \dot{\tilde{b}}_j(t) = -\dot{\hat{b}}_j(t),$$

applying Young's inequality, we get

$$\begin{aligned}
\dot{V}(t) \leq \frac{1}{1+\Omega(t)} &\left[-\frac{7}{8} \lambda_{\min}(Q) |X(t)|^2 \right. \\
&+ \frac{8}{\lambda_{\min}(Q)} |PB|^2 \hat{\eta}(0,t)^2 + 2 X^T P B_1 \tilde{d}(t) \\
&- 2\tilde{K}(t) X(t) B^T P X(t) + \frac{1}{2} q_2(l(t)) r_a e^{\delta L} \hat{\eta}(l(t),t)^2 \\
&- \frac{1}{2} q_2(0) r_a \hat{\eta}(0,t)^2 - \frac{1}{2} \delta r_a \int_0^{l(t)} e^{\delta x} q_2(x) \hat{\eta}(x,t)^2 dx \\
&- \frac{1}{2} r_a \int_0^{l(t)} e^{\delta x} q_2{}'(x) \hat{\eta}(x,t)^2 dx + r_a \int_0^{l(t)} c_4(x) e^{\delta x} \hat{\eta}(x,t)^2 dx \\
&- \frac{1}{2} (q_1(l(t)) - \dot{l}(t)) r_b e^{-\delta L} \hat{\alpha}(l(t),t)^2 + \frac{1}{2} q_1(0) r_b \hat{\alpha}(0,t)^2 \\
&- \frac{1}{2} \delta r_b \int_0^{l(t)} e^{-\delta x} q_1(x) \hat{\alpha}(x,t)^2 dx \\
&+ \frac{1}{2} r_b \int_0^{l(t)} e^{-\delta x} q_1'(x) \hat{\alpha}(x,t)^2 dx \\
&+ r_b \int_0^{l(t)} c_1(x) e^{-\delta x} \hat{\alpha}(x,t)^2 dx + r_a \int_0^{l(t)} e^{\delta x} \hat{\eta}(x,t) \\
&\times \left(H_b Z(t) - \dot{\hat{K}}(t) D_{\hat{K}(t)} X(t) - \hat{K}(t) R(x,t) \right) dx \\
&+ r_b \int_0^{l(t)} e^{-\delta x} \hat{\alpha}(x,t) \left(H_a Z(t) - \bar{\lambda}(x,0) q_1(0) C X(t) \right) dx
\end{aligned}$$

$$-r_a \int_0^{l(t)} e^{\delta x} \hat{\eta}(x,t) \Big(D(x;\hat{K}(t))B_1 \tilde{d}(t)$$

$$-D(x;\hat{K}(t))B\tilde{K}(t)X(t)\Big) dx\Bigg] - \tilde{K}(t)\gamma_c^{-1}\dot{\hat{K}}(t)^T$$

$$-\sum_{j=1}^{N} \frac{1}{\gamma_{aj}}\dot{\hat{a}}_j(t)\tilde{a}_j(t) - \sum_{j=1}^{N} \frac{1}{\gamma_{bj}}\dot{\hat{b}}_j(t)\tilde{b}_j(t). \tag{5.156}$$

Recalling (5.86) and applying the Young and Cauchy-Schwarz inequalities, one obtains

$$\dot{V}(t) \leq \frac{1}{1+\Omega(t)}\Bigg[-\left(\frac{6}{8}\lambda_{\min}(Q) - q_1(0)r_b\bar{D}^2\right.$$

$$\left.-\frac{q_1(0)^2}{2}Lr_b\bar{J}^2\left|C\right|^2\right)|X(t)|^2 + \frac{8}{\lambda_{\min}(Q)}|PB|^2\hat{\eta}(0,t)^2$$

$$+2X^T PB_1\tilde{d}(t) - 2\tilde{K}(t)X(t)B^T PX(t)$$

$$-\left(\frac{1}{2}q_2(0)r_a - q_1(0)r_b p_1^2\right)\hat{\eta}(0,t)^2$$

$$-\left(\frac{1}{2}\delta\underline{q_2}r_a - r_a\overline{c_4} - \frac{1}{2}r_a\overline{q_2'}\right)\int_0^{l(t)} e^{\delta x}\hat{\eta}(x,t)^2 dx$$

$$-\frac{1}{2}(q_1 - \dot{l}(t))r_b e^{-\delta L}\hat{\alpha}(l(t),t)^2$$

$$-\left(\frac{1}{2}\underline{q_1}\delta r_b - r_b\overline{c_1} - \frac{1}{2}r_b - \frac{1}{2}r_b\overline{q_1'}\right)\int_0^{l(t)} e^{-\delta x}\hat{\alpha}(x,t)^2 dx$$

$$+r_a \int_0^{l(t)} e^{\delta x}\hat{\eta}(x,t)\left(H_b Z(t) + \dot{\hat{K}}D_{\hat{K}(t)}X(t) - \dot{\hat{K}}R(x,t)\right) dx$$

$$+r_b \int_0^{l(t)} e^{-\delta x}\hat{\alpha}(x,t)H_a Z(t) dx$$

$$-r_a \int_0^{l(t)} e^{\delta x}\hat{\eta}(x,t)\big(D(x;\hat{K}(t))B_1\tilde{d}(t)$$

$$-\tilde{K}(t)X(t)B^T D(x;\hat{K}(t))^T\big) dx\Bigg] - \tilde{K}(t)\gamma_c^{-1}\dot{\hat{K}}(t)^T$$

$$-\sum_{j=1}^{N} \frac{1}{\gamma_{aj}}\dot{\hat{a}}_j(t)\tilde{a}_j(t) - \sum_{j=1}^{N} \frac{1}{\gamma_{bj}}\dot{\hat{b}}_j(t)\tilde{b}_j(t), \tag{5.157}$$

where

$$\bar{J} = \max_{0 \leq x \leq L} \{|\bar{\lambda}(x,0)|\}, \tag{5.158}$$

$$D = \max_{0 \leq x \leq L, \underline{k}_i \leq \hat{k}_i(t) \leq \bar{k}_i} \{|C - p_1 D(x;\hat{K}(t))|\}. \tag{5.159}$$

From (5.157), we arrive at (5.123).

5.8   NOTES

Inspired by a practical control problem in the mining cable elevator with a disturbed cage moving along a flexible guide, based on the basic boundary control design of coupled hyperbolic PDEs in [96], [48], we considered the adaptive control of coupled hyperbolic PDEs coupled with a disturbed and highly uncertain ODE anti-collocated with the PDE boundary control input in this chapter. Compared to the adaptive control of 2×2 hyperbolic systems in [5], [7], our plant was characterized by the time-varying PDE domain and an additional disturbed unstable ODE anti-collocated with the control input.

In this chapter we examined the transverse vibration instead of the axial vibration of the cable as in chapter 3, and thus the constant coefficients in the plant in chapter 3 are generalized to the spatially varying coefficients, considering the spatially varying tension in the transverse vibration cable dynamics. Moreover, the continuous-in-time controller in this chapter will be extended to a piecewise-constant one in chapter 13 to alleviate the adaptive learning transient and make the adaptive control law more implementable in string-actuated mechanisms with massive actuators.

Chapter Six

Deep-Sea Construction

In addition to control of the mining cable elevator dealt with in chapters 2–5, here we address the control problem for a deep-sea construction vessel (DCV), which is used to place the underwater parts of an offshore oil-drilling platform at designated locations on the seafloor. Similarly, cable actuation can be used in the construction of artificial reefs for enabling the development and growth of marine life, in laying communication cables on the seafloor, which is often uneven and rife with obstacles, and in other undersea applications. Additionally, the control design in this chapter can also be applied in unmanned aerial vehicles (UAVs), to achieve rapid suppression of oscillations of the cable and precisely place the suspended object at the target position via the regulation force provided by the UAV's rotor wings.

The key difference in the vibration control in this chapter relative to that in chapters 2–5 is the suppression of the lateral-longitudinal coupled vibrations of the cable instead of only one-dimensional axial vibrations.

By using the extended Hamilton's principle, in section 6.1 a nonlinear PDE system is derived governing the lateral-longitudinal coupled vibration dynamics of the DCV comprising a cable of time-varying length and an attached payload. Then this model is linearized at the steady state, generating a linear partial differential equation (PDE) model that is extended to a more general system including two coupled wave PDEs connected with two interacting ordinary differential equations (ODEs) at the uncontrolled boundary. Through a preliminary transformation, an equivalent reformulated plant is generated as a 4×4 coupled heterodirectional hyperbolic PDE-ODE system with spatially varying coefficients on a time-varying domain. For this cascade of a 4×4 PDE system into an ODE, a control design is then conducted.

In section 6.2, the state-feedback control design is presented, and the exponential stability result of the state-feedback closed-loop system is proved. In section 6.3, we design a state observer and prove the exponential convergence to zero of the observer errors. In section 6.4, we propose an observer-based output-feedback controller where the measurements are only placed at the actuated boundary of the PDE, namely, at the platform at the sea surface, and prove the exponential stability of the closed-loop system, as well as the boundedness and exponential convergence of the control inputs, via Lyapunov analysis. In section 6.5, the obtained theoretical result is tested on a nonlinear model with ocean disturbances, even though the design is developed in the absence of such real-world effects.

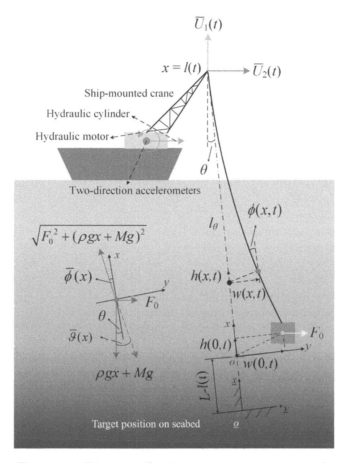

Figure 6.1. Diagram of a deep-sea construction vessel.

6.1   MODELING PROCESS AND LINEARIZATION

Modeling

DCVs are often used to install the underwater parts of an offshore drilling platform, such as a subsea manifold, a subsea pump station, a subsea distribution unit along with the associated foundations, flowlines, and umbilicals [167], [168]. A DCV is depicted in figure 6.1, where a crane mounted on a ship regulates a cable to install a piece of equipment (attached payload) at the target position on the seabed. The attached payload is subject to a constant drag force caused by a constant water stream velocity. We consider only the water-stream-caused drag force at the payload because the diameter of the cable is much smaller than that of the payload. For the suppression of cable oscillation/vibration, two-directional control forces implemented by two actuators (hydraulic cylinder and hydraulic motor) and measurements by accelerometers are applied/placed at the ship-mounted crane. The varying cable length, driven by an additional winch wound with the cable on the ship, can be considered a predefined time-varying function $l(t)$ in the cable vibration dynamics regulated by the cranes.

   Hamilton's principle [99] is applied to formulate the mathematical model of the two-dimensional vibrations of the DCV in figure 6.1, where we neglect the dynamics

Table 6.1. Physical parameters of the DCV

Parameters (units)	Values
Maximal cable length L (m)	1210
Initial cable length $l(0)$ (m)	250
Maximum descending velocity $\bar{M}$ (m/s)	10
Operation time t_f (s)	120
Cable cross-sectional area A_a (m^2)	0.47×10^{-3}
Cable effective Young's modulus E (N/m^2)	7.03×10^{10}
Cable linear density m_c (kg/m)	8.95
Payload mass M_L (kg)	8000
Payload volume V_p (m^3)	5
Gravitational acceleration g (m/s^2)	9.8
Drag coefficient C_d	1
Stream velocity V_s (m/s)	2
Seawater density ρ_w (kgm^{-3})	1024
Longitudinal damping coefficient in cable c_u	0.5
Lateral damping coefficient in cable c_v	0.3
Longitudinal damping coefficient at attached payload c_h	0.5
Lateral damping coefficient at attached payload c_w	0.3

of the ship because it can be kept at the desired position by the ship dynamic positioning system. The physical parameters of the DCV are shown in table 6.1, and the given values are used in the simulation. To describe the vibrations of the cable, the classical moving frame approach is used [36]. In figure 6.1, the xoy frame is a moving coordinate associated with cable motion—that is, the time-varying cable length $l(t)$, moving along line l_θ, where x ranges from $x = 0$ at the cable bottom to $x = l(t)$ at the top end. The other coordinate frame $\underline{xoy}$ is earth fixed. The buoyancy of the cable can easily be included by adjusting the cable linear density m_c as

$$\rho = m_c - \rho_w A_a$$

and changing the payload mass M_L as

$$M = M_L - \rho_w V_p$$

when calculating the gravity of the payload.

Denote the longitudinal and (in-plane) lateral dynamic deflections in the distributed position x in the cable as $h(x,t)$ and $w(x,t)$, respectively. The deformed position vector—that is, the vibration displacement vector—in the earth fixed coordinate frame $\underline{xoy}$ is

$$D(x,t) = [L - l(t) + x + h(x,t), w(x,t)]^T. \tag{6.1}$$

Thus, the velocity vector is

$$D_t(x,t) = [h_t(x,t) - \dot{l}(t), w_t(x,t)]^T. \tag{6.2}$$

According to the large displacement approach [66], the strain at the distributed positions x in the cable is given by

$$\varepsilon(x,t) = h_x(x,t) + \frac{1}{2}w_x(x,t)^2. \tag{6.3}$$

Therefore, the kinetic energy E_k and the potential energy E_p of the system are represented as

$$E_k = \frac{\rho}{2}\int_0^{l(t)} |D_t(x,t)|^2 dx + \frac{M_L}{2}|D_t(0,t)|^2, \tag{6.4}$$

$$E_p = \frac{1}{2}\int_0^{l(t)} EA_a\varepsilon(x,t)^2 dx + \int_0^{l(t)} T(x)\varepsilon(x,t)dx, \tag{6.5}$$

where $T(x)$ is the static tension, given by

$$T(x) = \rho g x + M g. \tag{6.6}$$

Because the long steel cable and the equipment at the bottom of the cable are very heavy and the resulting gravity is much larger than the water-stream-caused drag force

$$F_0 = \frac{\rho_w}{2}C_d V_s^2 \tag{6.7}$$

taken from [22], the angle

$$\theta = \arctan\frac{F_0}{\sqrt{\rho g L + M g}} \tag{6.8}$$

in figure 6.1 is small. Therefore, it is reasonable to assume that the static tension $T(x)$ only results from the mass of the payload and cable. The virtual work is

$$\delta W = \bar{U}_1\delta h(l(t),t) + \bar{U}_2\delta w(l(t),t) + c_h(h_t(0,t) - \dot{l}(t))\delta h(0,t)$$
$$+ c_w w_t(0,t)\delta w(0,t) + \int_0^{l(t)} c_v w_t(x,t)\delta w(x,t)dx$$
$$+ \int_0^{l(t)} c_u(h_t(x,t) - \dot{l}(t))\delta h(x,t)dx + F_0\delta w(0,t)$$
$$- M g\delta h(0,t) - \int_0^{l(t)} \rho g\delta h(x,t)dx, \tag{6.9}$$

where, due to the small θ, an approximation is adopted that the virtual work done by the gravity and U_1 are parallel to the x-direction and that the virtual work done by F_0 and U_2 are parallel to the y-direction.

Apply variations of (6.4), (6.5) into the extended Hamilton's principle

$$\int_{t_1}^{t_2} (\delta E_k - \delta E_p + \delta W)dt = 0, \tag{6.10}$$

where a difference [199] from the standard procedure due to the time-varying integration domain should be noted in (6.11), (6.12) that follow. The use of Leibnitz's rule gives

$$\rho \int_0^{l(t)} D_t(x,t)^T \delta D_t(x,t) dx$$

$$= \frac{\rho \int_0^{l(t)} D_t(x,t)^T \delta D(x,t) dx}{dt} - \dot{l}(t)\rho D_t(l(t),t)^T \delta D(l(t),t)$$

$$- \rho \int_0^{l(t)} D_{tt}(x,t)^T \delta D(x,t) dx. \tag{6.11}$$

Then, integrating (6.11) from t_1 to t_2 and imposing vanishing variations of δD yields

$$\int_{t_1}^{t_2} \rho \int_0^{l(t)} D_t(x,t)^T \delta D_t(x,t) dx dt$$

$$= -\int_{t_1}^{t_2} \dot{l}(t)\rho D_t(l(t),t)^T \delta D(l(t),t) dt$$

$$- \int_{t_1}^{t_2} \rho \int_0^{l(t)} D_{tt}(x,t)^T \delta D(x,t) dx dt. \tag{6.12}$$

Remark that $l(t)$ is treated as a pre-defined function, and thus it is not necessary to consider its variation in $\delta D(x,t), \delta D_t(x,t)$—that is, $\delta l(t) = 0$ and $\delta \dot{l}(t) = 0$.

Through a lengthy calculation for (6.10), the governing equations are then obtained as

$$-m_c(h_{tt}(x,t) - \ddot{l}(t)) + EA_a h_{xx}(x,t)$$

$$+ c_u(h_t(x,t) - \dot{l}(t)) + EA_a w_x(x,t)w_{xx}(x,t) = 0, \tag{6.13}$$

$$-m_c w_{tt}(x,t) + \frac{3}{2}EA_a w_x(x,t)^2 w_{xx}(x,t) + T(x)w_{xx}(x,t)$$

$$+ EA_a h_x(x,t)w_{xx}(x,t) + EA_a h_{xx}(x,t)w_x(x,t)$$

$$+ c_v w_t(x,t) + T'(x)w_x(x,t) = 0, \tag{6.14}$$

$$M_L(h_{tt}(0,t) - \ddot{l}(t)) + c_h(h_t(0,t) - \dot{l}(t))$$

$$+ EA_a h_x(0,t) + \frac{1}{2}EA_a w_x(0,t)^2 = 0, \tag{6.15}$$

$$M_L w_{tt}(0,t) + c_w w_t(0,t) + \frac{1}{2}EA_a w_x(0,t)^3$$

$$+ EA_a h_x(0,t)w_x(0,t) + T(0)w_x(0,t) + F_0 = 0, \tag{6.16}$$

$$-\left(EA_a(h_x(l(t),t) + \frac{1}{2}w_x(l(t),t)^2) + T(l(t))\right)w_x(l(t),t)$$

$$- \dot{l}(t)\rho w_t(l(t),t) + \bar{U}_2(t) = 0, \tag{6.17}$$

$$-EA_a\left(h_x(l(t),t) + \frac{1}{2}w_x(l(t),t)^2\right) - T(l(t))$$

$$- \dot{l}(t)\rho(h_t(l(t),t) - \dot{l}(t)) + \bar{U}_1(t) = 0. \tag{6.18}$$

Equations (6.13)–(6.18) are a strongly nonlinear system, so an approximated linear model that is suitable for control design should be constructed. The nonlinear PDE system (6.13)–(6.18) is linearized in the next subsection around a steady state, as in [22].

Linearization

The steady states of the distributed strain and pivot angle $\varepsilon(x,t), \phi(x,t)$ can be calculated analytically and expressed as

$$\bar{\varepsilon}(x) = \frac{1}{EA_a}\sqrt{(\rho g x + Mg)^2 + F_0^2}, \tag{6.19}$$

$$\bar{\phi}(x) = \bar{\vartheta}(x) - \theta = \arctan\left(\frac{F_0}{\rho g x + Mg}\right) - \theta. \tag{6.20}$$

In nonlinear terms in (6.13)–(6.18), replacing $h_x(x,t)$, which approximately describes the distributed strain in the cable, by $\bar{\varepsilon}(x)$, and replacing $-w_x(x,t)$, which is approximately equal to the pivot angle $\phi(x,t)$ in figure 6.1, by $\bar{\phi}(x)$, a linear model around the steady state is obtained as

$$-m_c u_{tt}(x,t) + EA_a u_{xx}(x,t) - EA_a w_x(x,t)\bar{\phi}'(x) + c_u u_t(x,t) = 0, \tag{6.21}$$

$$-m_c w_{tt}(x,t) + \left(\frac{3}{2}EA_a \bar{\phi}(x)^2 + T(x)\right)w_{xx}(x,t)$$
$$+(EA_a \bar{\varepsilon}'(x) + \rho g)w_x(x,t) + c_v w_t(x,t) - EA_a \bar{\phi}'(x)u_x(x,t) = 0, \tag{6.22}$$

$$M_L u_{tt}(0,t) + c_h u_t(0,t) + EA_a u_x(0,t) - \frac{EA_a}{2}\bar{\phi}(0)w_x(0,t) = 0, \tag{6.23}$$

$$M_L w_{tt}(0,t) + c_w w_t(0,t) + \frac{1}{2}EA_a \bar{\phi}(0)^2 w_x(0,t)$$
$$+EA_a \bar{\phi}(0)u_x(0,t) = 0, \tag{6.24}$$

$$w_x(l(t),t) - \frac{1}{EA_a \bar{\varepsilon}(l(t)) + \frac{EA_a}{2}\bar{\phi}(l(t))^2 + T(l(t))}U_2(t) = 0, \tag{6.25}$$

$$u_x(l(t),t) - \frac{1}{EA_a}U_1(t) = 0, \tag{6.26}$$

in which $u(x,t)$ denotes

$$u(x,t) = h(x,t) - l(t), \tag{6.27}$$

and

$$U_1(t) = \bar{U}_1(t) + \frac{EA_a}{2}\bar{\phi}(l(t))^2 - T(l(t)) - \dot{l}(t)\rho u_t(l(t),t), \tag{6.28}$$

$$U_2(t) = \bar{U}_2(t) - \dot{l}(t)\rho w_t(l(t),t). \tag{6.29}$$

We have used

$$T(0)w_x(0,t) + F_0 \approx 0 \tag{6.30}$$

in (6.16), which is obtained by replacing $w_x(0,t)$ with the steady state

$$-\bar{\phi}(0) \approx -\arctan\left(\frac{F_0}{Mg}\right) \approx \frac{-F_0}{Mg} \tag{6.31}$$

due to $\frac{F_0}{Mg}$ being small.

From a practical point of view, the available measurements are the acceleration signals $u_{tt}(l(t), t), w_{tt}(l(t), t)$ obtained by accelerometers placed at the crane because measuring vibrational acceleration rather than velocity/displacement is more viable in vibrational mechanical systems. The velocity signals $u_t(l(t), t), w_t(l(t), t)$ can then be obtained by integrations of the measured acceleration signals under known initial conditions.

Therefore, the vibration control design of the DCV corresponds to the boundary control of the above coupled wave PDEs (6.21)–(6.26), characterized by the spatially varying coefficients related to the steady states, the time-varying domain introduced by the time-varying length of the cable, and the second-order boundary conditions describing the dynamics of the attached payload. Using the known signals $u_t(l(t), t), w_t(l(t), t)$, by virtue of (6.6), (6.20), the designed control laws $U_1(t), U_2(t)$ are converted to the physical control forces $\bar{U}_1(t), \bar{U}_2(t)$ at the ship-mounted crane.

General Plant

We represent (6.21)–(6.26) in a more general form, which includes more couplings between two wave PDEs, in both the domain and at the dynamic boundary, and consider the boundary control problem for this general model. The obtained theoretical result is then applied back to the specific application problem of the DCV—that is, to (6.21)–(6.26)—in the simulation.

The class of plants we consider is

$$
\begin{aligned}
w_{tt}(x,t) = {} & d_1(x)w_{xx}(x,t) + d_2(x)w_x(x,t) + d_3(x)u_x(x,t) \\
& + d_4(x)w_t(x,t) + d_5(x)u_t(x,t),
\end{aligned} \tag{6.32}
$$

$$
\begin{aligned}
u_{tt}(x,t) = {} & d_6(x)u_{xx}(x,t) + d_7(x)w_x(x,t) + d_8(x)u_x(x,t) \\
& + d_9(x)w_t(x,t) + d_{10}(x)u_t(x,t),
\end{aligned} \tag{6.33}
$$

$$
w_{tt}(0,t) = d_{11}w_t(0,t) + d_{12}w_x(0,t) + d_{13}u_t(0,t) + d_{14}u_x(0,t), \tag{6.34}
$$

$$
u_{tt}(0,t) = d_{15}u_t(0,t) + d_{16}u_x(0,t) + d_{17}w_t(0,t) + d_{18}w_x(0,t), \tag{6.35}
$$

$$
u_x(l(t),t) = d_{19}(l(t))U_1(t), \tag{6.36}
$$

$$
w_x(l(t),t) = d_{20}(l(t))U_2(t), \tag{6.37}
$$

$\forall (x,t) \in [0, l(t)] \times [0, \infty)$. We assume the measurements of $u_t(l(t), t), w_t(l(t), t)$ according to the available measurements of the DCV mentioned in the last subsection. The wave PDEs w and u are coupled with each other both in the domain and at the dynamic boundary. The system coefficients $d_{11}, d_{12}, d_{13}, d_{14}, d_{15}, d_{16}, d_{17}, d_{18}$ are arbitrary constants, and $d_{19}(l(t)), d_{20}(l(t))$, which belong to $C^0([0, L])$, are nonzero, where the positive constant L is the maximal length of the varying spatial domain. The spatially varying coefficients $d_1(x), d_2(x), d_3(x), d_4(x), d_5(x), d_6(x), d_7(x), d_8(x), d_9(x), d_{10}(x)$ are subject to the following two assumptions.

Assumption 6.1. *The spatially varying coefficients $d_2(x), d_3(x), d_4(x), d_5(x), d_7(x), d_8(x), d_9(x), d_{10}(x)$ belong to $C^0([0, L])$.*

Assumption 6.2. *The spatially varying coefficients $d_1(x), d_6(x) \subset C^1$ are positive, and $d_1(x) \neq d_6(x), \forall x \in [0, L]$.*

The time-varying domain—that is, the moving boundary $l(t)$—is subject to the following two assumptions.

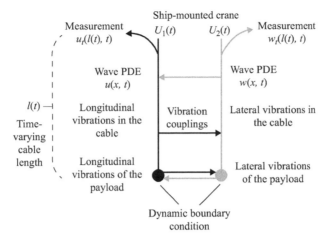

Figure 6.2. Diagram of the plant (6.32)–(6.37). The text in gray describes the physical meanings corresponding to (6.32)–(6.37) in the specific application of the DCV.

Assumption 6.3. *The function $l(t)$ is bounded—that is, $0 < l(t) \le L$, $\forall t \ge 0$.*

Assumption 6.4. *The function $\dot{l}(t)$ is bounded as*

$$\left| \dot{l}(t) \right| < \min_{0 \le x \le L} \{ \sqrt{d_1(x)}, \sqrt{d_6(x)} \}. \tag{6.38}$$

Assumption 6.4 ensures the well-posedness of the initial boundary value problem (6.32)–(6.37) according to [71, 72].

Assumptions 6.1–6.4 about the spatially varying coefficients and the time-varying spatial domain of (6.32)–(6.37) are fully satisfied in the application of the DCV, which can be easily checked by the specific expressions of $d_1, \ldots, d_{20}$ (6.215)–(6.221) and the parameter values in table 6.1 of the DCV in the simulation.

The general plant (6.32)–(6.37), whose diagram is shown in figure 6.2, covers the vibration dynamics of the DCV (6.21)–(6.26) considered in this chapter—namely, that (6.21)–(6.26) is a particular case of (6.32)–(6.37) by setting the coefficients $d_1, \ldots, d_{20}$ as the expressions (6.215)–(6.221) in the simulation.

Our objective is to exponentially stabilize (6.32)–(6.37) through designing the control inputs $U_1(t)$, $U_2(t)$ in (6.36)–(6.37) using only the boundary values $u_t(l(t), t)$, $w_t(l(t), t)$—that is, a collocated type output-feedback control system. The well-posedness of (6.32)–(6.37) can be seen based on an equivalently reformulated plant shown in the next subsection.

Reformulated Plant

A preliminary transformation, which allows one to convert the original plant (6.32)–(6.37) to a reformulated plant, where backstepping is readily applicable, is utilized. We introduce a set of the Riemann transformations

$$z(x,t) = w_t(x,t) + \sqrt{d_1(x)} w_x(x,t), \tag{6.39}$$

$$v(x,t) = w_t(x,t) - \sqrt{d_1(x)} w_x(x,t), \tag{6.40}$$

$$k(x,t) = u_t(x,t) + \sqrt{d_6(x)}u_x(x,t), \tag{6.41}$$

$$y(x,t) = u_t(x,t) - \sqrt{d_6(x)}u_x(x,t) \tag{6.42}$$

and define new variables as

$$X(t) = [w(0,t), w_t(0,t)], \quad Y(t) = [u(0,t), u_t(0,t)] \tag{6.43}$$

to reformulate (6.32)–(6.37) as

$$y_t(x,t) + \sqrt{d_6(x)}y_x(x,t)$$
$$= \left(\frac{d_7(x)}{2\sqrt{d_1(x)}} + \frac{d_9(x)}{2} \right) z(x,t)$$
$$+ \left(s_1(x) + \frac{d_{10}(x)}{2} \right) k(x,t) + \left(\frac{d_9(x)}{2} - \frac{d_7(x)}{2\sqrt{d_1(x)}} \right) v(x,t)$$
$$+ \left(\frac{d_{10}(x)}{2} - s_1(x) \right) y(x,t), \tag{6.44}$$

$$k_t(x,t) - \sqrt{d_6(x)}k_x(x,t)$$
$$= \left(\frac{d_7(x)}{2\sqrt{d_1(x)}} + \frac{d_9(x)}{2} \right) z(x,t)$$
$$+ \left(s_1(x) + \frac{d_{10}(x)}{2} \right) k(x,t) + \left(\frac{d_9(x)}{2} - \frac{d_7(x)}{2\sqrt{d_1(x)}} \right) v(x,t)$$
$$+ \left(\frac{d_{10}(x)}{2} - s_1(x) \right) y(x,t), \tag{6.45}$$

$$v_t(x,t) + \sqrt{d_1(x)}v_x(x,t)$$
$$= \left(s_2(x) + \frac{d_4(x)}{2} \right) z(x,t)$$
$$+ \left(\frac{d_3(x)}{2\sqrt{d_6(x)}} + \frac{d_5(x)}{2} \right) k(x,t) + \left(\frac{d_4(x)}{2} - s_2(x) \right) v(x,t)$$
$$+ \left(\frac{d_5(x)}{2} - \frac{d_3(x)}{2\sqrt{d_6(x)}} \right) y(x,t), \tag{6.46}$$

$$z_t(x,t) - \sqrt{d_1(x)}z_x(x,t)$$
$$= \left(s_2(x) + \frac{d_4(x)}{2} \right) z(x,t)$$
$$+ \left(\frac{d_3(x)}{2\sqrt{d_6(x)}} + \frac{d_5(x)}{2} \right) k(x,t) + \left(\frac{d_4(x)}{2} - s_2(x) \right) v(x,t)$$
$$+ \left(\frac{d_5(x)}{2} - \frac{d_3(x)}{2\sqrt{d_6(x)}} \right) y(x,t), \tag{6.47}$$

$$v(0,t) = 2C_2 X(t) - z(0,t), \tag{6.48}$$

$$y(0,t) = 2C_2 Y(t) - k(0,t), \tag{6.49}$$

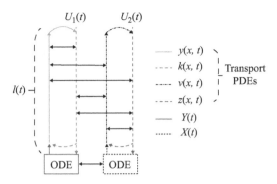

Figure 6.3. Diagram of the system (6.44)–(6.53).

$$\dot{X}(t) = \left(A_1 - \frac{B_1 d_{12} C_2}{\sqrt{d_1(0)}}\right) X(t) + \left(d_{13} B_1 C_2 - \frac{B_1 d_{14} C_2}{\sqrt{d_6(0)}}\right) Y(t)$$

$$+ \frac{B_1 d_{12}}{\sqrt{d_1(0)}} z(0,t) + \frac{B_1 d_{14}}{\sqrt{d_6(0)}} k(0,t), \tag{6.50}$$

$$\dot{Y}(t) = \left(A_2 - \frac{B_1 d_{16} C_2}{\sqrt{d_6(0)}}\right) Y(t) + \left(d_{17} B_1 C_2 - \frac{B_1 d_{18} C_2}{\sqrt{d_1(0)}}\right) X(t)$$

$$+ \frac{B_1 d_{18}}{\sqrt{d_1(0)}} z(0,t) + \frac{B_1 d_{16}}{\sqrt{d_6(0)}} k(0,t), \tag{6.51}$$

$$k(l(t),t) = 2\sqrt{d_6(l(t))} d_{19}(l(t)) U_1(t) + y(l(t),t), \tag{6.52}$$

$$z(l(t),t) = 2\sqrt{d_1(l(t))} d_{20}(l(t)) U_2(t) + v(l(t),t), \tag{6.53}$$

where

$$s_1(x) = \frac{d_8(x) - \sqrt{d_6(x)}\sqrt{d_6(x)}'}{2\sqrt{d_6(x)}}, \tag{6.54}$$

$$s_2(x) = \frac{d_2(x) - \sqrt{d_1(x)}\sqrt{d_1(x)}'}{2\sqrt{d_1(x)}}, \tag{6.55}$$

and

$$A_1 = \begin{bmatrix} 0 & 1 \\ 0 & d_{11} \end{bmatrix}, A_2 = \begin{bmatrix} 0 & 1 \\ 0 & d_{15} \end{bmatrix}, \tag{6.56}$$

$$B_1 = \begin{pmatrix} 0 \\ 1 \end{pmatrix}, C_2 = \begin{pmatrix} 0 & 1 \end{pmatrix}. \tag{6.57}$$

The diagram of the system (6.44)–(6.53) is depicted in figure 6.3. The system (6.44)–(6.53) is a 4×4 hyperbolic system coupled with ODEs, where $y(x,t), v(x,t)$ and $k(x,t), z(x,t)$ propagate in opposite directions, and where the four transport PDEs are coupled with each other in the time-varying spatial domain $[0, l(t)]$ and with the two ODEs $X(t), Y(t)$, which are coupled with each other at the uncontrolled boundary $x = 0$. Compared with the result in chapter 3, a set of two 2×2 pairs that are coupled not along the domain but at the boundary is extended to one full coupled 4×4 system because additional in-domain couplings between original wave

PDEs are considered. Moreover, the assumption in chapter 3 that some ODE states at the uncontrolled boundary are measurable is removed in this chapter.

In order to rewrite (6.44)–(6.53) in a compact form—that is, in a matrix representation—we define new variables as

$$p(x,t) = [y(x,t), v(x,t)]^T, \tag{6.58}$$

$$r(x,t) = [k(x,t), z(x,t)]^T, \tag{6.59}$$

$$W(t) = [X(t), Y(t)]^T. \tag{6.60}$$

The system (6.44)–(6.53) can then be rewritten as

$$p_t(x,t) + Q(x)p_x(x,t) = T_a(x)r(x,t) + T_b(x)p(x,t), \tag{6.61}$$

$$r_t(x,t) - Q(x)r_x(x,t) = T_a(x)r(x,t) + T_b(x)p(x,t), \tag{6.62}$$

$$p(0,t) = C_3 W(t) - r(0,t), \tag{6.63}$$

$$\dot{W}(t) = (\bar{A} - \bar{B}C_3)W(t) + 2\bar{B}r(0,t), \tag{6.64}$$

$$r(l(t),t) = R(l(t))U(t) + p(l(t),t), \tag{6.65}$$

where $U(t) = [U_1(t), U_2(t)]^T$, and

$$R(l(t)) = 2\mathrm{diag}\left\{ \sqrt{d_6(l(t))}d_{19}(l(t)), \sqrt{d_1(l(t))}d_{20}(l(t)) \right\}, \tag{6.66}$$

$$Q(x) = \mathrm{diag}\{Q_1(x), Q_2(x)\} = \mathrm{diag}\{\sqrt{d_6(x)}, \sqrt{d_1(x)}\}. \tag{6.67}$$

The matrices $\bar{A}, C_3, \bar{B}$, and

$$T_a(x) = \{T_{aij}(x)\}_{1 \le i,j \le 2}, \ \ T_b(x) = \{T_{bij}(x)\}_{1 \le i,j \le 2}$$

are given as

$$\bar{A} = \begin{bmatrix} A_1 & d_{13}B_1C_2 \\ d_{17}B_1C_2 & A_2 \end{bmatrix}, C_3 = 2 \begin{bmatrix} 0 & C_2 \\ C_2 & 0 \end{bmatrix}, \tag{6.68}$$

$$\bar{B} = \begin{bmatrix} \frac{B_1 d_{14}}{2\sqrt{d_6(0)}} & \frac{B_1 d_{12}}{2\sqrt{d_1(0)}} \\ \frac{B_1 d_{16}}{2\sqrt{d_6(0)}} & \frac{B_1 d_{18}}{2\sqrt{d_1(0)}} \end{bmatrix} = \begin{bmatrix} 0 & 0 \\ \frac{d_{14}}{2\sqrt{d_6(0)}} & \frac{d_{12}}{2\sqrt{d_1(0)}} \\ 0 & 0 \\ \frac{d_{16}}{2\sqrt{d_6(0)}} & \frac{d_{18}}{2\sqrt{d_1(0)}} \end{bmatrix}, \tag{6.69}$$

$$T_a(x) = \begin{bmatrix} s_1(x) + \frac{d_{10}(x)}{2} & \frac{d_7(x)}{2\sqrt{d_1(x)}} + \frac{d_9(x)}{2} \\ \frac{d_3(x)}{2\sqrt{d_6(x)}} + \frac{d_5(x)}{2} & s_2(x) + \frac{d_4(x)}{2} \end{bmatrix}, \tag{6.70}$$

$$T_b(x) = \begin{bmatrix} \frac{d_{10}(x)}{2} - s_1(x) & \frac{d_9(x)}{2} - \frac{d_7(x)}{2\sqrt{d_1(x)}} \\ \frac{d_5(x)}{2} - \frac{d_3(x)}{2\sqrt{d_6(x)}} & \frac{d_4(x)}{2} - s_2(x) \end{bmatrix}. \tag{6.71}$$

The following assumption is required for the stabilization of the ODE subsystem (6.64) in the state-feedback and observer design and is also satisfied in the DCV model by checking the system parameters in the simulation.

Assumption 6.5. *The pair $(\bar{A} - \bar{B}C_3, \bar{B})$ is controllable, and $(\bar{A} - \bar{B}C_3, C_3)$ is observable.*

The reformulated plant (6.61)–(6.65) obtained from (6.32)–(6.37) via the preliminary transformation in section 6.1 is well-posed because it is equivalent to the well-posed plant in [48] with setting $m = n = 2$, which indicates that the original plant (6.32)–(6.37) is also well-posed because of the invertible preliminary transformation.

In the following sections, we develop an observer-based output-feedback controller for (6.61)–(6.65) on a time-varying PDE domain and with only collocated boundary states assumed measurable. Then the control input and the stability result will be formulated back for the original plant (6.32)–(6.37).

6.2 BASIC CONTROL DESIGN USING FULL STATES

Backstepping Design

We seek an invertible transformation that converts the system (6.61)–(6.65), with variables $(p(x,t), r(x,t), W(t))$, into a so-called target system whose exponential stability is straightforward to prove.

We postulate the backstepping transformation in the form

$$\alpha(x,t) = p(x,t) - \int_0^x K(x,y)p(y,t)dy$$
$$- \int_0^x J(x,y)r(y,t)dy - \gamma(x)W(t), \tag{6.72}$$

$$\beta(x,t) = r(x,t) - \int_0^x F(x,y)p(y,t)dy$$
$$- \int_0^x N(x,y)r(y,t)dy - \lambda(x)W(t), \tag{6.73}$$

where

$$K(x,y) = \{K_{ij}(x,y)\}_{1 \le i,j \le 2}, J(x,y) = \{J_{ij}(x,y)\}_{1 \le i,j \le 2}, \tag{6.74}$$
$$F(x,y) = \{F_{ij}(x,y)\}_{1 \le i,j \le 2}, N(x,y) = \{N_{ij}(x,y)\}_{1 \le i,j \le 2}, \tag{6.75}$$

on the triangular domain $\mathcal{D} = \{0 \le y \le x \le l(t)\}$, and

$$\gamma(x) = \{\gamma_{ij}(x)\}_{1 \le i < 2, 1 \le j \le 4}, \ \lambda(x) = \{\lambda_{ij}(x)\}_{1 \le i \le 2, 1 \le j \le 4} \tag{6.76}$$

are to be determined.

The target system $(\alpha(x,t), \beta(x,t), W(t))$ is designed as

$$\dot{W}(t) = \hat{A}W(t) + 2\bar{B}\beta(0,t), \tag{6.77}$$
$$\alpha_t(x,t) = -Q(x)\alpha_x(x,t) + \bar{T}_b(x)\alpha(x,t) + g_1(x)\beta(0,t), \tag{6.78}$$
$$\beta_t(x,t) = Q(x)\beta_x(x,t) + \bar{T}_a(x)\beta(x,t) + g(x)\beta(0,t), \tag{6.79}$$
$$\alpha(0,t) = -\beta(0,t), \tag{6.80}$$
$$\beta(l(t),t) = 0, \tag{6.81}$$

where

$$\hat{A} = \bar{A} - \bar{B}C_3 + 2\bar{B}\kappa \tag{6.82}$$

is a Hurwitz matrix, by choosing

$$\kappa = \{\kappa_{ij}\}_{1\leq i\leq 2, 1\leq j\leq 4},\tag{6.83}$$

recalling assumption 6.5. The matrices $\bar{T}_a(x), \bar{T}_b(x)$ are diagonal ones consisting of the diagonal elements of (6.70), (6.71), denoted as $\bar{T}_a(x) = \mathrm{diag}(\bar{T}_{ai}(x))$ and $\bar{T}_b(x) = \mathrm{diag}(\bar{T}_{bi}(x))$ for $i = 1, 2$. Thus, all coupling terms in the PDE domain (6.61), (6.62) are removed. The functions $g(x), g_1(x)$ are in the form of

$$g(x) = \begin{pmatrix} 0 & 0 \\ g_a(x) & 0 \end{pmatrix}, \quad g_1(x) = \begin{pmatrix} 0 & 0 \\ g_b(x) & 0 \end{pmatrix},\tag{6.84}$$

where

$$g_a(x) = \sqrt{d_6(0)}F_{21}(x,0) - \frac{\lambda_{22}(x)d_{14}}{\sqrt{d_6(0)}} - \frac{\lambda_{24}(x)d_{16}}{\sqrt{d_6(0)}} + \sqrt{d_6(0)}N_{21}(x,0),\tag{6.85}$$

$$g_b(x) = \sqrt{d_6(0)}J_{21}(x,0) - \frac{\gamma_{22}(x)d_{14}}{\sqrt{d_6(0)}} - \frac{\gamma_{24}(x)d_{16}}{\sqrt{d_6(0)}} + \sqrt{d_6(0)}K_{21}(x,0).\tag{6.86}$$

The system (6.77)–(6.81) is exponentially stable, as we will see in the stability analysis via the Lyapunov function in theorem 6.1.

By matching the systems (6.61)–(6.65) and (6.77)–(6.81) through (6.72), (6.73), a lengthy but straightforward calculation leads to the conditions on the kernels in (6.72), (6.73) as follows. The functions $F(x,y)$, $N(x,y)$, $\lambda(x)$ should satisfy the matrix equations

$$Q(x)F(x,x) + F(x,x)Q(x) + T_b(x) = 0,\tag{6.87}$$

$$Q(x)N(x,x) - N(x,x)Q(x) - \bar{T}_a(x) + T_a(x) = 0,\tag{6.88}$$

$$N(x,0)Q(0) + F(x,0)Q(0) - 2\lambda(x)\bar{B} - g(x) = 0,\tag{6.89}$$

$$Q(x)N_x(x,y) + N_y(x,y)Q(y) + N(x,y)Q'(y)$$
$$-N(x,y)T_a(y) + \bar{T}_a(x)N(x,y) - F(x,y)T_a(y) = 0,\tag{6.90}$$

$$Q(x)F_x(x,y) - F_y(x,y)Q(y) - F(x,y)Q'(y)$$
$$-F(x,y)T_b(y) + \bar{T}_a(x)F(x,y) - N(x,y)T_b(y) = 0,\tag{6.91}$$

$$Q(x)\lambda'(x) - \lambda(x)(\bar{A} - \bar{B}C_3) + \bar{T}_a(x)\lambda(x)$$
$$-F(x,0)Q(0)C_3 + g(x)\lambda(0) = 0,\tag{6.92}$$

$$\lambda(0) - \kappa = 0,\tag{6.93}$$

and $K(x,y)$, $J(x,y)$, $\gamma(x)$ should satisfy

$$-Q(x)J(x,x) - J(x,x)Q(x) + T_a(x) = 0,\tag{6.94}$$

$$K(x,x)Q(x) - Q(x)K(x,x) - \bar{T}_b(x) + T_b(x) = 0,\tag{6.95}$$

$$K(x,0)Q(0) + J(x,0)Q(0) - 2\gamma(x)\bar{B} - g_1(x) = 0,\tag{6.96}$$

$$-Q(x)J_x(x,y) + J_y(x,y)Q(y) + J(x,y)Q'(y)$$
$$-J(x,y)T_a(y) + \bar{T}_b(x)J(x,y) - K(x,y)T_a(y) = 0,\tag{6.97}$$

$$-Q(x)K_x(x,y) - K_y(x,y)Q(y) - K(x,y)Q'(y)$$

$$-K(x,y)T_b(y) + \bar{T}_b(x)K(x,y) - J(x,y)T_b(y) = 0, \qquad (6.98)$$

$$Q(x)\gamma'(x) + \gamma(x)(\bar{A} - \bar{B}C_3) + \bar{T}_b(x)\gamma(x)$$
$$-K(x,0)Q(0)C_3 + g_1(x)\lambda(0) = 0, \qquad (6.99)$$

$$\gamma(0) - C_3 + \lambda(0) = 0. \qquad (6.100)$$

In order to ensure the existence of a unique solution of the above kernel equations, additional artificial boundary conditions should be imposed, which are shown in the following lemma about well-posedness of the kernel equations.

Lemma 6.1. *After adding two additional artificial boundary conditions N_{21} $(L,y) = 0$, $K_{21}(L,y) = 0$ for the subelements N_{21}, K_{21} of kernels N and K, respectively, the matrix equations (6.87)–(6.93) have a unique solution $F, N \in L^\infty(\mathcal{D})$, $\lambda \in L^\infty([0, l(t)])$, and (6.94)–(6.100) have a unique solution $K, J \in L^\infty(\mathcal{D}), \gamma \in L^\infty([0, l(t)])$.*

Proof. The proof is shown in appendix 6.6. $\square$

With similar derivations, one can show that the inverse transformations are defined as

$$p(x,t) = \alpha(x,t) - \int_0^x \bar{K}(x,y)\alpha(y,t)dy$$
$$- \int_0^x \bar{J}(x,y)\beta(y,t)dy - \bar{\gamma}(x)W(t), \qquad (6.101)$$

$$r(x,t) = \beta(x,t) - \int_0^x \bar{F}(x,y)\alpha(y,t)dy$$
$$- \int_0^x \bar{N}(x,y)\beta(y,t)dy - \bar{\lambda}(x)W(t), \qquad (6.102)$$

where kernels $\bar{K}(x,y), \bar{J}(x,y), \bar{\gamma}(x), \bar{F}(x,y), \bar{N}(x,y), \bar{\lambda}(x)$ can be proved well-posed, as in lemma 6.1.

Once the equations for the kernels are solved, the control input is obtained as

$$U(t) = -R(l(t))^{-1}\left[p(l(t),t) - \int_0^{l(t)} F(l(t),y)p(y,t)dy \right.$$
$$\left. - \int_0^{l(t)} N(l(t),y)r(y,t)dy - \lambda(l(t))W(t) \right] \qquad (6.103)$$

by matching boundary conditions (6.65) and (6.81) via (6.73).

Applying (6.39)–(6.42), (6.43), (6.58)–(6.60), we can expand (6.103) and rewrite it in terms of the original states as

$$U_1(t) = \frac{-1}{d_{19}(l(t))\sqrt{d_6(l(t))}}\left[u_t(l(t),t) \right.$$
$$- \int_0^{l(t)} \left[F_{11}(l(t),y)(u_t(y,t) - \sqrt{d_6(y)}u_x(y,t)) \right.$$
$$+ F_{12}(l(t),y)(w_t(y,t) - \sqrt{d_1(y)}w_x(y,t)) \big] dy$$
$$- \int_0^{l(t)} \left[N_{11}(l(t),y)(u_t(y,t) + \sqrt{d_6(y)}u_x(y,t)) \right.$$

$$+ N_{12}(l(t), y)(w_t(y, t) + \sqrt{d_1(y)}w_x(y, t))] dy$$
$$- \lambda_{11}(l(t))w(0, t) - \lambda_{12}(l(t))w_t(0, t)$$
$$- \lambda_{13}(l(t))u(0, t) - \lambda_{14}(l(t))u_t(0, t) \Big], \tag{6.104}$$

$$U_2(t) = \frac{-1}{d_{20}(l(t))\sqrt{d_1(l(t))}} \Big[w_t(l(t), t)$$
$$- \int_0^{l(t)} \big[F_{21}(l(t), y)(u_t(y, t) - \sqrt{d_6(y)}u_x(y, t))$$
$$+ F_{22}(l(t), y)(w_t(y, t) - \sqrt{d_1(y)}w_x(y, t))\big] dy$$
$$- \int_0^{l(t)} \big[N_{21}(l(t), y)(u_t(y, t) + \sqrt{d_6(y)}u_x(y, t))$$
$$+ N_{22}(l(t), y)(w_t(y, t) + \sqrt{d_1(y)}w_x(y, t))\big] dy$$
$$- \lambda_{21}(l(t))w(0, t) - \lambda_{22}(l(t))w_t(0, t)$$
$$- \lambda_{23}(l(t))u(0, t) - \lambda_{24}(l(t))u_t(0, t) \Big], \tag{6.105}$$

where we replace $u_x(l(t), t)$, $w_x(l(t), t)$ with (6.36), (6.37). With the above control laws $U_1(t), U_2(t)$, we obtain the stability results of the state-feedback closed-loop system, summarized in the following two theorems.

Stability Analysis under State-Feedback Control

The exponential stability result of the state-feedback closed-loop system is shown in the following theorem, which physically means that the vibrational energy of the cable, including kinetic energy and potential energy in two directions, bounded by $\xi(\|w_t(\cdot, t)\|^2 + \|w_x(\cdot, t)\|^2 + \|u_t(\cdot, t)\|^2 + \|u_x(\cdot, t)\|^2)$ with $\xi > 0$, is exponentially convergent to zero, where the decay rate of the vibrational energy is adjustable with the control parameters.

Theorem 6.1. *For all initial values* $(w(x, 0), w_t(x, 0)) \in H^2(0, L) \times H^1(0, L)$, $(u(x, 0), u_t(x, 0)) \in H^2(0, L) \times H^1(0, L)$, *the closed-loop system consisting of the plant* (6.32)–(6.37) *and the state-feedback control law* (6.104), (6.105) *is exponentially stable in the sense that there exist positive constants* Υ_1, σ_1 *such that*

$$\Big(\|w_t(\cdot, t)\|^2 + \|w_x(\cdot, t)\|^2 + \|u_t(\cdot, t)\|^2 + \|u_x(\cdot, t)\|^2$$
$$+ w(0, t)^2 + w_t(0, t)^2 + u(0, t)^2 + u_t(0, t)^2 \Big)^{1/2}$$
$$\leq \Upsilon_1 \Big(\|w_t(\cdot, 0)\|^2 + \|w_x(\cdot, 0)\|^2 + \|u_t(\cdot, 0)\|^2 + \|u_x(\cdot, 0)\|^2$$
$$+ w(0, 0)^2 + w_t(0, 0)^2 + u(0, 0)^2 + u_t(0, 0)^2 \Big)^{1/2} e^{-\sigma_1 t}, \tag{6.106}$$

where $\|u(\cdot, t)\|^2$ is a compact notation for $\int_0^{l(t)} u(x, t)^2 dx$. The convergence rate σ_1 is adjustable by the control parameters.

Proof. We start by studying the stability of the target system (6.77)–(6.81). The equivalent stability property between the target system (6.77)–(6.81) and the original system (6.32)–(6.37) is ensured via the definitions (6.39)–(6.42), (6.58)–(6.60) and the backstepping transformations (6.72), (6.73) and (6.101), (6.102).

Consider the following Lyapunov function for the target system (6.77)–(6.81):

$$V_1 = W^T(t)P_1 W(t) + \frac{1}{2}\int_0^{l(t)} e^{\delta_2 x}\beta(x,t)^T R_a Q(x)^{-1}\beta(x,t)dx$$

$$+ \frac{1}{2}\int_0^{l(t)} e^{-\delta_1 x}\alpha(x,t)^T R_b Q(x)^{-1}\alpha(x,t)dx, \tag{6.107}$$

where the positive definite matrix $P_1 = P_1^T$ is the solution to the Lyapunov equation $P_1\hat{A} + \hat{A}^T P_1 = -\hat{Q}_1$, for some $\hat{Q}_1 = \hat{Q}_1^T > 0$, and R_a, R_b are diagonal matrices given as

$$R_a = \text{diag}\{r_{a1}, r_{a2}\}, \quad R_b = \text{diag}\{r_{b1}, r_{b2}\}. \tag{6.108}$$

The positive parameters $r_{a1}, r_{a2}, r_{b1}, r_{b2}, \delta_1, \delta_2$ are to be chosen later.

According to (6.107), we get

$$\mu_1\Omega(t) \leq V_1(t) \leq \mu_2\Omega(t), \tag{6.109}$$

where

$$\Omega(t) = |W(t)|^2 + \|\beta(x,t)\|^2 + \|\alpha(x,t)\|^2 \tag{6.110}$$

and μ_1, μ_2 are some positive constants. The time derivative of $V_1(t)$ along (6.77)–(6.81) is obtained as

$$\dot{V}_1 = \dot{W}^T(t)P_1 W(t) + W^T(t)P_1\dot{W}(t)$$

$$+ \int_0^{l(t)} e^{\delta_2 x}\beta(x,t)^T R_a Q(x)^{-1}\beta_t(x,l)dx$$

$$+ \int_0^{l(t)} e^{-\delta_1 x}\alpha(x,t)^T R_b Q(x)^{-1}\alpha_t(x,t)dx$$

$$+ \frac{\dot{l}(t)}{2}e^{-\delta_1 l(t)}\alpha(l(t),t)^T R_b Q(l(t))^{-1}\alpha(l(t),t)$$

$$= W(t)^T(\hat{A}^T P_1 + P_1\hat{A})W(t) + 4W^T(t)P_1\bar{B}\beta(0,t)$$

$$+ \frac{\dot{l}(t)}{2}e^{-\delta_1 l(t)}\alpha(l(t),t)^T R_b Q(l(t))^{-1}\alpha(l(t),t)$$

$$+ \int_0^{l(t)} e^{\delta_2 x}\beta(x,t)^T R_a Q(x)^{-1}\bar{T}_a(x)\beta(x,t)dx$$

$$+ \int_0^{l(t)} e^{\delta_2 x}\beta(x,t)^T R_a Q(x)^{-1}g(x)\beta(0,t)dx$$

$$- \frac{1}{2}\beta(0,t)^T R_a\beta(0,t) - \frac{\delta_2}{2}\int_0^{l(t)} e^{\delta_2 x}\beta(x,t)^T R_a\beta(x,t)dx$$

$$- \frac{1}{2}e^{-\delta_1 l(t)}\alpha(l(t),t)^T R_b\alpha(l(t),t) + \frac{1}{2}\alpha(0,t)^T R_b\alpha(0,t)$$

$$-\frac{\delta_1}{2}\int_0^{l(t)} e^{-\delta_1 x}\alpha(x,t)^T R_b\alpha(x,t)dx$$

$$+\int_0^{l(t)} e^{-\delta_1 x}\alpha(x,t)^T R_b Q(x)^{-1}\bar{T}_b(x)\alpha(x,t)dx$$

$$+\int_0^{l(t)} e^{-\delta_1 x}\alpha(x,t)^T R_b Q(x)^{-1}g_1(x)\beta(0,t)dx. \tag{6.111}$$

Applying Young's inequality, by virtue of the boundedness of the elements $\frac{1}{\sqrt{d_1(x)}}$, $\frac{1}{\sqrt{d_6(x)}}$, $g_a(x)$, $g_b(x)$ in the matrices $Q(x)^{-1}$, $g(x)$, $g_1(x)$ yields $\xi > 0$ such that the following inequalities hold:

$$\int_0^{l(t)} e^{\delta_2 x}\beta(x,t)^T R_a Q(x)^{-1}g(x)\beta(0,t)dx$$

$$\leq \xi\int_0^{l(t)} e^{\delta_2 x}\beta(x,t)^T R_a\beta(x,t)dx + \xi\int_0^{l(t)} e^{\delta_2 L}\beta(0,t)^T\Lambda_a\beta(0,t)dx, \tag{6.112}$$

$$\int_0^{l(t)} e^{-\delta_1 x}\alpha(x,t)^T R_b Q(x)^{-1}g_1(x)\beta(0,t)dx$$

$$\leq \xi\int_0^{l(t)} e^{-\delta_1 x}\alpha(x,t)^T R_b\alpha(x,t)dx + \xi\int_0^{l(t)}\beta(0,t)^T\Lambda_b\beta(0,t)dx, \tag{6.113}$$

where

$$\Lambda_a = \begin{bmatrix} r_{a2} & 0 \\ 0 & 0 \end{bmatrix}, \Lambda_b = \begin{bmatrix} r_{b2} & 0 \\ 0 & 0 \end{bmatrix}. \tag{6.114}$$

Inserting (6.112), (6.113) and applying Young's inequality into (6.111), one obtains

$$\dot{V}_1(t) \leq -\frac{1}{2}\lambda_{\min}(Q_2)|W(t)|^2 - \beta(0,t)^T\left(\frac{R_a}{2} - \frac{R_b}{2}\right.$$

$$-\frac{8|P_1\bar{B}_1|^2}{\lambda_{\min}(Q_2)}I_2 - e^{\delta_2 L}\xi\Lambda_a - \xi\Lambda_b\right)\beta(0,t) - \int_0^{l(t)} e^{\delta_2 x}\beta(x,t)^T$$

$$\times R_a\left(\left(\frac{\delta_2}{2} - \xi\right)I_2 - Q(x)^{-1}\bar{T}_a(x)\right)\beta(x,t)dx - \int_0^{l(t)} e^{-\delta_1 x}$$

$$\times \alpha(x,t)^T R_b\left(\left(\frac{\delta_1}{2} - \xi\right)I_2 - Q(x)^{-1}\bar{T}_b(x)\right)\alpha(x,t)dx$$

$$-\frac{1}{2}e^{-\delta_1 l(t)}\alpha(l(t),t)^T R_b\left(I_2 - Q(l(t))^{-1}\dot{l}(t)\right)\alpha(l(t),t), \tag{6.115}$$

where I_2 is a 2×2 identity matrix.

The parameters $r_{a1}, r_{a2}, r_{b1}, r_{b2}, \delta_1, \delta_2$ are chosen to satisfy

$$r_{a1} > r_{b1} + \frac{16|P_1\bar{B}_1|^2}{\lambda_{\min}(Q_2)} + 2r_{a2}c^{\delta_2 L}\zeta + 2r_{b2}\xi, \tag{6.116}$$

$$r_{a2} > r_{b2} + \frac{16|P_1\bar{B}_1|^2}{\lambda_{\min}(Q_2)} \tag{6.117}$$

with sufficiently large δ_1, δ_2. The positive constants r_{b1}, r_{b2} can be arbitrary. We know that the elements in the diagonal matrix $Q(l(t))^{-1}l(t)$ are smaller than unity by recalling assumption 6.4. Additionally, the boundedness of all elements in the diagonal matrix $Q(x)^{-1}$, $\bar{T}_a(x), \bar{T}_b(x)$ is assured by recalling assumption 6.1. We thus arrive at

$$\dot{V}_1(t) \leq -\eta_1 V_1(t) \tag{6.118}$$

for some positive η_1. It then follows that

$$V_1(t) \leq V_1(0)e^{-\eta_1 t}, \tag{6.119}$$

and hence,

$$\Omega(t) \leq \frac{\mu_2}{\mu_1}\Omega(0)e^{-\eta_1 t} \tag{6.120}$$

by recalling (6.109).

In sum, we have obtained exponential stability in $\Omega(t)$. Establishing the relationship between the $\Omega(t)$ and the appropriate norm of the $u(x,t), w(x,t)$-system is the key to establishing exponential stability in the original variables.

Defining

$$\begin{aligned}
\Xi(t) = {} & \|u_x(\cdot,t)\|^2 + \|u_t(\cdot,t)\|^2 + |u(0,t)|^2 + |u_t(0,t)|^2 \\
& + \|w_x(\cdot,t)\|^2 + \|w_t(\cdot,t)\|^2 + |w(0,t)|^2 + |w_t(0,t)|^2,
\end{aligned} \tag{6.121}$$

recalling (6.39)–(6.42), (6.43)–(6.60), (6.101), (6.102), and applying the Cauchy-Schwarz inequality, the following inequality holds

$$\bar{\theta}_{1a}\Xi(t) \leq \Omega(t) \leq \bar{\theta}_{1b}\Xi(t) \tag{6.122}$$

for some positive $\bar{\theta}_{1a}$ and $\bar{\theta}_{1b}$. Therefore, we get

$$\Xi(t) \leq \frac{\mu_2\bar{\theta}_{1b}}{\mu_1\bar{\theta}_{1a}}\Xi(0)e^{-\eta_1 t}. \tag{6.123}$$

Thus, (6.106) is achieved with

$$\Upsilon_1 = \sqrt{\frac{\mu_2\bar{\theta}_{1b}}{\mu_1\bar{\theta}_{1a}}}, \quad \sigma_1 = \frac{\eta_1}{2}, \tag{6.124}$$

where the convergence rate σ_1 can be adjusted by the control parameter κ through affecting $\lambda_{\min}(\hat{Q}_1)$. With this, the proof of theorem 6.1 is complete. $\square$

Next, we prove the exponential convergence of the control input. We state first a lemma that shows the exponential stability result for the closed-loop system in the sense of the H^2 norm.

Lemma 6.2. *With arbitrary initial data $(w(x,0), w_t(x,0)) \in H^2(0,L) \times H^1(0,L)$, $(u(x,0), u_t(x,0)) \in H^2(0,L) \times H^1(0,L)$, the exponential stability of the closed-loop system consisting of the plant (6.32)–(6.37) and the state-feedback control law (6.104),*

(6.105) *holds in the sense that there exist positive constants* Υ_{1a} *and* σ_{1a} *such that*

$$\left(\|u_{xx}(\cdot,t)\|^2 + \|w_{xx}(\cdot,t)\|^2 + \|u_{tx}(\cdot,t)\|^2 + \|w_{tx}(\cdot,t)\|^2\right)^{\frac{1}{2}}$$

$$\leq \Upsilon_{1a}\Bigg(\|u_x(\cdot,0)\|^2 + \|w_x(\cdot,0)\|^2 + \|u_t(\cdot,0)\|^2 + \|w_t(\cdot,0)\|^2$$

$$+\|u_{xx}(\cdot,0)\|^2 + \|w_{xx}(\cdot,0)\|^2 + \|u_{tx}(\cdot,0)\|^2 + \|w_{tx}(\cdot,0)\|^2$$

$$+|w(0,0)|^2 + |w_t(0,0)|^2 + |u(0,0)|^2 + |u_t(0,0)|^2\Bigg)e^{-\sigma_{1a}t}. \tag{6.125}$$

Proof. Taking the spatial and time derivatives of (6.78)–(6.79) and (6.126), (6.129), (6.130), respectively, we get

$$\ddot{W}(t) = \hat{A}\dot{W}(t) + 2\bar{B}Q(0)\beta_x(0,t) + 2\bar{B}\bar{T}_a(0)\beta(0,t), \tag{6.126}$$

$$\alpha_{xt}(x,t) = -Q(x)\alpha_{xx}(x,t)$$
$$+ (\bar{T}_b(x) - Q'(x))\alpha_x(x,t) + \bar{T}_b'(x)\alpha(x,t), \tag{6.127}$$

$$\beta_{xt}(x,t) = Q(x)\beta_{xx}(x,t)$$
$$+ (\bar{T}_a(x) + Q'(x))\beta_x(x,t) + \bar{T}_a'(x)\beta(x,t), \tag{6.128}$$

$$Q(0)\alpha_x(0,t) = Q(0)\beta_x(0,t) + \bar{T}_a(0)\beta(0,t) + \bar{T}_b(0)\alpha(0,t), \tag{6.129}$$

$$\beta_x(l(t),t) = 0, \tag{6.130}$$

where (6.79), (6.81) have been used. Equation (6.130) results from

$$(\dot{l}(t) + Q(l(t)))\beta_x(l(t),t) = 0, \tag{6.131}$$

where the elements in the diagonal matrix $\dot{l}(t) + Q(l(t))$ are nonzero for all t by recalling assumption 6.4. Define new variables

$$\varpi(x,t) = \alpha_x(x,t), \ \ \zeta(x,t) = \beta_x(x,t), \ \ Z(t) = \dot{W}(t). \tag{6.132}$$

Consider a Lyapunov function as

$$V_2(t) = R_1 V_1(t) + Z(t)^T P_1 Z(t)$$

$$+ \frac{1}{2}\int_0^{l(t)} e^{\bar{\delta}_1 x}\zeta(x,t)^T \bar{R}_a Q(x)^{-1}\zeta(x,t)dx$$

$$+ \frac{1}{2}\int_0^{l(t)} e^{-\bar{\delta}_2 x}\varpi(x,t)^T \bar{R}_b Q(x)^{-1}\varpi(x,t)dx, \tag{6.133}$$

where $\bar{R}_a, \bar{R}_b$ are diagonal matrices given as

$$\bar{R}_a = \text{diag}\{\bar{r}_{a1}, \bar{r}_{a2}\}, \ \ \bar{R}_b = \text{diag}\{\bar{r}_{b1}, \bar{r}_{b2}\}.$$

The parameters $\bar{r}_{a1}, \bar{r}_{a2}, \bar{r}_{b1}, \bar{r}_{b2}, \bar{\delta}_1, \bar{\delta}_2$, and R_1 are positive.

Taking the derivative of (6.133) along (6.126)–(6.130), recalling (6.115), determining $\bar{r}_{a1}, \bar{r}_{a2}, \bar{r}_{b1}, \bar{r}_{b2}, \bar{\delta}_1, \bar{\delta}_2$ through a process similar to that in (6.111)–(6.119), and choosing large enough positive constant R_1, we arrive at

$$\dot{V}_2(t) \leq -\eta_2 V_2(t) \tag{6.134}$$

for some positive η_2. Recalling the backstepping transformations (6.101), (6.102), we obtain

$$\|p_x(\cdot,t)\|^2 + \|r_x(\cdot,t)\|^2 \leq \Upsilon_{1b}\big(W(0)^2 + \|p(\cdot,0)\|^2 + \|r(\cdot,0)\|^2$$
$$+ \|p_x(\cdot,0)\|^2 + \|r_x(\cdot,0)\|^2\big)e^{-\eta_2 t} \qquad (6.135)$$

for some positive Υ_{1b}. Applying (6.39)–(6.42), (6.58), (6.59), the proof of lemma 6.2 is complete. $\qquad\square$

Theorem 6.2. *In the closed-loop system (6.32)–(6.37), (6.104), (6.105), the state-feedback control signals $U_1(t)$, $U_2(t)$ in (6.104), (6.105) are bounded and exponentially convergent to zero in the sense that there exist positive constants σ_2 and Υ_2 such that*

$$|U_1(t)|^2 + |U_2(t)|^2 \leq \Upsilon_2 \bigg(\|u_x(\cdot,0)\|^2 + \|w_x(\cdot,0)\|^2 + \|u_t(\cdot,0)\|^2$$
$$+ \|w_t(\cdot,0)\|^2 + \|u_{xx}(\cdot,0)\|^2 + \|w_{xx}(\cdot,0)\|^2$$
$$+ \|u_{tx}(\cdot,0)\|^2 + \|w_{tx}(\cdot,0)\|^2 + |w(0,0)|^2 + |w_t(0,0)|^2$$
$$+ |u(0,0)|^2 + |u_t(0,0)|^2 \bigg)e^{\sigma_2 t}. \qquad (6.136)$$

Proof. According to (6.103) and the exponential stability result in theorem 6.1, we know that once

$$p(l(t),t) = [u_t(l(t),t) - \sqrt{d_6(l(t))}u_x(l(t),t), w_t(l(t),t) - \sqrt{d_1(l(t))}w_x(l(t),t)]^T$$

is established to be exponentially convergent to zero in the sense of $|p(l(t),t)|^2$, the exponential convergence of the control input is obtained.

Applying the Cauchy-Schwarz inequality and recalling (6.35), (6.36), we get

$$|p(l(t),t)| \leq 2|p(0,t)| + 2\sqrt{L}\|p_x(\cdot,t)\|$$
$$\leq 4|r(0,t)| + 4|C_3 W(t)| + 2\sqrt{L}\|p_x(\cdot,t)\|$$
$$\leq 8|r(l(t),t)| + 8\sqrt{L}\|r_x(\cdot,t)\| + 4|C_3 W(t)| + 2\sqrt{L}\|p_x(\cdot,t)\|. \qquad (6.137)$$

Recalling (6.102), (6.81) and the exponential convergence of $\|\alpha(\cdot,t)\|^2$, $\|\beta(\cdot,t)\|^2$, $|W(t)|^2$ proved in theorem 6.1, we have $|r(l(t),t)|$ as exponentially convergent to zero. Recalling lemma 6.2, we thus have $|p(l(t),t)|$ as exponentially convergent to zero. The proof of theorem 6.2 is complete. $\qquad\square$

6.3 OBSERVER FOR TWO-DIMENSIONAL OSCILLATIONS OF THE CABLE

Observer Structure

We consider only sensors placed at the actuated boundary (ship-mounted crane). An observer should then be designed to estimate the states w, u in the domain and at the uncontrolled boundary—that is, the lateral-longitudinal coupled vibration states of the cable and the attached payload. These estimates are required in the state-feedback control laws (6.104), (6.105).

The available measurements are $u_t(l(t), t)$, $w_t(l(t), t)$, which makes $p(l(t), t)$ known through the conversion

$$p(l(t), t) = [u_t(l(t), t) - \sqrt{d_6(l(t))} d_{19}(l(t)) U_1(t),$$
$$w_t(l(t), t) - \sqrt{d_1(l(t))} d_{20}(l(t)) U_2(t)] \tag{6.138}$$

by recalling (6.36), (6.37), (6.58), (6.39), and (6.42).

Using the known signal $p(l(t), t)$, the observer for the coupled wave PDE plant (6.32)–(6.37) is constructed as

$$\hat{w}_t(x, t) = \frac{1}{2}(\hat{z}(x, t) + \hat{v}(x, t)), \tag{6.139}$$

$$\hat{w}_x(x, t) = \frac{1}{2\sqrt{d_1(x)}}(\hat{z}(x, t) - \hat{v}(x, t)), \tag{6.140}$$

$$\hat{u}_t(x, t) = \frac{1}{2}(\hat{k}(x, t) + \hat{y}(x, t)), \tag{6.141}$$

$$\hat{u}_x(x, t) = \frac{1}{2\sqrt{d_6(x)}}(\hat{k}(x, t) - \hat{y}(x, t)), \tag{6.142}$$

$$\hat{p}_t(x, t) + Q(x)\hat{p}_x(x, t) = T_a(x)\hat{r}(x, t) + T_b(x)\hat{p}(x, t)$$
$$+ \Gamma_1(x, t)(p(l(t), t) - \hat{p}(l(t), t)), \tag{6.143}$$

$$\hat{r}_t(x, t) - Q(x)\hat{r}_x(x, t) = T_a(x)\hat{r}(x, t) + T_b(x)\hat{p}(x, t)$$
$$+ \Gamma_2(x, t)(p(l(t), t) - \hat{p}(l(t), t)), \tag{6.144}$$

$$\hat{p}(0, t) = C_3\hat{W}(t) - \hat{r}(0, t), \tag{6.145}$$

$$\dot{\hat{W}}(t) = (\bar{A} - \bar{B}C_3)\hat{W}(t) + 2\bar{B}\hat{r}(0, t)$$
$$+ \Gamma_3(t)(p(l(t), t) - \hat{p}(l(t), t)), \tag{6.146}$$

$$\hat{r}(l(t), t) = R(l(t))U(t) + p(l(t), t), \tag{6.147}$$

where

$$\hat{p} = [\hat{y}(x, t), \hat{v}(x, t)]^T, \tag{6.148}$$

$$\hat{r} = [\hat{k}(x, t), \hat{z}(x, t)]^T, \tag{6.149}$$

$$\hat{W}(t) = [\hat{X}(t), \hat{Y}(t)]^T = [\hat{w}(0, t), \hat{w}_t(0, t), \hat{u}(0, t), \hat{u}_t(0, t)]^T. \tag{6.150}$$

The observer consists of two parts:

1. The system (6.143)–(6.147), as a copy of plant (6.61)–(6.65) with the addition of output injection, is to estimate $p(x, t), r(x, t)$.
2. Once $p(x, t), r(x, t)$ are estimated successfully by (6.143)–(6.147), the estimates of the original plant are simply obtained as (6.139)–(6.142) by virtue of the Riemann transformations (6.39)–(6.42).

Next, the observer gains $\Gamma_1(x, t)$, $\Gamma_2(x, t)$, and $\Gamma_3(t)$ are determined, in order to achieve exponential stability of the observer error system, which is done in the next subsection. A difference from the traditional observer gain functions should be noted. The gains Γ_1, Γ_2 depend not only on the spatial variable x but also on time t because of the time-varying domain.

Observer Error System

The observer's task is to ensure that the observer errors (differences between the estimated and real states) are driven to zero, which is to be achieved by designing adequate observer gains. Denote the observer errors as

$$\tilde{w}_t(x,t) = w_t(x,t) - \hat{w}_t(x,t), \tag{6.151}$$

$$\tilde{w}_x(x,t) = w_x(x,t) - \hat{w}_x(x,t), \tag{6.152}$$

$$\tilde{u}_t(x,t) = u_t(x,t) - \hat{u}_t(x,t), \tag{6.153}$$

$$\tilde{u}_x(x,t) = u_x(x,t) - \hat{u}_x(x,t), \tag{6.154}$$

$$\begin{aligned}
\tilde{W}(x,t) &= W(x,t) - \hat{W}(t) \\
&= [X(t), Y(t)] - [\hat{X}(t), \hat{Y}(t)] \\
&= [w(0,t), w_t(0,t), u(0,t), u_t(0,t)]^T \\
&\quad - [\hat{w}(0,t), \hat{w}_t(0,t), \hat{u}(0,t), \hat{u}_t(0,t)]^T \\
&= [\tilde{X}(t), \tilde{Y}(t)] \\
&= [\tilde{w}(0,t), \tilde{w}_t(0,t), \tilde{u}(0,t), \tilde{u}_t(0,t)]^T,
\end{aligned} \tag{6.155}$$

$$\tilde{p}(x,t) = p(x,t) - \hat{p}(x,t) = [\tilde{y}(x,t), \tilde{v}(x,t)], \tag{6.156}$$

$$\tilde{r}(x,t) = r(x,t) - \hat{r}(x,t) = [\tilde{k}(x,l), \tilde{z}(x,t)]. \tag{6.157}$$

Recalling (6.61)–(6.65), (6.39)–(6.42), and (6.139)–(6.147), the resulting observer error dynamics are given by

$$\tilde{w}_t(x,t) = \frac{1}{2}(\tilde{z}(x,t) + \tilde{v}(x,t)), \tag{6.158}$$

$$\tilde{w}_x(x,t) = \frac{1}{2\sqrt{d_1(x)}}(\tilde{z}(x,t) - \tilde{v}(x,t)), \tag{6.159}$$

$$\tilde{u}_t(x,t) = \frac{1}{2}(\tilde{k}(x,t) + \tilde{y}(x,t)), \tag{6.160}$$

$$\tilde{u}_x(x,t) = \frac{1}{2\sqrt{d_6(x)}}(\tilde{k}(x,t) - \tilde{y}(x,t)), \tag{6.161}$$

$$\begin{aligned}
\tilde{p}_t(x,t) + Q(x)\tilde{p}_x(x,t) &= T_a(x)\tilde{r}(x,t) \\
&\quad + T_b(x)\tilde{p}(x,t) + \Gamma_1(x,t)\tilde{p}(l(t),t),
\end{aligned} \tag{6.162}$$

$$\begin{aligned}
\tilde{r}_t(x,t) - Q(x)\tilde{r}_x(x,t) &= T_a(x)\tilde{r}(x,t) \\
&\quad + T_b(x)\tilde{p}(x,t) + \Gamma_2(x,t)\tilde{p}(l(t),t),
\end{aligned} \tag{6.163}$$

$$\tilde{p}(0,t) = C_3\tilde{W}(t) - \tilde{r}(0,t), \tag{6.164}$$

$$\begin{aligned}
\dot{\tilde{W}}(t) &= (\bar{A} - \bar{B}C_3)\tilde{W}(t) + 2\bar{B}\tilde{r}(0,t) \\
&\quad + \Gamma_3(t)\tilde{p}(l(t),t),
\end{aligned} \tag{6.165}$$

$$\tilde{r}(l(t),t) = 0, \tag{6.166}$$

where the subsystem (6.162)–(6.166) describing the dynamics of the observer error of the system (6.61)–(6.65) determines the observer error of the plant (6.32)–(6.37) via (6.158)–(6.161). Therefore, the exponential stability of (6.162)–(6.166) is the key to making sure that the proposed observer is exponentially convergent to the actual states of the original plant (6.32)–(6.37).

Observer Backstepping Design

To find the observer gains $\Gamma_1(x,t), \Gamma_2(x,t), \Gamma_3(t)$ that guarantee that (6.162)–(6.166) is exponentially stable, we use a transformation to map (6.162)–(6.166) to a target observer error system whose exponential stability result is straightforward to obtain.

The transformation is introduced as

$$\tilde{p}(x,t) = \tilde{\alpha}(x,t) - \int_x^{l(t)} \bar{\varphi}(x,y)\tilde{\alpha}(y,t)dy, \tag{6.167}$$

$$\tilde{r}(x,t) = \tilde{\beta}(x,t) - \int_x^{l(t)} \bar{\psi}(x,y)\tilde{\alpha}(y,t)dy, \tag{6.168}$$

$$\tilde{W}(t) = \tilde{S}(t) + \int_0^{l(t)} \bar{K}(y)\tilde{\alpha}(y,t)dy, \tag{6.169}$$

where the kernels

$$\bar{\varphi}(x,y) = \{\bar{\varphi}_{ij}(x,y)\}_{1\leq i,j\leq 2}, \ \ \bar{\psi}(x,y) = \{\bar{\psi}_{ij}(x,y)\}_{1\leq i,j\leq 2}$$

on a triangular domain $\mathcal{D}_1 = \{0 \leq x \leq y \leq l(t)\}$ and

$$\bar{K}(y) = \{\bar{K}_{ij}(y)\}_{1\leq i\leq 4, 1\leq j\leq 2},$$

are to be determined.

The target observer error system is set up as

$$\tilde{\alpha}_t(x,t) + Q(x)\tilde{\alpha}_x(x,t) = T_a(x)\tilde{\beta}(x,t) + \bar{T}_b(x)\tilde{\alpha}(x,t)$$
$$+ \int_x^{l(t)} \bar{M}(x,y)\tilde{\beta}(y,t)dy, \tag{6.170}$$

$$\tilde{\beta}_t(x,t) - Q(x)\tilde{\beta}_x(x,t) = \int_x^{l(t)} \bar{N}(x,y)\tilde{\beta}(y,t)dy + T_a(x)\tilde{\beta}(x,t), \tag{6.171}$$

$$\tilde{\alpha}(0,t) = C_3\tilde{S}(t) - \tilde{\beta}(0,t) + \int_0^{l(t)} H(y)\tilde{\alpha}(y,t)dy, \tag{6.172}$$

$$\tilde{\beta}(l(t),t) = 0, \tag{6.173}$$

$$\dot{\tilde{S}}(t) = \check{A}\tilde{S}(t) + \check{E}\tilde{\beta}(0,t) + \int_0^{l(t)} G(y)\tilde{\beta}(y,t)dy, \tag{6.174}$$

where the matrix

$$\check{A} = \bar{A} - \bar{B}C_3 - L_0C_3$$

is made Hurwitz by choosing

$$L_0 = \{L_{0ij}\}_{1\leq i\leq 4, 1\leq j\leq 2}$$

and recalling assumption 6.5, and where $\bar{M}(x,y), \bar{N}(x,y)$ satisfy

$$\bar{M}(x,y) = \int_x^y \bar{\varphi}(x,z)\bar{M}(z,y)dz + \bar{\varphi}(x,y)T_a(y), \tag{6.175}$$

$$\bar{N}(x,y) = \int_x^y \bar{\psi}(x,z)\bar{M}(z,y)dz + \bar{\psi}(x,y)T_a(y). \tag{6.176}$$

The function $H(y) = \{h_{ij}(y)\}_{1 \le i,j \le 2}$ in (6.172) is a strictly lower triangular matrix given by

$$H(y) = \begin{pmatrix} 0 & 0 \\ \bar{\psi}_{2,1}(0,y) + \bar{\varphi}_{2,1}(0,y) + \bar{K}_{21}(y) & 0 \end{pmatrix}, \tag{6.177}$$

and $G(y), \check{E}$ in (6.174) are

$$G(y) = \{G_{ij}(y)\}_{1 \le i \le 4, 1 \le j \le 2} = -\bar{K}(0,y)\bar{T}_a(y) - \int_0^y \bar{K}(0,z)\bar{M}(z,y)dz,$$

$$\check{E} = \{\check{E}_{ij}\}_{1 \le i \le 4, 1 \le j \le 2} = L_0 + 2\bar{B}.$$

The exponential stability of the target system (6.170)–(6.174) will be given in lemma 6.4.

By matching (6.162)–(6.166) and (6.170)–(6.174) through the transformation (6.167)–(6.169), the conditions on the kernels in (6.167)–(6.169) and the observer gains in (6.143), (6.144), (6.146) are obtained as follows. Kernels $\bar{\varphi}(x,y)$, $\bar{\psi}(x,y)$, $\bar{K}(y)$ should satisfy the matrix equations

$$\begin{aligned} &- \bar{\varphi}_y(x,y)Q(y) - Q(x)\bar{\varphi}_x(x,y) - \bar{\varphi}(x,y)Q'(y) \\ &+ T_a(x)\bar{\psi}(x,y) + T_b(x)\bar{\varphi}(x,y) - \bar{\varphi}(x,y)\bar{T}_b(y) = 0, \end{aligned} \tag{6.178}$$

$$\begin{aligned} &- \bar{\psi}_y(x,y)Q(y) + Q(x)\bar{\psi}_x(x,y) - \bar{\psi}(x,y)Q'(y) \\ &+ T_a(x)\bar{\psi}(x,y) \quad \bar{\psi}(x,y)\bar{T}_b(y) + T_b(x)\bar{\varphi}(x,y) = 0, \end{aligned} \tag{6.179}$$

$$Q(x)\bar{\varphi}(x,x) - \bar{\varphi}(x,x)Q(x) - T_b(x) + \bar{T}_b(x) = 0, \tag{6.180}$$

$$Q(x)\bar{\psi}(x,x) + \bar{\psi}(x,x)Q(x) + T_b(x) = 0, \tag{6.181}$$

$$\bar{\psi}(0,y) + \bar{\varphi}(0,y) + C_3\bar{K}(y) \quad H(y) = 0, \tag{6.182}$$

$$\begin{aligned} &- \bar{K}'(y)Q(y) + (\bar{A} - \bar{B}C_3 - L_0C_3)\bar{K}(y) \\ &- \bar{K}(y)[Q'(y) + \bar{T}_b(y)] - L_0\bar{\varphi}(0,y) \\ &\qquad - (2\bar{B} + L_0)\bar{\psi}(0,y) = 0, \end{aligned} \tag{6.183}$$

$$\bar{K}(0) = L_0 Q(0)^{-1}, \tag{6.184}$$

and the observer gains are obtained as

$$\Gamma_1(x,t) = \dot{l}(t)\bar{\varphi}(x,l(t)) - \bar{\varphi}(x,l(t))Q(l(t)), \tag{6.185}$$

$$\Gamma_2(x,t) = \dot{l}(t)\bar{\psi}(x,l(t)) - \bar{\psi}(x,l(t))Q(l(t)), \tag{6.186}$$

$$\Gamma_3(t) = \dot{l}(t)\bar{K}(l(t)) - \bar{K}(l(t))Q(l(t)). \tag{6.187}$$

Lemma 6.3. *After adding an additional artificial boundary condition $\bar{\varphi}_{21}(x,L) = 0$ for the element $\bar{\varphi}_{21}$ in the matrix $\bar{\varphi}$, the matrix equations (6.178)–(6.184) have a unique solution $\bar{\varphi}, \bar{\psi} \in L^\infty(\mathcal{D}_1)$, $\bar{K} \in L^\infty([0,l(t)])$.*

Proof. After swapping the positions of the arguments in B.9–B.10 in [6]—that is, by changing the domain $\mathcal{D}_1$ to $\mathcal{D}$—(6.178)–(6.184) has the analogous form with kernels $F(x,y), N(x,y), \lambda(y)$ in (6.87)–(6.93). Following the steps in the proof of lemma 6.1, including the introduction of the extended domain $\mathcal{D}_0$ and the addition of the additional artificial boundary condition, lemma 6.3 is obtained. $\qquad \square$

Following similar steps as above, the inverse transformation of (6.167)–(6.169) is determined as

$$\tilde{\alpha}(x,t) = \tilde{p}(x,t) - \int_x^{l(t)} \check{\varphi}(x,y)\tilde{p}(y,t)dy, \qquad (6.188)$$

$$\tilde{\beta}(x,t) = \tilde{r}(x,t) - \int_x^{l(t)} \check{\psi}(x,y)\tilde{p}(y,t)dy, \qquad (6.189)$$

$$\tilde{S}(t) = \tilde{W}(t) + \int_0^{l(t)} \check{K}(y)\tilde{r}(y,t)dy, \qquad (6.190)$$

where $\check{\varphi}(x,y) \in R^{2\times 2}$, $\check{\psi}(x,y) \in R^{2\times 2}$, and $\check{K}(y) \in R^{4\times 2}$ are kernels on $\mathcal{D}_1$ and $0 \leq y \leq l(t)$, respectively.

Stability Analysis of Observer Error System

Before showing the performance of the proposed observer on tracking the actual states in the original plant (6.32)–(6.37) in the next theorem, the stability result of the observer error subsystem (6.162)–(6.166), which dominates the observer errors of the original plant (6.32)–(6.37), is given in the following lemma.

Lemma 6.4. *For the observer error subsystem* (6.162)–(6.166), *there exist positive constants* Υ_3, σ_3 *such that*

$$\left(\|\tilde{p}(\cdot,t)\|^2 + \|\tilde{r}(\cdot,t)\|^2 + \left|\tilde{W}(t)\right|^2 \right)^{\frac{1}{2}}$$

$$\leq \Upsilon_3 \left(\|\tilde{p}(\cdot,0)\|^2 + \|\tilde{r}(\cdot,0)\|^2 + \left|\tilde{W}(0)\right|^2 \right)^{\frac{1}{2}} e^{-\sigma_3 t}. \qquad (6.191)$$

Proof. Expanding (6.170)–(6.174) as $\tilde{\alpha} = [\tilde{\alpha}_1, \tilde{\alpha}_2]^T$, $\tilde{\beta} = [\tilde{\beta}_1, \tilde{\beta}_2]^T$, one obtains

$$\tilde{\alpha}_{it}(x,t) + Q_i(x)\tilde{\alpha}_{ix}(x,t) = \sum_{j=1}^2 T_{aij}(x)\tilde{\beta}_j(x,t) + \bar{T}_{bi}(x)\tilde{\alpha}_i(x,t)$$

$$+ \int_x^{l(t)} \sum_{j=1}^2 \bar{M}_{ij}(x,y)\tilde{\beta}_j(y,t)dy, \qquad (6.192)$$

$$\tilde{\beta}_{it}(x,t) - Q_i(x)\tilde{\beta}_{ix}(x,t) = \int_x^{l(t)} \sum_{j=1}^2 \bar{N}_{ij}(x,y)\tilde{\beta}_j(y,t)dy$$

$$+ \sum_{j=1}^2 T_{aij}(x)\tilde{\beta}_j(x,t), \qquad (6.193)$$

$$\tilde{\alpha}_i(0,t) = C_3\tilde{S}(t) - \tilde{\beta}_i(0,t)$$

$$+ (i-1)\int_0^{l(t)} h_{21}(y)\tilde{\alpha}_1(y,t)dy, \qquad (6.194)$$

$$\tilde{\beta}_i(l(t),t) = 0 \qquad (6.195)$$

for $i = 1, 2$, and $\tilde{S}(t)$ is governed by

$$\dot{\tilde{S}}(t) = \breve{A}\tilde{S}(t) + \breve{E}[\tilde{\beta}_1(0,t), \tilde{\beta}_2(0,t)]^T + \int_0^{l(t)} G(y)[\tilde{\beta}_1(y,t), \tilde{\beta}_2(y,t)]^T dy. \quad (6.196)$$

In (6.192)–(6.196), $\tilde{\beta}_i(\cdot, t)$ are independent, and $\tilde{\beta}_i(\cdot, t) \equiv 0$ after a finite time because of (6.195). Thus, $\tilde{S}(t)$ is exponentially convergent to zero because $\breve{A}$ is Hurwitz. The signals $\tilde{\alpha}_1(\cdot, t)$ are exponentially convergent to zero because of the exponential convergence of $\tilde{\alpha}_1(0,t)$ due to (6.194) for $i = 1$. The signals $\tilde{\alpha}_1(\cdot, t)$ flow into $\tilde{\alpha}_2(0,t)$ through the boundary (6.194), where the exponential convergence of $\tilde{\alpha}_2(0,t)$ is also obtained for $i = 2$ because all signals on the right-hand side of the equal sign are exponentially convergent to zero. It follows that $\tilde{\alpha}_2(\cdot, t)$ are exponentially convergent to zero as well.

The exponential stability result is seen more clearly by using the Lyapunov function

$$V_e(t) = \frac{\breve{r}_{b1}}{2} \int_0^{l(t)} e^{-\breve{\delta}_1 x} \tilde{\alpha}_1(x,t)^T Q_1(x)^{-1} \tilde{\alpha}_1(x,t) dx$$

$$+ \frac{\breve{r}_{a1}}{2} \int_0^{l(t)} e^{\breve{\delta}_2 x} \tilde{\beta}_1(x,t)^T Q_1(x)^{-1} \tilde{\beta}_1(x,t) dx$$

$$+ \frac{\breve{r}_{a2}}{2} \int_0^{l(t)} e^{\breve{\delta}_2 x} \tilde{\beta}_2(x,t)^T Q_2(x)^{-1} \tilde{\beta}_2(x,t) dx + \tilde{S}(t)^T P_2 \tilde{S}(t)$$

$$+ \frac{\breve{r}_{b2}}{2} \int_0^{l(t)} e^{-\breve{\delta}_1 x} \tilde{\alpha}_2(x,t)^T Q_2(x)^{-1} \tilde{\alpha}_2(x,t) dx, \quad (6.197)$$

where a positive definite matrix $P_2 = P_2^T$ is the solution to the Lyapunov equation

$$P_2 \breve{A} + \breve{A}^T P_2 = -\hat{Q}_2$$

for some $\hat{Q}_2 = \hat{Q}_2^T > 0$, and $\breve{r}_{a1}, \breve{r}_{a2}, \breve{r}_{b1}, \breve{r}_{b2}, \breve{\delta}_1, \breve{\delta}_2$ are positive constants. The following inequality holds

$$\mu_{e1}\Omega_e(t) \leq V_e(t) \leq \mu_{e2}\Omega_e(t) \quad (6.198)$$

for some positive μ_{e1}, μ_{e2}, where

$$\Omega_e(t) = \|\tilde{\alpha}(\cdot, t)\|^2 + \|\tilde{\beta}(\cdot, t)\|^2 + \left|\tilde{S}(t)\right|^2, \quad (6.199)$$

and

$$\|\tilde{\alpha}(\cdot, t)\|^2 = \int_0^{l(t)} \tilde{\alpha}_1(\cdot, t)^2 dx + \int_0^{l(t)} \tilde{\alpha}_2(\cdot, t)^2 dx. \quad (6.200)$$

Taking the derivative of (6.197) along (6.192)–(6.196) and choosing $\breve{r}_{a1}, \breve{r}_{a2}, \breve{r}_{b1}, \breve{r}_{b2}, \breve{\delta}_1, \breve{\delta}_2$ in a process similar to that in (6.111)–(6.119), we obtain

$$\dot{V}_e(t) \leq -\eta_e V_e(t) \quad (6.201)$$

for some positive η_e, which is associated with the choice of L_0. The exponential stability result follows in the sense of

$$\left(\|\tilde{\alpha}(x,t)\|^2 + \|\tilde{\beta}(x,t)\|^2 + |\tilde{S}(t)|^2 \right)^{\frac{1}{2}}$$

$$\leq \xi_e \left(\|\tilde{\alpha}(x,0)\|^2 + \|\tilde{\beta}(x,0)\|^2 + |\tilde{S}(0)|^2 \right)^{\frac{1}{2}} e^{-\eta_e t} \qquad (6.202)$$

for some positive ξ_e and η_e.

Recalling the direct and inverse backstepping transformations (6.167)–(6.169), (6.188)–(6.190), and applying the Cauchy-Schwarz inequality, the proof of lemma 6.4 is complete. $\qquad \square$

Applying the exponential stability result of the observer error subsystem (6.162)–(6.166) in lemma 6.4 and recalling the relationships (6.158)–(6.161), we acquire the following theorem about the performance of the observer on tracking the actual states in the original plant (6.32)–(6.37).

Theorem 6.3. *For the observer error system (6.158)–(6.166) with the observer gains $\Gamma_1(x,t)$ in (6.185), $\Gamma_2(x,t)$ in (6.186), and $\Gamma_3(t)$ in (6.187), with arbitrary initial data $(w(x,0), w_t(x,0)) \in H^2(0,L) \times H^1(0,L)$, $(u(x,0), u_t(x,0)) \in H^2(0,L) \times H^1(0,L)$, there exist positive constants Υ_4, σ_4 such that*

$$\Big(\|\tilde{u}_t(\cdot,t)\|^2 + \|\tilde{u}_x(\cdot,t)\|^2 + \|\tilde{w}_t(\cdot,t)\|^2 + \|\tilde{w}_x(\cdot,t)\|^2$$

$$+ \tilde{w}(0,t)^2 + \tilde{w}_t(0,t)^2 + \tilde{u}(0,t)^2 + \tilde{u}_t(0,t)^2 \Big)^{\frac{1}{2}}$$

$$\leq \Upsilon_4 \Big(\|\tilde{u}_t(\cdot,0)\|^2 + \|\tilde{u}_x(\cdot,0)\|^2 + \|\tilde{w}_t(\cdot,0)\|^2 + \|\tilde{w}_x(\cdot,0)\|^2$$

$$+ \tilde{w}(0,0)^2 + \tilde{w}_t(0,0)^2 + \tilde{u}(0,0)^2 + \tilde{u}_t(0,0)^2 \Big)^{\frac{1}{2}} e^{-\sigma_4 t}, \qquad (6.203)$$

which means that the observer states in (6.139)–(6.147) are exponentially convergent to the actual values in (6.32)–(6.37) according to (6.151)–(6.154).

Proof. Recalling lemma 6.4 and (6.155)–(6.157), the following inequality holds

$$\left(\|\tilde{y}(\cdot,t)\|^2 + \|\tilde{v}(\cdot,t)\|^2 + \|\tilde{k}(\cdot,t)\|^2 + \|\tilde{z}(\cdot,t)\|^2 + \left| \tilde{X}(t) \right|^2 + \left| \tilde{Y}(t) \right|^2 \right)^{\frac{1}{2}}$$

$$\leq \Upsilon_{4a} \Big(\|\tilde{y}(\cdot,0)\|^2 + \|\tilde{v}(\cdot,0)\|^2$$

$$+ \|\tilde{k}(\cdot,0)\|^2 + \|\tilde{z}(\cdot,0)\|^2 + \left| \tilde{X}(0) \right|^2 + \left| \tilde{Y}(0) \right|^2 \Big)^{\frac{1}{2}} e^{-\sigma_{4a} t}$$

for some positive constants $\Upsilon_{4a}, \sigma_{4a}$.

According to (6.158)–(6.161), with which $\tilde{u}_t(\cdot,t), \tilde{u}_x(\cdot,t), \tilde{w}_t(\cdot,t), \tilde{w}_x(\cdot,t)$ are represented by $\tilde{z}(\cdot,t), \tilde{v}(\cdot,t), \tilde{k}(\cdot,t), \tilde{y}(\cdot,t)$, the proof of theorem 6.3 is complete, recalling (6.155). $\qquad \square$

6.4   CONTROLLER WITH COLLOCATED BOUNDARY SENSING

The output-feedback control law, employing axial and lateral control forces at the ship-mounted crane with sensors placed only at that location, is obtained by combining the state-feedback controller from section 6.2 with the observer from section 6.3. After inserting the observer states into the state-feedback controller (6.104), (6.105) to replace the unmeasurable states, the control laws $U_1(t)$, $U_2(t)$ in (6.36), (6.37) assume the output-feedback form denoted as $U_{o1}(t)$, $U_{o2}(t)$ and given by

$$
\begin{aligned}
U_{o1}(t) = \frac{-1}{d_{19}(l(t))\sqrt{d_6(l(t))}} \Bigg[& u_t(l(t),t) - \int_0^{l(t)} \big[(F_{11}(l(t),y) + N_{11}(l(t),y))\hat{u}_t(y,t) \\
& + (F_{12}(l(t),y) + N_{12}(l(t),y))\hat{w}_t(y,t) \\
& + (N_{11}(l(t),y) - F_{11}(l(t),y))\sqrt{d_6(y)}\hat{u}_x(y,t) \\
& + (N_{12}(l(t),y) - F_{12}(l(t),y))\sqrt{d_1(y)}\hat{w}_x(y,t) \big] dy \\
& - \lambda_{11}(l(t))\hat{w}(0,t) - \lambda_{12}(l(t))\hat{w}_t(0,t) \\
& - \lambda_{13}(l(t))\hat{u}(0,t) - \lambda_{14}(l(t))\hat{u}_t(0,t) \Bigg],
\end{aligned}
\tag{6.204}
$$

$$
\begin{aligned}
U_{o2}(t) = \frac{-1}{d_{20}(l(t))\sqrt{d_1(l(t)}} \Bigg[& w_t(l(t),t) - \int_0^{l(t)} \big[(F_{21}(l(t),y) + N_{21}(l(t),y))\hat{u}_t(y,t) \\
& + (F_{22}(l(t),y) + N_{22}(l(t),y))\hat{w}_t(y,t) \\
& + (N_{21}(l(t),y) - F_{21}(l(t),y))\sqrt{d_6(y)}\hat{u}_x(y,t) \\
& + (N_{22}(l(t),y) - F_{22}(l(t),y))\sqrt{d_1(y)}\hat{w}_x(y,t) \big] dy \\
& - \lambda_{21}(l(t))\hat{w}(0,t) - \lambda_{22}(l(t))\hat{w}_t(0,t) \\
& - \lambda_{23}(l(t))\hat{u}(0,t) - \lambda_{24}(l(t))\hat{u}_t(0,t) \Bigg].
\end{aligned}
\tag{6.205}
$$

The control inputs $U_{o1}(t)$, $U_{o2}(t)$ are implemented based on the boundary measurements $u_t(l(t),t)$, $w_t(l(t),t)$ mentioned in section 6.1. To be exact, $u_t(l(t),t)$, $w_t(l(t),t)$ directly act as the first terms in the expressions (6.204), (6.205) and are also used to obtain the solutions of the observer system (6.139)–(6.147) that are required in the remaining terms in (6.204), (6.205), through the conversion (6.138).

Theorem 6.4. *For the closed-loop system consisting of the plant* (6.32)–(6.37), *the observer* (6.139)–(6.147), *and the output-feedback controller* (6.204), (6.205), *with arbitrary initial values* $(w(x,0), w_t(x,0)) \in H^2(0,L) \times H^1(0,L)$, $(u(x,0), u_t(x,0)) \in H^2(0,L) \times H^1(0,L)$, *one obtains*

1) there exist positive constants Υ_5 *and* σ_5 *such that*

$$
\left(\Xi(t) + \hat{\Xi}(t) \right)^{1/2} \leq \Upsilon_5 \left(\Xi(0) + \hat{\Xi}(0) \right)^{1/2} e^{-\sigma_5 t},
\tag{6.206}
$$

where $\Xi(t)$ *is given in* (6.121), *and* $\hat{\Xi}(t)$ *is defined as*

$$
\begin{aligned}
\hat{\Xi}(t) = & \|\hat{u}_x(\cdot,t)\|^2 + \|\hat{u}_t(\cdot,t)\|^2 + |\hat{u}(0,t)|^2 + |\hat{u}_t(0,t)|^2 \\
& + \|\hat{w}_x(\cdot,t)\|^2 + \|\hat{w}_t(\cdot,t)\|^2 + |\hat{w}(0,t)|^2 + |\hat{w}_t(0,t)|^2 ;
\end{aligned}
\tag{6.207}
$$

2) *the output-feedback control signals* (6.204), (6.205) *are bounded and exponentially convergent to zero.*

Proof. The output-feedback controller (6.204), (6.205) can be written as

$$[U_{o1}(t), U_{o2}(t)]^T = [U_{sf1}(t), U_{sf2}(t)]^T + \tilde{\delta}(t) \tag{6.208}$$

by virtue of (6.151)–(6.155), where $U_{sf1}(t), U_{sf2}(t)$ are the state-feedback laws given by (6.104), (6.105), and $\tilde{\delta}(t) \in R^{2 \times 1}$ is

$$\begin{aligned}
\tilde{\delta}(t) = 2R(l(t))^{-1} &\left[\int_0^{l(t)} F(l(t), y)[\tilde{u}_t(y,t) - \sqrt{d_6(y)}\tilde{u}_x(y,t), \right. \\
&\tilde{w}_t(y,t) - \sqrt{d_1(y)}\tilde{w}_x(y,t)]^T dy \\
&+ \int_0^{l(t)} N(l(t), y)[\tilde{u}_t(y,t) + \sqrt{d_6(y)}\tilde{u}_x(y,t), \\
&\tilde{w}_t(y,t) + \sqrt{d_1(y)}\tilde{w}_x(y,t)]^T dy \\
&\left. + \lambda(l(t))[\tilde{w}(0,t), \tilde{w}_t(0,t), \tilde{u}(0,t), \tilde{u}_t(0,t)]^T \right].
\end{aligned} \tag{6.209}$$

Applying the output-feedback controller (6.208) to the plant (6.32)–(6.37), that is, $U_1(t) = U_{1o}(t)$, $U_2(t) = U_{2o}(t)$, and recalling theorems 6.1 and 6.3, together with (6.151)–(6.155), we arrive at (6.206), which is property (1) in theorem 6.4. Moreover, by applying theorem 6.2, which shows that the state-feedback controllers $U_{sf1}(t)$, $U_{sf2}(t)$ are exponentially convergent to zero, and theorem 6.3, which guarantees the exponential convergence to zero of $\tilde{\delta}(t)$ (6.209), we obtain the results of boundedness and exponential convergence of the output-feedback control input in light of (6.208)—that is, property (2) of theorem 6.4 is established. Therefore, the proof of theorem 6.4 is complete. □

6.5 SIMULATION FOR A DEEP-SEA CONSTRUCTION SYSTEM

The simulation is conducted based both on the linear model (6.21)–(6.26) and on the actual nonlinear model (6.13)–(6.18) with unmodeled disturbances, where the simulation on the former model is performed to verify the theoretical results, and the second simulation is done to illustrate the effectiveness in the application of vibration control to a reasonably realistic model of a DCV.

The plant on the time-varying domain with pre-determined time-varying functions $l(t)$ and $\dot{l}(t)$ shown in figure 6.4 is converted to a plant on the fixed domain $\iota \in [0, 1]$, with time-varying coefficients related to $l(t), \dot{l}(t), \ddot{l}(t)$ by introducing

$$\iota = \frac{x}{l(t)} \tag{6.210}$$

that is, representing $u(x, t)$ by $u(\iota, t)$, as

$$u_x(x,t) = \frac{1}{l(t)}u_\iota(\iota, t), \tag{6.211}$$

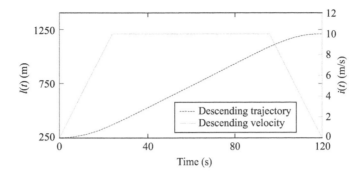

Figure 6.4. Descending trajectory and velocity—that is, the time-varying cable length $l(t)$ and the changing rate $\dot{l}(t)$.

$$u_{xx}(x,t) = \frac{1}{l(t)^2} u_{\iota\iota}(\iota,t), \tag{6.212}$$

$$u_t(x,t) = u_t(\iota,t) - \frac{\dot{l}(t)\iota}{l(t)} u_\iota(\iota,t), \tag{6.213}$$

$$u_{tt}(x,t) = u_{tt}(\iota,t) - \frac{2\dot{l}(t)\iota}{l(t)} u_{\iota t}(\iota,t) - \frac{\dot{l}(t)^2\iota^2}{l(t)^2} u_{\iota\iota}(\iota,t)$$

$$- \frac{(l(t)\ddot{l}(t) - 2\dot{l}(t)^2)\iota}{l(t)^2} u_\iota(\iota,t). \tag{6.214}$$

Then the simulation is conducted based on the finite-difference method with the time and space steps of 0.001 and 0.05, respectively. The observer (6.139) (6.147) is solved in the same way, and the following equations are used to obtain $\hat{u}, \hat{w}$ from $\hat{k}, \hat{y}, \hat{z}, \hat{v}$:

$$\hat{u}(\iota,t) = \int_0^\iota \frac{1}{2\sqrt{d_6(\bar{\iota})}} (\hat{k}(\bar{\iota},t) - \hat{y}(\bar{\iota},t))d\bar{\iota} + \bar{C}_1\hat{W}(t),$$

$$\hat{w}(\iota,t) = \int_0^\iota \frac{1}{2\sqrt{d_1(\bar{\iota})}} (\hat{z}(\bar{\iota},t) - \hat{v}(\bar{\iota},t))d\bar{\iota} + \bar{C}_2\hat{W}(t)$$

according to (6.139)–(6.142) and (6.150), where $\bar{C}_1 - [0,0,1,0]$, and $\bar{C}_2 = [1,0,0,0]$.

The initial conditions are defined according to the steady state, as $u_x(\cdot,0) = \bar{\varepsilon}(\cdot)$, $u_t(\cdot,0) = 0$, and $w_x(\cdot,0) = -\bar{\phi}(\cdot)$, $w_t(\cdot,0) = 0$. By defining $u(0,0) = 0$ and $w(l(0),0) = 0$, the initial conditions of (6.21)–(6.26) are thus defined completely in the finite-difference numerical calculation. All initial conditions $\hat{k}(\cdot,0)$, $\hat{y}(\cdot,0)$, $\hat{z}(\cdot,0)$, $\hat{v}(\cdot,0)$, $\hat{W}(0)$ of the observer (6.139)–(6.147) are set as zero.

Test on the Linear Model

Matching (6.32)–(6.37) with (6.21)–(6.26), we obtain the specific expressions of the coefficients in (6.32)–(6.37) as

$$d_1(x) = \frac{\frac{3}{2} E A_a \bar{\phi}(x)^2 + T(x)}{m_c}, \tag{6.215}$$

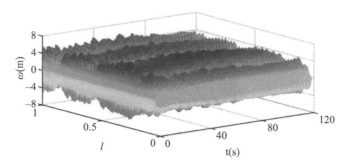

Figure 6.5. Responses of lateral vibrations $w(\iota, t)$ without control.

$$d_2(x) = \frac{EA_a\bar{\varepsilon}'(x) + \rho g}{m_c}, \quad d_3 = \frac{-EA_a\bar{\phi}'(x)}{m_c}, \tag{6.216}$$

$$d_4 = \frac{-c_v}{m_c}, \quad d_5 = 0, \quad d_6 = \frac{EA_a}{m_c}, \quad d_7(x) = \frac{-EA_a\bar{\phi}'(x)}{m_c}, \tag{6.217}$$

$$d_8 = d_9 = 0, \quad d_{10} = \frac{-c_u}{m_c}, \quad d_{11} = \frac{-c_w}{M_L}, \quad d_{12} = \frac{-EA_a\bar{\phi}(0)^2}{2M_L}, \tag{6.218}$$

$$d_{13} = 0, \quad d_{14} = \frac{-EA_a\bar{\phi}(0)}{M_L}, \quad d_{15} = \frac{-c_h}{M_L}, d_{16} = \frac{-EA_a}{M_L}, \tag{6.219}$$

$$d_{17} = 0, \quad d_{18} = \frac{EA_a\bar{\phi}(0)}{2M_L}, \quad d_{19} = \frac{1}{EA_a}, \tag{6.220}$$

$$d_{20}(l(t)) = \frac{1}{EA_a\bar{\varepsilon}(l(t)) + \frac{EA_a}{2}\bar{\phi}(l(t))^2 + T(l(t))}, \tag{6.221}$$

where $T(x)$, $\bar{\varepsilon}(x)$, and $\bar{\phi}(x)$ are given in (6.6), (6.19), (6.20), and the values of the physical parameters are shown in table 6.1. The variable x in (6.215)–(6.221) can be represented by ι via (6.210). We apply the proposed controllers (6.204), (6.205) into (6.21)–(6.26), where the approximate solution of the kernel equations (6.87)–(6.93) is also solved by the finite-difference method on a fixed triangular domain $\mathcal{D}_0 = \{0 \leq y \leq x \leq L\}$. Then we extract $F(l(t), y), N(l(t), y)$, which are used in the controller. The control parameters κ are chosen as

$$\begin{bmatrix} \kappa_{11} & \kappa_{12} & \kappa_{13} & \kappa_{14} \\ \kappa_{21} & \kappa_{22} & \kappa_{23} & \kappa_{24} \end{bmatrix} = \begin{bmatrix} 0.8 & 1.2 & 4.5 & 6 \\ 2.5 & 3 & 1.5 & 2 \end{bmatrix} \times 10^3, \tag{6.222}$$

which determines the kernel $\lambda(x)$ used in the controllers. The same process is used to get $\bar{\varphi}(x, l(t)), \bar{\psi}(x, l(t))$, which are used in the observer gains (6.185), (6.186). All elements in L_0 are defined as 1.

Figure 6.5 shows that the large lateral vibrations whose oscillation range is up to 10 m persist in the whole operation time 120 s in the case without control. Even though the longitudinal vibrations in figure 6.6 are decaying because of the material damping coefficient c_u of the cable in table 6.1, the longitudinal vibration at the top of the cable ($\iota = 1$ in figure 6.5) is excessive and decays slowly because this point bears the whole mass of the cable and payload, resulting in large elastic deflections. Applying the proposed output-feedback lateral and longitudinal vibration control forces at the ship-mounted crane, the lateral vibrations decay with a satisfied decay

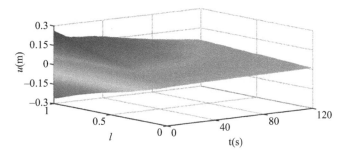

Figure 6.6. Responses of longitudinal vibrations $u(\iota, t)$ without control.

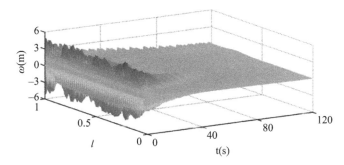

Figure 6.7. Closed-loop responses of lateral vibrations $w(\iota, t)$.

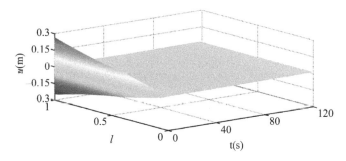

Figure 6.8. Closed-loop responses of longitudinal vibrations $u(\iota, t)$.

rate according to figure 6.7, and figure 6.8 shows that the longitudinal vibrations are suppressed very quickly. The output-feedback control forces at the ship-mounted crane are shown in figure 6.9, where the states of the proposed observer are used. The performance of the observer on tracking the actual states can be seen in figures 6.10 and 6.11, which show that the observer errors of both the lateral and the longitudinal vibrations are convergent to zero.

Test on the Actual Nonlinear Model with Ocean Current Disturbances

The nonlinear model (6.13)–(6.18) is converted to the one on the fixed domain $\iota \in [0, 1]$ through the process mentioned at the beginning of this section. The time

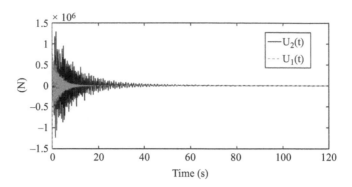

Figure 6.9. Control forces $U_1(t)$ and $U_2(t)$.

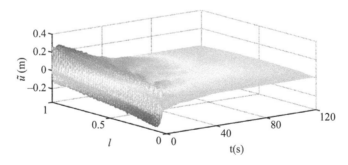

Figure 6.10. Observer error of lateral vibrations $\tilde{w}(\iota, t)$.

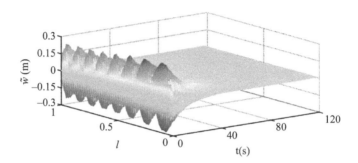

Figure 6.11. Observer error of longitudinal vibrations $\tilde{u}(\iota, t)$.

and space steps are changed to 0.0005 and 0.1, respectively, in the finite-difference method, to ensure numerical stability.

In practice, ocean current disturbances act as external lateral oscillating drag forces $f(\iota, t)$ on the cable. In the simulation, $f(\iota, t)$, which is added in (6.14) and (6.16), converted to a fixed domain, as above, is defined as follows. Consider the time-varying ocean surface current velocity $P(t)$ modeled by a first-order Gauss-Markov process [61], as follows:

$$\dot{P}(t) + \mu P(t) = \mathcal{G}(t), \quad P_{\min} \leq P(t) \leq P_{\max}, \tag{6.223}$$

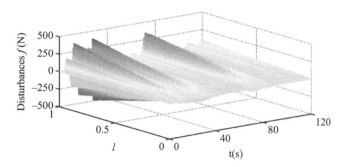

Figure 6.12. Lateral oscillation drag forces from ocean current disturbances.

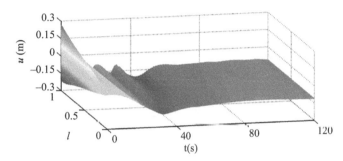

Figure 6.13. Closed-loop responses of longitudinal vibrations $u(\iota, t)$ in the actual nonlinear model with unmodeled disturbances.

where $\mathcal{G}(t)$ is Gaussian white noise. The constants $P_{\min}, P_{\max}$, and μ are chosen, respectively, as 1.6 ms^{-1}, 2.4 ms^{-1}, and 0 [95]. The function $f(\iota, t)$ is then given according to [95] as

$$f(\iota, t) = (0.9\iota + 0.1)\, \frac{1}{2}\rho_s C_d P(t)^2 R_D A_D \cos\left(4\pi \frac{S_t P(t)}{R_D} t + \varsigma\right), \qquad (6.224)$$

where $0.9\iota + 0.1$ means that the full disturbance load is applied at the top of the cable, at the ocean surface, and linearly declines to 0.1 at the bottom of the cable, at the payload. The constant $C_d = 1$ denotes the drag coefficient, and $\varsigma = \pi$ is the phase angle. The constant $A_D - 400$ denotes the amplitude of the oscillating drag force, and $S_t = 0.2$ is the Strouhal number [59]. The ocean disturbances f used in the simulation are shown in figure 6.12.

The control parameters κ_{11}, κ_{12}, κ_{13}, κ_{14} and κ_{21}, κ_{22}, κ_{23}, κ_{24} are increased to twice and ten times those in (6.222), respectively, due to the unmodeled disturbances. The observer parameters are kept the same as those in section 6.5. We apply the proposed output-feedback controller to the actual nonlinear model with the ocean current disturbances. Figures 6.13 and 6.14 show that the longitudinal vibrations and lateral vibrations are reduced as time goes on. It is particularly remarkable in figure 6.14 that while the control law is designed only for suppressing oscillations in response to initial lateral displacements and not for attenuating a lateral disturbance, like that in figure 6.12, the controller is effective in disturbance attenuation after an initial transient of about 40 seconds. From figure 6.15, the observer errors converge to a small range around zero. Simulation results in

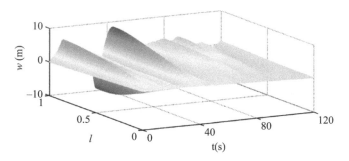

Figure 6.14. Closed-loop responses of lateral vibrations $w(\iota, t)$ in the actual nonlinear model with unmodeled disturbances.

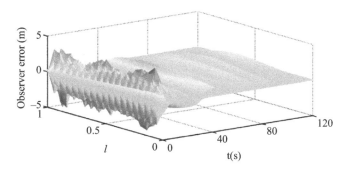

Figure 6.15. Observer errors $\tilde{u}(\cdot, t) + \tilde{w}(\cdot, t)$ in the actual nonlinear model with unmodeled disturbances.

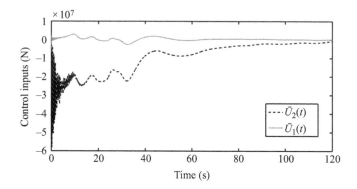

Figure 6.16. Control forces in the actual nonlinear model.

this section illustrate the effectiveness of the proposed control design when applied to vibration suppression of the DCV. The output-feedback control inputs in this actual model are shown in figure 6.16.

6.6 APPENDIX

Proof of lemma 6.1 The boundary conditions in (6.87)–(6.93) are along the lines $y = x$ and $y = 0$, and there are no conditions at the boundary $x = l(t)$ on the triangular domain $\mathcal{D}$. Therefore, we can extend the boundary $x = l(t)$ in $\mathcal{D}$ to $x = L$

due to $0 < l(t) \leq L$ in assumption 6.3 to solve F, N on a fixed triangular domain $\mathcal{D}_0 = \{0 < y < x < L\}$ and λ on a constant interval $0 \leq x \leq L$ in (6.87)–(6.93). The lines $y = x$ and $y = 0$ are overlapped boundaries of the triangular domains $\mathcal{D}_0$ and $\mathcal{D}$ and $\mathcal{D} \subseteq \mathcal{D}_0$. Once the solutions of (6.87)–(6.93) on $\mathcal{D}_0$ are obtained, those on the subset $\mathcal{D}$ are the required kernels F, N and λ. To ensure the well-posedness of the kernel equations (6.87)–(6.93) on $\mathcal{D}_0$, we add an additional artificial condition at $x = L$ for N_{21}, as done in [96]. Next, we will prove that there exists a unique solution of (6.87)–(6.93) on $\mathcal{D}_0$ by showing the system is in the form of a class of well-posed equations in [48], which ensures there exists a unique solution $F, N \in L^\infty(\mathcal{D})$, $\lambda \in L^\infty([0, l(t)])$.

Expanding (6.87)–(6.93) on $\mathcal{D}_0$, we know that $F_{ij}(x, y)_{1 \leq i, j \leq 2}$, $N_{ij}(x, y)_{1 \leq i, j \leq 2}$ should satisfy the following coupled hyperbolic PDEs:

$$\sqrt{d_6(x)} F_{11x}(x, y) - \sqrt{d_6(y)} F_{11y}(x, y)$$

$$= [\frac{d_{10}(y)}{2} - s_1(y)] N_{11}(x, y) + [\frac{d_5(y)}{2} - \frac{d_3(y)}{2\sqrt{d_6(y)}}] N_{12}(x, y)$$

$$+ [\frac{d_{10}(y)}{2} - s_1(y) + \sqrt{d_6(y)}' - \frac{d_{10}(x)}{2} + s_1(x)] F_{11}(x, y)$$

$$+ [\frac{d_5(y)}{2} - \frac{d_3(y)}{2\sqrt{d_6(y)}}] F_{12}(x, y), \tag{6.225}$$

$$\sqrt{d_6(x)} F_{12x}(x, y) - \sqrt{d_1(y)} F_{12y}(x, y)$$

$$= [\frac{d_9(y)}{2} - \frac{d_7(y)}{2\sqrt{d_1(y)}}] N_{11}(x, y) + [\frac{d_4(y)}{2} - s_2(y)] N_{12}(x, y)$$

$$+ [\frac{d_4(y)}{2} - s_2(y) + \sqrt{d_1(y)}' - \frac{d_{10}(x)}{2} + s_1(x)] F_{12}(x, y)$$

$$+ [\frac{d_9(y)}{2} - \frac{d_7(y)}{2\sqrt{d_1(y)}}] F_{11}(x, y), \tag{6.226}$$

$$\sqrt{d_1(x)} F_{21x}(x, y) - \sqrt{d_6(y)} F_{21y}(x, y)$$

$$= [\frac{d_{10}(y)}{2} - s_1(y)] N_{21}(x, y) + [\frac{d_5(y)}{2} - \frac{d_3(y)}{2\sqrt{d_6(y)}}] N_{22}(x, y)$$

$$+ [\frac{d_{10}(y)}{2} - s_1(y) + \sqrt{d_6(y)}' - \frac{d_4(x)}{2} + s_2(x)] F_{21}(x, y)$$

$$+ [\frac{d_5(y)}{2} - \frac{d_3(y)}{2\sqrt{d_6(y)}}] F_{22}(x, y), \tag{6.227}$$

$$\sqrt{d_1(x)} F_{22x}(x, y) - \sqrt{d_1(y)} F_{22y}(x, y)$$

$$= [\frac{d_9(y)}{2} - \frac{d_7(y)}{2\sqrt{d_1(y)}}] N_{21}(x, y) + [\frac{d_4(y)}{2} - s_2(y)] N_{22}(x, y)$$

$$+ [\frac{d_4(y)}{2} - s_2(y) + \sqrt{d_1(y)}' - \frac{d_4(x)}{2} + s_2(x)] F_{22}(x, y)$$

$$+ [\frac{d_9(y)}{2} - \frac{d_7(y)}{2\sqrt{d_1(y)}}] F_{21}(x, y), \tag{6.228}$$

$$\sqrt{d_6(x)}N_{11x}(x,y) + \sqrt{d_6(y)}N_{11y}(x,y)$$

$$= [s_1(y) + \frac{d_{10}(y)}{2}]F_{11}(x,y) + [\frac{d_3(y)}{2\sqrt{d_6(y)}} + \frac{d_5(y)}{2}]F_{12}(x,y)$$

$$+ [s_1(y) + \frac{d_{10}(y)}{2} - \sqrt{d_6(y)}' - s_1(x) - \frac{d_{10}(x)}{2}]N_{11}(x,y)$$

$$+ [\frac{d_3(y)}{2\sqrt{d_6(y)}} + \frac{d_5(y)}{2}]N_{12}(x,y), \tag{6.229}$$

$$\sqrt{d_6(x)}N_{12x}(x,y) + \sqrt{d_1(y)}N_{12y}(x,y)$$

$$= [\frac{d_7(y)}{2\sqrt{d_1(y)}} + \frac{d_9(y)}{2}]F_{11}(x,y) + [s_2(y) + \frac{d_4(y)}{2}]F_{12}(x,y)$$

$$+ [s_2(y) + \frac{d_4(y)}{2} - \sqrt{d_1(y)}' - s_1(x) - \frac{d_{10}(x)}{2}]N_{12}(x,y)$$

$$+ [\frac{d_7(y)}{2\sqrt{d_1(y)}} + \frac{d_9(y)}{2}]N_{11}(x,y), \tag{6.230}$$

$$\sqrt{d_1(x)}N_{21x}(x,y) + \sqrt{d_6(y)}N_{21y}(x,y)$$

$$= [s_1(y) + \frac{d_{10}(y)}{2}]F_{21}(x,y) + [\frac{d_3(y)}{2\sqrt{d_6(y)}} + \frac{d_5(y)}{2}]F_{22}(x,y)$$

$$+ [s_1(y) + \frac{d_{10}(y)}{2} - \sqrt{d_6(y)}' - s_2(x) - \frac{d_4(x)}{2}]N_{21}(x,y)$$

$$+ [\frac{d_3(y)}{2\sqrt{d_6(y)}} + \frac{d_5(y)}{2}]N_{22}(x,y), \tag{6.231}$$

$$\sqrt{d_1(x)}N_{22x}(x,y) + \sqrt{d_1(y)}N_{22y}(x,y)$$

$$= [\frac{d_7(y)}{2\sqrt{d_1(y)}} + \frac{d_9(y)}{2}]F_{21}(x,y) + [s_2(y) + \frac{d_4(y)}{2}]F_{22}(x,y)$$

$$+ [s_2(y) + \frac{d_4(y)}{2} - \sqrt{d_1(y)}' - s_2(x) - \frac{d_4(x)}{2}]N_{22}(x,y)$$

$$+ [\frac{d_7(y)}{2\sqrt{d_1(x)}} + \frac{d_9(y)}{2}]N_{21}(x,y) \tag{6.232}$$

along with the following set of boundary conditions:

$$F_{11}(x,x) = \frac{-d_{10}(x) + 2s_1(x)}{4\sqrt{d_6(x)}}, \tag{6.233}$$

$$F_{12}(x,x) = \frac{-d_9(x)\sqrt{d_1(x)} + d_7(x)}{2(\sqrt{d_6(x)} + \sqrt{d_1(x)})\sqrt{d_1(x)}}, \tag{6.234}$$

$$F_{21}(x,x) = \frac{-d_5(x)\sqrt{d_6(x)} + d_3(x)}{2(\sqrt{d_6(x)} + \sqrt{d_1(x)})\sqrt{d_6(x)}}, \tag{6.235}$$

$$F_{22}(x,x) = \frac{-d_4(x) + 2s_2(x)}{4\sqrt{d_1(x)}}, \tag{6.236}$$

$$N_{12}(x,x) = \frac{\frac{-d_7(x)}{2\sqrt{d_1(x)}} - \frac{d_9(x)}{2}}{\sqrt{d_6(x)} - \sqrt{d_1(x)}}, \tag{6.237}$$

$$N_{21}(x,x) = \frac{\frac{-d_3(x)}{2\sqrt{d_6(x)}} - \frac{d_5(x)}{2}}{\sqrt{d_1(x)} - \sqrt{d_6(x)}}, \tag{6.238}$$

$$N_{11}(x,0) = -F_{11}(x,0) + \frac{d_{14}\lambda_{12}(x)}{d_6(0)} + \frac{d_{16}\lambda_{14}(x)}{d_6(0)}, \tag{6.239}$$

$$N_{12}(x,0) = -F_{12}(x,0) + \frac{d_{12}\lambda_{12}(x)}{d_1(0)} + \frac{d_{18}\lambda_{14}(x)}{d_1(0)}, \tag{6.240}$$

$$N_{21}(x,0) = -F_{21}(x,0)$$
$$+ \frac{g_0(x)\sqrt{d_6(0)} + d_{14}\lambda_{22}(x) + d_{16}\lambda_{24}(x)}{d_6(0)}, \tag{6.241}$$

$$N_{22}(x,0) = -F_{22}(x,0) + \frac{d_{12}\lambda_{22}(x)}{d_1(0)} + \frac{d_{18}\lambda_{24}(x)}{d_1(0)}, \tag{6.242}$$

$$N_{21}(L,y) = 0, \tag{6.243}$$

where (6.243) is the artificial boundary condition, and assumption 6.2 ensures that the denominators of (6.237), (6.238) are nonzero. Additionally, $\lambda_{ij}(x)_{1 \le i \le 2, 1 \le j \le 4}$ should satisfy the following ODEs:

$$\sqrt{d_6(x)}\lambda_{11}'(x) + [s_1(x) + \frac{d_{10}(x)}{2}]\lambda_{11}(x) = 0, \tag{6.244}$$

$$\sqrt{d_6(x)}\lambda_{12}'(x) + [s_1(x) + \frac{d_{10}(x)}{2} - d_{11} + \frac{d_{12}}{\sqrt{d_1(0)}}]\lambda_{12}(x)$$
$$- \lambda_{11}(x) - (d_{17} - \frac{d_{18}}{\sqrt{d_1(0)}})\lambda_{14}(x) - 2\sqrt{d_1(0)}F_{12}(x,0) = 0, \tag{6.245}$$

$$\sqrt{d_6(x)}\lambda_{13}'(x) + [s_1(x) + \frac{d_{10}(x)}{2}]\lambda_{13}(x) = 0, \tag{6.246}$$

$$\sqrt{d_6(x)}\lambda_{14}'(x) + [s_1(x) + \frac{d_{10}(x)}{2} - d_{15} + \frac{d_{16}}{\sqrt{d_6(0)}}]\lambda_{14}(x)$$
$$- \lambda_{13}(x) - (d_{13} - \frac{d_{14}}{\sqrt{d_6(0)}})\lambda_{12}(x) - 2\sqrt{d_6(0)}F_{11}(x,0) = 0, \tag{6.247}$$

$$\sqrt{d_1(x)}\lambda_{21}'(x) + [s_2(x) + \frac{d_4(x)}{2}]\lambda_{21}(x) + g_0(x)\lambda_{11}(0) = 0, \tag{6.248}$$

$$\sqrt{d_1(x)}\lambda_{22}'(x) + [s_2(x) + \frac{d_4(x)}{2} - d_{11} + \frac{d_{12}}{\sqrt{d_1(0)}}]\lambda_{22}(x)$$
$$- \lambda_{21}(x) - (d_{17} - \frac{d_{18}}{\sqrt{d_1(0)}})\lambda_{24}(x)$$
$$- 2\sqrt{d_1(0)}F_{22}(x,0) + g_0(x)\lambda_{12}(0) = 0, \tag{6.249}$$

$$\sqrt{d_1(x)}\lambda_{23}'(x) + [s_2(x) + \frac{d_4(x)}{2}]\lambda_{23}(x) + g_0(x)\lambda_{13}(0) = 0, \tag{6.250}$$

$$\sqrt{d_1(x)}\lambda_{24}'(x) + [s_2(x) + \frac{d_4(x)}{2} - d_{15} + \frac{d_{16}}{\sqrt{d_6(0)}}]\lambda_{24}(x)$$

$$- (d_{13} - \frac{d_{14}}{\sqrt{d_6(0)}})\lambda_{22}(x) - \lambda_{23}(x)$$

$$- 2\sqrt{d_6(0)}F_{21}(x,0) + g_0(x)\lambda_{14}(0) = 0, \qquad (6.251)$$

with initial conditions

$$\begin{bmatrix} \lambda_{11}(0) & \lambda_{12}(0) & \lambda_{13}(0) & \lambda_{14}(0) \\ \lambda_{21}(0) & \lambda_{22}(0) & \lambda_{23}(0) & \lambda_{24}(0) \end{bmatrix}$$

$$= \begin{bmatrix} \kappa_{11} & \kappa_{12} & \kappa_{13} & \kappa_{14} \\ \kappa_{21} & \kappa_{22} & \kappa_{23} & \kappa_{24} \end{bmatrix}. \qquad (6.252)$$

The equation set (6.225)–(6.252) has the same structure as the kernel equations (17)–(24) in [48] with setting $m = n = 2$. More precisely, (6.225)–(6.228) corresponds to (17); (6.229)–(6.232) corresponds to (18); (6.233)–(6.236) corresponds to (19); (6.238) corresponds to (20); (6.239)–(6.242) corresponds to (21); the additional artificial boundary condition (6.243) corresponds to (24); the initial value problem (6.244)–(6.252) corresponds to (22), (23). Even though (17)–(24) in [48] are with constant coefficients, the well-posedness result still holds in (6.225)–(6.252) with spatially varying coefficients because (6.225)–(6.252) are in the form of a general class of hyperbolic PDEs (30), (31), whose well-posedness has been proved in [48]. Therefore, (6.225)–(6.252)—that is, (6.87)–(6.93)—are well-posed on the domain $\mathcal{D}_0$, which straightforwardly yields to a unique solution $F, N \in L^\infty(\mathcal{D})$, $\lambda \in L^\infty([0, l(t)])$ of (6.87)–(6.93) because of $\mathcal{D} \subseteq \mathcal{D}_0$, $l(t) \le L$. Kernel equations (6.94)–(6.100) of $J(x,y), K(x,y), \gamma(x)$ have the same structure as (6.87)–(6.93). Through a similar process as above, adding an additional artificial boundary condition $K_{21}(L, y) = 0$, one obtains a unique solution $J, K \in L^\infty(\mathcal{D})$, $\gamma \in L^\infty([0, l(t)])$ of (6.94)–(6.100). The proof of lemma 6.1 is complete.

6.7   NOTES

Chapters 2–5 presented control designs for one-dimensional cable vibrations, either axial or transversal. In this chapter, we dealt with axial-transversal coupled

Table 6.2. Comparison of previous results on the boundary vibration control of cables

	Multi-dimensional coupled vibrations in cables	Time-varying cable lengths	Number of controlled/fixed/ damped boundaries
[90, 131]	×	✓	2
[92]	✓	✓	2
[22]	✓	×	1
[178], [185]	×	✓	1
This chapter	✓	✓	1

vibrations of the cable, whose dynamics are linear wave PDEs with in-domain couplings obtained from linearizing the original nonlinear model around a steady state. After applying Riemann transformations, the plant becomes a 4×4 hyperbolic system. Compared to the 4×4 system in the dual-cable mining elevator in chapter 3, which is a set of two 2×2 pairs that are coupled not along the domain but at the boundary, the plant is a fully coupled 4×4 hyperbolic system in this chapter. Comparisons with some previous results on the boundary vibration control of cables are summarized in table 6.2.

Chapter Seven

Deep-Sea Construction with Event-Triggered Delay Compensation

A backstepping control law for the deep-sea construction vessel (DCV) was presented in chapter 6. One challenge to its practical implementation is that the massive actuator, the ship-mounted crane, is incapable of supporting the fast-changing control signal due to its low natural frequency. In chapter 6, we placed the sensors at the ship-mounted crane—that is, at the top of the cable. When the sensor is placed at the object, referred to as the payload, at the bottom of the cable, there exists another challenge: a sensor delay, due to the fact that the sensor signal is transmitted over a large distance from the seabed to the vessel on the ocean surface, through a set of acoustic devices. Such a sensor signal delay may result in information distortion or even make the control system lose stability.

In order to solve these problems, in this chapter we pursue a delay-compensated event-triggered control scheme, which compensates for the sensor delay of arbitrary length and reduces the changes in the actuator signal—that is, the control input employs piecewise-constant values.

The DCV model and the general model used for control are presented in section 7.1. The observer design is proposed in section 7.2, where the observer gains are determined in two transformations that convert the observer error system to a target observer error system whose exponential stability is straightforward to obtain. An observer-based output-feedback controller with delay compensation is designed in section 7.3 using the backstepping method. The dynamic event-triggering mechanism (ETM) is designed in section 7.4. The existence of a minimal dwell time between two successive triggering times and exponential convergence in the event-based closed-loop system are proved in section 7.5. The proposed control design is verified in the application of DCV control to seabed installation via a simulation in section 7.6.

7.1 PROBLEM FORMULATION

Model of DCV

We follow the modeling process in section 6.1 in chapter 6 but make the following three simplifications: 1) We take the cable length as constant, setting $l(t)$ as L; 2) The spatially dependent static tension of the cable, which considers the effect of the cable mass on the tension in the cable, is assumed to be constant as a result of the payload mass and buoyancy—that is, we set $T(x)$ as T_0; 3) We only focus on lateral vibrations instead of longitudinal-lateral coupled vibrations and model the

Table 7.1. Physical parameters of the DCV

Parameters (units)	Values
Cable length L (m)	1000
Cable diameter R_d (m)	0.2
Cable linear density ρ (kg/m)	8.02
Payload mass M_L (kg)	4.0×10^5
Gravitational acceleration g (m/s^2)	9.8
Cable material damping coefficient d_c (N·s/m)	0.5
Height of payload modeled as a cylinder h_c (m)	10
Diameter of payload modeled as a cylinder D_c (m)	5
Damping coefficient at payload d_L (N·s/m)	2.0×10^5
Seawater density ρ_s (kgm^{-3})	1024

dynamics of a DCV as

$$T_0 u_{\bar{x}}(0,t) = U(t), \tag{7.1}$$

$$\rho u_{tt}(\bar{x},t) = T_0 u_{\bar{x}\bar{x}}(\bar{x},t) - d_c u_t(\bar{x},t), \tag{7.2}$$

$$u(L,t) = b_L(t), \tag{7.3}$$

$$M_L \ddot{b}_L(t) = -d_L \dot{b}_L(t) + T_0 u_{\bar{x}}(L,t), \tag{7.4}$$

$\forall (\bar{x},t) \in [0,L] \times [0,\infty)$. The function $u(\bar{x},t)$ denotes the distributed transverse displacement along the cable. The function $b_L(t)$ represents the transverse displacement of the payload. Input $U(t)$ is the control force at the ship-mounted crane. The static tension T_0 is defined as

$$T_0 = M_L g - F_{\text{buoyant}}, \tag{7.5}$$

where the buoyancy F_{buoyant} is

$$F_{\text{buoyant}} = \frac{1}{4} \pi D_c^2 h_c \rho_s g. \tag{7.6}$$

The physical parameters of the DCV are given in table 7.1, whose values are taken from [95]. We simplify the DCV model derived in chapter 6 with the purpose of presenting the delay-compensated event-triggered control design in this chapter more clearly. This design can be applied to the model in chapter 6 with a modest extension.

In chapter 6, we placed the sensors at the ship-mounted crane, at the top of the cable, whereas the design in which the sensor is placed at the object, referred to as the payload, at the bottom of the cable, is considered here. As shown in figure 7.1, an accelerometer is placed at the payload to measure the lateral acceleration $\ddot{b}_L(t)$. However, there exists a sensor delay τ because the sensor signal is transmitted over a large distance from the seabed to the vessel on the ocean surface through a set of acoustic devices. Therefore, the acquisition of the wireless receiver at the vessel is the delayed measurement of the lateral acceleration of the payload, denoted as

$$y_{\text{out}}(t) = \dot{\zeta}(t-\tau), \quad t \in [\tau, \infty), \tag{7.7}$$

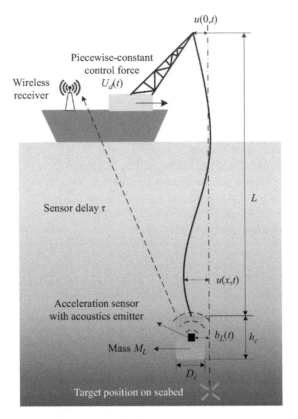

Figure 7.1. Diagram of a deep-sea construction vessel.

where a new variable is introduced as

$$\zeta(t) = \dot{b}_L(t). \tag{7.8}$$

In practice, the lateral velocity ζ of the payload is obtained by integrating the delayed acceleration sensing $y_{\text{out}}(t)$—that is,

$$\zeta(t - \tau) = \int_{\tau}^{t} y_{\text{out}}(h)dh + \zeta(0), \quad t \in [\tau, \infty), \tag{7.9}$$

which is a delayed signal of the lateral velocity where the initial value $\zeta(0)$ of the lateral velocity at the payload is used. Equation (7.9) only explains an acquisition method for ζ in practical DCVs via the acceleration sensor, and no restrictions are imposed on any initial conditions in the design and theory in this chapter. In the practical DCV there always exists, caused by ocean currents, a distributed oscillating drag force $f(\bar{x}, t)$ along the cable and a drag force $f_L(t)$ at the payload; that is, disturbances $f(\bar{x}, t)$, $f_L(t)$ would exist in (7.2), (7.4). Although we do not deal with the disturbances $f(\bar{x}, t)$, $f_L(t)$ in the control design in this chapter, they are considered in the simulation model for testing the robustness of the proposed controller.

Representing the Model in Riemann Coordinates

Applying the Riemann transformations

$$\bar{z}(\bar{x}, t) = u_t(\bar{x}, t) - \sqrt{\frac{T_0}{\rho}} u_{\bar{x}}(\bar{x}, t), \tag{7.10}$$

$$\bar{w}(\bar{x}, t) = u_t(\bar{x}, t) + \sqrt{\frac{T_0}{\rho}} u_{\bar{x}}(\bar{x}, t), \tag{7.11}$$

introducing a space normalization variable

$$x = \frac{\bar{x}}{L} \in [0, 1], \tag{7.12}$$

and defining

$$\bar{z}(\bar{x}, t) = z(x, t), \quad \bar{w}(\bar{x}, t) = w(x, t), \tag{7.13}$$

we can rewrite the equations (7.1)–(7.4) as the plant (7.16)–(7.20), to be shown shortly, with the coefficients

$$\bar{c} = 2\sqrt{\frac{1}{T_0\rho}}, \quad q_1 = q_2 = \frac{1}{L}\sqrt{\frac{T_0}{\rho}}, \quad c_1 = c_2 = c_3 = c_4 = \frac{d_c}{2\rho}, \tag{7.14}$$

$$q = -1, \quad p = 1, \quad c = 2, \quad a_1 = \frac{-d_L}{M_L} + \frac{\sqrt{T_0\rho}}{M_L}, \quad b_1 = \frac{\sqrt{T_0\rho}}{M_L}, \tag{7.15}$$

where $\bar{z}_{\bar{x}}(\bar{x}, t) = \frac{1}{L}z_x(x, t)$, $\bar{w}_{\bar{x}}(\bar{x}, t) = \frac{1}{L}w_x(x, t)$ have been used.

General Model

The considered plant is in a general form, given by

$$z(0, t) = pw(0, t) + \bar{c}U(t), \tag{7.16}$$

$$z_t(x, t) = -q_1 z_x(x, t) - c_1 w(x, t) - c_2 z(x, t), \tag{7.17}$$

$$w_t(x, t) = q_2 w_x(x, t) - c_3 w(x, t) - c_4 z(x, t), \tag{7.18}$$

$$w(1, t) = qz(1, t) + c\zeta(t), \tag{7.19}$$

$$\dot{\zeta}(t) = a_1\zeta(t) - b_1 z(1, t), \tag{7.20}$$

$\forall(x, t) \in [0, 1] \times [0, \infty)$, and the output measurement is

$$Y_{\text{out}}(t) = [\zeta(t - \tau), \dot{\zeta}(t - \tau)]^T, \quad t \in [\tau, \infty) \tag{7.21}$$

according to the obtained signals shown in (7.7)–(7.9). The scalar $\zeta(t)$ is an ODE state. The scalars $z(x, t), w(x, t)$ are states of the 2×2 coupled hyperbolic PDEs. The positive parameter τ is an arbitrary constant denoting the time delay in the measurement. The control input $U(t)$ in (7.16) is to be designed. The parameters $c_1, c_2, c_3, c_4, c \in \mathbb{R}$ are arbitrary. The positive constants q_1 and q_2 are transport speeds. The constants a_1 and $b_1 \neq 0$ are arbitrary. The parameters $q, p \in \mathbb{R}$ $(q \neq 0)$ satisfy assumption 7.1.

Assumption 7.1. *The plant parameters p, q satisfy*

$$|pq| < e^{\frac{c_3}{q_2} + \frac{c_2}{q_1}}.$$

One can readily check that assumption 7.1, which will be used in section 7.4, is satisfied in the DCV model with the parameters (7.14), (7.15), whose values are given in table 7.1.

Rewrite Delay as a Transport PDE

By defining

$$v(x, t) = \zeta(t - \tau(x - 1)), \tag{7.22}$$

we obtain from (7.21) a transport PDE

$$v(1, t) = \zeta(t), \tag{7.23}$$

$$v_t(x, t) = -\frac{1}{\tau} v_x(x, t), \quad x \in [1, 2], \tag{7.24}$$

$$Y_{\text{out}}(t) = [v(2, t), v_t(2, t)]^T \tag{7.25}$$

for all $(x, t) \in [1, 2] \times [0, \infty)$, to describe the time delay in the measurement (7.21). Replacing (7.21) with (7.23)–(7.25), we obtain a hyperbolic PDE-ODE system connecting with another transport PDE, the coupled hyperbolic PDE-ODE-transport PDE system (7.16)–(7.20), (7.23)–(7.25). The time delay is "removed" at a cost of adding a transport PDE into the plant (7.16)–(7.20). The continuous-in-time boundary control design for a PDE-ODE-PDE configuration was also studied in [2].

Now the task is equivalent to exponentially stabilizing the overall system (7.16)–(7.20), (7.23)–(7.25)—that is, the (z, w, v)-PDEs and ζ-ODE, by employing an event-triggered output-feedback control input $U(t)$ at the left boundary of the z PDE using only the right boundary state of the v-PDE. We adopt the following notation:

- The symbol $\mathbb{R}_-$ denotes the set of negative real numbers, whose complement of the real axis is $\mathbb{R}_+ := [0, +\infty)$. The symbol $\mathbb{Z}_+$ denotes the set of all nonnegative integers.
- Let $U \subseteq \mathbb{R}^n$ be a set with a nonempty interior, and let $\Omega \subseteq \mathbb{R}$ be a set. By $C^0(U; \Omega)$, we denote the class of continuous mappings on U, which take values in Ω. By $C^k(U; \Omega)$, where $k \geq 1$, we denote the class of continuous functions on U, which have continuous derivatives of order k on U and take values in Ω.
- We use the notation $L^2(0, 1)$ for the standard space of the equivalence class of square-integrable, measurable functions defined on $(0, 1)$, and $\|f\| = \left(\int_0^1 f(x)^2 dx\right)^{\frac{1}{2}} < +\infty$ for $f \in L^2(0, 1)$.
- For an $I \subseteq \mathbb{R}_+$, the space $C^0(I; L^2(0, 1))$ is the space of continuous mappings $I \ni t \mapsto u[t] \in L^2(0, 1)$.
- Let $u : \mathbb{R}_+ \times [0, 1] \to \mathbb{R}$ be given. We use the notation $u[t]$ to denote the profile of u at certain $t \geq 0$—that is, $(u[t])(x) = u(x, t)$ for all $x \in [0, 1]$.

7.2 OBSERVER DESIGN USING DELAYED MEASUREMENT

In order to build the observer-based event-triggered output-feedback controller of the plant (7.16)–(7.21), in this section we design a state observer to track the overall system (7.16)–(7.21) using only the delayed measurement $y_{out}(t)$. Through the reformulation in section 7.1, the estimation task is equivalent to designing a state observer to recover the overall system (7.16)–(7.20), (7.23)–(7.25) using only the measurements at the right boundary $x = 2$ of the v PDE.

The observer is built as a copy of the plant (7.16)–(7.20), (7.23)–(7.25) plus some output error injection terms, as follows:

$$\hat{z}(0,t) = p\hat{w}(0,t) + \bar{c}U(t), \tag{7.26}$$

$$\hat{z}_t(x,t) = -q_1\hat{z}_x(x,t) - c_1\hat{w}(x,t) - c_2\hat{z}(x,t)$$
$$+ H_2(x)(Y_{out}(t) - \hat{Y}_{out}(t)), \quad x \in [0,1], \tag{7.27}$$

$$\hat{w}_t(x,t) = q_2\hat{w}_x(x,t) - c_3\hat{w}(x,t) - c_4\hat{z}(x,t)$$
$$+ H_3(x)(Y_{out}(t) - \hat{Y}_{out}(t)), \quad x \in [0,1], \tag{7.28}$$

$$\hat{w}(1,t) = q\hat{z}(1,t) + c\hat{\zeta}(t) + H_4(Y_{out}(t) - \hat{Y}_{out}(t)), \tag{7.29}$$

$$\dot{\hat{\zeta}}(t) = a_1\hat{\zeta}(t) - b_1\hat{z}(1,t) + H_1(Y_{out}(t) - \hat{Y}_{out}(t)), \tag{7.30}$$

$$\hat{v}(1,t) = \hat{\zeta}(t), \tag{7.31}$$

$$\hat{v}_t(x,t) = -\frac{1}{\tau}\hat{v}_x(x,t) + H_5(x)(Y_{out}(t) - \hat{Y}_{out}(t)), \quad x \in [1,2], \tag{7.32}$$

where

$$\hat{Y}_{out}(t) = [\hat{v}(2,t), \hat{v}_t(2,t)]^T, \tag{7.33}$$

and the row vectors $H_1 = [H_{1a}, H_{1b}] \in \mathbb{R}^2$, $H_2(x) = [H_{2a}(x), H_{2b}(x)] \in \mathbb{R}^2$, $H_3(x) = [H_{3a}(x), H_{3b}(x)] \in \mathbb{R}^2$, $H_4 = [H_{4a}, H_{4b}] \in \mathbb{R}^2$, $H_5(x) = [H_{5a}(x), H_{5b}(x)] \in \mathbb{R}^2$ are observer gains to be determined. Defining the observer error states as

$$(\tilde{z}(x,t), \tilde{w}(x,t), \tilde{\zeta}(t), \tilde{v}(x,t))$$
$$= (z(x,t), w(x,t), \zeta(t), v(x,t)) - (\hat{z}(x,t), \hat{w}(x,t), \hat{\zeta}(t), \hat{v}(x,t)), \tag{7.34}$$

according to (7.16)–(7.20), (7.23)–(7.25), and (7.26)–(7.32), yields the observer error system

$$\tilde{z}(0,t) = p\tilde{w}(0,t), \tag{7.35}$$

$$\tilde{z}_t(x,t) = -q_1\tilde{z}_x(x,t) - c_1\tilde{w}(x,t) - c_2\tilde{z}(x,t) - H_2(x)[\tilde{v}(2,t), \tilde{v}_t(2,t)]^T,$$
$$x \in [0,1], \tag{7.36}$$

$$\tilde{w}_t(x,t) = q_2\tilde{w}_x(x,t) - c_3\tilde{w}(x,t) - c_4\tilde{z}(x,t) - H_3(x)[\tilde{v}(2,t), \tilde{v}_t(2,t)]^T,$$
$$x \in [0,1], \tag{7.37}$$

$$\tilde{w}(1,t) = q\tilde{z}(1,t) + c\tilde{\zeta}(t) - H_4[\tilde{v}(2,t), \tilde{v}_t(2,t)]^T, \tag{7.38}$$

$$\dot{\tilde{\zeta}}(t) = a_1\tilde{\zeta}(t) - b_1\tilde{z}(1,t) - H_1[\tilde{v}(2,t), \tilde{v}_t(2,t)]^T, \tag{7.39}$$

$$\tilde{v}(1,t) = \tilde{\zeta}(t), \tag{7.40}$$

$$\tilde{v}_t(x,t) = -\frac{1}{\tau}\tilde{v}_x(x,t) - H_5[\tilde{v}(2,t), \tilde{v}_t(2,t)]^T, \quad x \in [1,2]. \tag{7.41}$$

The determination of the observer gains H_1, $H_2(x)$, $H_3(x)$, H_4, $H_5(x)$ in (7.26)–(7.32) will be completed through two transformations presented next, which convert the observer error system (7.35)–(7.41) to a target observer error system whose exponential stability is straightforward to obtain.

First Transformation

We apply the transformation

$$\tilde{v}(x,t) = \tilde{\eta}(x,t) + \varphi(x)\tilde{\zeta}(t), \tag{7.42}$$

where $\varphi(x)$ satisfies

$$\varphi(x) = e^{-\tau a_1(x-1)}, \quad x \in [1,2], \tag{7.43}$$

with the purpose of converting (7.39)–(7.41) to

$$\dot{\tilde{\zeta}}(t) = -\bar{a}_1 \tilde{\zeta}(t) - b_1 \tilde{z}(1,t) - e^{\tau a_1} L_1 \tilde{\eta}(2,t), \tag{7.44}$$

$$\tilde{\eta}(1,t) = 0, \tag{7.45}$$

$$\tilde{\eta}_t(x,t) = -\frac{1}{\tau}\tilde{\eta}_x(x,t), \quad x \in [1,2], \tag{7.46}$$

where

$$\bar{a}_1 = L_1 - a_1 > 0, \tag{7.47}$$

by choosing the design parameter L_1. Matching (7.39) and (7.44) via (7.42), by virtue of (7.47), we obtain H_1 as

$$H_1 = [e^{\tau a_1} L_1, 0]. \tag{7.48}$$

Taking the time and spatial derivatives of (7.42), submitting the result into (7.41), and applying (7.39), (7.46), and (7.48), we obtain

$$\begin{aligned}
\tilde{v}_t(x,t) &+ \frac{1}{\tau}\tilde{v}_x(x,t) + H_5(x)[\tilde{v}(2,t), \tilde{v}_t(2,t)]^T \\
&= \tilde{\eta}_t(x,t) + \varphi(x)a_1\tilde{\zeta}(t) - \varphi(x)b_1\tilde{z}(1,t) - \varphi(x)e^{\tau a_1}L_1\tilde{v}(2,t) \\
&\quad + \frac{1}{\tau}\tilde{\eta}_x(x,t) + \frac{1}{\tau}\varphi'(x)\tilde{\zeta}(t) + H_5(x)[\tilde{v}(2,t), \tilde{v}_t(2,t)]^T \\
&= -\varphi(x)b_1\tilde{z}(1,t) - \varphi(x)e^{\tau a_1}L_1\tilde{v}(2,t) + H_5(x)[\tilde{v}(2,t), \tilde{v}_t(2,t)]^T \\
&\quad + \left(\varphi(x)a_1 + \frac{1}{\tau}\varphi'(x)\right)\tilde{\zeta}(t) = 0, \tag{7.49}
\end{aligned}$$

where $\varphi'(x) + \tau a_1 \varphi(x) = 0$ obtained from (7.43) has been used. The row vector $H_5(x)$ is determined to ensure

$$-\varphi(x)b_1\tilde{z}(1,t) - \varphi(x)e^{\tau a_1}L_1\tilde{v}(2,t) + H_5(x)[\tilde{v}(2,t), \tilde{v}_t(2,t)]^T = 0. \tag{7.50}$$

We solve the observer gains by using the Laplace transforms, considering the convenience of the algebraic relationships between the output estimation error injections and other observer error states in the frequency domain.

First, we derive the algebraic relationships between $\tilde{z}(1,s)$ and $\tilde{v}(2,s)$. Taking the Laplace transform of (7.44), we get

$$(s+\bar{a}_1)\tilde{\zeta}(s) = -b_1\tilde{z}(1,s) - e^{\tau a_1}L_1 e^{-\tau s}\tilde{\eta}(1,s). \tag{7.51}$$

For brevity, we consider all initial conditions to be zero when we take the Laplace transform. (Arbitrary initial conditions could be incorporated into the stability statement through an expanded analysis, which is routine but heavy on additional notation.) Recalling (7.45) as well as $\bar{a}_1 > 0$, which ensures that $(s+\bar{a}_1)$ does not have any zeros in the closed right-half plane, we use (7.51) to get

$$\tilde{\zeta}(s) = -(s+\bar{a}_1)^{-1}b_1\tilde{z}(1,s). \tag{7.52}$$

According to (7.42), (7.45), (7.46), applying (7.43), (7.52), we have

$$\tilde{v}(2,s) = e^{-\tau s}\tilde{\eta}(1,s) + \varphi(2)\tilde{\zeta}(s) = -r(s)\tilde{z}(1,s), \tag{7.53}$$

where

$$r(s) = e^{-\tau a_1}(s+\bar{a}_1)^{-1}b_1. \tag{7.54}$$

The function $r(s)$ is an asymptotically stable and strictly proper transfer function because $\bar{a}_1 > 0$.

After obtaining the algebraic relationship (7.53), we determine H_5 to make (7.50) hold. Taking the Laplace transform of (7.50) and inserting (7.43) and (7.53), we get

$$\begin{aligned}
&-\varphi(x)b_1\tilde{z}(1,s) - \varphi(x)e^{\tau a_1}L_1\tilde{v}(2,s) + H_5(x)[\tilde{v}(2,s), s\tilde{v}(2,s)]^T \\
&= -\varphi(x)b_1\tilde{z}(1,s) + \varphi(x)e^{\tau a_1}L_1 r(s)\tilde{z}(1,s) - H_{5a}(s;x)r(s)\tilde{z}(1,s) \\
&\quad - H_{5b}(s;x)sr(s)\tilde{z}(1,s) \\
&= -\varphi(x)b_1\tilde{z}(1,s) + \varphi(x)e^{\tau a_1}L_1 r(s)\tilde{z}(1,s) \\
&\quad - H_{5a}(s;x)r(s)\tilde{z}(1,s) - H_{5b}(s;x)(e^{-\tau a_1}b_1 - \bar{a}_1 r(s))\tilde{z}(1,s) \\
&= -(\varphi(x)b_1 + H_{5b}(s;x)e^{-\tau a_1}b_1)\tilde{z}(1,s) \\
&\quad + (\varphi(x)e^{\tau a_1}L_1 + H_{5b}(s;x)\bar{a}_1 - H_{5a}(s;x))r(s)\tilde{z}(1,s) \\
&= 0, \tag{7.55}
\end{aligned}$$

where

$$sr(s) = (s+\bar{a}_1-\bar{a}_1)e^{-\tau a_1}(s+\bar{a}_1)^{-1}b_1 = e^{-\tau a_1}b_1 - \bar{a}_1 r(s) \tag{7.56}$$

by recalling (7.54) has been used.

For (7.55) to hold, the transfer function $H_5(x)$ is then defined as

$$H_5(x) = [e^{-\tau a_1(x-2)}L_1 - e^{-\tau a_1(x-2)}\bar{a}_1, -e^{-\tau a_1(x-2)}]. \tag{7.57}$$

In the above derivation, we have completed the conversion between (7.39)–(7.41) and (7.44)–(7.46) through (7.42), under the designed $H_1, H_5(x)$.

In what follows, H_4 is determined to make $\tilde{w}(1,t)$ in the boundary condition (7.38) be zero—that is, to render

$$\tilde{w}(1,t) = q\tilde{z}(1,t) + c\tilde{\zeta}(t) - H_4[\tilde{v}(2,t), \tilde{v}_t(2,t)]^T = 0. \tag{7.58}$$

Taking the Laplace transform of (7.58) and inserting (7.52) and (7.53), we obtain

$$
\begin{aligned}
\tilde{w}(1,s) &= q\tilde{z}(1,s) + c\tilde{\zeta}(s) - H_{4a}(x)\tilde{v}(2,s) - H_{4b}s\tilde{v}(2,s) \\
&= q\tilde{z}(1,s) - c(s+\bar{a}_1)^{-1}b_1\tilde{z}(1,s) + H_{4a}r(s)\tilde{z}(1,s) + H_{4b}sr(s)\tilde{z}(1,s) \\
&= q\tilde{z}(1,s) - ce^{\tau a_1}r(s)\tilde{z}(1,s) \\
&\quad + H_{4a}r(s)\tilde{z}(1,s) + H_{4b}(e^{-\tau a_1}b_1 - \bar{a}_1 r(s))\tilde{z}(1,s) \\
&= (q + H_{4b}e^{-\tau a_1}b_1)\tilde{z}(1,s) + (H_{4a} - H_{4b}\bar{a}_1 - ce^{\tau a_1})r(s)\tilde{z}(1,s) \\
&= 0,
\end{aligned}
\tag{7.59}
$$

where (7.54), (7.56) have been used.

For (7.59) to hold, the row vector H_4 is chosen as

$$
H_4 = \left[-\frac{q\bar{a}_1}{e^{-\tau a_1}b_1} + ce^{\tau a_1}, \; -\frac{q}{e^{-\tau a_1}b_1} \right].
\tag{7.60}
$$

Through applying the first transformation (7.42), with H_4, $H_5(x)$, and H_1, (7.35)–(7.41) is converted to the intermediate system as

$$
\tilde{z}(0,t) = p\tilde{w}(0,t),
\tag{7.61}
$$

$$
\tilde{z}_t(x,t) = -q_1\tilde{z}_x(x,t) - c_1\tilde{w}(x,t) - c_2\tilde{z}(x,t) - H_2(x)[\tilde{v}(2,t),\tilde{v}_t(2,t)]^T,
\tag{7.62}
$$

$$
\tilde{w}_t(x,t) = q_2\tilde{w}_x(x,t) - c_3\tilde{w}(x,t) - c_4\tilde{z}(x,t) - H_3(x)[\tilde{v}(2,t),\tilde{v}_t(2,t)]^T,
\tag{7.63}
$$

$$
\tilde{w}(1,t) = 0,
\tag{7.64}
$$

$$
\dot{\tilde{\zeta}}(t) = -\bar{a}_1\tilde{\zeta}(t) - b_1\tilde{z}(1,t) - e^{\tau a_1}L_1\tilde{\eta}(2,t),
\tag{7.65}
$$

$$
\tilde{\eta}(1,t) = 0,
\tag{7.66}
$$

$$
\tilde{\eta}_t(x,t) = -\frac{1}{\tau}\tilde{\eta}_x(x,t).
\tag{7.67}
$$

Next, we introduce the second transformation to decouple the couplings in (7.62), (7.63).

Second Transformation

We now apply the second transformation [21]

$$
\tilde{w}(x,t) = \tilde{\beta}(x,t) - \int_x^1 \psi(x,y)\tilde{\alpha}(y,t)dy,
\tag{7.68}
$$

$$
\tilde{z}(x,t) = \tilde{\alpha}(x,t) - \int_x^1 \phi(x,y)\tilde{\alpha}(y,t)dy,
\tag{7.69}
$$

with the kernels $\psi(x,y),\phi(x,y)$ satisfying

$$
\psi(x,x) = \frac{c_4}{q_1 + q_2},
\tag{7.70}
$$

$$
\phi(0,y) = p\psi(0,y),
\tag{7.71}
$$

$$
-q_1\psi_y(x,y) + q_2\psi_x(x,y) - c_4\phi(x,y) + (c_2 - c_3)\psi(x,y) = 0,
\tag{7.72}
$$

$$
-q_1\phi_x(x,y) - q_1\phi_y(x,y) - c_1\psi(x,y) = 0.
\tag{7.73}
$$

The purpose of the transformations (7.68), (7.69) is to convert the intermediate system (7.61)–(7.67) to the target system (the subsystem $\tilde{\eta}(\cdot, t)$, given in (7.66), (7.67), is removed for brevity because $\tilde{\eta}(\cdot, t) \equiv 0$, $t \geq \tau$), as follows:

$$\tilde{\alpha}(0, t) = p\tilde{\beta}(0, t), \tag{7.74}$$

$$\tilde{\alpha}_t(x, t) = -q_1 \tilde{\alpha}_x(x, t) + \int_x^1 \bar{M}(x, y)\tilde{\beta}(y, t)dy - c_2\tilde{\alpha}(x, t) - c_1\tilde{\beta}(x, t), \tag{7.75}$$

$$\tilde{\beta}_t(x, t) = q_2\tilde{\beta}_x(x, t) + \int_x^1 \bar{N}(x, y)\tilde{\beta}(y, t)dy - c_3\tilde{\beta}(x, t), \tag{7.76}$$

$$\tilde{\beta}(1, t) = 0, \tag{7.77}$$

$$\dot{\tilde{\zeta}}(t) = -\bar{a}_1\tilde{\zeta}(t) - b_1\tilde{\alpha}(1, t) \tag{7.78}$$

for $t \geq \tau$, where the integral operator kernels $\bar{M}$ and $\bar{N}$ are defined as

$$\bar{M}(x, y) = \int_x^y \phi(x, \delta)\bar{M}(\delta, y)d\delta - c_1\phi(x, y), \tag{7.79}$$

$$\bar{N}(x, y) = \int_x^y \psi(x, \delta)\bar{M}(\delta, y)d\delta - c_1\psi(x, y). \tag{7.80}$$

In what follows, $\Pi_2(x)$, $H_3(x)$ are determined by matching the intermediate system (7.61)–(7.65) and the target system observer error system (7.74)–(7.78) via (7.68), (7.69).

Inserting (7.68), (7.69) into (7.63) along (7.75), (7.76) and applying (7.70)–(7.72), (7.80), we get

$$\tilde{w}_t(x, t) - q_2\tilde{w}_x(x, t) + c_4\tilde{z}(x, t) + c_3\tilde{w}(x, t) + \Pi_3(x)[\tilde{v}(2, t), \tilde{v}_t(2, t)]^T$$
$$= q_1\psi(x, 1)\tilde{\alpha}(1, t) + H_3(x)[\tilde{v}(2, t), \tilde{v}_t(2, t)]^T = 0, \tag{7.81}$$

where the detailed calculation is shown in (7.193) in appendix 7.7A. We thus know that the following equation needs to be satisfied:

$$q_1\psi(x, 1)\tilde{z}(1, t) + H_3(x)[\tilde{v}(2, t), \tilde{v}_t(2, t)]^T = 0, \tag{7.82}$$

where $\tilde{\alpha}(1, t) = \tilde{z}(1, t)$ according to (7.69) has been used. Rewriting (7.82) in the frequency domain and applying (7.53), (7.56), we obtain

$$q_1\psi(x, 1)\tilde{z}(1, s) + H_3(x)[\tilde{v}(2, s), s\tilde{v}(2, s)]^T$$
$$= q_1\psi(x, 1)\tilde{z}(1, s) - H_{3a}(x)r(s)\tilde{z}(1, s) - H_{3b}(x)sr(s)\tilde{z}(1, s)$$
$$= q_1\psi(x, 1)\tilde{z}(1, s) - H_{3a}(x)r(s)\tilde{z}(1, s) - H_{3b}(x)(e^{-\tau a_1}b_1 - \bar{a}_1 r(s))\tilde{z}(1, s)$$
$$= (q_1\psi(x, 1) - H_{3b}(x)e^{-\tau a_1}b_1)\tilde{z}(1, s)$$
$$+ (H_{3b}(x)\bar{a}_1 - H_{3a}(x))r(s)\tilde{z}(1, s) = 0. \tag{7.83}$$

The transfer function $H_3(x)$ is chosen as

$$H_3(x) = \left[\frac{\bar{a}_1 q_1\psi(x, 1)}{e^{-\tau a_1}b_1}, \frac{q_1\psi(x, 1)}{e^{-\tau a_1}b_1}\right]. \tag{7.84}$$

Inserting (7.68), (7.69) into (7.62) along (7.75), (7.76) and applying (7.73), (7.79), we get

$$\tilde{z}_t(x,t) + q_1\tilde{z}_x(x,t) + c_1\tilde{w}(x,t) + c_2\tilde{z}(x,t) + H_2(x)[\tilde{v}(2,t), \tilde{v}_t(2,t)]^T$$
$$= q_1\phi(x,1)\tilde{\alpha}(1,t) + H_2(x)[\tilde{v}(2,t), \tilde{v}_t(2,t)]^T = 0, \tag{7.85}$$

where the detailed calculation is shown in (7.194) in appendix 7.7A. Therefore, $H_2(x)[\tilde{v}(2,t), \tilde{v}_t(2,t)]^T$ should satisfy

$$q_1\phi(x,1)\tilde{z}(1,t) + H_2(x)[\tilde{v}(2,t), \tilde{v}_t(2,t)]^T = 0, \tag{7.86}$$

where $\tilde{\alpha}(1,t) = \tilde{z}(1,t)$ according to (7.69) has been used. Taking the Laplace transform of (7.86) and recalling (7.53), (7.56), we obtain

$$q_1\phi(x,1)\tilde{z}(1,s) + H_2(x)[\tilde{v}(2,s), s\tilde{v}(2,s)]^T$$
$$= q_1\phi(x,1)\tilde{z}(1,s) - H_{2a}(x)r(s)\tilde{z}(1,s) - H_{2b}(x)sr(s)\tilde{z}(1,s)$$
$$= q_1\phi(x,1)\tilde{z}(1,s) - H_{2a}(x)r(s)\tilde{z}(1,s) - H_{2b}(x)(e^{-\tau a_1}b_1 - \bar{a}_1 r(s))\tilde{z}(1,s)$$
$$= (q_1\phi(x,1) - H_{2b}(x)e^{-\tau a_1}b_1)\tilde{z}(1,s) + (H_{2b}(x)\bar{a}_1 - H_{2a}(x))r(s)\tilde{z}(1,s)$$
$$= 0. \tag{7.87}$$

The transfer function $H_2(x)$ is determined to be

$$H_2(x) = \left[\frac{\bar{a}_1 q_1\phi(x,1)}{e^{-\tau a_1}b_1}, \frac{q_1\phi(x,1)}{e^{-\tau a_1}b_1}\right]. \tag{7.88}$$

The boundary condition (7.74) follows directly from inserting $x = 0$ into (7.68), (7.69) and applying (7.61), (7.71). The second conversion is thus completed, and the two PDEs (7.62), (7.63) are decoupled, which can be seen in (7.75), (7.76).

After performing the above two transformations, we have converted the original observer error system (7.35)–(7.41) to the target observer error system (7.74)–(7.78) (for $t \in [\tau, \infty)$, $\tilde{\eta}(x,t) \equiv 0$ according to (7.66), (7.67) is removed for brevity). The exponential convergence of the observer error system is shown in the next subsection, which guarantees that the output estimation error injections in the observer (7.26)–(7.32) are exponentially convergent to zero.

Stability Analysis of the Observer Error System

Theorem 7.1. *For all initial data* $((\tilde{z}[0], \tilde{w}[0])^T, \tilde{v}[0], \tilde{\zeta}(0)) \in C^0([0,1]; \mathbb{R}^2) \times C^1([1,2]; \mathbb{R}) \times \mathbb{R}$ *and* $m(0) \in \mathbb{R}^-$*, the exponential convergence of the observer error system* (7.35)–(7.41) *holds in the sense of the norm*

$$\|\tilde{z}(\cdot,t)\|_\infty + \|\tilde{w}(\cdot,t)\|_\infty + \|\tilde{v}(\cdot,t)\|_\infty + \|\tilde{v}_t(\cdot,t)\|_\infty + \left|\tilde{\zeta}(t)\right| + \left|\dot{\tilde{\zeta}}(t)\right|, \tag{7.89}$$

where the decay rate is adjustable by L_1.

Proof. The stability of the original observer error system is obtained by analyzing the stability of the target observer error system (7.74)–(7.78) and using the invertibility of the transformations. With the method of characteristics, it is easy to show that $\tilde{\beta}[t] \equiv 0$ after $t = \tau + \frac{1}{q_2}$, considering (7.76), (7.77). According to (7.74), (7.75), we have $\tilde{\alpha}[t] \equiv 0$ after $t_0 = \tau + \frac{1}{q_1} + \frac{1}{q_2}$. Because $\bar{a}_1 > 0$, we then have $\tilde{\zeta}(t)$ exponentially convergent to zero after t_0. Recalling $\tilde{\eta}[t] \equiv 0$ for $t \geq \tau$, we obtain $\tilde{\eta}_t(x,t) \equiv 0$ for $t \geq \tau$. According to $\tilde{\alpha}(1,t) \equiv 0$ and the fact that $\tilde{\zeta}(t)$ is exponentially convergent to zero, after t_0, we find that $\dot{\tilde{\zeta}}(t)$ is exponentially convergent to zero after t_0 via (7.78). Therefore, we know that

$$\bar{\Omega}(t) = \|\tilde{\alpha}(\cdot,t)\|_\infty + \|\tilde{\beta}(\cdot,t)\|_\infty + \|\tilde{\eta}(\cdot,t)\|_\infty + \|\tilde{\eta}_t(\cdot,t)\|_\infty + \left|\tilde{\zeta}(t)\right| + \left|\dot{\tilde{\zeta}}(t)\right|$$

is bounded by an exponential decay with the decay rate λ_e for $t \geq t_0$. The decay rate λ_e depends on the decay rate of the ODE $\tilde{\zeta}(t)$. In other words, the decay rate λ_e depends on the choice of L_1 according to (7.47). It should be noted that the transient in the finite time $[0, t_0)$ can be bounded by an arbitrarily fast decay rate considering the trade-off between the decay rate and the overshoot coefficient— that is, the higher the decay rate, the higher the overshoot coefficient. Therefore, we conclude that there is exponential stability in the sense that $\bar{\Omega}(t)$ is bounded by an exponential decay rate λ_e with some overshoot coefficients for $t \geq 0$. Applying the transformations (7.42) and (7.68), (7.69), we respectively have

$$\|\tilde{v}(\cdot,t)\|_\infty \leq \Upsilon_{1a}\left(\|\tilde{\eta}(\cdot,t)\|_\infty + \left|\tilde{\zeta}(t)\right|\right),$$

$$\|\tilde{v}_t(\cdot,t)\|_\infty \leq \Upsilon_{1b}\left(\|\tilde{\eta}_t(\cdot,t)\|_\infty + \left|\dot{\tilde{\zeta}}(t)\right|\right),$$

$$\|\tilde{z}(\cdot,t)\|_\infty + \|\tilde{w}(\cdot,t)\|_\infty \leq \Upsilon_{1c}\left(\|\tilde{\alpha}(\cdot,t)\|_\infty + \|\tilde{\beta}(\cdot,t)\|_\infty\right),$$

for some positive $\Upsilon_{1a}, \Upsilon_{1b}, \Upsilon_{1c}$. The proof is complete. $\qquad\square$

7.3 DELAY-COMPENSATED OUTPUT-FEEDBACK CONTROLLER

In the last section, we designed an observer that compensates for the time delay in the output measurements of the distal ODE to track the states of the overall system (7.16)–(7.21). In this section, we design an output feedback control law $U(t)$ based on the observer (7.26)–(7.32) with output estimation error injection terms assumed absent, in accordance with the result of their convergence to zero in theorem 7.1. The separation principle is then verified and applied in the stability analysis of the resulting closed-loop system.

With the purpose of removing the coupling terms in (7.27), (7.28) and making the system parameter in the distal ODE (7.30) negative, a PDE backstepping transformation in the form [48]

$$\alpha(x,t) = \hat{z}(x,t) - \int_x^1 K_3(x,y)\hat{z}(y,t)dy$$

$$- \int_x^1 J_3(x,y)\hat{w}(y,t)dy - \gamma(x)\hat{\zeta}(t), \tag{7.90}$$

$$\beta(x,t) = \hat{w}(x,t) - \int_x^1 K_2(x,y)\hat{z}(y,t)dy$$

$$- \int_x^1 J_2(x,y)\hat{w}(y,t)dy - \lambda(x)\hat{\zeta}(t) \tag{7.91}$$

is introduced, where the kernels $K_3(x,y)$, $J_3(x,y)$, $\gamma(x)$, $K_2(x,y)$, $J_2(x,y)$, $\lambda(x)$ satisfy

$$q_1 K_3(x,1) = q_2 J_3(x,1)q - \gamma(x)b_1, \qquad (7.92)$$

$$J_3(x,x) = \frac{c_1}{q_2+q_1}, \qquad (7.93)$$

$$-q_1 J_{3x}(x,y) + q_2 J_{3y}(x,y)$$
$$+ c_1 K_3(x,y) - (c_2-c_3)J_3(x,y) = 0, \qquad (7.94)$$

$$-q_1 K_{3x}(x,y) - q_1 K_{3y}(x,y) + c_4 J_3(x,y) = 0, \qquad (7.95)$$

$$\gamma(1) = F_1, \qquad (7.96)$$

$$-q_1 \gamma'(x) - \gamma(x)(a_1+c_2) - q_2 J_3(x,1)c = 0, \qquad (7.97)$$

$$q_2 q J_2(x,1) = q_1 K_2(x,1) + \lambda(x)b_1, \qquad (7.98)$$

$$K_2(x,x) = \frac{-c_4}{q_1+q_2}, \qquad (7.99)$$

$$q_2 J_{2x}(x,y) + q_2 J_{2y}(x,y) + c_1 K_2(x,y) = 0, \qquad (7.100)$$

$$q_2 K_{2x}(x,y) - q_1 K_{2y}(x,y)$$
$$+ c_4 J_2(x,y) + c_2 K_2(x,y) - c_3 K_2(x,y) = 0, \qquad (7.101)$$

$$q_2 \lambda'(x) - \lambda(x)(a_1+c_3) - q_2 J_2(x,1)c = 0, \qquad (7.102)$$

$$\lambda(1) = q\gamma(1) + c \qquad (7.103)$$

in order to convert (7.26)–(7.32) into the target system:

$$\alpha(0,t) = p\beta(0,t), \qquad (7.104)$$

$$\alpha_t(x,t) = -q_1 \alpha_x(x,t) - c_2 \alpha(x,t), \qquad (7.105)$$

$$\beta_t(x,t) = q_2 \beta_x(x,t) - c_3 \beta(x,t), \qquad (7.106)$$

$$\beta(1,t) = q\alpha(1,t), \qquad (7.107)$$

$$\dot{\hat{\zeta}}(t) = -a_m \hat{\zeta}(t) - b_1 \alpha(1,t), \qquad (7.108)$$

$$\hat{v}(1,t) = c\hat{\zeta}(t), \qquad (7.109)$$

$$\hat{v}_t(x,t) = -\frac{1}{\tau}\hat{v}_x(x,t), \qquad (7.110)$$

where

$$a_m = b_1 F_1 - a_1 > 0 \qquad (7.111)$$

is imposed by choosing the design parameter F_1, which appears within the gain functions of the feedback law

$$U(t) = \frac{1}{c}\int_0^1 \bar{K}_1(x)\hat{z}(x,t)dx + \frac{1}{c}\int_0^1 \bar{K}_2(x)\hat{w}(x,t)dx + \frac{1}{c}\bar{K}_3\hat{\zeta}(t), \qquad (7.112)$$

where

$$\bar{K}_1(x) = K_3(0,x) - pK_2(0,x), \qquad (7.113)$$

$$\bar{K}_2(x) = J_3(0,x) - pJ_2(0,x), \qquad (7.114)$$

$$\bar{K}_3 = \gamma(0) - p\lambda(0). \qquad (7.115)$$

The conditions of the kernels (7.92)–(7.103) are obtained by matching (7.105)–(7.108) and (7.27)–(7.30), whose calculation details are given in steps 1–3 in appendix 7.7B. The equation sets (7.92)–(7.97) and (7.98)–(7.103) have an analogous structure with (19)–(24) in [183]. Following the proof of lemma 1 in [183], we obtain the well-posedness of (7.92)–(7.103).

Following section 2.4 in [183], we have that the inverse transformation of (7.90), (7.91) exists as

$$\hat{z}(x,t) = \alpha(x,t) - \int_x^1 \mathcal{M}(x,y)\alpha(y,t)dy$$

$$- \int_x^1 \mathcal{N}(x,y)\beta(y,t)dy - \mathcal{G}(x)\hat{\zeta}(t), \tag{7.116}$$

$$\hat{w}(x,t) = \beta(x,t) - \int_x^1 \mathcal{D}(x,y)\alpha(y,t)dy$$

$$- \int_x^1 \mathcal{T}(x,y)\beta(y,t)dy - \mathcal{P}(x)\hat{\zeta}(t), \tag{7.117}$$

where the well-posedness of the kernels $\mathcal{M}(x,y)$, $\mathcal{N}(x,y)$, $\mathcal{G}(x)$, $\mathcal{D}(x,y)$, $\mathcal{T}(x,y)$, $\mathcal{P}(x)$ can be obtained from the well-posedness of (7.92)–(7.103), according to chapter 9.9 of [175].

7.4   EVENT-TRIGGERING MECHANISM

In this section we introduce an event-triggered control scheme for the stabilization of the overall PDE-ODE-PDE system (7.16)–(7.20), (7.23)–(7.25). This scheme relies on the delay-compensated observer-based output-feedback continuous-in-time control $U(t)$ designed in the last section, and on a dynamic event-triggering mechanism (ETM) that is realized using the states from the delay-compensated observer. The event-triggered control signal $U_d(t)$ is the value of the continuous-in-time $U(t)$ at the time instants t_k but applied until time t_{k+1}—that is,

$$U_d(t) = U(t_k) = \frac{1}{\bar{c}} \int_0^1 \bar{K}_1(x)\hat{z}(x,t_k)dx$$

$$+ \frac{1}{\bar{c}} \int_0^1 \bar{K}_2(x)\hat{w}(x,t_k)dx + \frac{1}{\bar{c}}\bar{K}_3\hat{\zeta}(t_k), \quad t \in [t_k, t_{k+1}) \tag{7.118}$$

for $k \in \mathbb{Z}_+$. A deviation $d(t)$ between the continuous-in-time control signal and the event-based one is given as

$$d(t) = U(t) - U_d(t). \tag{7.119}$$

Inserting (7.118), the right boundary of the target system (7.104)–(7.110) becomes

$$\alpha(0,t) = p\beta(0,t) - \bar{c}d(t). \tag{7.120}$$

Taking the Laplace transform of (7.105)–(7.110), according to section 3.2 in [49], we obtain the following algebraic relationships between $\alpha(0,s)$ and $\beta(0,s)$ as

$$\beta(0,s) = qe^{\frac{-(c_3+s)}{q_2} - \frac{(c_2+s)}{q_1}}\alpha(0,s). \tag{7.121}$$

Inserting (7.121) into (7.120) in the frequency domain, we have

$$\alpha(0, s) = \frac{-\bar{c}}{(1 - h(s))} d(s), \tag{7.122}$$

where

$$h(s) = pq e^{-(\frac{c_3}{q_2} + \frac{c_2}{q_1})} e^{-(\frac{1}{q_2} + \frac{1}{q_1})s}. \tag{7.123}$$

From (7.122), (7.123), assumption 7.1 guarantees

$$|\alpha(0, t)| \leq |h(0)| \sup_{0 \leq \xi \leq t} |\alpha(0, \xi)| + |\bar{c}| \sup_{0 \leq \xi \leq t} |d(\xi)|,$$

where the constant $|h(0)|$ is strictly smaller than 1. By the small gain theorem, we obtain

$$\alpha(0, t)^2 \leq \lambda_d \sup_{0 \leq \xi \leq t} d(\xi)^2, \tag{7.124}$$

where

$$\lambda_d = \frac{\bar{c}^2}{(1 - |h(0)|)^2}.$$

Please note that we obtain (7.124), which will be used in the Lyapunov analysis, based on (7.105)–(7.110) (state-feedback loop)—that is, assuming the observer error injections absent—with the purpose of avoiding heavy additional notation. Incorporating these observer error injections will not change the stability result obtained in the Lyapunov analysis, which will be shown later, by virtue of exponential convergence of the observer errors in theorem 7.1.

Similarly to [57], the ETM to determine the triggering times is designed to be governed by the following dynamic triggering condition:

$$t_{k+1} = \inf\{t \in R^+ | t > t_k | d(t)^2 \geq \theta V(t) - \mu m(t)\}. \tag{7.125}$$

The internal dynamic variable $m(t)$ satisfies the ordinary differential equation:

$$\dot{m}(t) = -\eta m(t) - \kappa_1 \alpha(1, t)^2 - \kappa_2 \alpha(0, t)^2 - \kappa_3 \hat{v}(2, t)^2 - \kappa_4 \left| \tilde{Y}(t) \right|^2, \tag{7.126}$$

where

$$\tilde{Y}(t) = Y_{\text{out}}(t) - \hat{Y}_{\text{out}}(t),$$

and where the initial condition of m is $m(0) < 0$, which guarantees that

$$m(t) < 0. \tag{7.127}$$

Inequality (7.127) follows from the ODE in (7.126), the nonpositivity of the nonhomogeneous terms on its right-hand side, the strict negativity of $m(0)$, the variation-of-constants formula, and the comparison principle. Thus, the fact that $m(t) < 0$ would not be affected by (7.125). Introducing the internal dynamic variable $m(t)$ defined by (7.126) is used in proving the existence of a minimal dwell time, which will be seen clearly in the proof of lemma 7.2. The Lyapunov function $V(t)$ in (7.125) is defined as

$$V(t) = \frac{1}{2}\hat{\zeta}(t)^2 + \frac{1}{2}\int_0^1 e^{\delta_1 x}\beta(x,t)^2 dx$$
$$+ \frac{1}{2}r_b \int_0^1 e^{-\delta_2 x}\alpha(x,t)^2 dx + \frac{1}{2}r_c \int_1^2 e^{-x}\hat{v}(x,t)^2 dx. \tag{7.128}$$

The positive design parameters θ, μ, η, κ_1, κ_2, κ_3, δ_1, δ_2, r_c, r_b satisfy

$$\kappa_1 > 2\lambda_1, \tag{7.129}$$

$$\kappa_2 > 2\lambda_1, \tag{7.130}$$

$$\delta_1 > \frac{2|c_3|}{q_2}, \tag{7.131}$$

$$\delta_2 > \frac{2|c_2|}{q_1}, \tag{7.132}$$

$$r_c < \frac{a_m e\tau}{2}, \tag{7.133}$$

$$r_b > \frac{1}{q_1}q^2 q_2 e^{\delta_1+\delta_2} + \frac{b_1^2}{q_1 a_m}e^{\delta_2} + \frac{2}{q_1}\kappa_1 e^{\delta_2}, \tag{7.134}$$

$$\kappa_3 < \frac{1}{2\tau}r_c e^{-2}, \tag{7.135}$$

$$\mu < \frac{\sigma_a}{\frac{1}{2}q_1 r_b \lambda_d + \kappa_2 \lambda_d}, \tag{7.136}$$

$$\theta < \min\left\{\frac{\sigma_a}{\frac{1}{2}q_1 r_b \lambda_d + \kappa_2 \lambda_d}, \frac{\mu\tau}{r_c}\kappa_3 e^2, \frac{\mu}{2\lambda_{\alpha 1}}\kappa_1, \frac{\mu}{2\lambda_{\alpha 0}}\kappa_2, \frac{\mu}{2}\kappa_4\right\}, \tag{7.137}$$

where η is free and associated with the dwell time, which can be seen later, and λ_1 is a positive constant that will be given in lemma 7.1 and only depends on the parameters of the plant and the design parameters L_1, F_1 in the continuous-in-time control law, and where

$$\sigma_a = \min\left\{\lambda_a, \eta, \frac{2\kappa_4}{R_e}\right\} > 0, \tag{7.138}$$

$$\lambda_a = \frac{1}{\bar{\xi}_2}\min\left\{\frac{1}{2}\left(a_m - \frac{1}{\tau}r_c e^{-1}\right), \frac{1}{2}\delta_1 q_2 - |c_3|,\right.$$
$$\left.\frac{1}{2}\delta_2 q_1 r_b e^{-\delta_2} - r_b |c_2| e^{-\delta_2}, \frac{e^{-2}}{2\tau}r_c\right\}, \tag{7.139}$$

$$R_e > \frac{2\kappa_4}{\lambda_e}, \tag{7.140}$$

$$\lambda_{\alpha 1} = \frac{1}{2}|b_1| + \frac{1}{2}q_1 r_b e^{-\delta_2}, \tag{7.141}$$

$$\lambda_{\alpha 0} = \frac{1}{q^2}q_2 \max\{1, \bar{c}^2\}, \tag{7.142}$$

where λ_e is a positive constant depending on the design parameter L_1, as shown in proof of theorem 7.1.

The conditions of all the parameters $L_1, F_1, \kappa_1, \kappa_2, \delta_1, \delta_2, r_c, r_b, \kappa_3, \mu, \theta$ are given in (7.47), (7.111), (7.129)–(7.137), which are cascaded rather than coupled.

The event-triggering condition, including (7.125), (7.126), (7.128), only uses the observer states, recalling the transformations (7.90), (7.91).

Lemma 7.1. *For $d(t)$ defined in (7.119), there exists a positive constant λ_1 such that*

$$\dot{d}(t)^2 \leq \lambda_1 \left(\Omega(t) + \alpha(1,t)^2 + \alpha(0,t)^2 + d(t)^2 + \left| \tilde{Y}(t) \right|^2 \right) \tag{7.143}$$

for $t \in (t_k, t_{k+1})$, where

$$\Omega(t) = \|\alpha(\cdot,t)\|^2 + \|\beta(\cdot,t)\|^2 + \hat{\zeta}(t)^2 \tag{7.144}$$

and where λ_1 only depends on the parameters of the plant and the design parameters L_1, F_1 in the continuous-in-time control law.

Proof. Taking the time derivative of (7.119), we have

$$\dot{d}(t)^2 = \dot{U}(t)^2 \tag{7.145}$$

because $\dot{U}_d(t) = 0$ for $t \in (t_k, t_{k+1})$. Recalling (7.112), we have

$$
\begin{aligned}
\dot{d}(t)^2 &= \dot{U}(t)^2 \\
&= \frac{1}{\bar{c}^2} \Bigg[q_1 \bar{K}_1(0) \hat{z}(0,t) - q_2 \bar{K}_2(0) \hat{w}(0,t) \\
&\quad - \left(\bar{K}_1(1) q_1 - q \bar{K}_2(1) q_2 + \bar{K}_3 b_1 \right) \hat{z}(1,t) + (\bar{K}_3 a_1 + c \bar{K}_2(1) q_2) \hat{\zeta}(t) \\
&\quad + \int_0^L \left(\bar{K}_1'(x) q_1 - \bar{K}_1(x) c_2 - \bar{K}_2(x) c_4 \right) \hat{z}(x,t) dx \\
&\quad - \int_0^L \left(\bar{K}_2(x) c_3 - \bar{K}_2'(x) q_2 + \bar{K}_1(x) c_1 \right) \hat{w}(x,t) dx \\
&\quad + \left(\int_0^1 \bar{K}_1(x) H_2(x) dx + \int_0^1 \bar{K}_2(x) H_3(x) dx \right. \\
&\quad \left. + \bar{K}_3 H_1 + \bar{K}_2(1) q_2 H_4 \right) \tilde{Y}(t) \Bigg]^2 \\
&\leq \lambda_0 \Bigg[\hat{z}(1,t)^2 + \hat{w}(0,t)^2 + \hat{z}(0,t)^2 + \|\hat{z}(\cdot,t)\|^2 + \|\hat{w}(\cdot,t)\|^2 \\
&\quad + \hat{\zeta}(t)^2 + \left| \tilde{Y}(t) \right|^2 \Bigg] \tag{7.146}
\end{aligned}
$$

for some positive λ_0 depending only on the parameters of the plant and the design parameters L_1, F_1 in the continuous-in-time control law. Applying the inverse transformation and (7.120), we obtain (7.143). The proof of lemma 7.1 is complete. □

Remark 7.1. The target system corresponding to the output-feedback loop is obtained by applying the backstepping transformation (7.90), (7.91) into the observer (7.26)–(7.32) and replacing U in (7.26) by U_d defined in (7.118). The resulting system is (7.120), (7.105)–(7.110) with output estimation error injection terms

$G_2(x)\tilde{Y}(t)$, $G_3(x)\tilde{Y}(t)$, $H_4\tilde{Y}(t)$, $H_1\tilde{Y}(t)$, $H_5(x)\tilde{Y}(t)$ in (7.105), (7.106), (7.107), (7.108), (7.110), respectively, where the bounded functions G_2, G_3 depend on the observer gains obtained in section 7.2 and the kernels in the backstepping transformation (7.90), (7.91). In the following analysis in this chapter, we denote this system as the $\mathcal{S}$-system.

The following lemma proves the existence of a minimal dwell time independent of initial conditions.

Lemma 7.2. *For some $\kappa_1, \kappa_2, \theta$, there exists a minimal dwell time independent of initial conditions between any two successive triggering times—that is, $t_{k+1} - t_k \geq T$ for all $k \geq 0$ where T is a positive constant.*

Proof. Let us introduce the function

$$\psi(t) = \frac{d(t)^2 + \frac{\mu}{2}m(t)}{\theta V(t) - \frac{\mu}{2}m(t)}, \tag{7.147}$$

which is proposed in [57]. We see that $\psi(t_{k+1}) = 1$ because the event is triggered, and $\psi(t_k) < 0$ because of $m(t) < 0$ and $d(t_k) = 0$. The function $\psi(t)$ is continuous on $[t_k, t_{k+1}]$ for the given initial conditions $(\hat{z}(\cdot, 0), \hat{w}(\cdot, 0))^T \in C^0([0, 1]; \mathbb{R}^2)$, $\hat{v}(\cdot, 0) \in C^1([1, 2]; \mathbb{R})$, $\hat{\zeta}(0) \in \mathbb{R}$, $m(0) \in \mathbb{R}^-$, following the proof of proposition 14.1 in chapter 14 and recalling the backstepping transformations (7.90), (7.91). By the intermediate value theorem, there exists $t^* > t_k$ such that $\psi(t) \subset [0, 1]$ when $t \in [t^*, t_{k+1}]$. The lower bound of the minimal dwell time T can be defined as the minimal time it takes for $\psi(t)$ from 0 to 1.

For all $t \in [t^*, t_{i+1}]$, taking the time derivative of $V(t)$ (7.128) along the system defined in remark 7.1, applying integration by parts, one gets

$$\begin{aligned}
\dot{V}(t) = &- a_m \hat{\zeta}(t)^2 - \hat{\zeta}(t)b_1\alpha(1,t) + \frac{1}{2}q_2 e^{\delta_1}\beta(1,t)^2 - \frac{1}{2}q_2\beta(0,t)^2 \\
&- \frac{1}{2}\delta_1 q_2 \int_0^1 e^{\delta_1 x}\beta(x,t)^2 dx \quad c_3 \int_0^1 e^{\delta_1 x}\beta(x,t)^2 dx \\
&- \frac{1}{2}r_b q_1 e^{-\delta_2}\alpha(1,t)^2 + \frac{1}{2}q_1 r_b \alpha(0,t)^2 \\
&- \frac{1}{2}\delta_2 q_1 r_b \int_0^1 e^{-\delta_2 x}\alpha(x,t)^2 dx - c_2 r_b \int_0^1 e^{-\delta_2 x}\alpha(x,t)^2 dx \\
&- \frac{1}{2\tau}r_c \int_1^2 e^{-x}\hat{v}(x,t)^2 dx - \frac{1}{2\tau}r_c e^{-2}\hat{v}(2,t)^2 + \frac{1}{2\tau}r_c e^{-1}\hat{v}(1,t)^2 \\
&+ \left(\int_0^1 e^{\delta_1 x}\beta(x,t)H_3(x)dx + r_b \int_0^1 e^{-\delta_2 x}\alpha(x,t)H_2(x)dx \right. \\
&\left. + r_c \int_1^2 e^{-x}\hat{v}(x,t)H_5(x)dx + \hat{\zeta}(t)H_1 \right)\tilde{Y}(t). \tag{7.148}
\end{aligned}$$

We then have

$$\dot{V}(t) \geq - \mu_0 \Omega_1(t) - \lambda_{\alpha 1}\alpha(1,t)^2 - \lambda_{\alpha 0}(\alpha(0,t)^2 + d(t)^2) - \frac{1}{2\tau}r_c e^{-2}\hat{v}(2,t)^2 - \left|\tilde{Y}(t)\right|^2 \tag{7.149}$$

where

$$\Omega_1(t) = \|\alpha(\cdot, t)\|^2 + \|\beta(\cdot, t)\|^2 + \|\hat{v}(\cdot, t)\|^2 + \hat{\zeta}(t)^2 \tag{7.150}$$

and

$$\mu_0 = \max \left\{ \frac{1}{2} \delta_1 q_2 e^{\delta_1} + |c_3| e^{\delta_1} + e^{2\delta_1} \max_{x \in [0,1]} \{G_3(x)^2\}, \right.$$
$$\frac{1}{2} \delta_2 q_1 r_b + |c_2| r_b + r_b{}^2 \max_{x \in [0,1]} \{G_2(x)^2\},$$
$$\left. \frac{1}{2\tau} r_c e^{-1} + r_c{}^2 \max_{x \in [1,2]} \{H_5(x)^2\}, a_m + \frac{1}{2} |b_1| + H_1^2 \right\}, \tag{7.151}$$

and where the positive constants $\lambda_{\alpha 1}$, $\lambda_{\alpha 0}$ are shown in (7.141), (7.142).

Taking the derivative of (7.147), applying Young's inequality, using (7.143) in lemma 7.1, and inserting (7.149), we have

$$\dot{\psi} = \frac{(2d(t)\dot{d}(t) + \frac{\mu}{2}\dot{m}(t))}{\theta V(t) - \frac{\mu}{2}m(t)} - \frac{(\theta \dot{V}(t) - \frac{\mu}{2}\dot{m}(t))}{\theta V(t) - \frac{\mu}{2}m(t)} \psi$$
$$\leq \frac{1}{\theta V(t) - \frac{\mu}{2}m(t)} \left[r_1 \lambda_1 \Omega(t)^2 + r_1 \lambda_1 \alpha(1, t)^2 + r_1 \lambda_1 \alpha(0, t)^2 + \left(\frac{1}{r_1} + r_1 \lambda_1 \right) d(t)^2 \right.$$
$$\left. + r_1 \lambda_1 \left| \tilde{Y}(t) \right|^2 + \frac{\mu}{2}\dot{m}(t) \right] - \frac{1}{\theta V(t) - \frac{\mu}{2}m(t)} \left[\theta \left(-\mu_0 \Omega_1(t) - \lambda_{\alpha 1} \alpha(1, t)^2 \right. \right.$$
$$\left. \left. - \lambda_{\alpha 0}(\alpha(0, t)^2 + d(t)^2) - \left| \tilde{Y}(t) \right|^2 - \frac{1}{2\tau} r_c e^{-2} \hat{v}(2, t)^2 \right) - \frac{\mu}{2}\dot{m}(t) \right] \psi, \tag{7.152}$$

where the positive constant r_1 comes from applying Young's inequality. Inserting (7.126) to replace $\dot{m}(t)$, one obtains

$$\dot{\psi} \leq \frac{1}{\theta V(t) - \frac{\mu}{2}m(t)} \left[r_1 \lambda_1 \Omega(t) + r_1 \lambda_1 \alpha(1, t)^2 + r_1 \lambda_1 \alpha(0, t)^2 + \left(\frac{1}{r_1} + r_1 \lambda_1 \right) d(t)^2 \right.$$
$$\left. + r_1 \lambda_1 \left| \tilde{Y}(t) \right|^2 - \frac{\mu}{2}\eta m(t) - \frac{\mu}{2}\kappa_1 \alpha(1, t)^2 - \frac{\mu}{2}\kappa_2 \alpha(0, t)^2 - \frac{\mu}{2}\kappa_4 \left| \tilde{Y}(t) \right|^2 \right]$$
$$- \frac{1}{\theta V(t) - \frac{\mu}{2}m(t)} \left[-\theta \mu_0 \Omega_1(t) - \theta \lambda_{\alpha 1} \alpha(1, t)^2 \right.$$
$$- \theta \lambda_{\alpha 0} \alpha(0, t)^2 - \theta \lambda_{\alpha 0} d(t)^2 - \theta \left| \tilde{Y}(t) \right|^2 - \frac{\theta}{2\tau} r_c e^{-2} \hat{v}(2, t)^2$$
$$\left. + \frac{\mu}{2}\eta m(t) + \frac{\mu}{2}\kappa_1 \alpha(1, t)^2 + \frac{\mu}{2}\kappa_2 \alpha(0, t)^2 + \frac{\mu}{2}\kappa_3 \hat{v}(2, t)^2 + \frac{\mu}{2}\kappa_4 \left| \tilde{Y}(t) \right|^2 \right] \psi. \tag{7.153}$$

Recalling (7.128), (7.144), (7.150) and the fact that

$$\xi_1 \Omega_1(t) \leq V(t) \leq \xi_2 \Omega_1(t), \tag{7.154}$$

where the positive constants ξ_1, ξ_2 are

$$\xi_1 = \min \left\{ \frac{1}{2}, \frac{1}{2} r_b e^{-\delta_2}, \frac{1}{2} r_c e^{-2} \right\}, \tag{7.155}$$

$$\xi_2 = \max\left\{\frac{1}{2}, \frac{1}{2}e^{\delta_1}, \frac{1}{2}r_b, \frac{1}{2}r_c e^{-1}\right\}, \tag{7.156}$$

we have

$$\dot{\psi} \le \frac{1}{\theta V(t) - \frac{\mu}{2}m(t)}\left[\left(\frac{1}{r_1} + r_1\lambda_1\right)d(t)^2 + \left(r_1\lambda_1 - \frac{\mu}{2}\kappa_1\right)\alpha(1,t)^2\right.$$

$$+ \left(r_1\lambda_1 - \frac{\mu}{2}\kappa_2\right)\alpha(0,t)^2 + \left(r_1\lambda_1 - \frac{\mu}{2}\kappa_4\right)\left|\tilde{Y}(t)\right|^2 + r_1\frac{\lambda_1}{\xi_1}V(t) - \frac{\mu}{2}\eta m(t)\Big]$$

$$- \frac{1}{\theta V(t) - \frac{\mu}{2}m(t)}\left[-\theta\frac{\mu_0}{\xi_1}V(t) - \left(\theta\lambda_{\alpha 1} - \frac{\mu}{2}\kappa_1\right)\alpha(1,t)^2\right.$$

$$- \left(\theta\lambda_{\alpha 0} - \frac{\mu}{2}\kappa_2\right)\alpha(0,t)^2 - \theta\lambda_{\alpha 0}d(t)^2$$

$$- \left(\theta - \frac{\mu}{2}\kappa_4\right)\left|\tilde{Y}(t)\right|^2 - \left(\frac{\theta}{2\tau}r_c e^{-2} - \frac{\mu}{2}\kappa_3\right)\hat{v}(2,t)^2 + \frac{\mu}{2}\eta m(t)\Big]\psi, \tag{7.157}$$

where $\Omega_1(t) \ge \Omega(t)$ has been used. Choosing $r_1 = \mu$ and applying (7.129), (7.130), (7.137) (the last four terms) and the inequalities

$$-\frac{\frac{\mu}{2}\eta m(t)}{\theta V(t) - \frac{\mu}{2}m(t)} \le -\frac{\frac{\mu}{2}\eta m(t)}{-\frac{\mu}{2}m(t)} = \eta,$$

$$\frac{V(t)}{\theta V(t) - \frac{\mu}{2}m(t)} \le \frac{V(t)}{\theta V(t)} = \frac{1}{\theta},$$

$$\frac{d(t)^2}{\theta V(t) - \frac{\mu}{2}m(t)} = \frac{d(t)^2 + \frac{\mu}{2}m(t) - \frac{\mu}{2}m(t)}{\theta V(t) - \frac{\mu}{2}m(t)} \le \psi(t) + 1,$$

which hold because $m(t) < 0$, we see that (7.157) becomes

$$\dot{\psi}(t) \le \frac{(\frac{1}{\mu} + \mu\lambda_1)d(t)^2 + \mu\frac{\lambda_1}{\xi_1}V(t) - \frac{\mu}{2}\eta m(t)}{\theta V(t) - \frac{\mu}{2}m(t)} + \eta\psi + \frac{\mu_0}{\xi_1}\psi + \theta\lambda_{\alpha 0}\psi^2 + \theta\lambda_{\alpha 0}\psi$$

$$\le \theta\lambda_{\alpha 0}\psi^2 + \left(\frac{1}{\mu} + \mu\lambda_1 + \eta + \mu_0 + \theta\lambda_{\alpha 0}\right)\psi + \frac{1}{\mu} + \mu\lambda_1 + \frac{\mu\lambda_1}{\theta\xi_1} + \eta. \tag{7.158}$$

The differential inequality (7.158) has the form

$$\dot{\psi} \le n_1\psi^2 + n_2\psi + n_3, \tag{7.159}$$

where

$$n_1 = \theta\lambda_{\alpha 0}, \tag{7.160}$$

$$n_2 = \frac{1}{\mu} + \mu\lambda_1 + \eta + \mu_0 + \theta\lambda_{\alpha 0}, \tag{7.161}$$

$$n_3 = \frac{1}{\mu} + \mu\lambda_1 + \frac{\mu\lambda_1}{\theta\xi_1} + \eta \tag{7.162}$$

are positive constants.

It follows that the time needed by ψ to go from 0 to 1 is at least

$$T = \int_0^1 \frac{1}{n_1 + n_2\bar{s} + n_3\bar{s}^2} d\bar{s} > 0, \tag{7.163}$$

which is independent of initial conditions. $\square$

7.5   STABILITY ANALYSIS

In this section we state the main result of the chapter after we first establish an intermediate result.

Lemma 7.3. *With the arbitrary initial data* $(\alpha(x,0), \beta(x,0))^T \in C^0([0,1]; \mathbb{R}^2)$, $\hat{v}(x,0) \in C^1([1,2]; \mathbb{R})$, $\hat{\zeta}(0) \in \mathbb{R}$, $m(0) \in \mathbb{R}^-$, *for* (7.105)–(7.110), (7.120), (7.126), *the exponential convergence is achieved in the sense of the norm* $|\hat{\zeta}(t)| + \|\alpha(\cdot,t)\| + \|\beta(\cdot,t)\| + \|\hat{v}(\cdot,t)\| + |m(t)|$.

Proof. There exists a function

$$e(t) = \Upsilon_d e^{-\lambda_e t},$$

for some positive Υ_d, λ_e such that

$$\left| \tilde{Y}(t) \right| \leq e(t), \tag{7.164}$$

recalling the exponential convergence result proved in theorem 7.1. Let us consider the Lyapunov function

$$V_a(t) = V(t) - m(t) + \frac{1}{2}R_e e(t)^2 \tag{7.165}$$

where $m(t)$ is defined in (7.126) $(m(t) < 0)$, and $V(t)$ is given in (7.128).

By virtue of (7.150), we have

$$d_1(\Omega_1(t) + |m(t)| + e(t)^2) \leq V_a(t) \leq d_2(\Omega_1(t) + |m(t)| + e(t)^2) \tag{7.166}$$

for some positive d_1, d_2.

Taking the derivative of (7.165) along (7.105)–(7.110), (7.120) and recalling (7.126), we obtain

$$\dot{V}_a(t) = \dot{V} - \dot{m}(t) + R_e e(t)\dot{e}(t)$$

$$= -a_m\hat{\zeta}(t)^2 - \hat{\zeta}(t)b_1\alpha(1,t) + \frac{1}{2}q_2 e^{\delta_1}\beta(1,t)^2 - \frac{1}{2}q_2\beta(0,t)^2$$

$$- \frac{1}{2}\delta_1 q_2 \int_0^1 e^{\delta_1 x}\beta(x,t)^2 dx - c_3 \int_0^1 e^{\delta_1 x}\beta(x,t)^2 dx$$

$$- \frac{1}{2}r_b q_1 e^{-\delta_2}\alpha(1,t)^2 + \frac{1}{2}q_1 r_b\alpha(0,t)^2$$

$$- \frac{1}{2}\delta_2 q_1 r_b \int_0^1 e^{-\delta_2 x}\alpha(x,t)^2 dx - c_2 r_b \int_0^1 e^{-\delta_2 x}\alpha(x,t)^2 dx$$

$$- \frac{1}{2\tau}r_c \int_1^2 e^{-x}\hat{v}(x,t)^2 dx - \frac{1}{2\tau}r_c e^{-2}\hat{v}(2,t)^2 + \frac{1}{2\tau}r_c e^{-1}\hat{v}(1,t)^2$$

$$- R_e \lambda_e e(t)^2 + \eta m(t) + \kappa_1 \alpha(1,t)^2 + \kappa_2 \alpha(0,t)^2 + \kappa_3 \hat{v}(2,t)^2 + \kappa_4 \left| \tilde{Y}(t) \right|^2 . \tag{7.167}$$

Recalling (7.124), (7.31) and applying Young's inequality and the Cauchy-Schwarz inequality, we have

$$
\begin{aligned}
\dot{V}_a(t) \le & -\frac{1}{2} \left(a_m - \frac{1}{\tau} r_c e^{-1} \right) \hat{\zeta}(t)^2 - \left(\frac{1}{2} q_1 r_b e^{-\delta_2} - \frac{q^2}{2} q_2 e^{\delta_1} - \frac{b_1^2}{2a_m} - \kappa_1 \right) \alpha(1,t)^2 \\
& + \left(\frac{1}{2} q_1 r_b \lambda_d + \kappa_2 \lambda_d \right) \sup_{0 \le \xi \le t} d(\xi)^2 - \left(\frac{1}{2} \delta_1 q_2 - |c_3| \right) \int_0^1 e^{\delta_1 x} \beta(x,t)^2 dx \\
& - \left(\frac{1}{2} \delta_2 q_1 r_b - r_b |c_2| \right) \int_0^1 e^{-\delta_2 x} \alpha(x,t)^2 dx - \frac{1}{2\tau} r_c \int_1^2 e^{-x} \hat{v}(x,t)^2 dx \\
& - \left(\frac{1}{2\tau} r_c e^{-2} - \kappa_3 \right) \hat{v}(2,t)^2 - (R_e \lambda_e - \kappa_4) e(t)^2 + \eta m(t). \tag{7.168}
\end{aligned}
$$

Applying (7.131), (7.132), (7.133), (7.134), (7.135), we thus arrive at

$$\dot{V}_a(t) \le -\lambda_a V(t) + \eta m(t) - \kappa_4 e(t)^2 + \left(\frac{1}{2} q_1 r_b \lambda_d + \kappa_2 \lambda_d \right) \sup_{0 \le \xi \le t} d(\xi)^2, \tag{7.169}$$

where λ_a is given in (7.139). Recalling (7.165), we have

$$\dot{V}_a(t) \le -\sigma_a V_a(t) + \left(\frac{1}{2} q_1 r_b \lambda_d + \kappa_2 \lambda_d \right) \sup_{0 \le \xi \le t} d(\xi)^2, \tag{7.170}$$

where σ_a is given in (7.138).

Multiplying both sides of (7.170) by $e^{\sigma_a t}$, we have

$$e^{\sigma_a t} \dot{V}_a(t) + e^{\sigma_a t} \sigma_a V_a(t) \le e^{\sigma_a t} \left(\frac{1}{2} q_1 r_b \lambda_d + \kappa_2 \lambda_d \right) \sup_{0 \le \xi \le t} d(\xi)^2. \tag{7.171}$$

The left-hand side of (7.171) is $\frac{d(e^{\sigma_a t} V(t))}{dt}$. The integration of (7.171) from 0 to t yields

$$
\begin{aligned}
V_a(t) & \le V_a(0) e^{-\sigma_a t} + \frac{1}{\sigma_a} (1 - e^{-\sigma_a t}) \left(\frac{1}{2} q_1 r_b \lambda_d + \kappa_2 \lambda_d \right) \sup_{0 \le \xi \le t} d(\xi)^2 \\
& \le V_a(0) e^{-\sigma_a t} + \frac{1}{\sigma_a} \left(\frac{1}{2} q_1 r_b \lambda_d + \kappa_2 \lambda_d \right) \sup_{0 \le \xi \le t} d(\xi)^2. \tag{7.172}
\end{aligned}
$$

The triggering condition (7.125) guarantees

$$\sup_{0 \le \xi \le t} d(\xi)^2 \le \theta \sup_{0 \le \xi \le t} V(\xi) + \mu \sup_{0 \le \xi \le t} |m(\xi)|. \tag{7.173}$$

Inserting (7.173) into (7.172) and then recalling (7.165) yields

$$
\begin{aligned}
V_a(t) & \le V_a(0) e^{-\sigma_a t} + \frac{1}{\sigma_a} \left(\frac{1}{2} q_1 r_b \lambda_d + \kappa_2 \lambda_d \right) \sup_{0 \le \xi \le t} (\theta V(\xi) + \mu |m(\xi)|) \\
& \le V_a(0) e^{-\sigma_a t} + \bar{\Phi} \sup_{0 \le \xi \le t} V_a(\xi), \tag{7.174}
\end{aligned}
$$

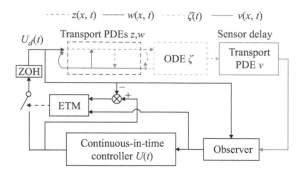

Figure 7.2. Event-based closed-loop system.

where

$$\bar{\Phi} = \frac{1}{\sigma_a}\left(\frac{1}{2}q_1 r_b \lambda_d + \kappa_2 \lambda_d\right)\max\{\theta,\mu\}. \tag{7.175}$$

Recalling (7.136), (7.137) (the first term), we find that

$$\bar{\Phi} < 1. \tag{7.176}$$

The following estimate then holds

$$\sup_{0\leq\xi\leq t}\left(V_a(\xi)e^{\sigma_a\xi}\right) \leq V_a(0) + \bar{\Phi}\sup_{0\leq\xi\leq t}\left(V_a(\xi)e^{\sigma_a\xi}\right) \tag{7.177}$$

as a consequence of (7.174). It follows that

$$\sup_{0\leq\xi\leq t} V_a(\xi) \leq \Upsilon_V V_a(0)e^{-\sigma_a t}, \tag{7.178}$$

where the constant

$$\Upsilon_V = \frac{1}{1-\bar{\Phi}} > 0 \tag{7.179}$$

by recalling (7.176). The choices of ETM parameters affect the overshoot coefficient in the exponential result according to (7.175), (7.179). Recalling (7.166), lemma 7.3 is obtained. $\qquad\square$

The block diagram of the event-based output-feedback closed-loop system is shown in figure 7.2, where an observer-based continuous-in-time control input $U(t)$ in (7.112) is updated at time instants t_k determined by ETM (7.125), (7.126) realized based on the observer (7.26)–(7.32). Between updates the input is held constant in a zero-order-hold (ZOH) fashion.

The properties of the output-feedback closed-loop system are shown in the following theorem.

Theorem 7.2. *For all initial data* $(z(x,0),w(x,0))^T \in C^0([0,1],\mathbb{R}^2)$, $v(x,0) \in C^1([1,2],\mathbb{R})$, $\zeta(0) \in \mathbb{R}$, $(\hat{z}(x,0),\hat{w}(x,0))^T \in C^0([0,1],\mathbb{R}^2)$, $\hat{v}(x,0) \in C^1([1,2],\mathbb{R})$, $\hat{\zeta}(0) \in \mathbb{R}$, $m(0) \in \mathbb{R}^-$, *choosing the design parameters to satisfy* (7.47), (7.111), (7.129)–(7.137), *the output-feedback closed-loop system—that is, the plant* (7.16)– (7.21) *under the event-based control input* $U_d(t)$ *in* (7.118), *which is realized using*

the observer (7.26)–(7.32), and the ETM (7.125), (7.126), has the following properties:

1) *No Zeno phenomenon occurs—that is,*

$$\lim_{i \to \infty} t_i = +\infty. \tag{7.180}$$

2) *The exponential convergence in the event-based output-feedback closed-loop sys-
tem is achieved in the sense of the norm* $|\zeta(t)| + \|z(\cdot,t)\| + \|w(\cdot,t)\| + \|v(\cdot,t)\| +
|\hat{\zeta}(t)| + \|\hat{z}(\cdot,t)\| + \|\hat{w}(\cdot,t)\| + \|\hat{v}(\cdot,t)\| + |m(t)|.$
3) *The event-triggered control input is convergent to zero in the sense of*

$$\lim_{t \to \infty} U_d(t) = 0. \tag{7.181}$$

Proof. 1) Recalling lemma 7.2, we have

$$t_i \geq iT, \quad i \in \mathbb{Z}_+. \tag{7.182}$$

Property (1) is thus obtained.
2) Through an additional analysis that is routine but heavy on additional notation,
we know that the stability result in lemma 7.3 still holds (by choosing a sufficiently large R_e) for the target system corresponding to the output-feedback
loop, that is, the $\mathcal{S}$-system defined in remark 7.1. Thus, the separation principle holds. Recalling the invertibility of the backstepping transformation (7.90),
(7.91) and applying theorem 7.1 as well as (7.34), property (2) is thus obtained.
3) Recalling (7.112) and the stability result proved in property (2), we have that
the continuous-in-time control input $U(t)$ is convergent to zero. According to
the definition (7.118), property (3) is obtained.   □

7.6   SIMULATION FOR DEEP-SEA CONSTRUCTION WITH SENSOR DELAY

Here we consider this simulation's applicability to the DCV when placing equipment to be installed on the seabed for offshore oil drilling. The equipment must be
installed accurately at a predetermined location with a tight tolerance. The permissible maximum tolerance for a typical subsea installation was 2.5 m, according
to [95]. Applying the design presented above, we obtain an output-feedback control
force employing piecewise-constant values at the crane to reduce oscillations of the
long cable and position the equipment in the target area while compensating for
the sensor delay.

Initial Conditions and Design Parameters

The DCV model in the simulation is (7.16)–(7.21), with coefficients in (7.14), (7.15)
whose values are shown in table 7.1. The obtained simulation results are represented
in the $(\bar{z}, \bar{w})$ and u models that are on the spatial domain $[0, L]$ via (7.10), (7.11),
(7.12), (7.13). The disturbances f, f_L, which will be formulated in the next subsection, are included (with multiplying by $\frac{1}{\rho}$ considering the conversion from the wave
PDE (7.1)–(7.4)) in the model (7.16)–(7.21). The initial conditions are chosen as

$$z(x,0) = 6\sin\left(5\pi x + \frac{\pi}{4}\right), \quad w(x,0) = 6\cos\left(5\pi x + \frac{\pi}{3}\right), \tag{7.183}$$

which gives the ODE initial conditions

$$\zeta(0) = 3\sin\left(\frac{21\pi}{4}\right) + 3\cos\left(\frac{16\pi}{3}\right) \tag{7.184}$$

by recalling (7.19), and the quantity is, physically, the initial oscillation velocity of the payload. According to the Riemann transformation (7.10), (7.11) and (7.12), (7.13), the initial conditions of $\bar{z}, \bar{w}$ and the initial oscillation velocity of the cable are hence

$$\bar{z}(\bar{x},0) = 6\sin\left(\frac{5\pi\bar{x}}{L} + \frac{\pi}{4}\right), \quad \bar{w}(\bar{x},0) = 6\cos\left(\frac{5\pi\bar{x}}{L} + \frac{\pi}{3}\right), \tag{7.185}$$

$$u_t(\bar{x},0) = \frac{1}{2}(\bar{z}(\bar{x},0) + \bar{w}(\bar{x},0)). \tag{7.186}$$

The initial distributed oscillation displacement of the cable is defined as $u(\bar{x},0) = 0$, which, recalling (7.1), (7.3), implies that the initial displacement of the payload is $b_L(0) = 0$. The initial values of the observer are defined as zero—that is, $\hat{z}[0] = 0, \hat{w}[0] = 0, \hat{\zeta}(0) = 0, \hat{v}[0] = 0$. The initial value of the internal dynamic variable $m(t)$ in ETM is set as $m(0) = -1.5 \times 10^5$. The design parameters are chosen as $L_1 = 1$, $F_1 = 50$, $\kappa_1 = 3000, \kappa_2 = 2000, \delta_1 = 0.017, \delta_2 = 0.01, r_c = 0.2, r_b = 40, \kappa_3 = 0.025, \mu = 0.2$, and $\theta = 0.08$ according to (7.47), (7.111), (7.129)–(7.137), and the free design parameter η is picked as 8.

Ocean Current Disturbances

The time-varying ocean surface current velocity is modeled by a first-order Gauss-Markov process [61]:

$$\dot{P}(t) + \bar{\mu}P(t) = \bar{\mathcal{G}}(t), \quad P_{\min} \leq P(t) \leq P_{\max}, \tag{7.187}$$

where $\bar{\mathcal{G}}(t)$ is Gaussian white noise. The constants $P_{\min}, P_{\max}$ and $\bar{\mu}$ are chosen as 1.6 ms^{-1}, 2.4 ms^{-1}, and 0 [95]. The full current load $P(t)$ is applied at the cable from $\bar{x} = 0$ to $\bar{x} = 300$ m and thereafter linearly declines to $0.1P(t)$ at the bottom of the cable, which is at $\bar{x} = 1000$ m [95]. The depth-dependent ocean current profile $P(\bar{x},t)$ is thus obtained as

$$P(\bar{x},t) = \begin{cases} P(t), & 0 \leq \bar{x} \leq 300 \\ \dfrac{970 - 0.9\bar{x}}{700}P(t), & 300 \leq \bar{x} \leq L, \end{cases} \tag{7.188}$$

which determines the ocean current disturbances $\bar{f}(\bar{x},t)$ and $f_L(t)$ next. The disturbance $\bar{f}(\bar{x},t)$ is the distributed oscillating drag force [95] modeled as

$$\bar{f}(\bar{x},t) = \frac{1}{2}\rho_s C_d P(\bar{x},t)^2 R_D A_D \cos\left(4\pi\frac{S_t P(\bar{x},t)}{R_D}t + \varsigma\right), \tag{7.189}$$

where $C_d = 1$ denotes the drag coefficient, $\varsigma = \pi$ is the phase angle, $A_D = 400$ denotes the amplitude of the oscillating drag force, and $S_t = 0.2$ is the Strouhal number [59].

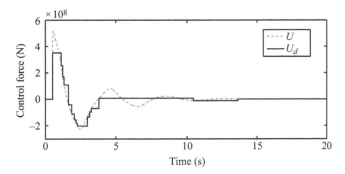

Figure 7.3. Control forces (continuous-in-time U and event-based U_d).

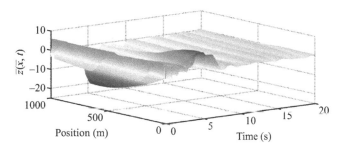

Figure 7.4. The evolution of $\bar{z}(\bar{x}, t)$ under the proposed event-based control input.

Then the distributed disturbance applied in the model (7.16)–(7.21) is $f(x,t) = \bar{f}(xL,t)$. The drag force $f_L(t)$ at the payload, which is considered to be a cylinder, is derived from Morison's equation [95] as

$$f_L(t) = \frac{1}{2} C_d \rho_s h_c D_c \left| P(L,t) \right| P(L,t). \tag{7.190}$$

Simulation Results

We concentrate on the end phase (20 s) of the descending process, when the payload is near the seabed, and the cable is at its fully extended total length L. Our task is to reduce the oscillations of the cable and place the payload at the bottom of the cable in the target area on the seabed—namely, within the permissible tolerance of 2.5 m around the predetermined location [95] by applying a piecewise-constant control input at the onboard crane driving the top of the cable, where the measurements are transmitted from the seabed and subject to a 0.5 s delay.

The event-based control input defined in (7.118) and the continuous-in-time control input defined in (7.112) are shown in figure 7.3, where the number of triggering times is 15, and the minimal dwell time is 0.097 s. The control input is zero at the beginning due to the sensor delay and the zero initial conditions of the observer based on which the control law is realized. With the event-based control input, we know from figures 7.4 and 7.5 that the PDE states $\bar{z}(\bar{x}, t)$ and $\bar{w}(\bar{x}, t)$ are regulated to a small range around zero, under the unknown ocean current disturbances (7.189), (7.190) and the sensor delay $\tau = 0.5$ s. Similar results from the

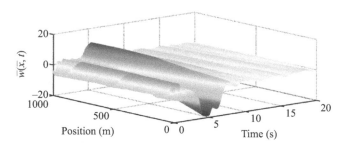

Figure 7.5. The evolution of $\bar{w}(\bar{x}, t)$ under the proposed event-based control input.

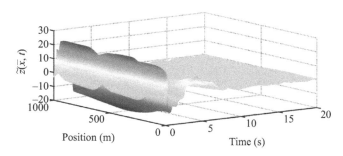

Figure 7.6. The evolution of the observer error $\tilde{z}(\bar{x}, t)$.

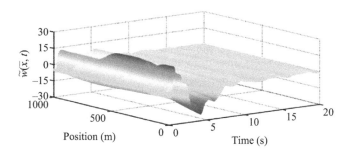

Figure 7.7. The evolution of the observer error $\tilde{w}(\bar{x}, t)$.

observer errors $\tilde{z}(\bar{x}, t) = \bar{z}(\bar{x}, t) - \hat{z}(\frac{\bar{x}}{L}, t)$, $\tilde{w}(\bar{x}, t) = \bar{w}(\bar{x}, t) - \hat{w}(\frac{\bar{x}}{L}, t)$ are observed in figures 7.6 and 7.7. Figure 7.8 shows that the internal dynamic variable $m(t)$ in ETM is less than zero all the time.

By virtue of (7.3), (7.8), the lateral displacement of the payload $b_L(t)$ is

$$b_L(t) = \int_0^t \zeta(\delta)d\delta + u(L, 0). \tag{7.191}$$

From figure 7.9, we know that the oscillation velocity of the payload, $\dot{b}_L(t) = \zeta(t)$, is convergent to zero, and the position error of the payload is -0.98 m from the desired location on the seabed, which satisfies the requirement of being within the permissible tolerance of 2.5 m given in [95]. Through (7.10), (7.11), the cable lateral oscillation energy, including the oscillation kinetic energy $\frac{\rho}{2}\|u_t(\bar{x}, t)\|^2$ and

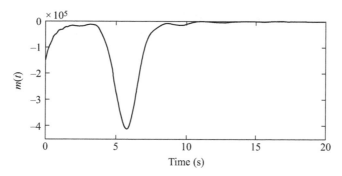

Figure 7.8. The evolution of the internal dynamic variable $m(t)$ in the ETM.

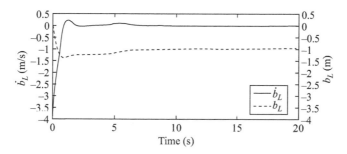

Figure 7.9. The oscillation velocity and displacement of the payload.

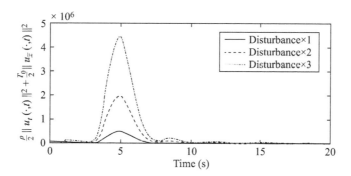

Figure 7.10. The cable oscillation energy.

potential energy $\frac{T_0}{2}\|u_{\bar{x}}(\bar{x},t)\|^2$, is represented by $\bar{z}(\bar{x},t), \bar{w}(\bar{x},t)$, shown in figures 7.4 and 7.5, as

$$\frac{\rho}{2}\|u_t(\cdot,t)\|^2 + \frac{T_0}{2}\|u_{\bar{x}}(\cdot,t)\|^2$$
$$= \frac{\rho}{8}\|\bar{w}(\cdot,t) + \bar{z}(\cdot,t)\|^2 + \frac{\rho}{8}\|\bar{w}(\cdot,t) - \bar{z}(\cdot,t)\|^2. \tag{7.192}$$

As shown in figure 7.10, the oscillation energy of the cable with the proposed control law in (7.118) is reduced to a low level after $t = 10$ s under the external ocean current disturbances (7.189), (7.190). This result, the solid line in figure 7.10, shows the robustness of the proposed control to small disturbances. However, as we continue to increase the amplitude of the disturbance (7.189), from the baseline value to its twice and to its triple, the results on cable oscillation energy are shown as the dashed line and dot-and-dash line in figure 7.10, where performance deterioration of the proposed controller is observed with the increase of disturbance amplitudes, and relatively large vibrations appear at the triple disturbance.

7.7   APPENDIX

A. Calculations of (7.81) and (7.85)

Details of calculating (7.81) are shown as follows:

$$
\tilde{w}_t(x,t) - q_2 \tilde{w}_x(x,t) + c_4 \tilde{z}(x,t) + c_3 \tilde{w}(x,t) + H_3(x)[\tilde{v}(2,t), \tilde{v}_t(2,t)]^T
$$

$$
= \tilde{\beta}_t(x,t) - \int_x^1 \psi(x,y) \tilde{\alpha}_t(y,t) dy
$$

$$
- q_2 \tilde{\beta}_x(x,t) + q_2 \int_x^1 \psi_x(x,y) \tilde{\alpha}(y,t) dy - q_2 \psi(x,x) \tilde{\alpha}(x,t)
$$

$$
+ c_4 \tilde{\alpha}(x,t) - c_4 \int_x^1 \phi(x,y) \tilde{\alpha}(y,t) dy + H_3(x)[\tilde{v}(2,t), \tilde{v}_t(2,t)]^T
$$

$$
+ c_3 \tilde{\beta}(x,t) - c_3 \int_x^1 \psi(x,y) \tilde{\alpha}(y,t) dy
$$

$$
= \int_x^1 \bar{N}(x,y) \tilde{\beta}(y,t) dy + c_2 \int_x^1 \psi(x,y) \tilde{\alpha}(y,t) dy
$$

$$
+ q_1 \int_x^1 \psi(x,y) \tilde{\alpha}_x(y,t) dy + \int_x^1 c_1 \psi(x,y) \tilde{\beta}(y,t) dy
$$

$$
- \int_x^1 \int_x^y \psi(x,\delta) \bar{M}(\delta,y) d\delta \tilde{\beta}(y,t) dy + q_2 \int_x^1 \psi_x(x,y) \tilde{\alpha}(y,t) dy
$$

$$
- q_2 \psi(x,x) \tilde{\alpha}(x,t) - c_3 \int_x^1 \psi(x,y) \tilde{\alpha}(y,t) dy
$$

$$
+ c_4 \tilde{\alpha}(x,t) - c_4 \int_x^1 \phi(x,y) \tilde{\alpha}(y,t) dy + H_3(x)[\tilde{v}(2,t), \tilde{v}_t(2,t)]^T
$$

$$
= (c_4 - (q_1 + q_2)\psi(x,x)) \tilde{\alpha}(x,t)
$$

$$
+ \int_x^1 (-q_1 \psi_y(x,y) + q_2 \psi_x(x,y) - c_4 \phi(x,y) + (c_2 - c_3)\psi(x,y)) \tilde{\alpha}(y,t) dy
$$

$$
+ \int_x^1 \left(c_1 \psi(x,y) + \bar{N}(x,y) - \int_x^y \psi(x,\delta) \bar{M}(\delta,y) d\delta \right) \tilde{\beta}(y,t) dy
$$

$$
+ q_1 \psi(x,1) \tilde{\alpha}(1,t) + H_3(x)[\tilde{v}(2,t), \tilde{v}_t(2,t)]^T
$$

$$
= q_1 \psi(x,1) \tilde{\alpha}(1,t) + H_3(x)[\tilde{v}(2,t), \tilde{v}_t(2,t)]^T = 0. \tag{7.193}
$$

Details of calculating (7.85) are shown as follows:

$$
\tilde{z}_t(x,t) + q_1 \tilde{z}_x(x,t) + c_1 \tilde{w}(x,t) + c_2 \tilde{z}(x,t) + H_2(x)[\tilde{v}(2,t), \tilde{v}_t(2,t)]^T
$$

$$= \tilde{\alpha}_t(x,t) - \int_x^1 \phi(x,y)\tilde{\alpha}_t(y,t)dy + q_1\tilde{\alpha}_x(x,t) - q_1 \int_x^1 \phi_x(x,y)\tilde{\alpha}(y,t)dy$$

$$+ q_1\phi(x,x)\tilde{\alpha}(x,t) + c_1\tilde{\beta}(x,t) - c_1 \int_x^1 \psi(x,y)\tilde{\alpha}(y,t)dy + H_2(x)[\tilde{v}(2,t), \tilde{v}_t(2,t)]^T$$

$$+ c_2\tilde{\alpha}(x,t) - c_2 \int_x^1 \phi(x,y)\tilde{\alpha}(y,t)dy$$

$$= -c_1\tilde{\beta}(x,t) + q_1 \int_x^1 \phi(x,y)\tilde{\alpha}_x(y,t)dy + \int_x^1 \left(-q_1\phi_x(x,y) - c_1\psi(x,y)\right)\tilde{\alpha}(y,t)dy$$

$$+ \int_x^1 \left(-\int_x^y \phi(x,\delta)\bar{M}(y,\delta)d\delta + \phi(x,y)c_1\right)\tilde{\beta}(y,t)dy$$

$$+ q_1\phi(x,x)\tilde{\alpha}(x,t) + c_1\tilde{\beta}(x,t) + \int_x^1 \bar{M}(x,y)\tilde{\beta}(y,t)dy + H_2(x)[\tilde{v}(2,t), \tilde{v}_t(2,t)]^T$$

$$= \int_x^1 \left(-q_1\phi_x(x,y) - c_1\psi(x,y) - q_1\phi_y(x,y)\right)\tilde{\alpha}(y,t)dy$$

$$+ \int_x^1 \left(\bar{M}(x,y) - \int_x^y \phi(x,\delta)\bar{M}(\delta,y)d\delta + c_1\phi(x,y)\right)\tilde{\beta}(y,t)dy$$

$$+ q_1\phi(x,1)\tilde{\alpha}(1,t) + H_2(x)[\tilde{v}(2,t), \tilde{v}_t(2,t)]^T$$

$$= q_1\phi(x,1)\tilde{\alpha}(1,t) + H_2(x)[\tilde{v}(2,t), \tilde{v}_t(2,t)]^T = 0. \tag{7.194}$$

B. Matching (7.27)–(7.30) and (7.105)–(7.108)

Step 1: Taking the time and spatial derivative of (7.91) along (7.26)–(7.30) and substituting the results into (7.106), we get

$$\beta_t(x,t) - q_2\beta_x(x,t) + c_3\beta(x,t)$$

$$= \hat{w}_t(x,t) - \int_x^1 K_2(x,y)\hat{z}_t(y,t)dy - \int_x^1 J_2(x,y)\hat{w}_t(y,t)dy - \lambda(x)\dot{\hat{\zeta}}(t)$$

$$- q_2\hat{w}_x(x,t) + q_2 \int_x^1 K_{2x}(x,y)\hat{z}(y,t)dy + q_2 \int_x^1 J_{2x}(x,y)\hat{w}(y,t)dy$$

$$+ q_2\lambda'(x)\hat{\zeta}(t) - q_2K_2(x,x)\hat{z}(x,t) - q_2J_2(x,x)\hat{w}(x,t)$$

$$+ c_3\hat{w}(x,t) - c_3 \int_x^1 K_2(x,y)\hat{z}(y,t)dy - c_3 \int_x^1 J_2(x,y)\hat{w}(y,t)dy - c_3\lambda(x)\hat{\zeta}(t)$$

$$= -c_4\hat{z}(x,t) + q_1 \int_x^1 K_2(x,y)\hat{z}_x(y,t)dy$$

$$+ \int_x^1 K_2(x,y)\left(c_2\hat{z}(y,t) + c_1\hat{w}(y,t)\right)dy - q_2 \int_x^1 J_2(x,y)\hat{w}_x(y,t)dy$$

$$+ \int_x^1 J_2(x,y)\left(c_4\hat{z}(y,t) + c_3\hat{w}(y,t)\right)dy$$

$$+ q_2 \int_x^1 K_{2x}(x,y)\hat{z}(y,t)dy + q_2 \int_x^1 J_{2x}(x,y)\hat{w}(y,t)dy$$

$$- q_2K_2(x,x)\hat{z}(x,t) - q_2J_2(x,x)\hat{w}(x,t)$$

$$- \lambda(x)a_1\hat{\zeta}(t) + \lambda(x)b_1\hat{z}(1,t) + q_2\lambda'(x)\hat{\zeta}(t)$$

$$- c_3\int_x^1 K_2(x,y)\hat{z}(y,t)dy - c_3\int_x^1 J_2(x,y)\hat{w}(y,t)dy - c_3\lambda(x)\hat{\zeta}(t)$$

$$= -\left(c_4 + (q_1+q_2)K_2(x,x)\right)\hat{z}(x,t)$$

$$+ \int_x^1 \left(c_1K_2(x,y) + q_2J_{2x}(x,y) + q_2J_{2y}(x,y)\right)\hat{w}(y,t)dy$$

$$+ \int_x^1 \left(c_4J_2(x,y) + c_2K_2(x,y) - c_3K_2(x,y) + q_2K_{2x}(x,y) - q_1K_{2y}(x,y)\right)\hat{z}(y,t)dy$$

$$+ \left(q_1K_2(x,1) - q_2J_2(x,1)q + \lambda(x)b_1\right)\hat{z}(1,t)$$

$$+ \left(q_2\lambda'(x) - \lambda(x)a_1 - c_3\lambda(x) - q_2J_2(x,1)c\right)\hat{\zeta}(t) = 0. \tag{7.195}$$

For (7.195) to hold, conditions (7.98)–(7.102) should be satisfied.

Step 2: Taking the time and spatial derivative of (7.90) along (7.26)–(7.30) and substituting the results into (7.105), we obtain

$$\alpha_t(x,t) + q_1\alpha_x(x,t) + c_2\alpha(x,t)$$

$$= \hat{z}_t(x,t) - \int_x^1 K_3(x,y)\hat{z}_t(y,t)dy - \int_x^1 J_3(x,y)\hat{w}_t(y,t)dy$$

$$- \gamma(x)\dot{\hat{\zeta}}(t) + q_1\hat{z}_x(x,t) - q_1\int_x^1 K_{3x}(x,y)\hat{z}(y,t)dy$$

$$- q_1\int_x^1 J_{3x}(x,y)\hat{w}(y,t)dy - q_1\gamma'(x)\hat{\zeta}(t) + q_1K_3(x,x)\hat{z}(x,t) + q_1J_3(x,x)\hat{w}(x,t)$$

$$+ c_2\hat{z}(x,t) - c_2\int_x^1 K_3(x,y)\hat{z}(y,t)dy - c_2\int_x^1 J_3(x,y)\hat{w}(y,t)dy - c_2\gamma(x)\hat{\zeta}(t)$$

$$= -c_1\hat{w}(x,t) + q_1\int_x^1 K_3(x,y)\hat{z}_x(y,t)dy + \int_x^1 K_3(x,y)(c_2\hat{z}(y,t) + c_1\hat{w}(y,t))dy$$

$$- q_2\int_x^1 J_3(x,y)\hat{w}_x(y,t)dy + \int_x^1 J_3(x,y)(c_4\hat{z}(y,t) + c_3\hat{w}(y,t))dy$$

$$- q_1\int_x^1 K_{3x}(x,y)\hat{z}(y,t)dy - q_1\int_x^1 J_{3x}(x,y)\hat{w}(y,t)dy + q_1K_3(x,x)\hat{z}(x,t)$$

$$+ q_1J_3(x,x)\hat{w}(x,t) - \gamma(x)a_1\hat{\zeta}(t) + \gamma(x)b_1\hat{z}(1,t) - q_1\gamma'(x)\hat{\zeta}(t)$$

$$- c_2\int_x^1 K_3(x,y)\hat{z}(y,t)dy - c_2\int_x^1 J_3(x,y)\hat{w}(y,t)dy - c_2\gamma(x)\hat{\zeta}(t)$$

$$= \left((q_2+q_1)J_3(x,x) - c_1\right)\hat{w}(x,t) + \int_x^1 \left(c_1K_3(x,y) - q_1J_{3x}(x,y)\right.$$

$$\left. + q_2J_{3y}(x,y) - (c_2-c_3)J_3(x,y)\right)\hat{w}(y,t)dy$$

$$+ \int_x^1 \left(c_4J_3(x,y) - q_1K_{3x}(x,y) - q_1K_{3y}(x,y)\right)\hat{z}(y,t)dy$$

$$+ \left(q_1K_3(x,1) - q_2J_3(x,1)q + \gamma(x)b_1\right)\hat{z}(1,t)$$

$$+ \left(-q_1\gamma'(x) - \gamma(x)a_1 - c_2\gamma(x) - q_2J_3(x,1)c\right)\hat{\zeta}(t) = 0. \tag{7.196}$$

For (7.196) to hold, conditions (7.92)–(7.95), (7.97) should be satisfied.

Step 3: Inserting (7.90), (7.91) into (7.108), (7.107), respectively, and applying (7.29), (7.30), we obtain

$$\dot{\hat{\zeta}}(t) + a_m \hat{\zeta}(t) + b_1 \alpha(1, t)$$

$$= \dot{\hat{\zeta}}(t) - a_1 \hat{\zeta}(t) + b_1 F_1 \hat{\zeta}(t) + b_1 \hat{z}(1, t) - b_1 \gamma(1) \hat{\zeta}(t)$$

$$= b_1 (F_1 - \gamma(1)) \hat{\zeta}(t) = 0, \tag{7.197}$$

$$\beta(1, t) - q\alpha(1, t)$$

$$= \hat{w}(1, t) - q\hat{z}(1, t) + (q\gamma(1) - \lambda(1)) \hat{\zeta}(t)$$

$$= (q\gamma(1) - \lambda(1) + c) \hat{\zeta}(t) = 0. \tag{7.198}$$

For (7.197), (7.198) to hold, conditions (7.96) and (7.103) should be satisfied.

7.8   NOTES

Most existing results [108, 115, 119, 180] on the delay-compensated control of PDEs are in a continuous-in-time form, while the control input in this chapter is in a piecewise-constant fashion.

In the control design in this chapter, we only considered the dynamics of the cable and payload and neglected the ship-mounted crane dynamics in the DCV. Incorporating the crane dynamics, the plant becomes an ODE-PDE-ODE sandwich structure, which is dealt with in chapter 10. Even though chapter 10 only provides a continuous-in-time control law, readers can refer to the result in chapter 11 where an event-triggered control design for sandwich systems is presented to develop an event-triggered-type controller.

Chapter Eight

Offshore Rotary Oil Drilling

Besides the control designs for the mining cable elevators and deep-sea construction vessels in chapters 2–6, here we present a torsional vibration control scheme for offshore rotary oil drilling.

We adopt the one-dimensional wave partial differential equation (PDE) [24, 25, 153] to describe the torsional vibration dynamics of the drill string, avoiding the spillover phenomenon [15] caused by a control design based on lumped parameter models, which neglects the distributed nature of the system. A spillover instability could potentially occur as a result of applying, to a long drill string, a feedback law designed using traditional ordinary differential equation (ODE) control strategies [68, 146] for an approximated mass-spring-damper system [156]. According to [24, 25, 26, 120, 153], the stick-slip instability between the bit and rock is characterized by a linear anti-damped term with a highly uncertain friction parameter (depending on the nature of the rock or soil that the drill is passing through) on the the wave PDE boundary opposite to the control input.

The external disturbance at the bit, which results from a wave-induced heaving motion of the drill rig [1], is usually described by a harmonic form with known frequencies and unknown amplitudes [128], because the amplitudes of the disturbance caused by the heaving motion may be affected by wind or ocean currents and are difficult to define in advance, while the dominant frequency components of the heaving motion in a specific sea area are usually accessible. In this chapter, the disturbance at the bit is modeled as the harmonic function with known frequencies and unknown amplitudes.

The control problem in this chapter is to design an adaptive output-feedback controller for the wave PDE with both high uncertainty and instability at the anti-collocated boundary. We begin this chapter by introducing the torsional vibration dynamics of oil drilling with stick-slip instability and a disturbance at the bit in section 8.1. The adaptive update laws for the unknown coefficients are presented in section 8.2, followed by the design of the output-feedback controller to adaptively cancel the disturbance and eliminate the destabilizing terms at the anti-collocated boundary in section 8.3. The asymptotic convergence to zero of the uncontrolled boundary states—that is, the oscillations of the angular displacement and velocity at the bit—and the boundedness of all states in the closed-loop system, are proved via Lyapunov analysis in section 8.4. Simulation tests in offshore rotary oil drilling are provided in section 8.5.

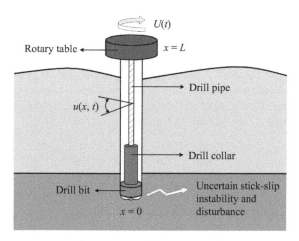

Figure 8.1. A drill string used in offshore oil drilling. The function $u(x,t)$ denotes the distributed elastic angular displacement of the drill string. The drill bit is subject to the uncertain stick-slip instability and disturbance, which are anti-collocated with the torque $U(t)$ at the rotary table.

8.1 DESCRIPTION OF OIL-DRILLING MODELS

A Wave PDE Model

The offshore oil-drilling system rotates around its vertical axis, penetrating through the rock on the seafloor (see figure 8.1). It consists of the assembly of a drill pipe, a drill collar, and a rock-cutting tool referred to as a drill bit. At the top of the drill string, the rotary table provides the necessary torque to push the system into a rotary motion [153]. The torsional vibration dynamic model of the rotary offshore oil drill string system [24] is given by

$$u_{tt}(x,t) = qu_{xx}(x,t), \tag{8.1}$$

$$u_x(L,t) = U(t), \tag{8.2}$$

$$I_b u_{tt}(0,t) = cu_t(0,t) - ku_x(0,t) - d(t), \tag{8.3}$$

where the state $u(x,t)$ denotes the distributed elastic angular displacement of the drill pipe, $x \in [0,L]$, with L the length of the drill pipe and $t \in [0,\infty)$ representing the time. The coefficients

$$q = GJ/I_d, \quad k = GJ, \tag{8.4}$$

with G, J, I_d representing the shear modulus of the drill pipe, drill pipe second moment of area, and drill pipe moment of inertia per unit of length, respectively, and I_b as the moment of inertia of the bottom-hole assembly (BHA). Input $U(t)$ is scalar, and $kU(t)$ represents the control torque. The function $cu_t(0,t)$ describes the stick-slip instability-introducing force between the bit and the rock. Moreover, $d(t)$ denotes the pressure oscillations on the drill bit caused by the wave-induced heaving motion of the drilling rig.

Two assumptions are made for the unknown coefficient c and the uncertain disturbance $d(t)$.

Assumption 8.1. *The anti-damping coefficient c in (8.3) is unknown but bounded by a known and arbitrary constant $\bar{c}$—that is, $c \in [0, \bar{c}]$.*

The coefficient c is treated as nonnegative without loss of generality. If, instead of anti-damping, damping were actually introduced by the rock cutting process, the control problem would be easier, the design developed here would still apply, and the resulting stability result would still hold.

Assumption 8.2. *The disturbance $d(t)$ is of the general harmonic form as*

$$d(t) = \sum_{j=1}^{N} [a_j \cos(\theta_j t) + b_j \sin(\theta_j t)], \tag{8.5}$$

where N is an arbitrary integer. The frequencies θ_j are known and arbitrary constants. The amplitudes a_j, b_j are unknown constants bounded by the known and arbitrary constants $\bar{a}_j$, $\bar{b}_j$—that is, $a_j \in [0, \bar{a}_j]$, $b_j \in [0, \bar{b}_j]$.

Equation (8.5) can model all periodic disturbance signals to an arbitrarily high degree of accuracy by choosing N sufficiently large. The frequency information requirement is reasonable in the wave-introduced disturbance considered in this chapter because the dominant frequency of the ocean waves is usually known for a particular area of the ocean [62].

The available boundary measurements are the torsional vibration acceleration $u_{tt}(0, t)$ measured by the acceleration sensor placed at the bit, and the torsional vibration velocity $u_t(L, t)$ obtained by the feedback signal from the actuator. We obtain the torsional vibration displacement and velocity at the bit $u(0, t), u_t(0, t)$ by twice integrating the measurement $u_{tt}(0, t)$ with known initial conditions $u(0, 0)$, $u_t(0, 0)$ because the installation of the acceleration sensor is more economical, reliable, and accurate. This paragraph only explains a signal acquisition method, imposing no restrictions on any initial conditions in the design and theory.

With the coefficient c of the anti-damping term unknown and the amplitudes a_j, b_j of the harmonic disturbance $d(t)$ unknown at the uncontrolled boundary $x = 0$, the control objective in this chapter is to design a control input $U(t)$ using the available boundary measurements to guarantee the asymptotic convergence to the origin of the torsional vibrational angular displacement $u(0, t)$ and velocity $u_t(0, t)$ of the drill bit at the downhole boundary. The uniform boundedness of all the states in the closed-loop system should be ensured as well.

A Wave PDE-ODE Model

We define

$$X(t) = [u(0, t), u_t(0, t)]^T, \tag{8.6}$$

and then the system (8.1)–(8.3) is written as a PDE-ODE coupled system

$$\dot{X}(t) = AX(t) + Bu_x(0, t) + \frac{1}{k}Bd(t), \tag{8.7}$$

$$u(0, t) = CX(t), \tag{8.8}$$

$$u_{tt}(x,t) = qu_{xx}(x,t), \tag{8.9}$$

$$u_x(L,t) = U(t), \tag{8.10}$$

where

$$A = \begin{bmatrix} 0 & 1 \\ 0 & \frac{c}{I_b} \end{bmatrix}, B = \frac{k}{I_b}\begin{bmatrix} 0 \\ -1 \end{bmatrix}, C = [1,0]. \tag{8.11}$$

The unknown anti-damping coefficient c is in the matrix A. The control objective is thus to ensure the asymptotic convergence of $X(t)$.

A 2×2 Transport PDEs-ODE Model

In order to reduce the plant (8.7)–(8.10) from second order to the conventional first order, we use the following Riemann coordinates:

$$z(x,t) = u_t(x,t) - \sqrt{q}u_x(x,t), \tag{8.12}$$

$$w(x,t) = u_t(x,t) + \sqrt{q}u_x(x,t) \tag{8.13}$$

to reversibly rewrite the system (8.7)–(8.10) as

$$\dot{X}(t) = \left(A - \frac{BC_1}{\sqrt{q}}\right)X(t) + \frac{B}{\sqrt{q}}w(0,t) + \frac{1}{k}Bd(t), \tag{8.14}$$

$$z(0,t) = 2C_1X(t) - w(0,t), \tag{8.15}$$

$$z_t(x,t) = -\sqrt{q}z_x(x,t), \tag{8.16}$$

$$w_t(x,t) = \sqrt{q}w_x(x,t), \tag{8.17}$$

$$w(L,t) = 2\sqrt{q}U(t) + z(L,t). \tag{8.18}$$

The constant q is positive according to the definition in (8.4), and the system matrix in ODE (8.14) is rewritten as

$$A - \frac{1}{\sqrt{q}}BC_1 = A_E + A_c - \frac{1}{\sqrt{q}}BC_1, \tag{8.19}$$

where

$$C_1 = [0,1] \tag{8.20}$$

by defining

$$A_E = \begin{bmatrix} 0 & 1 \\ 0 & 0 \end{bmatrix}, \tag{8.21}$$

$$A_c = A - A_E = \begin{bmatrix} 0 & 0 \\ 0 & \frac{c}{I_b} \end{bmatrix}, \tag{8.22}$$

which will be used in the following control design.

In what follows we design the adaptive controller $U(t)$ based on the plant (8.14)–(8.19) with the condition that $A_E - \frac{1}{\sqrt{q}}BC_1$ is controllable. This condition is satisfied in the oil-drilling model according to (8.11), (8.20), and (8.21).

8.2 ADAPTIVE UPDATE LAWS FOR UNKNOWN COEFFICIENTS

The objective in this section is to build adaptive update laws for the unknown coefficients c and a_j, b_j, respectively, where normalization and projection operators are used to guarantee boundedness, as is typical in adaptive control designs.

We propose adaptive update laws for the unknown coefficient c in the matrix A in (8.11) and the unknown coefficients a_j, b_j in $d(t)$ in (8.5) as follows:

$$\dot{\hat{c}}(t) = \gamma_c \text{Proj}_{[0,\bar{c}]} \left(\tau(t), \hat{c}(t) \right), \tag{8.23}$$

$$\dot{\hat{a}}_j(t) = \gamma_{aj} \text{Proj}_{[0,\bar{a}_j]} \left(\tau_{1j}(t), \hat{a}_j(t) \right), \tag{8.24}$$

$$\dot{\hat{b}}_j(t) = \gamma_{bj} \text{Proj}_{[0,\bar{b}_j]} \left(\tau_{2j}(t), \hat{b}_j(t) \right), \tag{8.25}$$

where the positive update gains $\gamma_c, \gamma_{aj}, \gamma_{bj}$ are tuning parameters to be determined. For all $m \leq M$ and all r, p, $\text{Proj}_{[m,M]}$ is the standard projection operator [116] given by

$$\text{Proj}_{[m,M]}(r,p) = \begin{cases} 0, & \text{if } p = m \text{ and } r < 0, \\ 0, & \text{if } p = M \text{ and } r > 0, \\ r, & \text{else.} \end{cases}$$

The role of the projection operator is to keep the parameter estimates bounded. The bounds $\bar{c}$ and $\bar{a}_j, \bar{b}_j$ are defined in assumption 8.1 and assumption 8.2, respectively. The functions $\tau, \tau_{1j}, \tau_{2j}$ in (8.23)–(8.25) are defined as

$$\tau(t) = \frac{1}{1+\Omega(t)} \left(X(t)^T P A_m - \lambda_a \int_0^L e^x \beta(x,t) \right.$$
$$\times \left(\bar{\kappa} + \left[0, \frac{\sqrt{q}\hat{c}(t)}{k} \right] \right) e^{\frac{1}{\sqrt{q}}(A_E + \hat{c}(t)A_m - \frac{1}{\sqrt{q}}BC_1)x} A_m dx \right) X(t), \tag{8.26}$$

$$\tau_{1j}(t) = \frac{1}{k(1+\Omega(t))} \left(X(t)^T P B \cos(\theta_j t) - \lambda_a \int_0^L e^x \beta(x,t) \right.$$
$$\times \left(\bar{\kappa} + \left[0, \frac{\sqrt{q}\hat{c}(t)}{k} \right] \right) e^{\frac{1}{\sqrt{q}}(A_E + \hat{A}_c(t) - \frac{1}{\sqrt{q}}BC_1)x} B \cos(\theta_j t) dx \right), \tag{8.27}$$

$$\tau_{2j}(t) = \frac{1}{k(1+\Omega(t))} \left(X(t)^T P B \sin(\theta_j t) - \lambda_a \int_0^L e^x \beta(x,t) \right.$$
$$\times \left(\bar{\kappa} + \left[0, \frac{\sqrt{q}\hat{c}(t)}{k} \right] \right) e^{\frac{1}{\sqrt{q}}(A_E + \hat{A}_c(t) - \frac{1}{\sqrt{q}}BC_1)x} B \sin(\theta_j t) dx \right). \tag{8.28}$$

The choices of $\tau(t), \tau_{1j}(t), \tau_{2j}(t)$ will be clear from Lyapunov analysis, which will be shown in section 8.4. The definition of the parameters and states used in (8.26)–(8.28) are shown as follows—that is, (8.29)–(8.41):

$$A_m = \begin{bmatrix} 0 & 0 \\ 0 & \frac{1}{I_b} \end{bmatrix}, \tag{8.29}$$

$$\hat{A}_c(t) = \begin{bmatrix} 0 & 0 \\ 0 & \frac{\hat{c}(t)}{I_b} \end{bmatrix}, \tag{8.30}$$

$$\Omega(t) = \frac{1}{2}\lambda_a \int_0^L e^x \beta(x,t)^2 dx + \frac{1}{2}\lambda_b \int_0^L e^{-x}\alpha(x,t)^2 dx + \frac{1}{2}X(t)^T P X(t). \quad (8.31)$$

The normalization $\Omega(t)$ is introduced in the denominator in (8.26)–(8.28) to limit the rates of changes of the parameter estimates—that is, in $\dot{c}(t)$ and $\dot{\hat{a}}_j(t), \dot{\hat{b}}_j(t)$. The normalization constants $\lambda_a > 0, \lambda_b > 0$ are tuning parameters to be determined.

The matrix $P = P^T > 0$, where the superscript T means transposition, is the unique solution to the Lyapunov equation

$$P\bar{A} + \bar{A}^T P = -Q \quad (8.32)$$

for some $Q = Q^T > 0$, and the known matrix

$$\bar{A} = A_E - \frac{1}{\sqrt{q}}BC_1 + \frac{1}{\sqrt{q}}B\bar{\kappa} \quad (8.33)$$

is made Hurwitz by appropriately choosing the control parameters $\bar{\kappa} = [\bar{k}_1, \bar{k}_2]$ later.

Next, disturbance-shifted state variables $\alpha(x,t), \beta(x,t)$ are introduced as

$$\alpha(x,t) = z(x,t) - \frac{\sqrt{q}}{k}[\hat{a}_1(t), \hat{b}_1(t), \ldots, \hat{a}_N(t), \hat{b}_N(t)]e^{-\frac{A_z}{\sqrt{q}}x}Z(t), \quad (8.34)$$

$$\beta(x,t) = w(x,t) + \frac{\sqrt{q}}{k}[\hat{a}_1(t), \hat{b}_1(t), \ldots, \hat{a}_N(t), \hat{b}_N(t)]e^{\frac{A_z}{\sqrt{q}}x}Z(t)$$
$$- \left(\kappa + \left[0, \frac{\sqrt{q}\hat{c}(t)}{k}\right]\right) e^{\frac{1}{\sqrt{q}}(A_E+\hat{A}_c(t)-\frac{1}{\sqrt{q}}BC_1)x}X(t)$$
$$- \int_0^x \frac{1}{q}\left(\bar{\kappa} + \left[0, \frac{\sqrt{q}\hat{c}(t)}{k}\right]\right) e^{\frac{1}{\sqrt{q}}(A_E+\hat{A}_c(t)-\frac{1}{\sqrt{q}}BC_1)(x-y)}B$$
$$\times \left(w(y,t) + \frac{\sqrt{q}}{k}[\hat{a}_1(t), \hat{b}_1(t), \ldots, \hat{a}_N(t), \hat{b}_N(t)]e^{\frac{A_z}{\sqrt{q}}y}Z(t)\right)dy, \quad (8.35)$$

where

$$A_z = \operatorname{diag}\left[\begin{pmatrix} 0 & -\theta_1 \\ \theta_1 & 0 \end{pmatrix}, \ldots, \begin{pmatrix} 0 & -\theta_N \\ \theta_N & 0 \end{pmatrix}\right], \quad (8.36)$$

and

$$Z(t) = [\cos(\theta_1 t), \sin(\theta_1 t), \ldots, \cos(\theta_N t), \sin(\theta_N t)]^T. \quad (8.37)$$

The signal $Z(t)$ in (8.37) will be used in constructing an important transformation in section 8.3. The functions $\alpha(x,t)$, $\beta(x,t)$ in (8.34), (8.35) are related to transformations (8.42), (8.43) and (8.62), (8.63), which will be shown later.

The states $z(x,t)$, $w(x,t)$ in (8.34) and (8.35) are calculated by means of the available boundary states through the transport PDEs (8.14)–(8.18), as follows:

$$w(x,t) = w\left(L, t - \frac{1}{\sqrt{q}}(L-x)\right), \quad (8.38)$$

$$z(x,t) = 2C_1 X\left(t - \frac{1}{\sqrt{q}}x\right) - w\left(0, t - \frac{1}{\sqrt{q}}x\right)$$
$$= 2C_1 X\left(t - \frac{1}{\sqrt{q}}x\right) - w\left(L, t - \frac{L}{\sqrt{q}} - \frac{1}{\sqrt{q}}x\right). \quad (8.39)$$

Applying (8.6), (8.13), (8.20), we write the states $w(x,t), z(x,t)$ as

$$w(x,t) = u_t\left(L, t - \frac{1}{\sqrt{q}}(L-x)\right) + \sqrt{q}u_x\left(L, t - \frac{1}{\sqrt{q}}(L-x)\right), \tag{8.40}$$

$$z(x,t) = 2u_t\left(0, t - \frac{1}{\sqrt{q}}x\right) - u_t\left(L, t - \frac{L}{\sqrt{q}} - \frac{1}{\sqrt{q}}x\right)$$
$$- \sqrt{q}u_x\left(L, t - \frac{L}{\sqrt{q}} - \frac{1}{\sqrt{q}}x\right). \tag{8.41}$$

The functions $u_t(L, t - \frac{L}{\sqrt{q}} - \frac{1}{\sqrt{q}}x)$, $u_t(L, t - \frac{1}{\sqrt{q}}(L-x))$ are the measurement $u_t(L,t)$ at previous time moments, and recalling (8.2), $u_x(L, t - \frac{L}{\sqrt{q}} - \frac{1}{\sqrt{q}}x)$ and $u_x(L, t - \frac{1}{\sqrt{q}}(L-x))$ can be replaced by $U(t - \frac{L}{\sqrt{q}} - \frac{1}{\sqrt{q}}x)$ and $U(t - \frac{1}{\sqrt{q}}(L-x))$ which are control inputs at previous time moments, and are, therefore, known quantities. Hence, according to (8.34)–(8.37), (8.40), (8.41), and (8.6), we have that $\alpha(x,t), \beta(x,t)$, and $X(t)$ can be obtained by the available boundary measurements proposed in section 8.1. Therefore, the adaptive update laws can be expressed using the available boundary states proposed in section 8.1.

8.3   OUTPUT-FEEDBACK CONTROL DESIGN

In this section, we design an output-feedback controller to compensate for the uncertain stick-slip instability and cancel the external disturbance at the ODE, which is anti-collocated with the controller. The controller employs the adaptive laws presented in section 8.2, using the measurements mentioned in section 8.1. To develop our output-feedback design, we first propose a transformation to make the unmatched external disturbance collocated with the control input, and then the disturbance is easily canceled via control design.

The Transformation Making Unmatched Disturbances Collocated with Control

We now transform the system (8.14)–(8.18) into an intermediate system so that the control input at $x = L$ and the anti-collocated disturbance at the ODE become collocated.

We introduce the invertible transformations $(w, z) \to (v, s)$ as

$$v(x,t) = w(x,t) + \Gamma(x,t)Z(t), \tag{8.42}$$
$$s(x,t) = z(x,t) + \Gamma_1(x,t)Z(t), \tag{8.43}$$

where $\Gamma(x,t), \Gamma_1(x,t)$ are to be determined.

Through (8.42), (8.43), we convert the system (8.14)–(8.18) into the following system:

$$\dot{X}(t) = \left(A - \frac{BC_1}{\sqrt{q}}\right)X(t) + \frac{B}{\sqrt{q}}v(0,t) + \frac{1}{k}B\tilde{d}(t), \tag{8.44}$$

$$s(0,t) + v(0,t) = 2C_1X(t), \tag{8.45}$$

$$s_t(x,t) = -\sqrt{q}s_x(x,t) + \Gamma_{1t}(x,t)Z(t), \tag{8.46}$$

$$v_t(x,t) = \sqrt{q}v_x(x,t) + \Gamma_t(x,t)Z(t), \tag{8.47}$$

$$v(L,t) = 2\sqrt{q}U(t) + s(L,t) + (\Gamma(L,t) - \Gamma_1(L,t))Z(t), \tag{8.48}$$

where $\tilde{d}(t)$ is defined as

$$\tilde{d}(t) = \sum_{j=1}^{N} \left[\tilde{a}_j(t)\cos(\theta_j t) + \tilde{b}_j(t)\sin(\theta_j t) \right], \tag{8.49}$$

with $\tilde{a}_j(t), \tilde{b}_j(t)$, $j = 1, \ldots, N$ defined as

$$\tilde{a}_j(t) = a_j - \hat{a}_j(t), \tag{8.50}$$

$$\tilde{b}_j(t) = b_j - \hat{b}_j(t). \tag{8.51}$$

Recalling (8.36), (8.37), we immediately have

$$\dot{Z}(t) = A_z Z(t). \tag{8.52}$$

Taking the time and spatial derivatives of (8.42) and substituting the result into (8.47), we get

$$\begin{aligned}
&v_t(x,t) - \sqrt{q}v_x(x,t) - \Gamma_t(x,t)Z(t) \\
&= w_t(x,t) - \sqrt{q}w_x(x,t) + \Gamma_t(x,t)Z(t) + \Gamma(x,t)A_z Z(t) \\
&\quad - \sqrt{q}\Gamma_x(x,t)Z(t) - \Gamma_t(x,t)Z(t) \\
&= (\Gamma(x,t)A_z - \sqrt{q}\Gamma_x(x,t))Z(t) = 0.
\end{aligned} \tag{8.53}$$

For (8.53) to hold, we obtain the sufficient condition

$$\Gamma(x,t)A_z - \sqrt{q}\Gamma_x(x,t) = 0. \tag{8.54}$$

By mapping (8.14) and (8.44) through the transformation (8.42), we get the condition

$$\Gamma(0,t) = \frac{\sqrt{q}}{k}[\hat{a}_1(t), \hat{b}_1(t), \ldots, \hat{a}_N(t), \hat{b}_N(t)]. \tag{8.55}$$

According to (8.54), (8.55), we obtain the solution

$$\Gamma(x,t) = \frac{\sqrt{q}}{k}[\hat{a}_1(t), \hat{b}_1(t), \ldots, \hat{a}_N(t), \hat{b}_N(t)]e^{\frac{A_z}{\sqrt{q}}x}, \tag{8.56}$$

where $e^{\frac{A_z}{\sqrt{q}}x}$ is written as

$$e^{\frac{A_z}{\sqrt{q}}x} = \mathrm{diag}\left[\begin{pmatrix} \cos(\frac{\theta_1}{\sqrt{q}}x), & -\sin(\frac{\theta_1}{\sqrt{q}}x) \\ \sin(\frac{\theta_1}{\sqrt{q}}x), & \cos(\frac{\theta_1}{\sqrt{q}}x) \end{pmatrix}, \cdots, \right.$$

$$\left. \begin{pmatrix} \cos(\frac{\theta_N}{\sqrt{q}}x), & -\sin(\frac{\theta_N}{\sqrt{q}}x) \\ \sin(\frac{\theta_N}{\sqrt{q}}x), & \cos(\frac{\theta_N}{\sqrt{q}}x) \end{pmatrix} \right], \tag{8.57}$$

according to (8.36).

Similarly, mapping (8.15), (8.16) and (8.45), (8.46) through (8.42), (8.43), we obtain

$$\Gamma_1(x,t) = -\frac{\sqrt{q}}{k}[\hat{a}_1(t), \hat{b}_1(t), \dots, \hat{a}_N(t), \hat{b}_N(t)]e^{-\frac{A_s}{\sqrt{q}}x}. \tag{8.58}$$

The functions $|\Gamma(x,t)|$ and $|\Gamma_1(x,t)|$ are bounded as

$$\max_{x\in[0,L],t\in[0,\infty)} \{|\Gamma(x,t)|, |\Gamma_1(x,t)|\} \leq \frac{\sqrt{2Nq}}{k} \max_{j=1,\dots,N}\{\bar{a}_j, \bar{b}_j\}, \tag{8.59}$$

where $|\cdot|$ is a Euclidean norm, recalling (8.57) being a rotation matrix and the adaptive estimates $\hat{a}_j(t), \hat{b}_j(t)$ bounded by $[0, \bar{a}_j]$ and $[0, \bar{b}_j]$ via the projection operator in section 8.2.

Thus, through the transformation (8.42), (8.43) with (8.56)–(8.58), we complete the conversion from the plant (8.14)–(8.18) to the intermediate system (8.44)–(8.48) where the control input and the unmatched disturbance information are collocated.

Backstepping Control Design

In addition to making the cancellation of the disturbance term $Z(t)$ at (8.48) straightforward, the objective in this section is to compensate for the uncertain anti-damped term $cu_t(0,t)$ included in (8.44) by designing the control input $U(t)$ at (8.48).

The PDE backstepping method [114] is used to design the controller. Recalling (8.44) and (8.19), through a backstepping transformation, we assign the poles of $A_E - \frac{1}{\sqrt{q}}BC_1$ to form a Hurwitz matrix and compensate for the anti-damping term $A_cX(t)$ by $\hat{A}_c(t)X(t)$. Recalling (8.22) and (8.30), we obtain

$$\tilde{A}_c(t) = A_c - \hat{A}_c(t) = \begin{bmatrix} 0 & 0 \\ 0 & \frac{\tilde{c}(t)}{I_b} \end{bmatrix}, \tag{8.60}$$

where $\tilde{c}(t)$ is the estimation error of c, given by

$$\tilde{c}(t) = c - \hat{c}(t). \tag{8.61}$$

The backstepping transformation is defined as

$$\alpha(x,t) = s(x,t), \tag{8.62}$$

$$\beta(x,t) = v(x,t) - \int_0^x \phi(x,y,t)v(y,t)dy - \gamma(x,t)X(t), \tag{8.63}$$

where the time-varying kernel functions $\phi(x,y,t), \gamma(x,t)$ are to be determined later.

From the spatially causal structure, the inverse of the transformation (8.62), (8.63) is written as

$$s(x,t) = \alpha(x,l), \tag{8.64}$$

$$v(x,t) = \beta(x,t) - \int_0^x \psi(x,y,t)\beta(y,t)dy - \chi(x,t)X(t), \tag{8.65}$$

where $\psi(x,y,t)$ and $\chi(x,t)$ are kernel functions to be determined.

Remark 8.1. The fact that the kernel functions are time-varying in the backstepping transformation (8.63) is due to the adaptive estimate $\hat{c}(t)$ included. Because $\hat{c}(t)$ and $\dot{\hat{c}}(t)$ are bounded according to the designed update laws, $\hat{c}(t)$ is continuously differentiable.

Through the backstepping transformation (8.62), (8.63) and (8.64), (8.65), we convert the intermediate system (8.44)–(8.48) into the following target system:

$$\dot{X}(t) = \bar{A}X(t) + \tilde{A}_c(t)X(t) + B\frac{1}{\sqrt{q}}\beta(0,t) + B\frac{1}{k}\tilde{d}(t), \qquad (8.66)$$

$$\beta_t(x,t) = \sqrt{q}\beta_x(x,t) + \Gamma_t(x,t)Z(t) - \int_0^x \phi(x,y,t)\Gamma_t(y,t)dyZ(t)$$

$$- (\gamma_t(x,t) + \gamma(x,t)\tilde{A}_c)X(t) - \int_0^x \phi_t(x,y,t)\beta(y,t)dy$$

$$+ \int_0^x \phi_t(x,y,t)\int_0^y \psi(y,\omega,t)\beta(\omega,t)d\omega dy$$

$$+ \int_0^x \phi_t(x,y,t)\chi(y,t)X(t)dy - \gamma(x,t)\frac{B}{k}\tilde{d}(t), \qquad (8.67)$$

$$\alpha_t(x,t) = -\sqrt{q}\alpha_x(x,t) + \Gamma_{1t}(x,t)Z(t), \qquad (8.68)$$

$$\beta(0,t) = -\alpha(0,t) + (\gamma(0,t) + 2C_1)X(t), \qquad (8.69)$$

$$\beta(L,t) = 0, \qquad (8.70)$$

where

$$\bar{A} = A_F - \frac{1}{\sqrt{q}}BC_1 + \frac{1}{\sqrt{q}}B\bar{\kappa} = \frac{k}{\sqrt{q}I_b}\begin{bmatrix} 0 & \frac{\sqrt{q}I_b}{k} \\ -\bar{\kappa}_1 & (1-\bar{\kappa}_2) \end{bmatrix} \qquad (8.71)$$

is Hurwitz by choosing the control parameters $\bar{\kappa}_1, \bar{\kappa}_2$ to satisfy

$$\bar{\kappa}_2 > 1, \quad 0 < \bar{\kappa}_1 < \frac{k(1-\bar{\kappa}_2)^2}{4\sqrt{q}I_b}. \qquad (8.72)$$

The kernel functions are determined as follows. Taking the time and spatial derivatives of (8.63) gives

$$\beta_t(x,t) = v_t(x,t) - \int_0^x \phi(x,y,t)v_t(y,t)dy$$

$$- \int_0^x \phi_t(x,y,t)v(y,t)dy - \gamma(x,t)\dot{X}(t) - \gamma_t(x,t)X(t), \qquad (8.73)$$

$$\beta_x(x,t) = v_x(x,t) - \int_0^x \phi_x(x,y,t)v(y,t)dy$$

$$- \phi(x,x,t)v(x,t) - \gamma_x(x,t)X(t). \qquad (8.74)$$

According to (8.44)–(8.48), the following equality holds

$$\beta_t(x,t) - \sqrt{q}\beta_x(x,t) - \Gamma_t(x,t)Z(t)$$

$$+ \int_0^x \phi(x,y,t)\Gamma_t(y,t)dyZ(t) + (\gamma_t(x,t) + \gamma(x,t)\tilde{A}_c)X(t)$$

$$+ \int_0^x \phi_t(x,y,t)v(y,t)dy + \gamma(x,t)\frac{B}{k}\tilde{d}(t)$$

$$= -\sqrt{q} \int_0^x \phi(x,y,t)v_x(y,t)dy + \sqrt{q} \int_0^x \phi_x(x,y,t)v(y,t)dy$$

$$- \gamma(x,t)\dot{X}(t) + \sqrt{q}\gamma_x(x,t)X(t)$$

$$+ \sqrt{q}\phi(x,x,t)v(x,t) + \gamma(x,t)\tilde{A}_c X(t) + \gamma(x,t)\frac{B}{k}\tilde{d}(t)$$

$$= \left(\sqrt{q}\phi(x,0,t) - \frac{1}{\sqrt{q}}\gamma(x,t)B \right)v(0,t)$$

$$+ \int_0^x \left(\sqrt{q}\phi_x(x,y,t) + \sqrt{q}\phi_y(x,y,t) \right)v(y,t)dy$$

$$+ \left(\sqrt{q}\gamma_x(x,t) - \gamma(x,t)\left(A_E + \hat{A}_c - \frac{1}{\sqrt{q}}BC_1 \right) \right)X(t). \tag{8.75}$$

For (8.75) to be zero, which ensures that (8.67) holds by applying (8.65), together with mapping (8.66) with (8.44) via (8.63), we obtain the following kernel conditions:

$$q\phi(x,0,t) = \gamma(x,t)B, \tag{8.76}$$

$$\phi_x(x,y,t) + \phi_y(x,y,t) = 0, \tag{8.77}$$

$$\sqrt{q}\gamma_x(x,t) - \gamma(x,t)\left(A_E + \hat{A}_c(t) - \frac{1}{\sqrt{q}}BC_1 \right) = 0, \tag{8.78}$$

$$B\frac{1}{\sqrt{q}}\gamma(0,t) = \frac{1}{\sqrt{q}}B\bar{\kappa} - \hat{A}_c(t). \tag{8.79}$$

We thus obtain the unique kernel solutions as

$$\gamma(x,t) = \left(\bar{\kappa} + \left[0, \frac{\sqrt{q}\hat{c}(t)}{k} \right] \right) e^{\frac{1}{\sqrt{q}}(A_E + \hat{A}_c(t) - \frac{1}{\sqrt{q}}BC_1)x}, \tag{8.80}$$

$$\phi(x,y,t) = \frac{1}{q}\left(\bar{\kappa} + \left[0, \frac{\sqrt{q}\hat{c}(t)}{k} \right] \right) e^{\frac{1}{\sqrt{q}}(A_E + \hat{A}_c(t) - \frac{1}{\sqrt{q}}BC_1)(x-y)}B, \tag{8.81}$$

where $\hat{A}_c(t) = -\frac{1}{k}B[0,\hat{c}(t)]$ has been used. The adaptive estimate $\hat{c}(t)$, which is continuous and included in the kernel functions (8.80), (8.81), ensures that $\gamma(x,t)$ and $\phi(x,y,t)$ are continuous.

The kernels $\psi(x,y,t), \chi(x,t)$ in the inverse transformation (8.65) also exist and are continuous, which is shown as follows. Rewrite (8.63) as

$$v(x,t) - \int_0^x \phi(x,y,t)v(y,t)dy = \beta(x,t) + \gamma(x,t)X(t).$$

Because $\phi(x,y,t)$ is continuous, recalling remark 8.1, we obtain a unique continuous $\eta(x,y,t)$ existing on $\{(x,y)|0 \le y \le x \le L\}$ such that

$$v(x,t) = \beta(x,t) + \gamma(x,t)X(t) + \int_0^x \eta(x,y,t)\left(\beta(y,t) + \gamma(y,t)X(t) \right)dy$$

$$= \beta(x,t) + \int_0^x \eta(x,y,t)\beta(y,t)dy$$

$$+ \left(\int_0^x \eta(x,y,t)\gamma(y,t)dy + \gamma(x,t) \right)X(t),$$

according to [169]. The kernels in the inverse transformation (8.64), (8.65) are

$$\psi(x,y,t) = -\eta(x,y,t),$$

$$\chi(x,t) = -\int_0^x \eta(x,y,t)\gamma(y,t)dy - \gamma(x,t),$$

which are also continuous, by recalling the continuity of $\gamma(x,t)$ and $\eta(x,y,t)$.

The two transformations (8.42), (8.43) and (8.62), (8.63) with the kernels (8.56)–(8.58) and (8.80), (8.81) are used to obtain (8.34) and (8.35). According to the boundary condition (8.70), (8.48) and the transformation (8.63), we derive the controller as

$$U(t) = \frac{1}{2\sqrt{q}}\bigg(-s(L,t) - (\Gamma(L,t) - \Gamma_1(L,t))Z(t)$$

$$+ \int_0^L \phi(L,y,t)v(y,t)dy + \gamma(L,t)X(t) \bigg). \tag{8.82}$$

Using (8.42), (8.43) and (8.12), (8.13), the control law (8.82) is rewritten as

$$U(t) = \frac{1}{2\sqrt{q}}\bigg(-u_t(L,t) + \sqrt{q}u_x(L,t) - \Gamma(L,t)Z(t)$$

$$+ \int_0^L \phi(L,y,t)\Gamma(y,t)dyZ(t) + \gamma(L,t)X(t)$$

$$+ \int_0^L \phi(L,y,t)\bigg[u_t\bigg(L,t-\frac{1}{\sqrt{q}}(L-y)\bigg)$$

$$+ \sqrt{q}u_x\bigg(L,t-\frac{1}{\sqrt{q}}(L-y)\bigg)\bigg]dy\bigg), \tag{8.83}$$

where

$$v(y,t) = w\bigg(L,t-\frac{1}{\sqrt{q}}(L-y)\bigg) + \Gamma(y,t)Z(t) \tag{8.84}$$

has been used, which is obtained by (8.40), (8.42).

Substituting (8.56)–(8.58), (8.80), (8.81), and (8.6), the controller (8.83) is written as

$$U(t) = \frac{1}{2\sqrt{q}}\bigg[-u_t(L,t) + \sqrt{q}u_x(L,t)$$

$$- \frac{\sqrt{q}}{k}[\hat{a}_1(t),\hat{b}_1(t),\ldots,\hat{a}_N(t),\hat{b}_N(t)]e^{\frac{A_z}{\sqrt{q}}L}Z(t)$$

$$+ \frac{1}{q}\bigg(\bar{\kappa} + \bigg[0,\frac{\sqrt{q}\hat{c}(t)}{k}\bigg]\bigg)\int_0^L e^{\frac{1}{\sqrt{q}}(A_E+\hat{c}(t)A_m-\frac{1}{\sqrt{q}}BC_1)(L-y)}B$$

$$\times \frac{\sqrt{q}}{k}[\hat{a}_1(t),\hat{b}_1(t),\ldots,\hat{a}_N(t),\hat{b}_N(t)]e^{\frac{A_z}{\sqrt{q}}y}dyZ(t)$$

$$+ \left(\bar{\kappa} + \left[0, \frac{\sqrt{q}\hat{c}(t)}{k} \right] \right) e^{\frac{1}{\sqrt{q}}(A_E + \hat{c}(t)A_m - \frac{1}{\sqrt{q}}BC_1)L} [u(0,t), u_t(0,t)]^T$$

$$+ \frac{1}{q} \left(\bar{\kappa} + \left[0, \frac{\sqrt{q}\hat{c}(t)}{k} \right] \right)$$

$$\times \int_{t-\frac{L}{\sqrt{q}}}^{t} e^{(A_E + \hat{c}(t)A_m - \frac{1}{\sqrt{q}}BC_1)(t-\delta)} B \times [u_t(L,\delta) + \sqrt{q}u_x(L,\delta)]d\delta \Bigg], \quad (8.85)$$

where $\hat{A}_c(t) = \hat{c}(t)A_m$, implied from (8.29) and (8.30), is used. The function $Z(t)$ is defined as (8.37), and $\hat{c}(t), \hat{a}_j(t), \hat{b}_j(t), j = 1, \ldots, N$ are calculated from the adaptive update laws (8.23) and (8.26), (8.24) and (8.27), (8.25) and (8.28) proposed in section 8.2. The signals $u(0,t), u_t(0,t), u_t(L,t)$ are obtained from the measurements discussed in section 8.1. The function $u_x(L,t)$ can be replaced as $U(t)$ by applying (8.2), and the final form of the control input (8.85) is shown in the following paragraph.

We choose the controller to be activated at $t = \frac{2L}{\sqrt{q}}$ to make sure the second argument of $w(L, t - \frac{L}{\sqrt{q}} - \frac{x}{\sqrt{q}})$ in (8.39) is positive. In other words, the system runs as open loop and $U(t) = 0$ when $t \in [0, \frac{2L}{\sqrt{q}})$. However, if we have a history of the states and the input prior to $t = 0$, we can start the feedback law at $t = 0$. Substituting (8.2) into (8.85) to replace $u_x(L,t)$ as $U(t)$, the control input is rewritten as

$$U(t) = \frac{1}{\sqrt{q}} \Bigg[-u_t(L,t) - \frac{\sqrt{q}}{k} [\hat{a}_1(t), \hat{b}_1(t), \ldots, \hat{a}_N(t), \hat{b}_N(t)] e^{\frac{A_z}{\sqrt{q}}L} Z(t)$$

$$+ \frac{1}{q} \left(\bar{\kappa} + \left[0, \frac{\sqrt{q}\hat{c}(t)}{k} \right] \right) \int_0^L e^{\frac{1}{\sqrt{q}}(A_E + \hat{c}(t)A_m - \frac{1}{\sqrt{q}}BC_1)(L-y)} B$$

$$\times \frac{\sqrt{q}}{k} [\hat{a}_1(t), \hat{b}_1(t), \ldots, \hat{a}_N(t), \hat{b}_N(t)] e^{\frac{A_z}{\sqrt{q}}y} dy Z(t)$$

$$+ \left(\bar{\kappa} + \left[0, \frac{\sqrt{q}\hat{c}(t)}{k} \right] \right) e^{\frac{1}{\sqrt{q}}(A_E + \hat{c}(t)A_m - \frac{1}{\sqrt{q}}BC_1)L} [u(0,t), u_t(0,t)]^T$$

$$+ \frac{1}{q} \left(\bar{\kappa} + \left[0, \frac{\sqrt{q}\hat{c}(t)}{k} \right] \right) \int_{t-\frac{L}{\sqrt{q}}}^{t} e^{(A_E + \hat{c}(t)A_m - \frac{1}{\sqrt{q}}BC_1)(t-\delta)} B$$

$$\times [u_t(L,\delta) + \sqrt{q}U(\delta)]d\delta \Bigg], \quad t \geq \frac{2L}{\sqrt{q}}, \quad (8.86)$$

where the value $U(\delta)$ of the control input in the previous time interval $\delta \in [t - \frac{L}{\sqrt{q}}, t)$ is used. In addition, $U(\bar{\delta}_1)$ with $\bar{\delta}_1 \in [t - \frac{2L}{\sqrt{q}}, t)$ are also used in (8.31), (8.34), (8.35) and (8.40), (8.41) in implementing the update laws (8.23)–(8.28) for the estimates $\hat{c}(t), \hat{a}_j(t), \hat{b}_j(t)$. Because $U(t)$ in the previous time interval $t \in [0, \frac{2L}{\sqrt{q}})$ is zero, $U(\delta)$, $U(\bar{\delta}_1)$ can be regarded as known quantities in calculating $U(t)$ in (8.86) at $t = \frac{2L}{\sqrt{q}}$. By that analogy, $U(t)$ at each time point in $t \geq \frac{2L}{\sqrt{q}}$ can be calculated by the measurements $u(0,t), u_t(0,t), u_t(L,t)$ and $U(\delta), U(\bar{\delta}_1)$ in the previous steps. Therefore, the proposed output-feedback controller (8.86) is entirely computable with the available

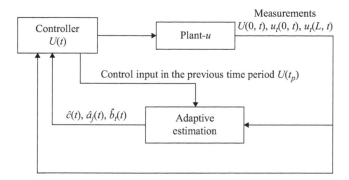

Figure 8.2. Block diagram of the closed-loop system.

boundary measurements $u(0,t), u_t(0,t), u_t(L,t)$. The block diagram of the closed-loop system is shown in figure 8.2.

8.4   STABILITY ANALYSIS

Before proving the main result of this chapter, we state the following lemma.

Lemma 8.1. *For all initial values $s(\cdot,0), v(\cdot,0)$ in $L^2(0,L)$, the system (8.44)–(8.48) under the control (8.82) is uniformly bounded and asymptotically stable in the sense of the norm*

$$\left(\|s(\cdot,t)\|^2 + \|v(\cdot,t)\|^2 + |X(t)|^2 \right)^{1/2}. \tag{8.87}$$

Proof. Define

$$\Theta(t) = \|\beta(\cdot,t)\|^2 + \|\alpha(\cdot,t)\|^2 + |X(t)|^2, \tag{8.88}$$

where $\|\cdot\|$ denotes the L_2 norm. Recalling (8.31), we get

$$\mu_1 \Theta(t) \leq \Omega(t) \leq \mu_2 \Theta(t), \tag{8.89}$$

with positive μ_1, μ_2 as

$$\mu_1 = \frac{1}{2} \min\{\lambda_a, \lambda_b e^{-L}, \lambda_{\min}(P)\}, \tag{8.90}$$

$$\mu_2 = \frac{1}{2} \min\{\lambda_a e^{L}, \lambda_b, \lambda_{\max}(P)\}, \tag{8.91}$$

where $\lambda_{\min}$ and $\lambda_{\max}$ denote the minimum and maximum eigenvalues of the corresponding matrix.

Choose a Lyapunov function as

$$V(t) = \ln\left(1 + \Omega(t)\right) + \sum_{j=1}^{N} \frac{1}{2\gamma_{aj}} \tilde{a}_j(t)^2 + \sum_{j=1}^{N} \frac{1}{2\gamma_{bj}} \tilde{b}_j(t)^2 + \frac{1}{2\gamma_c} \tilde{c}(t)^2, \tag{8.92}$$

where $\tilde{a}_j(t), \tilde{b}_j(t), \tilde{c}(t)$ are defined in (8.50), (8.51), and (8.61), respectively.

By virtue of (8.89) and the fact that the function $\ln(\cdot)$ is monotonically increasing, the following inequality holds

$$0 \leq \ln\left(1 + \mu_1 \Theta(t)\right) + \bar{\mu}_1 \left(\sum_{j=1}^{N} \left(\tilde{a}_j(t)^2 + \tilde{b}_j(t)^2\right) + \tilde{c}(t)^2\right)$$

$$\leq V(t) \leq \ln\left(1 + \mu_2 \Theta(t)\right) + \bar{\mu}_2 \left(\sum_{j=1}^{N} \left(\tilde{a}_j(t)^2 + \tilde{b}_j(t)^2\right) + \tilde{c}(t)^2\right), \qquad (8.93)$$

where

$$\bar{\mu}_1 = \frac{1}{2 \max_{j \in \{1 \cdots N\}} \{\gamma_{aj}, \gamma_{bj}, \gamma_c\}} > 0,$$

$$\bar{\mu}_2 = \frac{1}{2 \min_{j \in \{1 \cdots N\}} \{\gamma_{aj}, \gamma_{bj}, \gamma_c\}} > 0. \qquad (8.94)$$

Taking the derivative of (8.92) and inserting (8.66), we obtain

$$\dot{V}(t) = \frac{1}{1 + \Omega(t)} \left(-X(t)^T Q X(t) + \frac{1}{\sqrt{q}} X^T P B \beta(0, t) \right.$$

$$+ X(t)^T P \frac{B}{k} \sum_{j=1}^{N} [\tilde{a}_j(t) \cos(\theta_j t) + \tilde{b}_j(t) \sin(\theta_j t)]$$

$$+ X(t)^T P \tilde{c}(t) A_{\mathrm{m}} X(t) + \lambda_a \int_0^L e^x \beta(x, t) \beta_t(x, t) dx$$

$$\left. + \lambda_b \int_0^L e^{-x} \alpha(x, t) \alpha_t(x, t) dx \right) - \sum_{j=1}^{N} \frac{1}{\gamma_{aj}} \dot{\tilde{a}}_j(t) \tilde{a}_j(t)$$

$$- \sum_{j=1}^{N} \frac{1}{\gamma_{bj}} \dot{\tilde{b}}_j(t) \tilde{b}_j(t) - \frac{1}{\gamma_c} \dot{\tilde{c}}(t) \tilde{c}(t), \qquad (8.95)$$

where $\tilde{A}(t) = \tilde{c}(t) A_{\mathrm{m}}$ implied by (8.29) and (8.60) is used.

Inserting (8.67)–(8.70) into (8.95) and applying the Cauchy-Schwarz inequality, we obtain

$$\dot{V}(t) \leq \frac{1}{1 + \Omega(t)} \left[-\left(\frac{3}{4} \lambda_{\min}(Q) - \sqrt{q} \lambda_b (\gamma(0, t) + 2C_1)^2\right) |X(t)|^2 \right.$$

$$- \frac{1}{2} \sqrt{q} \lambda_a \int_0^L e^x \beta(x, t)^2 dx - \lambda_a \int_0^L e^x \beta(x, t) \gamma_t(x, t) dx X(t)$$

$$- \lambda_a \int_0^L e^x \beta(x, t) \left(\int_0^x \phi_t(x, y, t) \beta(y, t) dy \right.$$

$$- \int_0^x \phi_t(x, y, t) \int_0^y \psi(y, \omega, t) \beta(\omega, t) d\omega dy$$

$$\left. - \int_0^x \phi_t(x, y, t) \chi(y, t) X(t) dy \right) dx - \frac{1}{2} \sqrt{q} \lambda_b e^{-L} \alpha(L, t)^2$$

$$- \left(\frac{1}{2}\sqrt{q}\lambda_a - \frac{|PB|^2}{q\lambda_{\min}(Q)} - \sqrt{q}\lambda_b \right) \beta(0,t)^2$$

$$- \frac{1}{2}\sqrt{q}\lambda_b \int_0^L e^{-x}\alpha(x,t)^2 dx + \lambda_b \int_0^L e^{-x}\alpha(x,t)\Gamma_{1t}(x,t)Z(t)dx$$

$$+ \lambda_a \int_0^L e^x \beta(x,t)\Gamma_t(x,t)Z(t)dx - \lambda_a \int_0^L e^x \beta(x,t) \int_0^x \phi(x,y,t)\Gamma_t(y,t)dy Z(t)dx \Bigg]$$

$$+ \sum_{j=1}^N \left(\frac{X(t)^T PB\cos(\theta_j t)}{k(1+\Omega(t))} - \frac{\lambda_a \int_0^L e^x \beta(x,t)\gamma(x,t)B\cos(\theta_j t)dx}{k(1+\Omega(t))} - \frac{1}{\gamma_a}\dot{\tilde{a}}_j(t) \right)\tilde{a}_j(t)$$

$$+ \sum_{j=1}^N \left(\frac{X(t)^T PB\sin(\theta_j t)}{k(1+\Omega(t))} - \frac{\lambda_a \int_0^L e^x \beta(x,t)\gamma(x,t)B\sin(\theta_j t)dx}{k(1+\Omega(t))} - \frac{1}{\gamma_b}\dot{\tilde{b}}_j(t) \right)\tilde{b}_j(t)$$

$$+ \left(\frac{X(t)^T PA_{\mathrm{m}}X(t)}{1+\Omega(t)} - \frac{\lambda_a \int_0^L e^x \beta(x,t)\gamma(x,t)A_m dx X(t)}{1+\Omega(t)} - \frac{1}{\gamma_c}\dot{\tilde{c}}(t) \right)\tilde{c}(t). \tag{8.96}$$

Step 1. Consider the third and fourth terms in the square bracket in (8.96).

Because $\hat{c}(t) \in [0, \bar{c}]$, one can see that $\gamma(x,t), \phi(x,y,t)$ are bounded according to (8.80), (8.81). Here we define

$$\bar{\gamma} = \max\{|\gamma(x,t)|; x \in [0,L], t \in [0,\infty)\}, \tag{8.97}$$

$$\bar{\phi} = \max\{|\phi(x,y,t)|; x \in [0,L], y \in [0,L], t \in [0,\infty)\}. \tag{8.98}$$

Similarly, the boundedness of $\chi(x,t), \psi(x,y,t)$ can also be obtained, defining

$$\bar{\chi} = \max\{|\chi(x,t)|; x \in [0,L], t \in [0,\infty)\}, \tag{8.99}$$

$$\bar{\psi} = \max\{|\psi(x,y,t)|; x \subset [0,L], y \in [0,L], t \in [0,\infty)\}. \tag{8.100}$$

Applying the Cauchy-Schwarz and Young's inequalities to the numerator of (8.26), we have that the absolute value of the numerator is less than or equal to $m_1(|X(t)|^2 + \|\beta(\cdot,t)\|^2)$ for some positive m_1. Recalling the form of $\Omega(t)$, given by (8.31), which appears in the denominator of (8.26), we also have

$$1 + \Omega(t) > \frac{1}{2}\min\{\lambda_a, \lambda_b e^{-L}, \lambda_{\min}(P)\}(|X(t)|^2 + \|\beta(\cdot,t)\|^2).$$

It follows that

$$\left| \dot{\hat{c}}(t) \right| \leq \frac{2\gamma_c m_1(|X(t)|^2 + \|\beta(\cdot,t)\|^2)}{\min\{\lambda_a, \lambda_b e^{-L}, \lambda_{\min}(P)\}(|X(t)|^2 + \|\beta(\cdot,t)\|^2)}$$

$$= \frac{2\gamma_c m_1}{\min\{\lambda_a, \lambda_b e^{-L}, \lambda_{\min}(P)\}} \tag{8.101}$$

via (8.23).

Recalling (8.80), (8.81), we obtain the bounds on $\gamma_t(x,t)$ and $\phi_t(x,t)$ as

$$|\gamma_t(x,t)| = \left| \left[0, \frac{\sqrt{q}\dot{\hat{c}}(t)}{k} \right] + \left(\bar{\kappa} + \left[0, \frac{\sqrt{q}\dot{\hat{c}}(t)}{k} \right] \right) \frac{\dot{\hat{c}}(t)}{\sqrt{q}} A_{\mathrm{m}} \right| m_e$$

$$\leq \frac{\gamma_c m_c m_e}{\min\{\lambda_a, \lambda_b e^{-L}, \lambda_{\min}(P)\}}, \tag{8.102}$$

$$|\phi_t(x,t)| \leq \frac{\gamma_c m_c m_e |B|}{q \min\{\lambda_a, \lambda_b e^{-L}, \lambda_{\min}(P)\}}, \tag{8.103}$$

where the constants m_c, m_e are

$$m_c = \frac{2\sqrt{q}m_1}{k} + \frac{2}{I_b\sqrt{q}}\sqrt{|\bar{\kappa}|^2 + \frac{q\bar{c}^2}{k^2}}m_1, \tag{8.104}$$

$$m_e = \max_{\hat{c}(t)\in[0,\bar{c}], x\in[0,L]} \left\{ \bar{\sigma}\left(e^{\frac{1}{\sqrt{q}}(A_E + \hat{c}(t)A_m - \frac{1}{\sqrt{q}}BC_1)x}\right) \right\}. \tag{8.105}$$

The symbol $\bar{\sigma}\left(e^{\frac{1}{\sqrt{q}}(A_E + \hat{c}(t)A_m - \frac{1}{\sqrt{q}}BC_1)x}\right)$ in (8.105) stands for the largest singular value of $e^{\frac{1}{\sqrt{q}}(A_E + \hat{c}(t)A_m - \frac{1}{\sqrt{q}}BC_1)x}$ at $c(t)$ and x.

Applying the Young and Cauchy-Schwarz inequalities into the third and fourth terms in the square bracket in (8.96), using (8.99)–(8.103), we get

$$\lambda_a \int_0^L e^x \beta(x,t)\gamma_t(x,t)dx X(t)$$

$$\leq \frac{\gamma_c \bar{M}_0}{\min\{\lambda_a, \lambda_b e^{-L}, \lambda_{\min}(P)\}}\left(\|\beta(\cdot,t)\|^2 + |X(t)|^2\right), \tag{8.106}$$

$$\lambda_a \int_0^L e^x \beta(x,t)\left(\int_0^x \phi_t(x,y,t)\beta(y,t)dy\right.$$

$$-\int_0^x \phi_t(x,y,t)\int_0^y \psi(y,\omega,t)\beta(\omega,t)d\omega dy$$

$$\left.-\int_0^x \phi_t(x,y,t)\chi(y,t)X(t)dy\right)dx$$

$$\leq \frac{\gamma_c \bar{M}_1}{\min\{\lambda_a, \lambda_b e^{-L}, \lambda_{\min}(P)\}}\left(\|\beta(\cdot,t)\|^2 + |X(t)|^2\right), \tag{8.107}$$

where

$$\bar{M}_0 = 2\lambda_a e^L m_c m_e \max\{L,1\},$$

$$\bar{M}_1 = \lambda_a e^L L\frac{1}{q}m_c m_e |B|\max\{1 + L\bar{\psi} + 2\bar{\chi}, 2\bar{\chi}L\}.$$

Step 2. Consider the eighth to tenth terms in the square bracket in (8.96).

By virtue of the numerator of (8.27), (8.28), together with $1 + \Omega(t) \geq 1$ in the denominator, we straightforwardly show that

$$\max_{j=1,\ldots,N}\{|\tau_{1j}(t)|, |\tau_{2j}(t)|\} \leq m_2\left(|X(t)| + \sqrt{L}\|\beta(\cdot,t)\|\right) \tag{8.108}$$

for all t, where

$$m_2 = \frac{2}{k}\max\{|PB|, e^L \lambda_a \sqrt{L}\bar{\gamma}|B|\} > 0. \tag{8.109}$$

Therefore, we obtain

$$\max_{j=1,\ldots,N}\left\{\left|\dot{\hat{a}}_j(t)\right|, \left|\dot{\hat{b}}_j(t)\right|\right\} \leq m_2 \max_{j=1,\ldots,N}\{\gamma_{aj}, \gamma_{bj}\}\left(|X(t)| + \sqrt{L}\|\beta(\cdot,t)\|\right), \tag{8.110}$$

establishing the boundedness of (8.24), (8.25).

According to (8.56)–(8.58), we obtain

$$\max_{x\in[0,L],t\in[0,\infty)} \{|\Gamma_t(x,t)|, |\Gamma_{1t}(x,t)|\}$$

$$\leq \frac{\sqrt{q}}{k}\left|\dot{\hat{a}}_1(t), \dot{\hat{b}}_1(t), \ldots, \dot{\hat{a}}_N(t), \dot{\hat{b}}_N(t)\right|$$

$$\leq \frac{\sqrt{2qN}}{k}m_2 \max_{j=1,\ldots,N}\{\gamma_{aj}, \gamma_{bj}\}\left(|X(t)| + \sqrt{L}\|\beta(\cdot,t)\|\right). \tag{8.111}$$

Recalling (8.111) and applying Young's inequality to the eighth, ninth, and tenth terms in the square bracket in (8.96), we obtain

$$\lambda_b \int_0^L e^{-x}\alpha(x,t)\Gamma_{1t}(x,t)Z(t)dx$$

$$\leq \max_{j=1,\ldots,N}\{\gamma_{aj}, \gamma_{bj}\}\bar{M}_2\left(\|\alpha(\cdot,t)\|^2 + \|\beta(\cdot,t)\|^2 + |X(t)|^2\right), \tag{8.112}$$

$$\lambda_a \int_0^L e^{x}\beta(x,t)\Gamma_t(x,t)Z(t)dx$$

$$-\lambda_u \int_0^L e^{x}\beta(x,t)\int_0^x \phi(x,y,t)\Gamma_t(y,t)dy\,Z(t)dx$$

$$\leq \max_{j=1,\ldots,N}\{\gamma_{aj}, \gamma_{bj}\}\bar{M}_3\left(\|\beta(\cdot,t)\|^2 + |X(t)|^2\right), \tag{8.113}$$

for some positive constants $\bar{M}_2, \bar{M}_3$, where $|Z(t)| = \sqrt{N}$ and (8.98) are used.

Step 3. Substituting (8.106), (8.107), (8.112), (8.113), (8.23)–(8.25), and (8.80) into (8.96) and again applying Young's inequality, we obtain

$$\dot{V}(t) \leq \frac{1}{1+\Omega(t)}\left(-h_1|X(t)|^2 - h_2\|\beta(\cdot,t)\|^2\right.$$

$$\left.-h_3\|\alpha(\cdot,t)\|^2 - h_4\alpha(L,t)^2 - h_5\beta(0,t)^2\right),$$

where

$$h_1 = \frac{3}{4}\lambda_{\min}(Q) - \sqrt{q}\lambda_b(\gamma(0,t) + 2C_1)^2$$

$$-\frac{\gamma_c\bar{M}_0 + \gamma_c\bar{M}_1}{\min\{\lambda_a, \lambda_b c^{-L}, \lambda_{\min}(P)\}} - \max_{j=1,\ldots,N}\{\gamma_{aj}, \gamma_{bj}\}(\bar{M}_2 + \bar{M}_3),$$

$$h_2 = \frac{1}{2}\sqrt{q}\lambda_a - \frac{\gamma_c\bar{M}_0 + \gamma_c\bar{M}_1}{\min\{\lambda_a, \lambda_b e^{-L}, \lambda_{\min}(P)\}}$$

$$-\max_{j=1,\ldots,N}\{\gamma_{aj}, \gamma_{bj}\}(\bar{M}_2 + \bar{M}_3),$$

$$h_3 = \frac{1}{2}\sqrt{q}\lambda_b e^{-L} - \max_{j=1,\ldots,N}\{\gamma_{aj}, \gamma_{bj}\}\bar{M}_2,$$

$$h_4 = \frac{1}{2}\sqrt{q}\lambda_b e^{-L} > 0,$$

$$h_5 = \frac{1}{2}\sqrt{q}\lambda_a - \frac{|PB|^2}{q\lambda_{\min}(Q)} - \sqrt{q}\lambda_b.$$

Choosing

$$\lambda_a > \frac{2|PB|^2}{q\sqrt{q}\lambda_{\min}(Q)} + 2\lambda_b$$

to guarantee $h_5 > 0$ and using sufficiently small positive constants $\lambda_b, \gamma_{aj}, \gamma_{bj}, \gamma_c$ to make $h_1 > 0$, $h_2 > 0$, and $h_3 > 0$, we get

$$\dot{V}(t) \leq \frac{-\xi}{1+\Omega(t)}\left(|X(t)|^2 + \|\beta(\cdot,t)\|^2 + \|\alpha(\cdot,t)\|^2 + \alpha(L,t)^2 + \beta(0,t)^2\right) \quad (8.114)$$

with a positive $\xi = \min\{h_1, h_2, h_3, h_4, h_5\}$, and hence

$$V(t) \leq V(0), \quad \forall t \geq 0. \quad (8.115)$$

Step 4. Recalling (8.93), one easily gets that $\tilde{c}(t), \tilde{a}_j(t), \tilde{b}_j(t), j = 1, \ldots, N$, and $\Theta(t)$ are uniformly bounded. Therefore, together with (8.88), we find that $\|\beta(\cdot,t)\|, \|\alpha(\cdot,t)\|, |X(t)|$ are uniformly bounded. It follows that $\|v(\cdot,t)\|, \|s(\cdot,t)\|$ are uniformly bounded via (8.64), (8.65). By recalling (8.49) and (8.60), we also know that $\tilde{d}(t)$ and $|\tilde{A}_c(t)|$ are uniformly bounded. According to (8.111), $\Gamma_t(x,t)$ and $\Gamma_{1t}(x,t)$ are bounded as well.

According to (8.66)–(8.70), we further get

$$\frac{d}{dt}|X(t)|^2 = 2X^T(t)\left(\bar{A}X(t) + \tilde{A}_c(t)X(t) + B\frac{1}{\sqrt{q}}\beta(0,t) + B\frac{1}{k}\tilde{d}(t)\right), \quad (8.116)$$

$$\frac{d}{dt}\|\beta(\cdot,t)\|^2 = -\sqrt{q}\beta(0,t)^2 + 2\int_0^L \beta(x,t)\left(\Gamma_t(x,t)Z(t)\right.$$

$$- \int_0^x \phi(x,y,t)\Gamma_t(y,t)dy Z(t) - (\gamma_t(x,t) + \gamma(x,t)\tilde{A}_c(t))X(t)$$

$$- \int_0^x \phi_t(x,y,t)\beta(y,t)dy + \int_0^x \phi_t(x,y,t)\int_0^y \psi(y,\omega,t)\beta(\omega,t)d\omega dy$$

$$\left. + \int_0^x \phi_t(x,y,t)\chi(y,t)X(t)dy - \gamma(x,t)\frac{B}{k}\tilde{d}(t)\right)dx, \quad (8.117)$$

$$\frac{d}{dt}\|\alpha(\cdot,t)\|^2 = -\sqrt{q}\alpha(L,t)^2 + \sqrt{q}\alpha(0,t)^2 + 2\int_0^L \alpha(x,t)\Gamma_{1t}(x,t)Z(t)dx. \quad (8.118)$$

Recalling (8.63) and (8.70), we have $v(L,t)$ as uniformly bounded. According to (8.42) and the boundedness of $\Gamma(x,t)$, we find that $w(L,t)$ is uniformly bounded. Because $w(0,t) = w(L,t - \frac{L}{\sqrt{q}})$, $w(0,t)$ is uniformly bounded for $t > \frac{L}{\sqrt{q}}$. Therefore, $z(0,t)$ is uniformly bounded via (8.15). Because $z(L,t) = z(0,t - \frac{L}{\sqrt{q}})$, $z(L,t)$ is uniformly bounded for $t > \frac{L}{\sqrt{q}}$. According to (8.42), (8.43) together with the boundedness of $\Gamma(x,t), \Gamma_1(x,t)$ and (8.62), (8.63), we have $\beta(0,t), \beta(L,t), \alpha(L,t), \alpha(0,t)$ as uniformly bounded.

Applying the Cauchy-Schwarz inequality, we get

$$\frac{d}{dt}|X(t)|^2 \leq \mu_3\left(|X(t)|^2 + \beta(0,t)^2 + \tilde{d}(t)^2\right),$$

$$\frac{d}{dt}\|\beta(\cdot,t)\|^2 \leq \mu_4\left(\|\beta(\cdot,t)\|^2 + |X(t)|^2 + \beta(0,t)^2 + \tilde{d}(t)^2\right),$$

$$\frac{d}{dt}\|\alpha(\cdot,t)\|^2 \le \mu_5 \Big(\|\alpha(\cdot,t)\|^2 + \|\beta(\cdot,t)\|^2 + |X(t)|^2 + \alpha(0,t)^2 + \alpha(L,t)^2 \Big),$$

with some positive constants μ_3, μ_4, μ_5. Thus, $\frac{d}{dt}|X(t)|^2$, $\frac{d}{dt}\|\beta(\cdot,t)\|^2$, and $\frac{d}{dt}\|\alpha(\cdot,t)\|^2$ are uniformly bounded thanks to the boundedness results established above.

Finally, integrating (8.114) from 0 to ∞, it follows that $|X(t)|$, $\|\alpha(\cdot,t)\|$, $\|\beta(\cdot,t)\|$ are square integrable. Following Barbalat's lemma, we conclude that $|X(t)|$, $\|\alpha(\cdot,t)\|$, $\|\beta(\cdot,t)\|$ tend to zero as $t \to \infty$.

Due to the invertibility and continuity of the backstepping transformations (8.62), (8.63) and (8.64), (8.65), the proof of lemma 8.1 is complete. □

The closed-loop system is

$$u_{tt}(x,t) = q u_{xx}(x,t), \tag{8.119}$$

$$u_x(L,t) = U(t), \tag{8.120}$$

$$I_b u_{tt}(0,t) = c u_t(0,t) - k u_x(0,t) - d(t), \tag{8.121}$$

$$d(t) = \sum_{j=1}^{N} [a_j \cos(\theta_j t) + b_j \sin(\theta_j t)], \tag{8.122}$$

$$\dot{\hat{c}}(t) = \gamma_c \mathrm{Proj}_{[0,\bar{c}]}\{\tau(t), \hat{c}(t)\}, \tag{8.123}$$

$$\dot{\hat{a}}_j(t) = \gamma_{aj} \mathrm{Proj}_{[0,\bar{a}_j]}\{\tau_{1j}(t), \hat{a}_j(t)\}, \tag{8.124}$$

$$\dot{\hat{b}}_j(t) = \gamma_{bj} \mathrm{Proj}_{[0,\bar{b}_j]}\{\tau_{2j}(t), \hat{b}_j(t)\}, \tag{8.125}$$

where the control input $U(t)$ is defined in (8.86). The functions $\tau(t)$, $\tau_{1j}(t)$, $\tau_{2j}(t)$, which are defined in (8.26)–(8.28), can be represented as the original state u by applying (8.34)–(8.37), (8.40), (8.41), and (8.6).

Define

$$\mathcal{H} = H^2(0,L) \times H^1(0,L), \tag{8.126}$$

where

$$H^1(0,L) = \{u | u(\cdot,t) \in L^2(0,L), u_x(\cdot,t) \in L^2(0,L)\},$$

$$H^2(0,L) = \{u | u(\cdot,t) \in L^2(0,L), u_x(\cdot,t) \in L^2(0,L), u_{xx}(\cdot,t) \in L^2(0,L)\}$$

and let $u(\cdot,t) \in L^2(0,L)$ mean that $u(\cdot,t)$ is square integrable in x. The main result is presented in the following theorem.

Theorem 8.1. *For all initial values $(u(\cdot,0), u_t(\cdot,0)) \in \mathcal{H}$, the closed-loop system consisting of the plant (8.119)–(8.121) and the controller (8.86) with the adaptive update laws (8.123)–(8.125) has the following properties:*

1. *The outputs $u(0,t), u_t(0,t)$ of the closed-loop system are asymptotically convergent to zero—that is,*

$$\lim_{t\to\infty} u(0,t) = 0, \quad \lim_{t\to\infty} u_t(0,t) = 0.$$

2. *Distributed states in the closed-loop system are uniformly ultimately bounded in the sense of the norm*

$$\left(\|u_x(\cdot,t)\|^2 + \|u_t(\cdot,t)\|^2\right)^{\frac{1}{2}}.$$

Proof. According to the asymptotic stability result in lemma 8.1, we know that $X(t) = [u(0,t), u_t(0,t)]^T$ is asymptotically convergent to zero, and thus property 1 of theorem 8.1 is proved.

Recalling (8.42), (8.43) and applying the Cauchy-Schwarz inequality, we obtain

$$\|w(\cdot,t)\|^2 \leq 2\|v(\cdot,t)\|^2 + 2N\|\Gamma(\cdot,t)\|^2, \tag{8.127}$$

$$\|z(x,t)\|^2 \leq 2\|s(\cdot,t)\|^2 + 2N\|\Gamma_1(\cdot,t)\|^2, \tag{8.128}$$

where $Z(t)^2 = N$ is used. The functions $\|\Gamma(\cdot,t)\|^2$, $\|\Gamma_1(\cdot,t)\|^2$ are bounded by a positive constant according to (8.59). Together with the convergence to zero and the uniform boundedness of $\|v(\cdot,t)\|^2 + \|s(\cdot,t)\|^2$ proved in lemma 8.1, we obtain the uniform ultimate boundedness of $\|w(\cdot,t)\|^2$, $\|z(\cdot,t)\|^2$. Recalling (8.12) and (8.13), we get

$$u_t(x,t) = \frac{1}{2}(z(x,t) + w(x,t)), \tag{8.129}$$

$$u_x(x,t) = \frac{1}{2\sqrt{q}}(w(x,t) - z(x,t)). \tag{8.130}$$

Applying the Cauchy-Schwarz inequality, we obtain

$$\|u_t(\cdot,t)\|^2 + \|u_x(\cdot,t)\|^2$$
$$\leq \frac{L}{2}(\|z(\cdot,t)\|^2 + \|w(\cdot,t)\|^2) + \frac{L}{2q}(\|z(\cdot,t)\|^2 + \|w(\cdot,t)\|^2)$$
$$\leq \frac{q+1}{2q}L(\|z(\cdot,t)\|^2 + \|w(\cdot,t)\|^2).$$

Property 2 of theorem 8.1 is thus proved. □

Summary of the implementation of the control algorithm supported by the above theorem: According to theorem 8.1, by applying the control torque $GJU(t)$ to the rotary table of the oil-drilling system with the uncertain stick-slip instability and external disturbance at the drill bit as shown in figure 8.1, the torsional vibration displacement $u(0,t)$ and velocity $u_t(0,t)$ of the drill bit are driven toward zero as time goes on. The constants G, J are physical parameters given in section 8.1. Input $U(t)$ in (8.86) includes the adaptive estimates $\hat{c}(t)$, $\hat{a}_j(t)$, $\hat{b}_j(t)$ defined in section 8.2 and is constructed by measuring the signals $u(0,t), u_t(0,t), u_t(L,t)$, which are obtained from the acceleration sensor placed at the bit and the feedback signal of the actuator at the rotary table, as mentioned in section 8.1. All signals in the controller are obtained from the direct measurements or integrals without using derivatives, which avoids the measurement noise amplification.

8.5   SIMULATION FOR OFFSHORE OIL DRILLING

The Oil-Drilling Model

The offshore oil-drilling model tested in the simulation is (8.1)–(8.3), with the physical parameters shown in table 8.1, which are borrowed from [24, 153]. The disturbance at the drill bit is given as a harmonic form of

$$d(t) = 2\cos(2t) + \sin(2t). \tag{8.131}$$

Table 8.1. Physical parameters of the oil-drilling system

Parameters (units)	Values
Length of the drill pipe L (m)	2000
Shear modulus of the drill pipe G (N/m^2)	7.96×10^{11}
Drill pipe moment of inertia	
per unit of length I_d (kg·m)	0.095
Moment of inertia of the BHA I_b (kg·m^2)	311
Drill pipe second moment of area J (m^4)	1.19×10^{-5}
Anti-damping parameter c (N·m·s/rad)	1

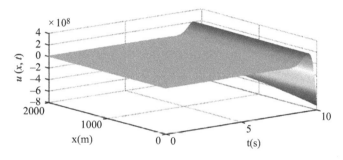

Figure 8.3. Open-loop responses of $u(x,t)$.

Therefore, the unknown amplitudes a_1, b_1 in the disturbance are $2, 1$, respectively. We only know that their upper bounds $\bar{a}_1, \bar{b}_1$ are $4, 2$. The unknown anti-damping coefficient c is 1, and the upper bound is known as $\bar{c} = 2$.

According to section 8.1, the system (8.1)–(8.3) is written as (8.14)–(8.18), where $X(t) = [u(0,t), u_t(0,t)]^T$. The main simulation is conducted based on (8.14)–(8.18), and then the responses z, w are converted to the responses u of the system (8.1)–(8.3) through

$$u(x,t) = \frac{1}{2\sqrt{q}} \int_0^x (w(y,t) - z(y,t))dy + u(0,t) \qquad (8.132)$$

by recalling (8.12), (8.13). The finite-difference method is adopted to conduct the simulation with a time step and space step of 0.0005 and 0.05, respectively. According to the physical parameters in table 8.1, the coefficients in (8.14)–(8.18) are obtained as $q = 9.971 \times 10^7$, $k = 9.472 \times 10^6$ through the definitions in section 8.1. Consider the initial conditions in (8.1)–(8.3) to be $u(x,0) = 0.15$, $u_t(x,0) = \sin(\frac{2\pi}{L}x)$. Then the according initial conditions in (8.14)–(8.18) are $w(x,0) = z(x,0) = \sin(\frac{2\pi}{L}x)$ by virtue of (8.12), (8.13), (8.132).

Open-Loop Responses

In the open-loop case, it is shown that the plant (8.1)–(8.3) is unstable in figures 8.3 and 8.4 because of the effect of the anti-damping term at the bit $x = 0$. The expected diverging results for $z(x,t), w(x,t)$ (8.14)–(8.18) are seen in figure 8.5. To be exact, the diverging phenomenon starts at $X(t) = [u(0,t), u_t(0,t)]^T$ in (8.14) (shown in figure 8.4), flowing into $z(0,t)$ via (8.15) and giving rise to the increase

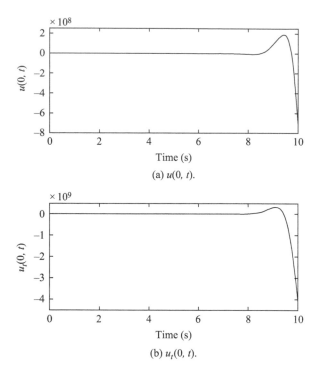

(a) $u(0, t)$.

(b) $u_t(0, t)$.

Figure 8.4. Open-loop responses of $u(0, t), u_t(0, t)$.

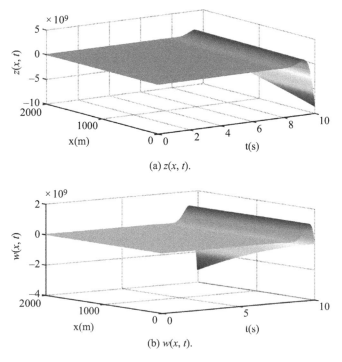

(a) $z(x, t)$.

(b) $w(x, t)$.

Figure 8.5. Open-loop responses of $z(x, t), w(x, t)$.

of $|z(0,t)|$, following which the diverging response develops for $z(x,t)$ via (8.16), with instability traveling up to the boundary $x = L$, which is shown in figure 8.5(a). According to (8.18), diverging performance then appears at $w(L,t)$, which leads to the instability of $w(x,t)$ via (8.17) and causes the increase of $|w(0,t)|$, which is shown in figure 8.5(b).

Closed-Loop Responses

We apply the control law (8.86) into (8.14)–(8.18), with the control parameters chosen as $\bar{\kappa} = [0.1, 1.5]$, $\gamma_a = 0.005$, $\gamma_b = 0.008$, and $\gamma_c = 0.006$. The constant N in (8.86) is 1. The function u_t in (8.86) is represented by the states in (8.14)–(8.18) via $u_t = \frac{1}{2}(w + z)$. The adaptive estimates $\hat{c}(t)$, $\hat{a}_1(t)$, $\hat{b}_1(t)$ in (8.86) are calculated from (8.23) and (8.26), (8.24) and (8.27), and (8.25) and (8.28). Recalling (8.62), (8.63), (8.42), (8.43), (8.38), (8.39), the integrals $\int_0^L e^{-x}\alpha(x,t)^2 dx$, $\int_0^L e^x \beta(x,t)^2 dx$ in the weight norm $\Omega(t)$ defined in (8.31) and appearing in the adaptive update laws for the estimates are also represented by the states in (8.14)–(8.18) as

$$
\int_0^L e^{-x}\alpha(x,t)^2 dx
$$
$$
= \int_0^L e^{-x}(z(x,t) + \Gamma_1(x,t)Z(t))^2 dx
$$
$$
= -\int_{t-\frac{L}{\sqrt{q}}}^t e^{-\sqrt{q}(t-\delta_1)}\left(2C_1 X(\delta_1) - w(L, \delta_1 - \frac{L}{\sqrt{q}})\right.
$$
$$
\left. - \frac{\sqrt{q}}{k}[\hat{a}_1(t), \hat{b}_1(t)]e^{-A_z(t-\delta_1)}Z(t)\right)^2 d\delta_1, \tag{8.133}
$$

and

$$
\int_0^L c^x \beta(x,t)^2 dx
$$
$$
= \int_0^L e^x \left(w(x,t) + \Gamma(x,t)Z(t)\right.
$$
$$
\left. - \int_0^x \phi(x,y,t)(w(y,t) + \Gamma(y,t)Z(t))dy - \gamma(x,t)X(t)\right)^2 dx
$$
$$
= \int_{t-\frac{L}{\sqrt{q}}}^t e^{L-\sqrt{q}(t-\delta_1)}\left[w(L, \delta_1) + \frac{\sqrt{q}}{k}[\hat{a}_1(t), \hat{b}_1(t)]c^{\frac{A_z}{\sqrt{q}}(L-\sqrt{q}(t-\delta_1))}Z(t)\right.
$$
$$
- \int_{t-\frac{L}{\sqrt{q}}}^{\delta_1} \frac{1}{q}\left(\bar{\kappa} + \left[0, \frac{\sqrt{q}\hat{c}(t)}{k}\right]\right)e^{(A_E + \hat{A}_c(t) - \frac{1}{\sqrt{q}}BC_1)(\delta_1 - \delta_2)}B
$$
$$
\times \left(w(L, \delta_2) + \frac{\sqrt{q}}{k}[\hat{a}_1(t), \hat{b}_1(t)]e^{\frac{A_z}{\sqrt{q}}(L-\sqrt{q}(t-\delta_2))}Z(t)\right)d\delta_2
$$
$$
\left. - \left(\bar{\kappa} + \left[0, \frac{\sqrt{q}\hat{c}(t)}{k}\right]\right)e^{\frac{1}{\sqrt{q}}(A_E + \hat{A}_c(t) - \frac{1}{\sqrt{q}}BC_1)(L-\sqrt{q}(t-\delta_1))}X(t)\right]^2 d\delta_1. \tag{8.134}
$$

The same process is adopted to calculate $\int_0^L \beta(x,t)dx$ used in (8.26)–(8.28).

The controller is activated at $t = \frac{2L}{\sqrt{q}} = 0.4$ s, ensuring that $\delta_1 - \frac{L}{\sqrt{q}}$ with $\delta_1 \in [t - \frac{L}{\sqrt{q}}, t]$ in $w(L, \delta_1 - \frac{L}{\sqrt{q}})$ in (8.133) is nonnegative.

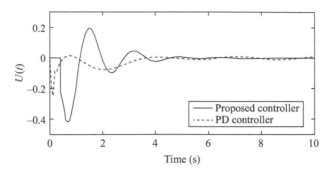

Figure 8.6. The proposed control input and the PD control input.

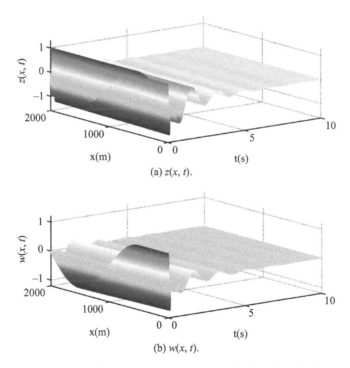

(a) $z(x, t)$.

(b) $w(x, t)$.

Figure 8.7. Closed-loop responses of $z(x, t), w(x, t)$.

We compare the proposed controller with the classical proportional-derivative (PD) controller, which uses the signal $X(t) = [u(0, t), u_t(0, t)]^T$ and is given by

$$U_{\mathrm{PD}}(t) = k_p u(0, t) + k_d u_t(0, t). \tag{8.135}$$

The best regulating PD performance is achieved with $k_p = 0.13$ and $k_d = 1.2$ in (8.14)–(8.18).

The evolution of the proposed controller, activated at $t = 0.4$ s, and the PD controller, activated at $t = 0$, is shown in figure 8.6.

According to figure 8.7, we know that the responses of $z(x, t), w(x, t)$ under the proposed controller are convergent to a small neighborhood of zero. From figure

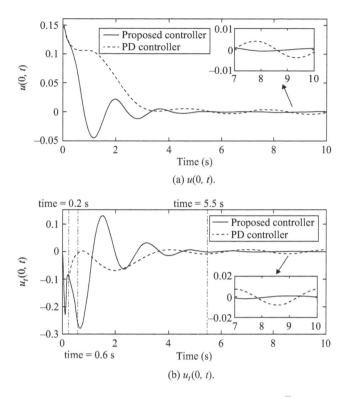

(a) $u(0, t)$.

(b) $u_t(0, t)$.

Figure 8.8. Closed-loop responses of $X(t) = [u(0,t), u_t(0,t)]^T$ under the proposed adaptive controller and the PD controller, which physically represents the torsional vibration angular displacement and velocity at the bit (the reasons for the phenomenon of the PD controller's better transient performance before about $t = 5.5$ s are discussed in remark 8.2).

8.8, even though the transient performance of $u(0,t), u_t(0,t)$—that is, of the vector state $X(t)$—under the proposed controller is worse than that under the PD controller before about $t = 5.5$ s (the reasons for this phenomenon are discussed in remark 8.2), the responses of $u(0,t), u_t(0,t)$ under the proposed controller are convergent to a smaller neighborhood of zero than those under the PD controller as time goes on. It physically means that the proposed controller achieves better performance in the suppression of the torsional vibration displacement and velocity at the bit.

Recalling (8.132) and the closed-loop responses of $z(x,t), w(x,t), u(0,t)$, the response of $u(x,t)$ under the proposed controller is obtained and shown in figure 8.9, which indicates that the torsional vibrations of the oil-drilling pipe have been suppressed. The norm $(\|u_t(\cdot,t)\|^2 + \|u_x(\cdot,t)\|^2)^{\frac{1}{2}}$ obtained from (8.129), (8.130) denotes torsional vibration energy consisting of kinetic energy and potential energy. The responses of $(\|u_t(\cdot,t)\|^2 + \|u_x(\cdot,t)\|^2)^{\frac{1}{2}}$ under the proposed controller and the PD controller are shown in figure 8.10. We see that, even though the PD controller has a better transient performance before about $t = 5.5$ s (the two reasons for this phenomenon are explained in remark 8.2), the proposed controller reduces the vibration to a smaller range around zero as time goes on, which verifies that the proposed adaptive controller performs better at suppressing vibrations in the offshore oil-drilling platform.

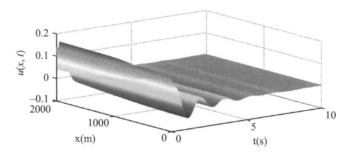

Figure 8.9. Closed-loop response of $u(x,t)$, which physically represents the torsional vibrations of the oil-drilling pipe under the proposed controller.

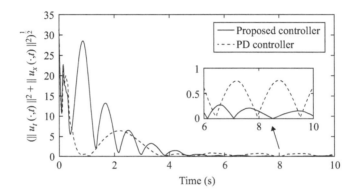

Figure 8.10. Closed-loop response of the norm $(\|u_t(\cdot,t)\|^2 + \|u_x(\cdot,t)\|^2)^{\frac{1}{2}}$ under the proposed adaptive controller and the PD controller, which physically represents torsional vibration energy, including kinetic energy and potential energy of the oil-drilling pipe (the reasons for the phenomenon of the PD controller's better transient performance before about $t = 5.5$ s are discussed in remark 8.2).

The adaptive estimation action is activated at $t = 0.4$ s. From figure 8.11, which shows the adaptive estimation errors of the constants c, a_1, b_1 in the above regulation process, we know that the estimates $\hat{c}(t), \hat{a}_1(t), \hat{b}_1(t)$ converge to values that are close to the actual c, a_1, b_1 as time goes on. Even though the estimates do not exactly arrive at their actual values, the state convergence is achieved, which is typical in adaptive control in the absence of persistence of excitation.

Remark 8.2. There are two reasons for the phenomenon that the simpler PD controller has a better transient performance than the proposed adaptive controller before $t = 5.5$ s. First, the proposed model-based adaptive controller is activated later than the PD controller with a 0.4 s delay. The proposed adaptive controller is activated at $t = 0.4$ s and the regulation action reaches the ODE at $x = 0$ until $t = 0.6$ s because the propagation time from $x = L$ to $x = 0$ is $\frac{L}{\sqrt{q}} = 0.2$ s in (8.14)–(8.18), while the PD controller is activated at $t = 0$ and the regulation action reaches the ODE at $t = 0.2$ s. This is shown in figure 8.8(b), where the PD controller starts regulating the ODE states toward zero after $t = 0.2$ s, and the response under the proposed adaptive controller continues to deteriorate until about $t = 0.6$ s because of

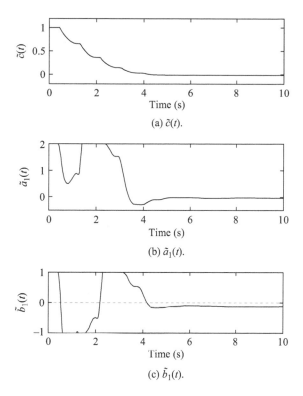

(a) $\tilde{c}(t)$.

(b) $\tilde{a}_1(t)$.

(c) $\tilde{b}_1(t)$.

Figure 8.11. Adaptive estimation errors of the anti-damping coefficient c and the disturbance amplitudes a_1, b_1.

no regulation action to stabilize the anti-stable ODE. Second, the PD controller is running under the very best parameters that we have chosen over many simulation tests (which is equivalent to knowing the model parameters), to be more than fair to the model-free PD controller and to give it a maximal advantage. In contrast, the proposed adaptive controller is operating with a poor knowledge of the anti-damping parameter—that is, with very bad initial gains, and a poor knowledge of the disturbance parameters, where the adaptive estimation of these parameters introduces an adaptive learning transient until as late as about $t = 5.5$ s, which is shown in figure 8.11. In summary, as soon as our model-based adaptive output-feedback disturbance-canceling controller completes, or nearly completes, the learning process, this controller vastly outperforms the PD controller in subsequent performance, as expected.

8.6 NOTES

Chapters 2–6 dealt with the longitudinal or lateral vibrations of cables, while the torsional vibrations in the "drill string," which is not a cable but a kilometers-long thin cylinder, were dealt with in this chapter, along with stick-slip instability and external disturbances resulting from the wave-induced heaving motion of the drilling rig, at the bit. The plant is a one-dimensional wave PDE system with an uncontrolled dynamic boundary, which includes an anti-damping term with an unknown coefficient and a harmonic disturbance with unknown amplitudes. The control design incorporates adaptive control designs in [24, 25, 26, 117, 124], which were developed for a one-dimensional wave PDE with an actuator on one boundary and an anti-damping instability with an unknown coefficient on the other boundary, as well as the idea of adaptive cancellation of anti-collocated uncertain disturbances for wave PDEs in [85].

Part II

Generalizations

Chapter Nine

Basic Control of Sandwich Hyperbolic PDEs

In part I of this book we neglected the actuator dynamics, assuming that the control input directly flows into the partial differential equation (PDE) boundary. In many situations, however, the actuator dynamics must be considered in the control design, especially when the dominant time constant of the actuator is close to that of the plant. When the model incorporates the actuator dynamics in the input channel of the string-payload model considered in part I, such as the hydraulic cylinder and head sheaves in chapters 2–5, the ship-mounted crane in chapter 6, or the rotary table in chapter 8, the plant becomes a hyperbolic PDE sandwiched between two ordinary differential equations (ODEs).

We address a new theoretical problem: boundary control of sandwich PDEs, starting from a basic design presented in this chapter, moving on to delay-compensated control in chapter 10, developing an event-triggered controller in chapter 11, and dealing with nonlinearities in chapter 12.

In this chapter, we solve the problem of stabilization of 2×2 coupled linear first-order hyperbolic PDEs sandwiched between two ODEs by combining PDE backstepping and ODE backstepping. In section 9.2, we seek a PDE backstepping transformation that maps the plant into a stable system where the in-domain couplings between the hyperbolic PDEs are removed, and the system matrix of the ODE at the left boundary is Hurwitz. The resulting right boundary condition is an ODE with a number of perturbation given in terms of the PDE states. We then deal with this ODE in the input channel via the ODE backstepping method in section 9.3. In section 9.4, a controller is built, and the exponential stability of the overall closed-loop system is proved by Lyapunov analysis, where some control parameters, which exist in the ODE backstepping design procedure, are determined, with the goal of tolerating the PDE perturbations. The boundedness and exponential convergence of the controller in the closed-loop system are proved in section 9.5. In section 9.6, we extend the proposed method and the associated proofs to a more general case where the input ODE is of arbitrary order. The simulation results are provided in section 9.7.

9.1 PROBLEM FORMULATION

The sandwich hyperbolic PDE system considered in this chapter is

$$\dot{X}(t) = AX(t) + Bv(0, t), \tag{9.1}$$

$$u_t(x, t) = -pu_x(x, t) + c_1 v(x, t), \tag{9.2}$$

$$v_t(x, t) = pv_x(x, t) + c_2 u(x, t), \tag{9.3}$$

$$u(0,t) = qv(0,t) + CX(t), \tag{9.4}$$

$$v(1,t) = z(t), \tag{9.5}$$

$$\ddot{z}(t) = c_0 \dot{z}(t) + ru(1,t) + U(t), \tag{9.6}$$

$\forall (x,t) \in [0,1] \times [0,\infty)$, where $X(t) \in \mathbb{R}^{n \times 1}$, and $Z^T(t) = [z(t), \dot{z}(t)] = [z_1(t), z_2(t)] \in \mathbb{R}^{2 \times 1}$ are ODE states. The scalars $u(x,t) \in \mathbb{R}, v(x,t) \in \mathbb{R}$ are states of the PDEs. The matrices $A \in \mathbb{R}^{n \times n}$, $B \in \mathbb{R}^{n \times 1}$ satisfy that the pair $[A;B]$ is controllable. The matrix $C \in \mathbb{R}^{1 \times n}$ and constants $c_0, c_1, c_2, r, q \in \mathbb{R}$ are arbitrary. The positive constant p denotes the arbitrary transport speed. We consider the transport speed of (9.2), (9.3) to be equal in this chapter but with a possible extension to the case where the transport speeds of the two transport PDEs are different. The control input $U(t)$ is to be designed. The full relative degree in (9.5), (9.6) is assumed for the design. The objective here is to exponentially stabilize all ODE states $Z(t), X(t)$ and PDE states $u(x,t), v(x,t)$ by the control input $U(t)$. Moreover, the result is extended to a more general system with arbitrary-order ODEs sandwiching a PDE.

9.2   BACKSTEPPING FOR THE PDE-ODE CASCADE

Backstepping Transformations and the Target System

We consider the infinite-dimensional backstepping transformation of the PDE state $u(x,t), v(x,t)$ as follows:

$$\alpha(x,t) \equiv u(x,t), \tag{9.7}$$

$$\beta(x,t) = v(x,t) - \int_0^x \psi(x,y)u(y,t)dy - \int_0^x \phi(x,y)v(y,t)dy - \gamma(x)X(t). \tag{9.8}$$

The kernel equations for $\psi(x,y), \phi(x,y), \gamma(x)$ are determined later. The reason why we only apply the backstepping transformation on v is that given the difference in the propagation directions in the u and v subsystems, only the u-term in the v-subsystem in (9.3) acts as a potentially destabilizing feedback term, whereas the v-term in the u-subsystem in (9.2) acts as a feedforward term, which disturbs the u system but cannot destabilize it, once the presence of the u-term in (9.3) is eliminated. Using the partial backstepping transformation (9.7), (9.8) results in a requirement of the calculation of fewer kernels and a simpler structure of the controller.

   The inverse of (9.7), (9.8) is postulated as

$$u(x,t) \equiv \alpha(x,t), \tag{9.9}$$

$$v(x,t) = \beta(x,t) - \int_0^x \psi^I(x,y)\alpha(y,t)dy - \int_0^x \phi^I(x,y)\beta(y,t)dy - \gamma^I(x)X(t), \tag{9.10}$$

where $\phi^I(x,y), \psi^I(x,y), \gamma^I(x)$ are the kernels of the inverse transformation (9.10). The well-posedness of these kernels is shown later.

   Our aim is to convert the original system (9.1)–(9.5) to the following target system:

$$\dot{X}(t) = (A + B\kappa)X(t) + B\beta(0,t), \tag{9.11}$$

$$\alpha_t(x,t) = -p\alpha_x(x,t) + c_1\beta(x,t) - c_1\int_0^x \psi^I(x,y)\alpha(y,t)dy$$

$$- c_1\int_0^x \phi^I(x,y)\beta(y,t)dy - c_1\gamma^I(x)X(t), \tag{9.12}$$

$$\beta_t(x,t) = p\beta_x(x,t), \tag{9.13}$$

$$\alpha(0,t) = q\beta(0,t) + C_0X(t), \tag{9.14}$$

where

$$C_0 = C + q\gamma(0). \tag{9.15}$$

Since the pair $[A; B]$ is controllable, there exists indeed κ such that $A + B\kappa$ is Hurwitz. The system (9.13) contains no feedback connection, as we had indicated a few lines above, whereas the system (9.12) contains only feedforward connections from β in (9.13) and X in (9.11). While less obvious, the integral of $\alpha(y,t)$ from $y=0$ to $y=x$ is also a feedforward connection, given the transport direction in the α-PDE from $y=0$ toward $y=1$.

Let us now consider the boundary state $\beta(1,t)$. It is easily seen that

$$\beta_{tt}(1,t) = v_{tt}(1,t) + p\psi(1,1)u_t(1,t) - p\psi(1,0)u_t(0,t)$$

$$+ p\Big(p\psi_y(1,1) + c_2\phi(1,1)\Big)u(1,t)$$

$$- p\Big(p\psi_y(1,0) + c_2\phi(1,0)\Big)u(0,t) - p\phi(1,1)v_t(1,t)$$

$$+ \Big(p\phi(1,0) - \gamma(1)B\Big)v_t(0,t)$$

$$- \gamma(1)A^2X(t) - p\Big(c_1\psi(1,1) - p\phi_y(1,1)\Big)v(1,t)$$

$$+ \Big(pc_1\psi(1,0) - p^2\phi_y(1,0) - \gamma(1)AB\Big)v(0,t)$$

$$- \int_0^1 \Big(p^2\psi_{yy}(1,y) + c_1c_2\psi(1,y)\Big)u(y,t)dy$$

$$- \int_0^1 \Big(p^2\phi_{yy}(1,y) + c_1c_2\phi(1,y)\Big)v(y,t)dy. \tag{9.16}$$

By virtue of (9.5), (9.6), we obtain

$$v_{tt}(1,t) = c_0v_t(1,t) + ru(1,t) + U(t). \tag{9.17}$$

Plugging (9.13) and the inverse transformations (9.9), (9.10) into (9.16), after a lengthy calculation, which involves a change of the order of integration in a double integral, we get

$$\beta_{tt}(1,t) = h_1\beta_t(1,t) + h_5\beta(1,t) + U(t) + h_2\alpha_t(1,t) + h_3\beta_t(0,t)$$

$$+ h_4\alpha_t(0,t) + (h_6 + r)\alpha(1,t) + h_7\beta(0,t) + h_8\alpha(0,t)$$

$$+ \int_0^1 h_9(y)\beta(y,t)dy + \int_0^1 h_{10}(y)\alpha(y,t)dy + H_{11}X(t), \tag{9.18}$$

which also belongs to the target system, where h_1, h_2, h_3, h_4, h_5, h_6, h_7, h_8, $h_9(y)$, $h_{10}(y)$, H_{11} are shown in appendix 9.8. It should be noted that (9.18) is a second-order ODE system $(\beta(1,t), \beta_t(1,t))$ with a number of PDE state perturbation terms.

We can obtain well-posedness of the closed-loop system through analyzing the well-posedness of the target system (9.11)–(9.14), (9.18). It can be expected that well-posedness of the target system depends on that of the 2×2 coupled linear hyperbolic PDE-ODE (9.11)–(9.14), which can be obtained similarly to the proof of lemma 9.1 which will be shown later.

Kernel Equations

Taking the derivative of (9.8) with respect to x and t, respectively, along the solution of (9.1)–(9.4) and substituting the results to (9.13), we get

$$\beta_t(x,t) - p\beta_x(x,t)$$

$$= v_t(x,t) - \int_0^x \psi(x,y)u_t(y,t)dy - \int_0^x \phi(x,y)v_t(y,t)dy$$

$$- pv_x(x,t) + p\int_0^x \psi_x(x,y)u(y,t)dy + p\int_0^x \phi_x(x,y)v(y,t)dy$$

$$+ p\psi(x,x)u(x,t) + p\phi(x,x)v(x,t) - \gamma(x)\dot{X}(t) + p\gamma'(x)X(t)$$

$$= c_2 u(x,t) + \int_0^x p\psi(x,y)u_x(y,t)dy - \int_0^x c_1\psi(x,y)v(y,t)dy$$

$$- \int_0^x p\phi(x,y)v_x(y,t)dy - \int_0^x c_2\phi(x,y)u(y,t)dy$$

$$+ \int_0^x p\psi_x(x,y)u(y,t)dy + \int_0^x p\phi_x(x,y)v(y,t)dy$$

$$+ p\psi(x,x)u(x,t) + p\phi(x,x)v(x,t) - \gamma(x)\dot{X}(t) + p\gamma'(x)X(t)$$

$$= \left(c_2 + 2p\psi(x,x)\right)u(x,t) + \left(p\phi(x,0) - \gamma(x)B - p\psi(x,0)q\right)v(0,t)$$

$$+ \int_0^x \left(-c_1\psi(x,y) + p\phi_x(x,y) + p\phi_y(x,y)\right)v(y,t)dy$$

$$- \int_0^x \left(c_2\phi(x,y) - p\psi_x(x,y) + p\psi_y(x,y)\right)u(y,t)dy$$

$$+ \left(p\gamma'(x) - \gamma(x)A - p\psi(x,0)C\right)X(t) = 0. \tag{9.19}$$

For (9.19) to hold and matching (9.11), (9.14) with (9.1), (9.4) via the transformations (9.8), we obtain the following kernel equations:

$$c_2 + 2p\psi(x,x) = 0, \tag{9.20}$$

$$p\phi(x,0) = \gamma(x)B + p\psi(x,0)q, \tag{9.21}$$

$$p\phi_x(x,y) + p\phi_y(x,y) - c_1\psi(x,y) = 0, \tag{9.22}$$

$$-p\psi_x(x,y) + p\psi_y(x,y) + c_2\phi(x,y) = 0, \tag{9.23}$$

$$p\gamma'(x) - \gamma(x)A - p\psi(x,0)C = 0, \tag{9.24}$$

$$\gamma(0) = \kappa, \tag{9.25}$$

for $0 \le y \le x \le 1$.

Well-Posedness of the Kernel Equations

We show the well-posedness of the kernel equations (9.20)–(9.25) by using the methods of characteristics and successive approximations [50].

Lemma 9.1. *The kernel equations* (9.20)–(9.25) *have a unique solution* $(\psi, \phi) \in C^1(D) \times C^1(D)$, *where* $D = \{(x,y)|0 \le y \le x \le 1\}$.

Proof. The proof of this lemma is presented in appendix 9.8. □

Inverse Transformation

In order to ensure the invertibility of the transformation (9.8), we search for the inverse transformation of (9.8), which can convert the target system (9.11)–(9.14) into the original system (9.1)–(9.4).

Recalling the transformation (9.8) and rewriting it as

$$v(x,t) - \int_0^x \phi(x,y)v(y,t)dy = \beta(x,t) + \int_0^x \psi(x,y)u(y,t)dy + \gamma(x)X(t). \tag{9.26}$$

According to lemma 9.1, $\phi(x,y)$ is continuous, and we obtain a unique continuous $\chi(x,y)$ existing on $D = \{(x,y)|0 \le y \le x \le 1\}$ such that (see, e.g., [169])

$$v(x,t) = \beta(x,t) + \int_0^x \psi(x,y)u(y,t)dy + \gamma(x)X(t)$$
$$+ \int_0^x \chi(x,y)\left(\beta(y,t) + \int_0^y \psi(y,z)u(z,t)dz + \gamma(y)X(t)\right)dy, \tag{9.27}$$

whose proof can be seen in chapter 9.9 of [175].

Equation (9.27) is rewritten in the form of (9.10), as follows:

$$v(x,t) = \beta(x,t) + \int_0^x \chi(x,y)\beta(y,t)dy$$
$$+ \int_0^x \left(\int_y^x \chi(x,z)\psi(z,y)dz + \psi(x,y)\right)\alpha(y,t)dy$$
$$+ \left(\gamma(x) + \int_0^x \chi(x,y)\gamma(y)dy\right)X(t). \tag{9.28}$$

Comparing (9.28) with (9.10), we obtain

$$\psi^I(x,y) = -\int_y^x \chi(x,z)\psi(z,y)dz - \psi(x,y), \tag{9.29}$$

$$\phi^I(x,y) = -\chi(x,y), \tag{9.30}$$

$$\gamma^I(x) = -\gamma(x) - \int_0^x \chi(x,y)\gamma(y)dy. \tag{9.31}$$

According to the well-posedness of $\psi(x,y), \chi(x,y), \gamma(y)$ in $D = \{(x,y)|0 \leq y \leq x \leq 1\}$, we obtain the well-posedness of the kernels $\psi^I(x,y), \phi^I(x,y), \gamma^I(x)$ on D in (9.10), which shows the invertibility between the target system $(\alpha(x,t), \beta(x,t))$ and the original one $(u(x,t), v(x,t))$.

9.3   BACKSTEPPING FOR THE INPUT ODE

The following backstepping transformation for the $(\beta(1,t), \beta_t(1,t))$ system (9.18) is made:

$$y_1(t) = \beta(1,t), \tag{9.32}$$
$$y_2(t) = \beta_t(1,t) + \tau_1[\beta(1,t)], \tag{9.33}$$

where the function τ_1, which is to be defined in the following steps, is the virtual control law in the ODE backstepping method.

Step 1. We consider a Lyapunov function candidate as

$$V_{y1} = \frac{1}{2}y_1(t)^2. \tag{9.34}$$

Taking the derivative of (9.34), we obtain

$$\dot{V}_{y1} = y_1(t)\dot{y}_1(t) = y_1(t)(y_2(t) - \tau_1). \tag{9.35}$$

Define

$$\tau_1(y_1) = \bar{c}_1 y_1, \tag{9.36}$$

where $\bar{c}_1$ is a positive constant to be determined later.

Substituting (9.36) into (9.35) yields

$$\dot{V}_{y1} = -\bar{c}_1 y_1(t)^2 + y_1(t)y_2(t). \tag{9.37}$$

Step 2. Similarly, a Lyapunov function candidate is considered as

$$V_y = V_{y1} + \frac{1}{2}y_2(t)^2 = \frac{1}{2}y_1(t)^2 + \frac{1}{2}y_2(t)^2. \tag{9.38}$$

Taking the derivative of (9.38), we get

$$\dot{V}_y = -\bar{c}_1 y_1(t)^2 + y_1(t)y_2(t) + y_2(t)(\beta_{tt}(1,t) + \dot{\tau}_1). \tag{9.39}$$

Recalling (9.18), we obtain

$$\begin{aligned}
\dot{V}_y = {} & -\bar{c}_1 y_1(t)^2 + y_1(t)y_2(t) + y_2(t)\bigg(U(t) + h_1\beta_t(1,t) + h_5\beta(1,t) \\
& + h_2\alpha_t(1,t) + h_3\beta_t(0,t) + h_4\alpha_t(0,t) + (h_6 + r)\alpha(1,t) \\
& + h_7\beta(0,t) + h_8\alpha(0,t) + \int_0^1 h_9(y)\beta(y,t)dy \\
& + \int_0^1 h_{10}(y)\alpha(y,t)dy + H_{11}X(t) + \dot{\tau}_1 \bigg),
\end{aligned} \tag{9.40}$$

where the gains $h_1, \ldots, h_{10}, H_{11}$ shown in appendix 9.8 are related to the kernel functions $(\psi(x,y), \phi(x,y)) \in W^{2,1}(D)$.

Choosing

$$\begin{aligned}
U(t) &= -\bar{c}_2 y_2(t) - y_1(t) - \dot{\tau}_1 - h_1 \beta_t(1,t) - h_5 \beta(1,t) \\
&\quad - h_2 \alpha_t(1,t) - h_3 \beta_t(0,t) - h_4 \alpha_t(0,t) \\
&= -(\bar{c}_2 + \bar{c}_1 + h_1)\beta_t(1,t) - (\bar{c}_1 \bar{c}_2 + 1 + h_5)\beta(1,t) \\
&\quad - h_2 \alpha_t(1,t) - h_3 \beta_t(0,t) - h_4 \alpha_t(0,t),
\end{aligned} \tag{9.41}$$

where $\bar{c}_2$ is a positive constant to be determined later, we get

$$\begin{aligned}
\dot{V}_y = {}&-\bar{c}_1 y_1(t)^2 - \bar{c}_2 y_2(t)^2 + y_2(t)\bigg((h_6 + r)\alpha(1,t) + h_7 \beta(0,t) + h_8 \alpha(0,t) \\
&+ \int_0^1 h_9(y)\beta(y,t)dy + \int_0^1 h_{10}(y)\alpha(y,t)dy + H_{11}X(t) \bigg).
\end{aligned} \tag{9.42}$$

9.4 CONTROLLER AND STABILITY ANALYSIS

Control Law

Substituting the PDE transformation (9.7), (9.8) into (9.41), we get the controller expressed by the original states, as follows:

$$\begin{aligned}
U(t) = {}&-n_1 v_t(1,t) + n_2 v(1,t) - h_2 u_t(1,t) - n_3 u(1,t) - h_3 v_t(0,t) \\
&- n_4 v(0,t) - h_4 u_t(0,t) + n_5 u(0,t) + N_8 X(t) \\
&+ \int_0^1 n_6(y)u(y,t)dy + \int_0^1 n_7(y)v(y,t)dy,
\end{aligned} \tag{9.43}$$

where

$$n_1 = \bar{c}_2 + \bar{c}_1 + h_1, \tag{9.44}$$
$$n_2 = (\bar{c}_2 + \bar{c}_1 + h_1)\phi(1,1) - (\bar{c}_1 \bar{c}_2 + 1 + h_5), \tag{9.45}$$
$$n_3 = (\bar{c}_2 + \bar{c}_1 + h_1)\psi(1,1), \tag{9.46}$$
$$n_4 = (\bar{c}_2 + \bar{c}_1 + h_1)(\phi(1,0) - \gamma(1)B) - h_3\gamma(0)B, \tag{9.47}$$
$$n_5 = (\bar{c}_2 + \bar{c}_1 + h_1)\psi(1,0), \tag{9.48}$$
$$\begin{aligned}
n_6(y) = {}&(\bar{c}_2 + \bar{c}_1 + h_1)(\psi_y(1,y) + c_2\phi(1,y)) \\
&+ (\bar{c}_1 \bar{c}_2 + 1 + h_5)\psi(1,y),
\end{aligned} \tag{9.49}$$
$$n_7(y) = (\bar{c}_2 + \bar{c}_1 + h_1)(c_1\psi(1,y) - \phi_y(1,y)) + (\bar{c}_1 \bar{c}_2 + 1 + h_5)\phi(1,y), \tag{9.50}$$
$$N_8 = h_3\gamma(0)A + (\bar{c}_2 + \bar{c}_1 + h_1)\gamma(1)A + (\bar{c}_1 \bar{c}_2 + 1 + h_5)\gamma(1). \tag{9.51}$$

The pending control parameters $\bar{c}_1, \bar{c}_2$ will be determined in the following stability analysis.

By substituting (9.2), (9.3) at $x=0$ and $x=1$ into (9.43), the controller is rewritten as

$$\begin{aligned}
U(t) = {}&-n_1 p v_x(1,t) + (n_2 - h_2 c_1)v(1,t) \\
&+ h_2 p u_x(1,t) - (n_3 + n_1 c_2)u(1,t) - h_3 p v_x(0,t)
\end{aligned}$$

$$- (n_4 + h_4 c_1) v(0, t) + h_4 p u_x(0, t) + (n_5 - h_3 c_2) u(0, t)$$

$$+ \int_0^1 n_6(y) u(y, t) dy + \int_0^1 n_7(y) v(y, t) dy + N_8 X(t). \tag{9.52}$$

Stability Analysis of States

Theorem 9.1. *For all initial values* $(u(x, 0), v(x, 0)) \in W^{2,2}(0, 1)$, *with some* $\bar{c}_1, \bar{c}_2$, *the closed-loop system consisting of the plant* (9.1)–(9.6) *and the control law* (9.52) *is exponentially stable at the origin in the sense of the norm*

$$\left(\int_0^1 u^2(x, t) dx + \int_0^1 v^2(x, t) dx + |X(t)|^2 + z_1(t)^2 + z_2(t)^2 \right)^{1/2}, \tag{9.53}$$

where $|\cdot|$ *denotes the Euclidean norm.*

Proof. We start from studying the stability of the target system. The equivalent stability property between the target system and the original system is ensured due to the invertibility of the PDE backstepping transformation (9.7), (9.8) and the ODE backstepping transformation (9.32), (9.33).

First, we study the stability proof of the target system via Lyapunov analysis of the PDE-ODE system. Second, with the Lyapunov analysis of the input ODE in section 9.3, Lyapunov analysis of the whole ODE-PDE-ODE system is provided, where the control parameters $\bar{c}_1, \bar{c}_2$ in the control law (9.52) are determined.

1) LYAPUNOV ANALYSIS FOR THE PDE-ODE SYSTEM

Define

$$\Omega_1(t) = \|\beta(\cdot, t)\|^2 + \|\alpha(\cdot, t)\|^2 + |X(t)|^2, \tag{9.54}$$

where $\|\beta(\cdot, t)\|^2$ is a compact notation for $\int_0^1 \beta(x, t)^2 dx$.

Now consider a Lyapunov function

$$V_1(t) = X^T(t) P_1 X(t) + \frac{a_1}{2} \int_0^1 e^{\delta_1 x} \beta(x, t)^2 dx + \frac{b_1}{2} \int_0^1 e^{-\delta_1 x} \alpha(x, t)^2 dx, \tag{9.55}$$

where there exists a matrix $P_1 = P_1^T > 0$, which is the unique solution to the Lyapunov equation

$$P_1(A + B\kappa) + (A + B\kappa)^T P_1 = -Q_1, \tag{9.56}$$

for some $Q_1 = Q_1^T > 0$, by recalling that $A + B\kappa$ is Hurwitz. The positive parameters a_1, b_1, δ_1 are to be chosen later.

From (9.54), we obtain

$$\theta_{11} \Omega_1(t) \leq V_1(t) \leq \theta_{12} \Omega_1(t), \tag{9.57}$$

where

$$\theta_{11} = \min \left\{ \lambda_{\min}(P_1), \frac{a_1}{2}, \frac{b_1 e^{-\delta_1}}{2} \right\} > 0, \tag{9.58}$$

$$\theta_{12} = \max\left\{\lambda_{\max}(P_1), \frac{a_1 e^{\delta_1}}{2}, \frac{b_1}{2}\right\} > 0. \tag{9.59}$$

The time derivative of $V_1(t)$ along (9.11)–(9.14) is obtained as

$$
\begin{aligned}
\dot{V}_1(t) \leq{}& -\lambda_{\min}(Q_1)|X(t)|^2 + 2X^T P_1 B\beta(0,t) + \frac{p}{2}a_1 e^{\delta_1}\beta(1,t)^2 - \frac{p}{2}a_1\beta(0,t)^2 \\
&- \frac{p}{2}\delta_1 a_1 \int_0^1 e^{\delta_1 x}\beta(x,t)^2 dx - \frac{p}{2}\delta_1 b_1 \int_0^1 e^{-\delta_1 x}\alpha(x,t)^2 dx \\
&- \frac{p}{2}b_1 e^{-\delta_1}\alpha(1,t)^2 + \frac{p}{2}b_1\alpha(0,t)^2 \\
&+ b_1 \int_0^1 e^{-\delta_1 x}\alpha(x,t)c_1\left(\beta(x,t) - \int_0^x \psi^I(x,y)\alpha(y,t)dy\right. \\
&\left. - \int_0^x \phi^I(x,y)\beta(y,t)dy - \gamma^I(x)X(t)\right)dx. \tag{9.60}
\end{aligned}
$$

Let us consider the final part in (9.60) first. Using Young's inequality and the Cauchy-Schwarz inequality for the the final part in (9.60) yields the existence of $\xi > 0$ such that

$$
\begin{aligned}
\int_0^1 e^{-\delta_1 x}\alpha(x,t)c_1\beta(x,t)dx <{}& \xi \int_0^1 e^{-\delta_1 x}\alpha(x,t)^2 dx \\
&+ \xi \int_0^1 e^{\delta_1 x}\beta(x,t)^2 dx, \tag{9.61}
\end{aligned}
$$

$$\int_0^1 e^{-\delta_1 x}\alpha(x,t)c_1\int_0^x \psi^I(x,y)\alpha(y,t)dydx < \frac{\xi}{\delta_1}\int_0^1 e^{-\delta_1 x}\alpha(x,t)^2 dx, \tag{9.62}$$

$$
\begin{aligned}
\int_0^1 e^{-\delta_1 x}\alpha(x,t)c_1\int_0^x \phi^I(x,y)\beta(y,t)dydx <{}& \frac{\xi}{\delta_1}\int_0^1 e^{-\delta_1 x}\alpha(x,t)^2 dx \\
&+ \frac{\xi}{\delta_1}\int_0^1 e^{\delta_1 x}\beta(x,t)^2 dx, \tag{9.63}
\end{aligned}
$$

$$
\begin{aligned}
\int_0^1 e^{-\delta_1 x}\alpha(x,t)c_1\gamma^I(x)X(t)dx <{}& \frac{\lambda_{\min}(Q_1)}{4b_1}|X(t)|^2 \\
&+ \frac{\xi^2 b_1}{\lambda_{\min}(Q_1)}\int_0^1 e^{-\delta_1 x}\alpha(x,t)^2 dx. \tag{9.64}
\end{aligned}
$$

Recalling (9.14), applying Young's inequality, and substituting (9.61)–(9.64) into (9.60), we obtain

$$
\begin{aligned}
\dot{V}_1(t) \leq{}& -\left(\frac{1}{2}\lambda_{\min}(Q_1) - pb_1|C_0|^2\right)|X(t)|^2 - \left(\frac{p}{2}a_1 - pb_1 q^2 - \frac{4|PB|}{\lambda_{\min}(Q_1)}\right)\beta(0,t)^2 \\
&- \left(\frac{p}{2}\delta_1 a_1 - b_1\xi - b_1\frac{\xi}{\delta_1}\right)\int_0^1 \beta(x,t)^2 dx \\
&- \left(\frac{p}{2}\delta_1 b_1 - \frac{2b_1\xi}{\delta_1} - \frac{\xi^2 b_1^2}{\lambda_{\min}(Q_1)} - b_1\xi\right)e^{-\delta_1}\int_0^1 \alpha(x,t)^2 dx \\
&- \frac{p}{2}b_1 e^{-\delta_1}\alpha(1,t)^2 + \frac{p}{2}a_1 e^{\delta_1}\beta(1,t)^2. \tag{9.65}
\end{aligned}
$$

Choose parameters b_1, δ_1, a_1 in sequence to satisfy

$$0 < b_1 < \frac{\lambda_{\min}(Q_1)}{2p\,|C_0|^2}, \tag{9.66}$$

$$\delta_1 > \max\left\{1, \frac{2}{p}\left(3\xi + \frac{\xi^2 b_1}{\lambda_{\min}(Q_1)}\right)\right\}, \tag{9.67}$$

$$a_1 > \max\left\{\frac{8\,|PB|}{p\lambda_{\min}(Q_1)} + 2q^2 b_1, \ \frac{2b_1\xi}{p\delta_1} + \frac{2b_1\xi}{p\delta_1^2}\right\} \tag{9.68}$$

to make

$$\eta_1 = \frac{1}{2}\lambda_{\min}(Q_1) - pb_1\,|C_0|^2 > 0, \tag{9.69}$$

$$\eta_2 = \frac{p}{2}a_1 - pb_1 q^2 - \frac{4\,|PB|}{\lambda_{\min}(Q_1)} > 0, \tag{9.70}$$

$$\eta_3 = \frac{p}{2}\delta_1 a_1 - b_1\xi - b_1\frac{\xi}{\delta_1} > 0, \tag{9.71}$$

$$\eta_4 = \left(\frac{p}{2}\delta_1 b_1 - \frac{2b_1\xi}{\delta_1} - \frac{\xi^2 b_1^2}{\lambda_{\min}(Q_1)} - b_1\xi\right)e^{-\delta_1} > 0. \tag{9.72}$$

Defining

$$\eta_5 = \frac{p}{2}b_1 e^{-\delta_1} > 0, \quad \eta_6 = \frac{p}{2}a_1 e^{\delta_1} > 0, \tag{9.73}$$

we arrive at

$$\dot{V}_1(t) \le -\eta_1|X(t)|^2 - \eta_2\beta(0,t)^2 - \eta_3\int_0^1 \beta(x,t)^2 dx$$
$$-\eta_4\int_0^1 \alpha(x,t)^2 dx - \eta_5\alpha(1,t)^2 + \eta_6\beta(1,t)^2. \tag{9.74}$$

2) Lyapunov analysis for the whole ODE-PDE-ODE system

Recalling (9.38), we define a Lyapunov function

$$V(t) = V_1(t) + V_y(t). \tag{9.75}$$

Denoting the norm

$$\Omega_2(t) = \|\beta(\cdot,t)\|^2 + \|\alpha(\cdot,t)\|^2 + |X(t)|^2 + y_1(t)^2 + y_2(t)^2, \tag{9.76}$$

we get

$$\theta_{21}\Omega_2(t) \le V(t) \le \theta_{22}\Omega_2(t), \tag{9.77}$$

where

$$\theta_{21} = \min\left\{\lambda_{\min}(P_1), \frac{a_1}{2}, \frac{b_1 e^{-\delta_1}}{2}, \frac{1}{2}\right\} > 0, \tag{9.78}$$

$$\theta_{22} = \max\left\{\lambda_{\max}(P_1), \frac{a_1 e^{\delta_1}}{2}, \frac{b_1}{2}, \frac{1}{2}\right\} > 0. \tag{9.79}$$

Taking the derivative of (9.75) and using (9.74) and (9.42), we get

$$\dot{V} \leq -\eta_1 |X(t)|^2 - \eta_2 \beta(0,t)^2 - \eta_3 \int_0^1 \beta(x,t)^2 dx - \eta_4 \int_0^1 \alpha(x,t)^2 dx - \eta_5 \alpha(1,t)^2$$
$$+ \eta_6 \beta(1,t)^2 - \bar{c}_1 y_1(t)^2 - \bar{c}_2 y_2(t)^2 + y_2(t)\Big((h_6 + r)\alpha(1,t) + h_7\beta(0,t)$$
$$+ h_8(q\beta(0,t) + C_0 X(t)) + \int_0^1 h_9(y)\beta(y,t)dy$$
$$+ \int_0^1 h_{10}(y)\alpha(y,t)dy + H_{11}X(t)\Big), \tag{9.80}$$

where we have used (9.14).

Applying Young's inequality, the Cauchy-Schwarz inequality, and (9.32) to (9.80), we obtain

$$\dot{V} \leq -\left(\eta_1 - r_1 |H_{11}|^2 - r_7 h_8{}^2 |C_0|^2\right) |X(t)|^2 - (\eta_2 - h_7^2 r_3 - r_6 h_8{}^2 q^2)\beta(0,t)^2$$
$$- (\eta_3 - r_5 h_{9\,\max}^2) \int_0^1 \beta(x,t)^2 dx - (\eta_4 - r_2 h_{10\,\max}^2) \int_0^1 \alpha(x,t)^2 dx$$
$$- (\eta_5 - (h_6 + r)^2 r_4)\alpha(1,t)^2 - (\bar{c}_1 - \eta_6)y_1(t)^2$$
$$- \left(\bar{c}_2 - \left(\frac{1}{4r_1} + \frac{1}{4r_2} + \frac{1}{4r_3} + \frac{1}{4r_4} + \frac{1}{4r_5} + \frac{1}{4r_6} + \frac{1}{4r_7}\right)\right)y_2(t)^2. \tag{9.81}$$

We choose the positive constants $r_1, r_2, r_3, r_4, r_5, r_6, r_7$ as

$$r_1 < \frac{\eta_1}{|H_{11}|^2}, \quad r_2 < \frac{\eta_4}{h_{10\,\max}^2}, \quad r_3 < \frac{\eta_2}{h_7^2}, \quad r_4 < \frac{\eta_5}{(h_6 + r)^2},$$
$$r_5 < \frac{\eta_3}{h_{9\,\max}^2}, \quad r_6 < \frac{\eta_2 - h_7^2 r_3}{h_8^2 q^2}, \quad r_7 < \frac{\eta_1 - r_1|H_{11}|^2}{h_8^2|C_0|^2}, \tag{9.82}$$

where

$$h_{9\,\max} = \max_{x \in [0,1]}\{|h_9(x)|\}, \quad h_{10\,\max} = \max_{x \in [0,1]}\{|h_{10}(x)|\} \tag{9.83}$$

and choose the control parameters $\bar{c}_1, \bar{c}_2$ as

$$\bar{c}_1 > \eta_6, \tag{9.84}$$
$$\bar{c}_2 > \frac{1}{4}\left(\frac{1}{r_1} + \frac{1}{r_2} + \frac{1}{r_3} + \frac{1}{r_4} + \frac{1}{r_5} + \frac{1}{r_6} + \frac{1}{r_7}\right) \tag{9.85}$$

such that

$$\dot{V} \leq -\lambda \theta_{22}\Omega_2 - \hat{\eta}_0 \beta(0,t)^2 - \hat{\eta}_1 \alpha(1,t)^2$$
$$\leq -\lambda V - \hat{\eta}_0 \beta(0,t)^2 - \hat{\eta}_1 \alpha(1,t)^2 \tag{9.86}$$

for some positive λ, where

$$\hat{\eta}_0 = \eta_2 - h_7^2 r_3 - r_6 h_8{}^2 q^2 > 0$$

and

$$\hat{\eta}_1 = \eta_5 - (h_6 + r)^2 r_4 > 0.$$

From (9.77) and (9.86), we conclude that the target system $(\beta(x,t), \alpha(x,t), X(t), y_1(t), y_2(t))$ is exponentially stable in the sense of the norm

$$\left(\int_0^1 \alpha^2(x,t)dx + \int_0^1 \beta^2(x,t)dx + |X(t)|^2 + y_1(t)^2 + y_2(t)^2 \right)^{1/2}. \qquad (9.87)$$

Using the invertibility between the target $(\alpha(x,t), \beta(x,t))$ system and the original $(u(x,t), v(x,t))$ system via the transformation (9.8) and its inverse (9.10) and the invertibility between $(y_1(t), y_2(t))$ and $(\beta(1,t), \beta_t(1,t))$ via the invertible transformations (9.32), (9.33), together with (9.5), we obtain that the $(v(x,t), u(x,t), X(t), z_1(t), \beta_t(1,t))$ system is exponentially stable in the sense of the norm

$$\left(\int_0^1 u^2(x,t)dx + \int_0^1 v^2(x,t)dx + |X(t)|^2 + z_1(t)^2 + \beta_t(1,t)^2 \right)^{1/2}. \qquad (9.88)$$

Taking the derivative of the inverse transformation (9.10) and setting $x = 1$, together with (9.5), we get

$$\begin{aligned}
\dot{z}(t) &= v_t(1,t) \\
&= \beta_t(1,t) + p\psi^I(1,1)\alpha(1,t) - p\phi^I(1,1)\beta(1,t) \\
&\quad - p\psi^I(1,0)\alpha(0,t) + \left(p\phi^I(1,0) - \gamma^I(1)B \right)\beta(0,t) \\
&\quad + \int_0^1 \left(\int_y^1 c_1\psi^I(1,\sigma)\psi^I(\sigma,y)d\sigma - p\psi_y{}^I(1,y) \right)\alpha(y,t)dy \\
&\quad + \int_0^1 \left(\int_y^1 c_1\psi^I(1,\sigma)\phi^I(\sigma,y)d\sigma - c_1\psi^I(1,y) + p\phi_y{}^I(1,y) \right)\beta(y,t)dy \\
&\quad + \left(\int_0^1 c_1\psi^I(1,y)\gamma^I(y)dy - \gamma^I(1)(A + B\kappa) \right)X(t).
\end{aligned} \qquad (9.89)$$

Applying the Cauchy-Schwarz inequality into (9.89), with the exponential stability results in terms of $\|\alpha(\cdot,t)\|^2 + \|\beta(\cdot,t)\|^2 + |X(t)|^2 + |z(t)|^2 + |\beta_t(1,t)|^2$, shown in (9.87) and (9.88), we obtain the exponential convergence results in terms of $|\dot{z}(t)|^2$, namely $z_2(t)^2$.

With this, the proof of theorem 9.1 is complete. $\qquad \square$

9.5 BOUNDEDNESS AND EXPONENTIAL CONVERGENCE OF THE CONTROLLER

In the last section, we proved that all states of PDEs and ODEs are exponentially stable in the closed-loop system, which consists of the plant (9.1)–(9.6) and the controller (9.52). In this section, we prove the exponential convergence and bound-edness of the controller $U(t)$ (9.52) in the closed-loop system.

Theorem 9.2. *In the closed-loop system, which consists of the plant* (9.1)–(9.6) *and the controller $U(t)$* (9.52), *there exist positive constants λ_2 and Υ_0 such that*

$$|U(t)| \leq \Upsilon_0 e^{-\frac{\lambda_2}{2}t}, \quad \forall t \geq 0, \tag{9.90}$$

namely, ensuring that $U(t)$ is bounded and exponentially convergent to zero.

According to (9.52) and theorem 9.1, we know that if we want to show the exponential convergence of the controller (9.52), the exponential convergence of eight signals $u(1,t), u_x(1,t), v(0,t), v_x(0,t), v(1,t), v_x(1,t), u(0,t), u_x(0,t)$ in (9.52) needs to be proved, which can be obtained through producing L_2 estimates of $u_x(x,t), v_x(x,t), u_{xx}(x,t), v_{xx}(x,t)$.

Before the proof of theorem 9.2, we present two lemmas. The first shows the exponential stability estimates in terms of $\|u_x(x,t)\|^2 + \|v_x(x,t)\|^2$. The second gives the exponential stability estimates in terms of $\|u_{xx}(x,t)\|^2 + \|v_{xx}(x,t)\|^2$.

Lemma 9.2. *For all initial data $(u(x,0), v(x,0)) \in H^1(0,1)$, the closed-loop system $(u(x,t), v(x,t))$, comprising* (9.1)–(9.6) *with the controller* (9.52), *is exponentially stable in the sense of*

$$\|u_x(\cdot, t)\|^2 + \|v_x(\cdot, t)\|^2. \tag{9.91}$$

Proof. Differentiating (9.12) and (9.13) with respect to x and differentiating (9.14) with respect to t, we obtain

$$\begin{aligned}
\alpha_{xt}(x,t) = &-p\alpha_{xx}(x,t) + c_1\beta_x(x,t) - c_1\gamma^{I\prime}(x)X(t) \\
&- c_1\psi^I(x,x)\alpha(x,t) - c_1\phi^I(x,x)\beta(x,t) \\
&- c_1 \int_0^x \psi_x{}^I(x,y)\alpha(y,t)dy - c_1 \int_0^x \phi_x{}^I(x,y)\beta(y,t)dy,
\end{aligned} \tag{9.92}$$

$$\beta_{xt}(x,t) = p\beta_{xx}(x,t), \tag{9.93}$$

$$-\alpha_x(0,t) = q\beta_x(0,t) + \frac{1}{p}\left(C_0(A+B\kappa) + c_1\gamma^I(0)\right)X(t)$$

$$+ \frac{1}{p}(C_0B - c_1)\beta(0,t). \tag{9.94}$$

Let us define

$$A_1 = \frac{1}{2} \int_0^1 b_2 e^{-\delta_2 x}\alpha_x(x,t)^2 dx, \tag{9.95}$$

$$A_2 = \frac{1}{2} \int_0^1 a_2 e^{\delta_2 x}\beta_x(x,t)^2 dx, \tag{9.96}$$

where b_2 is an arbitrary positive constant, which can adjust the convergence rate, and the positive constants δ_2, a_2 shall be chosen later.

Taking the derivative of (9.95) along (9.92), (9.93), we obtain

$$\dot{A}_1 = -\frac{p}{2}b_2 e^{-\delta_2}\alpha_x(1,t)^2 + \frac{p}{2}b_2\alpha_x(0,t)^2 - \frac{p}{2}b_2\delta_2 \int_0^1 e^{-\delta_2 x}\alpha_x(x,t)^2 dx$$

$$-\int_0^1 b_2 e^{-\delta_2 x}\alpha_x(x,t)c_1\psi^I(x,x)\alpha(x,t)dx$$

$$-\int_0^1 b_2 e^{-\delta_2 x}\alpha_x(x,t)c_1\phi^I(x,x)\beta(x,t)dx$$

$$-\int_0^1 b_2 e^{-\delta_2 x}\alpha_x(x,t)c_1\int_0^x \psi_x^{I}(x,y)\alpha(y,t)dydx$$

$$-\int_0^1 b_2 e^{-\delta_2 x}\alpha_x(x,t)c_1\int_0^x \phi_x^{I}(x,y)\beta(y,t)dydx$$

$$+\int_0^1 b_2 e^{-\delta_2 x}\alpha_x(x,t)c_1\beta_x(x,t)dx$$

$$-\int_0^1 b_2 e^{-\delta_2 x}\alpha_x(x,t)c_1\gamma^{I'}(x)X(t)dx. \tag{9.97}$$

Let us consider the last six terms in (9.97) first. Using Young's inequality and the Cauchy-Schwarz inequality yields the existence of $\xi_2 > 0$ such that

$$\int_0^1 e^{-\delta_2 x}\alpha_x(x,t)c_1\psi^I(x,x)\alpha(x,t)dx$$

$$< \xi_2 \int_0^1 e^{-\delta_2 x}\alpha_x(x,t)^2 dx + \xi_2 \int_0^1 e^{-\delta_2 x}\alpha(x,t)^2 dx, \tag{9.98}$$

$$\int_0^1 e^{-\delta_2 x}\alpha_x(x,t)c_1\phi^I(x,x)\beta(x,t)dx$$

$$< \xi_2 \int_0^1 e^{-\delta_2 x}\alpha_x(x,t)^2 dx + \xi_2 \int_0^1 e^{\delta_2 x}\beta(x,t)^2 dx, \tag{9.99}$$

$$\int_0^1 e^{-\delta_2 x}\alpha_x(x,t)c_1\int_0^x (\psi_x^{I}(x,y)+\psi^I(x,y))\alpha(y,t)dydx$$

$$< \frac{\xi_2}{\delta_2} \int_0^1 e^{-\delta_2 x}\alpha_x(x,t)^2 dx + \frac{\xi_2}{\delta_2} \int_0^1 e^{-\delta_2 x}\alpha(x,t)^2 dx, \tag{9.100}$$

$$\int_0^1 e^{-\delta_2 x}\alpha_x(x,t)c_1\int_0^x (\phi_x^{I}(x,y)+\phi^I(x,y))\beta(y,t)dydx$$

$$< \frac{\xi_2}{\delta_2} \int_0^1 e^{-\delta_2 x}\alpha_x(x,t)^2 dx + \frac{\xi_2}{\delta_2} \int_0^1 e^{\delta_2 x}\beta(x,t)^2 dx, \tag{9.101}$$

$$\int_0^1 e^{-\delta_2 x}\alpha_x(x,t)c_1\beta_x(x,t)dx$$

$$< \xi_2 \int_0^1 e^{-\delta_2 x}\alpha_x(x,t)^2 dx + \xi_2 \int_0^1 e^{\delta_2 x}\beta_x(x,t)^2 dx, \tag{9.102}$$

$$\int_0^1 e^{-\delta_2 x}\alpha_x(x,t)c_1\gamma^{I'}(x)X(t)dx$$

$$< \xi_2 |X(t)|^2 + \xi_2 \int_0^1 e^{-\delta_2 x}\alpha_x(x,t)^2 dx. \tag{9.103}$$

Substituting (9.98)–(9.103) into (9.97), we obtain

$$\dot{A}_1 \le -\frac{p}{2}b_2 e^{-\delta_2}\alpha_x(1,t)^2 + \frac{p}{2}b_2\alpha_x(0,t)^2$$
$$- \left(\frac{p}{2}b_2\delta_2 - 4b_2\xi_2 - \frac{2b_2\xi_2}{\delta_2}\right)\int_0^1 e^{-\delta_2 x}\alpha_x(x,t)^2 dx$$
$$+ \left(\xi_2 b_2 + \frac{\xi_2 b_2}{\delta_2}\right)\int_0^1 e^{-\delta_2 x}\alpha(x,t)^2 dx + b_2\xi_2|X(t)|^2$$
$$+ \left(\xi_2 b_2 + \frac{\xi_2 b_2}{\delta_2}\right)\int_0^1 e^{\delta_2 x}\beta(x,t)^2 dx + \xi_2 b_2 \int_0^1 e^{\delta_2 x}\beta_x(x,t)^2 dx. \quad (9.104)$$

Taking the derivative of (9.96), we get

$$\dot{A}_2 = \frac{p}{2}a_2 e^{\delta_2}\beta_x(1,t)^2 - \frac{p}{2}a_2\beta_x(0,t)^2 - \frac{p}{2}a_2\delta_2 \int_0^1 e^{\delta_2 x}\beta_x(x,t)^2 dx. \quad (9.105)$$

Defining

$$\bar{A} = A_1 + A_2, \quad (9.106)$$

taking the derivative of $\bar{A}$, and using (9.104), (9.105), we obtain

$$\dot{\bar{A}}(t) \le -\frac{p}{2}b_2 e^{-\delta_2}\alpha_x(1,t)^2 + \frac{p}{2}b_2\alpha_x(0,t)^2 + \frac{p}{2}a_2 e^{\delta_2}\beta_x(1,t)^2 - \frac{p}{2}a_2\beta_x(0,t)^2$$
$$- \left(\frac{p}{2}b_2\delta_2 - 4\xi_2 b_2 - \frac{2\xi_2 b_2}{\delta_2}\right)\int_0^1 e^{-\delta_2 x}\alpha_x(x,t)^2 dx$$
$$- \left(\frac{p}{2}a_2\delta_2 - \xi_2 b_2\right)\int_0^1 e^{\delta_2 x}\beta_x(x,t)^2 dx$$
$$+ \left(\xi_2 b_2 + \frac{\xi_2 b_2}{\delta_2}\right)\int_0^1 e^{-\delta_2 x}\alpha(x,t)^2 dx$$
$$+ \left(\xi_2 b_2 + \frac{\xi_2 b_2}{\delta_2}\right)\int_0^1 e^{\delta_2 x}\beta(x,t)^2 dx + \xi_2 b_2|X(t)|^2. \quad (9.107)$$

Recalling (9.106), (9.75), we propose a Lyapunov function

$$V_2(t) = \bar{A} + R_1 V. \quad (9.108)$$

Denoting

$$\Omega_3(t) = \|\beta_x(\cdot,t)\|^2 + \|\alpha_x(\cdot,t)\|^2 + \|\beta(\cdot,t)\|^2$$
$$+ \|\alpha(\cdot,t)\|^2 + |X(t)|^2 + \beta(1,t)^2 + y_2(t)^2, \quad (9.109)$$

we have

$$\theta_{31}\Omega_3(t) \le V_2(t) \le \theta_{32}\Omega_3(t), \quad (9.110)$$

where

$$\theta_{31} = \min\left\{R_1\theta_{21}, \frac{a_2}{2}, \frac{b_2 e^{-\delta_2}}{2}\right\} > 0, \quad (9.111)$$

$$\theta_{32} = \max\left\{ R_1\theta_{22}, \frac{a_2 e^{\delta_2}}{2}, \frac{b_2}{2} \right\} > 0. \tag{9.112}$$

Taking the derivative of (9.108), recalling (9.107), (9.86), and applying the Cauchy-Schwarz inequality into (9.94) to rewrite $\alpha_x(0,t)^2$ in (9.107) as

$$\alpha_x(0,t)^2 \le 3q^2 \beta_x(0,t)^2 + \frac{3}{p^2}\left| C_0(A+B\kappa) + c_1 \gamma^I(0) \right|^2 |X(t)|^2$$

$$+ \frac{3}{p^2}(C_0 B - c_1)^2 \beta(0,t)^2, \tag{9.113}$$

we get

$$\dot{V}_2(t) = \dot{\tilde{A}} + R_1 \dot{V}$$

$$\le -\frac{p}{2}b_2 e^{-\delta_2}\alpha_x(1,t)^2$$

$$-\left(\frac{p}{2}a_2 - \frac{3pb_2 q^2}{2}\right)\beta_x(0,t)^2$$

$$-\left(\frac{p}{2}b_2\delta_2 - 4\xi_2 b_2 - \frac{2\xi_2 b_2}{\delta_2}\right)\int_0^1 \alpha_x(x,t)^2 dx$$

$$-\left(\frac{p}{2}a_2\delta_2 - \xi_2 b_2\right)\int_0^1 e^{\delta_2 x}\beta_x(x,t)^2 dx$$

$$-\left(R_1\hat{\eta}_0 - \frac{3b_2}{2p}(c_1 + C_0 B)^2\right)\beta(0,t)^2$$

$$-\left(R_1\theta_{22}\lambda - \frac{a_2 e^{\delta_2}}{p}\right)y_2(t)^2 - \left(R_1\theta_{22}\lambda - \frac{a_2 e^{\delta_2}}{p}\bar{c}_1^2\right)\beta(1,t)^2$$

$$-\left(R_1\theta_{22}\lambda - \xi_2 b_2 - \frac{\xi_2 b_2}{\delta_2}\right)\int_0^1 \alpha(x,t)^2 dx$$

$$-\left(R_1\theta_{22}\lambda - \xi_2 b_2 e^{\delta_2} - \frac{\xi_2 b_2}{\delta_2}e^{\delta_2}\right)\int_0^1 \beta(x,t)^2 dx$$

$$-\left(R_1\theta_{22}\lambda - \xi_2 b_2 - \frac{3b_2}{2p}|c_1\gamma^I(0) + C_0(A+B\kappa)|^2\right)|X(t)|^2 - R_1\hat{\eta}_1\alpha(1,t)^2$$

$$\le -\lambda_1\theta_{32}\Omega_3(t) - R_1\hat{\eta}_1\alpha(1,t)^2 - \hat{\eta}_2\beta(0,t)^2 - \hat{\eta}_3\beta_x(0,t)^2$$

$$\le -\lambda_1 V_2(t) - R_1\hat{\eta}_1\alpha(1,t)^2 - \hat{\eta}_2\beta(0,t)^2 - \hat{\eta}_3\beta_x(0,t)^2 \tag{9.114}$$

for some positive λ_1, where we have chosen

$$\delta_2 > \max\left\{1, \frac{12\xi_2}{p}\right\}, \quad a_2 > \max\left\{\frac{2\xi_2}{\delta_2}, 3b_2 q^2\right\} \tag{9.115}$$

and sufficiently large R_1. The coefficients $\hat{\eta}_1, \hat{\eta}_2, \hat{\eta}_3$ are

$$\hat{\eta}_2 = R_1\hat{\eta}_0 - \frac{3pb_2}{2p^2}(c_1 + C_0 B)^2 > 0, \tag{9.116}$$

$$\hat{\eta}_3 = \frac{p}{2}a_2 - \frac{3pb_2 q^2}{2} > 0. \tag{9.117}$$

Recalling (9.110), (9.114), we obtain the exponential stability estimates in the sense of

$$\|\alpha_x(\cdot, t)\|^2 + \|\beta_x(\cdot, t)\|^2. \tag{9.118}$$

Differentiating (9.9), (9.10) with respect to x, we get

$$u_x(x, t) = \alpha_x(x, t), \tag{9.119}$$

$$v_x(x, t) = \beta_x(x, t) - \int_0^x \psi_x^I(x, y)\alpha(y, t)dy - \int_0^x \phi_x^I(x, y)\beta(y, t)dy$$
$$- \dot{\gamma}^I(x)X(t) - \psi^I(x, x)\alpha(x, t) - \phi^I(x, x)\beta(x, t). \tag{9.120}$$

Using the Young and Cauchy-Schwarz inequalities, we get the inequalities

$$\|v_x(x, t)\|^2 \le 6\Big(\|\beta_x(x, t)\|^2 + K_\infty\|\alpha(x, t)\|^2$$
$$+ L_\infty\|\beta(x, t)\|^2 + \max_{x\in[0,1]}\{|\gamma^{I'}(x)|^2\}|X(t)|^2\Big), \tag{9.121}$$

where

$$K_\infty = \max_{(x,y)\in D}\{|\psi_x(x, y)|^2\} + \max_{x\in[0,1]}\{|\psi^I(x, x)|^2\}, \tag{9.122}$$

$$L_\infty = \max_{(x,y)\in D}\{|\phi_x(x, y)|^2\} + \max_{x\in[0,1]}\{|\phi^I(x, x)|^2\}. \tag{9.123}$$

Based on the exponential stability estimates in terms of $\|\alpha_x(\cdot, t)\|^2 + \|\beta_x(\cdot, t)\|^2$ proved above, together with the exponential stability results in terms of the norm that includes $\|\alpha(\cdot, t)\|^2 + \|\beta(\cdot, t)\|^2 + |X(t)|^2$, provided in theorem 9.1, we obtain the exponential stability estimates in terms of $\|u_x(\cdot, t)\|^2 + \|v_x(\cdot, t)\|^2$.

The proof of lemma 9.2 is complete. □

Lemma 9.3. *For all initial data $(u(x, 0), v(x, 0)) \in H^2(0, 1)$, the closed-loop system $(u(x, t), v(x, t))$ (9.1)–(9.6) with the controller (9.52) is exponentially stable in the sense of*

$$\|v_{xx}(\cdot, t)\|^2 + \|u_{xx}(\cdot, t)\|^2. \tag{9.124}$$

Proof. Differentiating (9.12) and (9.13) twice with respect to x and differentiating (9.14) twice with respect to t, we obtain

$$\alpha_{xxt}(x, t) = -p\alpha_{xxx}(x, t) + c_1\beta_{xx}(x, t) - c_1\gamma^{I''}(x)X(t)$$
$$- c_1\int_0^x \psi_{xx}^I(x, y)\alpha(y, t)dy - c_1\psi^I(x, x)\alpha_x(x, t)$$
$$- c_1\int_0^x \phi_{xx}^I(x, y)\beta(y, t)dy - c_1\phi^I(x, x)\beta_x(x, t)$$
$$- \Big(2c_1\psi_x^I(x, x) + c_1\psi_y^I(x, x)\Big)\alpha(x, t)$$
$$- \Big(2c_1\phi_x^I(x, x) + c_1\phi_y^I(x, x)\Big)\beta(x, t), \tag{9.125}$$

$$\beta_{xxt}(x, t) = p\beta_{xxx}(x, t), \tag{9.126}$$

and

$$\alpha_{xx}(0,t) = q\beta_{xx}(0,t) + \frac{1}{p}C_0 B\beta_x(0,t) - \frac{1}{p}c_1\psi^I(0,0)\alpha(0,t)$$

$$- \frac{1}{p^2}\left(pc_1\gamma^{I\prime}(0) - C_0(A+B\kappa)^2 - c_1\gamma(0)(A+B\kappa)\right)X(t)$$

$$- \frac{1}{p^2}\left(C_0(A+B\kappa)B + pc_1\phi^I(0,0) - c_1\gamma(0)B\right)\beta(0,t). \qquad (9.127)$$

We denote

$$B_1 = \frac{1}{2}\int_0^1 b_3 e^{-\delta_3 x}\alpha_{xx}(x,t)^2 dx, \qquad (9.128)$$

$$B_2 = \frac{1}{2}\int_0^1 a_3 e^{\delta_3 x}\beta_{xx}(x,t)^2 dx, \qquad (9.129)$$

where the positive constant b_3 can be chosen arbitrarily to adjust the convergence rate, and the the positive constants δ_3, a_3 shall be defined later.

Taking the derivative of (9.128) along (9.125), (9.126), we get

$$\dot{B}_1(t) = -\frac{p}{2}b_3 e^{-\delta_3}\alpha_{xx}(1,t)^2 + \frac{p}{2}b_3\alpha_{xx}(0,t)^2 - \frac{p}{2}b_3\delta_3\int_0^1 e^{-\delta_3 x}\alpha_{xx}(x,t)^2 dx$$

$$+ \int_0^1 b_3 e^{-\delta_3 x}\alpha_{xx}(x,t)c_1\beta_{xx}(x,t)dx$$

$$- \int_0^1 b_3 e^{-\delta_3 x}\alpha_{xx}(x,t)c_1\gamma^{I\prime\prime}(x)X(t)dx$$

$$- \int_0^1 b_3 e^{-\delta_3 x}\alpha_{xx}(x,t)c_1\int_0^x \psi_{xx}{}^I(x,y)\alpha(y,t)dydx$$

$$- \int_0^1 b_3 e^{-\delta_3 x}\alpha_{xx}(x,t)c_1\int_0^x \phi_{xx}{}^I(x,y)\beta(y,t)dydx$$

$$- \int_0^1 b_3 e^{-\delta_3 x}\alpha_{xx}(x,t)\left(2c_1\psi_x{}^I(x,x) + c_1\psi_y{}^I(x,x)\right)\alpha(x,t)dx$$

$$- \int_0^1 b_3 e^{-\delta_3 x}\alpha_{xx}(x,t)\left(2c_1\phi_x{}^I(x,x) + c_1\phi_y{}^I(x,x)\right)\beta(x,t)dx$$

$$- \int_0^1 b_3 e^{-\delta_3 x}\alpha_{xx}(x,t)c_1\psi^I(x,x)\alpha_x(x,t)dx$$

$$- \int_0^1 b_3 e^{-\delta_3 x}\alpha_{xx}(x,t)c_1\phi^I(x,x)\beta_x(x,t)dx. \qquad (9.130)$$

Now let us deal with the last eight terms in (9.130) by using Young's inequality and the Cauchy-Schwarz inequality. Similarly to (9.98)–(9.103), there exists a $\xi_3 > 0$ such that

$$\int_0^1 e^{-\delta_3 x}\alpha_{xx}(x,t)c_1\beta_{xx}(x,t)dx$$

$$\leq \xi_3 \int_0^1 e^{-\delta_3 x}\alpha_{xx}(x,t)^2 dx + \xi_3 \int_0^1 e^{\delta_3 x}\beta_{xx}(x,t)^2 dx, \qquad (9.131)$$

$$\int_0^1 e^{-\delta_3 x}\alpha_{xx}(x,t)c_1\ddot{\gamma}^I(x)X(t)dx$$

$$\leq \xi_3\int_0^1 e^{-\delta_3 x}\alpha_{xx}(x,t)^2 dx + \xi_3\left|X(t)\right|^2, \tag{9.132}$$

$$\int_0^1 e^{-\delta_3 x}\alpha_{xx}(x,t)c_1\int_0^x \psi_{xx}{}^I(x,y)\alpha(y,t)dydx$$

$$\leq \frac{\xi_3}{\delta_3}\int_0^1 e^{-\delta_3 x}\alpha_{xx}(x,t)^2 dx + \frac{\xi_3}{\delta_3}\int_0^1 e^{-\delta_3 x}\alpha(x,t)^2 dx, \tag{9.133}$$

$$\int_0^1 e^{-\delta_3 x}\alpha_{xx}(x,t)c_1\int_0^x \phi_{xx}{}^I(x,y)\beta(y,t)dydx$$

$$\leq \frac{\xi_3}{\delta_3}\int_0^1 e^{-\delta_3 x}\alpha_{xx}(x,t)^2 dx + \frac{\xi_3}{\delta_3}\int_0^1 e^{-\delta_3 x}\beta(x,t)^2 dx, \tag{9.134}$$

$$\int_0^1 e^{-\delta_3 x}\alpha_{xx}(x,t)\left(2c_1\psi_x{}^I(x,x)+c_1\psi_y{}^I(x,x)\right)\alpha(x,t)dx$$

$$\leq \xi_3\int_0^1 e^{-\delta_3 x}\alpha_{xx}(x,t)^2 dx + \xi_3\int_0^1 e^{\delta_3 x}\alpha(x,t)^2 dx, \tag{9.135}$$

$$\int_0^1 e^{-\delta_3 x}\alpha_{xx}(x,t)\left(2c_1\phi_x{}^I(x,x)+c_1\phi_y{}^I(x,x)\right)\beta(x,t)dx$$

$$\leq \xi_3\int_0^1 e^{-\delta_3 x}\alpha_{xx}(x,t)^2 dx + \xi_3\int_0^1 e^{\delta_3 x}\beta(x,t)^2 dx, \tag{9.136}$$

$$\int_0^1 e^{-\delta_3 x}\alpha_{xx}(x,t)c_1\psi^I(x,x)\alpha_x(x,t)dx$$

$$\leq \xi_3\int_0^1 e^{-\delta_3 x}\alpha_{xx}(x,t)^2 dx + \xi_3\int_0^1 e^{-\delta_3 x}\alpha_x(x,t)^2 dx, \tag{9.137}$$

$$\int_0^1 e^{-\delta_3 x}\alpha_{xx}(x,t)c_1\phi^I(x,x)\beta_x(x,t)dx$$

$$\leq \xi_3\int_0^1 e^{-\delta_3 x}\alpha_{xx}(x,t)^2 dx + \xi_3\int_0^1 e^{\delta_3 x}\beta_x(x,t)^2 dx. \tag{9.138}$$

Substituting (9.131)–(9.138) into (9.130), we obtain

$$\dot{B}_1(t) \leq -\frac{p}{2}b_3 e^{-\delta_3}\alpha_{xx}(1,t)^2 + \frac{p}{2}b_3\alpha_{xx}(0,t)^2$$

$$-\left(\frac{p}{2}b_3\delta_3 - 6\xi_3 b_3 - 2\frac{\xi_3 b_3}{\delta_3}\right)\int_0^1 e^{-\delta_3 x}\alpha_{xx}(x,t)^2 dx$$

$$+\xi_3 b_3\int_0^1 e^{\delta_3 x}\beta_{xx}(x,t)^2 dx + \xi_3 b_3\int_0^1 e^{\delta_3 x}\beta_x(x,t)^2 dx$$

$$+b_3\left(\xi_3 + \frac{\xi_3}{\delta_3}\right)\int_0^1 e^{\delta_3 x}\beta(x,t)^2 dx + b_3\left(\xi_3 + \frac{\xi_3}{\delta_3}\right)\int_0^1 e^{-\delta_3 x}\alpha(x,t)^2 dx$$

$$+\xi_3 b_3\int_0^1 e^{-\delta_3 x}\alpha_x(x,t)^2 dx + \xi_3 b_3\left|X(t)\right|^2. \tag{9.139}$$

Taking the derivative of (9.129) along (9.125), (9.126), we get

$$\dot{B}_2(t) = \int_0^1 a_3 e^{\delta_3 x} \beta_{xx}(x,t) \beta_{xxt}(x,t) dx$$

$$= p \int_0^1 a_3 e^{\delta_3 x} \beta_{xx}(x,t) \beta_{xxx}(x,t) dx$$

$$= \frac{p}{2} a_3 e^{\delta_3} \beta_{xx}(1,t)^2 - \frac{p}{2} a_3 \beta_{xx}(0,t)^2 - \frac{p}{2} a_3 \delta_3 \int_0^1 e^{\delta_3 x} \beta_{xx}(x,t)^2 dx. \quad (9.140)$$

Let us define

$$\bar{B} = B_1 + B_2. \quad (9.141)$$

Applying the Cauchy-Schwarz inequality into (9.127) yields

$$\alpha_{xx}(0,t)^2 \le 5q^2 \beta_{xx}(0,t)^2 + \frac{5}{p^2} |C_0 B|^2 \beta_x(0,t)^2 + \frac{5}{p^2} c_1^2 \psi^I(0,0)^2 \alpha(0,t)^2$$

$$+ \frac{5}{p^4} \left(pc_1 \gamma^{I'}(0) - C_0(A+B\kappa)^2 - c_1\gamma(0)(A+B\kappa) \right)^2 |X(t)|^2$$

$$+ \frac{5}{p^4} \left(C_0(A+B\kappa)B + pc_1\phi^I(0,0) - c_1\gamma(0)B \right)^2 \beta(0,t)^2, \quad (9.142)$$

which is used to replace $\alpha_{xx}(0,t)^2$ in (9.139). Then by recalling (9.139), (9.140), the inequality of the derivative of (9.141) is obtained as

$$\dot{\bar{B}} = \dot{B}_1 + \dot{B}_2$$

$$\le -\left(\frac{p}{2} b_3 \delta_3 - 6\xi_3 b_3 - 2\frac{\xi_3 b_3}{\delta_3} \right) \int_0^1 e^{-\delta_3 x} \alpha_{xx}(x,t)^2 dx$$

$$- \left(\frac{p}{2} a_3 \delta_3 - \xi_3 b_3 \right) \int_0^1 e^{\delta_3 x} \beta_{xx}(x,t)^2 dx - \left(\frac{p}{2} a_3 - \frac{5pb_3 q^2}{2} \right) \beta_{xx}(0,t)^2$$

$$- \frac{p}{2} b_3 e^{-\delta_3} \alpha_{xx}(1,t)^2 + \xi_3 b_3 \int_0^1 e^{\delta_3 x} \beta_x(x,t)^2 dx$$

$$+ b_3 \left(\xi_3 + \frac{\xi_3}{\delta_3} \right) \int_0^1 e^{\delta_3 x} \beta(x,t)^2 dx$$

$$+ b_3 \left(\xi_3 + \frac{\xi_3}{\delta_3} \right) \int_0^1 e^{-\delta_3 x} \alpha(x,t)^2 dx$$

$$+ \xi_3 b_3 \int_0^1 e^{-\delta_3 x} \alpha_x(x,t)^2 dx + \frac{p}{2} a_3 e^{\delta_3} \beta_{tt}(1,t)^2$$

$$+ \left(\frac{5b_3}{2p^3} \left(pc_1 \gamma^{I'}(0) - C_0(A+B\kappa)^2 - c_1\gamma(0)(A+B\kappa) \right)^2 + \xi_3 b_3 \right) |X(t)|^2$$

$$+ \frac{5b_3}{2p^3} \left(C_0(A+B\kappa)B + pc_1\phi^I(0,0) - c_1\gamma(0)B \right)^2 \beta(0,t)^2$$

$$+ \frac{5b_3}{2p} c_1^2 \psi^I(0,0)^2 \alpha(0,t)^2 + \frac{5|C_0 B|^2 b_3}{2p} \beta_x(0,t)^2. \quad (9.143)$$

Equation (9.143) includes a positive term $\beta_{tt}(1,t)^2$ ("positive term" here means a quadratic term with a plus sign) which is dealt with next. Substituting (9.41) into (9.18) yields

$$\begin{aligned}
\beta_{tt}(1,t) = &-(\bar{c}_2 + \bar{c}_1)\beta_t(1,t) - (\bar{c}_1\bar{c}_2 + 1)\beta(1,t) \\
&+ (h_6 + r)\alpha(1,t) + h_7\beta(0,t) + h_8\alpha(0,t) \\
&+ \int_0^1 h_9(y)\beta(y,t)dy + \int_0^1 h_{10}(y)\alpha(y,t)dy + H_{11}X(t).
\end{aligned} \tag{9.144}$$

Applying the Cauchy-Schwarz inequality in (9.144), we get

$$\begin{aligned}
\beta_{tt}(1,t)^2 \leq &8(\bar{c}_2 + \bar{c}_1)^2\beta_t(1,t)^2 + 8(\bar{c}_1\bar{c}_2 + 1)^2\beta(1,t)^2 \\
&+ 8(h_6 + r)^2\alpha(1,t)^2 + 8h_7{}^2\beta(0,t)^2 + 8h_8{}^2\alpha(0,t)^2 \\
&+ 8h_{9\,\max}\|\beta(\cdot,t)\|^2 + 8h_{10\,\max}\|\alpha(\cdot,t)\|^2 + 8H_{11}{}^2\,|X(t)|^2.
\end{aligned} \tag{9.145}$$

Replacing $\beta_{tt}(1,t)^2$ in (9.143) with (9.145) will be used in the following Lyapunov analysis.

Defining a Lypunov function

$$V_u = R_2V_2 + \bar{B} \tag{9.146}$$

and

$$\begin{aligned}
\Omega_4(t) = &\|\beta_{xx}(\cdot,t)\|^2 + \|\alpha_{xx}(\cdot,t)\|^2 + \|\beta_x(\cdot,t)\|^2 \\
&+ \|\alpha_x(\cdot,t)\|^2 + \|\beta(\cdot,t)\|^2 + \|\alpha(\cdot,t)\|^2 \\
&+ |X(t)|^2 + \beta(1,t)^2 + y_2(t)^2,
\end{aligned} \tag{9.147}$$

we obtain

$$\theta_{41}\Omega_4(t) \leq V_u(t) \leq \theta_{42}\Omega_4(t), \tag{9.148}$$

where

$$\theta_{41} = \min\left\{R_2\theta_{31}, \frac{a_3}{2}, \frac{b_3e^{-\delta_3}}{2}\right\} > 0, \tag{9.149}$$

$$\theta_{42} = \max\left\{R_2\theta_{32}, \frac{a_3e^{\delta_3}}{2}, \frac{b_3}{2}\right\} > 0. \tag{9.150}$$

By virtue of (9.32), (9.33), (9.36), (9.112), (9.114), (9.143), (9.145), taking the derivative of V_u, we get

$$\begin{aligned}
\dot{V}_u = &R_2\dot{V}_2 + \dot{\bar{B}} \\
\leq &-\frac{1}{2}R_2\lambda_1V_2 - \left(\frac{p}{2}b_3\delta_3 - 6\xi_3b_3 - 2\frac{\xi_3b_3}{\delta_3}\right)\int_0^1 \alpha_{xx}(x,t)^2dx \\
&- \left(\frac{p}{2}a_3\delta_3 - \xi_3b_3\right)\int_0^1 e^{\delta_3 x}\beta_{xx}(x,t)^2dx - \left(\frac{p}{2}a_3 - \frac{5pb_3q^2}{2}\right)\beta_{xx}(0,t)^2 \\
&- \frac{p}{2}b_3e^{-\delta_3}\alpha_{xx}(1,t)^2 - \left(\frac{1}{2}R_2\lambda_1\theta_{32} - \xi_3b_3e^{\delta_3}\right)\int_0^1 \beta_x(x,t)^2dx
\end{aligned}$$

$$- \left(\frac{1}{2} R_2 \lambda_1 \theta_{32} - \xi_3 b_3 e^{\delta_3} - \frac{\xi_3 b_3}{\delta_3} e^{\delta_3} - 4pa_3 e^{\delta_3} h_{9 \max} \right) \int_0^1 \beta(x,t)^2 dx$$

$$- \left(\frac{1}{2} R_2 \lambda_1 \theta_{32} - \xi_3 b_3 - \frac{\xi_3 b_3}{\delta_3} - 4pa_3 e^{\delta_3} h_{10 \max} \right) \int_0^1 \alpha(x,t)^2 dx$$

$$- \left(\frac{1}{2} R_2 \lambda_1 \theta_{32} - \xi_3 b_3 \right) \int_0^1 \alpha_x(x,t)^2 dx$$

$$- \left[\frac{1}{2} R_2 \lambda_1 \theta_{32} - \left(\frac{5b_3}{2p^3} \left(pc_1 \gamma^{I'}(0) - C_0(A + B\kappa)^2 - c_1\gamma(0)(A + B\kappa) \right)^2 + \xi_3 b_3 \right. \right.$$

$$\left. + 4pa_3 e^{\delta_3} {H_{11}}^2 \right) - 2 \left(\frac{5b_3}{2p} c_1^2 \psi(0,0)^2 + 4pa_3 e^{\delta_3} {h_8}^2 \right) |C_0|^2 \bigg] |X(t)|^2$$

$$- \left[R_2 \hat{\eta}_2 - \left(\frac{5b_3}{2p^3} \left(C_0(A + B\kappa)B + pc_1 \phi^I(0,0) - c_1\gamma(0)B \right)^2 + 4pa_3 e^{\delta_3} {h_7}^2 \right) \right.$$

$$\left. - 2q^2 \left(\frac{5b_3}{2p} c_1^2 \psi(0,0)^2 + 4pa_3 e^{\delta_3} {h_8}^2 \right) \right] \beta(0,t)^2$$

$$- \left(R_2 \hat{\eta}_3 - \frac{5|C_0 B|^2 b_3}{2p} \right) \beta_x(0,t)^2$$

$$- \left(\frac{1}{2} R_2 \lambda_1 \theta_{32} - 4pa_3 e^{\delta_3} (\bar{c}_2 + \bar{c}_1)^2 \right) \beta_t(1,t)^2$$

$$- \left(\frac{1}{2} R_2 \lambda_1 \theta_{32} - 4pa_3 e^{\delta_3} (\bar{c}_1 \bar{c}_2 + 1)^2 \right) \beta(1,t)^2$$

$$- \left(R_2 R_1 \hat{\eta}_1 - 4pa_3 e^{\delta_3} (h_6 + r)^2 \right) \alpha(1,t)^2. \tag{9.151}$$

Choosing

$$\delta_3 > \max \left\{ 1, \frac{16\xi_3}{p} \right\}, \quad a_3 > \max \left\{ \frac{2\xi_3 b_3}{p\delta_3}, 5b_3 q^2 \right\}$$

and sufficiently large R_2, we obtain

$$\dot{V}_u(t) \leq -\frac{1}{2} R_2 \lambda_1 V_2 - \sigma_2 \bar{B}, \tag{9.152}$$

with the positive constant

$$\sigma_2 = \frac{2 \min\{\frac{p}{2} b_3 \delta_3 - 6\xi_3 b_3 - 2\frac{\xi_3 b_3}{\delta_3}, \frac{p}{2} a_3 \delta_3 - \xi_3 b_3\}}{\max\{a_3, b_3\} e^{\delta_3}}. \tag{9.153}$$

Then we arrive at

$$\dot{V}_u(t) \leq -\lambda_2 V_u(t), \tag{9.154}$$

where

$$\lambda_2 = \min \left\{ \frac{1}{2} \lambda_1, \sigma_2 \right\}. \tag{9.155}$$

Hence, we obtain

$$V_u(t) \le e^{-\lambda_2 t} V_u(0), \quad \forall t \ge 0. \tag{9.156}$$

Therefore, we obtain the exponential stability estimates in terms of $\|\alpha_{xx}(\cdot, t)\|^2 + \|\beta_{xx}(\cdot, t))\|^2$.

Differentiating (9.9), (9.10) twice with respect to x, we get

$$u_{xx}(x, t) = \alpha_{xx}(x, t),$$

$$\begin{aligned}
v_{xx}(x, t) &= \beta_{xx}(x, t) - \gamma^{I''}(x) X(t) \\
&\quad - \left(2\psi^I_x(x, x) + \psi^I_y(x, x) \right) \alpha(x, t) - \left(2\phi_x^I(x, x) + \phi_y^I(x, x) \right) \beta(x, t) \\
&\quad - \int_0^x \psi^I_{xx}(x, y) \alpha(y, t) dy - \int_0^x \phi_{xx}^I(x, y) \beta(y, t) dy \\
&\quad - \psi^I(x, x) \alpha_x(x, t) - \phi^I(x, x) \beta_x(x, t).
\end{aligned} \tag{9.157}$$

Through a similar calculation as in (9.121), with the exponential stability estimates in terms of $\|\alpha_{xx}(\cdot, t)\|^2 + \|\beta_{xx}(\cdot, t)\|^2$ proved above, and recalling the exponential stability estimates in terms of $\|\alpha_x(\cdot, t)\|^2 + \|\beta_x(\cdot, t)\|^2$ shown in lemma 9.2, together with the exponential stability results in terms of the norm including $\|\alpha(\cdot, t)\|^2 + \|\beta(\cdot, t)\|^2 + |X(t)|^2$ provided in theorem 9.1, we obtain the exponential stability estimates in terms of $\|u_{xx}(\cdot, t)\|^2 + \|v_{xx}(\cdot, t)\|^2$.

The proof of lemma 9.3 is thus complete. $\qquad\square$

Using lemmas 9.2 and 9.3, we prove theorem 9.2 as follows.

Proof. Recalling (9.52) and using the Cauchy-Schwarz inequality, we obtain

$$\begin{aligned}
&|U(t)|^2 \\
&\le \Big(\bar{\xi}_1 v_x(1, t)^2 + \xi_2 v(1, t)^2 + \bar{\xi}_3 u_x(1, t)^2 + \bar{\xi}_4 u(1, t)^2 + \bar{\xi}_5 v_x(0, t)^2 + \bar{\xi}_6 v(0, t)^2 \\
&\quad + \bar{\xi}_7 u_x(0, t)^2 + \bar{\xi}_8 u(0, t)^2 + \bar{\xi}_9 |X(t)|^2 + \bar{\xi}_{10} \|u(\cdot, t)\|^2 + \bar{\xi}_{11} \|v(\cdot, t)\|^2 \Big)
\end{aligned} \tag{9.158}$$

for some positive constants $\bar{\xi}_1, \bar{\xi}_2, \bar{\xi}_3, \bar{\xi}_4, \bar{\xi}_5, \bar{\xi}_6, \bar{\xi}_7, \bar{\xi}_8, \bar{\xi}_9, \bar{\xi}_{10}, \bar{\xi}_{11}$.

Recalling the exponential estimates in terms of the norms $\|u(\cdot, t)\|_{H_2} + \|v(\cdot, t)\|_{H_2}$ proved in lemmas 9.2 and 9.3, using the Sobolev inequality, we obtain the exponential estimate in terms of the norm $\|u(\cdot, t)\|_{C^1} + \|v(\cdot, t)\|_{C^1}$, which gives the exponential convergence of $U(t)$ by recalling (9.158) and theorem 9.1. The upper boundedness Υ_0 of $|U(t)|$ depends on the initial values of the terms in (9.158).

The proof of theorem 9.2 is complete. $\qquad\square$

9.6   EXTENSION TO ODES OF ARBITRARY ORDER

In this section, we allow the possibility that the input ODE is not of second-order but of arbitrary order m and provide a sketch of the design and analysis for this general case. To avoid repetition, we omit some detailed calculations that can be developed by relying on calculations in sections 9.2–9.5.

Replace (9.5), (9.6) with

$$v(1, t) = C_z Z(t), \tag{9.159}$$

$$\dot{Z}(t) = A_z Z(t) + B_z U(t) + \phi_z, \tag{9.160}$$

where $Z(t) \in \mathbb{R}^{m \times 1}$, and

$$A_z = [0, 1, 0, \ldots, 0; 0, 0, 1, 0, \ldots, 0; \cdots; 0, \ldots, 0, 1; a_{z1}, \ldots, a_{zm}] \in \mathbb{R}^{m \times m}, \tag{9.161}$$

with $a_{z1}, \ldots, a_{zm}$ as arbitrary constants. The matrices B_z and C_z are

$$B_z = [0, 0, \ldots, 1]^T \in \mathbb{R}^{m \times 1}, \ \ C_z = [1, 0, \ldots, 0] \in \mathbb{R}^{1 \times m},$$

and ϕ_z is

$$\phi_z = R_{z1} v(0, t) + R_{z2} u(1, t) + \int_0^1 R_{z3}(x) u(x, t) dx$$

$$+ \int_0^1 R_{z4}(x) v(x, t) dx + R_{z5} X(t), \tag{9.162}$$

where the constants R_{z1}, R_{z2}, the bounded functions $R_{z3}(x)$, $R_{z4}(x)$, and the constant matrix R_{z5} are arbitrary. It is assumed that the system (9.159), (9.160) has the full relative degree.

Control Design

The equation (9.160), with the state variable choices $Z(t) = [z_1(t), \ldots, z_m(t)]^T$, is in the form of a chain of m integrators given by

$$\dot{z}_1(t) = z_2(t), \tag{9.163}$$

$$\dot{z}_2(t) = z_3(t), \tag{9.164}$$

$$\vdots \tag{9.165}$$

$$\dot{z}_{m-1}(t) = z_m(t), \tag{9.166}$$

$$\dot{z}_m(t) = a_{z1} z_1(t) + \cdots + a_{zm} z_m(t) + U(t) + \phi_z. \tag{9.167}$$

The form of system (9.18) extended to order m is obtained as

$$\partial_t^m \beta(1, t)$$
$$= U(t) + [q_m \partial_t^{m-1} \beta(1, t) + q_{m-1} \partial_t^{m-2} \beta(1, t) + \cdots + q_2 \beta_t(1, t) + q_1 \beta(1, t)]$$
$$+ [p_m \partial_t^{m-1} \alpha(1, t) + p_{m-1} \partial_t^{m-2} \alpha(1, t) + \cdots + p_2 \alpha_t(1, t) + p_1 \alpha(1, t)]$$
$$+ [\bar{q}_m \partial_t^{m-1} \beta(0, t) + \bar{q}_{m-1} \partial_t^{m-2} \beta(0, t) + \cdots + \bar{q}_2 \beta_t(0, t) + \bar{q}_1 \beta(0, t)]$$
$$+ [\bar{p}_m \partial_t^{m-1} \alpha(0, t) + \bar{p}_{m-1} \partial_t^{m-2} \alpha(0, t) + \cdots + \bar{p}_2 \alpha_t(0, t) + \bar{p}_1 \alpha(0, t)]$$
$$+ \int_0^1 \bar{Q}(y) \beta(y, t) dy + \int_0^1 \bar{P}(y) \alpha(y, t) dy + H_z X(t), \tag{9.168}$$

where $q_m, \ldots, q_1, p_m, \ldots, p_1, \bar{q}_m, \ldots, \bar{q}_1, \bar{p}_m, \ldots, \bar{p}_1, \bar{Q}(y), \bar{P}(y), H_z$ are coefficients consisting of the kernel functions in the backstepping transformation (9.8), (9.10) and the plant parameters in (9.1)–(9.4), (9.159)–(9.162).

The following backstepping transformation for the $(\beta(1,t), \beta_t(1,t), \ldots, \partial_t^{m-1}\beta(1,t))$ system (9.168) is postulated as

$$y_1(t) = \beta(1,t), \tag{9.169}$$
$$y_2(t) = \beta_t(1,t) + \tau_1[\beta(1,t)], \tag{9.170}$$

$$\vdots \tag{9.171}$$

$$y_m(t) = \partial_t^{m-1}\beta(1,t) + \tau_{m-1}[\beta(1,t), \ldots, \partial_t^{m-2}\beta(1,t)], \tag{9.172}$$

where $\tau_1, \ldots, \tau_{m-1}$ are determined in the following steps as the virtual control laws in the ODE backstepping method.

Step 1. We consider a Lyapunov function candidate

$$V_{y1} = \frac{1}{2}y_1(t)^2. \tag{9.173}$$

Taking the derivative of (9.173), we obtain

$$\dot{V}_{y1} = -\hat{c}_1 y_1(t)^2 + y_1(t)y_2(t), \tag{9.174}$$

with the choice of

$$\tau_1 = \hat{c}_1 y_1(t), \tag{9.175}$$

where $\hat{c}_1$ is a positive constant to be determined later.

Step 2. A Lyapunov function candidate is considered as

$$V_{y2} = V_{y1} + \frac{1}{2}y_2(t)^2 = \frac{1}{2}y_1(t)^2 + \frac{1}{2}y_2(t)^2. \tag{9.176}$$

Taking the derivative of (9.176), we obtain

$$\dot{V}_{y2} = -\hat{c}_1 y_1(t)^2 + y_1(t)y_2(t) + y_2(t)(y_3(t) - \tau_2 + \dot{\tau}_1). \tag{9.177}$$

Choosing

$$\tau_2 = \dot{\tau}_1 + y_1(t) + \hat{c}_2 y_2(t), \tag{9.178}$$

where

$$\dot{\tau}_1 = \hat{c}_1 \dot{y}_1 = \hat{c}_1 \beta_t(1,t) \tag{9.179}$$

according to (9.175), (9.169), we get

$$\dot{V}_{y2} = -\hat{c}_1 y_1(t)^2 - \hat{c}_2 y_2(t)^2 + y_2(t)y_3(t). \tag{9.180}$$

Step 3. through **step m-1.** We make an induction hypothesis that for a Lyapunov function candidate

$$V_{yi} = V_{yi-1} + \frac{1}{2}y_i(t)^2$$
$$= \frac{1}{2}y_1(t)^2 + \frac{1}{2}y_2(t)^2 + \cdots + \frac{1}{2}y_{i-1}(t)^2 + \frac{1}{2}y_i(t)^2, \quad \forall\, 2 \le i \le m-2, \tag{9.181}$$

we have

$$\dot{V}_{yi} = -\hat{c}_1 y_1(t)^2 - \hat{c}_2 y_2(t)^2 - \cdots - \hat{c}_{i-1} y_{i-1}(t)^2 - \hat{c}_i y_i(t)^2$$
$$+ y_i(t) y_{i+1}(t), \quad \forall i \geq 2 \tag{9.182}$$

by choosing

$$\tau_i = \dot{\tau}_{i-1} + y_{i-1}(t) + \hat{c}_i y_i(t). \tag{9.183}$$

Consider a Lyapunov function candidate

$$V_{yi+1} = V_{yi} + \frac{1}{2} y_{i+1}(t)^2$$
$$= \frac{1}{2} y_1(t)^2 + \frac{1}{2} y_2(t)^2 + \ldots + \frac{1}{2} y_i(t)^2 + \frac{1}{2} y_{i+1}(t)^2. \tag{9.184}$$

Taking the derivative of (9.184), using (9.182) and (9.169)–(9.172), we obtain

$$\dot{V}_{yi+1} = \dot{V}_{yi} + y_{i+1}(t) \dot{y}_{i+1}(t)$$
$$= -\hat{c}_1 y_1(t)^2 - \hat{c}_2 y_2(t)^2 - \ldots - \hat{c}_{i-1} y_{i-1}(t)^2 - \hat{c}_i y_i(t)^2$$
$$+ y_i(t) y_{i+1}(t) + y_{i+1}(t)(\dot{\tau}_i + y_{i+2} - \tau_{i+1})$$
$$= -\hat{c}_1 y_1(t)^2 - \hat{c}_2 y_2(t)^2 - \ldots - \hat{c}_{i-1} y_{i-1}(t)^2 - \hat{c}_i y_i(t)^2$$
$$- \hat{c}_{i+1} y_{i+1}(t)^2 + y_{i+1}(t) y_{i+2}(t) \tag{9.185}$$

by choosing

$$\tau_{i+1} = \dot{\tau}_i + y_i(t) + \hat{c}_{i+1} y_{i+1}(t). \tag{9.186}$$

Therefore, (9.181)–(9.183) holds for $i+1$. Recalling steps 1 and 2, it follows that for a Lyapunov function candidate

$$V_{ym-1} = \frac{1}{2} y_1(t)^2 + \frac{1}{2} y_2(t)^2 + \cdots + \frac{1}{2} y_{m-1}(t)^2, \tag{9.187}$$

we have

$$\dot{V}_{ym-1} = -\hat{c}_1 y_1(t)^2 - \hat{c}_2 y_2(t)^2 - \ldots - \hat{c}_{m-2} y_{m-2}(t)^2$$
$$- \hat{c}_{m-1} y_{m-1}(t)^2 + y_{m-1}(t) y_m(t) \tag{9.188}$$

by choosing

$$\tau_{m-1} = \dot{\tau}_{m-2} + y_{m-2}(t) + \hat{c}_{m-1} y_{m-1}(t). \tag{9.189}$$

Step m. A Lyapunov function candidate is considered as

$$V_{ym} = V_{ym-1} + \frac{1}{2} y_m(t)^2$$
$$= \frac{1}{2} y_1(t)^2 + \frac{1}{2} y_2(t)^2 + \cdots + \frac{1}{2} y_{m-1}(t)^2 + \frac{1}{2} y_m(t)^2. \tag{9.190}$$

Taking the derivative of (9.190), recalling (9.188), we obtain

$$\dot{V}_{ym} = -\hat{c}_1 y_1(t)^2 - \hat{c}_2 y_2(t)^2 - \cdots - \hat{c}_{m-1} y_{m-1}(t)^2$$
$$+ y_{m-1}(t) y_m(t) + y_m(t) \dot{y}_m(t). \tag{9.191}$$

According to (9.168) and (9.172), (9.191) is rewritten as

$$\dot{V}_{ym}$$

$$= -\hat{c}_1 y_1(t)^2 - \hat{c}_2 y_2(t)^2 - \cdots - \hat{c}_{m-1} y_{m-1}(t)^2 + y_{m-1}(t) y_m(t) + y_m(t) \Bigg[U(t)$$
$$+ \big[q_m \partial_t^{m-1} \beta(1,t) + q_{m-1} \partial_t^{m-2} \beta(1,t) + \cdots + q_2 \beta_t(1,t) + q_1 \beta(1,t) \big]$$
$$+ \big[p_m \partial_t^{m-1} \alpha(1,t) + p_{m-1} \partial_t^{m-2} \alpha(1,t) + \cdots + p_2 \alpha_t(1,t) + p_1 \alpha(1,t) \big]$$
$$+ \big[\bar{q}_m \partial_t^{m-1} \beta(0,t) + \bar{q}_{m-1} \partial_t^{m-2} \beta(0,t) + \cdots + \bar{q}_2 \beta_t(0,t) + \bar{q}_1 \beta(0,t) \big]$$
$$+ \big[\bar{p}_m \partial_t^{m-1} \alpha(0,t) + \bar{p}_{m-1} \partial_t^{m-2} \alpha(0,t) + \cdots + \bar{p}_2 \alpha_t(0,t) + \bar{p}_1 \alpha(0,t) \big]$$
$$+ \int_0^1 \bar{Q}(y) \beta(y,t) dy + \int_0^1 \bar{P}(y) \alpha(y,t) dy + H_z X(t) + \dot{\tau}_{m-1} \Bigg], \tag{9.192}$$

where

$$\dot{\tau}_{m-1} = \hat{c}_1 y_1^{m-1}(t) + \sum_{i=1}^{i=m-2} \big(y_i^{m-1-i}(t) + \hat{c}_{i+1} y_{i+1}^{m-1-i}(t) \big) \tag{9.193}$$

for $m-1 \geq 2$, by virtue of (9.189), and $y_i^n(t)$ denotes the n order derivative of $y_i(t)$, $\forall i = 1, \ldots, m$. For $m-1 = 1$, $\dot{\tau}_1$ is given in (9.179).

Design the controller as

$$U(t) = - \big[q_m \partial_t^{m-1} \beta(1,t) + q_{m-1} \partial_t^{m-2} \beta(1,t) + \cdots + q_2 \beta_t(1,t) + q_1 \beta(1,t) \big]$$
$$- \big[p_m \partial_t^{m-1} \alpha(1,t) + p_{m-1} \partial_t^{m-2} \alpha(1,t) + \cdots + p_2 \alpha_t(1,t) \big]$$
$$- \big[\bar{q}_m \partial_t^{m-1} \beta(0,t) + \bar{q}_{m-1} \partial_t^{m-2} \beta(0,t) + \cdots + \bar{q}_2 \beta_t(0,t) \big]$$
$$- \big[\bar{p}_m \partial_t^{m-1} \alpha(0,t) + \bar{p}_{m-1} \partial_t^{m-2} \alpha(0,t) + \cdots + \bar{p}_2 \alpha_t(0,t) \big]$$
$$- y_{m-1}(t) - \dot{\tau}_{m-1} - \hat{c}_m y_m(t)$$

$$= - \Bigg[\Bigg(q_m + \sum_{i=1}^{m-2} \hat{c}_i \Bigg) \partial_t^{m-1} \beta(1,t)$$
$$+ \Bigg(q_{m-1} + m - 2 + \sum_{i=2}^{m-2} \hat{c}_i \hat{c}_{i-1} \Bigg) \partial_t^{m-2} \beta(1,t) + \cdots + q_1 \beta(1,t) \Bigg]$$
$$- \big[p_m \partial_t^{m-1} \alpha(1,t) + p_{m-1} \partial_t^{m-2} \alpha(1,t) + \cdots + p_2 \alpha_t(1,t) \big]$$
$$- \big[\bar{q}_m \partial_t^{m-1} \beta(0,t) + \bar{q}_{m-1} \partial_t^{m-2} \beta(0,t) + \cdots + \bar{q}_2 \beta_t(0,t) \big]$$
$$- \big[\bar{p}_m \partial_t^{m-1} \alpha(0,t) + \bar{p}_{m-1} \partial_t^{m-2} \alpha(0,t) + \cdots + \bar{p}_2 \alpha_t(0,t) \big]$$
$$- y_{m-1}(t) - \hat{c}_{m-1} \dot{y}_{m-1}(t) - \hat{c}_m y_m(t). \tag{9.194}$$

Using the transformations (9.169)–(9.172), (9.7), (9.8), the controller (9.194) is expressed as a function of the original state $u(x,t), v(x,t)$ as

$$U(t) = \hat{n}_{m-1} \partial_x^{m-1} v(1,t) + \hat{n}_{m-2} \partial_x^{m-2} v(1,t), \cdots + \hat{n}_0 v(1,t)$$
$$+ \hat{k}_{m-1} \partial_x^{m-1} u(1,t) + \hat{k}_{m-2} \partial_x^{m-2} u(1,t) + \cdots + \hat{k}_0 u(1,t)$$

$$+ \hat{h}_{m-1}\partial_x^{m-1}v(0,t) + \hat{h}_{m-2}\partial_x^{m-2}v(0,t) + \cdots + \hat{h}_0 v(0,t)$$

$$+ \hat{l}_{m-1}\partial_x^{m-1}u(0,t) + \hat{l}_{m-2}\partial_x^{m-2}u(0,t) + \cdots + \hat{l}_0 u(0,t)$$

$$+ \hat{D}X(t) + \int_0^1 \hat{N}(y)u(y,t)dy + \int_0^1 \hat{L}(y)v(y,t)dy, \qquad (9.195)$$

where the equations (9.2), (9.3) have been used to replace the time derivatives arising in (9.194) with the spatial derivatives in (9.195).

The gains $\hat{n}_{m-1}, \ldots, \hat{n}_0$, $\hat{k}_{m-1}, \ldots, \hat{k}_0$, $\hat{h}_{m-1}, \ldots, \hat{h}_0$, $\hat{l}_{m-1}, \ldots, \hat{l}_0$, and $\hat{D}, \hat{N}(y), \hat{L}(y)$ consist of the kernels in the backstepping transformations (9.8), (9.10), the plant parameters in (9.1)–(9.4), (9.159), (9.160), and the control parameters $\hat{c}_1, \ldots, \hat{c}_m, \kappa$.

Now we get

$$\begin{aligned}
\dot{V}_{ym} = {}& -\hat{c}_1 y_1(t)^2 - \hat{c}_2 y_2(t)^2 - \cdots - \hat{c}_m y_m(t)^2 \\
& + y_m(t)\Big(\bar{q}_1\beta(0,t) + \bar{p}_1\alpha(0,t) + p_1\alpha(1,t) \\
& \quad + \int_0^1 \bar{Q}(y)\beta(y,t)dy + \int_0^1 \bar{P}(y)\alpha(y,t)dy + H_z X(t) \Big),
\end{aligned} \qquad (9.196)$$

where $\hat{c}_1, \ldots, \hat{c}_m$ are positive constants to be determined later.

Stability Analysis

Theorem 9.3. *For all initial values $(u(x,0), v(x,0)) \in W^{m,2}(0,1)$, the closed-loop system consisting of the plant (9.1)–(9.4), (9.159), (9.160) and the control law (9.195) is exponentially stable at the origin in the sense of the norm*

$$\left(\int_0^1 u^2(x,t)dx + \int_0^1 v^2(x,t)dx + |X(t)|^2 + z_1(t)^2 + \cdots + z_m(t)^2 \right)^{1/2}. \qquad (9.197)$$

Proof. Recalling (9.190) and (9.55), we define a Lyapunov function as

$$V_m(t) = V_1(t) + V_{ym}(t). \qquad (9.198)$$

Denoting

$$\Omega_{2m}(t) = \|\beta(\cdot,t)\|^2 + \|\alpha(\cdot,t)\|^2 + |X(t)|^2 + y_1(t)^2 + \cdots + y_m(t)^2, \qquad (9.199)$$

we get

$$\theta_{21}\Omega_{2m}(t) \le V_m(t) \le \theta_{22}\Omega_{2m}(t), \qquad (9.200)$$

where

$$\theta_{21} = \min\left\{ \lambda_{\min}(P_1), \frac{a_1}{2}, \frac{b_1 e^{-\delta_1}}{2}, \frac{1}{2} \right\} > 0, \qquad (9.201)$$

$$\theta_{22} = \max\left\{ \lambda_{\max}(P_1), \frac{a_1 e^{\delta_1}}{2}, \frac{b_1}{2}, \frac{1}{2} \right\} > 0. \qquad (9.202)$$

Taking the derivative of (9.198) and using (9.74) and (9.196), we get

$$\dot{V}_m \leq -\eta_1 |X(t)|^2 - \eta_2 \beta(0,t)^2 - \eta_3 \int_0^1 \beta(x,t)^2 dx - \eta_4 \int_0^1 \alpha(x,t)^2 dx - \eta_5 \alpha(1,t)^2$$

$$+ \eta_6 y_1(t)^2 - \hat{c}_1 y_1(t)^2 - \hat{c}_2 y_2(t)^2 - \cdots - \hat{c}_{m-1} y_{m-1}(t)^2 - \hat{c}_m y_m(t)^2$$

$$+ y_m(t) \bigg(\bar{q}_1 \beta(0,t) + \bar{p}_1 \alpha(0,t) + p_1 \alpha(1,t) + \int_0^1 Q(y)\beta(y,t)dy$$

$$+ \int_0^1 P(y)\alpha(y,t)dy + H_z X(t) \bigg). \tag{9.203}$$

Recalling (9.14) and applying Young's inequality, the Cauchy-Schwarz inequality, and (9.169) into (9.203), we obtain

$$\dot{V}_m \leq - \left(\eta_1 - \hat{r}_1 |H_z|^2 - \hat{r}_7 \bar{p}_1^2 |C_0|^2 \right) |X(t)|^2 - (\eta_2 - \bar{q}_1^2 \hat{r}_3 - \hat{r}_6 \bar{p}_1^2 q^2)\beta(0,t)^2$$

$$- (\eta_3 - \hat{r}_5 \bar{Q}_{\max}^2) \int_0^1 \beta(x,t)^2 dx - (\eta_4 - \hat{r}_2 \bar{P}_{\max}^2) \int_0^1 \alpha(x,t)^2 dx$$

$$- (\eta_5 - p_1^2 \hat{r}_4)\alpha(1,t)^2 - (\hat{c}_1 - \eta_6)y_1(t)^2 - \hat{c}_2 y_2(t)^2 - \cdots - \hat{c}_{m-1} y_{m-1}(t)^2$$

$$- \left(\hat{c}_m - \left(\frac{1}{4\hat{r}_1} + \frac{1}{4\hat{r}_2} + \frac{1}{4\hat{r}_3} + \frac{1}{4\hat{r}_4} + \frac{1}{4\hat{r}_5} + \frac{1}{4\hat{r}_6} + \frac{1}{4\hat{r}_7} \right) \right) y_m(t)^2. \tag{9.204}$$

We choose the positive constants $\hat{r}_1, \hat{r}_2, \hat{r}_3, \hat{r}_4, \hat{r}_5, \hat{r}_6, \hat{r}_7$ as

$$\hat{r}_1 < \frac{\eta_1}{|H_z|^2}, \quad \hat{r}_2 < \frac{\eta_4}{\bar{P}_{\max}^2}, \quad \hat{r}_3 < \frac{\eta_2}{\bar{q}_1^2}, \quad \hat{r}_4 < \frac{\eta_5}{p_1^2},$$

$$\hat{r}_5 < \frac{\eta_3}{\bar{Q}_{\max}^2}, \quad \hat{r}_6 < \frac{\eta_2 - \bar{q}_1^2 \hat{r}_3}{\bar{p}_1^2 q^2}, \quad \hat{r}_7 < \frac{\eta_1 - \hat{r}_1 |H_z|^2}{\bar{p}_1^2 |C_0|^2}, \tag{9.205}$$

where

$$\bar{Q}_{\max} = \max_{x \in [0,1]} \{|\bar{Q}(x)|\}, \quad \bar{P}_{\max} = \max_{x \in [0,1]} \{|\bar{P}(x)|\} \tag{9.206}$$

and choose the control parameters $\hat{c}_1, \hat{c}_m$ as

$$\hat{c}_1 > \eta_6, \tag{9.207}$$

$$\hat{c}_m > \frac{1}{4} \left(\frac{1}{\hat{r}_1} + \frac{1}{\hat{r}_2} + \frac{1}{\hat{r}_3} + \frac{1}{\hat{r}_4} + \frac{1}{\hat{r}_5} + \frac{1}{\hat{r}_6} + \frac{1}{\hat{r}_7} \right). \tag{9.208}$$

The positive control parameters $\hat{c}_2, \ldots, \hat{c}_{m-1}$ can be chosen arbitrarily to adjust the exponential decay rate of the closed-loop system.

Finally, we arrive at

$$\dot{V}_m(t) \leq -\hat{\lambda} V_m(t) - \hat{g}_0 \beta(0,t)^2 - \hat{g}_1 \alpha(1,t)^2 \tag{9.209}$$

for some positive $\hat{\lambda}$, and

$$\hat{g}_0 = \eta_2 - \bar{q}_1^2 \hat{r}_3 - \hat{r}_6 \bar{p}_1^2 q^2 > 0, \quad \hat{g}_1 = \eta_5 - p_1^2 \hat{r}_4 > 0.$$

Through a process similar to (9.87)–(9.89), we arrive at theorem 9.3.　　　□

Boundedness and Exponential Convergence of the Controller

In this section, we prove the exponential convergence and boundedness of the controller $U(t)$ (9.195) in the closed-loop system including the input ODE of order m.

Theorem 9.4. *In the closed-loop system including the plant* (9.1)–(9.4), (9.159), (9.160) *and the controller* $U(t)$ (9.195), *there exist positive constants* λ_m *and* Υ_{0m} *such that*

$$|U(t)| \leq \Upsilon_{0m} e^{-\frac{\lambda_m}{2}t}, \quad \forall t \geq 0, \tag{9.210}$$

namely, ensuring that $U(t)$ *is bounded and exponentially convergent to zero.*

Before showing the proof of theorem 9.4, we present two lemmas where we produce and analyze L_2 estimates of $u_x(x,t), v_x(x,t), \cdots, \partial_x^{m-1}u(x,t), \partial_x^{m-1}v(x,t), \partial_x^m u(x,t), \partial_x^m v(x,t)$ in order to prove the exponential convergence of the bounds of signals $\partial_x^{m-1}v(1,t), \partial_x^{m-1}u(1,t), \partial_x^{m-1}v(0,t), \partial_x^{m-1}u(0,t), \partial_x^{m-2}u(1,t), \partial_x^{m-2}v(0,t), \partial_x^{m-2}u(0,t), \cdots, v(1,t), u(1,t), v(0,t), u(0,t)$ in the controller (9.195).

Lemma 9.4. *For all initial data* $(u(x,0), v(x,0)) \in H^{m-1}(0,1)$, *the closed-loop* $(u(x,t), v(x,t))$ *system* (9.1)–(9.4), (9.159), (9.160) *with the controller* (9.195) *is exponentially stable in the sense of*

$$\|\partial_x^{m-1}u(\cdot,t)\|^2 + \|\partial_x^{m-1}v(\cdot,t)\|^2. \tag{9.211}$$

Proof. We consider the Lyapunov function

$$\bar{B}_{m-1}(t) = \frac{1}{2}\int_0^1 b_{m-1}e^{-\delta_{m-1}x}\partial_x^{m-1}\alpha(x,t)^2 dx$$
$$+ \frac{1}{2}\int_0^1 a_{m-1}e^{\delta_{m-1}x}\partial_x^{m-1}\beta(x,t)^2 dx, \tag{9.212}$$

where the positive constant b_{m-1} can be chosen arbitrarily to adjust the convergence rate, and the positive constants a_{m-1}, δ_{m-1} shall be defined later.

Taking the derivative of (9.212) along the system obtained from differentiating (9.12), (9.13) $m-1$ times with respect to x and differentiating (9.14) $m-1$ times with respect to t, using the Cauchy-Schwarz inequality, we can choose positive a_{m-1} and δ_{m-1} (see remark 9.1) such that

$$\dot{\bar{B}}_{m-1}(t)$$
$$\leq -\bar{M}_1 \int_0^1 e^{-\delta_{m-1}x}\partial_x^{m-1}\alpha(x,t)^2 dx - \bar{M}_2 \int_0^1 e^{\delta_{m-1}x}\partial_x^{m-1}\beta(x,t)^2 dx$$
$$- \bar{M}_3\partial_x^{m-1}\beta(0,t)^2 - \frac{1}{2}b_{m-1}e^{-\delta_{m-1}}\partial_x^{m-1}\alpha(1,t)^2 + \frac{1}{2}a_{m-1}e^{\delta_{m-1}}\partial_x^{m-1}\beta(1,t)^2$$
$$+ \bar{M}_4 \int_0^1 e^{\delta_{m-1}x}\left[\partial_x^{m-2}\beta(x,t)^2 + \cdots + \beta_x(x,t)^2 + \beta(x,t)^2\right] dx$$
$$+ \bar{M}_5 \int_0^1 e^{-\delta_{m-1}x}\left[\partial_x^{m-2}\alpha(x,t)^2 + \cdots + \alpha_x(x,t)^2 + \alpha(x,t)^2\right] dx$$
$$+ \bar{M}_6|X(t)|^2 + \bar{M}_7\left[\partial_x^{m-2}\beta(0,t)^2 + \cdots + \beta_x(0,t)^2 + \beta(0,t)^2\right], \tag{9.213}$$

where $\bar{M}_1, \bar{M}_2, \bar{M}_3, \bar{M}_4, \bar{M}_5, \bar{M}_6, \bar{M}_7$ are positive constants.

Remark 9.1. As in (9.143), we can choose a_{m-1}, δ_{m-1} to make sure that the terms of order $m-1$, namely the terms $\|\partial_x^{m-1}\alpha(\cdot,t)\|^2$, $\|\partial_x^{m-1}\beta(\cdot,t)\|^2$, $\partial_x^{m-1}\beta(0,t)^2$, are with negative signs in $\dot{\bar{B}}_{m-1}$, except for $\partial_x^{m-1}\beta(1,t)^2$. The positive term $\partial_x^{m-1}\beta(1,t)^2$ ("positive term" here means a quadratic term with a plus sign) bounded by $\frac{1}{p^2}\partial_t^{m-1}$ $\beta(1,t)^2$ can be accommodated by the exponential results in terms of $y_1(t)^2, \ldots,$ $y_m(t)^2$ provided in theorem 9.3. As in (9.142), the positive term $\partial_x^{m-1}\alpha(0,t)^2$ has been written as the positive terms $\partial_x^{m-1}\beta(0,t)^2, \partial_x^{m-2}\beta(0,t)^2, \cdots, \beta(0,t)^2$, and $|X(t)|^2$ by using the Cauchy-Schwarz inequality in the time derivative of (9.14) of $m-1$ order and using (9.12). As in (9.143), the positive term $\partial_x^{m-1}\beta(0,t)^2$ is "canceled" by choosing a_{m-1}, and other positive terms with coefficients $\bar{M}_6, \bar{M}_7$ are kept in (9.213). The remaining positive terms will be dealt with in the following steps.

All positive terms of order $m-2, \ldots, 1$ can be accommodated by the exponential estimates in the sense of $\|\partial_x^{m-2}\alpha(\cdot,t)\|^2 + \|\partial_x^{m-2}\beta(\cdot,t)\|^2, \ldots, \|\alpha_x(\cdot,t)\|^2 + \|\beta_x(\cdot,t)\|^2$, which can be obtained according to lemma 9.2 and lemma 9.3. Together with the exponential results in the sense of $\|\alpha(\cdot,t)\|^2 + \|\beta(\cdot,t)\|^2 + |X(t)|^2 + y_1(t)^2 + \cdots + y_m(t)^2$ provided in theorem 9.3, we define a Lyapunov function

$$V_{u(m-1)}(t) = R_{m-1}\left[\prod_{i=1}^{m-2} R_i V_m(t) + \prod_{i=2}^{m-2} R_i \bar{A}(t) + \prod_{i=3}^{m-2} R_i \bar{B}(t) \right.$$
$$\left. + \prod_{i=4}^{m-2} R_i \bar{B}_3(t) + \cdots + \bar{B}_{m-2}(t) \right] + \bar{B}_{m-1}(t), \qquad (9.214)$$

where $\bar{B}_i(t)$, $\forall i = 3, \ldots, m-2$ are Lyapunov functions similar to (9.212) by replacing $m-1$ with i.

Taking the derivative of (9.214) and choosing sufficiently large $R_i > 0$, we get

$$\dot{V}_{u(m-1)}(t) = -\lambda_{m-1} V_{u(m-1)}(t) - \hat{g}_2 \alpha(1,t)^2$$
$$- \hat{g}_3[\partial_x^{m-1}\beta(0,t)^2 + \partial_x^{m-2}\beta(0,t)^2 + \cdots + \beta_x(0,t) + \beta(0,t)^2] \quad (9.215)$$

for some positive $\lambda_{m-1}, \hat{g}_2, \hat{g}_3$.

Thus, we obtain the exponential stability estimates in terms of $\|\partial_t^{m-1}\alpha(\cdot,t)\|^2 + \|\partial_t^{m-1}\beta(\cdot,t)\|^2$.

Through a similar process with (9.119)–(9.121), we obtain lemma 9.4. $\qquad \square$

Lemma 9.5. *For all initial data $(u(x,0), v(x,0)) \in H^m(0,1)$, the closed-loop system $(u(x,t), v(x,t))$ (9.1)–(9.4), (9.159), (9.160) with the controller (9.195) is exponentially stable in the sense of*

$$\|\partial_x^m u(\cdot,t)\|^2 + \|\partial_x^m v(\cdot,t)\|^2. \qquad (9.216)$$

Proof. Define a Lyapunov function

$$\bar{B}_m(t) = \frac{1}{2}\int_0^1 b_m e^{-\delta_m x} \partial_x^m \alpha(x,t)^2 dx + \frac{1}{2}\int_0^1 a_m e^{\delta_m x} \partial_x^m \beta(x,t)^2 dx, \qquad (9.217)$$

where the positive constant b_m can be chosen arbitrarily to adjust the convergence rate, and the positive constants a_m, δ_m will be defined later.

Taking the derivative of (9.217) along the system obtained from differentiating (9.12), (9.13) m times with respect to x and differentiating (9.14) m times with respect to t, using the Cauchy-Schwarz inequality, we can choose positive a_m and δ_m (see remark 9.2) such that

$$
\begin{aligned}
\dot{B}_m(t) \leq & -M_1 \int_0^1 e^{-\delta_m x} \partial_x^m \alpha(x,t)^2 dx - M_2 \int_0^1 e^{\delta_3 x} \partial_x^m \beta(x,t)^2 dx \\
& - M_3 \partial_x^m \beta(0,t)^2 - \frac{1}{2} b_m e^{-\delta_m} \partial_x^m \alpha(1,t)^2 + \frac{1}{2} a_m e^{\delta_m} \partial_x^m \beta(1,t)^2 \\
& + M_4 \int_0^1 e^{\delta_m x} \left[\partial_x^{m-1} \beta(x,t)^2 + \cdots + \beta_x(x,t)^2 + \beta(x,t)^2 \right] dx \\
& + M_5 \int_0^1 e^{-\delta_3 x} \left[\partial_x^{m-1} \alpha(x,t)^2 + \cdots + \alpha_x(x,t)^2 + \alpha(x,t)^2 \right] dx \\
& + M_6 |X(t)|^2 + M_7 \left[\partial_x^{m-1} \beta(0,t)^2 + \cdots + \beta_x(0,t)^2 + \beta(0,t)^2 \right],
\end{aligned}
\tag{9.218}
$$

where $M_1, M_2, M_3, M_4, M_5, M_6, M_7, M_8$ are positive constants.

Remark 9.2. As in (9.143), we can choose a_m, δ_m to make sure that all terms of order m, namely, $\|\partial_x^m \alpha(\cdot,t)\|^2$, $\|\partial_x^m \beta(\cdot,t)\|^2$, $\partial_x^m \beta(0,t)^2$, are with negative signs in $\dot{B}_m$, except for $\partial_x^m \beta(1,t)^2$. As in (9.142), the positive term $\partial_x^m \alpha(0,t)^2$ ("positive term" here means a quadratic term with a plus sign) has been written as positive terms $\partial_x^m \beta(0,t)^2$, $\partial_x^{m-1} \beta(0,t)^2$, $\cdots$, $\beta(0,t)^2$, and $|X(t)|^2$ via applying the Cauchy-Schwarz inequality in the time derivative of (9.14) of m order and using (9.12). As in (9.143), the positive term $\partial_x^m \beta(0,t)^2$ is "canceled" by choosing a_m, and other positive terms are kept in (9.218), with coefficients M_6, M_7.

In (9.218), the function $\partial_x^m \beta(1,t)^2$ can be written as the positive terms

$$
\begin{aligned}
M_9 \Bigg[& \left[\partial_t^{m-1} \beta(1,t)^2 + \partial_t^{m-2} \beta(1,t)^2 + \cdots + \beta_t(1,t) + \beta(1,t)^2 \right] \\
& + \beta(0,t)^2 + \alpha(1,t)^2 + \int_0^1 \beta(y,t)^2 dy + \int_0^1 \alpha(y,t)^2 dy + |X(t)|^2 \Bigg],
\end{aligned}
\tag{9.219}
$$

where M_9 is a positive constant, obtained by using (9.13), substituting (9.194) into (9.168), and applying the Cauchy-Schwarz inequality, as well as by recalling that $\alpha(0,t)^2$ is bounded by $\beta(0,t)^2$ and $|X(t)|^2$ via applying the Cauchy-Schwarz inequality in (9.14).

According to lemma 9.2, lemma 9.3, and lemma 9.4, we obtain the exponential estimates in the sense of $\|\partial_x^{m-1} \alpha(\cdot,t)\|^2 + \|\partial_x^{m-1} \beta(\cdot,t)\|^2 + \cdots + \|\alpha_x(\cdot,t)\|^2 + \|\beta_x(\cdot,t)\|^2$, which accommodate the positive terms of order $m-1,\ldots,1$ in $\dot{B}_m$ (9.218). Together with the exponential results in terms of $\|\alpha(\cdot,t)\|^2 + \|\beta(\cdot,t)\|^2 + |X(t)|^2 + y_1(t)^2 + \cdots + y_m(t)^2$ provided in theorem 9.3, we define a Lyapunov function

$$
V_{um} = R_m V_{u(m-1)} + \bar{B}_m.
\tag{9.220}
$$

Taking the derivative of V_{um} and choosing sufficiently large R_m, we arrive at

$$
\dot{V}_{um}(t) \leq -\lambda_m V_{um}(t)
\tag{9.221}
$$

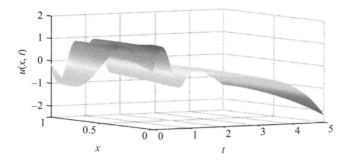

Figure 9.1. Response of $u(x,t)$ in the plant (9.1)–(9.6) without control.

for some positive λ_m.

Through a similar process with (9.156), (9.157), we arrive at lemma 9.5. □

Now we give the proof of theorem 9.4.

Proof. Using the exponential estimates in terms of $\|u(\cdot,t)\|^2 + \|v(\cdot,t)\|^2 + \|u_x(\cdot,t)\|^2 + \|v_x(\cdot,t)\|^2 + \cdots + \|\partial_x^m u(\cdot,t)\|^2 + \|\partial_x^m v(\cdot,t)\|^2 + z_1(t)^2 + z_2(t)^2 + \cdots + z_m(t)^2 + |X(t)|^2$ provided in lemma 9.2, lemma 9.3, lemma 9.4, lemma 9.5, and theorem 9.3, through a similar process with proof of theorem 9.2, we obtain theorem 9.4. □

9.7 SIMULATION

We use the finite-difference method to conduct the simulation with a time step of 0.00025 and a spatial step of 0.005. The solutions of the kernel equations (9.20)–(9.25), which are coupled linear hyperbolic PDEs on $D = \{(x,y)|0 \leq y \leq x \leq 1\}$, are also solved by the finite-difference method. We define the plant parameters in (9.1)–(9.6) as

$$[A, B, c_1, c_2, q, p, C, c_0, r] = [0.5, 1, 0.5, 0.5, 1, 1, 1, 1, 1],$$

and the control parameters are chosen as

$$[\kappa, \bar{c}_1, \bar{c}_2] = [-2, 5, 13].$$

The initial conditions of $v(x,t)$ and $u(x,t)$ are defined as

$$v(x,0) = u(x,0) = \sin(2\pi x),$$

and the initial conditions of $X(t)$ and $z(t)$ are

$$X(0) = u(0,0) - v(0,0) = 0, \quad z(0) = v(1,0) = 0$$

according to (9.4) and (9.5).

Comparing figure 9.1, which shows the open-loop response of $u(x,t)$, and figure 9.2, which gives the closed-loop response of $u(x,t)$, one can observe that in the latter case the convergence to zero is achieved, whereas the states grow unbounded

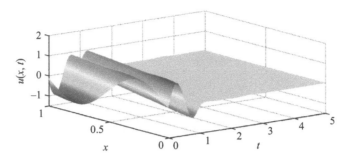

Figure 9.2. Response of $u(x,t)$ in the plant (9.1)–(9.6) with the controller (9.52).

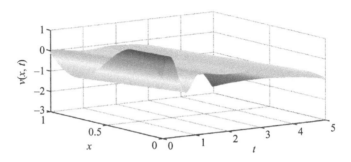

Figure 9.3. Response of $v(x,t)$ in the plant (9.1)–(9.6) without control.

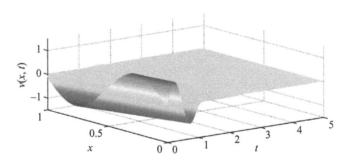

Figure 9.4. Response of $v(x,t)$ in the plant (9.1)–(9.6) with the controller (9.52).

in the former case. Similar observations are made by comparing figure 9.3 and figure 9.4, which show the open-loop and closed-loop responses of $v(x,t)$, respectively.

In figure 9.5, we show the responses of the input ODE states $z(t), \dot{z}(t)$ in both open-loop (*left side*) and closed-loop (*right side*) cases. We observe that the states $z(t), \dot{z}(t)$ grow unbounded in the open-loop case and converge to zero under control. Similar results are observed in figure 9.6, which shows both the open-loop and closed-loop responses of the ODE state $X(t)$. In figure 9.7, we show the response of the control input $U(t)$ (9.52) in the closed-loop system. As one can observe, $U(t)$ converges to zero.

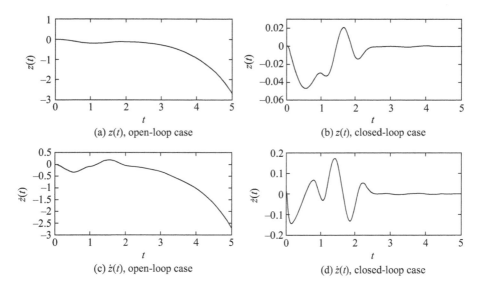

Figure 9.5. Response of input ODE states $z(t)$, $\dot{z}(t)$ in the open-loop and closed-loop cases.

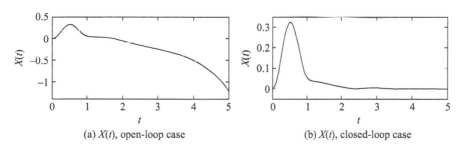

Figure 9.6. Response of ODE state $X(t)$ in the open-loop and closed-loop cases.

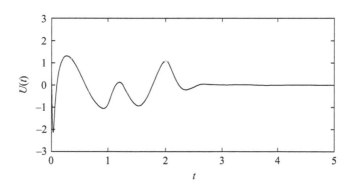

Figure 9.7. Control input $U(t)$ (9.52) in the closed-loop system.

9.8 APPENDIX

A. The quantities h_1, h_2, h_3, h_4, h_5, h_6, h_7, h_8, $h_9(y)$, $h_{10}(y)$, H_{11}

The quantities $h_1, h_2, h_3, h_4, h_5, h_6, h_7, h_8, h_9(y), h_{10}(y), H_{11}$ are given by

$$h_1 = c_0 - p\phi(1,1),$$

$$h_2 = p\psi(1,1),$$

$$h_3 = p\phi(1,0) - \gamma(1)B,$$

$$h_4 = -p\psi(1,0),$$

$$h_5 = -pc_1\psi(1,1) + p^2\phi_y(1,1) + p^2\phi(1,1)\phi^I(1,1) - c_0\phi^I(1,1)p,$$

$$h_6 = p^2\psi_y(1,1) + pc_2\phi(1,1) + c_0\psi^I(1,1)p - p^2\phi(1,1)\psi^I(1,1),$$

$$h_7 = pc_1\psi(1,0) - p^2\phi_y(1,0) - \gamma(1)AB - p^2\phi(1,1)\phi^I(1,0) - c_0\gamma^I(1)B$$
$$\quad + p\phi(1,1)\gamma^I(1)B + c_0\phi^I(1,0)p - (p\phi(1,0) - \gamma(1)B)\gamma^I(0)B,$$

$$h_8 = p^2\phi(1,1)\psi^I(1,0) - c_0\psi^I(1,0)p - p(p\psi_y(1,0) + c_2\phi(1,0)),$$

$$h_9(y) = p\phi(1,1)\psi^I(1,y)c_1 - \int_y^1 p\phi(1,1)\psi^I(1,\delta)c_1\phi^I(\delta,y)d\delta$$

$$\quad + \int_y^1 \left(p^2\phi_{yy}(1,\delta) + c_1c_2\phi(1,\delta)\right)\phi^I(\delta,y)d\delta$$

$$\quad - \left(p^2\phi_{yy}(1,y) + c_1c_2\phi(1,y)\right)$$

$$\quad + p\left(c_1\psi(1,1) - p\phi_y(1,1)\right)\phi^I(1,y) + c_0\phi_y^I(1,y)p$$

$$\quad + \int_y^1 c_0\psi^I(1,\delta)c_1\phi^I(\delta,y)d\delta - c_0\psi^I(1,y)c_1 - p^2\phi(1,1)\phi^I{}_y(1,y),$$

$$h_{10}(y) = \int_y^1 c_0\psi^I(1,\delta)c_1\psi^I(\delta,y)d\delta - c_0p\psi^I{}_y(1,y)$$

$$\quad - \left(p^2\psi_{yy}(1,y) + c_2c_1\psi(1,y)\right) + p\left(c_1\psi(1,1) - p\phi_y(1,1)\right)\psi^I(1,y)$$

$$\quad + \int_y^1 \left(p^2\phi_{yy}(1,\delta) + c_1c_2\phi(1,\delta)\right)\psi^I(\delta,y)d\delta$$

$$\quad + p^2\phi(1,1)\psi_y^I(1,y) - \int_y^1 p\phi(1,1)\psi^I(1,\delta)c_1\psi^I(\delta,y)d\delta,$$

$$H_{11} = p\phi(1,1)\gamma^I(1)A + \int_0^1 \left(p^2\phi_{yy}(1,\delta) + c_1c_2\phi(1,\delta)\right)\gamma^I(\delta)d\delta$$

$$\quad + p\left(c_1\psi(1,1) - p\phi_y(1,1)\right)\gamma^I(1)$$

$$\quad - \left(pc_1\psi(1,0) - p^2\phi_y(1,0) - \gamma(1)AB\right)\gamma^I(0)$$

$$\quad + c_0\gamma^I(1)B\gamma^I(0) - p\phi(1,1)\gamma^I(1)B\gamma^I(0)$$

$$\quad - \gamma(1)A^2 + (p\phi(1,0) - \gamma(1)B)\gamma^I(0)^2B - (p\phi(1,0) - \gamma(1)B)\gamma^I(0)A$$

$$\quad + \int_0^1 c_0\psi^I(1,y)c_1\gamma^I(y)dy - \int_0^1 p\phi(1,1)\psi^I(1,y)c_1\gamma^I(y)dy - c_0\gamma^I(1)A.$$

B. Proof of lemma 9.1

First, we transform the kernel equations (9.20)–(9.25) into integral equations using the method of characteristics.

By virtue of (9.24), (9.25), we express the solution of $\gamma(x)$ in terms of ψ, as follows:

$$\gamma(x) = \kappa e^{\frac{1}{p}Ax} + C \int_0^x e^{\frac{1}{p}A(x-\tau)}\psi(\tau,0)d\tau. \tag{9.222}$$

Then we use the method of characteristic lines, as in [50], to give the successive approximations of $\psi(x,y), \phi(x,y)$. Along the line $x = -y + \bar{a}_1$, according to (9.23), (9.20), we obtain

$$\frac{d\psi(s)}{ds} = -c_2\phi(\bar{a}_1 - s, s), \tag{9.223}$$

$$\psi(\frac{\bar{a}_1}{2}) = -\frac{c_2}{2p}, \tag{9.224}$$

with the characteristic $(-s+\bar{a}_1, s)$ that reaches to (x, y). According to the ODE (9.223), (9.224), we obtain

$$\psi(x,y) = -\frac{c_2}{2p} - \int_{\frac{x+y}{2}}^y \frac{c_2}{p}\phi(x+y-\tau,\tau)d\tau. \tag{9.225}$$

Then (9.225) is rewritten as the integral form

$$\psi(x,y) = G_0(x,y) + G[\phi](x,y), \tag{9.226}$$

where,

$$G_0(x,y) = -\frac{c_2}{2p}, \tag{9.227}$$

$$G[\phi](x,y) = -\int_{\frac{x+y}{2}}^y \frac{c_2}{p}\phi(x+y-\tau,\tau)d\tau. \tag{9.228}$$

Substituting (9.222) into (9.21), with (9.21), (9.22) along the line $x = y + \bar{a}_2$, we get

$$\frac{d\phi(s)}{s} = \frac{c_1}{p}\psi(\bar{a}_2 + s, s), \tag{9.229}$$

$$\phi(0) = \kappa e^{\frac{1}{p}A\bar{a}_2}B + \frac{1}{p}C\int_0^{\bar{a}_2} e^{\frac{1}{p}A(\bar{a}_2-\tau)}\psi(\tau,0)d\tau B + \psi(\bar{a}_2,0)q, \tag{9.230}$$

with the characteristic $(s+\bar{a}_2, s)$ that reaches to (x, y). According to the ODE (9.229), (9.230), we obtain

$$\phi(x,y) = \kappa e^{\frac{1}{p}A(x-y)}B + \frac{1}{p}C\int_0^{x-y} e^{\frac{1}{p}A(x-y-\tau)}\psi(\tau,0)d\tau B$$

$$+ \psi(x-y,0)q + \frac{1}{p}\int_0^y c_1\psi(x-y+\tau,\tau)d\tau. \tag{9.231}$$

Substituting (9.225) into (9.231) yields

$$\phi(x,y) = \kappa e^{\frac{1}{p}A(x-y)}B + \frac{1}{p}C\int_0^{x-y} e^{\frac{1}{p}A(x-y-\tau)}\psi(\tau,0)d\tau B$$

$$-\frac{qc_2}{2p} + q\int_0^{\frac{x-y}{2}} \frac{c_2}{p}\phi(x-y-\tau,\tau)d\tau - \frac{c_1c_2}{2p^2}y$$

$$-\int_0^y \int_{\frac{x-y+2\tau}{2}}^\tau \frac{c_1c_2}{p^2}\phi(x-y+2\tau-\mu,\mu)d\mu d\tau, \tag{9.232}$$

which is rewritten as the integral form

$$\phi(x,y) = F_0(x,y) + F[\psi,\phi](x,y), \tag{9.233}$$

where

$$F_0(x,y) = \kappa e^{\frac{1}{p}A(x-y)}B - \frac{qc_2}{2p} - \frac{c_1c_2}{2p^2}y, \tag{9.234}$$

$$F[\psi,\phi](x,y) = \frac{1}{p}C\int_0^{x-y} e^{\frac{1}{p}A(x-y-\tau)}\psi(\tau,0)d\tau B + q\int_0^{\frac{x-y}{2}} \frac{c_2}{p}\phi(x-y-\tau,\tau)d\tau$$

$$-\int_0^y \int_{\frac{x-y+2\tau}{2}}^\tau \frac{c_1c_2}{p^2}\phi(x-y+2\tau-\mu,\mu)d\mu d\tau. \tag{9.235}$$

Second, we use the method of successive approximations to construct a solution to the integral equations (9.226), (9.233) in the form of a converging series.

Setting

$$\psi^0(x,y) = 0, \tag{9.236}$$

$$\phi^0(x,y) = 0, \tag{9.237}$$

$$\psi^{n+1}(x,y) = G_0(x,y) + G[\phi^n](x,y), \tag{9.238}$$

$$\phi^{n+1}(x,y) = F_0(x,y) + F[\psi^n,\phi^n](x,y), \tag{9.239}$$

for $n = 0,1...$, with the definition of increments

$$\Delta\psi^{n+1} = \psi^{n+1} - \psi^n$$

and

$$\Delta\phi^{n+1} = \phi^{n+1} - \phi^n,$$

where

$$\Delta\psi^0 = G_0(x,y), \quad \Delta\phi^0 = F_0(x,y),$$

it is easy to see that

$$\Delta\psi^{n+1} = G[\Delta\phi^n](x,y), \tag{9.240}$$

$$\Delta\phi^{n+1} = F[\Delta\psi^n,\Delta\phi^n](x,y). \tag{9.241}$$

Define

$$\bar{a} = \max_{x\in[0,1]} \left\{ Ce^{\frac{1}{p}Ax}B, \kappa e^{\frac{1}{p}Ax}B \right\}, \tag{9.242}$$

$$\bar{b} = \frac{1}{p} \max \left\{ c_2, q c_2, \frac{1}{p} c_1 c_2 \right\}, \tag{9.243}$$

$$\eta = 2\bar{a} + 4\bar{b}. \tag{9.244}$$

According to the definition of $\Delta \psi^0, \Delta \phi^0$, we observe

$$\Delta \psi^0 < \eta$$

and

$$\Delta \phi^0 < \eta.$$

Assume now that

$$|\Delta \psi^n| \leq \eta^{n+1} \frac{(x-y)^n}{n!}, \tag{9.245}$$

$$|\Delta \phi^n| \leq \eta^{n+1} \frac{(x-y)^n}{n!} \tag{9.246}$$

are true for some $n \in \mathbb{N}^*$.

Substituting (9.245), (9.246) into (9.240), (9.241) expressed in (9.228) and (9.235), through a straightforward calculation we obtain

$$\left| \Delta \psi^{n+1} \right| \leq \eta^{n+2} \frac{(x-y)^{n+1}}{(n+1)!}, \tag{9.247}$$

$$\left| \Delta \phi^{n+1} \right| \leq \eta^{n+2} \frac{(x-y)^{n+1}}{(n+1)!}. \tag{9.248}$$

Therefore, the series

$$\psi(x, y) = \sum_{n=0}^{\infty} \Delta \psi^n (x, y), \tag{9.249}$$

$$\phi(x, y) - \sum_{n=0}^{\infty} \Delta \phi^n (x, y) \tag{9.250}$$

uniformly converges to the solution $(\psi(x, y), \phi(x, y))$ of the kernel equations (9.20)–(9.25) on

$$D = \{(x, y) | 0 \leq y \leq x \leq 1\},$$

and then the solution $\gamma(x)$ is obtained via (9.222).

Now we show the continuity of the sum (9.249), (9.250). First, it is straightforward to show that for $n \in \mathbb{N}^*$,

$$\Delta \psi^0 = G_0(x, y), \quad \Delta \phi^0 = F_0(x, y)$$

are continuous on D. Indeed, $\Delta \psi^0, \Delta \phi^0$ are continuous on D as a composition of continuous functions. Besides, if we assume that $\Delta \psi^n$ and $\Delta \phi^n$ are continuous, then $\Delta \psi^{n+1}$ and $\Delta \phi^{n+1}$ are continuous as the integral (with continuous limits of integration) of continuous functions times $\Delta \psi^n$ and $\Delta \phi^n$ composed of continuous functions. Finally, the normal convergence ensures continuity of the solutions $\psi(x, y), \phi(x, y)$ on D [50].

The proof of uniqueness of the solutions, which directly relies on the linearity of the kernel equations, is identical to [32]. We do not get into detail here for the sake of brevity.

The proof of lemma 9.1 is complete.

9.9 NOTES

In chapters 2–8, we presented control designs for PDE-ODE systems. In this chapter, we dealt with ODE-PDE-ODE sandwich configurations. In these configurations the control input is directly applied not at the PDE boundary but at an ODE in the input channel of the PDE-ODE system. As mentioned at the beginning of the chapter, the ODE-PDE-ODE sandwich configuration physically represents the incorporation of the dynamics of the actuators (the hydraulic cylinder and head sheaves, the ship-mounted crane, and the rotary table) whose inertias are considerable, into the mining cable elevator, deep-sea construction vessel, and oil-drilling systems considered in chapters 2–8. The control design in this chapter is also applicable in the unmanned aerial vehicle (UAV) for the rapid suppression of oscillations of the cable and suspended object through a control force provided by the UAV's rotor wings, where the plant is structured as a UAV-cable-payload (ODE-PDE-ODE), and in control of an overhead crane [34] that consists of a motorized platform (ODE) driving a cable (PDE) connecting a payload (ODE) at the bottom. The basic control design of sandwich systems was presented in this chapter, and more additional effects: delay, sampled (or event-triggered) sensing, and nonlinearities are dealt with in chapters 10–12. A summary of theoretical results on the boundary control of sandwich systems will be given in the notes section of chapter 12.

Chapter Ten

Delay-Compensated Control of Sandwich Hyperbolic Systems

Compared with the basic control of sandwich hyperbolic partial differential equations (PDEs) in chapter 9, here we develop the delay-compensated boundary control of sandwich hyperbolic PDEs and propose an observer-based output-feedback controller. We solve the problem by employing a framework where the sensor delay is represented as a transport PDE, and by estimating the delay value as the reciprocal of the convection speed in the transport PDE, generating an ODE (ordinary differential equation)-PDE-ODE-PDE plant. Moreover, the restriction on the proximal ODE structure that is a chain of integrators in chapter 9 is relaxed.

The control design is applied in the deep-sea construction vessel (DCV), to build output-feedback control forces at the ship-mounted crane to reduce oscillations of the long cable and place the equipment at the predetermined location on the seafloor while compensating for the sensor delay that arises from the long-distance transmission (from the seabed to the ship at the ocean surface) of the sensing signal via acoustic devices.

We start this chapter by introducing the model of the system and the control task in section 10.1. The observer design is proposed in section 10.2. Therein, three transformations are used to convert the observer error system to a target observer error system whose exponential stability is straightforward to obtain, where all the output injection terms required for constructing the observer are determined. An observer-based output-feedback control design is proposed in section 10.3, where two transformations are applied to transform the observer into a so-called target system in a stable form, except for the proximal ODE that is influenced by perturbations originating from PDEs and the distal ODE. After representing this target system in the frequency domain to obtain the relationships between the states of the proximal ODE and those perturbation states, the proximal ODE is reformulated as a new ODE without external perturbations in the frequency domain, and then the stabilizing control input is designed. The exponential stability of the closed-loop system and the boundedness and exponential convergence to zero of the control input are proved in section 10.4. In simulation, the obtained theoretical result is applied to the oscillation suppression and position control of a DCV in section 10.5.

10.1 PROBLEM FORMULATION

Model Description

The plant considered in this chapter is

$$\dot{X}(t) = A_0 X(t) + E_0 w(0,t) + B_0 U(t), \tag{10.1}$$

$$z(0,t) = pw(0,t) + C_0 X(t), \tag{10.2}$$

$$z_t(x,t) = -q_1 z_x(x,t) - c_1 w(x,t) - c_1 z(x,t), \tag{10.3}$$

$$w_t(x,t) = q_2 w_x(x,t) - c_2 w(x,t) - c_2 z(x,t), \tag{10.4}$$

$$w(1,t) = qz(1,t) + C_1 Y(t), \tag{10.5}$$

$$\dot{Y}(t) = A_1 Y(t) + B_1 z(1,t), \tag{10.6}$$

$$y_{\text{out}}(t) = C_1 Y(t-\tau), \tag{10.7}$$

$\forall (x,t) \in [0,1] \times [0,\infty)$. The block diagram of (10.1)–(10.7) is shown in figure 10.1. The vectors $X(t) \in \mathbb{R}^{n \times 1}$, $Y(t) \in \mathbb{R}^{m \times 1}$ are ODE states. The scalars $z(x,t) \in \mathbb{R}$, $w(x,t) \in \mathbb{R}$ are states of the 2×2 coupled hyperbolic PDEs with initial conditions $(z(x,0), w(x,0)) \in L^2(0,1) \times L^2(0,1)$. The parameter τ is an arbitrary constant denoting the time delay in the measurement. Input $U(t)$ is the control input to be designed. The parameters $c_1, c_2 \in \mathbb{R}$ and $E_0 \in \mathbb{R}^{n \times 1}$ are arbitrary. The positive constants q_1 and q_2 are transport velocities. The parameters $q, p \in \mathbb{R}$ satisfy assumption 10.1. The matrices $A_0 \in \mathbb{R}^{n \times n}$, $B_0 \in \mathbb{R}^{n \times 1}$, $C_0 \in \mathbb{R}^{1 \times n}$, $A_1 \in \mathbb{R}^{m \times m}$, $B_1 \in \mathbb{R}^{m \times 1}$, $C_1 \in \mathbb{R}^{1 \times m}$ satisfy assumptions 10.2 and 10.3.

Assumption 10.1. *The plant parameters p, q satisfy*

$$|pq| < e^{\frac{c_2}{q_2} + \frac{c_1}{q_1}} \tag{10.8}$$

and $q \neq 0$.

This assumption will be used in the output-feedback control design in section 10.3.

Assumption 10.2. *The pairs (A_0, B_0), (A_1, B_1) are stabilizable, and (A_0, C_0), (A_1, C_1) are detectable.*

According to assumption 10.2, there exist constant matrices L_0, L_1, F_0, F_1 to make the following matrices Hurwitz:

$$\bar{A}_0 = A_0 - L_0 C_0, \tag{10.9}$$

$$\bar{A}_1 = A_1 - e^{\tau A_1} L_1 C_1 e^{-\tau A_1}, \tag{10.10}$$

$$\hat{A}_0 = A_0 - B_0 F_0, \tag{10.11}$$

$$\hat{A}_1 = A_1 - B_1 F_1, \tag{10.12}$$

where $A_1 - e^{\tau A_1} L_1 C_1 e^{-\tau A_1}$ has the same eigenvalues as $A_1 - L_1 C_1$ [125].

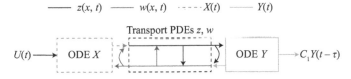

Figure 10.1. Block diagram of the plant (10.1)–(10.7).

Assumption 10.3. *The matrices C_0, A_0, B_0 satisfy*

$$\det\left(\begin{bmatrix} sI - A_0 & B_0 \\ C_0 & 0 \end{bmatrix}\right) \neq 0 \tag{10.13}$$

for all $s \in \mathbb{C}, \Re(s) \geq 0$.

Assumption 10.3 is about matrices of the proximal ODE $X(t)$, namely actuator dynamics. Even though zeros in the closed right-half plane are excluded here while zeros are allowed in some previous results on the control of sandwich PDE systems, such as [43], this assumption relaxes some restrictions on the structure of the proximal ODE in the existing literature (such as A_0, B_0, C_0 being scalar in [8, 49], B_0 being invertible in [151], $\det(C_0 B_0) \neq 0$ in [43], or a form of a chain of integrators in [179, 183]). This assumption, which is first used in sandwich systems in [152], is equal to the existence of a stable left inversion system [135] of (10.1) and is used in the control input design in section 10.3.

Assumption 10.4. *The matrices C_1, A_1, B_1 satisfy*

$$\det\left(\begin{bmatrix} sI - A_1 & B_1 \\ C_1 e^{-\tau A_1} & 0 \end{bmatrix}\right) \neq 0 \tag{10.14}$$

for all $s \in \mathbb{C}, \Re(s) \geq 0$.

Assumption 10.4 is about matrices of the distal ODE $Y(t)$ with a sensor delay τ in the measurement output state. This assumption also prohibits the zeros of the ODE subsystem (C_1, A_1, B_1) from being located in the closed right-half plane. It is not particularly restrictive and (C_1, A_1, B_1) is still quite general, covering many application cases. This assumption is used in the observer design for the overall sandwich system with the delayed measurement in section 10.2. If the sensor delay is zero, this assumption has the same form as assumption 10.3.

The design in this chapter is also suitable for collocated control, namely, when the measurement is the output state of the proximal ODE $X(t)$ with a time delay τ, provided the matrix triple (C_0, A_0, E_0) satisfies

$$\det\left(\begin{bmatrix} sI - A_0 & E_0 \\ C_0 e^{-\tau A_0} & 0 \end{bmatrix}\right) \neq 0$$

for all $s \in \mathbb{C}, \Re(s) \geq 0$.

The control objective of this chapter is to exponentially stabilize the overall sandwich system—that is, the ODE states $Y(t), X(t)$ and the PDE states $u(x,t)$, $v(x,t)$ by constructing an output-feedback control input $U(t)$ applied at the proximal ODE $X(t)$ using the delayed measurement $y_{\text{out}}(t)$.

Rewrite Delay as a Transport PDE

By defining

$$v(x,t) = C_1 Y(t - \tau(x-1)), \quad \forall (x,t) \in [1,2] \times [\tau, \infty), \tag{10.15}$$

we obtain a transport PDE

$$v(1,t) = C_1 Y(t), \tag{10.16}$$

$$v_t(x,t) = -\frac{1}{\tau}v_x(x,t), \tag{10.17}$$

$$y_{\text{out}}(t) = v(2,t) \tag{10.18}$$

for all $(x,t) \in [1,2] \times [\tau,\infty)$ to describe the time delay in the measurement (10.7).

Replacing (10.7) by (10.16)–(10.18), we obtain a sandwich hyperbolic PDE-ODE system connecting with another transport PDE—that is, the following ODE-coupled hyperbolic PDEs-ODE-transport PDE system:

$$\dot{X}(t) = A_0 X(t) + E_0 w(0,t) + B_0 U(t), \tag{10.19}$$

$$z(0,t) = pw(0,t) + C_0 X(t), \tag{10.20}$$

$$z_t(x,t) = -q_1 z_x(x,t) - c_1 w(x,t) - c_1 z(x,t), \ x \in [0,1], \tag{10.21}$$

$$w_t(x,t) = q_2 w_x(x,t) - c_2 w(x,t) - c_2 z(x,t), \ x \in [0,1], \tag{10.22}$$

$$w(1,t) = qz(1,t) + C_1 Y(t), \tag{10.23}$$

$$\dot{Y}(t) = A_1 Y(t) + B_1 z(1,t), \tag{10.24}$$

$$v(1,t) = C_1 Y(t), \tag{10.25}$$

$$v_t(x,t) = -\frac{1}{\tau}v_x(x,t), \ x \in [1,2], \tag{10.26}$$

$$y_{\text{out}}(t) = v(2,t) \tag{10.27}$$

for $t \geq \tau$.

The time delay is "removed" at a cost of adding a transport PDE into the plant (10.1)–(10.7). Now, the control task is equivalent to exponentially stabilizing the overall system (10.19)–(10.27)—that is, ODE(X)-PDE(z,w)-ODE(Y)-PDE(v)—by constructing an output-feedback control input $U(t)$ at the first ODE (10.19) using the right boundary state of the last PDE (10.27).

10.2   OBSERVER DESIGN

In order to build the observer-based output-feedback controller of the plant (10.1)–(10.7), in this section we design a state observer to track the overall system (10.1)–(10.7) using only the delayed measurement $y_{\text{out}}(t)$. Through the reformulation in section 10.1, the estimation task is equivalent to designing a state observer to recover the overall system (10.19)–(10.27) using only measurements at the right boundary $x = 2$ of the last transport PDE v.

The observer is built as a copy of the plant (10.19)–(10.27) plus some dynamic output error injection terms, as follows:

$$\dot{\hat{X}}(t) = A_0 \hat{X}(t) + E_0 \hat{w}(0,t) + B_0 U(t) + h_1(y_{\text{out}}(t) - \hat{v}(2,t)), \tag{10.28}$$

$$\hat{z}(0,t) = p\hat{w}(0,t) + C_0 \hat{X}(t), \tag{10.29}$$

$$\hat{z}_t(x,t) = -q_1 \hat{z}_x(x,t) - c_1 \hat{w}(x,t) - c_1 \hat{z}(x,t) + h_2(y_{\text{out}}(t) - \hat{v}(2,t);x), \tag{10.30}$$

$$\hat{w}_t(x,t) = q_2 \hat{w}_x(x,t) - c_2 \hat{w}(x,t) - c_2 \hat{z}(x,t) + h_3(y_{\text{out}}(t) - \hat{v}(2,t);x), \tag{10.31}$$

$$\hat{w}(1,t) = q\hat{z}(1,t) + C_1 \hat{Y}(t) + h_4(y_{\text{out}}(t) - \hat{v}(2,t)), \tag{10.32}$$

$$\dot{\hat{Y}}(t) = A_1 \hat{Y}(t) + B_1 \hat{z}(1,t) + \Gamma_1(y_{\text{out}}(t) - \hat{v}(2,t)), \tag{10.33}$$

$$\hat{v}(1,t) = C_1 \hat{Y}(t), \tag{10.34}$$

$$\hat{v}_t(x,t) = -\frac{1}{\tau}\hat{v}_x(x,t) + h_5(y_{\text{out}}(t) - \hat{v}(2,t); x), \tag{10.35}$$

where a constant matrix Γ_1 and the dynamics h_1, h_2, h_3, h_4, h_5 are to be determined. Initial conditions are taken as $(\hat{z}(x,0), \hat{w}(x,0), \hat{v}(x,0)) \in L^2(0,1) \times L^2(0,1) \times L^2(1,2)$. Defining the observer error states as

$$\begin{aligned}
(\tilde{z}(x,t), \tilde{w}(x,t), \tilde{X}(t), \tilde{Y}(t), \tilde{v}(x,t)) \\
= (z(x,t), w(x,t), X(t), Y(t), v(x,t)) \\
- (\hat{z}(x,t), \hat{w}(x,t), \hat{X}(t), \hat{Y}(t), \hat{v}(x,t)),
\end{aligned} \tag{10.36}$$

according to (10.19)–(10.27) and (10.28)–(10.35), we obtain the observer error system

$$\dot{\tilde{X}}(t) = A_0\tilde{X}(t) + E_0\tilde{w}(0,t) - h_1(\tilde{v}(2,t)), \tag{10.37}$$

$$\tilde{z}(0,t) = p\tilde{w}(0,t) + C_0\tilde{X}(t), \tag{10.38}$$

$$\tilde{z}_t(x,t) = -q_1\tilde{z}_x(x,t) - c_1\tilde{w}(x,t) - c_1\tilde{z}(x,t) - h_2(\tilde{v}(2,t); x), \tag{10.39}$$

$$\tilde{w}_t(x,t) = q_2\tilde{w}_x(x,t) - c_2\tilde{w}(x,t) - c_2\tilde{z}(x,t) - h_3(\tilde{v}(2,t); x), \tag{10.40}$$

$$\tilde{w}(1,t) = q\tilde{z}(1,t) + C_1\tilde{Y}(t) - h_4(\tilde{v}(2,t)), \tag{10.41}$$

$$\dot{\tilde{Y}}(t) - A_1 Y(t) + B_1\tilde{z}(1,t) - \Gamma_1\tilde{v}(2,t), \tag{10.42}$$

$$\tilde{v}(1,t) = C_1\tilde{Y}(t), \tag{10.43}$$

$$\tilde{v}_t(x,t) = -\frac{1}{\tau}\tilde{v}_x(x,t) - h_5(\tilde{v}(2,t); x), \tag{10.44}$$

where $\Gamma_1\tilde{v}(2,t)$ is an output injection, and $h_1(\tilde{v}(2,t))$, $h_2(\tilde{v}(2,t); x)$, $h_3(\tilde{v}(2,t); x)$, $h_4(\tilde{v}(2,t))$, $h_5(\tilde{v}(2,t); x)$ are dynamic output injections, which are defined as

$$h_1(\tilde{v}(2,t)) = \mathcal{L}^{-1}[H_1(s)\tilde{v}(2,s)], \tag{10.45}$$

$$h_2(\tilde{v}(2,t); x) = \mathcal{L}^{-1}[H_2(s; x)\tilde{v}(2,s)], \tag{10.46}$$

$$h_3(\tilde{v}(2,t); x) = \mathcal{L}^{-1}[H_3(s; x)\tilde{v}(2,s)], \tag{10.47}$$

$$h_4(\tilde{v}(2,t)) = \mathcal{L}^{-1}[H_4(s)\tilde{v}(2,s)], \tag{10.48}$$

$$h_5(\tilde{v}(2,t); x) = \mathcal{L}^{-1}[H_5(s; x)v(2,s)], \tag{10.49}$$

where $\mathcal{L}^{-1}$ denotes the inverse Laplace transform, and the transfer functions $H_1(s)$, $H_2(s; x)$, $H_3(s; x)$, $H_4(s)$, $H_5(s; x)$ are to be determined later. It should be noted that x in H_2, H_3, H_5 is only a parameter.

Introducing (10.45)–(10.49) is helpful in constructing the dynamics $h_i(\cdot)$ in (10.28)–(10.35) because the algebraic relationships between $\tilde{v}(2,s)$ and other states can be obtained by using the Laplace transform, and the transfer functions in (10.45)–(10.49) can be solved in algebraic equations after rewriting the conditions required to achieve an exponentially stable observer error system in the frequency domain.

The determination of $H_1(s)$, $H_2(s; x)$, $H_3(s; x)$, $H_4(s)$, $H_5(s; x)$, and Γ_1 in the observer (10.28)–(10.35) will be completed through three transformations presented next, which convert the observer error system (10.37)–(10.44) to a target observer error system whose exponential stability is straightforward to obtain.

First Transformation

Applying the transformation

$$\tilde{v}(x,t) = \tilde{\eta}(x,t) + \varphi(x)\tilde{Y}(t), \tag{10.50}$$

where $\varphi(x)$ is to be determined, we convert (10.42)–(10.44) to a stable form as

$$\dot{\tilde{Y}}(t) = \bar{A}_1\tilde{Y}(t) + B_1\tilde{z}(1,t) - \Gamma_1\tilde{\eta}(2,t), \tag{10.51}$$

$$\tilde{\eta}(1,t) = 0, \tag{10.52}$$

$$\tilde{\eta}_t(x,t) = -\frac{1}{\tau}\tilde{\eta}_x(x,t), \quad x \in [1,2], \tag{10.53}$$

where $\bar{A}_1$ is a Hurwitz matrix defined in (10.10). In what follows, $\varphi(x)$, Γ_1, $H_5(s;x)$ are determined by matching (10.42)–(10.44) and (10.51)–(10.53) via (10.50).

Inserting the transformation (10.50) into (10.42), we obtain

$$\dot{\tilde{Y}}(t) = (A_1 - \Gamma_1\varphi(2))\tilde{Y}(t) + B_1\tilde{z}(1,t) - \Gamma_1\tilde{\eta}(2,t). \tag{10.54}$$

By virtue of (10.10), (10.51), Γ_1 should satisfy

$$\Gamma_1\varphi(2) = e^{\tau A_1}L_1C_1e^{-\tau A_1}. \tag{10.55}$$

Evaluating (10.50) at $x = 1$ and applying (10.43), (10.52), we get

$$\tilde{v}(1,t) = \tilde{\eta}(1,t) + \varphi(1)\tilde{Y}(t) = \varphi(1)\tilde{Y}(t) = C_1\tilde{Y}(t). \tag{10.56}$$

Therefore,

$$\varphi(1) = C_1. \tag{10.57}$$

Taking the time and spatial derivatives of (10.50) and submitting the result into (10.44), we obtain

$$\tilde{v}_t(x,t) + \frac{1}{\tau}\tilde{v}_x(x,t) + h_5(\tilde{v}(2,t);x)$$

$$= \tilde{\eta}_t(x,t) + \varphi(x)A_1\tilde{Y}(t) + \varphi(x)B_1\tilde{z}(1,t) - \varphi(x)\Gamma_1\tilde{v}(2,t)$$

$$+ \frac{1}{\tau}\tilde{\eta}_x(x,t) + \frac{1}{\tau}\varphi'(x)\tilde{Y}(t) + h_5(\tilde{v}(2,t);x)$$

$$= \varphi(x)B_1\tilde{z}(1,t) - \varphi(x)\Gamma_1\tilde{v}(2,t) + h_5(\tilde{v}(2,t);x)$$

$$+ \left[\varphi(x)A_1 + \frac{1}{\tau}\varphi'(x)\right]\tilde{Y}(t) = 0, \tag{10.58}$$

where (10.42), (10.53) are used, and $\varphi(x)$ should satisfy

$$\varphi'(x) = -\tau A_1\varphi(x) \tag{10.59}$$

to make $[\varphi(x)\bar{A}_1 + \frac{1}{\tau}\varphi'(x)]\tilde{Y}(t)$ zero. The transfer function $H_5(s; x)$ that determines the signal $h_5(\tilde{v}(2,t); x)$ via (10.49) should be defined to ensure the remainder of (10.58) is zero—that is,

$$\varphi(x)B_1\tilde{z}(1,t) - \varphi(x)\Gamma_1\tilde{v}(2,t) + h_5(\tilde{v}(2,t); x) = 0. \tag{10.60}$$

Before determining $H_5(s; x)$, we solve conditions (10.55), (10.57), (10.59) to obtain $\varphi(x), \Gamma_1$ as

$$\varphi(x) = C_1 e^{-\tau A_1(x-1)}, \quad x \in [1, 2], \tag{10.61}$$

$$\Gamma_1 = e^{\tau A_1} L_1. \tag{10.62}$$

With (10.52), (10.53), we know that

$$\tilde{\eta}(2,t) = 0, \ t \geq \tau. \tag{10.63}$$

Thus, (10.51) is written as

$$\dot{\tilde{Y}}(t) = \bar{A}_1\tilde{Y}(t) + B_1\tilde{z}(1,t) \tag{10.64}$$

for $t \geq \tau$. Taking the Laplace transform of (10.64), we get

$$(sI - \bar{A}_1)\tilde{Y}(s) = B_1\tilde{z}(1,s), \tag{10.65}$$

where I is an identity matrix with appropriate dimensions. For brevity, we consider all initial conditions to be zero when we take the Laplace transform. (Arbitrary initial conditions could be incorporated into the stability statement through an expanded analysis which is routine but heavy on additional notation.)

Recalling that $\bar{A}_1$ is Hurwitz, $\det(sI - \bar{A}_1)$ does not have any zeros in the closed right-half plane. Then the matrix $sI - \bar{A}_1$ is invertible for any $s \in \mathbb{C}$, $\Re(s) \geq 0$. Multiplying both sides of (10.65) by $(sI - \bar{A}_1)^{-1}$, we obtain

$$\tilde{Y}(s) = (sI - \bar{A}_1)^{-1}B_1\tilde{z}(1,s). \tag{10.66}$$

According to (10.50) and (10.63), we get

$$\tilde{v}(2,t) = \varphi(2)\tilde{Y}(t), \quad t \geq \tau. \tag{10.67}$$

Writing (10.67) in the frequency domain and inserting (10.61), (10.66), we obtain

$$\tilde{v}(2,s) = \varphi(2)\tilde{Y}(s) - r(s)\tilde{z}(1,s), \tag{10.68}$$

where

$$r(s) = C_1 e^{-\tau A_1}(sI - \bar{A}_1)^{-1}B_1. \tag{10.69}$$

Notice $r(s) \in \mathbb{R}$ due to $C_1 \in \mathbb{R}^{1 \times m}$ and $B_1 \in \mathbb{R}^{m \times 1}$.

Lemma 10.1. *The function $r(s) = C_1 e^{-\tau A_1}(sI - \bar{A}_1)^{-1}B_1$ is nonzero for any $s \in \mathbb{C}$ with $\Re(s) \geq 0$ under assumptions 10.2 and 10.4, or, in plain words, $r(s)$ is a stable, strictly proper transfer function.*

Proof. Using (10.10) in assumption 10.2, we get

$$\begin{bmatrix} I & e^{\tau A_1} L_1 \\ 0 & I \end{bmatrix} \begin{bmatrix} sI - A_1 & B_1 \\ C_1 e^{-\tau A_1} & 0 \end{bmatrix} \begin{bmatrix} I & -(sI - \bar{A}_1)^{-1} B_1 \\ 0 & I \end{bmatrix}$$

$$= \begin{bmatrix} sI - \bar{A}_1 & 0 \\ C_1 e^{-\tau A_1} & -C_1 e^{-\tau A_1} (sI - \bar{A}_1)^{-1} B_1 \end{bmatrix}. \tag{10.70}$$

Recalling assumption 10.4, we know that

$$\det\left(\begin{bmatrix} sI - \bar{A}_1 & 0 \\ C_1 e^{-\tau A_1} & -C_1 e^{-\tau A_1} (sI - \bar{A}_1)^{-1} B_1 \end{bmatrix} \right) \neq 0$$

for any $s \in \mathbb{C}$, $\Re(s) \geq 0$. Therefore, $C_1 e^{-\tau A_1} (sI - \bar{A}_1)^{-1} B_1 \neq 0$. The proof of the lemma is complete. $\qquad\square$

According to lemma 10.1, we have the existence of

$$r(s)^{-1} = \frac{1}{r(s)},$$

which is a stable though improper transfer function.

Let us now go back to (10.60) to determine $H_5(s; x)$. Taking the Laplace transform of (10.60), recalling (10.49), and inserting (10.61) and (10.68), we obtain

$$\varphi(x) B_1 \tilde{z}(1, s) - \varphi(x) \Gamma_1 \tilde{v}(2, s) + H_5(s; x) \tilde{v}(2, s)$$

$$= \left[C_1 e^{-\tau A_1 (x-1)} B_1 - \left(C_1 e^{-\tau A_1 (x-1)} \Gamma_1 - H_5(s; x) \right) r(s) \right] \tilde{z}(1, s)$$

$$= 0. \tag{10.71}$$

The transfer function $H_5(s; x)$ is then chosen as

$$H_5(s; x) = C_1 e^{-\tau A_1 (x-1)} \Gamma_1 - C_1 e^{-\tau A_1 (x-1)} B_1 r(s)^{-1}$$

$$= C_1 e^{-\tau A_1 (x-1)} \Gamma_1 - \frac{C_1 e^{-\tau A_1 (x-1)} B_1}{C_1 e^{-\tau A_1} (sI - \bar{A}_1)^{-1} B_1}, \tag{10.72}$$

where lemma 10.1 has been used to ascertain that this transfer function is stable. This transfer function, as well as all the other transfer functions $H_i(s; x)$, yet to be determined, is improper. In remark 10.1, we discuss how to implement these transfer functions despite their improperness.

Therefore, (10.71) holds. Then (10.60) holds by rewriting (10.71) in the time domain. Together with (10.59), then (10.58) holds for $t \geq \tau$. The function $h_5(\tilde{v}(2, t); x)$ can then be defined via (10.72) and (10.49).

In the above derivation, we have completed the conversion between (10.42)–(10.44) and (10.51)–(10.53) through (10.50) and determined $\Gamma_1, h_5(\tilde{v}(2, t); x)$ needed in the observer.

In what follows, $\Pi_4(s)$ is determined to make the boundary condition (10.41) to be zero—that is, to render

$$\tilde{w}(1, t) = q \tilde{z}(1, t) + C_1 \tilde{Y}(t) - h_4(\tilde{v}(2, t)) = 0. \tag{10.73}$$

Taking the Laplace transform of (10.73), recalling (10.48), and inserting (10.66) and (10.68), we obtain

$$\tilde{w}(1,s) = q\tilde{z}(1,s) + C_1\tilde{Y}(s) - H_4(s)\tilde{v}(2,s)$$
$$= \left(q + C_1(sI - \bar{A}_1)^{-1}B_1 - H_4(s)r(s)\right)\tilde{z}(1,s)$$
$$= 0. \tag{10.74}$$

The transfer function $H_4(s)$ is chosen as

$$H_4(s) = [q + C_1(sI - \bar{A}_1)^{-1}B_1]r(s)^{-1}$$
$$= \frac{q + C_1(sI - \bar{A}_1)^{-1}B_1}{C_1 e^{-\tau A_1}(sI - \bar{A}_1)^{-1}B_1} \tag{10.75}$$

to make (10.74) hold. This transfer function is stable, thanks to lemma 10.1, but it is improper. We discuss its implementation in remark 10.1. We then get $\tilde{w}(1,t) = 0$ in (10.73) by rewriting $\tilde{w}(1,s) = 0$ in the time domain. The function $h_4(\tilde{v}(2,t))$ is thus determined by (10.48), (10.75).

Therefore, through the first transformation (10.50), with determining the dynamic output injection terms $h_4(\tilde{v}(2,t))$, $h_5(\tilde{v}(2,t);x)$, (10.37)–(10.44) is converted to the first intermediate system as

$$\dot{\tilde{X}}(t) = A_0\tilde{X}(t) + E_0\tilde{w}(0,t) - h_1(\tilde{v}(2,t)), \tag{10.76}$$
$$\tilde{z}(0,t) = p\tilde{w}(0,t) + C_0\tilde{X}(t), \tag{10.77}$$
$$\tilde{z}_t(x,t) = -q_1\tilde{z}_x(x,t) - c_1\tilde{w}(x,t) - c_1\tilde{z}(x,t) - h_2(\tilde{v}(2,t);x), \tag{10.78}$$
$$\tilde{w}_t(x,t) = q_2\tilde{w}_x(x,t) - c_2\tilde{w}(x,t) - c_2\tilde{z}(x,t) - h_3(\tilde{v}(2,t);x), \tag{10.79}$$
$$\tilde{w}(1,t) = 0, \tag{10.80}$$
$$\dot{\tilde{Y}}(t) = \bar{A}_1\tilde{Y}(t) + B_1\tilde{z}(1,t), \tag{10.81}$$
$$\tilde{\eta}(1,t) = 0, \tag{10.82}$$
$$\tilde{\eta}_t(x,t) = -\frac{1}{\tau}\tilde{\eta}_x(x,t) \tag{10.83}$$

for $t \geq \tau$, where (10.80)–(10.83) are in a stable form while coupling terms exist in the domain $x \in [0,1]$—that is, (10.78), (10.79). Next, we introduce the second transformation to decouple the couplings in (10.78), (10.79).

Second Transformation

We now apply the second transformation [21]

$$\tilde{w}(x,t) = \tilde{\beta}(x,t) - \int_x^1 \psi(x,y)\tilde{\alpha}(y,t)dy, \tag{10.84}$$

$$\tilde{z}(x,t) = \tilde{\alpha}(x,t) - \int_x^1 \phi(x,y)\tilde{\alpha}(y,t)dy \tag{10.85}$$

with the kernels $\psi(x,y)$, $\phi(x,y)$ satisfying

$$\psi(x,x) = \frac{c_2}{q_1 + q_2}, \tag{10.86}$$

$$\phi(0,y) = p\psi(0,y) - C_0 K_1(y), \tag{10.87}$$

$$-q_1\psi_y(x,y) + q_2\psi_x(x,y) = (c_2 - c_1)\psi(x,y) + c_2\phi(x,y), \tag{10.88}$$

$$-q_1\phi_x(x,y) - q_1\phi_y(x,y) = c_1\psi(x,y), \tag{10.89}$$

where the function $K_1(y)$ will be defined later and then the well-posedness of (10.86)–(10.89) will be shown. The purpose of the transformations (10.84), (10.85) is to convert the first intermediate system (10.76)–(10.81) to a second intermediate system, which is given by

$$\dot{\tilde{X}}(t) = A_0\tilde{X}(t) + E_0\tilde{\beta}(0,t) - E_0\int_0^1 \psi(0,y)\tilde{\alpha}(y,t)dy + h_1(\tilde{v}(2,t)), \tag{10.90}$$

$$\tilde{\alpha}(0,t) = p\tilde{\beta}(0,t) + C_0\tilde{X}(t) - \int_0^1 C_0 K_1(y)\tilde{\alpha}(y,t)dy, \tag{10.91}$$

$$\tilde{\alpha}_t(x,t) = -q_1\tilde{\alpha}_x(x,t) + \int_x^1 \bar{M}(x,y)\tilde{\beta}(y,t)dy$$
$$- c_1\tilde{\alpha}(x,t) - c_1\tilde{\beta}(x,t), \tag{10.92}$$

$$\tilde{\beta}_t(x,t) = q_2\tilde{\beta}_x(x,t) + \int_x^1 \bar{N}(x,y)\tilde{\beta}(y,t)dy - c_2\tilde{\beta}(x,t), \tag{10.93}$$

$$\tilde{\beta}(1,t) = 0, \tag{10.94}$$

$$\dot{\tilde{Y}}(t) = \bar{A}_1\tilde{Y}(t) + B_1\tilde{\alpha}(1,t) \tag{10.95}$$

for $t \geq \tau$, where the integral operator kernels $\bar{M}$ and $\bar{N}$ are defined as

$$\bar{M}(x,y) = \int_x^y \phi(x,\delta)\bar{M}(\delta,y)d\delta - c_1\phi(x,y), \tag{10.96}$$

$$\bar{N}(x,y) = \int_x^y \psi(x,\delta)\bar{M}(\delta,y)d\delta - c_1\psi(x,y). \tag{10.97}$$

The subsystem $\tilde{\eta}(\cdot,t)$, given in (10.82), (10.83), is removed from the second intermediate system (10.90)–(10.95) for brevity because $\tilde{\eta}(\cdot,t) \equiv 0$, $t \geq \tau$.

In what follows, $H_2(s;x)$, $H_3(s;x)$ are determined in matching the first intermediate system (10.76)–(10.81) and the second intermediate system (10.90)–(10.95) via (10.84), (10.85).

Inserting (10.84), (10.85) into (10.79) along (10.92), (10.93) and applying (10.86)–(10.88), (10.97), we get

$$\tilde{w}_t(x,t) - q_2\tilde{w}_x(x,t) + c_2\tilde{z}(x,t) + c_2\tilde{w}(x,t) + h_3(\tilde{v}(2,t);x)$$
$$= q_1\psi(x,1)\tilde{\alpha}(1,t) + h_3(\tilde{v}(2,t);x) = 0, \tag{10.98}$$

where the detailed calculation is shown in (10.254) in appendix 10.6B. We thus know that the following equation needs to be satisfied:

$$q_1\psi(x,1)\tilde{z}(1,t) + h_3(\tilde{v}(2,t);x) = 0, \tag{10.99}$$

where $\tilde{\alpha}(1,t) = \tilde{z}(1,t)$ according to (10.85) is used. Writing (10.99) in the frequency domain and applying (10.47), (10.68), we obtain

$$q_1\psi(x,1)\tilde{z}(1,s) + H_3(s;x)\tilde{v}(2,s)$$
$$= (q_1\psi(x,1) + H_3(s;x)r(s))\,\tilde{z}(1,s) = 0. \tag{10.100}$$

The transfer function $H_3(s;x)$ is chosen as

$$H_3(s;x) = -q_1\psi(x,1)r(s)^{-1} = \frac{-q_1\psi(x,1)}{C_1 e^{-\tau \bar{A}_1}(sI - \bar{A}_1)^{-1}B_1} \tag{10.101}$$

to make (10.100) hold. This transfer function is stable, thanks to lemma 10.1, but it is improper. We discuss its implementation in remark 10.1. We get that (10.98) holds by rewriting (10.100) in the time domain. The function $h_3(\tilde{v}(2,t);x)$ can then be obtained by (10.47), (10.101).

Inserting (10.84), (10.85) into (10.78) along (10.92), (10.93) and applying (10.89), (10.96), we get

$$\tilde{z}_t(x,t) + q_1\tilde{z}_x(x,t) + c_1\tilde{w}(x,t) + c_1\tilde{z}(x,t) + h_2(\tilde{v}(2,t);x)$$
$$= q_1\phi(x,1)\tilde{\alpha}(1,t) + h_2(\tilde{v}(2,t);x) = 0, \tag{10.102}$$

where the detailed calculation is shown in (10.255) in appendix 10.6B. Therefore, $h_2(\tilde{v}(2,t);x)$ should satisfy

$$q_1\phi(x,1)\tilde{z}(1,t) + h_2(\tilde{v}(2,t);x) = 0, \tag{10.103}$$

where $\tilde{\alpha}(1,t) = \tilde{z}(1,t)$ according to (10.85) is used. Taking the Laplace transform of (10.103) and recalling (10.46), (10.68), we obtain

$$q_1\phi(x,1)\tilde{z}(1,s) + H_2(s;x)\tilde{v}(2,s)$$
$$= (q_1\phi(x,1) + H_2(s;x)r(s))\,\tilde{z}(1,s) = 0. \tag{10.104}$$

The transfer function $H_2(s;x)$ is obtained as

$$H_2(s;x) = -q_1\phi(x,1)r(s)^{-1} = \frac{q_1\phi(x,1)}{C_1 e^{-\tau \bar{A}_1}(sI - \bar{A}_1)^{-1}B_1} \tag{10.105}$$

to ensure (10.102) holds. This transfer function is stable, thanks to lemma 10.1, but it is improper. We discuss its implementation in remark 10.1. The function $h_2(\tilde{v}(2,t);x)$ can thus be defined by (10.105), (10.46).

The boundary conditions (10.77), (10.80) follow directly from inserting $x=0$, $x=1$ into (10.84), (10.85) and applying (10.87), (10.91), (10.94). The ODEs (10.76), (10.81) are obtained directly from (10.90), (10.95) via (10.84), (10.85), respectively.

The second conversion is thus completed, and two PDEs (10.78), (10.79) are decoupled now, which can be seen in (10.92), (10.93).

Third Transformation

In order to decouple the ODE (10.90) with the PDEs and rebuild the ODE in a stable form, we intend to convert the second intermediate system (10.90)–(10.95) to the following target observer error system:

$$\dot{\tilde{Z}}(t) = \bar{A}_0\tilde{Z}(t), \tag{10.106}$$

$$\tilde{\alpha}(0,t) = C_0 \tilde{Z}(t), \tag{10.107}$$

$$\tilde{\alpha}_t(x,t) = -q_1 \tilde{\alpha}_x(x,t) - c_1 \tilde{\alpha}(x,t), \tag{10.108}$$

$$\dot{\tilde{Y}}(t) = \bar{A}_1 \tilde{Y}(t) + B_1 \tilde{\alpha}(1,t) \tag{10.109}$$

for $t \geq t_0 = \tau + \frac{1}{q_2}$, where $\bar{A}_0$ is a Hurwitz matrix defined in (10.9). According to (10.93), (10.94), $\tilde{\beta}(x,t) \equiv 0$ after $t_0 = \tau + \frac{1}{q_2}$ and $\tilde{\beta}(x,t)$ are removed for brevity. Equations (10.90)–(10.95) are thus rewritten as

$$\dot{\tilde{X}}(t) = A_0 \tilde{X}(t) - E_0 \int_0^1 \psi(0,y)\tilde{\alpha}(y,t)dy + h_1(\tilde{v}(2,t)), \tag{10.110}$$

$$\tilde{\alpha}(0,t) = C_0 \tilde{X}(t) - \int_0^1 C_0 K_1(y)\tilde{\alpha}(y,t)dy, \tag{10.111}$$

$$\tilde{\alpha}_t(x,t) = -q_1 \tilde{\alpha}_x(x,t) - c_1 \tilde{\alpha}(x,t), \tag{10.112}$$

$$\dot{\tilde{Y}}(t) = \bar{A}_1 \tilde{Y}(t) + B_1 \tilde{\alpha}(1,t) \tag{10.113}$$

for $t \geq t_0$. Equations (10.112), (10.113) are the same as (10.108), (10.109). We thus only need to convert (10.110), (10.111) to (10.106), (10.107).

The transformation

$$\tilde{Z}(t) = \tilde{X}(t) - \int_0^1 K_1(y)\tilde{\alpha}(y,t)dy \tag{10.114}$$

is applied to complete the conversion, where $K_1(y)$ satisfies

$$L_0 - q_1 K_1(0) = 0, \tag{10.115}$$

$$(\bar{A}_0 + c_1)K_1(y) - q_1 K_1'(y) - E_0\psi(0,y) + L_0 C_0 K_1(y) = 0. \tag{10.116}$$

The equation set (10.86)–(10.89) and (10.115), (10.116) is a 2×2 hyperbolic PDE-ODE system, which is a scalar case of the well-posed kernel equations (17)–(23) in [48] (setting dimensions in [48] to 1). Therefore, the conditions of the kernels $\psi(x,y), \phi(x,y)$ in (10.84), (10.85) and $K_1(y)$ in (10.114)—that is, (10.86)–(10.89), (10.115), (10.116), are well-posed.

In what follows, $H_1(s)$ is determined by matching (10.110), (10.111) and (10.106), (10.107) via (10.114). Substituting (10.114) into (10.106) and applying (10.110)–(10.112), (10.115), (10.116), we obtain

$$\dot{\tilde{Z}}(t) - \bar{A}_0 \tilde{Z}(t)$$

$$= A_0 \tilde{X}(t) - E_0 \int_0^1 \psi(0,y)\tilde{\alpha}(y,t)dy + h_1\tilde{v}(2,t)$$

$$+ q_1 K_1(1)\tilde{\alpha}(1,t) - q_1 K_1(0)\tilde{\alpha}(0,t)$$

$$- q_1 \int_0^1 K_1'(y)\tilde{\alpha}(y,t)dy + c_1 \int_0^1 K_1(y)\tilde{\alpha}(y,t)dy$$

$$- A_0 \tilde{X}(t) + L_0 C_0 \tilde{X}(t) + \bar{A}_0 \int_0^1 K_1(y)\tilde{\alpha}(y,t)dy$$

$$= h_1\tilde{v}(2,t) + q_1 K_1(1)\tilde{\alpha}(1,t) + [L_0 - q_1 K_1(0)]\tilde{\alpha}(0,t)$$

$$+ \int_0^1 \left[\bar{A}_0 K_1(y) + c_1 K_1(y) - q_1 K_1{}'(y) - E_0 \psi(0,y) \right.$$

$$\left. + L_0 C_0 K_1(y) \right] \tilde{\alpha}(y,t) dy$$

$$= h_1(\tilde{v}(2,t)) + q_1 K_1(1) \tilde{\alpha}(1,t) = 0, \quad t \geq t_0. \tag{10.117}$$

The function $H_1(s)$, which defines $h_1(\tilde{v}(2,t))$ by (10.45), is solved from

$$h_1(\tilde{v}(2,t)) + q_1 K_1(1) \tilde{z}(1,t) = 0, \tag{10.118}$$

where

$$\tilde{\alpha}(1,t) = \tilde{z}(1,t), \tag{10.119}$$

according to (10.85), has been used.

Writing (10.118) in the frequency domain and applying (10.45), (10.68) yields

$$H_1(s)\tilde{v}(2,s) + q_1 K_1(1)\tilde{z}(1,s)$$
$$= (H_1(s)r(s) + q_1 K_1(1)) \, \tilde{z}(1,s) = 0. \tag{10.120}$$

Solving for $H_1(s)$, we get it as

$$H_1(s) = -q_1 K_1(1) r(s)^{-1} = \frac{-q_1 K_1(1)}{C_1 e^{-\tau A_1} (sI - \bar{A}_1)^{-1} B_1}. \tag{10.121}$$

This transfer function is stable, thanks to lemma 10.1, but it is improper. We discuss its implementation in remark 10.1. We get that (10.117) holds by rewriting (10.120) in the time domain. The function $h_1(\tilde{v}(2,t))$ can then be defined via (10.45), (10.121).

Inserting (10.114) into (10.111), it is straightforward to obtain (10.107). Therefore, (10.106), (10.107) is converted from (10.110), (10.111) through (10.114) for $t \geq t_0$. The third transformation is completed, and the ODE (10.106) is independent and exponentially stable now.

After the above three transformations, we have converted the original observer error system (10.37)–(10.44) to the target observer error system (10.106)–(10.109) (for $t \in [t_0, \infty)$, $\tilde{\eta}(x,t) \equiv 0$, according to (10.82), (10.83), and $\tilde{\beta}(x,t) \equiv 0$, according to (10.93), (10.94), are removed for brevity). Because the original observer error system (10.37)–(10.44) is bounded in the finite time $t \in [0, t_0)$, we prove the exponential stability of (10.37)–(10.44) for $t \in [t_0, \infty)$ in the next subsection.

The spatially dependent transfer functions H_i determined, whose outputs are dynamic output injections in the observer (10.28)–(10.35), are employed as follows:

$$y_1(t) = h_1(\tilde{v}(2,t)), \tag{10.122}$$
$$y_2(x,t) = h_2(\tilde{v}(2,t); x), \tag{10.123}$$
$$y_3(x,t) = h_3(\tilde{v}(2,t); x), \tag{10.124}$$
$$y_4(t) = h_4(\tilde{v}(2,t)), \tag{10.125}$$
$$y_5(x,t) = h_5(\tilde{v}(2,t); x). \tag{10.126}$$

The signals y_i are proved exponentially convergent to zero in the next subsection.

Remark 10.1. While stable, thanks to lemma 10.1, the transfer functions $H_i(s)$ are improper and, therefore, the signals $H_i(s)\tilde{v}(2,s)$ contain time derivatives of $\tilde{v}(2,t)$. In practice, one way to avoid taking the time derivatives, which may lead to measurement noise amplification, is by measuring n-order time-derivative states $\partial_t^n v(2,t)$ and calculating $\tilde{v}(2,t)$ by n times integrations of $\partial_t^n \tilde{v}(2,t)$, which amounts to multiplying $H_i(s)$ by $\frac{1}{s^n}$ to make $H_i(s)$ proper. Measuring $\partial_t^n v(2,t)$ is not viable in general, but in the case of the control application to the DCV in section 10.5 the payload oscillation acceleration is measured, and the velocity is calculated by integration starting from the known initial conditions, as mentioned in chapter 7. Measuring acceleration is a prevalent method in many mechanical systems because the acceleration sensor is cheaper and far easier to manufacture and install [17].

Stability Analysis of the Observer Error System

Theorem 10.1. *For all initial data* $(\tilde{z}(x,0), \tilde{w}(x,0), \tilde{v}(x,0), \tilde{X}(0), \tilde{Y}(0)) \in L^2(0,1) \times L^2(0,1) \times L^2(1,2) \times \mathbb{R}^n \times \mathbb{R}^m$, *the internal exponential stability of the observer error system* (10.37)–(10.44) *holds in the sense of the norm*

$$\|\tilde{z}(\cdot,t)\|_\infty + \|\tilde{w}(\cdot,t)\|_\infty + \|\tilde{v}(\cdot,t)\|_\infty + \left|\tilde{X}(t)\right| + \left|\tilde{Y}(t)\right| + |y_1(t)|$$
$$+ |y_4(t)| + \|y_2(\cdot,t)\|_\infty + \|y_3(\cdot,t)\|_\infty + \|y_5(\cdot,t)\|_\infty, \tag{10.127}$$

with the decay rate adjustable by L_0, L_1.

Proof. The stability of the original observer error system is obtained by analyzing the stability of the target observer error system (10.106)–(10.109) and using the invertibility of the transformations. Equations (10.106)–(10.109) are a cascade of $\tilde{Z}(t)$ into $\tilde{\alpha}(\cdot,t)$ into $\tilde{Y}(t)$. From (10.106), $\tilde{Z}(t)$ is exponentially convergent to zero because $\bar{A}_0$ is Hurwitz. With the method of characteristics, as in [43] it is easy to show that $\tilde{\alpha}(x,t)$ in the PDE subsystem (10.106), (10.107) is exponentially convergent to zero. Because $\bar{A}_1$ is Hurwitz, $\tilde{Y}(t)$ is exponentially convergent to zero. The decay rate λ_e of (10.106)–(10.109) depends on the decay rate of the ODEs $\tilde{Z}(t), \tilde{Y}(t)$. In other words, the decay rate λ_e is adjustable by L_0, L_1 according to (10.9), (10.10). Recalling $\tilde{\eta}(x,t) \equiv 0$ and $\tilde{\beta}(x,t) \equiv 0$ after $t_0 = \frac{1}{q_2} + \tau$, we find that

$$\bar{\Omega}(t) = \|\tilde{\alpha}(\cdot,t)\|_\infty + \|\tilde{\beta}(\cdot,t)\|_\infty + \|\tilde{\eta}(\cdot,t)\|_\infty + \left|\tilde{Z}(t)\right| + \left|\tilde{Y}(t)\right|$$

is bounded by an exponential decay with the decay rate λ_e for $t \geq t_0$. It should be noted that the transient in the finite time $[0, t_0)$ can be bounded by an arbitrarily fast decay rate considering the trade-off between the decay rate and the overshoot coefficient—that is, the higher the decay rate, the higher the overshoot coefficient. Therefore, we conclude that the exponential stability in the sense of $\bar{\Omega}(t)$ is bounded by an exponential decay rate λ_e with some overshoot coefficients for $t \geq 0$. Applying the transformation (10.50), (10.114) and (10.84), (10.85), we respectively have that

$$\|\tilde{v}(\cdot,t)\|_\infty \leq \Upsilon_{1a} \left(\|\tilde{\eta}(\cdot,t)\|_\infty + \left|\tilde{Y}(t)\right|\right),$$
$$\left|\tilde{X}(t)\right| \leq \Upsilon_{1b} \left(\|\tilde{\alpha}(\cdot,\iota)\|_\infty + \left|\tilde{Z}(\iota)\right|\right),$$
$$\|\tilde{z}(\cdot,t)\|_\infty + \|\tilde{w}(\cdot,t)\|_\infty \leq \Upsilon_{1c} \left(\|\tilde{\alpha}(\cdot,t)\|_\infty + \|\tilde{\beta}(\cdot,t)\|_\infty\right)$$

for some positive $\Upsilon_{1a}, \Upsilon_{1b}, \Upsilon_{1c}$.

According to (10.45)–(10.49), (10.72), (10.75), (10.101), (10.105), (10.121), we know that the output injection states $y_1(t)$, $y_2(x,t)$, $y_3(x,t)$, $y_4(t)$, $y_5(x,t)$ are the output states of the following dynamical systems given by their spatially dependent transfer functions:

$$H_1(s) = \frac{-q_1 K_1(1)}{C_1 e^{-\tau A_1}(sI - \bar{A}_1)^{-1} B_1}, \tag{10.128}$$

$$H_2(s;x) = \frac{-q_1 \phi(x,1)}{C_1 e^{-\tau A_1}(sI - \bar{A}_1)^{-1} B_1}, \tag{10.129}$$

$$H_3(s;x) = \frac{-q_1 \psi(x,1)}{C_1 e^{-\tau A_1}(sI - \bar{A}_1)^{-1} B_1}, \tag{10.130}$$

$$H_4(s) = \frac{q + C_1(sI - \bar{A}_1)^{-1} B_1}{C_1 e^{-\tau A_1}(sI - \bar{A}_1)^{-1} B_1}, \tag{10.131}$$

$$H_5(s;x) = C_1 e^{-\tau A_1(x-1)} \Gamma_1 - \frac{C_1 e^{-\tau A_1(x-1)} B_1}{C_1 e^{-\tau A_1}(sI - \bar{A}_1)^{-1} B_1}, \tag{10.132}$$

whose input signal is $\tilde{v}(2,t)$, which is exponentially convergent to zero. Recalling lemma 10.1, we know that there is no pole in the closed right-half plane in the transfer function (10.128)–(10.132). The exponential convergence of $|y_1(t)|$, $\|y_2(\cdot,t)\|_\infty$, $\|y_3(\cdot,t)\|_\infty$, $|y_4(t)|$, $\|y_5(\cdot,t)\|_\infty$ is thus obtained. It should be noted that $x \in [0,1]$ is just a parameter in the numerators of the transfer functions (10.129), (10.130), (10.132), and the stability result is not affected. $\qquad \square$

10.3   OUTPUT-FEEDBACK CONTROL DESIGN

In the last section, we obtained the observer that compensates for the time delay in the output measurement $y_{\text{out}}(t)$ of the distal ODE, which is the only measurement used in the observer, to track the states of the overall sandwich PDE system (10.1)–(10.7). In this section, we design an output-feedback control law $U(t)$ based on the observer (10.28)–(10.35) by using backstepping transformations and frequency-domain designs.

First, two transformations are introduced to transform the observer (10.28)–(10.35) into a target system (10.173)–(10.180), which is in a stable form except for the proximal ODE influenced by perturbations originating from the PDEs and the distal ODE. Representing this "target system" in the frequency domain by using the Laplace transform, the algebraic relationships (10.196)–(10.202) between the states of the proximal ODE and the states of the PDEs and the distal ODE are obtained. Inserting these algebraic relationships to rewrite the perturbations in the proximal ODE, a new ODE (10.208) without external perturbations is obtained in the frequency domain, where the control input to exponentially stabilize this ODE is to be designed.

First Transformation

The aim of the first transformation is to remove the source terms in the PDE domain $x \in [0,1]$—that is, the couplings in (10.30), (10.31)—and to make the system matrix of the distal ODE (10.33) Hurwitz. A PDE backstepping transformation in the form [48]

$$\alpha(x,t) = \hat{z}(x,t) - \int_x^1 K_3(x,y)\hat{z}(y,t)dy$$

$$- \int_x^1 J_3(x,y)\hat{w}(y,t)dy - \gamma(x)\hat{Y}(t), \tag{10.133}$$

$$\beta(x,t) = \hat{w}(x,t) - \int_x^1 K_2(x,y)\hat{z}(y,t)dy$$

$$- \int_x^1 J_2(x,y)\hat{w}(y,t)dy - \lambda(x)\hat{Y}(t) \tag{10.134}$$

is introduced, where the kernels $K_3(x,y)$, $J_3(x,y)$, $\gamma(x)$, $K_2(x,y)$, $J_2(x,y)$, $\lambda(x)$ are to be determined later, to convert (10.28)–(10.35) into the following intermediate system:

$$\dot{\hat{X}}(t) = A_0\hat{X}(t) + E_0\beta(0,t) + \int_0^1 \bar{K}_4(x)\alpha(x,t)dx$$

$$+ \int_0^1 \bar{K}_5(x)\beta(x,t)dx + \bar{K}_6\hat{Y}(t) + B_0U(t) + h_1(\tilde{v}(2,t)), \tag{10.135}$$

$$\alpha(0,t) = p\beta(0,t) + C_0\hat{X}(t) + \int_0^1 \bar{K}_1(x)\alpha(x,t)dx$$

$$+ \bar{K}_3\hat{Y}(t) + \int_0^1 \bar{K}_2(x)\beta(x,t)dx, \tag{10.136}$$

$$\alpha_t(x,t) = -q_1\alpha_x(x,t) - c_1\alpha(x,t) - \gamma(x)\Gamma_1\tilde{v}(2,t)$$

$$- \int_x^1 J_2(x,y)h_3(\tilde{v}(2,t);y)dy$$

$$- \int_x^1 K_3(x,y)h_2(\tilde{v}(2,t);y)dy$$

$$+ h_2(\tilde{v}(2,t);x) - q_2 J_3(x,1)h_4(\tilde{v}(2,t)), \tag{10.137}$$

$$\beta_t(x,t) = q_2\beta_x(x,t) - c_2\beta(x,t)$$

$$- \lambda(x)\Gamma_1\tilde{v}(2,t) - \int_x^1 J_2(x,y)h_3(\tilde{v}(2,t);y)dy$$

$$- \int_x^1 K_2(x,y)h_2(\tilde{v}(2,t);y)dy$$

$$+ h_3(\tilde{v}(2,t);x) - q_2 J_2(x,1)h_4(\tilde{v}(2,t)), \tag{10.138}$$

$$\beta(1,t) = q\alpha(1,t) + h_4(\tilde{v}(2,t)), \tag{10.139}$$

$$\dot{\hat{Y}}(t) = \hat{A}_1\hat{Y}(t) + B_1\alpha(1,t) + \Gamma_1\tilde{v}(2,t), \tag{10.140}$$

$$\hat{v}(1,t) = C_1\hat{Y}(t), \tag{10.141}$$

$$\hat{v}_t(x,t) = -\frac{1}{\tau}\hat{v}_x(x,t) + h_5(\tilde{v}(2,t);x), \tag{10.142}$$

where the matrix $\hat{A}_1$ is made Hurwitz by choosing the control parameter F_1 according to assumption 10.2. The functions $\bar{K}_1(x)$, $\bar{K}_2(x)$, $\bar{K}_3$, $\bar{K}_4(x)$, $\bar{K}_5(x)$, $\bar{K}_6$ satisfy

$$\bar{K}_1(x) = pK_2(0, x) - K_3(0, x) + \int_0^x \bar{K}_1(y)K_3(y, x)dy$$

$$+ \int_0^x \bar{K}_2(y)K_2(y, x)dy, \tag{10.143}$$

$$\bar{K}_2(x) = -pJ_2(0, x) + J_3(0, x) + \int_0^x \bar{K}_1(y)J_3(y, x)dy$$

$$+ \int_0^x \bar{K}_2(y)J_2(y, x)dy, \tag{10.144}$$

$$\bar{K}_3 = \int_0^1 \bar{K}_2(x)\lambda(x)dx + \int_0^1 \bar{K}_1(x)\gamma(x)dx$$

$$+ p\lambda(0) - \gamma(0), \tag{10.145}$$

$$\bar{K}_4(x) = \int_0^x \bar{K}_4(y)K_3(y, x)dy + \int_0^x \bar{K}_5(y)K_2(y, x)dy$$

$$- E_0 K_2(0, x), \tag{10.146}$$

$$\bar{K}_5(x) = \int_0^x \bar{K}_4(y)J_3(y, x)dy + \int_0^x \bar{K}_5(y)J_2(y, x)dy + E_0 J_2(0, x), \tag{10.147}$$

$$\bar{K}_6 = \int_0^1 \bar{K}_5(x)\lambda(x)dx + \int_0^1 \bar{K}_4(x)\gamma(x)dx + E_0\lambda(0), \tag{10.148}$$

which are obtained by matching (10.135), (10.136) and (10.28), (10.29) via (10.133), (10.134) (the details are given in step 4 in appendix 10.6A). The following conditions of the kernels in the transformations (10.133), (10.134) are obtained by matching (10.137)–(10.140) and (10.30)–(10.33), with the details given in steps 1–3 in appendix 10.6A:

$$q_1 K_3(x, 1) = q_2 J_3(x, 1)q + \gamma(x)B_1, \tag{10.149}$$

$$J_3(x, x) = \frac{c_1}{q_2 + q_1}, \tag{10.150}$$

$$-q_1 J_{3x}(x, y) + q_2 J_{3y}(x, y)$$
$$+ (c_2 - c_1)J_3(x, y) + c_1 K_3(x, y) = 0, \tag{10.151}$$

$$-q_1 K_{3x}(x, y) - q_1 K_{3y}(x, y) + c_2 J_3(x, y) = 0, \tag{10.152}$$

$$\gamma(1) = -F_1, \tag{10.153}$$

$$-q_1\gamma'(x) - \gamma(x)(A_1 + c_1) - q_2 J_3(x, 1)C_1 = 0, \tag{10.154}$$

$$q_2 q J_2(x, 1) = q_1 K_2(x, 1) - \lambda(x)B_1, \tag{10.155}$$

$$K_2(x, x) = \frac{-c_2}{q_1 + q_2}, \tag{10.156}$$

$$q_2 J_{2x}(x, y) + q_2 J_{2y}(x, y) + c_1 K_2(x, y) = 0, \tag{10.157}$$

$$q_2 K_{2x}(x, y) - q_1 K_{2y}(x, y)$$
$$+ (c_1 - c_2)K_2(x, y) + c_2 J_2(x, y) = 0, \tag{10.158}$$

$$q_2\lambda'(x) - \lambda(x)(A_1 + c_2) - q_2 J_2(x, 1)C_1 = 0, \tag{10.159}$$

$$\lambda(1) = q\gamma(1) + C_1. \tag{10.160}$$

The well-posedness of (10.149)–(10.160) is established in the following lemma.

Lemma 10.2. *The kernel equations* (10.149)–(10.154) *have a unique solution* K_3, $J_3 \in C^1(D)$, $\gamma \in C^1([0,1])$, *and the kernel equations* (10.155)–(10.160) *have a unique solution* $K_2, J_2 \in C^1(D)$, $\lambda \in C^1([0,1])$ *on* $D = \{(x,y) | 0 \le x \le y \le 1\}$.

Proof. The equation sets (10.149)–(10.154) and (10.155)–(10.160) have a structure analogous to (9.20)–(9.25) in chapter 9. Following the proof of lemma 1 in chapter 9, we obtain this lemma. □

Similarly, the inverse of the transformation (10.133)–(10.134) is obtained as

$$\hat{z}(x,t) = \alpha(x,t) - \int_x^1 \mathcal{M}(x,y)\alpha(y,t)dy$$

$$- \int_x^1 \mathcal{N}(x,y)\beta(y,t)dy - \mathcal{G}(x)\hat{Y}(t), \tag{10.161}$$

$$\hat{w}(x,t) = \beta(x,t) - \int_x^1 \mathcal{D}(x,y)\alpha(y,t)dy$$

$$- \int_x^1 \mathcal{T}(x,y)\beta(y,t)dy - \mathcal{P}(x)\hat{Y}(t), \tag{10.162}$$

where $\mathcal{M}(x,y), \mathcal{N}(x,y), \mathcal{G}(x), \mathcal{D}(x,y), \mathcal{T}(x,y), \mathcal{P}(x)$ are kernels that can be determined through a process similar to that in appendix 10.6A. The first transformation in the control design is completed.

Second Transformation

In order to remove the last three terms in the boundary condition (10.136) and render Hurwitz the system matrix of the proximal ODE (10.135), we introduce the second transformation

$$\hat{Z}(t) = \hat{X}(t) + C_0^+ \int_0^1 \bar{K}_1(x)\alpha(x,t)dx$$

$$+ C_0^+ \int_0^1 \bar{K}_2(x)\beta(x,t)dx + C_0^+ \bar{K}_3\hat{Y}(t), \tag{10.163}$$

where C_0^+ denotes the Moore-Penrose right inverse of C_0. Because C_0 is full-row rank (with rank equal to 1), a right inverse exists for C_0—that is, $C_0 C_0^+ = I$. A choice of C_0^+ is

$$C_0^+ = C_0^T (C_0 C_0^T)^{-1}.$$

Using (10.163), we convert (10.135), (10.136) to

$$\dot{\hat{Z}}(t) = \hat{A}_0 \hat{Z}(t) + q_1 C_0^+ \bar{K}_1(0) C_0 \hat{Z}(t) + B_0 \bar{U}(t)$$

$$+ M_Y \hat{Y}(t) + \int_0^1 M_\alpha(x)\alpha(x,t)dx + \int_0^1 M_\beta(x)\beta(x,t)dx$$

$$+ N_1\alpha(1,t) + N_2\beta(0,t) + \mathcal{H}[h_1(\tilde{v}(2,t)), h_2(\tilde{v}(2,t);x),$$

$$h_3(\tilde{v}(2,t);x), h_4(\tilde{v}(2,t)), h_5(\tilde{v}(2,t);x), \tilde{v}(2,t)], \tag{10.164}$$

$$\alpha(0,t) = p\beta(0,t) + C_0\hat{Z}(t), \tag{10.165}$$

where

$$\mathcal{H}[h_1(\tilde{v}(2,t)), h_2(\tilde{v}(2,t);x), h_3(\tilde{v}(2,t);x), h_4(\tilde{v}(2,t)), h_5(\tilde{v}(2,t);x), \tilde{v}(2,t)]$$
$$= h_1(\tilde{v}(2,t)) + C_0^+ \bar{K}_3 \Gamma_1 \tilde{v}(2,t) + q_2 C_0^+ \bar{K}_2(1) q h_4(\tilde{v}(2,t))$$
$$+ C_0^+ \int_0^1 \bar{K}_1(x) \int_x^1 J_2(x,y) h_3(\tilde{v}(2,t);y) dy dx$$
$$- C_0^+ \int_0^1 \bar{K}_1(x) \int_x^1 K_3(x,y) h_2(\tilde{v}(2,t);y) dy dx$$
$$+ C_0^+ \int_0^1 \bar{K}_1(x) h_2(\tilde{v}(2,t);x) dx$$
$$- C_0^+ \int_0^1 \bar{K}_1(x) q_2 J_3(x,1) dx h_4(\tilde{v}(2,t))$$
$$- C_0^+ \int_0^1 \bar{K}_2(x) \lambda(x) dx \Gamma_1 \tilde{v}(2,t)$$
$$- C_0^+ \int_0^1 \bar{K}_2(x) \int_x^1 J_2(x,y) h_3(\tilde{v}(2,t);y) dy dx$$
$$- C_0^+ \int_0^1 \bar{K}_2(x) \int_x^1 K_2(x,y) h_2(\tilde{v}(2,t);y) dy dx$$
$$+ C_0^+ \int_0^1 \bar{K}_2(x) h_3(\tilde{v}(2,t);x) dx$$
$$- C_0^+ \int_0^1 \bar{K}_2(x) q_2 J_2(x,1) dx h_4(\tilde{v}(2,t)), \tag{10.166}$$

and

$$\bar{U}(t) = U(t) - F_0 \hat{Z}(t). \tag{10.167}$$

The matrix $\hat{A}_0$ is made Hurwitz by choosing the control parameter F_0 recalling assumption 10.2, and N_1, N_2, $M_\alpha(x)$, $M_\beta(x)$, M_Y in (10.164) are

$$N_1 = C_0^+ \bar{K}_3 B_1 - q_1 C_0^+ \bar{K}_1(1) + q_2 C_0^+ \bar{K}_2(1) q, \tag{10.168}$$
$$N_2 = E_0 - q_2 C_0^+ \bar{K}_2(0) + q_1 C_0^+ \bar{K}_1(0) p, \tag{10.169}$$
$$M_\alpha(x) = \bar{K}_4(x) + q_1 C_0^+ \bar{K}_1'(x) - (\hat{A}_0 + c_1) C_0^+ \bar{K}_1(x), \tag{10.170}$$
$$M_\beta(x) = \bar{K}_5(x) - q_2 C_0^+ \bar{K}_2'(x) - (\hat{A}_0 + c_2) C_0^+ \bar{K}_2(x), \tag{10.171}$$
$$M_Y = C_0^+ \bar{K}_3 \hat{A}_1 + \bar{K}_6 - \hat{A}_0 C_0^+ \bar{K}_3. \tag{10.172}$$

We thus arrive at the target system consisting of (10.137)–(10.142), (10.164), (10.165), which includes dynamic output injections in (10.166). Considering theorem 10.1 and (10.122)–(10.126), we know that $h_1(\tilde{v}(2,t))$, $h_2(\tilde{v}(2,t);x)$, $h_3(\tilde{v}(2,t);x)$, $h_4(\tilde{v}(2,t))$, $h_5(\tilde{v}(2,t);x)$, and $\Gamma_1 \tilde{v}(2,t)$ in the target system (10.137)–(10.142), (10.164), (10.165) can be regarded as zero, at least after the time gets large—that is, $\mathcal{H}[h_1, h_2, h_3, h_4, h_5, \tilde{v}] = 0$—for brevity. Therefore, the target system (10.137)–(10.142), (10.164), (10.165) is rewritten as

$$\dot{\hat{Z}}(t) = \hat{A}_0 \hat{Z}(t) + q_1 C_0^+ \bar{K}_1(0) C_0 \hat{Z}(t)$$

$$+ M_Y \hat{Y}(t) + \int_0^1 M_\alpha(x)\alpha(x,t)dx + \int_0^1 M_\beta(x)\beta(x,t)dx$$

$$+ N_1\alpha(1,t) + N_2\beta(0,t) + B_0\bar{U}(t), \tag{10.173}$$

$$\alpha(0,t) = p\beta(0,t) + C_0\hat{Z}(t), \tag{10.174}$$

$$\alpha_t(x,t) = -q_1\alpha_x(x,t) - c_1\alpha(x,t), \quad x \in [0,1], \tag{10.175}$$

$$\beta_t(x,t) = q_2\beta_x(x,t) - c_2\beta(x,t), \quad x \in [0,1], \tag{10.176}$$

$$\beta(1,t) = q\alpha(1,t), \tag{10.177}$$

$$\dot{\hat{Y}}(t) = \hat{A}_1\hat{Y}(t) + B_1\alpha(1,t), \tag{10.178}$$

$$\hat{v}(1,t) = C_1\hat{Y}(t), \tag{10.179}$$

$$\hat{v}_t(x,t) = -\frac{1}{\tau}\hat{v}_x(x,t), \quad x \in [1,2]. \tag{10.180}$$

Control Design in the Frequency Domain

In the last two subsections, the system (10.28)–(10.35) is converted to the target system (10.173)–(10.180) through the two transformations, (10.133), (10.134) and (10.163). In this subsection, the control $\bar{U}(t)$ in (10.173) of the target system (10.173)–(10.180) will be designed in the frequency domain by using the Laplace transform.

Taking the Laplace transform of (10.173)–(10.180), we obtain

$$(sI - \hat{A}_0)\hat{Z}(s) = q_1 C_0{}^+\bar{K}_1(0)C_0\hat{Z}(s) + M_Y\hat{Y}(s)$$

$$+ \int_0^1 M_\alpha(x)\alpha(x,s)dx + \int_0^1 M_\beta(x)\beta(x,s)dx$$

$$+ N_1\alpha(1,s) + N_2\beta(0,s) + B_0\bar{U}(s), \tag{10.181}$$

$$\alpha(0,s) = p\beta(0,s) + C_0\hat{Z}(s), \tag{10.182}$$

$$s\alpha(x,s) = -q_1\alpha_x(x,s) - c_1\alpha(x,s), \tag{10.183}$$

$$s\beta(x,s) = q_2\beta_x(x,s) - c_2\beta(x,s), \tag{10.184}$$

$$\beta(1,s) = q\alpha(1,s), \tag{10.185}$$

$$(sI - \hat{A}_1)\hat{Y}(s) = B_1\alpha(1,s), \tag{10.186}$$

$$\hat{v}(1,s) = C_1\hat{Y}(s), \tag{10.187}$$

$$s\hat{v}(x,s) = -\frac{1}{\tau}\hat{v}_x(x,s). \tag{10.188}$$

For brevity, we consider all initial conditions to be zero when we take the Laplace transform. (Arbitrary initial conditions could be incorporated into the stability statement through an expanded analysis which is routine but heavy on additional notation.)

We use the definition of a (strictly) proper transfer function for infinite-dimensional systems from [33].

Definition 10.1. *The function G is said to be proper if, for sufficiently large ρ,*

$$\sup_{\text{Re}(s) \geq 0 \bigcap |s| > \rho} |G(s)| < \infty. \tag{10.189}$$

If the limit of $G(s)$ at infinity exists and is 0, we say that G is strictly proper.

According to [33], the definition of a stable transfer function for infinite-dimensional systems in this book is given next.

Definition 10.2. *An irrational transfer function $G(s)$ appearing in this book is said to be stable if it satisfies*

$$\sup_{\mathrm{Re}(s) \geq 0} |G(s)| < \infty. \tag{10.190}$$

Definition 10.2 indicates there is no pole in the closed right-half plane.

Defining

$$h(s) = 1 - pq e^{-\left(\frac{c_2}{q_2} + \frac{c_1}{q_1}\right)} e^{-\left(\frac{1}{q_2} + \frac{1}{q_1}\right)s}, \tag{10.191}$$

according to (10.182)–(10.188) and section 3.2 in [49], we obtain the following algebraic relationships between $C_0 \hat{Z}(s)$ and other states in (10.182)–(10.188) as

$$h(s)\alpha(x,s) = e^{\frac{-(c_1+s)}{q_1}x} C_0 \hat{Z}(s), \tag{10.192}$$

$$h(s)\beta(x,s) = q e^{\frac{-(c_2+s)}{q_2}(1-x) - \frac{(c_1+s)}{q_1}} C_0 \hat{Z}(s), \tag{10.193}$$

$$h(s)\hat{v}(x,s) = C_1(sI - \hat{A}_1)^{-1} B_1 e^{\frac{-(c_1+s)}{q_1} - \tau(x-1)s} C_0 \hat{Z}(s), \tag{10.194}$$

$$h(s)\hat{v}(1,s) = C_1(sI - \hat{A}_1)^{-1} B_1 e^{\frac{-(c_1+s)}{q_1}} C_0 \hat{Z}(s), \tag{10.195}$$

$$h(s)\alpha(0,s) = C_0 \hat{Z}(s), \tag{10.196}$$

$$h(s)\beta(1,s) = q e^{\frac{-(c_1+s)}{q_1}} C_0 \hat{Z}(s), \tag{10.197}$$

$$h(s)\beta(0,s) = q c^{\frac{-(c_2+s)}{q_2} - \frac{(c_1+s)}{q_1}} C_0 \hat{Z}(s), \tag{10.198}$$

$$h(s)\alpha(1,s) = e^{\frac{-(c_1+s)}{q_1}} C_0 \hat{Z}(s), \tag{10.199}$$

$$h(s)\hat{Y}(s) = (sI - \hat{A}_1)^{-1} B_1 e^{\frac{-(c_1+s)}{q_1}} C_0 \hat{Z}(s), \tag{10.200}$$

$$h(s)\int_0^1 M_\beta(y)\beta(y,s)dy = \int_0^1 M_\beta(y) q e^{\frac{-(c_2+s)}{q_2}(1-y) - \frac{(c_1+s)}{q_1}} dy C_0 \hat{Z}(s), \tag{10.201}$$

$$h(s)\int_0^1 M_\alpha(y)\alpha(y,s)dy = \int_0^1 M_\alpha(y) e^{\frac{-(c_1+s)}{q_1}y} dy C_0 \hat{Z}(s). \tag{10.202}$$

Multiplying both sides of (10.181) by scalar $h(s)$ and substituting (10.196)–(10.202) therein yields

$$h(s)(sI - \hat{A}_0)\hat{Z}(s)$$
$$= h(s)q_1 C_0{}^+ \bar{K}_1(0) C_0 \hat{Z}(s)$$
$$\quad + M_Y(sI - \hat{A}_1)^{-1} B_1 e^{\frac{-(c_1+s)}{q_1}} C_0 \hat{Z}(s)$$
$$\quad + \int_0^1 M_\alpha(y) e^{\frac{-(c_1+s)}{q_1}y} dy C_0 \hat{Z}(s)$$
$$\quad + \int_0^1 M_\beta(y) q e^{\frac{-(c_2+s)}{q_2}(1-y) - \frac{(c_1+s)}{q_1}} dy C_0 \hat{Z}(s)$$
$$\quad + N_1 e^{\frac{-(c_1+s)}{q_1}} C_0 \hat{Z}(s) + N_2 q e^{\frac{-(c_2+s)}{q_2} - \frac{(c_1+s)}{q_1}} C_0 \hat{Z}(s) + h(s) B_0 \bar{U}(s). \tag{10.203}$$

Recalling assumption 10.1, we know that $h(s)$ is nonzero for any $s \in \mathbb{C}$, $\Re(s) \geq 0$, and then $h(s)$ has an inverse $h(s)^{-1}$ that is a stable, proper transfer function according to definitions 10.1 and 10.2. Multiplying both sides of (10.203) by $h(s)^{-1}$ and defining

$$\hat{\xi}(t) = C_0 \hat{Z}(t), \tag{10.204}$$

we rewrite (10.203) as

$$(sI - \hat{A}_0)\hat{Z}(s)$$
$$= q_1 C_0{}^+ \bar{K}_1(0)\hat{\xi}(s) + h(s)^{-1} M_Y (sI - \hat{A}_1)^{-1} B_1 e^{\frac{-(c_1+s)}{q_1}} \hat{\xi}(s)$$
$$+ h(s)^{-1} \int_0^1 M_\alpha(y) e^{\frac{-(c_1+s)}{q_1} y} dy \hat{\xi}(s)$$
$$+ h(s)^{-1} \int_0^1 M_\beta(y) q e^{\frac{-(c_2+s)}{q_2}(1-y) - \frac{(c_1+s)}{q_1}} dy \hat{\xi}(s)$$
$$+ h(s)^{-1} N_1 e^{\frac{-(c_1+s)}{q_1}} \hat{\xi}(s)$$
$$+ h(s)^{-1} N_2 q e^{\frac{-(c_2+s)}{q_2} - \frac{(c_1+s)}{q_1}} \hat{\xi}(s) + B_0 \bar{U}(s)$$

for any $s \in \mathbb{C}$, $\Re(s) \geq 0$. Defining

$$G(s) = q_1 C_0{}^+ \bar{K}_1(0) + h(s)^{-1} \left[M_Y (sI - \hat{A}_1)^{-1} B_1 e^{\frac{-(c_1+s)}{q_1}} \right.$$
$$+ \int_0^1 M_\alpha(y) e^{\frac{-(c_1+s)}{q_1} y} dy + \int_0^1 M_\beta(y) q e^{\frac{-(c_2+s)}{q_2}(1-y) - \frac{(c_1+s)}{q_1}} dy$$
$$\left. + N_1 e^{\frac{-(c_1+s)}{q_1}} + N_2 q e^{\frac{-(c_2+s)}{q_2} - \frac{(c_1+s)}{q_1}} \right], \tag{10.205}$$

which is a stable, proper transfer matrix according to definitions 10.1 and 10.2, we get

$$(sI - \hat{A}_0)\hat{Z}(s) = G(s)\hat{\xi}(s) + B_0 \bar{U}(s). \tag{10.206}$$

Recalling $\hat{A}_0$ being Hurwitz, $\det(sI - \hat{A}_0)$ does not have any zeros in the closed right-half plane. Then the matrix $(sI - \hat{A}_0)$ is invertible for any $s \in \mathbb{C}$, $\Re(s) \geq 0$. Multiplying both sides of (10.206) with $C_0(sI - \hat{A}_0)^{-1}$ from the left, we obtain

$$C_0 \hat{Z}(s) = C_0(sI - \hat{A}_0)^{-1} G(s)\hat{\xi}(s) + C_0(sI - \hat{A}_0)^{-1} B_0 \bar{U}(s). \tag{10.207}$$

That is,

$$\hat{\xi}(s) = C_0(sI - \hat{A}_0)^{-1} G(s)\hat{\xi}(s) + W_0 \bar{U}(s), \tag{10.208}$$

where

$$W_0(s) = C_0(sI - \hat{A}_0)^{-1} B_0.$$

Recall assumption 10.3, which is equivalent to the existence of a right inverse for W_0. A possible choice is given by the Moore-Penrose right inverse $W_0^+(s) = W_0^T(s)(W_0(s)W_0^T(s))^{-1}$ [152].

Choose $\bar{U}(s)$ in (10.208) as

$$\bar{U}(s) = -W_0^+(s)\Omega(s)C_0(sI - \hat{A}_0)^{-1}G(s)\hat{\xi}(s) = F(s)\hat{\xi}(s), \qquad (10.209)$$

where a single-input single-output (SISO) low-pass filter $\Omega(s)$ satisfying

$$|1 - \Omega(j\omega)| < \frac{1}{\sup_{\omega \in R} \bar{\sigma}(G(j\omega))\bar{\sigma}(C_0(j\omega I - \hat{A}_0)^{-1})}, \forall \omega \in R \qquad (10.210)$$

is adopted to make sure that $F(s)$ is strictly proper. Because $G(s)$ is uniformly bounded in the closed right-half plane, $\sup_{\omega \in R} \bar{\sigma}(G(j\omega))$ is bounded where $\bar{\sigma}$ stands for the largest singular value. A low-pass filter $\Omega(s)$ can always be chosen to ensure $F(s)$ strictly proper and to satisfy (10.210) concurrently, because there exists a ω_1 to make the right-hand side of (10.210) larger than 1 at $\omega \geq \omega_1$ ($\sup_{\omega \in R} \bar{\sigma}(G(j\omega))$ is bounded, and $\bar{\sigma}(C_0(j\omega - \hat{A}_0)^{-1})$ can be small enough at sufficiently high frequencies), and thus (10.210) still holds even if the gain $|\Omega(j\omega)|$ of the low-pass filter is close to zero at $\omega \geq \omega_1$. It means that a choice of the cutoff frequency of the low-pass filter $\Omega(s)$ is ω_1.

The control $\bar{U}$ has been chosen as strictly proper by introducing the low-pass filter $\Omega(s)$, which means that the controller is robust to small input delays [152].

Substituting (10.209) into (10.208), we obtain

$$\hat{\xi}(s) = (1 - \Omega(s))C_0(sI - \hat{A}_0)^{-1}G(s)\hat{\xi}(s) = \Phi(s)\hat{\xi}(s), \qquad (10.211)$$

that is,

$$(1 - \Phi(s))\hat{\xi}(s) = 0, \qquad (10.212)$$

where

$$\bar{\sigma}(\Phi(j\omega)) \leq |1 - \Omega(j\omega)|\bar{\sigma}\left(C_0(j\omega I - \hat{A}_0)^{-1}\right)\sup_{\omega \in R}\bar{\sigma}(G(j\omega)) < 1 \qquad (10.213)$$

by recalling (10.210), which is a sufficient condition for exponential convergence to zero of $\hat{\xi}$. By virtue of (10.209), (10.167), $U(s)$ is written as

$$U(s) = \bar{U}(s) + F_0\hat{Z}(s)$$
$$= \left[F_0 - W_0^+(s)\Omega(s)C_0(sI - \hat{A}_0)^{-1}G(s)C_0\right]\hat{Z}(s), \qquad (10.214)$$

where the inverse Laplace transform is required to represent $U(s)$ in the time domain for the implementation of the controller and where, moreover, $\hat{Z}$ can be replaced in terms of the observer states by (10.163), (10.133), (10.134).

10.4   STABILITY ANALYSIS OF THE CLOSED-LOOP SYSTEM

The closed-loop system includes the plant (10.1)–(10.7), the observer (10.28)–(10.35), and the controller (10.214). The block diagram of the closed-loop system is shown in figure 10.2. We have provided theorem 10.1, which shows that the observer error states between the plant and the observer are exponentially convergent to zero in the sense of the norm (10.127). By virtue of (10.36), in order to prove the

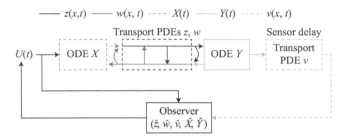

Figure 10.2. Diagram of the closed-loop system.

exponential stability result of the closed-loop system, we present the next lemma to establish the exponential stability of the $(\hat{z}(x,t), \hat{w}(x,t), \hat{v}(x,t), \hat{X}(t), \hat{Y}(t))$-system (10.28)–(10.35) under the controller (10.214).

Lemma 10.3. *For all initial values* $(\hat{z}(x,0), \hat{w}(x,0), \hat{v}(x,0), \hat{X}(0), \hat{Y}(0)) \in L^2(0,1) \times L^2(0,1) \times L^2(1,2) \times \mathbb{R}^n \times \mathbb{R}^m$, *the exponential stability of the system* (10.28)–(10.35) *under the controller* (10.214) *holds in the sense of the norm*

$$\|\hat{z}(\cdot,t)\|_\infty + \|\hat{w}(\cdot,t)\|_\infty + \|\hat{v}(\cdot,t)\|_\infty + \left|\hat{X}(t)\right| + \left|\hat{Y}(t)\right|, \qquad (10.215)$$

with the convergence rate adjustable by F_0, F_1.

Proof. We prove the exponential convergence of the states in the overall system based on the exponential convergence of $\hat{\xi}(t)$ by applying their algebraic relationships obtained in section 10.3.

According to the exponential convergence to zero of $\hat{\xi}(t) = C_0 \hat{Z}(t)$, which is obtained from (10.212), (10.213), recalling (10.192)–(10.194) and (10.200)–(10.202), we show that $\alpha(x,t)$, $\beta(x,t)$, $\hat{v}(x,t)$, $\|\alpha(\cdot,t)\|$, $\|\beta(\cdot,t)\|$, $|\hat{Y}(t)|$ are exponentially convergent to zero, where the convergence rate is adjustable by F_0, F_1 with (10.11), (10.12).

Substituting (10.209) into (10.206), we get

$$\hat{Z}(s) = (sI - \hat{A}_0)^{-1} \left[G(s) - B_0 W_0^+ \Omega(s) C_0 (sI - \hat{A}_0)^{-1} G(s) \right] \hat{\xi}(s). \qquad (10.216)$$

Because

$$(sI - \hat{A}_0)^{-1} \left[G(s) - B_0 W_0^+ \Omega(s) C_0 (sI - \hat{A}_0)^{-1} G(s) \right] \qquad (10.217)$$

is a (stable) proper transfer matrix, using the exponential convergence result of $\hat{\xi}$, we also obtain the exponential convergence to zero of $|\hat{Z}|$ via (10.216).

Applying the Cauchy-Schwarz inequality into the inverse transformations (10.161), (10.162) and the transformation (10.163), we obtain

$$|\hat{z}(x,t)| + |\hat{w}(x,t)| \leq \Upsilon_{2a} \Big(|\alpha(x,t)| + |\beta(x,t)| + \|\alpha(\cdot,t)\|$$

$$+ \|\beta(\cdot,t)\| + \left|\hat{Y}(t)\right| \Big), \qquad (10.218)$$

$$\left|\hat{X}(t)\right| \leq \Upsilon_{2b}\left(\|\alpha(\cdot,t)\| + \|\beta(\cdot,t)\| + \left|\hat{Y}(t)\right| + \left|\hat{Z}(t)\right|\right) \qquad (10.219)$$

for some positive $\Upsilon_{2a}, \Upsilon_{2b}$. Recalling the obtained exponential convergence of $\alpha(x,t)$, $\beta(x,t)$, $\|\alpha(\cdot,t)\|$, $\|\beta(\cdot,t)\|$, $|\hat{Y}(t)|$, $|\hat{Z}(t)|$, we thus obtain the exponential convergence to zero of $\hat{z}(x,t) + \hat{w}(x,t) + |\hat{X}(t)|$. Recalling the exponential convergence to zero of $|\hat{Y}(t)|$ and $\hat{v}(x,t)$, we obtain lemma 10.3. $\qquad\square$

Theorem 10.2. *For all initial values* $(z(x,0), w(x,0), v(x,0), X(0), Y(0)) \in L^2(0,1) \times L^2(0,1) \times L^2(1,2) \times \mathbb{R}^n \times \mathbb{R}^m$, *the closed-loop system including the plant* (10.1)– (10.7), *the observer* (10.28)–(10.35), *and the controller* (10.214) *has the following properties:*

1) *The exponential stability holds in the sense of the norm*

$$\|z(\cdot,t)\|_\infty + \|w(\cdot,t)\|_\infty + \|v(\cdot,t)\|_\infty + |X(t)| + |Y(t)|$$

$$+ \|\hat{z}(\cdot,t)\|_\infty + \|\hat{w}(\cdot,t)\|_\infty + \|\hat{v}(\cdot,t)\|_\infty + \left|\hat{X}(t)\right| + \left|\hat{Y}(t)\right|$$

$$+ |y_1(t)| + |y_4(t)| + \|y_2(\cdot,t)\|_\infty + \|y_3(\cdot,t)\|_\infty + \|y_5(\cdot,t)\|_\infty \qquad (10.220)$$

with the convergence rate adjustable by L_0, L_1, F_0, F_1.
2) *There exist the positive constants* Γ_c *and* λ_c *making the dynamic feedback control* $U(t)$ *bounded and exponentially convergent to zero in the sense of*

$$|U(t)| \leq \Gamma_c e^{-\lambda_c t}.$$

Proof. 1) Applying (10.36) and the Cauchy-Schwarz inequality, recalling theorem 10.1 and lemma 10.3, we obtain property (1).
2) According to the control design in section 10.3, we know that

$$F(s) = W_0^+ \Omega(s) C_0 (sI - \hat{A}_0)^{-1} G(s) C_0 \qquad (10.221)$$

in (10.209) is strictly proper. It follows that

$$F_0 - W_0^+ \Omega(s) C_0 (sI - \hat{A}_0)^{-1} G(s) C_0 \qquad (10.222)$$

in (10.214) is a (stable) proper transfer function because F_0 is a constant matrix. Recalling (10.214) and the exponential convergence of $|\hat{Z}|$ proved in lemma 10.3, we obtain the exponential convergence to zero of the dynamic feedback control $U(t)$, which is a dynamic extension generated by utilizing the frequency-domain design approach.

The proof of theorem 10.2 is complete. $\qquad\square$

10.5 APPLICATION IN DEEP-SEA CONSTRUCTION

A DCV is used to place equipment to be installed on the seafloor for offshore oil drilling, which is shown in figure 10.3. The equipment, referred to as the payload, has to be installed accurately at the predetermined location with a tight tolerance, such as the permissible maximum tolerance of 2.5 m for a typical subsea installation, according to [95] and chapter 7. We consider only one-dimensional oscillations of the

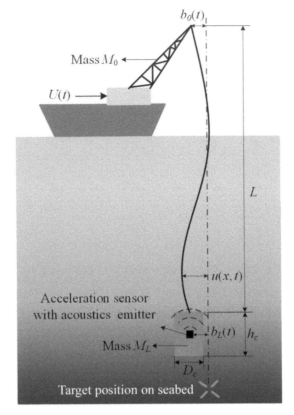

Figure 10.3. Schematic of a DCV used in seafloor installation.

DCV and the end phase of the descending process, during which the cable length can be deemed constant. Control problems of two-dimensional coupled oscillations of the DCV during the entire descending/ascending process, with a time-varying cable length, were addressed in chapter 6. However, chapter 6 does not focus on a sandwich system—that is, neglecting the crane dynamics, and does not include delay compensation.

Modeling

A nonlinear PDE model of the DCV consisting of the vessel, crane, cable, and payload, with transverse oscillations, is presented in [95].

a) Simplifications of the nonlinear model of the vessel crane

We impose the following simplifications of the nonlinear model of the DCV in the equations (1), (2), (6)–(8) of [95]:

- The nonuniform distributed tension $T(z, t)$ in equation (2) in [95], which introduces a nonlinearity into the model, is simplified to a uniform distributed tension T_0, which is defined by an equation of static equilibrium.
- The dynamics of the ocean surface vessel $y_s(t)$ (equation (1) in [95]) are neglected, and only the regulation of the dynamics of the crane-cable-payload is pursued

because the vessel can be kept at the desired position by the ship dynamic positioning system.
- The time-varying damping coefficients $d_0(t)$, $d_L(t)$ of the onboard crane and the payload in [95] are taken as constant.

Thus, the nonlinear model of the DCV (equations (1), (2), (6)–(8) in [95]) is simplified to the following linear model (in [95] $u_0(t)$, $u_L(t)$ in (7), (8) are two control inputs that are not required in our design):

$$M_0 \ddot{b}_0(t) = -d_0 \dot{b}_0(t) + T_0 u_x(0, t) + U(t), \tag{10.223}$$

$$u(0, t) = b_0(t), \tag{10.224}$$

$$\rho u_{tt}(x, t) = T_0 u_{xx}(x, t) - d_c u_t(x, t) + f(x, t), x \in [0, L], \tag{10.225}$$

$$u(L, t) = b_L(t), \tag{10.226}$$

$$M_L \ddot{b}_L(t) = -d_L \dot{b}_L(t) + T_0 u_x(L, t) + f_L(t). \tag{10.227}$$

The static tension T_0 is defined as

$$T_0 = M_L g - F_{\text{buoyant}} \tag{10.228}$$

with

$$F_{\text{buoyant}} = \frac{1}{4} \pi D_c^2 h_c \rho_s g. \tag{10.229}$$

The functions $u(x, t)$ denote distributed transverse displacements along the cable. The functions $b_0(t)$ and $b_L(t)$ represent transverse displacements of the onboard crane and the payload. The functions $f(z, t)$, $f_L(t)$ are ocean current disturbances— that is, external drag forces at the cable and payload. The physical parameters of the DCV in simulation are from [95] and are shown in table 10.1.

Even though the DCV model here is linear, while that in [95] is a more complicated nonlinear model, some boundedness assumptions in [95] are not required here. Moreover, only one control input at the onboard crane is required here, while one more control input for the payload is required in the control system in [95].

b) Reformulation of the linear model

Applying the Riemann transformations

$$z(x, t) = u_t(x, t) - \sqrt{\frac{T_0}{\rho}} u_x(x, t), \tag{10.230}$$

$$w(x, t) = u_t(x, t) + \sqrt{\frac{T_0}{\rho}} u_x(x, t) \tag{10.231}$$

and defining the new variables

$$X(t) = \dot{b}_0(t), \ Y(t) = \dot{b}_L(t), \tag{10.232}$$

we can write the equations (10.223)–(10.227) as

$$\dot{X}(t) = A_0 X(t) + E_0 w(0, t) + B_0 U(t), \tag{10.233}$$

$$z(0, t) = pw(0, t) + C_0 X(t), \tag{10.234}$$

$$z_t(x,t) = -q_1 z_x(x,t) - c_1(z(x,t) + w(x,t)) + f(x,t), \qquad (10.235)$$

$$w_t(x,t) = q_2 w_x(x,t) - c_2(z(x,t) + w(x,t)) + f(x,t), \qquad (10.236)$$

$$w(L,t) = qz(L,t) + C_1 Y(t), \qquad (10.237)$$

$$\dot{Y}(t) = A_1 Y(t) + B_1 z(1,t) + f_L(t), \qquad (10.238)$$

$$y_{\text{out}}(t) = C_1 Y(t-\tau), \qquad (10.239)$$

where $y_{\text{out}}(t)$ is the delayed measurement output, and $y_{\text{out}}(t) = 0, t \in [0, \tau)$ because the sensing signal has not yet been received before $t = \tau$. The observer and controller designs simulated are based on (10.233)–(10.239) except for the disturbances $f(x,t), f_L(t)$, which are regarded as model uncertainties in the simulation to test the robustness of the controller. The sensor delay τ is considered to be 0.1 s, and

$$q_1 = q_2 = \sqrt{\frac{T_0}{\rho}}, \ \ c_1 = c_2 = \frac{d_c}{2\rho}. \qquad (10.240)$$

The constants p, q satisfy assumption 10.1 ($|pq| = 1 < e^{\frac{c_2}{q_2} + \frac{c_1}{q_1}} = 1.0014$), and

$$A_0 = \frac{-d_0}{M_0} - \frac{\sqrt{T_0 \rho}}{M_0}, \ \ E_0 = \frac{\sqrt{T_0 \rho}}{M_0}, \ \ B_0 = \frac{1}{M_0}, C_0 = 2, \qquad (10.241)$$

$$A_1 = \frac{-d_L}{M_L} + \frac{\sqrt{T_0 \rho}}{M_L}, \ \ B_1 = -\frac{\sqrt{T_0 \rho}}{M_L}, \ \ C_1 = 2 \qquad (10.242)$$

satisfy assumptions 10.2–10.4.

The initial conditions are chosen as

$$z(x,0) = 4\sin\left(\frac{\pi x}{L}\right), \ \ w(x,0) = 4\cos\left(\frac{\pi x}{L}\right), \qquad (10.243)$$

which gives the ODE initial conditions

$$X(0) = 2, \ Y(0) = -2 \qquad (10.244)$$

Table 10.1. Physical parameters of the DCV

Parameters (units)	Values
Cable length L (m)	1000
Cable diameter R_D (m)	0.2
Cable effective Young's modulus E (N/m^2)	4.0×10^9
Cable linear density ρ (kg/m)	8.02
Crane mass M_0 (kg)	1.0×10^6
Payload mass M_L (kg)	4.0×10^5
Gravitational acceleration g (m/s^2)	9.8
Cable material damping coefficient d_c (N·s/m)	0.5
Height of payload modeled as a cylinder h_c (m)	10
Diameter of payload modeled as a cylinder D_c (m)	5
Damping coefficient at payload d_L (N·s/m)	2.0×10^5
Damping coefficient at crane d_0 (N·m·s/rad)	8.0×10^5
Seawater density ρ_s (kgm^{-3})	1024

by recalling (10.234), (10.237), and those two quantities are, physically, the initial oscillation velocities of the crane and payload. The initial conditions of (10.223)–(10.227) are determined based on the initial conditions of (10.233)–(10.239)—that is, from $z(x,0), w(x,0)$. The initial oscillation velocity of the cable is hence

$$u_t(x,0) = \frac{1}{2}(z(x,0) + w(x,0)) = 2\sin\left(\frac{\pi x}{L}\right) + 2\cos\left(\frac{\pi x}{L}\right). \tag{10.245}$$

The initial distributed oscillation displacement of the cable is defined as $u(x,0) = 0$, which, recalling (10.224), (10.226), implies that the initial displacement of the payload is $b_L(0) = 0$, and the initial displacement of the crane is $b_0(0) = 0$.

The ocean current disturbances, the distributed oscillating drag force $f(x,t)$ on the cable, and the drag force $f_L(t)$ at the payload are modeled as (7.189), (7.190) in chapter 7.

Simulation Results

Our task is to reduce the oscillations of the cable and place the payload in the target area, namely, within the permissible tolerance of 2.5 m around the predetermined location [95] by applying the observer-based output-feedback control force at the onboard crane. We concentrate on the end phase (20 s) of the descending process, when the payload is near the seafloor, and the cable is at the fully extended total length L. This is the most important and challenging phase because the cable is long, and the oscillations are large. The simulation is based on (10.233)–(10.239) using the finite-difference method with the time and space steps of 0.001 s and 0.1 m, respectively. With the sensor delay $\tau = 0.1$ s, the measurement output is delayed by 100 time steps. Applying the proposed observer and controller designs to the DCV (10.233)–(10.242), obtains the following simulation results.

a) Responses of z, w, X, Y

In figure 10.4, we note that the oscillations appear in the open-loop responses of $w(x,t), z(x,t)$, which is the result of the property of the long cable and the external disturbances (7.189), (7.190) in chapter 7. From figure 10.5, we observe that the designed control input reduces the oscillation amplitudes even though the plant is subject to external disturbances. The moving velocity of the controlled crane and the oscillation velocity of the payload—namely, $X(t)$ and $Y(t)$—are shown in figure 10.6, from which we know that $X(t)$ and $Y(t)$ converge to zero. Figure 10.7 shows that the observer errors $\tilde{w}(x,t), \tilde{z}(x,t)$ converge to a small range around zero under the unknown external disturbances and the sensor delay τ.

b) Representing the obtained responses as u, b_L in the DCV

The physical meaning of the responses z, w in figures 10.5, 10.6 will be clear after representing them as the responses of the cable oscillation and position error— that is, as u and b_L in (10.223)–(10.227). Through (10.230), (10.231), the cable transverse oscillation energy, including the oscillation kinetic energy $\frac{\rho}{2}\|u_t(\cdot,t)\|^2$ and the potential energy $\frac{T_0}{2}\|u_x(\cdot,t)\|^2$, is represented by $z(x,t), w(x,t)$ as

$$\frac{\rho}{2}\|u_t(\cdot,t)\|^2 + \frac{T_0}{2}\|u_x(\cdot,t)\|^2$$
$$= \frac{\rho}{8}\|w(\cdot,t) + z(\cdot,t)\|^2 + \frac{\rho}{8}\|w(\cdot,t) - z(\cdot,t)\|^2, \tag{10.246}$$

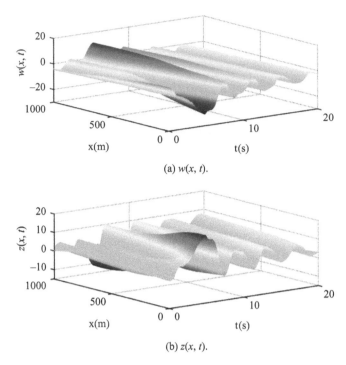

(a) $w(x, t)$.

(b) $z(x, t)$.

Figure 10.4. Responses of $w(x, t)$, $z(x, t)$ without control.

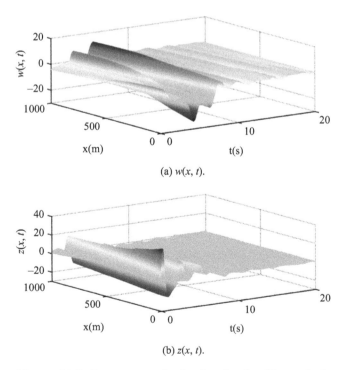

(a) $w(x, t)$.

(b) $z(x, t)$.

Figure 10.5. Responses of $w(x, t)$, $z(x, t)$ with control.

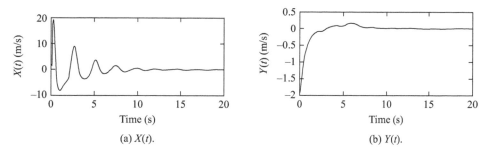

Figure 10.6. Responses of $X(t)$, $Y(t)$ with control.

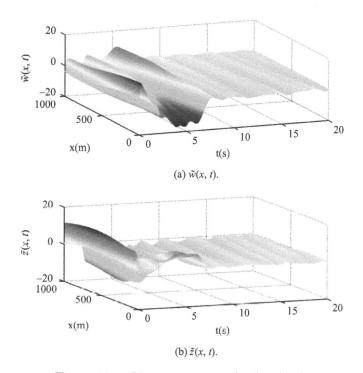

Figure 10.7. Observer errors $\tilde{w}(x,t)$, $\tilde{z}(x,t)$.

where $\|u_t(\cdot,t)\|^2$ denotes $\int_0^L u_t(\cdot,t)^2 dx$. The transverse displacement of the payload $b_L(t)$ is obtained as

$$b_L(t) = u(L,t) = \frac{1}{2}\int_0^t (z(L,\delta) + w(L,\delta))d\delta + b_L(0). \tag{10.247}$$

As shown in figure 10.8, the oscillation energy of the cable with the proposed control law is reduced faster and to a level below the uncontrolled case, after $t = 5.5$ s, under the external disturbances (7.189), (7.190) in chapter 7. This result shows the robustness of the proposed control to small disturbances. However, similar to the simulation results in chapter 7, as we continue to increase the amplitude of the disturbance given in (7.189) by gradually raising A_D, the amplitude of the oscillating drag force, from its baseline value 400, the dot-and-dash line in figure 10.8

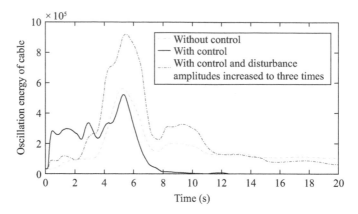

Figure 10.8. Cable transverse oscillation energy $\frac{\rho}{2}\|u_t(\cdot,t)\|^2 + \frac{T_0}{2}\|u_x(\cdot,t)\|^2$.

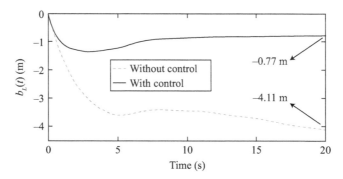

Figure 10.9. Transverse displacement $b_L(t)$ of the payload. The end point at $t = 20$ s means the position error on the seafloor. The permissible tolerance of this typical model is 2.5 m [95].

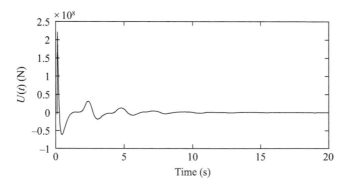

Figure 10.10. Control force of the onboard crane.

shows that the controller fails to achieve effective vibration suppression once A_D reaches three times the baseline value—that is, $A_D = 1200$. From figure 10.9, the position error of the payload is -0.77 m from the desired location on the seafloor, satisfying the requirement of being within the permissible tolerance of 2.5 m, while the position error is -4.11 m in the case without control, exceeding the tolerance. The control signal shown in figure 10.10 is bounded and convergent.

10.6   APPENDIX

A. Matching (10.28)–(10.33) and (10.135)–(10.140)

Step 1. Taking the time and spatial derivative of (10.134) along (10.28)–(10.33) and substituting the results into (10.138), we get

$$\beta_t(x,t) - q_2\beta_x(x,t) + c_2\beta(x,t) + \lambda(x)\Gamma_1\tilde{v}(2,t) + \int_x^1 J_2(x,y)h_3(\tilde{v}(2,t);y)dy$$

$$+ \int_x^1 K_2(x,y)h_2(\tilde{v}(2,t);y)dy - h_3(\tilde{v}(2,t);x) + q_2J_2(x,1)h_4(\tilde{v}(2,t))$$

$$= \hat{w}_t(x,t) - \int_x^1 K_2(x,y)\hat{z}_t(y,t)dy - \int_x^1 J_2(x,y)\hat{w}_t(y,t)dy - \lambda(x)\dot{\hat{Y}}(t)$$

$$- q_2\hat{w}_x(x,t) + q_2\int_x^1 K_{2x}(x,y)\hat{z}(y,t)dy + q_2\int_x^1 J_{2x}(x,y)\hat{w}(y,t)dy$$

$$+ q_2\lambda'(x)\hat{Y}(t) - q_2K_2(x,x)\hat{z}(x,t) - q_2J_2(x,x)\hat{w}(x,t)$$

$$+ c_2\hat{w}(x,t) - c_2\int_x^1 K_2(x,y)\hat{z}(y,t)dy - c_2\int_x^1 J_2(x,y)\hat{w}(y,t)dy$$

$$- c_2\lambda(x)\hat{Y}(t) + \lambda(x)\Gamma_1\tilde{v}(2,t) + \int_x^1 J_2(x,y)h_3(\tilde{v}(2,t);y)dy$$

$$+ \int_x^1 K_2(x,y)h_2(\tilde{v}(2,t);y)dy - h_3(\tilde{v}(2,t);x) + q_2J_2(x,1)h_4(\tilde{v}(2,t))$$

$$= -c_2\hat{z}(x,t) + h_3(\tilde{v}(2,t);x) + q_1\int_x^1 K_2(x,y)\hat{z}_x(y,t)dy$$

$$+ \int_x^1 c_1K_2(x,y)[\hat{z}(y,t) + \hat{w}(y,t)]dy$$

$$- \int_x^1 K_2(x,y)h_2(\tilde{v}(2,t);y)dy - q_2\int_x^1 J_2(x,y)\hat{w}_x(y,t)dy$$

$$+ \int_x^1 c_2J_2(x,y)[\hat{z}(y,t) + \hat{w}(y,t)]dy - \int_x^1 J_2(x,y)h_3(\tilde{v}(2,t);y)dy$$

$$+ q_2\int_x^1 K_{2x}(x,y)\hat{z}(y,t)dy + q_2\int_x^1 J_{2x}(x,y)\hat{w}(y,t)dy$$

$$- q_2K_2(x,x)\hat{z}(x,t) - q_2J_2(x,x)\hat{w}(x,t)$$

$$- \lambda(x)A_1\hat{Y}(t) - \lambda(x)B_1\hat{z}(1,t) - \lambda(x)\Gamma_1\tilde{v}(2,t) + q_2\lambda'(x)\hat{Y}(t)$$

$$- c_2\int_x^1 K_2(x,y)\hat{z}(y,t)dy - c_2\int_x^1 J_2(x,y)\hat{w}(y,t)dy - c_2\lambda(x)\hat{Y}(t)$$

$$+ \lambda(x)\Gamma_1\tilde{v}(2,t) + \int_x^1 J_2(x,y)h_3(\tilde{v}(2,t);y)dy$$

$$+ \int_x^1 K_2(x,y)h_2(\tilde{v}(2,t);y)dy - h_3(\tilde{v}(2,t);x) + q_2 J_2(x,1)h_4(\tilde{v}(2,t))$$

$$= -[c_2 + (q_1 + q_2)K_2(x,x)]\hat{z}(x,t)$$

$$+ \int_x^1 [c_1 K_2(x,y) + q_2 J_{2x}(x,y) + q_2 J_{2y}(x,y)]\,\hat{w}(y,t)dy$$

$$+ \int_x^1 [c_2 J_2(x,y) + c_1 K_2(x,y) - c_2 K_2(x,y) + q_2 K_{2x}(x,y) - q_1 K_{2y}(x,y)]\hat{z}(y,t)dy$$

$$+ [q_1 K_2(x,1) - q_2 J_2(x,1)q - \lambda(x)B_1]\,\hat{z}(1,t)$$

$$+ [q_2\lambda'(x) - \lambda(x)A_1 - c_2\lambda(x) - q_2 J_2(x,1)C_1]\,\hat{Y}(t). \qquad (10.248)$$

For (10.248) to hold, conditions (10.155)–(10.159) should be satisfied.

Step 2. Taking the time and spatial derivative of (10.133) along (10.28)–(10.33) and substituting the results into (10.137), we obtain

$$\alpha_t(x,t) + q_1\alpha_x(x,t) + c_1\alpha(x,t) + \gamma(x)\Gamma_1\tilde{v}(2,t) + \int_x^1 J_3(x,y)h_3(\tilde{v}(2,t);y)dy$$

$$+ \int_x^1 K_3(x,y)h_2(\tilde{v}(2,t);y)dy - h_2(\tilde{v}(2,t);x) + q_2 J_3(x,1)h_4(\tilde{v}(2,t))$$

$$= \hat{z}_t(x,t) - \int_x^1 K_3(x,y)\hat{z}_t(y,t)dy - \int_x^1 J_3(x,y)\hat{w}_t(y,t)dy$$

$$- \gamma(x)\dot{\hat{Y}}(t) + q_1\hat{z}_x(x,t) - q_1\int_x^1 K_{3x}(x,y)\hat{z}(y,t)dy$$

$$- q_1\int_x^1 J_{3x}(x,y)\hat{w}(y,t)dy - q_1\gamma'(x)\hat{Y}(t)$$

$$+ q_1 K_3(x,x)\hat{z}(x,t) + q_1 J_3(x,x)\hat{w}(x,t)$$

$$+ c_1\hat{z}(x,t) - c_1\int_x^1 K_3(x,y)\hat{z}(y,t)dy - c_1\int_x^1 J_3(x,y)\hat{w}(y,t)dy - c_1\gamma(x)\hat{Y}(t)$$

$$+ \gamma(x)\Gamma_1\tilde{v}(2,t) + \int_x^1 J_3(x,y)h_3(\tilde{v}(2,t);y)dy$$

$$+ \int_x^1 K_3(x,y)h_2(\tilde{v}(2,t);y)dy - h_2(\tilde{v}(2,t);x) + q_2 J_3(x,1)h_4(\tilde{v}(2,t))$$

$$= -c_1\hat{w}(x,t) + h_2(\tilde{v}(2,t);x) + q_1\int_x^1 K_3(x,y)\hat{z}_x(y,t)dy$$

$$+ \int_x^1 c_1 K_3(x,y)[\hat{z}(y,t) + \hat{w}(y,t)]dy - \int_x^1 K_3(x,y)h_2(\tilde{v}(2,t);y)dy$$

$$- q_2\int_x^1 J_3(x,y)\hat{w}_x(y,t)dy + \int_x^1 c_2 J_3(x,y)[\hat{z}(y,t) + \hat{w}(y,t)]dy$$

$$- \int_x^1 J_3(x,y)h_3(\tilde{v}(2,t);y)dy - q_1\int_x^1 K_{3x}(x,y)\hat{z}(y,t)dy$$

$$- q_1\int_x^1 J_{3x}(x,y)\hat{w}(y,t)dy + q_1 K_3(x,x)\hat{z}(x,t) + q_1 J_3(x,x)\hat{w}(x,t)$$

$$-\gamma(x)A_1\hat{Y}(t) - \gamma(x)B_1\hat{z}(1,t) - \gamma(x)\Gamma_1\tilde{v}(2,t) - q_1\gamma'(x)\hat{Y}(t)$$

$$-c_1\int_x^1 K_3(x,y)\hat{z}(y,t)dy - c_1\int_x^1 J_3(x,y)\hat{w}(y,t)dy - c_1\gamma(x)\hat{Y}(t)$$

$$+\gamma(x)\Gamma_1\tilde{v}(2,t) + \int_x^1 J_3(x,y)h_3(\tilde{v}(2,t);y)dy$$

$$+\int_x^1 K_3(x,y)h_2(\tilde{v}(2,t);y)dy - h_2(\tilde{v}(2,t);x) + q_2 J_3(x,1)h_4(\tilde{v}(2,t))$$

$$= [(q_2+q_1)J_3(x,x) - c_1]\hat{w}(x,t) + \int_x^1 \big[c_1 K_3(x,y) - q_1 J_{3x}(x,y)$$

$$+ q_2 J_{3y}(x,y) + (c_2-c_1)J_3(x,y)\big]\hat{w}(y,t)dy$$

$$+\int_x^1 [c_2 J_3(x,y) - q_1 K_{3x}(x,y) - q_1 K_{3y}(x,y)]\,\hat{z}(y,t)dy$$

$$+ [q_1 K_3(x,1) - q_2 J_3(x,1)q - \gamma(x)B_1]\,\hat{z}(1,t)$$

$$+ [-q_1\gamma'(x) - \gamma(x)A_1 - c_1\gamma(x) - q_2 J_3(x,1)C_1]\,\hat{Y}(t). \tag{10.249}$$

For (10.249) to hold, conditions (10.149)–(10.152), (10.154) should be satisfied.

Step 3. Inserting (10.133), (10.134) into (10.140), (10.139), respectively, and applying (10.32), (10.33), we obtain

$$\dot{\hat{Y}}(t) - \bar{A}_1\hat{Y}(t) - B_1\alpha(1,t) - \Gamma_1\tilde{v}(2,t)$$

$$= \dot{\hat{Y}}(t) - A_1\hat{Y}(t) + B_1 F_1\hat{Y}(t) - B_1\hat{z}(1,t) + B_1\gamma(1)\hat{Y}(t) - \Gamma_1\tilde{v}(2,t)$$

$$= B_1(F_1 + \gamma(1))\hat{Y}(t) = 0, \tag{10.250}$$

$$\beta(1,t) - q\alpha(1,t) - h_4(\tilde{v}(2,t))$$

$$= \hat{w}(1,t) - q\hat{z}(1,t) + (q\gamma(1) - \lambda(1))\hat{Y}(t) - h_4(\tilde{v}(2,t))$$

$$= (q\gamma(1) - \lambda(1) + C_1)\hat{Y}(t) = 0. \tag{10.251}$$

For (10.250), (10.251) to hold, conditions (10.153) and (10.160) should be satisfied.

Step 4. Inserting (10.133), (10.134) into (10.136) and (10.135) and applying (10.29), (10.28) we get

$$\alpha(0,t) - p\beta(0,t) - \int_0^1 \bar{K}_1(x)\alpha(x,t)dx - \bar{K}_3\hat{Y}(t)$$

$$-\int_0^1 \bar{K}_2(x)\beta(x,t)dx - C_0\hat{X}(t)$$

$$= \int_0^1 \bigg[pK_2(0,x) - K_3(0,x) + \int_0^x \bar{K}_1(y)K_3(y,x)dy$$

$$+\int_0^x \bar{K}_2(y)K_2(y,x)dy - \bar{K}_1(x)\bigg]z(x,t)dx$$

$$-\int_0^1 \bigg[pJ_2(0,x) - J_3(0,x) - \int_0^x \bar{K}_1(y)J_3(y,x)dy$$

$$-\int_0^x \bar{K}_2(y)J_2(y,x)dy + \bar{K}_2(x)\bigg]w(x,t)dx$$

$$+ \left[\int_0^1 \bar{K}_2(x)\lambda(x)dx + \int_0^1 \bar{K}_1(x)\gamma(x)dx + p\lambda(0) - \gamma(0) - \bar{K}_3 \right] Y(t) = 0,$$

$$(10.252)$$

$$\dot{\hat{X}}(t) - A_0\hat{X}(t) - E_0\beta(0,t) - \int_0^1 \bar{K}_4(x)\alpha(x,t)dx$$

$$- \int_0^1 \bar{K}_5(x)\beta(x,t)dx - \bar{K}_6\hat{Y}(t) - B_0U(t) - h_1(\tilde{v}(2,t))$$

$$= -\int_0^1 \left[\int_0^x \bar{K}_4(y)K_3(y,x)dy + \int_0^x \bar{K}_5(y)K_2(y,x)dy \right.$$

$$\left. - E_0K_2(0,x) - \bar{K}_4(x) \right] z(x,t)dx$$

$$- \int_0^1 \left[\int_0^x \bar{K}_4(y)J_3(y,x)dy + \int_0^x \bar{K}_5(y)J_2(y,x)dy \right.$$

$$\left. + E_0J_2(0,x) - \bar{K}_5(x) \right] w(x,t)dx$$

$$- \left[\int_0^1 \bar{K}_5(x)\lambda(x)dx + \int_0^1 \bar{K}_4(x)\gamma(x)dx + E_0\lambda(0) - \bar{K}_6 \right] Y(t) = 0. \quad (10.253)$$

For (10.252), (10.253) to hold, conditions (10.143)–(10.148) are obtained.

B. Calculations of (10.98) and (10.102)

The details of calculating (10.98) are shown as follows:

$$\tilde{w}_t(x,t) - q_2\tilde{w}_x(x,t) + c_2\tilde{z}(x,t) + c_2\tilde{w}(x,t) + h_3(\tilde{v}(2,t);x)$$

$$= \tilde{\beta}_t(x,t) - \int_x^1 \psi(x,y)\tilde{\alpha}_t(y,t)dy$$

$$- q_2\tilde{\beta}_x(x,t) + q_2\int_x^1 \psi_x(x,y)\tilde{\alpha}(y,t)dy - q_2\psi(x,x)\tilde{\alpha}(x,t)$$

$$+ c_2\tilde{\alpha}(x,t) - c_2\int_x^1 \phi(x,y)\tilde{\alpha}(y,t)dy + h_3(\tilde{v}(2,t);x)$$

$$+ c_2\tilde{\beta}(x,t) - c_2\int_x^1 \psi(x,y)\tilde{\alpha}(y,t)dy$$

$$= \int_x^1 \bar{N}(x,y)\tilde{\beta}(y,t)dy + c_1\int_x^1 \psi(x,y)\tilde{\alpha}(y,t)dy$$

$$+ q_1\int_x^1 \psi(x,y)\tilde{\alpha}_x(y,t)dy + \int_x^1 c_1\psi(x,y)\tilde{\beta}(y,t)dy$$

$$- \int_x^1 \int_x^y \psi(x,\delta)\bar{M}(\delta,y)d\delta\tilde{\beta}(y,t)dy + q_2\int_x^1 \psi_x(x,y)\tilde{\alpha}(y,t)dy$$

$$- q_2\psi(x,x)\tilde{\alpha}(x,t) - c_2\int_x^1 \psi(x,y)\tilde{\alpha}(y,t)dy$$

$$+ c_2\tilde{\alpha}(x,t) - c_2\int_x^1 \phi(x,y)\tilde{\alpha}(y,t)dy + h_3(\tilde{v}(2,t);x)$$

$$= [c_2 - (q_1 + q_2)\psi(x,x)]\tilde{\alpha}(x,t)$$

$$+ \int_x^1 [-q_1\psi_y(x,y) + q_2\psi_x(x,y) - c_2\phi(x,y) - (c_2 - c_1)\psi(x,y)]\tilde{\alpha}(x,t)dy$$

$$+ \int_x^1 \left[c_1\psi(x,y) + \bar{N}(x,y) - \int_x^y \psi(x,\delta)\bar{M}(\delta,y)d\delta \right] \tilde{\beta}(y,t)dy$$

$$+ q_1\psi(x,1)\tilde{\alpha}(1,t) + h_3(\tilde{v}(2,t);x)$$

$$= q_1\psi(x,1)\tilde{\alpha}(1,t) + h_3(\tilde{v}(2,t);x) = 0. \tag{10.254}$$

The details of calculating (10.102) are shown as follows:

$$\tilde{z}_t(x,t) + q_1\tilde{z}_x(x,t) + c_1\tilde{w}(x,t) + c_1\tilde{z}(x,t) + h_2(\tilde{v}(2,t);x)$$

$$= \tilde{\alpha}_t(x,t) - \int_x^1 \phi(x,y)\tilde{\alpha}_t(y,t)dy + q_1\tilde{\alpha}_x(x,t) - q_1 \int_x^1 \phi_x(x,y)\tilde{\alpha}(y,t)dy$$

$$+ q_1\phi(x,x)\tilde{\alpha}(x,t) + c_1\tilde{\beta}(x,t) - c_1 \int_x^1 \psi(x,y)\tilde{\alpha}(y,t)dy + h_2(\tilde{v}(2,t);x)$$

$$+ c_1\tilde{\alpha}(x,t) - c_1 \int_x^1 \phi(x,y)\tilde{\alpha}(y,t)dy$$

$$= -c_1\tilde{\beta}(x,t) + q_1 \int_x^1 \phi(x,y)\tilde{\alpha}_x(y,t)dy + \int_x^1 [-q_1\phi_x(x,y) - c_1\psi(x,y)]\tilde{\alpha}(y,t)dy$$

$$+ \int_x^1 \left[- \int_x^y \phi(x,\delta)\bar{M}(y,\delta)d\delta + \phi(x,y)c_1 \right] \tilde{\beta}(y,t)dy$$

$$+ q_1\phi(x,x)\tilde{\alpha}(x,t) + c_1\tilde{\beta}(x,t) + \int_x^1 \bar{M}(x,y)\tilde{\beta}(y,t)dy + h_2(\tilde{v}(2,t);x)$$

$$= \int_x^1 [-q_1\phi_x(x,y) - c_1\psi(x,y) - q_1\phi_y(x,y)]\tilde{\alpha}(y,t)dy$$

$$+ \int_x^1 \left[\bar{M}(x,y) - \int_x^y \phi(x,\delta)\bar{M}(\delta,y)d\delta + c_1\phi(x,y) \right] \tilde{\beta}(y,t)dy$$

$$+ q_1\phi(x,1)\tilde{\alpha}(1,t) + h_2(\tilde{v}(2,t);x)$$

$$= q_1\phi(x,1)\tilde{\alpha}(1,t) + h_2(\tilde{v}(2,t);x) = 0. \tag{10.255}$$

10.7 NOTES

In this chapter, an observer-based output-feedback controller was designed for heterodirectional coupled hyperbolic PDEs sandwiched between two general ODEs using backstepping and frequency-domain design methods, with compensation for arbitrarily long sensor delays. The proximal ODE structure in chapter 9 was restricted in a chain of integrators, which is relieved in this chapter. Unlike the delay-robust stabilization feedback control designs for sandwich systems in [49, 152], which achieve robustness to a small delay, a delay compensation technique originally developed in [116, 125] was adopted in the control design in this chapter to compensate for arbitrarily long delays.

Chapter Eleven

───

Event-Triggered Control of Sandwich Hyperbolic PDEs

In order to avoid too-frequent actions of the massive actuators of the string-actuated mechanisms in various applications, such as the hydraulic cylinder and head sheave in the mining cable elevator, the ship-mounted crane in the deep-sea construction vessel, or the rotary table in the oil-drilling system, starting from the continuous-in-time control laws derived in chapters 9 and 10, we develop event-triggered output-feedback backstepping boundary controllers for hyperbolic partial differential equation (PDE) sandwich systems, so that the actuators only move at the necessary times determined by a dynamic event-triggering criterion, also referred to as an *event-triggering mechanism.*

The content of the chapter is organized as follows. The model considered is described in section 11.1, and a state observer is designed in section 11.2. The control design is in two steps. The first step is the design of a continuous-in-time output-feedback boundary stabilization law in section 11.3, which is similar to the control design in chapter 10. The second step is the design of a dynamic event-triggering mechanism in section 11.4, where the existence of a minimal dwell time between two triggering times is proved. The exponential convergence result of the event-based closed-loop system is proved via Lyapunov analysis in section 11.5. The controller is applied into the axial vibration control of a mining cable elevator in section 11.6.

11.1 PROBLEM FORMULATION

The plant considered in this chapter is

$$\dot{Y}(t) = A_0 Y(t) + E_0 w(0,t) + B_0 U(t), \tag{11.1}$$

$$\hat{z}(0,t) = pw(0,t) + C_0 Y(t), \tag{11.2}$$

$$z_t(x,t) = -q_1 z_x(x,t) - c_1 w(x,t) - c_1 z(x,t), \tag{11.3}$$

$$w_t(x,t) = q_2 w_x(x,t) - c_2 w(x,t) - c_2 z(x,t), \tag{11.4}$$

$$w(1,t) = qz(1,t) + CX(t), \tag{11.5}$$

$$\dot{X}(t) = AX(t) + Bz(1,t) \tag{11.6}$$

for $\forall (x,t) \in [0,1] \times [0,\infty)$. The vectors $Y(t) \in \mathbb{R}^{\bar{n} \times 1}$, $X(t) \in \mathbb{R}^{n \times 1}$ are ordinary differential equation (ODE) states, where positive constants $\bar{n}, n$ are the dimensions of the two ODEs. The scalars $z(x,t) \in \mathbb{R}$, $w(x,t) \in \mathbb{R}$ are the states of the 2×2 coupled hyperbolic PDEs. The positive constants q_1 and q_2 are the transport velocities. The constants c_1, c_2 are arbitrary (the design in this chapter is straightforwardly suitable for the case that has four arbitrary coefficients of in-domain couplings, like

c_1, c_2, c_3, c_4 in (7.17), (7.18)), and p, q satisfy assumption 11.3 shown below. The vector $E_0 \in \mathbb{R}^{\bar{n} \times 1}$ is arbitrary. The control input $U(t) \in \mathbb{R}$ is to be designed. The plant initial conditions are taken as

$$(Y(0), z(x,0), w(x,0), X(0)) \in \chi = \mathbb{R}^{\bar{n}} \times H^{n_r+1}(0,1)^2 \times \mathbb{R}^n,$$

where n_r is the relative degree of (11.1). The matrices in the ODE subsystems $A \in \mathbb{R}^{n \times n}$, $A_0 \in \mathbb{R}^{\bar{n} \times \bar{n}}$, $B \in \mathbb{R}^{n \times 1}$, $B_0 \in \mathbb{R}^{\bar{n} \times 1}$, $C \in \mathbb{R}^{1 \times n}$, $C_0 \in \mathbb{R}^{1 \times \bar{n}}$ are subject to assumptions 11.1 and 11.2.

We adopt the following notation in this chapter. The symbol R^- denotes the set of negative real numbers, whose complement of the real axis is $\mathbb{R}_+ := [0, +\infty)$. For an interval $\bar{U} \subseteq \mathbb{R}_+$ and a set $\bar{\Omega} \subseteq \mathbb{R}$, by $C^0(\bar{U}; \bar{\Omega})$, we denote the class of continuous mappings on $\bar{U}$, which take values in $\bar{\Omega}$. For an integer $k \geq 1$, $H^k(0,1)$ denotes the Sobolev space of functions in $L^2(0,1)$ with all its weak derivatives up to order k in $L^2(0,1)$. The usual Euclidean norm is denoted by $|\cdot|$. For an interval $I \subseteq \mathbb{R}_+$, the space $C^0(I; L^2(0,1))$ is the space of continuous mappings $I \ni t \mapsto u(x.t) \in L^2(0,1)$.

Assumption 11.1. *The pairs (A, B) and (A_0, B_0) are controllable. The pairs (A, C) and (A_0, C_0) are observable.*

Based on this assumption, there exist constant matrices K, L, K_0, L_0 to make the following matrices Hurwitz:

$$\hat{A} = A - BK, \tag{11.7}$$
$$\bar{A} = A - LC, \tag{11.8}$$
$$\hat{A}_0 = A_0 - B_0 K_0, \tag{11.9}$$
$$\bar{A}_0 = A_0 - L_0 C_0. \tag{11.10}$$

The following assumption for the ODE in the input channel ensures a stable left inversion of the system (A_0, B_0, C_0) [135].

Assumption 11.2. *The matrices C_0, A_0, B_0 satisfy*

$$\det\left(\begin{bmatrix} sI - A_0 & B_0 \\ C_0 & 0 \end{bmatrix} \right) \neq 0 \tag{11.11}$$

for all $s \in \mathbb{C}, \Re(s) \geq 0$.

The following assumption about the PDE subsystem parameters, which will be used in the low-pass filter design, is satisfied in the mining cable elevator model.

Assumption 11.3. *The plant parameters p, q satisfy*

$$|pq| e^{\left(\frac{c_2}{q_2} + \frac{c_1}{q_1}\right)} < 1 \tag{11.12}$$

and $q \neq 0$.

The control objective of this chapter is to exponentially stabilize (11.1)–(11.6) through designing an event-triggered boundary control input using only a collocated measurement $w(0,t)$.

11.2   OBSERVER

In order to estimate the distributed states, which usually cannot be measured but are required in the controller, a state observer for the plant (11.1)–(11.6) is designed in this section using only the measurement $w(0, t)$ at the actuated boundary—that is, using only the sensing of signals at the head sheaves of the mining cable elevator. The observer is postulated with the same structure as in [43], as follows:

$$\dot{\hat{Y}}(t) = A_0\hat{Y}(t) + E_0w(0,t) + B_0U(t)$$
$$+ \Psi_0(w(0,t) - \hat{w}(0,t)), \qquad (11.13)$$

$$\hat{z}(0,t) = pw(0,t) + C_0\hat{Y}(t), \qquad (11.14)$$

$$\hat{z}_t(x,t) = -q_1\hat{z}_x(x,t) - c_1\hat{w}(x,t) - c_1\hat{z}(x,t)$$
$$+ \Psi_1(x)(w(0,t) - \hat{w}(0,t)), \qquad (11.15)$$

$$\hat{w}_t(x,t) = q_2\hat{w}_x(x,t) - c_2\hat{w}(x,t) - c_2\hat{z}(x,t)$$
$$+ \Psi_2(x)(w(0,t) - \hat{w}(0,t)), \qquad (11.16)$$

$$\hat{w}(1,t) = q\hat{z}(1,t) + C\hat{X}(t), \qquad (11.17)$$

$$\dot{\hat{X}}(t) = A\hat{X}(t) + B\hat{z}(1,t) + \Psi_3(w(0,t) - \hat{w}(0,t)), \qquad (11.18)$$

where $\Psi_0, \Psi_1(x), \Psi_2(x), \Psi_3$ are the observer gains to be determined. The control signal $U(t)$ can be replaced by an event-triggered control law $U_d(t)$ presented later in the event-based closed-loop system.

Defining observer error states as

$$[\tilde{z}(x,t), \tilde{w}(x,t), \tilde{X}(t), \tilde{Y}(t)]$$
$$= [z(x,t), w(x,t), X(t), Y(t)] - [\hat{z}(x,t), \hat{w}(x,t), \hat{X}(t), \hat{Y}(t)], \qquad (11.19)$$

we obtain the observer error system:

$$\dot{\tilde{Y}}(t) = A_0\tilde{Y}(t) - \Psi_0\tilde{w}(0,t), \qquad (11.20)$$

$$\tilde{z}(0,t) = C_0\tilde{Y}(t), \qquad (11.21)$$

$$\tilde{z}_t(x,t) = -q_1\tilde{z}_x(x,t) - c_1\tilde{w}(x,t) - c_1\tilde{z}(x,t) - \Psi_1(x)\tilde{w}(0,t), \qquad (11.22)$$

$$\tilde{w}_t(x,t) = q_2\tilde{w}_x(x,t) - c_2\tilde{w}(x,t) - c_2\tilde{z}(x,t) - \Psi_2(x)\tilde{w}(0,t), \qquad (11.23)$$

$$\tilde{w}(1,t) = q\tilde{z}(1,t) + C\tilde{X}(t), \qquad (11.24)$$

$$\dot{\tilde{X}}(t) = A\tilde{X}(t) + B\tilde{z}(1,t) - \Psi_3\tilde{w}(0,t). \qquad (11.25)$$

The following backstepping transformation [96] is made:

$$\tilde{z}(x,t) = \tilde{\alpha}(x,t) - \int_0^x \bar{\phi}(x,y)\tilde{\alpha}(y,t)dy - \int_0^x \bar{\phi}_1(x,y)\tilde{\beta}(y,t)dy - \bar{\gamma}(x)\tilde{Y}(t), \quad (11.26)$$

$$\tilde{w}(x,t) = \tilde{\beta}(x,t) - \int_0^x \bar{\psi}(x,y)\tilde{\alpha}(y,t)dy - \int_0^x \bar{\psi}_1(x,y)\tilde{\beta}(y,t)dy - \bar{\varphi}(x)\tilde{Y}(t) \quad (11.27)$$

to convert the observer error dynamics (11.20)–(11.25) to the intermediate observer error system as

$$\dot{\tilde{Y}}(t) = \bar{A}_0 \tilde{Y}(t) - \Psi_0 \tilde{\beta}(0, t), \tag{11.28}$$

$$\tilde{\alpha}(0, t) = 0, \tag{11.29}$$

$$\tilde{\alpha}_t(x, t) = -q_1 \tilde{\alpha}_x(x, t) - c_1 \tilde{\alpha}(x, t), \tag{11.30}$$

$$\tilde{\beta}_t(x, t) = q_2 \tilde{\beta}_x(x, t) - c_2 \tilde{\beta}(x, t), \tag{11.31}$$

$$\tilde{\beta}(1, t) = q\tilde{\alpha}(1, t) + C\tilde{X}(t) - \int_0^1 (q\bar{\phi}(1, y) - \bar{\psi}(1, y))\tilde{\alpha}(y, t)dy$$

$$- \int_0^1 (q\bar{\phi}_1(1, y) - \bar{\psi}_1(1, y))\tilde{\beta}(y, t)dy + (\bar{\varphi}(1) - q\bar{\gamma}(1))\tilde{Y}(t), \tag{11.32}$$

$$\dot{\tilde{X}}(t) = A\tilde{X}(t) + B\left(\tilde{\alpha}(1, t) - \int_0^1 \bar{\phi}(1, y)\tilde{\alpha}(y, t)dy - \int_0^1 \bar{\phi}_1(1, y)\tilde{\beta}(y, t)dy\right)$$

$$- \Psi_3 \tilde{\beta}(0, t) + (\Psi_3 \bar{\varphi}(0) - B\bar{\gamma}(1))\tilde{Y}(t). \tag{11.33}$$

By matching (11.20)–(11.25) and (11.28)–(11.33), the conditions on the kernel functions $\bar{\phi}(x, y)$, $\bar{\phi}_1(x, y)$, $\bar{\psi}(x, y)$, $\bar{\psi}_1(x, y)$ on $\mathcal{D} = \{0 \leq y \leq x \leq 1\}$ and the observer gains Ψ_0, $\Psi_1(x)$, $\Psi_2(x)$ are obtained as follows. The conditions on the kernels are

$$-q_1 \bar{\phi}_x(x, y) - q_1 \bar{\phi}_y(x, y) - c_1 \bar{\psi}(x, y) = 0, \tag{11.34}$$

$$q_2 \bar{\psi}_x(x, y) - q_1 \bar{\psi}_y(x, y)$$
$$-(c_2 - c_1)\bar{\psi}(x, y) - c_2 \bar{\phi}(x, y) = 0, \tag{11.35}$$

$$\bar{\phi}(1, y) = \frac{1}{q}CK_1(y) + \frac{1}{q}\bar{\psi}(1, y), \tag{11.36}$$

$$\bar{\psi}(x, x) = -\frac{c_2}{q_1 + q_2}, \tag{11.37}$$

$$q_2 \bar{\phi}_{1y}(x, y) - q_1 \bar{\phi}_{1x}(x, y)$$
$$+(c_2 - c_1)\bar{\phi}_1(x, y) - c_1 \bar{\psi}_1(x, y) = 0, \tag{11.38}$$

$$q_2 \bar{\psi}_{1y}(x, y) + q_2 \bar{\psi}_{1x}(x, y) - c_2 \bar{\phi}_1(x, y) = 0, \tag{11.39}$$

$$\bar{\psi}_1(1, y) = q\bar{\phi}_1(1, y) - CK_2(y), \tag{11.40}$$

$$\bar{\phi}_1(x, x) = \frac{c_1}{q_1 + q_2}, \tag{11.41}$$

$$-q_2 \bar{\varphi}'(x) + \bar{\varphi}(x)\bar{A}_0 + c_2 \bar{\gamma}(x) + c_2 \bar{\varphi}(x) = 0, \tag{11.42}$$

$$q_1 \bar{\gamma}'(x) + \bar{\gamma}(x)\bar{A}_0 + c_1 \bar{\gamma}(x) + c_1 \bar{\varphi}(x) = 0, \tag{11.43}$$

$$\bar{\varphi}(0) = -C_0, \tag{11.44}$$

$$\bar{\gamma}(0) = -C_0. \tag{11.45}$$

The choices of (11.36), (11.40) will be evident later, where the definition of $K_1(y)$ and $K_2(y)$ will be given and the well-posedness of (11.34)–(11.45) will be shown. The matrix $\bar{A}_0$ defined in (11.10) is made Hurwitz by choosing L_0, in accordance with assumption 11.1.

The observer gains Ψ_0, $\Psi_1(x)$, $\Psi_2(x)$ are

$$\Psi_1(x) = -q_2 \bar{\phi}_1(x, 0) - \bar{\gamma}(x)L_0, \tag{11.46}$$

$$\Psi_2(x) = -q_2 \bar{\psi}_1(x, 0) - \bar{\phi}(x)L_0, \tag{11.47}$$

$$\Psi_0 = L_0. \tag{11.48}$$

We next apply another transformation, as in chapter 10, with the purpose of removing the two integral terms in the boundary condition (11.32) and rendering Hurwitz the system matrix of the ODE (11.33). This second transformation is

$$\tilde{Z}(t) = \tilde{X}(t) - \int_0^1 K_1(y)\tilde{\alpha}(y,t)dy - \int_0^1 K_2(y)\tilde{\beta}(y,t)dy, \qquad (11.49)$$

where $K_1(y), K_2(y)$ are the solutions of the ODEs

$$-q_1 K_1{}'(y) + AK_1(y) + c_1 K_1(y) - B\bar{\phi}(1,y) = 0, \qquad (11.50)$$

$$K_1(1) = \frac{Lq - B}{q_1}, \qquad (11.51)$$

$$q_2 K_2{}'(y) + AK_2(y) + c_2 K_2(y) - B\bar{\phi}_1(1,y) = 0, \qquad (11.52)$$

$$K_2(1) = \frac{L}{q_2}. \qquad (11.53)$$

Then the equations (11.32), (11.33) become

$$\tilde{\beta}(1,t) = q\tilde{\alpha}(1,t) + C\tilde{Z}(t) + (\bar{\varphi}(1) - q\bar{\gamma}(1))\tilde{Y}(t), \qquad (11.54)$$

$$\dot{\tilde{Z}}(t) = \bar{A}\tilde{Z}(t) + (\Psi_3 \bar{\varphi}(0) - B\bar{\gamma}(1) - L(\bar{\varphi}(1) - q\bar{\gamma}(1)))\tilde{Y}(t), \qquad (11.55)$$

where matrix $\bar{A}$, defined in (11.8), is made Hurwitz by choosing L in accordance with assumption 11.1.

By matching (11.32), (11.33) and (11.54), (11.55) via (11.49), the final observer gain Ψ_3 is obtained as

$$\Psi_3 = K_2(0)q_2. \qquad (11.56)$$

Regarding the conditions (11.34)–(11.45), (11.50)–(11.53), the equation sets (11.34)–(11.37), (11.50), (11.51) and (11.38)–(11.41), (11.52), (11.53) are two independent 2×2 hyperbolic PDE-ODE systems whose well-posedness can be obtained following the proof of lemma 9.1 in chapter 9. Solving the equation set (11.42)–(11.45) is an initial value problem, and explicit solutions can be obtained. Therefore, the equations (11.34)–(11.45), (11.50)–(11.53) are well-posed.

The following lemma holds for the observer error system.

Lemma 11.1. *Consider the observer* (11.13)–(11.18) *with the observer gains* (11.46)–(11.48), (11.56) *applied to the system* (11.1)–(11.6). *The resulting observer error system* (11.20)–(11.25) *is exponentially stable in the sense that there exist positive constants* Υ_e, λ_e *such that, for all initial values* $(\tilde{z}(x,0), \tilde{w}(x,0), \tilde{X}(0), \tilde{Y}(0)) \in L^2(0,1) \times L^2(0,1) \times \mathbb{R}^n \times \mathbb{R}^{\bar{n}}$,

$$\left|\tilde{X}(t)\right|^2 + \left|\tilde{Y}(t)\right|^2 + \|\tilde{z}(\cdot,t)\|^2 + \|\tilde{w}(\cdot,t)\|^2$$

$$\leq \Upsilon_e \left(\left|\tilde{X}(0)\right|^2 + \left|\tilde{Y}(0)\right|^2 + \|\tilde{z}(\cdot,0)\|^2 + \|\tilde{w}(\cdot,0)\|^2 \right) e^{-\lambda_e t}, \qquad (11.57)$$

where the decay rate λ_e *depends on the choices of* L, L_0 *in* (11.8), (11.10).

Proof. The target observer error system obtained from the original observer error system via the transformations (11.26), (11.27), (11.49) is (11.28)–(11.31), (11.54),

(11.55), where the ODEs (11.28), (11.55) are organized as

$$
\begin{pmatrix} \dot{\tilde{Y}}(t) \\ \dot{\tilde{Z}}(t) \end{pmatrix} = \begin{pmatrix} \bar{A}_0 & 0 \\ I_a & \bar{A} \end{pmatrix} \begin{pmatrix} \tilde{Y}(t) \\ \tilde{Z}(t) \end{pmatrix} - \begin{pmatrix} \Psi_0 \\ 0 \end{pmatrix} \tilde{\beta}(0,t) \tag{11.58}
$$

with

$$
I_a = \Psi_3 \bar{\varphi}(0) - B\bar{\gamma}(1) - L(\bar{\varphi}(1) - q\bar{\gamma}(1)).
$$

The system matrix in (11.58) is Hurwitz because $\bar{A}_0, \bar{A}$ are Hurwitz matrices.

Through a similar Lyapunov analysis in the proof of theorem 6.1 in chapter 6 for the target observer error system (11.29)–(11.31), (11.54), (11.58), the exponential stability result is obtained, and the exponential convergence rate is determined by the eigenvalue assignment for $\bar{A}, \bar{A}_0$—that is, the choices of L, L_0. Applying the invertibility of the backstepping transformation (11.26), (11.27), and the transformation (11.49), one obtains this lemma. $\qquad\square$

11.3   CONTINUOUS-IN-TIME CONTROL LAW

Before designing the event-triggered controller $U_d(t)$, a continuous-in-time feedback controller $U(t)$ is developed to exponentially stabilize the sandwich plant in this section. The proximal reflection term is compensated by the control input going through the ODE at the input channel. As a result, the continuous-in-time control law includes n_r-order time derivatives of $w(0,t)$, where n_r is the relative degree of the ODE. This high-order term would cause difficulties in designing the subsequent event-triggering mechanism guaranteeing the existence of a minimal dwell time between two triggering times. A low-pass-filter-based modification of the backstepping control design is presented to address this problem.

The continuous-in-time control law is derived by conducting a state-feedback design based on the observer (11.13)–(11.18) with the output estimation error injection terms assumed absent, in accordance to the result on their asymptotic convergence to zero in lemma 11.1. The separation principle is then verified and applied in the stability analysis of the resulting closed-loop system.

First Transformation

In order to remove the in-domain couplings between the PDEs and make the system matrix of the ODE at the right boundary Hurwitz, we introduce a PDE backstepping transformation

$$
\alpha(x,t) = \hat{z}(x,t) - \int_x^1 M(x,y)\hat{z}(y,t)dy
$$
$$
\qquad\qquad - \int_x^1 N(x,y)\hat{w}(y,t)dy - \gamma(x)\hat{X}(t), \tag{11.59}
$$
$$
\beta(x,t) = \hat{w}(x,t) - \int_x^1 H(x,y)\hat{z}(y,t)dy
$$
$$
\qquad\qquad - \int_x^1 J(x,y)\hat{w}(y,t)dy - \lambda(x)\hat{X}(t), \tag{11.60}
$$

with $M(x,y)$, $N(x,y)$, $\gamma(x)$, $H(x,y)$, $J(x,y)$, $\lambda(x)$ satisfying conditions (11.205)–(11.216) in appendix 11.7A, whose well-posedness is proved in lemma 9.1 of chapter 9. With this transformation, the system (11.13)–(11.18) with the observer errors assumed absent is converted to

$$\dot{\hat{Y}}(t) = A_0\hat{Y}(t) + E_0\left(\beta(0,t) - \int_0^1 \bar{K}_4(0,y)\alpha(y,t)dy\right.$$

$$\left. - \int_0^1 \bar{K}_5(0,y)\beta(y,t)dy - \bar{K}_6\hat{X}(t)\right) + B_0U(t), \tag{11.61}$$

$$\alpha(0,t) = p\beta(0,t) + \int_0^1 \bar{K}_1(x)\alpha(x,t)dx + C_0\hat{Y}(t)$$

$$+ \int_0^1 \bar{K}_2(x)\beta(x,t)dx + \bar{K}_3\hat{X}(t), \tag{11.62}$$

$$\alpha_t(x,t) = -q_1\alpha_x(x,t) - c_1\alpha(x,t), \tag{11.63}$$

$$\beta_t(x,t) = q_2\beta_x(x,t) - c_2\beta(x,t), \tag{11.64}$$

$$\beta(1,t) = q\alpha(1,t), \tag{11.65}$$

$$\dot{\hat{X}}(t) = \hat{A}\hat{X}(t) + B\alpha(1,t), \tag{11.66}$$

where $\hat{A} = A - BK$ is Hurwitz in light of assumption 11.1, and the gains $\bar{K}_1(x)$, $\bar{K}_2(x)$, $\bar{K}_3$, $\bar{K}_4(x)$, $\bar{K}_5(x)$, $\bar{K}_6$ are shown in (11.217)–(11.222) in appendix 11.7B.

Second Transformation

Similar to chapter 10, a transformation is used to remove the additional terms introduced by the transformation (11.59), (11.60) at the left boundary, so that they can be compensated by the design of the control input. This transformation is given by

$$\hat{Z}(t) = \hat{Y}(t) + C_0{}^+\int_0^1 \bar{K}_1(x)\alpha(x,t)dx$$

$$+ C_0{}^+\int_0^1 \bar{K}_2(x)\beta(x,t)dx + C_0{}^+\bar{K}_3\hat{X}(t), \tag{11.67}$$

where $C_0{}^+$ is a right inverse matrix of C_0. Because C_0 is full-row rank according to assumption 11.2, a right inverse exists for C_0.

Then (11.61), (11.62) is converted to

$$\dot{\hat{Z}}(t) = \hat{A}_0\hat{Z}(t) + q_1C_0{}^+\bar{K}_1(0)C_0\hat{Z}(t) + B_0\bar{U}(t)$$

$$+ M_Y\hat{X}(t) + \int_0^1 M_\alpha(x)\alpha(x,t)dx$$

$$+ \int_0^1 M_\beta(x)\beta(x,t)dx + N_1\alpha(1,t) + N_2\beta(0,t), \tag{11.68}$$

$$\alpha(0,t) = p\beta(0,t) + C_0\hat{Z}(t), \tag{11.69}$$

where

$$\bar{U}(t) = U(t) - K_0 \hat{Z}(t), \tag{11.70}$$

and K_0 is chosen to make $\hat{A}_0$ Hurwitz according to assumption 11.1. The scalars $N_1, N_2, M_\alpha, M_\beta, M_X$ are shown in (11.223)–(11.227).

Third Transformation with Frequency-Domain Design

Taking the Laplace transform of (11.63)–(11.66), (11.68), (11.69), we obtain

$$(sI - \hat{A}_0)\hat{Z}(s) = q_1 C_0{}^+ \bar{K}_1(0) C_0 \hat{Z}(s) + M_X \hat{X}(s)$$
$$+ \int_0^1 M_\alpha(x)\alpha(x,s)dx + \int_0^1 M_\beta(x)\beta(x,s)dx$$
$$+ N_1\alpha(1,s) + N_2\beta(0,s) + B_0\bar{U}(s), \tag{11.71}$$

$$\alpha(0,s) = p\beta(0,s) + C_0\hat{Z}(s), \tag{11.72}$$

$$s\alpha(x,s) = -q_1\alpha_x(x,s) - c_1\alpha(x,s), \tag{11.73}$$

$$s\beta(x,s) = q_2\beta_x(x,s) - c_2\beta(x,s), \tag{11.74}$$

$$\beta(1,s) = q\alpha(1,s), \tag{11.75}$$

$$(sI - \hat{A})\hat{X}(s) = B\alpha(1,s). \tag{11.76}$$

For brevity, we consider all the initial conditions to be zero while taking the Laplace transform (arbitrary initial conditions could be incorporated into the stability statement through a routine expanded analysis).

Recalling that $\hat{A}_0$ is Hurwitz, $\det(sI - \hat{A}_0)$ does not have any zeros in the closed right half-plane. Multiplying both sides of (11.71) with $(sI - \hat{A}_0)^{-1}$, we get

$$\hat{Z}(s) = (sI - \hat{A}_0)^{-1}\left[q_1 C_0{}^+ \bar{K}_1(0) C_0 \hat{Z}(s) + M_X \hat{X}(s) \right.$$
$$+ \int_0^1 M_\alpha(x)\alpha(x,s)dx + \int_0^1 M_\beta(x)\beta(x,s)dx$$
$$\left. + N_1\alpha(1,s) + N_2\beta(0,s) + B_0\bar{U}(s) \right]. \tag{11.77}$$

In order to cancel the proximal reflection term $\beta(0,s)$ in (11.72), the transformation

$$\hat{W}(s) = C_0{}^+ p\beta(0,s) + \hat{Z}(s) \tag{11.78}$$

is applied to (11.77) and (11.72), yielding

$$\hat{W}(s) = (sI - \hat{A}_0)^{-1}\left[q_1 C_0{}^+ \bar{K}_1(0) C_0 \hat{W}(s) + M_X \hat{X}(s) \right.$$
$$+ \int_0^1 M_\alpha(x)\alpha(x,s)dx + \int_0^1 M_\beta(x)\beta(x,s)dx + N_1\alpha(1,s)$$
$$\left. + \bar{N}_2\beta(0,s) \right] + (sI - \hat{A}_0)^{-1}B_0\bar{U}(s) + C_0{}^+ p\beta(0,s), \tag{11.79}$$

$$\alpha(0,s) = C_0\hat{W}(s), \tag{11.80}$$

where

$$\bar{N}_2 = N_2 - q_1 C_0{}^+ \bar{K}_1(0)p. \tag{11.81}$$

Together with (11.73)–(11.76), according to section 3.2 in [49], we obtain the following relationships:

$$\alpha(x,s) = e^{\left(\frac{c_1-s}{q_1}\right)x} C_0 \hat{W}(s), \tag{11.82}$$

$$\beta(x,s) = qe^{\left(\frac{c_2-s}{q_2}\right)(1-x)+\left(\frac{c_1-s}{q_1}\right)} C_0 \hat{W}(s), \tag{11.83}$$

$$\alpha(0,s) = C_0 \hat{W}(s), \tag{11.84}$$

$$\beta(1,s) = qe^{\left(\frac{c_1-s}{q_1}\right)} C_0 \hat{W}(s), \tag{11.85}$$

$$\beta(0,s) = qe^{\left(\frac{c_2-s}{q_2}\right)+\left(\frac{c_1-s}{q_1}\right)} C_0 \hat{W}(s), \tag{11.86}$$

$$\alpha(1,s) = e^{\left(\frac{c_1-s}{q_1}\right)} C_0 \hat{W}(s), \tag{11.87}$$

$$\hat{X}(s) = (sI - \hat{A})^{-1} B e^{\left(\frac{c_1-s}{q_1}\right)} C_0 \hat{W}(s), \tag{11.88}$$

$$\int_0^1 M_\beta(y)\beta(y,s)dy = \int_0^1 M_\beta(y)qe^{\left(\frac{c_2-s}{q_2}\right)(1-y)+\left(\frac{c_1-s}{q_1}\right)}dy C_0 \hat{W}(s), \tag{11.89}$$

$$\int_0^1 M_\alpha(y)\alpha(y,s)dy = \int_0^1 M_\alpha(y)e^{\left(\frac{c_1-s}{q_1}\right)y}dy C_0 \hat{W}(s). \tag{11.90}$$

Inserting (11.82)–(11.90) into (11.79) and multiplying by C_0 on both sides, we get

$$C_0 \hat{W}(s) = C_0(sI - \hat{A}_0)^{-1}\bigg(q_1 C_0{}^+ \bar{K}_1(0)C_0 \hat{W}(s)$$

$$+ M_X(sI - \hat{A})^{-1} B e^{\left(\frac{c_1-s}{q_1}\right)} C_0 \hat{W}(s)$$

$$+ \int_0^1 M_\alpha(y)e^{\left(\frac{c_1-s}{q_1}\right)y}dy C_0 \hat{W}(s)$$

$$+ \int_0^1 M_\beta(y)qe^{\left(\frac{c_2-s}{q_2}\right)(1-y)+\left(\frac{c_1-s}{q_1}\right)}dy C_0 \hat{W}(s)$$

$$+ N_1 e^{\left(\frac{c_1-s}{q_1}\right)} C_0 \hat{W}(s) + \bar{N}_2 qe^{\left(\frac{c_2-s}{q_2}\right)+\left(\frac{c_1-s}{q_1}\right)} C_0 \hat{W}(s) \bigg)$$

$$+ C_0(sI - \hat{A}_0)^{-1}B_0\bar{U}(s) + pqe^{\left(\frac{c_2-s}{q_2}\right)+\left(\frac{c_1-s}{q_1}\right)} C_0 \hat{W}(s). \tag{11.91}$$

Define a new variable as

$$\xi(s) = C_0 \hat{W}(s), \tag{11.92}$$

and thus (11.91) is rewritten as

$$\xi(s) = \left[C_0(sI - \hat{A}_0)^{-1}G(s) + pqe^{\left(\frac{c_2-s}{q_2}+\frac{c_1-s}{q_1}\right)} \right] \xi(s)$$

$$+ C_0(sI - \hat{A}_0)^{-1}B_0\bar{U}(s), \tag{11.93}$$

where

$$G(s) = q_1 C_0 + \bar{K}_1(0) + M_X(sI - \hat{A})^{-1} Be^{\left(\frac{c_1 - s}{q_1}\right)}$$

$$+ \int_0^1 M_\alpha(y) e^{\left(\frac{c_1 - s}{q_1}\right)y} dy + \int_0^1 M_\beta(y) q e^{\left(\frac{c_2 - s}{q_2}\right)(1-y) + \left(\frac{c_1 - s}{q_1}\right)} dy$$

$$+ N_1 e^{\left(\frac{c_1 - s}{q_1}\right)} + \bar{N}_2 q e^{\left(\frac{c_2 - s}{q_2} + \frac{c_1 - s}{q_1}\right)} \tag{11.94}$$

is a vector of stable, proper transfer functions because there is no pole in the closed right half-plane (since $\hat{A}$ is Hurwitz). The definitions of proper, stable properties of irrational transfer functions in this book are given in definitions 10.1 and 10.2.

Using assumption 11.2, according to [135], we get

$$C_0(sI - \hat{A}_0)^{-1} B_0 \neq 0$$

for all $s \in \mathbb{C}, \Re(s) \geq 0$. We now choose $\bar{U}(s)$ in (11.93) as

$$\bar{U}(s) = -F(s)\xi(s), \tag{11.95}$$

where

$$F(s) = \frac{\Omega(s)}{C_0(sI - \hat{A}_0)^{-1} B_0} \left[C_0(sI - \hat{A}_0)^{-1} G(s) + pqc^{\left(\frac{c_2 - s}{q_2}\right) + \left(\frac{c_1 - s}{q_1}\right)} \right]. \tag{11.96}$$

The transfer function $\Omega(s)$ is a stable single-input single-output (SISO) low-pass filter of sufficient order to be designed, which makes $F(s)$ (strictly) proper and satisfies

$$|\Phi(s)| < 1 \tag{11.97}$$

for all $s \in \mathbb{C}, \Re(s) \geq 0$, where

$$\Phi(s) = (1 - \Omega(s)) \left(C_0(sI - \hat{A}_0)^{-1} G(s) + pqe^{\left(\frac{c_2 - s}{q_2}\right) + \left(\frac{c_1 - s}{q_1}\right)} \right). \tag{11.98}$$

The matrix $\hat{A}_0$ being Hurwitz and $G(s)$ in (11.94) being a stable, proper transfer matrix, along with assumption 11.3, ensures the existence of the low-pass filter $\Omega(s)$ satisfying (11.97) when s approaches infinity along the imaginary axis, so a low-pass filter $\Omega(s)$ exists with the desired properties for all $s \in \mathbb{C}, \Re(s) \geq 0$. The transfer function $F(s)$ (11.96) is stable and proper because there is no pole in the closed right half-plane.

Inserting $\bar{U}(s)$ defined by (11.95), we see that (11.93) becomes

$$(1 - \Phi(s))\xi(s) = 0, \tag{11.99}$$

where $\Phi(s)$ is given in (11.98) and satisfies (11.97).

In the continuous-in-time control design in this section, (11.97) and (11.99) ensure the exponential convergence to zero of ξ. Recalling (11.79), with (11.82)–(11.90), (11.92) and (11.95) incorporated, we get

$$
\begin{aligned}
\hat{W}(s) = \Bigg[& (sI - \hat{A}_0)^{-1} \bigg(q_1 C_0{}^+ \bar{K}_1(0) + M_X (sI - \hat{A})^{-1} B e^{\left(\frac{c_1 - s}{q_1}\right)} \\
& + \int_0^1 M_\alpha(y) e^{\left(\frac{c_1 - s}{q_1}\right) y} dy + \int_0^1 M_\beta(y) q e^{\left(\frac{c_2 - s}{q_2}\right)(1-y) + \left(\frac{c_1 - s}{q_1}\right)} dy \\
& + N_1 e^{\left(\frac{c_1 - s}{q_1}\right)} + \bar{N}_2 q e^{\left(\frac{c_2 - s}{q_2}\right) + \left(\frac{c_1 - s}{q_1}\right)} \bigg) \\
& - (sI - \hat{A}_0)^{-1} B_0 F(s) + C_0{}^+ p q e^{\left(\frac{c_2 - s}{q_2}\right) + \left(\frac{c_1 - s}{q_1}\right)} \Bigg] \xi(s). \quad (11.100)
\end{aligned}
$$

The exponential convergence to zero of $\hat{W}$ is thus obtained because the transfer function before $\xi(s)$ is stable, where $F(s)$ is defined in (11.96). The exponential convergence to zero of the signals in the α, β PDEs and the distal $\hat{X}$ ODE is also obtained by recalling the relationships (11.82)–(11.90).

Recalling (11.70), (11.95), the continuous-in-time control law, in s domain, is obtained as

$$
U(s) = K_0 \hat{Z}(s) + \bar{U}(s) = K_0 \hat{Z}(s) - F(s)\xi(s), \quad (11.101)
$$

on the basis of which the event-triggering mechanism in the next section is designed.

11.4   EVENT-TRIGGERING MECHANISM

In this section we introduce an event-triggered control scheme for stabilization of the 2×2 hyperbolic PDE sandwich system (11.1)–(11.6). The scheme relies on both the continuous-in-time control $U(t)$, designed in the last section, and a dynamic event-triggering mechanism (ETM), which determines triggering times. The event-triggered control signal $U_d(t)$ is the value of the continuous-in-time $U(t)$ at the time instants t_k but applied until time t_{k+1}—that is,

$$
U_d(t) = U(t_k), \quad t \in [t_k, t_{k+1}), \quad (11.102)
$$

where integer $k \geq 0$ and $t_0 = 0$. A deviation $d(t)$ between the continuous-in-time control signal and the event-based one is given as

$$
d(t) = U(t) - U_d(t). \quad (11.103)
$$

Recalling (11.101), we know that

$$
U_d(s) = U(s) - d(s) = K_0 \hat{Z}(s) - F(s)\xi(s) - d(s). \quad (11.104)
$$

Replacing $U(s)$ by $U_d(s)$ in the control design in the last section and applying (11.104), the relation (11.99) becomes

$$
(1 - \Phi(s))\xi(s) = C_0 (sI - \hat{A}_0)^{-1} B_0 d(s). \quad (11.105)
$$

Multiplying both sides of (11.79) with $(sI - \hat{A}_0)$, where $\bar{U}(s) = F(s)\xi(s) - d(s)$ now, and then inserting (11.82)–(11.90), (11.105), yields

$$
s\hat{W}(s) = \hat{A}_0 \hat{W}(s) + D(s)d(s), \quad (11.106)
$$

where

$$D(s) = \left(G(s) - B_0 F(s) + (sI - \hat{A}_0) C_0^+ p q e^{\left(\frac{c_2 - s}{q_2}\right) + \left(\frac{c_1 - s}{q_1}\right)} \right)$$
$$\times \frac{C_0 (sI - \hat{A}_0)^{-1} B_0}{1 - \Phi(s)} - B_0 \qquad (11.107)$$

is an $\bar{n}$-dimensional column vector of stable, proper transfer functions. The components of $D(s)$ are stable—that is, they have no unstable poles, due to (11.94), (11.96), $\hat{A}_0$ being Hurwitz, and

$$1 - \Phi(s) \neq 0$$

for all $s \in \mathbb{C}, \Re(s) \geq 0$. Let us now denote

$$\mathcal{D}(d(t))_{\bar{n} \times 1} = \mathcal{L}^{-1}(D(s) d(s)), \qquad (11.108)$$

which is the inverse Laplace transform of $D(s) d(s)$—that is, of the outputs of the system $D(s)$ under the input $d(s)$. The stable, proper $D(s)$ guarantees a BIBO (bounded-input bounded-output) relationship so that the following estimate holds

$$\mathcal{D}(d(t))^2 \leq \bar{d} \sup_{0 \leq \zeta \leq t} d(\zeta)^2, \qquad (11.109)$$

where the constant $\bar{d} > 0$ is associated with the L_1 norm of the impulse response (chapter 2 in [52], appendix B of [116]) of the entries of $D(s)$ in (11.107). It follows that $\bar{d}$ depends only on the parameters of the plant and of the low-pass-filter-based backstepping continuous-in-time control law.

Recalling (11.63)–(11.66), (11.80), and (11.106), we obtain the event-based target system in the time domain:

$$\dot{\hat{W}}(t) = \hat{A}_0 \hat{W}(t) + \mathcal{D}(d(t)), \qquad (11.110)$$
$$\alpha(0, t) = C_0 \hat{W}(t), \qquad (11.111)$$
$$\alpha_t(x, t) = -q_1 \alpha_x(x, t) - c_1 \alpha(x, t), \qquad (11.112)$$
$$\beta_t(x, t) = q_2 \beta_x(x, t) - c_2 \beta(x, t), \qquad (11.113)$$
$$\beta(1, t) = q \alpha(1, t), \qquad (11.114)$$
$$\dot{\hat{X}}(t) = \hat{A} \hat{X}(t) + B \alpha(1, t). \qquad (11.115)$$

The ETM to determine the triggering times is designed to be governed by the following dynamic triggering condition:

$$t_{k+1} = \inf\{t \in R^+ | t > t_k | d(t)^2 \geq \theta \hat{W}^T(t) P_0 \hat{W}(t) - \mu m(t)\}, \qquad (11.116)$$

where the internal dynamic variable $m(t)$ satisfies the ordinary differential equation

$$\dot{m}(t) = -\eta m(t) - \mu_W \sup_{0 \leq \zeta \leq t} \hat{W}(\zeta)^2 - \mu_d \sup_{0 \leq \zeta \leq t} d(\zeta)^2 \qquad (11.117)$$

with initial condition $m(0) < 0$, which guarantees that

$$m(t) < 0. \qquad (11.118)$$

Inequality (11.118) follows from the ODE (11.117), the nonpositivity of the non-homogeneous terms on its right-hand side, the strict negativity of $m(0)$, the variation-of-constants formula, and the comparison principle.

The positive definite matrix $P_0 = P_0^T$ in (11.116) is the unique solution to the Lyapunov equation

$$\hat{A}_0^T P_0 + P_0 \hat{A}_0 = -Q_0 \tag{11.119}$$

for some $Q_0 = Q_0{}^T > 0$. The positive constants θ, μ, η, μ_W, μ_d in the ETM are to be determined later.

The observer-based event-triggering condition (11.116) uses the transformed ODE state $\hat{W}(t)$ because it contains the estimated states of the overall ODE-PDE-ODE system through the transformations (11.59), (11.60), (11.67), (11.78).

As will be seen in lemma 11.2, $\dot{d}(t)^2$, on which the minimal dwell time relies, is bounded by $\sup_{0 \leq \zeta \leq t} |\hat{W}(\zeta)|^2$, $\sup_{0 \leq \zeta \leq t} d(\zeta)^2$ (instead of $|\hat{W}(t)|^2$, $d(t)^2$ in (11.116)). In order to avoid the Zeno phenomenon, namely, to ensure that

$$\lim_{k \to \infty} t_k = +\infty, \tag{11.120}$$

the internal dynamic variable $m(t)$ is introduced in (11.116) to offset $\sup_{0 \leq \zeta \leq t} |\hat{W}(\zeta)|^2$, $\sup_{0 \leq \zeta \leq t} d(\zeta)^2$ when proving the existence of a minimal dwell time, which will be seen clearly in the proof of lemma 11.3.

Proposition 11.1. *For the given* $(z(\cdot, t_k), w(\cdot, t_k))^T \in L^2((0,1); \mathbb{R}^2)$, $X(t_k) \in \mathbb{R}^n$, $Y(t_k) \in \mathbb{R}^{\bar{n}}$ *and* $(\hat{z}(\cdot, t_k), \hat{w}(\cdot, t_k))^T \in L^2((0,1); \mathbb{R}^2)$, $\hat{X}(t_k) \in \mathbb{R}^n$, $\hat{Y}(t_k) \in \mathbb{R}^{\bar{n}}$, $m(t_k) \in \mathbb{R}^-$, *there exist unique (weak) solutions* $((z, w)^T, X, Y) \in C^0([t_k, t_{k+1}]; L^2(0,1); \mathbb{R}^2)$ $\times C^0([t_k, t_{k+1}]; \mathbb{R}^n) \times C^0([t_k, t_{k+1}]; \mathbb{R}^{\bar{n}})$ *and* $((\hat{z}, \hat{w})^T, \hat{X}, \hat{Y}) \in C^0([t_k, t_{k+1}]; L^2(0,1);$ $\mathbb{R}^2) \times C^0([t_k, t_{k+1}]; \mathbb{R}^n) \times C^0([t_k, t_{k+1}]; \mathbb{R}^{\bar{n}})$, $m \in C^0([t_k, t_{k+1}]; \mathbb{R}^-)$ *to the systems* (11.1)–(11.6) *and* (11.13)–(11.18), (11.117) *with the event-based control input* $U_d(t)$ *applied in* (11.1) *and* (11.13), *respectively, between two time instants* t_k *and* t_{k+1}.

Proof. Adopting the definition of the weak solution for linear hyperbolic PDE-ODE systems, which will be shown in chapter 14 as definition 14.1, for the given $(z(\cdot, t_k), w(\cdot, t_k))^T \in L^2((0,1); \mathbb{R}^2)$, $X(t_k) \in \mathbb{R}^n$, and $Y(t_k) \in \mathbb{R}^{\bar{n}}$, with the results in [35, 141], we can show that there exist unique (weak) solutions $((z, w)^T, X, Y) \in C^0([t_k, t_{k+1}]; L^2(0,1); \mathbb{R}^2) \times C^0([t_k, t_{k+1}]; \mathbb{R}^n) \times C^0([t_k, t_{k+1}]; \mathbb{R}^{\bar{n}})$ to the system (11.1)–(11.6) under the event-based control input $U_d(t)$. Similarly, the unique solution of the observer error system also can be obtained by recalling the target observer error system (11.28)–(11.31), (11.54), (11.55), as well as the backstepping transformation (11.26), (11.27), and (11.49). According to (11.19), it can then be shown that there exist unique (weak) solutions $((\hat{z}, \hat{w})^T, \hat{X}, \hat{Y}) \in C^0([t_k, t_{k+1}];$ $L^2(0,1); \mathbb{R}^2) \times C^0([t_k, t_{k+1}]; \mathbb{R}^n) \times C^0([t_k, t_{k+1}]; \mathbb{R}^{\bar{n}})$ and $m \in C^0([t_k, t_{k+1}]; \mathbb{R}^-)$ to the system (11.13)–(11.18), (11.117) under the event-based control input $U_d(t)$. Proposition 11.1 is thus obtained. $\square$

Lemma 11.2. *For* $d(t)$ *defined in* (11.103), *there exist the positive constants* λ_W, λ_d *such that*

$$\dot{d}(t)^2 \leq \lambda_W \sup_{0 \leq \zeta \leq t} \hat{W}(\zeta)^2 + \lambda_d \sup_{0 \leq \zeta \leq t} d(\zeta)^2 \tag{11.121}$$

for $t \in (t_k, t_{k+1})$, where λ_W, λ_d depend only on the parameters of the plant and the low-pass-filter-based backstepping continuous-in-time control law.

Proof. Taking the time derivative of (11.103), we obtain

$$\dot{d}(t)^2 = \dot{U}(t)^2 \tag{11.122}$$

because $\dot{U}_d(t) = 0$ for $t \in (t_k, t_{k+1})$. Recalling (11.78), (11.86), (11.101), (11.106), we get

$$sU(s) = R(s)\hat{W}(s) + R_d(s)d(s), \tag{11.123}$$

where

$$R(s) = \left(K_0 - \left(K_0 C_0^+ pqe^{\left(\frac{c_2-s}{q_2}\right)+\left(\frac{c_1-s}{q_1}\right)} + F(s) \right) C_0 \right) \hat{A}_0 \tag{11.124}$$

is an $\bar{n}$-dimensional row vector of stable, proper transfer functions, and

$$R_d(s) = \left(K_0 - \left(K_0 C_0^+ pqe^{\left(\frac{c_2-s}{q_2}\right)+\left(\frac{c_1-s}{q_1}\right)} + F(s) \right) C_0 \right) D(s) \tag{11.125}$$

is a stable, proper transfer function ($F(s)$ and $D(s)$ are stable and proper). We thus have BIBO with the estimate

$$\dot{U}(t)^2 \leq \lambda_W \sup_{0 \leq \zeta \leq t} \hat{W}(\zeta)^2 + \lambda_d \sup_{0 \leq \zeta \leq t} d(\zeta)^2, \tag{11.126}$$

where λ_W is associated with the $\| \cdot \|_1$ norm of the impulse response of the entries of the transfer function vector $R(s)$, and λ_d is, likewise, associated with $\|R_d\|_1$, which depend only on the plant and the design of the continuous-in-time control law. The proof is complete. $\square$

The following lemma proves the existence of a minimal dwell time between two triggering times (independent of initial conditions). It ensures the Zeno phenomenon does not occur and a reduction of changes in the value in the actuator signal compared with the continuous-in-time control.

Lemma 11.3. *For some μ_W, μ_d, θ, there exists a $\tau > 0$, independent of initial conditions, such that $t_{k+1} - t_k \geq \tau$ for all $k \geq 0$.*

Proof. Let us introduce a function

$$\psi(t) = \frac{d(t)^2 + \frac{\mu}{2}m(t)}{\theta \hat{W}^T(t) P_0 \hat{W}(t) - \frac{\mu}{2}m(t)} \tag{11.127}$$

according to [57]. We have

$$\psi(t_{k+1}) = 1 \tag{11.128}$$

because the event is triggered, and

$$\psi(t_k) < 0 \tag{11.129}$$

because $m(t) < 0$ and $d(t_k) = 0$. The function $\psi(t)$ is continuous on $[t_k, t_{k+1}]$ due to proposition 11.1. By the intermediate value theorem, there exists $t^* > t_k$ such that $\psi(t) \in [0, 1]$ when $t \in [t^*, t_{k+1}]$. The lower bound τ of the minimal dwell time can be defined as the minimal time it takes for $\psi(t)$ from 0 to 1.

Recalling (11.109), (11.110), we obtain

$$\frac{\hat{W}^T(t) P_0 \hat{W}(t)}{dt} \geq -\frac{3}{2} \lambda_{\max}(Q_0) \hat{W}(t)^2 - \frac{2 \bar{d} \lambda_{\max}(P_0)^2}{\lambda_{\max}(Q_0)} \sup_{0 \leq \zeta \leq t} d(\zeta)^2. \qquad (11.130)$$

Taking the derivative of $\psi(t)$ (11.127) and using (11.121), (11.130), we have

$$\dot{\psi}(t) = \frac{2 d(t) \dot{d}(t) + \frac{\mu}{2} \dot{m}(t)}{\theta \hat{W}^T(t) P_0 \hat{W}(t) - \frac{\mu}{2} m(t)} - \frac{(\theta \frac{\hat{W}^T(t) P_0 \hat{W}(t)}{dt} - \frac{\mu}{2} \dot{m}(t))}{\theta \hat{W}^T(t) P_0 \hat{W}(t) - \frac{\mu}{2} m(t)} \psi(t)$$

$$\leq \frac{1}{\theta \hat{W}^T(t) P_0 \hat{W}(t) - \frac{\mu}{2} m(t)} \times \left[r_1 \lambda_W \sup_{0 \leq \zeta \leq t} \hat{W}(\zeta)^2 \right.$$

$$\left. + r_1 \lambda_d \sup_{0 \leq \zeta \leq t} d(\zeta)^2 + \frac{1}{r_1} d(t)^2 + \frac{\mu}{2} \dot{m}(t) \right]$$

$$- \frac{1}{\theta \hat{W}^T(t) P_0 \hat{W}(t) - \frac{\mu}{2} m(t)} \left[\theta \left(-\frac{3}{2} \lambda_{\max}(Q_0) \hat{W}(t)^2 \right.\right.$$

$$\left.\left. - \frac{2 \bar{d} \lambda_{\max}(P_0)^2}{\lambda_{\max}(Q_0)} \sup_{0 \leq \zeta \leq t} d(\zeta)^2 \right) - \frac{\mu}{2} \dot{m}(t) \right] \psi(t), \qquad (11.131)$$

where r_1 is a positive constant from Young's inequality. Inserting (11.117), one obtains

$$\dot{\psi}(t) \leq \frac{1}{\theta \hat{W}^T(t) P_0 \hat{W}(t) - \frac{\mu}{2} m(t)} \left[r_1 \lambda_W \sup_{0 \leq \zeta \leq t} \hat{W}(\zeta)^2 \right.$$

$$\left. + r_1 \lambda_d \sup_{0 \leq \zeta \leq t} d(\zeta)^2 + \frac{1}{r_1} d(t)^2 - \frac{\mu}{2} \eta m(t) \right.$$

$$\left. - \frac{\mu}{2} \mu_W \sup_{0 \leq \zeta \leq t} \hat{W}(\zeta)^2 - \frac{\mu}{2} \mu_d \sup_{0 \leq \zeta \leq t} d(\zeta)^2 \right]$$

$$- \frac{1}{\theta \hat{W}^T(t) P_0 \hat{W}(t) - \frac{\mu}{2} m(t)} \left[-\frac{3\theta}{2} \lambda_{\max}(Q_0) \hat{W}(t)^2 \right.$$

$$- \frac{2 \theta \bar{d} \lambda_{\max}(P_0)^2}{\lambda_{\max}(Q_0)} \sup_{0 \leq \zeta \leq t} d(\zeta)^2 + \frac{\mu}{2} \eta m(t)$$

$$\left. + \frac{\mu}{2} \mu_W \sup_{0 \leq \zeta \leq t} \hat{W}(\zeta)^2 + \frac{\mu}{2} \mu_d \sup_{0 \leq \zeta \leq t} d(\zeta)^2 \right] \psi(t)$$

$$\leq \frac{1}{\theta \hat{W}^T(t) P_0 \hat{W}(t) - \frac{\mu}{2} m(t)} \left[\left(r_1 \lambda_W - \frac{\mu}{2} \mu_W \right) \sup_{0 < \zeta < t} \hat{W}(\zeta)^2 \right.$$

$$\left. + \left(r_1 \lambda_d - \frac{\mu}{2} \mu_d \right) \sup_{0 \leq \zeta \leq t} d(\zeta)^2 + \frac{1}{r_1} d(t)^2 - \frac{\mu}{2} \eta m(t) \right]$$

$$+ \frac{1}{\theta \hat{W}^T(t) P_0 \hat{W}(t) - \frac{\mu}{2} m(t)} \left[-\frac{\mu}{2} \eta m(t) \right.$$

$$+ \left(\frac{3\theta}{2} \lambda_{\max}(Q_0) - \frac{\mu}{2} \mu_W \right) \sup_{0 \leq \zeta \leq t} \hat{W}(\zeta)^2$$

$$\left. + \left(\frac{2\theta \bar{d} \lambda_{\max}(P_0)^2}{\lambda_{\max}(Q_0)} - \frac{\mu}{2} \mu_d \right) \sup_{0 \leq \zeta \leq t} d(\zeta)^2 \right] \psi(t). \tag{11.132}$$

Choosing positive constants μ_W, μ_d, θ in the ETM such that they satisfy

$$\mu_W \geq \frac{2 r_1 \lambda_W}{\mu}, \tag{11.133}$$

$$\mu_d \geq \frac{2 r_1 \lambda_d}{\mu}, \tag{11.134}$$

$$\theta \leq \min \left\{ \frac{\mu \mu_W}{3 \lambda_{\max}(Q_0)}, \frac{\mu_d \mu \lambda_{\max}(Q_0)}{4 \bar{d} \lambda_{\max}(P_0)^2} \right\}, \tag{11.135}$$

we get

$$\dot{\psi}(t) \leq \frac{\frac{1}{r_1} d(t)^2 - \frac{\mu}{2} \eta m(t)}{\theta \hat{W}^T(t) P_0 \hat{W}(t) - \frac{\mu}{2} m(t)}$$

$$+ \frac{-\frac{\mu}{2} \eta m(t)}{\theta \hat{W}^T(t) P_0 \hat{W}(t) - \frac{\mu}{2} m(t)} \psi(t). \tag{11.136}$$

Applying the inequalities

$$\frac{-\frac{\mu}{2} \eta m(t)}{\theta \hat{W}^T(t) P_0 \hat{W}(t) - \frac{\mu}{2} m(t)} < \frac{-\frac{\mu}{2} \eta m(t)}{-\frac{\mu}{2} m(t)} = \eta,$$

$$\frac{d(t)^2}{\theta \hat{W}^T(t) P_0 \hat{W}(t) - \frac{\mu}{2} m(t)} = \frac{d(t)^2 + \frac{\mu}{2} m(t) - \frac{\mu}{2} m(t)}{\theta \hat{W}^T(t) P_0 \hat{W}(t) - \frac{\mu}{2} m(t)}$$

$$\leq \psi(t) + 1, \tag{11.137}$$

which hold because $m(t) < 0$, inequality (11.136) becomes

$$\dot{\psi}(t) \leq \frac{1}{r_1} + \eta + \left(\frac{1}{r_1} + \eta \right) \psi(t). \tag{11.138}$$

It follows that the time needed by $\psi(t)$ to go from 0 to 1 is at least

$$\tau = \int_0^1 \frac{1}{(\frac{1}{r_1} + \eta) \bar{s} + \frac{1}{r_1} + \eta} d\bar{s} > 0, \tag{11.139}$$

which is independent of initial conditions, where η is a free design parameter appearing in (11.117), and the condition on the positive constant r_1 from Young's inequality will be determined later. $\qquad \square$

11.5   STABILITY ANALYSIS OF THE EVENT-BASED CLOSED-LOOP SYSTEM

In the event-based output-feedback closed-loop system, a low-pass-filter-based back-stepping control law $U(t)$ in (11.101), using the states from the observer, is updated at time instants t_k determined by the ETM (11.116), (11.117) implemented based on the observer, to regulate the PDE plant (11.1)–(11.6).

Lemma 11.4. *With arbitrary initial data $(\hat{Y}(0), \hat{z}(x,0), \hat{w}(x,0), \hat{X}(0)) \in \chi$, in the event-based state-feedback loop, the exponential convergence is achieved in the sense that there exist positive constants Υ_f, λ_f such that*

$$\hat{\Xi}(t) \leq \Upsilon_f \hat{\Xi}(0) e^{-\lambda_f t}, \tag{11.140}$$

where

$$\hat{\Xi}(t) = \left| \hat{X}(t) \right|^2 + \left| \hat{Y}(t) \right|^2 + \|\hat{z}(\cdot,t)\|^2 + \|\hat{w}(\cdot,t)\|^2 + |m(t)| + \bar{U}(t)^2. \tag{11.141}$$

Proof. The state of the low-pass filter is $\bar{U}(t)$ in (11.95). According to (11.92), we get

$$\bar{U}(s) = -F(s)C_0\hat{W}(s)$$

where $F(s)$ (11.96) is a stable, proper transfer function guaranteeing the BIBO property with the estimate as

$$\bar{U}(t)^2 \leq \lambda_{lp}|C_0|^2 \sup_{0 \leq \zeta \leq t} \hat{W}(\zeta)^2, \tag{11.142}$$

where the positive constant λ_{lp} is associated with the L_1 norm of the impulse response of $F(s)$.

Let us consider the Lyapunov function

$$V(t) = r_w \hat{X}(t)^T P \hat{X}(t) + \hat{W}(t)^T P_0 \hat{W}(t) + \frac{1}{2} r_a \int_0^1 e^{\delta_1 x} \beta(x,t)^2 dx$$

$$+ \frac{1}{2} r_b \int_0^1 e^{-\delta_2 x} \alpha(x,t)^2 dx - m(t), \tag{11.143}$$

where a positive definite matrix $P = P^T$ is the solution to the Lyapunov equation

$$\hat{A}^T P + P\hat{A} = -Q \tag{11.144}$$

for some $Q = Q^T > 0$. The positive constants $r_a, r_b, \delta_1, \delta_2, r_w$ are to be determined later. The Lyapunov function (11.143) is positive definite because $m(t) < 0$.

Defining

$$\Omega_0(t) = \|\alpha(\cdot,t)\|^2 + \|\beta(\cdot,t)\|^2 + \left| \hat{X}(t) \right|^2 + \left| \hat{W}(t) \right|^2 + |m(t)|, \tag{11.145}$$

recalling (11.143)–(11.145), the following inequality holds

$$\mu_1 \Omega_0(t) \leq V(t) \leq \mu_2 \Omega_0(t) \tag{11.146}$$

for positive constants

$$\mu_1 = \min\left\{r_w\lambda_{\min}(P), \lambda_{\min}(P_0), \frac{1}{2}r_a, \frac{1}{2}r_b e^{-\delta_2}, 1\right\}, \tag{11.147}$$

$$\mu_2 = \max\left\{r_w\lambda_{\max}(P), \lambda_{\max}(P_0), \frac{1}{2}r_a e^{\delta_1}, \frac{1}{2}r_b, 1\right\}. \tag{11.148}$$

Taking the derivative of (11.143) along (11.110)–(11.115), recalling (11.117), one obtains

$$\begin{aligned}
\dot{V}(t) =& -r_w\hat{X}(t)^T Q\hat{X}(t) + 2r_w\hat{X}^T PB\alpha(1,t) - \hat{W}(t)^T Q_0\hat{W}(t) + 2\hat{W}^T P_0\mathcal{D}(d(t)) \\
&+ q_2 r_a \int_0^1 e^{\delta_1 x}\beta(x,t)\beta_x(x,t)dx - q_1 r_b \int_0^1 e^{-\delta_2 x}\alpha(x,t)\alpha_x(x,t)dx \\
&- r_a c_2 \int_0^1 e^{\delta_1 x}\beta(x,t)^2 dx - r_b c_1 \int_0^1 e^{-\delta_2 x}\alpha(x,t)^2 dx \\
&+ \eta m(t) + \mu_W \sup_{0\le\zeta\le t}\hat{W}(\zeta)^2 + \mu_d \sup_{0\le\zeta\le t}d(\zeta)^2 \\
=& -r_w\hat{X}(t)^T Q\hat{X}(t) + 2r_w\hat{X}^T PB\alpha(1,t) - \hat{W}(t)^T Q_0\hat{W}(t) + 2\hat{W}^T P_0\mathcal{D}(d(t)) \\
&+ \frac{1}{2}q_2 r_a e^{\delta_1}\beta(1,t)^2 - \frac{1}{2}q_2 r_a\beta(0,t)^2 - \frac{1}{2}\delta_1 q_2 r_a \int_0^1 e^{\delta_1 x}\beta(x,t)^2 dx \\
&- \frac{1}{2}q_1 r_b e^{-\delta_2}\alpha(1,t)^2 + \frac{1}{2}q_1 r_b(C_0\hat{W}(t))^2 - \frac{1}{2}\delta_2 q_1 r_b \int_0^1 e^{-\delta_2 x}\alpha(x,t)^2 dx \\
&- r_a c_2 \int_0^1 e^{\delta_1 x}\beta(x,t)^2 dx - r_b c_1 \int_0^1 e^{-\delta_2 x}\alpha(x,t)^2 dx \\
&+ \eta m(t) + \mu_W \sup_{0\le\zeta\le t}\hat{W}(\zeta)^2 + \mu_d \sup_{0\le\zeta\le t}d(\zeta)^2. \tag{11.149}
\end{aligned}$$

Applying Young's inequality and the Cauchy-Schwarz inequality and recalling (11.109), we obtain

$$\dot{V}(t) \le -\frac{r_w}{2}\lambda_{\min}(Q)|\hat{X}(t)|^2 - \left(\frac{1}{2}\lambda_{\min}(Q_0) - \frac{1}{2}q_1 r_b|C_0|^2\right)|\hat{W}(t)|^2$$

$$- \left(\frac{1}{2}q_1 r_b e^{-\delta_2} - \frac{2r_w|PB|^2}{\lambda_{\min}(Q)} - \frac{1}{2}q_2 r_a e^{\delta_1}q^2\right)\alpha(1,t)^2 \tag{11.150}$$

$$- \frac{1}{2}q_2 r_a\beta(0,t)^2 + \frac{2\bar{d}\lambda_{\max}(P_0)^2}{\lambda_{\min}(Q_0)}\sup_{0\le\zeta\le t}d(\zeta)^2$$

$$- \left(\frac{1}{2}\delta_1 q_2 r_a - r_a|c_2|\right)\int_0^1 e^{\delta_1 x}\beta(x,t)^2 dx \tag{11.151}$$

$$- \left(\frac{1}{2}\delta_2 q_1 r_b - r_b|c_1|\right)\int_0^1 e^{-\delta_2 x}\alpha(x,t)^2 dx \tag{11.152}$$

$$- \eta|m(t)| + \mu_W \sup_{0\le\zeta\le t}\hat{W}(\zeta)^2 + \mu_d \sup_{0\le\zeta\le t}d(\zeta)^2. \tag{11.153}$$

Choosing $\delta_1, \delta_2, r_a, r_b, r_w$ as

$$\delta_1 > \frac{2|c_2|}{q_2}, \tag{11.154}$$

$$\delta_2 > \frac{2|c_1|}{q_1}, \tag{11.155}$$

$$r_b < \frac{\lambda_{\min}(Q_0)}{q_1|C_0|^2}, \tag{11.156}$$

$$r_w < \frac{q_1 r_b e^{-\delta_2} \lambda_{\min}(Q)}{8|PB|^2}, \tag{11.157}$$

$$r_a < \frac{q_1 r_b}{2 q_2 q^2} e^{-(\delta_2 + \delta_1)}, \tag{11.158}$$

we thus arrive at

$$\dot{V}(t) \le -\sigma_a V(t) + \mu_W \sup_{0 \le \zeta \le t} \hat{W}(\zeta)^2 + \left(\mu_d + \frac{2\bar{d}\lambda_{\max}(P_0)^2}{\lambda_{\min}(Q_0)} \right) \sup_{0 \le \zeta \le t} d(\zeta)^2, \tag{11.159}$$

where

$$\sigma_a = \frac{1}{\mu_2} \min \left\{ \frac{r_w}{2} \lambda_{\min}(Q), \frac{1}{2}\lambda_{\min}(Q_0) - \frac{1}{2}q_1 r_b |C_0|^2, \right.$$
$$\left. \frac{1}{2}\delta_1 q_2 r_a - r_a |c_2|, \left(\frac{1}{2}\delta_2 q_1 r_b - r_b |c_1| \right) e^{-\delta_2}, \eta \right\} > 0. \tag{11.160}$$

Multiplying both sides of (11.159) by $e^{\sigma_a t}$, we get

$$e^{\sigma_a t} \dot{V}(t) + e^{\sigma_a t} \sigma_a V(t)$$
$$\le e^{\sigma_a t} \mu_W \sup_{0 \le \zeta \le t} \hat{W}(\zeta)^2 + e^{\sigma_a t} \left(\mu_d + \frac{2\bar{d}\lambda_{\max}(P_0)^2}{\lambda_{\min}(Q_0)} \right) \sup_{0 \le \zeta \le t} d(\zeta)^2. \tag{11.161}$$

The left side of (11.161) is $\frac{d(e^{\sigma_a t} V(t))}{dt}$. Integrating (11.161) from 0 to t yields

$$V(t) \le V(0)e^{-\sigma_a t} + \frac{1}{\sigma_a}(1 - e^{-\sigma_a t}) \left[\mu_W \sup_{0 \le \zeta \le t} \hat{W}(\zeta)^2 \right.$$
$$\left. + \left(\mu_d + \frac{2\bar{d}\lambda_{\max}(P_0)^2}{\lambda_{\min}(Q_0)} \right) \sup_{0 \le \zeta \le t} d(\zeta)^2 \right]$$
$$\le V(0)e^{-\sigma_a t} + \frac{1}{\sigma_a} \left[\mu_W \sup_{0 \le \zeta \le t} \hat{W}(\zeta)^2 \right.$$
$$\left. + \left(\mu_d + \frac{2\bar{d}\lambda_{\max}(P_0)^2}{\lambda_{\min}(Q_0)} \right) \sup_{0 \le \zeta \le t} d(\zeta)^2 \right]. \tag{11.162}$$

The triggering condition (11.116) guarantees

$$\sup_{0 \le \zeta \le t} d(\zeta)^2 \le \theta \lambda_{\max}(P_0) \sup_{0 \le \zeta \le t} \hat{W}(\zeta)^2 + \mu \sup_{0 \le \zeta \le t} |m(\zeta)|. \tag{11.163}$$

Inserting (11.163) into (11.162) and then recalling (11.145), (11.146) yields

$$V(t) \le V(0)e^{-\sigma_a t} + \frac{1}{\sigma_a} \left[\mu_W \sup_{0 \le \zeta \le t} \hat{W}(\zeta)^2 + \left(\mu_d + \frac{2\bar{d}\lambda_{\max}(P_0)^2}{\lambda_{\min}(Q_0)} \right) \right.$$
$$\left. \times \sup_{0 \le \zeta \le t} \left(\theta \lambda_{\max}(P_0) \hat{W}(\zeta)^2 + \mu|m(\zeta)| \right) \right]$$

$$\leq V(0)e^{-\sigma_a t} + \frac{1}{\sigma_a}\left[\mu_W + \left(\mu_d + \frac{2\bar{d}\lambda_{\max}(P_0)^2}{\lambda_{\min}(Q_0)}\right)\theta\lambda_{\max}(P_0)\right]$$

$$\times \sup_{0\leq\zeta\leq t} \hat{W}(\zeta)^2 + \frac{1}{\sigma_a}\left(\mu_d + \frac{2\bar{d}\lambda_{\max}(P_0)^2}{\lambda_{\min}(Q_0)}\right)\mu \sup_{0\leq\zeta\leq t} |m(\zeta)|$$

$$\leq V(0)e^{-\sigma_a t} + \max\left\{\frac{1}{\sigma_a}\left[\mu_W + \left(\mu_d + \frac{2\bar{d}\lambda_{\max}(P_0)^2}{\lambda_{\min}(Q_0)}\right)\theta\lambda_{\max}(P_0)\right],\right.$$

$$\left.\frac{1}{\sigma_a}\left(\mu_d + \frac{2\bar{d}\lambda_{\max}(P_0)^2}{\lambda_{\min}(Q_0)}\right)\mu\right\} \sup_{0\leq\zeta\leq t} \Omega_0(\zeta)$$

$$\leq V(0)e^{-\sigma_a t} + \bar{\Phi} \sup_{0\leq\zeta\leq t} V(\zeta), \tag{11.164}$$

where

$$\bar{\Phi} = \frac{1}{\mu_1}\max\left\{\frac{1}{\sigma_a}\mu_W + \frac{1}{\sigma_a}\left(\mu_d + \frac{2\bar{d}\lambda_{\max}(P_0)^2}{\lambda_{\min}(Q_0)}\right)\theta\lambda_{\max}(P_0),\right.$$

$$\left.\frac{\mu_d\mu}{\sigma_a} + \frac{2\mu\bar{d}\lambda_{\max}(P_0)^2}{\sigma_a\lambda_{\min}(Q_0)}\right\}. \tag{11.165}$$

In order to ensure that

$$\bar{\Phi} < 1, \tag{11.166}$$

along with combining the conditions (11.133)–(11.135) used to avoid the Zeno phenomenon, the design parameters $\mu, \mu_d, \mu_W, \theta$ are chosen according to the following guidelines.

1) Choose μ as

$$\mu < \frac{\mu_1\sigma_a\lambda_{\min}(Q_0)}{4\bar{d}\lambda_{\max}(P_0)^2} \tag{11.167}$$

to ensure that

$$\frac{2\mu\bar{d}\lambda_{\max}(P_0)^2}{\sigma_a\lambda_{\min}(Q_0)} < \frac{\mu_1}{2} \tag{11.168}$$

in (11.165). Before showing the guidelines for the other design parameters, we define the analysis parameter r_1, which is from Young's inequality applied in (11.131), as

$$r_1 < \min\left\{\frac{\mu_1\sigma_a}{4\lambda_d}, \frac{\mu\mu_1\sigma_a}{4\lambda_W}\right\}. \tag{11.169}$$

The choice of r_1 comes from the need to guarantee that both (11.166) and (11.133), (11.134) hold, which will be clearly seen later.

2) Choose μ_d to satisfy

$$\frac{2r_1\lambda_d}{\mu} \leq \mu_d < \frac{\mu_1\sigma_a}{2\mu} \tag{11.170}$$

in order to ensure that

$$\frac{\mu_d\mu}{\sigma_a} < \frac{\mu_1}{2} \tag{11.171}$$

in (11.165) (with the right inequality of (11.170)). Recalling (11.167), the final term in (11.165) is less than μ_1. The condition (11.134) is incorporated as the left inequality of (11.170).

3) Choose μ_W to satisfy

$$\frac{2r_1\lambda_W}{\mu} \leq \mu_W < \frac{\mu_1\sigma_a}{2} \tag{11.172}$$

in order to ensure that

$$\frac{1}{\sigma_a}\mu_W < \frac{\mu_1}{2} \tag{11.173}$$

in (11.165) (with the right inequality of (11.172)), where the condition (11.133) is incorporated as the left inequality of (11.172).

The parameter r_1 (11.169) is chosen to ensure that the far-left terms of (11.170), (11.172) are less than the far-right ones, where the far-right terms are from ensuring (11.166), and the far-left terms are from the conditions (11.133), (11.134) on avoiding the Zeno phenomenon.

4) Choose θ to satisfy

$$\theta < \min\left\{ \frac{\sigma_a\mu_1}{2\left(\mu_d + \frac{2\bar{d}\lambda_{\max}(P_0)^2}{\lambda_{\min}(Q_0)}\right)\lambda_{\max}(P_0)}, \frac{\mu\mu_W}{3\lambda_{\max}(Q_0)}, \frac{\mu_d\mu\lambda_{\max}(Q_0)}{4\bar{d}\lambda_{\max}(P_0)^2} \right\} \tag{11.174}$$

in order to ensure that

$$\frac{1}{\sigma_a}\left(\mu_d + \frac{2\bar{d}\lambda_{\max}(P_0)^2}{\lambda_{\min}(Q_0)}\right)\theta\lambda_{\max}(P_0) < \frac{\mu_1}{2} \tag{11.175}$$

in (11.165). Recalling (11.172), the first term in (11.165) is thus less than μ_1. The condition (11.135) is incorporated into (11.174) as the last two terms of (11.174). Because both terms in (11.165) are less than μ_1, ensured by steps 1–4, the inequality (11.166) is assured.

The following estimate then holds

$$\sup_{0\leq\zeta\leq t}\left(V(\zeta)e^{\sigma_a\zeta}\right) \leq V(0) + \bar{\Phi}\sup_{0\leq\zeta\leq t}\left(V(\zeta)e^{\sigma_a\zeta}\right) \tag{11.175}$$

as a consequence of (11.164). It follows that

$$\sup_{0\leq\zeta\leq t}V(\zeta) \leq \Upsilon_V V(0)e^{-\sigma_a t}, \tag{11.176}$$

where the constant

$$\Upsilon_V = \frac{1}{1-\bar{\Phi}} > 0 \tag{11.177}$$

by recalling (11.166). The choice of the low-pass filter and the ETM parameters affect the overshoot coefficient in the exponential result according to (11.165).

Since, according to definitions 10.1 and 10.2, the transfer function between $\beta(0,s)$ and $\hat{W}(s)$ is stable and proper, (11.86) leads to

$$\beta(0,t)^2 \leq \gamma_\beta \sup_{0\leq\zeta\leq t} \hat{W}(\zeta)^2, \tag{11.178}$$

where the positive constant γ_β only depends on the plant parameters. Recalling (11.78), the following inequality holds

$$\hat{Z}(t)^2 = 2 \sup_{0\leq\zeta\leq t} \hat{W}(\zeta)^2 + 2|C_0^+|^2 p^2 \sup_{0\leq\zeta\leq t} \beta(0,\zeta)^2$$

$$\leq \sup_{0\leq\zeta\leq t} \gamma_Z \hat{W}(\zeta)^2, \tag{11.179}$$

where the positive constant

$$\gamma_Z = \max\{2, 2|C_0^+|^2 p^2 \gamma_\beta\}$$

depends only on the plant parameters.

According to (11.179), (11.176), (11.142), we obtain

$$\left(\left|\hat{X}(t)\right|^2 + \left|\hat{Z}(t)\right|^2 + \|\alpha(\cdot,t)\|^2 + \|\beta(\cdot,t)\|^2 + |m(t)| + \bar{U}(t)^2 \right)$$

$$\leq \bar{\Upsilon}_f \left(\left|\hat{X}(0)\right|^2 + \left|\hat{Z}(0)\right|^2 + \|\alpha(\cdot,0)\|^2 + \|\beta(\cdot,0)\|^2 \right.$$

$$\left. + |m(0)| + \bar{U}(0)^2 \right) e^{-\bar{\lambda}_f t} \tag{11.180}$$

for some positive $\bar{\Upsilon}_f, \bar{\lambda}_f$, which are associated with Υ_V (11.177) and σ_a (11.160), respectively. Applying the invertibility of the backstepping transformations (11.59), (11.60), (11.67), we arrive at (11.140), where Υ_f, λ_f are associated with Υ_V (11.177) and σ_a (11.160), respectively. $\qquad\square$

The guidelines for the choices of all the parameters are given by (11.7)–(11.10), (11.97), (11.98), (11.154)–(11.158), (11.167)–(11.174), which are cascaded rather than coupled. The optimal choices of these parameters are not studied in this book, but in future work, the trade-off between the convergence rate and the lower bound of the minimal dwell time is worth studying.

Theorem 11.1. *For all initial data* $(Y(0), z(x,0), w(x,0), X(0)) \in \chi$ *and* $(\hat{Y}(0), \hat{z}(x,0), \hat{w}(x,0), \hat{X}(0)) \in \chi$, $m(0) \in \mathbb{R}^-$, *choosing the design parameters to satisfy* (11.7)–(11.10), (11.97), (11.98), (11.167), (11.170)–(11.174), *the output-feedback closed-loop system—that is, the plant* (11.1)–(11.6) *under the event-based control input* $U_d(t)$ *in* (11.102), *which is realized using the observer* (11.13)–(11.18), *the low-pass filter* $\Omega(s)$ *and the event-triggering mechanism* (11.116), (11.117), *has the following properties:*

1) No Zeno phenomenon occurs—that is,

$$\lim_{k\to\infty} t_k = +\infty. \tag{11.181}$$

2) The closed-loop system has unique (weak) solutions $((z,w)^T, X, Y) \in C^0([0,\infty); L^2(0,1); \mathbb{R}^2) \times C^0([0,\infty); \mathbb{R}^n) \times C^0([0,\infty); \mathbb{R}^{\bar{n}}), ((\hat{z},\hat{w})^T, \hat{X}, \hat{Y}) \in C^0([0,\infty); L^2(0,1); \mathbb{R}^2) \times C^0([0,\infty); \mathbb{R}^{\bar{n}}) \times C^0([0,\infty); \mathbb{R}^{\bar{n}}),$ *and* $m \in C^0([0,\infty); \mathbb{R}^-).$

3) The exponential convergence in the closed-loop system is achieved in the sense that there exist the positive constants Υ_a, λ_a such that

$$\Xi(t) + \hat{\Xi}(t) \le \Upsilon_a \left(\Xi(0) + \hat{\Xi}(0) \right) e^{-\lambda_a t}, \tag{11.182}$$

where

$$\Xi(t) = |X(t)|^2 + |Y(t)|^2 + \|z(\cdot,t)\|^2 + \|w(\cdot,t)\|^2,$$

and $\hat{\Xi}(t)$ is defined in (11.141), which includes $|m(t)|$.

4) The event-triggered control input is convergent to zero in the sense of

$$\lim_{t \to \infty} U_d(t) = 0. \tag{11.183}$$

Proof. 1) Recalling lemma 11.3, property (1) is obtained.

2) By virtue of proposition 11.1 and lemma 11.3, through iterative constructions between successive triggering times, property (2) is obtained.

3) Rewriting the observer states in the output-feedback control input as a sum of the plant states and the observer errors according to (11.19) and inserting the result into the plant (11.1)–(11.6), through the same steps as in the above state-feedback control designs, it follows that the closed-loop dynamics are a cascade of the observer error dynamics feeding into the target system dynamics in the form of (11.110)–(11.115) (the state-feedback loop). Because the stability of the observer error dynamics (which depends on the choices of L, L_0) and the stability of the state-feedback loop dynamics (which depends on the choices of K, K_0) are independent, the separation principle holds. Equation (11.182) is thus obtained recalling lemma 11.1 and lemma 11.4, where the overshoot Υ_a is associated with Υ_e, Υ_f, and the decay rate σ_a is associated with λ_e, λ_f. Property (3) is obtained.

4) Recalling (11.101) and the stability result proved in property (3), we have that the continuous-in-time control input $U(t)$ is convergent to zero. According to the definition (11.102), property (4) is obtained. □

11.6   APPLICATION IN THE MINING CABLE ELEVATOR

In this section, the proposed event-triggered backstepping boundary control design is applied to the axial vibration control of a mining cable elevator that is 2000 m deep and whose dynamics include a hydraulic actuator, mining cable, and cage.

Model

Figure 11.1 shows that the axial vibration dynamics of the mining cable elevator consisting of the hydraulic actuator, mining cable, and cage are described by a wave PDE sandwich system. The figure also shows that this system can be transformed, using a Riemann transformation, into a 2×2 coupled transport PDE sandwich system, considered in this chapter, and for which assumptions 11.1–11.3 are satisfied.

The vibration model of the mining cable elevator, which includes the hydraulic actuator dynamics, is described (as in chapter 2) by a wave PDE that is sandwiched between two ODEs, as follows:

$$M_h \ddot{b}_0(t) = -c_h \dot{b}_0(t) + \frac{\pi R_d^2}{4} E u_x(0,t) + U_d(t), \tag{11.184}$$

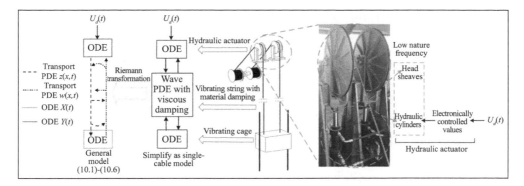

Figure 11.1. The relationship between the sandwich ODE-PDE-ODE hyperbolic system and the mining cable elevators consisting of a hydraulic-driven head sheave, mining cable, and cage.

$$u(0,t) = b_0(t), \tag{11.185}$$

$$\rho u_{tt}(x,t) = \frac{\pi R_d^2}{4} E u_{xx}(x,t) - d_c u_t(x,t), \ x \in [0, \bar{L}], \tag{11.186}$$

$$u(L,t) = b_L(t), \tag{11.187}$$

$$M_c \ddot{b}_L(t) = -c_L \dot{b}_L(t) + \frac{\pi R_d^2}{4} E u_x(\bar{L}, t), \tag{11.188}$$

where $U_d(t)$ is the event-triggered backstepping control input of the electronically controlled valves that regulates the hydraulic actuator to suppress vibration in the mining cable elevator. The PDE state $u(x,t)$ denotes the distributed axial vibration dynamics along the cable. The ODE state $b_0(t)$ represents the displacement of the hydraulic actuator, and $b_L(t)$ is the vibration displacement of the cage. The physical parameters in (11.184)–(11.188) of the mining cable elevator are shown in table 11.1. We apply the Riemann transformations

$$z(x,t) = u_t(x,t) - \sqrt{\frac{E\pi}{\rho}} \frac{R_d}{2} u_x(x,t), \tag{11.189}$$

$$w(x,t) = u_t(x,l) + \sqrt{\frac{E\pi}{\rho}} \frac{R_d}{2} u_x(x,t) \tag{11.190}$$

and define the new variables

$$Y(t) = \dot{b}_0(t), \ X(t) = \dot{b}_L(t), \tag{11.191}$$

which allows us to rewrite (11.184)–(11.188) as (11.1)–(11.6), with

$$q_1 = q_2 = \sqrt{\frac{E\pi}{\rho}} \frac{R_d}{2}, \quad c_1 = c_2 = \frac{-d_c}{2\rho}, \tag{11.192}$$

$$q = p = -1, \quad C_0 = C_1 = 2, \tag{11.193}$$

$$A_0 = \frac{-c_h}{M_h} - \frac{R_d\sqrt{E\pi\rho}}{2M_h}, \quad E_0 = \frac{R_d\sqrt{E\pi\rho}}{2M_h}, \quad B_0 = \frac{1}{M_h}, \tag{11.194}$$

Table 11.1. Physical parameters of the mining cable elevator

Parameters (units)	Values
Depth $\bar{L}$ (m)	2000
Cable diameter R_d (m)	0.2
Cable effective Young's modulus E (N/m^2)	1.02×10^9
Cable linear density ρ (kg/m)	8.1
Mass of hydraulic actuator M_h (kg)	300
Mass of cage M_c (kg)	15000
Damping coefficient of hydraulic actuator c_h	0.4
Damping coefficient of cage c_L	0.4
Cable material damping coefficient d_c	0.5
Gravitational acceleration g (m/s^2)	9.8

$$A = \frac{-c_L}{M_c} + \frac{R_d\sqrt{E\pi\rho}}{2M_c}, \ \ B = -\frac{R_d\sqrt{E\pi\rho}}{2M_c}, \tag{11.195}$$

which satisfy assumptions 11.1–11.3. The initial conditions of $z(x,t)$ and $w(x,t)$ are defined as

$$z(x,0) = 0.01\sin(2\pi(\bar{L}-x)/\bar{L} + \pi/6), \tag{11.196}$$
$$w(x,0) = 0.01\sin(3\pi(\bar{L}-x)/\bar{L}) \tag{11.197}$$

and

$$X(0) = \frac{1}{2}(w(\bar{L},0) - qz(\bar{L},0)), \tag{11.198}$$

$$Y(0) = \frac{1}{2}(z(0,0) - pw(0,0)) \tag{11.199}$$

according to (11.5). The observer initial conditions are defined as

$$\hat{z}(x,0) = z(x,0) + 0.2, \tag{11.200}$$
$$\hat{w}(x,0) = w(x,0) + 0.2, \tag{11.201}$$

where 0.2 is an initial observer error, and

$$\hat{X}(0) = \frac{1}{2}(\hat{w}(\bar{L},0) - q\hat{z}(\bar{L},0)), \tag{11.202}$$

$$\hat{Y}(0) = \frac{1}{2}(\hat{z}(0,0) - p\hat{w}(0,0)) \tag{11.203}$$

according to (11.17). We pick the initial value of $m(t)$ as $m(0) = -0.001$. The simulation is conducted based on the finite-difference method with the time step of 0.0015 s and the space step of 0.5 m.

Determining Design Parameters

The free design parameter η in (11.117) is selected as $\eta = 0.11$. The parameters affecting the decay rate of the states in the closed-loop system are determined next. According to A_0, A, B_0, B, C_0, C in (11.193)–(11.195) and the parameter values in table 11.1, recalling (11.7)–(11.10), the control gains and observer gains are chosen as

$$K_0 = 1, \quad K = 1.5$$

and

$$L_0 = 1, \quad L = 2,$$

respectively, yielding

$$\hat{A}_0 = -106.7, \ \ \hat{A} = -1.067, \ \ \bar{A} = -2.9, \ \ \bar{A}_0 = -55.4.$$

Defining $P = P_0 = \frac{1}{2}$, we then have

$$\lambda_{\min}(Q_0) = 106.7, \ \ \lambda_{\min}(Q) = 1.067$$

via (11.119), (11.144). According to (11.97), (11.98) and

$$C_0(Is - \hat{A}_0)^{-1} B_0 = \frac{2}{300(s + 106.7)}$$

in $F(s)$ in (11.96), the low-pass filter is chosen as the first-order type

$$\Omega(s) = \frac{1}{1 + 0.0011s},$$

which can be implemented with a resistor-capacitor (RC) circuit. Next, choosing

$$\delta_1 = 0.5, \quad \delta_2 = 0.5$$

according to (11.154), (11.155) and then determining

$$r_b = 0.013, \ \ r_w = 7.3, \ \ r_a = 0.0023$$

from (11.156)–(11.158) leads to

$$\mu_2 = 3.65, \ \ \mu_1 = 0.0011$$

according to the formulae (11.147), (11.148). Therefore, the estimate of the decay rate σ_a obtained from (11.160) is 0.108.

The parameters of the ETM are determined next. According to the transfer functions (11.107), (11.124), (11.125), the plant parameters, and the choices of K_0, K, a group of conservative estimates of $\bar{d}, \lambda_W, \lambda_d$ is

$$\bar{d} = 5, \ \ \lambda_W = 250, \ \ \lambda_d = 600.$$

Recalling (11.167), (11.169), μ is defined as

$$\mu = 0.0024$$

and r_1 as

$$r_1 = 0.22 \times 10^{-9}.$$

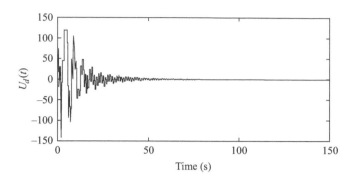

Figure 11.2. Output-feedback event-based control input $U_d(t)$.

Then μ_W, μ_d are determined by (11.170), (11.172) as

$$\mu_d = 0.01, \quad \mu_W = 0.5 \times 10^{-4}.$$

Finally, pick

$$\theta = 0.36 \times 10^{-9}$$

via (11.174). Recalling (11.139) for the highly conservative minimal dwell-time estimate τ, we get

$$\tau = 0.15 \times 10^{-9} \text{ s}.$$

Substituting the above parameters into (11.165), we arrive at

$$\bar{\Phi} = 0.6754.$$

The approximate solutions of the kernels $M(x,y)$, $N(x,y)$, $\gamma(x)$, $H(x,y)$, $J(x,y)$, $\lambda(x)$ are obtained from the conditions (11.205)–(11.216), which are two groups of coupled linear hyperbolic PDE-ODE systems on the domain $\{(x,y)|0 \leq x \leq y \leq \bar{L}\}$. The finite-difference method is employed with a step length of 1 m for y running from x to $\bar{L}$. The approximate solutions of $\bar{K}_1(x)$, $\bar{K}_2(x)$, $\bar{K}_3$, $\bar{K}_4(x)$, $\bar{K}_5(x), \bar{K}_6$ are obtained from conditions (11.217)–(11.222) by the finite-difference method with respect to $x \in [0, \bar{L}]$ with a step length of 1 m as well. Based on the above approximate solutions, N_1, N_2, $M_\alpha(x)$, $M_\beta(x)$, M_X are obtained using (11.223)–(11.227).

Closed-Loop Responses

Figure 11.2 shows the event-triggered control input, where the minimal dwell time is 0.297 s, which is much larger than the conservative estimate $\tau = 0.15 \times 10^{-9}$ s. If the design parameter η is picked as a smaller one of 0.106 (other design parameters are not changed), compared with the first value $\eta = 0.11$ defined at the beginning of this subsection, the number of update times of the control input decreases from 373 to 361, further reducing the actuation frequency. However, the control performance is slightly degraded because η also affects the convergence rate of the closed-loop system.

Figure 11.3 shows the convergence of the ODE states $X(t), Y(t)$—that is, the suppression of the axial vibration velocity of the cage and the regulation of the

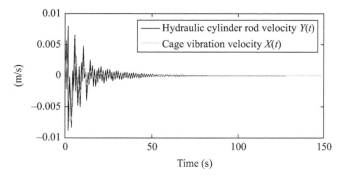

Figure 11.3. Axial vibration velocity of the cage $X(t)$ and moving velocity of the hydraulic rod $Y(t)$ in the hydraulic cylinder.

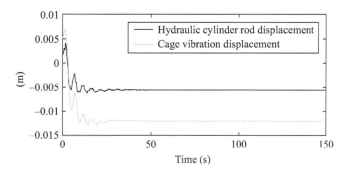

Figure 11.4. Axial vibration displacement of the cage (initial elastic displacement 0.005 m) and movement of the hydraulic rod in the hydraulic cylinder (initial position 0.001 m).

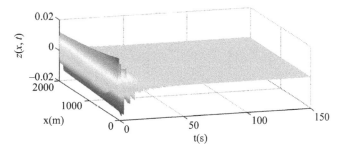

Figure 11.5. Response of $z(x,t)$.

moving velocity of the hydraulic rod in the hydraulic cylinder at the head sheaves. Integrations of $X(t), Y(t)$—the axial vibration displacement of the cage and the movement of the hydraulic rod in the hydraulic cylinder—are shown in figure 11.4 under the initial elastic displacement of 0.005 m at the cable-cage connection point and the initial position at 0.001 m of the hydraulic rod. Figures 11.5 and 11.6 show the convergence of the PDE states $z(x,t), w(x,t)$. The axial vibration energy of the

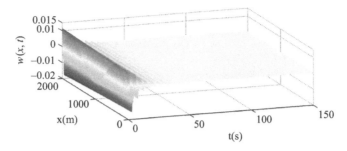

Figure 11.6. Response of $w(x,t)$.

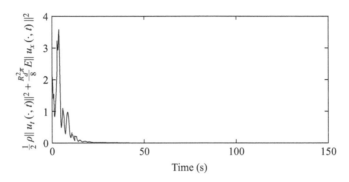

Figure 11.7. Axial vibration energy V_E of the cable.

cable,

$$V_E = \frac{1}{2}\rho\|u_t(\cdot,t)\|^2 + \frac{R_d^2\pi E}{8}\|u_x(\cdot,t)\|^2, \tag{11.204}$$

is converted into

$$V_E = \frac{\rho}{8}\|z(\cdot,t) - w(\cdot,t)\|^2 + \frac{\rho}{8}\|z(\cdot,t) + w(\cdot,t)\|^2$$

using (11.189), (11.190). Figure 11.7 shows that the axial vibration energy of the cable is reduced with the help of the proposed event-based vibration control system.

11.7   APPENDIX

A. The conditions on the kernels $M(x,y)$, $N(x,y)$, $\gamma(x)$, $H(x,y)$, $J(x,y)$, $\lambda(x)$

The conditions on the kernels $M(x,y)$, $N(x,y)$, $\gamma(x)$, $H(x,y)$, $J(x,y)$, $\lambda(x)$ in (11.59), (11.60) are given by

$$q_1 M(x,1) - q_2 N(x,1)q - \gamma(x)B = 0, \tag{11.205}$$

$$(q_2 + q_1)N(x,x) - c_1 = 0, \tag{11.206}$$

$$c_1 M(x,y) - q_1 N_x(x,y) + q_2 N_y(x,y)$$
$$+ (c_2 - c_1) N(x,y) = 0, \tag{11.207}$$
$$c_2 N(x,y) - q_1 M_x(x,y) - q_1 M_y(x,y) = 0, \tag{11.208}$$
$$\gamma(1) = -K, \tag{11.209}$$
$$-q_1 \gamma'(x) - \gamma(x)A - c_1 \gamma(x) - q_2 N(x,1)C = 0, \tag{11.210}$$
$$q_1 H(x,1) - q_2 J(x,1)q - \lambda(x)B = 0, \tag{11.211}$$
$$-c_2 - (q_1 + q_2)H(x,x) = 0, \tag{11.212}$$
$$c_1 H(x,y) + q_2 J_x(x,y) + q_2 J_y(x,y) = 0, \tag{11.213}$$
$$c_2 J(x,y) + (c_1 - c_2)H(x,y)$$
$$+ q_2 H_x(x,y) - q_1 H_y(x,y) = 0, \tag{11.214}$$
$$q_2 \lambda'(x) - \lambda(x)A - c_2 \lambda(x) - q_2 J(x,1)C = 0, \tag{11.215}$$
$$q\gamma(1) - \lambda(1) + C = 0. \tag{11.216}$$

B. The conditions of $\bar{K}_1(x)$, $\bar{K}_2(x)$, $\bar{K}_3(x)$, $\bar{K}_4(x)$, $\bar{K}_5(x)$, $\bar{K}_6$

In (11.61), (11.62), $\bar{K}_1(x), \bar{K}_2(x), \bar{K}_3, \bar{K}_4(x), \bar{K}_5(x), \bar{K}_6$ are the solutions of the following linear Volterra integral equations of the second kind:

$$\bar{K}_1(x) = pH(0,x) - M(0,x) + \int_0^x \bar{K}_1(y)M(y,x)dy$$
$$+ \int_0^x \bar{K}_2(y)H(y,x)dy, \tag{11.217}$$

$$\bar{K}_2(x) = -pJ(0,x) + N(0,x) + \int_0^x \bar{K}_1(y)N(y,x)dy$$
$$+ \int_0^x \bar{K}_2(y)J(y,x)dy, \tag{11.218}$$

$$\bar{K}_3 = \int_0^1 \bar{K}_2(x)\lambda(x)dx + \int_0^1 \bar{K}_1(x)\gamma(x)dx + p\lambda(0) - \gamma(0), \tag{11.219}$$

$$\bar{K}_4(x) = \int_0^x \bar{K}_4(y)M(y,x)dy + \int_0^x \bar{K}_5(y)H(y,x)dy - E_0 H(0,x), \tag{11.220}$$

$$\bar{K}_5(x) = \int_0^x \bar{K}_4(y)N(y,x)dy + \int_0^x \bar{K}_5(y)J(y,x)dy + E_0 J(0,x), \tag{11.221}$$

$$\bar{K}_6 = \int_0^1 \bar{K}_5(x)\lambda(x)dx + \int_0^1 \bar{K}_4(x)\gamma(x)dx + E_0\lambda(0). \tag{11.222}$$

C. Expressions for N_1, N_2, M_α, M_β, M_X

Expressions for $N_1, N_2, M_\alpha, M_\beta, M_X$ are

$$N_1 = C_0^+ \bar{K}_3 B - q_1 C_0^+ \bar{K}_1(1) + q_2 C_0^+ \bar{K}_2(1)q, \tag{11.223}$$
$$N_2 = E_0 - q_2 C_0^+ \bar{K}_2(0) + q_1 C_0^+ \bar{K}_1(0)p, \tag{11.224}$$
$$M_\alpha(x) = \bar{K}_4(x) + q_1 C_0^+ \bar{K}_1'(x) - (\hat{A}_0 + c_1)C_0^+ \bar{K}_1(x), \tag{11.225}$$
$$M_\beta(x) = \bar{K}_5(x) - q_2 C_0^+ \bar{K}_2'(x) - (\hat{A}_0 + c_2)C_0^+ \bar{K}_2(x), \tag{11.226}$$
$$M_X = C_0^+ \bar{K}_3 \hat{A} + \bar{K}_6 - \hat{A}_0 C_0^+ \bar{K}_3. \tag{11.227}$$

11.8 NOTES

The sandwich system control designs in chapters 9 and 10 used continuous-in-time control input signals, whereas the control input in this chapter is piecewise-constant by designing an event-triggering mechanism. As a result, changes in the actuator signal are reduced, which physically facilitates the practical implementation of many industrial string-actuated mechanisms, where the control inputs are provided by massive actuators. The first result of the event-triggered backstepping control of a 2×2 hyperbolic system is given in [57]. In this chapter, besides two additional ODEs, the proximal reflection term was required to be compensated by the event-based control input that goes through the input ODE which is stabilized meanwhile, making the control design more challenging.

Chapter Twelve

Sandwich Hyperbolic PDEs with Nonlinearities

In this chapter, we extend the results in chapter 9 for linear sandwich partial differential equations (PDEs) to a more challenging case: the control of sandwich hyperbolic PDEs with nonlinearities, motivated by the brake control of cable mining elevators, where the dynamics consist of a brake, a shock absorber (described by a nonlinear ordinary differential equation (ODE)), a cable of time-varying length, and a cage.

We begin by describing, in section 12.1, a particular class of coupled hyperbolic PDEs sandwiched between a nonlinear ODE on the actuated side and a linear ODE on the distal side, with a PDE domain that is time-varying. A state-feedback controller entering a single ODE state is designed to exponentially stabilize the overall system through several backstepping transformations in section 12.2. An observer that only uses the boundary values at the actuated side is constructed to recover all the states of the overall system in section 12.3. The global exponential stability of the observer-based output-feedback closed-loop control system, as well as the boundedness and exponential convergence of the control input, are proved via Lyapunov analysis in section 12.4. The performance is investigated via numerical simulation in section 12.5.

12.1 PROBLEM FORMULATION

The plant considered in this chapter is

$$\dot{X}(t) = AX(t) + Bv(0,t), \tag{12.1}$$

$$u_t(x,t) = -p_1 u_x(x,t) + c_1 v(x,t), \tag{12.2}$$

$$v_t(x,t) = p_2 v_x(x,t) + c_2 u(x,t), \tag{12.3}$$

$$u(0,t) = qv(0,t) + CX(t), \tag{12.4}$$

$$v(l(l),t) = s_1(t), \tag{12.5}$$

$$\dot{s}_1(t) = c_3 s_2(t) + f_1\left(s_1(t), \int_0^{l(t)} u(x,t)dx\right), \tag{12.6}$$

$$\dot{s}_2(t) = f_2(s_1(t), s_2(t), u(l(t),t)) + z(t), \tag{12.7}$$

$$\dot{z}(t) = c_4 z(t) + ru(l(t),t) + U(t), \tag{12.8}$$

$\forall (x,t) \in [0, l(t)] \times [0, \infty)$, where $X(t) \in \mathbb{R}^{n \times 1}$, $z(t) \in \mathbb{R}$ are ODE states, which describe the vibration dynamics of the cage and drum. The nonlinear ODE with the vector state $S(t) = [s_1(t), s_2(t)]^T \in \mathbb{R}^{2 \times 1}$ represents the shock absorber dynamics, where $s_1(t)$ represents the displacement of the connection point with the cable, and $s_2(t)$ means the velocity of the connection point with the drum. The functions $u(x,t) \in \mathbb{R}, v(x,t) \in \mathbb{R}$ are the states of the 2×2 coupled hyperbolic PDEs, which model the vibration states of the cable. The matrices $A \in \mathbb{R}^{n \times n}$, $B \in \mathbb{R}^{n \times 1}$, $C \in \mathbb{R}^{1 \times n}$ are assumed to be such that the pair (A, B) is controllable, and (A, C) is observable. The constants $c_1, c_2, c_3, c_4, r, q \in \mathbb{R}$ are arbitrary, whereas p_1 and p_2 are arbitrary positive transport velocities. The signal $U(t)$ is the control input to be designed.

Functions f_1, f_2, and $l(t)$ satisfy the following assumptions.

Assumption 12.1. *The functions f_1 and f_2 vanish at the origin—that is, $f_1(0,0) = 0$ and $f_2(0,0,0) = 0$.*

Assumption 12.2. *The functions $f_1(x_1, x_2)$ and $f_2(x_1, x_2, x_3)$ are continuously differentiable and globally Lipschitz in (x_1, x_2) and (x_1, x_2, x_3), respectively.*

Assumption 12.3. *The function $l : [0, \infty) \to C^2$ is uniformly bounded—that is, $0 < l(t) \leq L$, $\forall t \geq 0$, where L is a positive constant.*

Assumption 12.4. *The speed of the moving boundary is lower than the transport speed in the PDEs (12.2), (12.3)—that is,*

$$\left| \dot{l}(t) \right| < \min\{p_1, p_2\}, \qquad \forall t \geq 0. \tag{12.9}$$

Equations (12.1)–(12.5) can be regarded as reversibly converted from a wave PDE with in-domain damping through a Riemann coordinate transformation (as in chapter 3). Therefore, according to the conclusions in [71, 72], the fact that the speed of the moving boundary $l(t)$ is smaller than the wave speed—that is, assumption 12.4—ensures the well-posedness of the initial boundary value problem (12.1)–(12.5).

The signal flow of the plant (12.1)–(12.8) is configured as follows. The control input $U(t)$ goes through a linear ODE (12.8), which acts as a filter and whose output signal $z(t)$ drives a nonlinear ODE (12.6), (12.7), which includes the states $s_2(t)$ and $s_1(t)$, the latter of which, in turn, flows into the right boundary $x = l(t)$ (12.5) of the transport PDE (12.3), which is coupled with another transport PDE (12.2), (12.4) and connected with a linear ODE (12.1) at the left boundary $x = 0$. The reflected signal u flows back to the ODEs (12.6)–(12.8) via the transport PDE (12.2).

The control objective is to exponentially stabilize all the ODE states $S(t)$, $X(t)$, $z(t)$ and the PDE states $u(x,t), v(x,t)$ by designing a control input $U(t)$ applied at the first ODE (12.8) using the measurements $v(l(t), t), u(l(t), t), z(t)$.

12.2 STATE-FEEDBACK CONTROL DESIGN

In this section, a PDE backstepping transformation is used to convert the coupled hyperbolic PDE-ODE subsystem to a stable intermediate system where the in-domain couplings between the hyperbolic PDEs are removed, and the system matrix of the ODE at the left boundary is Hurwitz. The right boundary condition of the intermediate system can be regarded as a cascade of a nonlinear ODE and a linear ODE under some perturbations from the PDE states in the time-varying domain and the left boundary. The right boundary condition will be dealt with by using an ODE backstepping procedure. The global exponential stability of the closed-loop system is then proved, where control parameters are determined that come from the ODE backstepping design procedure and whose purpose is to retain stability in the face of the PDE perturbations. Moreover, the boundedness and exponential convergence of the designed control input are proved as well.

Backstepping for a PDE-ODE Subsystem

We use the infinite-dimensional backstepping transformation in chapter 9 for the PDE states $u(x,t), v(x,t)$, as follows:

$$\alpha(x,t) = u(x,t), \tag{12.10}$$

$$\beta(x,t) = v(x,t) - \int_0^x \psi(x,y)u(y,t)dy$$

$$- \int_0^x \phi(x,y)v(y,t)dy - \gamma(x)X(t). \tag{12.11}$$

The kernels $\psi(x,y), \phi(x,y)$ on $\mathcal{D} = \{0 \le y \le x \le l(t)\}$ and the row vector $\gamma(x)$ on $\{0 \le x \le l(t)\}$ satisfy

$$\psi(x,x) = \frac{-c_2}{p_1 + p_2}, \tag{12.12}$$

$$\phi(x,0) - \frac{1}{p_2}\gamma(x)B + \frac{p_1 q}{p_2}\psi(x,0), \tag{12.13}$$

$$p_2\phi_x(x,y) + p_2\phi_y(x,y) - c_1\psi(x,y) = 0, \tag{12.14}$$

$$-p_2\psi_x(x,y) + p_1\psi_y(x,y) + c_2\phi(x,y) = 0, \tag{12.15}$$

$$p_2\gamma'(x) - \gamma(x)A - p_1\psi(x,0)C = 0, \tag{12.16}$$

$$\gamma(0) = \kappa, \tag{12.17}$$

where κ is a row vector such that $A + B\kappa$ is Hurwitz since the pair (A, B) is controllable. Please refer to lemma 9.1 of chapter 9 for the well-posedness of (12.12)–(12.17).

As in chapter 9, the inverse of (12.10), (12.11) is

$$u(x,t) = \alpha(x,t), \tag{12.18}$$

$$v(x,t) = \beta(x,t) - \int_0^x \mathcal{D}(x,y)\alpha(y,t)dy$$

$$- \int_0^x \mathcal{M}(x,y)\beta(y,t)dy - \mathcal{J}(x)X(t), \tag{12.19}$$

where $\mathcal{D}(x,y), \mathcal{M}(x,y)$ and the row vector $\mathcal{J}(x)$ are the kernels of the inverse transformation (12.19), whose well-posedness is shown in section 9.2 in chapter 9.

Applying the above backstepping transformations, the original system (12.1)–(12.5) is converted to the following intermediate system (without the right boundary condition):

$$\dot{X}(t) = (A + B\kappa)X(t) + B\beta(0,t), \tag{12.20}$$

$$\alpha_t(x,t) = -p_1\alpha_x(x,t) + c_1\beta(x,t) - c_1 \int_0^x \mathcal{D}(x,y)\alpha(y,t)dy$$

$$- c_1 \int_0^x \mathcal{M}(x,y)\beta(y,t)dy - c_1\mathcal{J}(x)X(t), \tag{12.21}$$

$$\beta_t(x,t) = p_2\beta_x(x,t), \tag{12.22}$$

$$\alpha(0,t) = q\beta(0,t) + C_0 X(t), \tag{12.23}$$

where the row vector

$$C_0 = C + q\gamma(0). \tag{12.24}$$

Let us now consider the right boundary condition. Inserting $x = l(t)$ into (12.11) and taking the derivative with respect to t, we obtain

$$\dot{\beta}(l(t),t) = \dot{v}(l(t),t) - \dot{l}(t)\psi(l(t),l(t))u(l(t),t)$$

$$- \dot{l}(t)\phi(l(t),l(t))v(l(t),t) - \dot{l}(t)\int_0^{l(t)} \psi_x(l(t),y)u(y,t)dy$$

$$- \dot{l}(t)\int_0^{l(t)} \phi_x(l(t),y)v(y,t)dy$$

$$- \dot{l}(t)\gamma'(l(t))X(t) - \int_0^{l(t)} \psi(l(t),y)u_t(y,t)dy$$

$$- \int_0^{l(t)} \phi(l(t),y)v_t(y,t)dy - \gamma(l(t))\dot{X}(t). \tag{12.25}$$

Using (12.5), (12.6) to replace $\dot{v}(l(t),t)$ in (12.25) and then plugging the inverse transformations (12.18), (12.19) into (12.25) to replace u, v with α, β, through a change of the order of integration in a double integral, we get $\dot{\beta}(l(t),t)$ as

$$\dot{\beta}(l(t),t) = c_3 s_2(t) + f_1\bigg(\beta(l(t),t) - \int_0^{l(t)} \mathcal{D}(l(t),y)\alpha(y,t)dy$$

$$- \int_0^{l(t)} \mathcal{M}(l(t),y)\beta(y,t)dy - \mathcal{J}(l(t))X(t), \int_0^{l(t)} \alpha(y,t)dy\bigg)$$

$$+ \mathcal{F}(\beta(l(t),t), \beta(0,t), \alpha(l(t),t), \alpha(0,t), \beta(x,t), \alpha(x,t), X(t)), \tag{12.26}$$

where $\mathcal{F}$ is a perturbation that includes $\beta(l(t),t)$, $\beta(0,t)$, $\alpha(l(t),t)$, $\alpha(0,t)$, $\beta(x,t)$, $\alpha(x,t)$, and $X(t)$. The complete expression of $\mathcal{F}$ is shown in appendix 12.6A. Recalling (12.5), (12.7), (12.8) and (12.18), (12.19) yields

$$\dot{s}_2(t) = f_2\bigg(\beta(l(t),t) - \int_0^{l(t)} \mathcal{D}(l(t),y)\alpha(y,t)dy$$

$$- \int_0^{l(t)} \mathcal{M}(l(t),y)\beta(y,t)dy - \mathcal{J}(l(t))X(t), s_2(t), \alpha(l(t),t)\bigg) + z(t), \tag{12.27}$$

$$\dot{z}(t) = c_4 z(t) + r\alpha(l(t),t) + U(t). \tag{12.28}$$

The equation set (12.26)–(12.28) is the right boundary condition of the intermediate system in the form of several ODEs regulated by the control input $U(t)$.

The equations (12.26)–(12.28) governing the variables $(\beta(l(t),t), s_2(t), z(t))$ are a cascade of ODEs converted from the equations (12.5)–(12.8) for $(s_1(t), s_2, z(t))$ via transformation (12.10), (12.11). The system (12.26), (12.27) is a second-order nonlinear ODE for $(\beta(l(t),t), s_2(t))$ with perturbations $\mathcal{F}$. Equation (12.28) is a first-order linear ODE $z(t)$ with a perturbation $\alpha(l(t),t)$.

Through the backstepping transformations (12.10), (12.11), the original $(u(x,t)$, $v(x,t)$, $X(t)$, $s_1(t)$, $s_2(t)$, $z(t))$-system in (12.1)–(12.8) is converted to the intermediate $(\alpha(x,t)$, $\beta(x,t)$, $X(t)$, $\beta(l(t),t)$, $s_2(t)$, $z(t))$-system in (12.20)–(12.23), (12.26)–(12.28). Next, we propose a backstepping design for the ODEs for $(\beta(l(t),t)$, $s_2(t)$, $z(t))$ in (12.26)–(12.28) at the right boundary of the intermediate system.

Backstepping for ODEs in the Input Channel

The following backstepping transformation for the $(\beta(l(t),t)$, $s_2(t))$-system (12.26), (12.27) is introduced:

$$y_1(t) = \beta(l(t),t), \tag{12.29}$$
$$y_2(t) = s_2(t) + \tau_1(t), \tag{12.30}$$

where $\tau_1(t)$ is to be chosen in the following steps as the virtual control law in the ODE backstepping method.

Step 1. We consider the Lyapunov function candidate

$$V_{y1} = \frac{1}{2}y_1(t)^2.$$

Taking the derivative of V_{y1}, recalling (12.22), (12.26), and (12.30), we obtain

$$\dot{V}_{y1} = y_1(t)\dot{y}_1(t) = y_1(t)\dot{\beta}(l(t),t)$$
$$= y_1(t)\big(c_3y_2(t) - c_3\tau_1(t) + f_1 \mid \mathcal{F}\big). \tag{12.31}$$

The arguments of f_1 and $\mathcal{F}$ are omitted in (12.31), which are the same as those in (12.26).

We choose

$$\tau_1(t) = \frac{\bar{c}_1}{c_3}y_1(t), \tag{12.32}$$

where $\bar{c}_1$ is a positive constant to be determined later.

Substituting (12.32) into (12.31) yields

$$\dot{V}_{y1} = -\bar{c}_1y_1(t)^2 + c_3y_1(t)y_2(t) + y_1(t)f_1 + y_1(t)\mathcal{F}. \tag{12.33}$$

Step 2. A Lyapunov function candidate for $y_1(t), y_2(t)$ is considered to be

$$V_y = V_{y1} + \frac{1}{2}y_2(t)^2 = \frac{1}{2}y_1(t)^2 + \frac{1}{2}y_2(t)^2. \tag{12.34}$$

Taking the derivative of (12.34), we get

$$\dot{V}_y = -\bar{c}_1y_1(t)^2 + c_3y_1(t)y_2(t) + y_1(t)f_1$$
$$+ y_1(t)\mathcal{F} + y_2(t)\big(f_2 + z(t) + \dot{\tau}_1\big), \tag{12.35}$$

where (12.30) and (12.27) are used, and the argument omitted in f_2 is the same as that in (12.27).

Step 3. Define a new variable $\mathcal{E}(t)$ as

$$\mathcal{E}(t) = z(t) + \bar{c}_2y_2(t) + c_3y_1(t), \tag{12.36}$$

where the positive constant $\bar{c}_2$ is to be determined later.

Inserting (12.36) into (12.35) to replace $z(t)$, we obtain

$$\dot{V}_y = -\bar{c}_1 y_1(t)^2 - \bar{c}_2 y_2(t)^2 + y_1(t)f_1 + y_1(t)\mathcal{F}$$
$$+ y_2(t)\mathcal{E}(t) + y_2(t)f_2 + \frac{\bar{c}_1}{c_3} y_2(t)\dot{y}_1(t). \tag{12.37}$$

Using (12.36), we write (12.28) as

$$\dot{\mathcal{E}} = c_4 \mathcal{E}(t) + r\alpha(l(t), t) + \bar{c}_2 \dot{y}_2(t) + c_3 \dot{y}_1(t)$$
$$- c_4 \bar{c}_2 y_2(t) - c_4 c_3 y_1(t) + U(t). \tag{12.38}$$

Choosing $U(t)$ in (12.38) as

$$U(t) = -\bar{a}_0 \mathcal{E}(t) - r\alpha(l(t), t) + c_4 \bar{c}_2 y_2(t) + c_4 c_3 y_1(t), \tag{12.39}$$

we then have

$$\dot{\mathcal{E}}(t) = -k_{\mathcal{E}} \mathcal{E}(t) + \bar{c}_2 \dot{y}_2(t) + c_3 \dot{y}_1(t), \tag{12.40}$$

where

$$k_{\mathcal{E}} = \bar{a}_0 - c_4 > 0 \tag{12.41}$$

by choosing the control gain $\bar{a}_0$.

Through the transformations (12.10), (12.11), (12.29), (12.30), and (12.36), the original $(u(x,t),\, v(x,t),\, X(t),\, s_1(t),\, s_2(t),\, z(t))$-system is converted to the target $(\alpha(x,t),\, \beta(x,t),\, X(t),\, y_1(t),\, y_2(t),\, \mathcal{E}(t))$-system. The exponential stability of the target system will be ensured in the following Lyapunov analysis by appropriately choosing the control parameters $\bar{c}_1, \bar{c}_2, \bar{a}_0$.

Stability Analysis of the State-Feedback Closed-Loop System

Substituting (12.36), (12.32), (12.29), (12.30), (12.10), (12.11) into (12.39), we express the controller in terms of the original states, as

$$U(t) = -\bar{a}_0 z(t) + (c_4 - \bar{a}_0)\bar{c}_2 s_2(t) - ru(l(t), t)$$
$$+ (c_4 - \bar{a}_0)\left(\frac{\bar{c}_1 \bar{c}_2}{c_3} + c_3\right)\left(s_1(t) - \int_0^{l(t)} \psi(l(t), y)u(y, t)dy\right)$$
$$- \int_0^{l(t)} \phi(l(t), y)v(y, t)dy - \gamma(l(t))X(t)\Bigg). \tag{12.42}$$

The control parameters $\bar{c}_1, \bar{c}_2, \bar{a}_0$ will be determined in the stability analysis. Because the control law (12.42) uses the signal $u(l(t), t)$, in order to ensure that the control law is sufficiently regular, we will require the initial value $u(x, 0)$ to be in $H^1(0, L)$, where the positive constant L, given in assumption 12.3, is the maximum value of the time-varying PDE domain.

Theorem 12.1. *For all initial values* $(u(x, 0), v(x, 0)) \in H^1(0, L)$, *with some* $\bar{c}_1$, $\bar{c}_2, \bar{a}_0$, *the closed-loop system consisting of the plant* (12.1)–(12.8) *and the control law* (12.42) *is exponentially stable in the sense that there exist the positive constants*

Υ_1, λ_1 *such that*

$$\Omega_a(t) \leq \Upsilon_1 \Omega_a(0) e^{-\lambda_1 t}, \tag{12.43}$$

where

$$\Omega_a(t) = \|u(\cdot, t)\|^2 + \|v(\cdot, t)\|^2 + |X(t)|^2 + s_1(t)^2 + s_2(t)^2 + z(t)^2. \tag{12.44}$$

Proof. We start by studying the stability of the target system. The equivalent stability property between the target system and the original system is ensured due to the invertibility of the transformations (12.10), (12.11), (12.29), (12.30), and (12.36).

First, we produce the stability proof of the target system via Lyapunov analysis of the PDE-ODE subsystem. Second, combining the Lyapunov analysis for the ODEs under the input channel in the backstepping design, Lyapunov analysis of the overall system is provided, through which the control parameters $\bar{c}_1, \bar{c}_2, \bar{a}_0$ in the control law (12.42) are determined.

a) Lyapunov analysis for the PDE-ODE subsystem-$(\alpha(x,t), \beta(x,t), X(t))$: Consider the Lyapunov function

$$V_1(t) = X^T(t) P_1 X(t) + \frac{a_1}{2} \int_0^{l(t)} e^{\delta_1 x} \beta(x,t)^2 dx$$

$$+ \frac{b_1}{2} \int_0^{l(t)} e^{-\delta_1 x} \alpha(x,t)^2 dx, \tag{12.45}$$

where $P_1 = P_1^T > 0$ is the solution to the Lyapunov equation

$$P_1(A + B\kappa) + (A + B\kappa)^T P_1 = -Q_1$$

for some $Q_1 = Q_1^T > 0$. The positive parameters a_1, b_1, δ_1 shall be chosen later.

Taking the derivative of $V_1(t)$, we arrive at

$$\dot{V}_1(t) \leq -\eta_1 |X(t)|^2 - \eta_2 \beta(0,t)^2 - \eta_3 \int_0^{l(t)} \beta(x,t)^2 dx$$

$$- \eta_4 \int_0^{l(t)} \alpha(x,t)^2 dx - \eta_5 \alpha(l(t),t)^2 + \eta_6 \beta(l(t),t)^2, \tag{12.46}$$

where the detailed process of calculating $\dot{V}_1(t)$ is shown in appendix 12.6B. In this appendix, the choices of a_1, b_1, δ_1 and the expressions of positive constants η_1, η_2, η_3, η_4 are also given. Defining

$$v_{\max} = \max_{t \in [0,\infty)} \left\{ \left| \dot{l}(t) \right| \right\},$$

we know, by recalling assumption 12.4, that

$$\eta_5 = (p_1 - v_{\max}) \frac{b_1}{2} e^{-\delta_1 L} > 0,$$

as well as that

$$\eta_6 = (p_2 + v_{\max}) \frac{a_1}{2} e^{\delta_1 L} > 0.$$

b) Lyapunov analysis for the overall system: Consider the Lyapunov function

$$V(t) = V_1(t) + V_y(t) + \frac{1}{2}\mathcal{E}(t)^2, \tag{12.47}$$

where V_y was introduced in (12.34). Defining

$$\begin{aligned}
\Omega_1(t) = &\|\beta(\cdot,t)\|^2 + \|\alpha(\cdot,t)\|^2 + |X(t)|^2 \\
&+ y_1(t)^2 + y_2(t)^2 + \mathcal{E}(t)^2,
\end{aligned} \tag{12.48}$$

we get

$$\theta_{1a}\Omega_1(t) \leq V(t) \leq \theta_{1b}\Omega_1(t) \tag{12.49}$$

for some positive constants θ_{1a} and θ_{1b}.

Taking the derivative of (12.47), using (12.46), (12.40) and (12.37) with (12.149)–(12.156) in appendix 12.6A, recalling assumptions 12.1, 12.2, 12.4, we show that

$$\dot{V}(t) \leq -\lambda V(t) - \hat{\eta}_0 \beta(0,t)^2 - \hat{\eta}_1 \alpha(l(t),t)^2 \tag{12.50}$$

for some positive λ, where $\hat{\eta}_0$, $\hat{\eta}_1$ are positive constants given as (12.181), (12.182) in appendix. The detailed process of calculating $\dot{V}(t)$ is shown in appendix 12.6C, where the choices of the control parameters $\bar{c}_1$, $\bar{c}_2$, a_0 in the ODE backstepping are chosen to dominate the PDE perturbations.

We thus have

$$V(t) \leq V(0)e^{-\lambda t}. \tag{12.51}$$

It then follows that

$$\Omega_1(t) \leq \frac{\theta_{1b}}{\theta_{1a}}\Omega_1(0)e^{-\lambda t}$$

by recalling (12.49).

By defining

$$\Xi(t) = \|u(\cdot,t)\|^2 + \|v(\cdot,t)\|^2 + |X(t)|^2 + s_1(t)^2 + s_2(t)^2 + z(t)^2 \tag{12.52}$$

and applying the Cauchy-Schwarz inequality and transformations (12.10), (12.11), (12.18), (12.19), (12.29), (12.30), and (12.36), it is straightforward to obtain

$$\bar{\theta}_{1a}\Xi(t) \leq \Omega_1(t) \leq \bar{\theta}_{1b}\Xi(t) \tag{12.53}$$

for some positive $\bar{\theta}_{1a}$ and $\bar{\theta}_{1b}$. Therefore, we get

$$\Xi(t) \leq \frac{\theta_{1b}\bar{\theta}_{1b}}{\theta_{1a}\bar{\theta}_{1a}}\Xi(0)e^{-\lambda t}. \tag{12.54}$$

Thus, (12.43) is achieved with

$$\Upsilon_1 = \frac{\theta_{1b}\bar{\theta}_{1b}}{\theta_{1a}\bar{\theta}_{1a}}, \quad \lambda_1 = \lambda. \tag{12.55}$$

The proof of theorem 12.1 is complete.   □

In theorem 12.1, we proved that all PDEs and ODEs are exponentially stable in the closed-loop system including the plant (12.1)–(12.8) and the controller (12.42). Moreover, next we prove that the controller $U(t)$ (12.42) in the closed-loop system is also bounded and exponentially convergent to zero.

According to (12.42) and the exponential stability result proved in theorem 12.1, the exponential convergence of the control input requires, additionally, the exponential convergence of the signal $u(l(t), t)$, which can be obtained by proving the exponential stability estimate of $\|u_x(\cdot, t)\| + \|v_x(\cdot, t)\|$. Before proving the exponential convergence of the control input, we state a lemma first.

Lemma 12.1. *With arbitrary initial data* $(u(x, 0), v(x, 0)) \in H^1(0, L)$, *the exponential stability estimate of the closed-loop system* $(u(x, t), v(x, t))$ *is obtained in the sense that there exist the positive constants* Υ_{1a} *and* λ_{1a} *such that*

$$\|u_x(\cdot, t)\|^2 + \|v_x(\cdot, t)\|^2$$
$$\leq \Upsilon_{1a} \left(\Xi(0) + \|u_x(\cdot, 0)\|^2 + \|v_x(\cdot, 0)\|^2 \right) e^{-\lambda_{1a} t}, \tag{12.56}$$

where $\Xi(t)$ *is given in* (12.52).

The proof of lemma 12.1 is shown in appendix 12.6D. Lemma 12.1 will be used in proving the exponential convergence and boundedness of the controller (12.42) in the following theorem.

Theorem 12.2. *In the closed-loop system, which consists of the plant* (12.1)–(12.8) *and the controller given by* (12.42), *there exist the positive constants* λ_2 *and* Υ_2, *which ensure that* $U(t)$ *is bounded and exponentially convergent to zero in the sense of*

$$|U(t)| \leq \Upsilon_2 \bigg(\|u(\cdot, 0)\|^2 + \|v(\cdot, 0)\|^2 + |X(0)|^2 + s_1(0)^2$$
$$+ s_2(0)^2 + z(0)^2 + \|u_x(\cdot, 0)\|^2 + \|v_x(\cdot, 0)\|^2 \bigg)^{\frac{1}{2}} e^{-\lambda_2 t}. \tag{12.57}$$

Proof. The proof is shown in appendix 12.6E. □

12.3 OBSERVER DESIGN AND STABILITY ANALYSIS

In section 12.2, a state-feedback controller that requires distributed states is designed to stabilize the original system exponentially. However, it is difficult to measure distributed states in practice. We propose an output-feedback control law, which only requires measurements $u(l(t), t), v(l(t), t), z(t)$ at the controlled boundary of the PDE—that is, a "collocated" output-feedback law—based on a state observer designed in this section. The observer gains are determined in two transformation steps from the observer error system to an intermediate observer error system, and then to a target observer error system. The exponential stability of the observer error system is inferred from the stability of the target system and the invertibility of the transformations.

Observer Design

Using the measurements $u(l(t),t)$, $v(l(t),t)$, $z(t)$, the observer is designed as

$$\dot{\hat{X}}(t) = A\hat{X}(t) + B\hat{v}(0,t) + \Gamma_0(t)(u(l(t),t) - \hat{u}(l(t),t)), \tag{12.58}$$

$$\hat{u}_t(x,t) = -p_1\hat{u}_x(x,t) + c_1\hat{v}(x,t) + \Gamma_1(x,t)(u(l(t),t) - \hat{u}(l(t),t)), \tag{12.59}$$

$$\hat{v}_t(x,t) = p_2\hat{v}_x(x,t) + c_2\hat{u}(x,t) + \Gamma_2(x,t)(u(l(t),t) - \hat{u}(l(t),t)), \tag{12.60}$$

$$\hat{u}(0,t) = q\hat{v}(0,t) + C\hat{X}(t), \tag{12.61}$$

$$\hat{v}(l(t),t) = v(l(t),t), \tag{12.62}$$

$$\dot{\hat{s}}_1(t) = c_3\hat{s}_2(t) + f_1\left(\hat{s}_1(t), \int_0^{l(t)} \hat{u}(y,t)dy\right) + \mu_1(v(l(t),t) - \hat{s}_1(t))), \tag{12.63}$$

$$\dot{\hat{s}}_2(t) = f_2(\hat{s}_1(t), \hat{s}_2(t), \hat{u}(l(t),t)) + z(t) + \mu_2(v(l(t),t) - \hat{s}_1(t))), \tag{12.64}$$

$$\dot{\hat{z}}(t) = c_4\hat{z}(t) + ru(l(t),t) + \mu_3(z(t) - \hat{z}(t)) + U(t), \tag{12.65}$$

where $\Gamma_0(t), \Gamma_1(x,t), \Gamma_2(x,t), \mu_1, \mu_2, \mu_3$ are observer gains to be determined later. The initial values $\hat{u}(x,0), \hat{v}(x,0)$ are required to be in $H_1(0,L)$ to be consistent with section 12.2. Define observer errors as

$$[\tilde{X}(t), \tilde{u}(x,t), \tilde{v}(x,t), \tilde{s}_1(t), \tilde{s}_2(t), \tilde{z}(t)]$$
$$= [X(t), u(x,t), v(x,t), s_1(t), s_2(t), z(t)]$$
$$- [\hat{X}(t), \hat{u}(x,t), \hat{v}(x,t), \hat{s}_1(t), \hat{s}_2(t), \hat{z}(t)]. \tag{12.66}$$

According to (12.58)–(12.65) and (12.1)–(12.8), the observer error dynamics can be obtained as

$$\dot{\tilde{X}}(t) = A\tilde{X}(t) + B\tilde{v}(0,t) - \Gamma_0(t)\tilde{u}(l(t),t), \tag{12.67}$$

$$\tilde{u}_t(x,t) = -p_1\tilde{u}_x(x,t) + c_1\tilde{v}(x,t) - \Gamma_1(x,t)\tilde{u}(l(t),t), \tag{12.68}$$

$$\tilde{v}_t(x,t) = p_2\tilde{v}_x(x,t) + c_2\tilde{u}(x,t) - \Gamma_2(x,t)\tilde{u}(l(t),t), \tag{12.69}$$

$$\tilde{u}(0,t) = q\tilde{v}(0,t) + C\tilde{X}(t), \tag{12.70}$$

$$\tilde{v}(l(t),t) = 0, \tag{12.71}$$

$$\dot{\tilde{s}}_1(t) = c_3\tilde{s}_2(t) + \tilde{f}_1 - \mu_1\tilde{s}_1(t), \tag{12.72}$$

$$\dot{\tilde{s}}_2(t) = \tilde{f}_2 - \mu_2\tilde{s}_1(t), \tag{12.73}$$

$$\dot{\tilde{z}}(t) = -k_z\tilde{z}(t), \tag{12.74}$$

where

$$k_z = \mu_3 - c_4 > 0$$

by choosing the control parameter μ_3, and where

$$\tilde{f}_1 = f_1\left(s_1(t), \int_0^{l(t)} u(y,t)dy\right) - f_1\left(\hat{s}_1(t), \int_0^{l(t)} \hat{u}(y,t)dy\right), \tag{12.75}$$

$$\tilde{f}_2 = f_2\left(s_1(t), s_2(t), u(l(t), t)\right) - f_2\left(\hat{s}_1(t), \hat{s}_2(t), \hat{u}(l(t), t)\right). \tag{12.76}$$

Defining

$$\tilde{S}(t) = [\tilde{s}_1(t), \tilde{s}_2(t)]^T, \tag{12.77}$$

we rewrite the system (12.72), (12.73) as

$$\dot{\tilde{S}}(t) = (A_s - \mathcal{B}C_2)\tilde{S}(t) + \left[\tilde{f}_1, \tilde{f}_2\right]^T, \tag{12.78}$$

where

$$A_s = \begin{pmatrix} 0 & c_3 \\ 0 & 0 \end{pmatrix}, \quad C_2 = [1, 0], \quad \mathcal{B} = [\mu_1, \mu_2]^T. \tag{12.79}$$

The matrix $A_s - \mathcal{B}C_2$ can be made Hurwitz by choosing $\mathcal{B}$ because (A_s, C_2) is observable. In order to remove the potentially destabilizing feedback terms—that is, the $\tilde{u}$-terms in $\tilde{v}$ in (12.69), as mentioned in chapter 9—we apply, as in [96], the invertible backstepping transformation for the PDE states $(\tilde{u}, \tilde{v})$,

$$\tilde{u}(x, t) = \tilde{\alpha}(x, t) - \int_x^{l(t)} \bar{\phi}(x, y)\tilde{\alpha}(y, t)dy, \tag{12.80}$$

$$\tilde{v}(x, t) = \tilde{\beta}(x, t) - \int_x^{l(t)} \bar{\psi}(x, y)\tilde{\alpha}(y, t)dy, \tag{12.81}$$

to convert (12.67)–(12.74) to the intermediate observer error system as

$$\dot{\tilde{X}}(t) = A\tilde{X}(t) + B\tilde{\beta}(0, t) - B\int_0^{l(t)} \bar{\psi}(0, y)\tilde{\alpha}(y, t)dy - \Gamma_0(t)\tilde{\alpha}(l(t), t), \tag{12.82}$$

$$\tilde{\alpha}_t(x, t) = -p_1\tilde{\alpha}_x(x, t) + \int_x^{l(t)} \bar{M}(x, y)\tilde{\beta}(y, t)dy + c_1\tilde{\beta}(x, t), \tag{12.83}$$

$$\tilde{\beta}_t(x, t) = p_2\tilde{\beta}_x(x, t) + \int_x^{l(t)} \bar{N}(x, y)\tilde{\beta}(y, t)dy, \tag{12.84}$$

$$\tilde{\alpha}(0, t) = q\tilde{\beta}(0, t) + C\tilde{X}(t) + \int_0^{l(t)} (\bar{\phi}(0, y) - q\bar{\psi}(0, y))\tilde{\alpha}(y, t)dy, \tag{12.85}$$

$$\tilde{\beta}(l(t), t) = 0, \tag{12.86}$$

$$\dot{\tilde{S}}(t) = (A_s - \mathcal{B}C_2)\tilde{S}(t) + [\tilde{f}_1, \tilde{f}_2]^T, \tag{12.87}$$

$$\dot{\tilde{z}}(t) = -k_z\tilde{z}(t). \tag{12.88}$$

By matching (12.67)–(12.71) and (12.82)–(12.86), the kernel functions $\bar{\phi}, \bar{\psi}$ on $\mathcal{D}_1 = \{0 \leq x \leq y \leq l(t)\}$ should satisfy

$$-p_1\bar{\phi}_x(x, y) - p_1\bar{\phi}_y(x, y) - c_1\bar{\psi}(x, y) = 0, \tag{12.89}$$

$$\bar{\psi}(x, x) = \frac{c_2}{p_1 + p_2}, \tag{12.90}$$

$$-p_1\bar{\psi}_y(x, y) + p_2\bar{\psi}_x(x, y) - c_2\bar{\phi}(x, y) = 0. \tag{12.91}$$

The boundary condition of $\bar{\phi}$ is set as

$$\bar{\phi}(0, y) = q\bar{\psi}(0, y) - CK_0(y), \tag{12.92}$$

where $K_0(x)$ is shown later. The choice of (12.92) will be clear later.

The functions $\bar{M}(x, y), \bar{N}(x, y)$ in (12.82)–(12.88) satisfy

$$\bar{M}(x, y) = \int_x^y \bar{\phi}(x, z)\bar{M}(z, y)dz + c_1\bar{\phi}(x, y), \tag{12.93}$$

$$\bar{N}(x, y) = \int_x^y \bar{\psi}(x, z)\bar{M}(z, y)dz + c_1\bar{\psi}(x, y). \tag{12.94}$$

The observer gains $\Gamma_1(x, t)$ and $\Gamma_2(x, t)$ are obtained as

$$\Gamma_1(x, t) = \dot{l}(t)\bar{\phi}(x, l(t)) - p_1\bar{\phi}(x, l(t)), \tag{12.95}$$

$$\Gamma_2(x, t) = \dot{l}(t)\bar{\psi}(x, l(t)) - p_1\bar{\psi}(x, l(t)). \tag{12.96}$$

In order to decouple the ODE (12.82) from the PDE state $\tilde{\alpha}$ ($\tilde{\beta}$ reaches to zero after a finite time because of (12.84), (12.86)) and to make the system matrix in the ODE (12.82) Hurwitz, with the observer gain $\Gamma_0(t)$ in (12.82) still to be defined, we apply another transformation as

$$\tilde{Y}(t) = \tilde{X}(t) - \int_0^{l(t)} K_0(x)\tilde{\alpha}(x, t)dx - \int_0^{l(t)} K_1(x)\tilde{\beta}(x, t)dx \tag{12.97}$$

to convert (12.82) into

$$\dot{\tilde{Y}}(t) = (A - L_0 C)\tilde{Y}(t) - \int_0^{l(t)} \left[\int_0^x K_0(y)\bar{M}(y, x)dy \right.$$
$$\left. + \int_0^x K_1(y)\bar{N}(y, x)dy \right] \tilde{\beta}(x, t)dx, \tag{12.98}$$

where the matrix $A - L_0 C$ is made Hurwitz by recalling that (A, C) is observable and suitably choosing L_0, and where $K_0(x)$, $K_1(x)$ are determined next.

Substituting (12.97) into (12.98), with (12.82)–(12.86), using integration by parts and a change of the order of integration in a double integral, we obtain

$$\left[K_0(l(t))p_1 - \dot{l}(t)K_0(l(t)) - \Gamma_0(t) \right] \tilde{\alpha}(l(t), t)$$

$$- \int_0^{l(t)} \left[K_0'(x)p_1 - AK_0(x) + B\bar{\psi}(0, x) \right] \tilde{\alpha}(x, t)dx$$

$$+ (L_0 - K_0(0)p_1)\tilde{\alpha}(0, t) + \int_0^{l(t)} \left[- K_0(x)c_1 + K_1'(x)p_2 \right.$$
$$\left. + (A - L_0 C)K_1(x) \right] \tilde{\beta}(x, t)dx$$

$$+ [K_1(0)p_2 - L_0 q + B]\tilde{\beta}(0, t) = 0. \tag{12.99}$$

For (12.99) to hold, $K_0(x)$, $K_1(x)$ should satisfy

$$K_0'(x)p_1 - AK_0(x) + B\bar{\psi}(0, x) = 0, \tag{12.100}$$

$$K_0(0) = \frac{L_0}{p_1}, \tag{12.101}$$

$$K_1'(x)p_2 + (A - L_0C)K_1(x) - K_0(x)c_1 = 0, \tag{12.102}$$

$$K_1(0) = \frac{L_0 q - B}{p_2}. \tag{12.103}$$

Equations (12.89)–(12.92), (12.100)–(12.103) for the kernels $\bar{\phi}(x,y)$, $\bar{\psi}(x,y)$, $K_0(x)$, $K_1(x)$ are well-posed, which is established using the fact that, after swapping positions of arguments, like B.9–B.10 in [6]—that is, by changing the domain $\mathcal{D}_1$ to $\mathcal{D}$— the conditions (12.89)–(12.92), (12.100), (12.101) on $\bar{\phi}, \bar{\psi}, K_0$ have the same form as the conditions (12.12)–(12.17) on the kernels ϕ, ψ, γ, which have been proved to be well-posed in chapters 6 and 9. The explicit solution of $K_1(x)$ is then easy to obtain for the initial value problem (12.102), (12.103).

The observer gain $\Gamma_0(t)$ is obtained as

$$\Gamma_0(t) = -\dot{l}(t)K_0(l(t)) + K_0(l(t))p_1. \tag{12.104}$$

The target observer error system thus can be written as

$$\dot{\tilde{Y}}(t) = (A - L_0C)\tilde{Y}(t) - \int_0^{l(t)} \left[\int_0^x K_0(y)\bar{M}(y,x)dy \right.$$
$$\left. + \int_0^x K_1(y)\bar{N}(y,x)dy \right] \tilde{\beta}(x,t)dx, \tag{12.105}$$

$$\tilde{\alpha}_t(x,t) = -p_1\tilde{\alpha}_x(x,t) + \int_x^{l(t)} \bar{M}(x,y)\tilde{\beta}(y,t)dy + c_1\tilde{\beta}(x,t), \tag{12.106}$$

$$\tilde{\beta}_t(x,l) = p_2\tilde{\beta}_x(x,t) + \int_x^{l(t)} \bar{N}(x,y)\tilde{\beta}(y,t)dy, \tag{12.107}$$

$$\tilde{\alpha}(0,t) = q\tilde{\beta}(0,t) + C\tilde{Y}(t) + \int_0^{l(t)} CK_1(y)\tilde{\beta}(y,t)dy, \tag{12.108}$$

$$\tilde{\beta}(l(t),t) = 0, \tag{12.109}$$

$$\dot{\tilde{S}}(t) = (A_s - \mathcal{B}C_2)\tilde{S}(t) + \left[\tilde{f}_1, \tilde{f}_2\right]^T, \tag{12.110}$$

$$\dot{\tilde{z}}(t) = -k_z\tilde{z}(t), \tag{12.111}$$

where (12.92) has been used in forming (12.108).

The following theorem shows the exponential stability of the observer error system (12.67)–(12.74), which is obtained through the stability analysis of the target observer error system (12.105)–(12.111) and by applying the invertibility of the transformations. The initial data $(\tilde{u}(x,0), \tilde{v}(x,0))$ of the observer error system, defined by the initial conditions of the plant and the observer via (12.66), belong to $H^1(0,L)$.

Stability Analysis of the Observer Error System

Theorem 12.3. *Consider the observer system* (12.58)–(12.65) *with observer gains* $\Gamma_0(t)$ (12.104), $\Gamma_1(x,t)$ (12.95), $\Gamma_2(x,t)$ (12.96). *The observer error system* (12.67)– (12.74) *is exponentially stable in the sense that there exist the positive constants* Υ_e, λ_e *such that*

$$\Omega_e(t) \leq \Upsilon_e \Omega_e(0) e^{-\lambda_e t}, \tag{12.112}$$

where

$$\Omega_e(t) = \|\tilde{u}(\cdot, t)\|^2 + \|\tilde{v}(\cdot, t)\|^2 + \left|\tilde{X}(t)\right|^2 + \tilde{s}_1(t)^2 + \tilde{s}_2(t)^2 + \tilde{z}(t)^2. \tag{12.113}$$

Proof. 1) Analysis for the observer error subsystems of $(\tilde{u}(x,t), \tilde{v}(x,t), \tilde{X}(t), \tilde{z}(t))$: The ODE (12.110) with the state $\tilde{z}(t)$ is exponentially stable because $k_z > 0$. From (12.107), (12.109), the $\tilde{\beta}$-dynamics are independent of $\tilde{\alpha}$ and $\tilde{\beta}(x,t) \equiv 0$ after $t_{f0} = \frac{L}{p_2}$—that is, when the boundary condition (12.109) has propagated through the whole domain. The subsystem (12.105)–(12.109) becomes

$$\dot{\tilde{Y}}(t) = (A - L_0 C)\tilde{Y}(t), \tag{12.114}$$

$$\tilde{\alpha}_t(x,t) = -p_1 \tilde{\alpha}_x(x,t), \tag{12.115}$$

$$\tilde{\alpha}(0,t) = C\tilde{Y}(t) \tag{12.116}$$

for $t \geq t_{f0}$. The signal $\left|\tilde{Y}(t)\right|$ is exponentially convergent to zero because $A - L_0 C$ in the ODE (12.114) is Hurwitz. Define

$$V_a(t) = \tilde{Y}(t)^T P_a \tilde{Y}(t) + \frac{b_a}{2} \int_0^{l(t)} e^{-x} \tilde{\alpha}(x,t)^2 dx, \tag{12.117}$$

where b_a is a positive constant, and $P_a = P_a^T > 0$ is the solution to the Lyapunov equation

$$P_a(A - L_0 C) + (A - L_0 C)^T P_a = -Q_a$$

for some $Q_a = Q_a^T > 0$.

Taking the derivative of $V_a(t)$ along (12.114)–(12.116), we get

$$\dot{V}_a(t) \leq -\lambda_{\min}(Q_a)\tilde{Y}(t)^2 - p_1 b_a \int_0^{l(t)} e^{-x} \tilde{\alpha}(x,t)\tilde{\alpha}_x(x,t)dx$$

$$+ \frac{b_a}{2}\dot{l}(t)e^{-l(t)}\tilde{\alpha}((t),t)^2$$

$$\leq -\left(\lambda_{\min}(Q_a) - \frac{1}{2}p_1 b_a |C|^2\right)\tilde{Y}(t)^2 - \frac{1}{2}b_a(p_1 - \dot{l}(t))e^{-l(t)}\tilde{\alpha}(l(t),t)^2$$

$$- \frac{1}{2}p_1 b_a \int_0^{l(t)} e^{-x}\tilde{\alpha}(x,t)^2 dx. \tag{12.118}$$

Choosing

$$b_a < \frac{2\lambda_{\min}(Q_a)}{p_1 |C|^2}$$

and recalling assumption 12.4 yields

$$\dot{V}_a(t) \leq -\lambda_a V_a(t) - \lambda_{a1}\tilde{\alpha}(l(t),t)^2 \tag{12.119}$$

for some positive λ_a, λ_{a1}. The exponential stability result in the sense of $|\tilde{Y}(t)|^2 + \|\tilde{\alpha}(\cdot, t)\|^2 + \|\tilde{\beta}(\cdot, t)\|^2$ is obtained.

Even though $\tilde{\beta}(x,t) \equiv 0$ and (12.114)–(12.116) holds for $t \geq t_{f0}$ (if $\tilde{\beta}(x,0) = 0$, $\tilde{\beta}(x,t) \equiv 0$ and (12.114)–(12.116) hold at $t=0$, and the obtained exponential stability begins from $t=0$), the obtained exponential stability also holds from the beginning at $t=0$, because any transient in the finite time $[0, t_{f0})$ can be bounded by an exponentially decaying signal with an arbitrary decay rate and an appropriate overshoot coefficient.

According to the invertible transformation (12.80), (12.81), (12.97), we obtain the exponential stability in the sense of $|\tilde{X}(t)|^2 + \|\tilde{u}(\cdot,t)\|^2 + \|\tilde{v}(\cdot,t)\|^2$.

2) Analysis for the observer error subsystem of $\tilde{S}(t)$: Next we conduct the stability analysis for the $\tilde{S}(t)$-ODE in (12.110). Consider the Lyapunov function

$$V_s(t) = \tilde{S}(t)^T P_0 \tilde{S}(t), \tag{12.120}$$

where P_0 is a positive definite and symmetric solution of

$$(A_s - \mathcal{B}C_2)^T P_0 + P_0(A_s - \mathcal{B}C_2) + \bar{\gamma}^2 P_0^T P_0 + I^T < 0 \tag{12.121}$$

with

$$\bar{\gamma}^2 = \gamma_1^2 + 2\gamma_2^2, \tag{12.122}$$

and γ_1, γ_2 are positive Lipschitz constants shown in appendix 12.6F. The existence of the solution P_0 of the above equation (12.121) and the procedure to define the observer gain $\mathcal{B}$ are shown in section 4 in [144].

Taking the derivative of $V_s(t)$ (12.120), through the calculation process presented in appendix 12.6F, where assumption 12.2 is revoked, we achieve

$$\dot{V}_s(t) \leq -\sigma_s V_s(t) + \frac{\hat{\gamma}_1^2}{\gamma_1^2 + 2\gamma_2^2} \|\tilde{\alpha}(\cdot,t)\|^2 + \frac{\gamma_2^2}{\gamma_1^2 + 2\gamma_2^2} \tilde{\alpha}(l(t),t)^2 \tag{12.123}$$

for some positive σ_s.

Consider the Lyapunov function

$$V_e(t) = V_s(t) + R_{\tilde{\alpha}} V_a(t). \tag{12.124}$$

Taking the derivative of (12.124), recalling (12.123), (12.119), and choosing large enough positive constant $R_{\tilde{\alpha}}$, we obtain

$$\dot{V}_e(t) \leq -\sigma_s V_s(t) - \frac{1}{2} R_{\tilde{\alpha}} \lambda_a V_a(t) - \left(\frac{1}{4} R_{\tilde{\alpha}} \lambda_a b_a e^{-L} - \frac{\hat{\gamma}_1^2}{\gamma_1^2 + 2\gamma_2^2} \right) \|\tilde{\alpha}(\cdot,t)\|^2$$
$$- \left(R_{\tilde{\alpha}} \lambda_{a1} - \frac{\gamma_2^2}{\gamma_1^2 + 2\gamma_2^2} \right) \tilde{\alpha}(l(t),t)^2$$
$$\leq -\sigma_e V_e(t) \tag{12.125}$$

for some positive σ_e. Recalling (12.124), (12.120), we obtain that $|\tilde{S}(t)|^2$ is exponentially convergent to zero.

Finally, using the obtained exponential stability result in the sense of $\|\tilde{u}(\cdot,t)\|^2 + \|\tilde{v}(\cdot,t)\|^2 + |\tilde{X}(t)|^2 + |\tilde{z}(t)|^2$ in portion (1) and the exponential stability result of $|\tilde{S}(t)|^2$ in portion (2) with (12.77), the proof of theorem 12.3 is complete. □

Next, we state a lemma that shows the exponential stability estimates in the sense of $\|\tilde{v}_x(\cdot,t)\|^2 + \|\tilde{u}_x(\cdot,t)\|^2$.

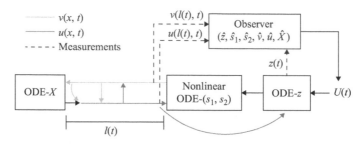

Figure 12.1. Block diagram of the output-feedback closed-loop system consisting of the plant (12.1)–(12.8), the observer (12.58)–(12.65), and the controller (12.127).

Lemma 12.2. *The exponential stability estimate of the observer error system* $(\tilde{u}(x,t), \tilde{v}(x,t))$ *is obtained in the sense that there exist the positive constants* $\Upsilon_{3a}, \lambda_{3a}$ *such that*

$$\|\tilde{v}_x(\cdot,t)\|^2 + \|\tilde{u}_x(\cdot,t)\|^2$$
$$\leq \Upsilon_{3a} \left(\Omega_e(0) + \|\tilde{u}_x(\cdot,0)\|^2 + \|\tilde{v}_x(\cdot,0)\|^2 \right) e^{-\lambda_{3a}t}. \tag{12.126}$$

Proof. Take the spatial derivative of (12.106), (12.107) and the time derivative of (12.105), (12.108), (12.109), where

$$\dot{\tilde{\beta}}(l(t),t) = \dot{l}(t)\tilde{\beta}_x(l(t),t) + \tilde{\beta}_t(l(t),t) = \left(\dot{l}(t) + p_2\right)\tilde{\beta}_x(l(t),t) = 0$$

is used. This identity results in $\tilde{\beta}_x(l(t),t) = 0$ because $\dot{l}(t) + p_2 \neq 0$, which follows from assumption 12.4. Through similar steps as in portion (1) of the proof of theorem 12.3, using the spatial derivative of the transformation (12.80), (12.81), recalling theorem 12.3, we obtain lemma 12.2. ☐

12.4 STABILITY OF THE OUTPUT-FEEDBACK CLOSED-LOOP SYSTEM

Replacing all the original states in the state-feedback controller (12.42) by the observer states, the output-feedback controller can be written as

$$U_{of}(t) = -\bar{a}_0\hat{z}(t) + (c_4 - \bar{a}_0)\bar{c}_2\hat{s}_2(t) - r\hat{u}(l(t),t)$$
$$+ (c_4 - \bar{a}_0)\left(\frac{\bar{c}_1\bar{c}_2}{c_3} + c_3\right)\left(\hat{s}_1(t) - \int_0^{l(t)} \psi(l(t),y)\hat{u}(y,t)dy\right)$$
$$- \int_0^{l(t)} \phi(l(t),y)\hat{v}(y,t)dy - \gamma(l(t))\hat{X}(t)\bigg). \tag{12.127}$$

The output-feedback closed-loop system consists of the plant (12.1)–(12.8), the observer (12.58)–(12.65), and the output-feedback controller (12.127). The block diagram of the output-feedback closed-loop system is shown in figure 12.1. The following theorem shows the exponential stability of the output-feedback closed-loop system and the exponential convergence of the controller (12.127).

Theorem 12.4. *The closed-loop system consisting of the plant* (12.1)–(12.8), *the observer* (12.58)–(12.65), *and the output-feedback controller* (12.127), *with arbitrary initial values* $(u(\cdot,0), v(\cdot,0)) \in H^1(0,L)$, $(\hat{u}(\cdot,0), \hat{v}(\cdot,0)) \in H^1(0,L)$, *has the following properties:*

1) There exist the positive constants Υ_4 *and* λ_4 *such that*

$$\Omega(t) \le \Upsilon_4 \Omega(0) e^{-\lambda_4 t}, \tag{12.128}$$

where

$$\begin{aligned}
\Omega(t) &= \|\hat{v}(\cdot,t)\|^2 + \|\hat{u}(\cdot,t)\|^2 + \left|\hat{X}(t)\right|^2 + \hat{s}_1(t)^2 + \hat{s}_2(t)^2 \\
&\quad + \hat{z}(t)^2 + \|v(\cdot,t)\|^2 + \|u(\cdot,t)\|^2 + |X(t)|^2 \\
&\quad + s_1(t)^2 + s_2(t)^2 + z(t)^2.
\end{aligned} \tag{12.129}$$

2) The output-feedback signal in (12.127) *is bounded and exponentially convergent to zero.*

Proof. 1) Rewrite (12.58)–(12.65) as

$$\dot{\hat{X}}(t) - A\hat{X}(t) + B\hat{v}(0,t) + \Gamma_0(t)\tilde{u}(l(t),t), \tag{12.130}$$

$$\hat{u}_t(x,t) = -p_1 \hat{u}_x(x,t) + c_1 \hat{v}(x,t) + \Gamma_1(x,t)\tilde{u}(l(t),t), \tag{12.131}$$

$$\hat{v}_t(x,t) = p_2 \hat{v}_x(x,t) + c_2 \hat{u}(x,t) + \Gamma_2(x,t)\tilde{u}(l(t),t), \tag{12.132}$$

$$\hat{u}(0,t) = q\hat{v}(0,t) + C\hat{X}(t), \tag{12.133}$$

$$\hat{v}(l(t),t) = \hat{s}_1(t) + \tilde{s}_1(t), \tag{12.134}$$

$$\dot{\hat{s}}_1(t) = c_3 \hat{s}_2(t) + f_1\left(\hat{s}_1(t), \int_0^{l(t)} \hat{u}(y,t)dy\right) + \mu_1 \tilde{s}_1(t), \tag{12.135}$$

$$\dot{\hat{s}}_2(t) = f_2(\hat{s}_1(t), \hat{s}_2(t), \hat{u}(l(t),t)) + \ddot{z}(t) + z(t) + \mu_2 \tilde{s}_1(t), \tag{12.136}$$

$$\dot{\hat{z}}(t) = c_4 \hat{z}(t) + r\hat{u}(l(t),t) + r\tilde{u}(l(t),t) + \mu_3 \tilde{z}(t) + U_{of}(t), \tag{12.137}$$

which has the same structure as the original system (12.1)–(12.8) plus the injections $\tilde{u}(l(t),t)$, $\tilde{s}_1(t)$, $\tilde{z}(t)$. Applying transformations (12.10), (12.11), (12.18), (12.19), (12.29), (12.30), and (12.36) (all states in the transformation should have a "hat" added on top, such as in "$\hat{u}$"), through the same steps as in section 12.2, we arrive at the target system of $(\hat{\alpha}, \hat{\beta}, \hat{X}, \hat{y}_1, \hat{y}_2, \hat{\mathcal{E}}, \tilde{u}(l(t),t), \tilde{z}(t), \tilde{s}_1(t), \dot{\tilde{s}}_1(t))$, whose portion $(\hat{\alpha}, \hat{\beta}, \hat{X}, \hat{y}_1, \hat{y}_2, \hat{\mathcal{E}})$ has the same structure as the exponentially stable target system in the state-feedback design but with several observer errors $\tilde{u}(l(t),t), \tilde{z}(t), \tilde{s}_1(t), \dot{\tilde{s}}_1(t)$ injected. Recalling theorem 12.3 and lemma 12.2, we show that $\tilde{u}(l(t),t), \tilde{z}(t), \tilde{s}_1(t)$ are exponentially convergent to zero. According to (12.72), (12.204), we obtain

$$\begin{aligned}
\dot{\tilde{s}}_1(t)^2 &\le 3c_3^2 \tilde{s}_2(t)^2 + 3\tilde{f}_1^2 + 3\mu_1^2 \tilde{s}_1(t)^2 \\
&\le 3c_3^2 \tilde{s}_2(t)^2 + 3\hat{\gamma}_1^2 \|\tilde{\alpha}(\cdot,t)\|^2 + 3(\mu_1^2 + \gamma_1^2)\tilde{s}_1(t)^2.
\end{aligned} \tag{12.138}$$

Thus, $\dot{\tilde{s}}_1(t)$ is also exponentially convergent to zero in light of theorem 12.3.

Define the Lyapunov function

$$V_{of} = \hat{X}(t)^T P_2 \hat{X}(t) + \frac{\bar{a}_1}{2} \int_0^{l(t)} e^{\bar{\delta}_1 x} \hat{\beta}(x,t)^2 dx + \frac{1}{2}\hat{\mathcal{E}}(t)^2$$
$$+ \frac{\bar{b}_1}{2} \int_0^{l(t)} e^{-\bar{\delta}_1 x} \hat{\alpha}(x,t)^2 dx + \frac{1}{2}\hat{y}_1(t)^2 + \frac{1}{2}\hat{y}_2(t)^2$$
$$+ R_{Ve} V_e(t) + R_z \tilde{z}(t)^2, \tag{12.139}$$

where $\bar{a}_1$, $\bar{b}_1$, $\bar{\delta}_1$, R_{Ve}, R_z are positive constants, and $P_2 = P_2^T > 0$ is the solution to the Lyapunov equation

$$P_2(A + B\kappa) + (A + B\kappa)^T P_2 = -Q_2$$

for some $Q_2 = Q_2^T > 0$.

Through the same steps as in theorem 12.1, using (12.138), (12.125), (12.111), we obtain

$$\dot{V}_{of} \le -\lambda_{of} V_{of}(t)$$

for some positive λ_{of}. We then obtain

$$\Omega_4(t) \le \Upsilon_{4a}\Omega_4(0)e^{-\lambda_{of}t}$$

for some positive Υ_{4a}, where

$$\Omega_4(t) = \|\hat{\alpha}(\cdot,t)\|^2 + \|\hat{\beta}(\cdot,t)\|^2 + \left|\hat{X}(t)\right|^2 + \hat{y}_1(t)^2 + \hat{y}_2(t)^2 + \hat{\mathcal{E}}(t)^2$$
$$+ \left|\tilde{S}(t)\right|^2 + \left|\tilde{Y}(t)\right|^2 + \|\tilde{\alpha}(\cdot,t)\|^2 + \|\tilde{\beta}(\cdot,t)\|^2 + \tilde{z}(t)^2. \tag{12.140}$$

Applying all transformations and their inverses, through the same steps as in (12.52)–(12.54), we get

$$\bar{\Omega}(t) \le \Upsilon_{4b}\bar{\Omega}(0)e^{-\lambda_{of}t}, \tag{12.141}$$

where Υ_{4b} is a positive constant, and

$$\bar{\Omega}(t) = \|\hat{u}(\cdot,t)\|^2 + \|\hat{v}(\cdot,t)\|^2 + \left|\hat{X}(t)\right|^2$$
$$+ \hat{s}_1(t)^2 + \hat{s}_2(t)^2 + \hat{z}(t)^2 + \|\tilde{u}(\cdot,t)\|^2 + \|\tilde{v}(\cdot,t)\|^2$$
$$+ \left|\tilde{X}(t)\right|^2 + \tilde{s}_1(t)^2 + \tilde{s}_2(t)^2 + \tilde{z}(t)^2.$$

Then recalling (12.66) and applying the Cauchy-Schwarz inequality, we obtain (12.128).

2) In order to prove the boundedness and exponential convergence of the output-feedback controller (12.127), having proved the above exponential stability results in property (1), we need to additionally prove the exponential convergence of $\hat{u}(l(t),t)$ to zero. This result can be obtained with an exponential stability estimate in the sense of $\|\hat{u}_x(\cdot,t)\| + \|\hat{v}_x(\cdot,t)\|$, which can be obtained through the same steps as lemma 12.1, with the aid of lemma 12.2 and theorem 12.3. Then, through the same steps as in theorem 12.2, we show that the output-feedback controller (12.127) is bounded and exponentially convergent to zero as well.

The proof of theorem 12.4 is complete. $\square$

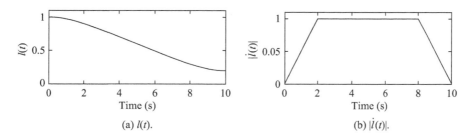

Figure 12.2. Moving boundary and its velocity.

12.5 SIMULATION

Consider the system given by

$$\dot{X}(t) = 0.4X(t) + v(0, t), \tag{12.142}$$

$$u_t(x, t) = -u_x(x, t) + 0.5v(x, t), \tag{12.143}$$

$$v_t(x, t) = v_x(x, t) + 0.5u(x, t), \tag{12.144}$$

$$u(0, t) = v(0, t) + X(t), \quad v(l(t), t) = s_1(t), \tag{12.145}$$

$$\dot{s}_1(t) = s_2(t) + s_1(t)^2 + \int_0^{l(t)} u(x, t)dx, \tag{12.146}$$

$$\dot{s}_2(t) = s_1(t)s_2(t) + u(l(t), t) + z(t), \tag{12.147}$$

$$\dot{z}(t) = 0.5z(t) + u(l(t), t) + U(t), \tag{12.148}$$

for $x \in [0, l(t)]$. The time-varying function $l(t)$ is known ahead of time and is decreasing from $l(0) = 1$ to 0.2 over a span of 10 s, as shown in figure 12.2. The initial values are given as

$$u(x, 0) = 3\sin(4\pi x),$$
$$v(x, 0) = 3\sin(4\pi x),$$
$$X(0) = u(0, 0) - v(0, 0),$$
$$s_1(0) = v(l(0), 0),$$
$$s_2(0) = z(0) = 0.$$

The initial values of the observer are given as

$$\hat{u}(x, 0) = u(x, 0) + 0.2\sin(2\pi(l(0) - x)),$$
$$\hat{v}(x, 0) = v(x, 0) + 0.2\sin(2\pi(l(0) - x)),$$
$$\hat{X}(0) = \hat{u}(0, 0) - \hat{v}(0, 0),$$
$$\hat{s}_1(0) = \hat{v}(l(0), 0),$$
$$\hat{s}_2(0) = s_2(0) + 0.5,$$
$$\hat{Z}(0) = Z(0) + 0.5$$

where the additional terms are the initial observer errors.

The simulation is performed by the finite-difference method for the discretization in time and space after converting the time-varying PDE domain to a fixed PDE

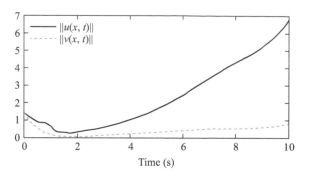

Figure 12.3. Open-loop responses of $\|u(\cdot,t)\|$ and $\|v(\cdot,t)\|$.

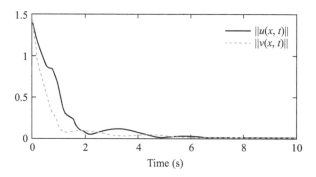

Figure 12.4. Responses of $\|u(\cdot,t)\|$ and $\|v(\cdot,t)\|$ under the proposed output-feedback controller.

domain by introducing $\tilde{\xi} = \frac{x}{l(t)}$, and then the time step and space step are chosen as 0.001 and 0.02, respectively. The kernels (12.12)–(12.17), (12.89)–(12.92), (12.100), (12.101), used in the control input, are also solved by the finite-difference method. The control parameters are chosen as

$$c_1 = 80, \quad c_2 = 150, \quad \bar{a}_0 = 350, \quad \kappa = -10,$$
$$L_0 = 10, \quad \mu_1 = \mu_2 = \mu_3 = 5.$$

The simulation results are shown next.

Comparing figure 12.3, which shows the open-loop responses of $\|u(\cdot,t)\|$, $\|v(\cdot,t)\|$, and figure 12.4, which gives the closed-loop responses of $\|u(\cdot,t)\|$, $\|v(\cdot,t)\|$, one can observe that in the latter case the convergence to zero is achieved, whereas the states grow unbounded in the former case. According to figure 12.5, we see that the responses of the $z(t)$-ODE, the nonlinear $(s_1(t), s_2(t))$-ODE, and the $X(t)$-ODE at the opposite boundary converge to zero under the proposed output-feedback controller. Moreover, figures 12.6 and 12.7 show that the proposed observer converges to the actual plant for both PDE and ODE states. Because $v(l(t),t)$ and $z(t)$ are the measurements, $\tilde{s}_1(t)$ and $\tilde{z}(t)$ are of small magnitude and converge to zero fast, so their plots are omitted here to avoid repetition. Figure 12.8 shows that the observer-based output-feedback control input is bounded and convergent to zero.

We limited ourselves in this chapter to known nonlinearities, with no uncertainties. In future work, it might be of interest to extend the control design to a

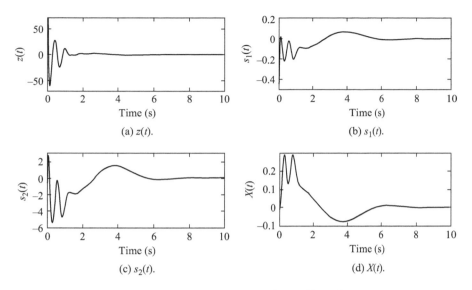

Figure 12.5. Responses of ODE states $z(t), s_1(t), s_2(t), X(t)$ under the proposed output-feedback controller.

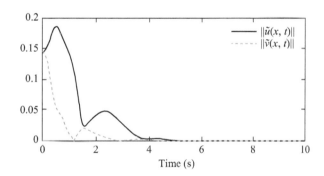

Figure 12.6. Observer errors of $\|\tilde{u}(\cdot, t)\|$, $\|\tilde{v}(\cdot, t)\|$.

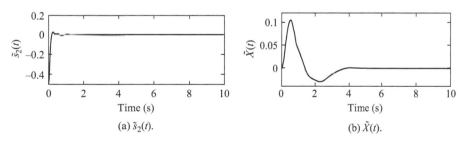

Figure 12.7. Observer errors of $\tilde{s}_2(t), \tilde{X}(t)$.

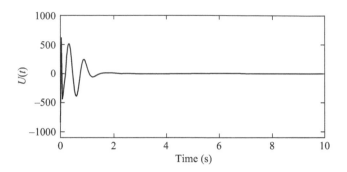

Figure 12.8. Output-feedback control input.

more complicated and practical case with some system parameters unknown and an adaptive design developed.

12.6   APPENDIX

A. The expression of $\mathcal{F}$

$$\mathcal{F}(\beta(l(t),\beta(0,t),\alpha(l(t),t),\alpha(0,t),\beta(x,t),\alpha(x,t),X(t))$$
$$= h_1(l(t))\beta(l(t),t) + h_2(l(t))\beta(0,t)$$
$$+ h_3(l(t))\alpha(l(t),t) + h_4(l(t))\alpha(0,t)$$
$$+ \int_0^{l(t)} h_5(l(t),y)\beta(y,t)dy$$
$$+ \int_0^{l(t)} h_6(l(t),y)\alpha(y,t)dy + H_7(l(t))X(t), \tag{12.149}$$

where

$$h_1(l(t)) = -p_2\phi(l(t),l(t)) - \dot{l}(t)\phi(l(t),l(t)), \tag{12.150}$$
$$h_2(l(t)) = p_2\phi(l(t),0) - \gamma(l(t))B, \tag{12.151}$$
$$h_3(l(t)) = p_1\psi(l(t),l(t)) - \dot{l}(t)\psi(l(t),l(t)), \tag{12.152}$$
$$h_4(l(t)) = -p_1\psi(l(t),0), \tag{12.153}$$
$$h_5(l(t),y) = \big(p_2\phi(l(t),l(t)) + \dot{l}(t)\phi(l(t),l(t))\big)\mathcal{M}(l(t),y)$$
$$+ p_2\phi_y(l(t),y) - c_1\psi(l(t),y) - \dot{l}(t)\phi_x(l(t),y)$$
$$- \int_y^{l(t)} \big(p_2\phi_y(l(t),z) - c_1\psi(l(t),z)$$
$$- \dot{l}(t)\phi_x(l(t),z)\big)\mathcal{M}(z,y)dz, \tag{12.154}$$
$$h_6(l(t),y) = p_1\psi_y(l(t),y) + c_2\phi(l(t),y) + \dot{l}(t)\psi_x(l(t),y)$$
$$- \big(p_2\phi(l(t),l(t)) + \dot{l}(t)\phi(l(t),l(t))\big)\mathcal{D}(l(t),y)$$
$$- \int_y^{l(t)} \big(p_2\phi_y(l(t),z) - c_1\psi(l(t),z)$$
$$- \dot{l}(t)\phi_x(l(t),z)\big)\mathcal{D}(z,y)dz, \tag{12.155}$$

$$H_7(l(t)) = \big(p_2\phi(l(t), l(t)) + \dot{l}(t)\phi(l(t), l(t))\big)\mathcal{J}(l(t))$$
$$- \gamma(l(t))A - \dot{l}(t)\gamma'(l(t)) - \big(p_2\phi(l(t), 0) - \gamma(l(t))B\big)\mathcal{J}(0)$$
$$- \int_0^{l(t)} \big(p_2\phi_y(l(t), y) - c_1\psi(l(t), y)$$
$$- \dot{l}(t)\phi_x(l(t), y)\big)\mathcal{J}(y)dy. \tag{12.156}$$

B. Calculation of $\dot{V}_1(t)$

Taking the time derivative of (12.45) along (12.20)–(12.23), we obtain

$$\dot{V}_1(t) \leq -\lambda_{\min}(Q_1)|X(t)|^2 + 2X^T P_1 B\beta(0, t)$$
$$+ \frac{p_2}{2}a_1 e^{\delta_1 l(t)}\beta(l(t), t)^2 - \frac{p_2}{2}a_1\beta(0, t)^2$$
$$+ \frac{a_1\dot{l}(t)}{2}e^{\delta_1 l(t)}\beta(l(t), t)^2 + \frac{b_1\dot{l}(t)}{2}e^{-\delta_1 l(t)}\alpha(l(t), t)^2$$
$$- \frac{p_2}{2}\delta_1 a_1 \int_0^{l(t)} e^{\delta_1 x}\beta(x, t)^2 dx - \frac{p_1}{2}\delta_1 b_1 \int_0^{l(t)} e^{-\delta_1 x}\alpha(x, t)^2 dx$$
$$- \frac{p_1}{2}b_1 e^{-\delta_1 l(t)}\alpha(l(t), t)^2 + \frac{p_1}{2}b_1\alpha(0, t)^2$$
$$+ b_1 c_1 \int_0^{l(t)} e^{-\delta_1 x}\alpha(x, t)\left(\beta(x, t) - \int_0^x \mathcal{D}(x, y)\alpha(y, t)dy\right.$$
$$\left. - \int_0^x \mathcal{M}(x, y)\beta(y, t)dy - \mathcal{J}(x)X(t)\right)dx. \tag{12.157}$$

Recalling (12.23), using Young's inequality and the Cauchy-Schwarz inequality for the last part in (12.157) yields the existence of $\xi > 0$ such that

$$\dot{V}_1(t) \leq -\left(\frac{1}{2}\lambda_{\min}(Q_1) - p_1 b_1 |C_0|^2\right)|X(t)|^2$$
$$- \left(\frac{p_2}{2}a_1 - p_1 b_1 q^2 - \frac{4|P_1 B|}{\lambda_{\min}(Q_1)}\right)\beta(0, t)^2$$
$$- \left(\frac{p_2}{2}\delta_1 a_1 - b_1\xi - b_1\frac{\xi}{\delta_1}\right)\int_0^{l(t)}\beta(x, t)^2 dx$$
$$- \left(\frac{p_2\delta_1 b_1}{2} - \frac{b_1\xi}{\delta_1} - \frac{\xi b_1^2}{\lambda_{\min}(Q_1)} - b_1\xi\right)e^{-\delta_1 L}\int_0^{l(t)}\alpha(x, t)^2 dx$$
$$- (p_1 - \dot{l}(t))\frac{b_1}{2}e^{-\delta_1 l(t)}\alpha(l(t), t)^2$$
$$+ (p_2 + \dot{l}(t))\frac{a_1}{2}e^{\delta_1 l(t)}\beta(l(t), t)^2, \tag{12.158}$$

where

$$\xi = \max\left\{\frac{c_1}{4}[\bar{\mathcal{D}}(1 + L) + \bar{\mathcal{M}}], L\bar{\mathcal{J}}^2 c_1^2, \frac{1}{2}c_1, \frac{c_1}{4}\bar{\mathcal{D}}L\right\} \tag{12.159}$$

and

$$\bar{\mathcal{D}} = \max_{0 \le y \le x \le L} \{|\mathcal{D}(x,y)|\}, \tag{12.160}$$

$$\bar{\mathcal{M}} = \max_{0 \le y \le x \le L} \{|\mathcal{M}(x,y)|\}, \tag{12.161}$$

$$\bar{\mathcal{J}} = \max_{0 \le x \le L} \{|\mathcal{J}(x)|\}. \tag{12.162}$$

Choosing the parameters b_1, δ_1, a_1 to satisfy

$$0 < b_1 < \frac{\lambda_{\min}(Q_1)}{2p_1 |C_0|^2}, \tag{12.163}$$

$$\delta_1 > \max \left\{ 1, \frac{2}{p_2} \left(2\xi + \frac{\xi b_1}{\lambda_{\min}(Q_1)} \right) \right\}, \tag{12.164}$$

$$a_1 > \max \left\{ \frac{8 |P_1 B|}{p_2 \lambda_{\min}(Q_1)} + 2q^2 b_1 \frac{p_1}{p_2}, \frac{2b_1 \xi}{p_2 \delta_1} + \frac{2b_1 \xi}{p_2 \delta_1^2} \right\}, \tag{12.165}$$

we arrive at (12.46), where

$$\eta_1 = \frac{1}{2} \lambda_{\min}(Q_1) - p_1 b_1 |C_0|^2 > 0, \tag{12.166}$$

$$\eta_2 = \frac{p_2}{2} a_1 - p_1 b_1 q^2 - \frac{4 |P_1 B|}{\lambda_{\min}(Q_1)} > 0, \tag{12.167}$$

$$\eta_3 = \frac{p_2}{2} \delta_1 a_1 - b_1 \xi - b_1 \frac{\xi}{\delta_1} > 0, \tag{12.168}$$

$$\eta_4 = \left(\frac{p_2 \delta_1 b_1}{2} - \frac{b_1 \xi}{\delta_1} - \frac{\xi b_1^2}{\lambda_{\min}(Q_1)} - b_1 \xi \right) e^{-\delta_1 L} > 0. \tag{12.169}$$

C. Calculation of $\dot{V}(t)$

Taking the time derivative of (12.47) and recalling (12.46), (12.40) and (12.37) with (12.149)–(12.156), we obtain

$$\begin{aligned}
\dot{V} \le & -\eta_1 |X(t)|^2 - \eta_2 \beta(0,t)^2 - \eta_3 \int_0^{l(t)} \beta(x,t)^2 dx \\
& -\eta_4 \int_0^{l(t)} \alpha(x,t)^2 dx - \eta_5 \alpha(l(t),t)^2 + \eta_6 \beta(l(t),t)^2 \\
& -\bar{c}_1 y_1(t)^2 - \bar{c}_2 y_2(t)^2 + y_1(t) f_1 \\
& + y_1(t) \Big(h_1(l(t))\beta(l(t),t) + h_2(l(t))\beta(0,t) \\
& + h_3(l(t))\alpha(l(t),t) + h_4(l(t))\alpha(0,t) \\
& + \int_0^{l(t)} h_5(l(t),y)\beta(y,t)dy + \int_0^{l(t)} h_6(l(t),y)\alpha(y,t)dy \\
& + H_7(l(t))X(t) \Big) + y_2(t)\mathcal{E}(t) + y_2(t) f_2 + \frac{\bar{c}_1}{c_3} y_2(t)\dot{y}_1(t) \\
& - k_{\mathcal{E}} \mathcal{E}(t)^2 + \mathcal{E}(t)\bar{c}_2 \dot{y}_2(t) + \mathcal{E}(t)c_3 \dot{y}_1(t).
\end{aligned} \tag{12.170}$$

Recalling assumptions 12.1 and 12.2 and (12.29), (12.30), (12.32), we obtain

$$f_1^2 \leq \gamma_{f1}\left(4y_1(t)^2 + 4\|\beta(\cdot,t)\|^2 + 5\|\alpha(\cdot,t)\|^2 + 4|X(t)|^2\right), \tag{12.171}$$

$$f_2^2 \leq \gamma_{f2}\left(\left(4 + \frac{2\bar{c}_1^2}{c_3^2}\right)y_1(t)^2 + 4\|\beta(\cdot,t)\|^2 + 4\|\alpha(\cdot,t)\|^2 \right.$$
$$\left. + 4|X(t)|^2 + 2y_2(t)^2 + \alpha(l(t),t)^2\right), \tag{12.172}$$

where γ_{f1}, γ_{f2} are positive constants associated with the kernels $\mathcal{D}, \mathcal{M}, \mathcal{J}$ and the Lipschitz constants. The omitted arguments of f_1, f_2 are the same as those in (12.26), (12.27).

Applying Young's inequality and the Cauchy-Schwarz inequality into, respectively, the ninth and tenth terms, and $y_2(t)\mathcal{E}(t) + y_2(t)f_2 + \frac{\bar{c}_1}{c_3}y_2(t)\dot{y}_1(t) + \mathcal{E}(t)\bar{c}_2\dot{y}_2(t) + \mathcal{E}(t)c_3\dot{y}_1(t)$ in (12.170), where (12.30), (12.27), (12.32), (12.36) are used to rewrite $\dot{y}_2(t)$ as

$$\dot{y}_2(t) = f_2 + \mathcal{E}(t) - \bar{c}_2 y_2(t) - c_3 y_1(t) + \frac{\bar{c}_1}{c_3}\dot{y}_1(t),$$

using (12.171), (12.172) to replace the resulting f_1^2, f_2^2, and recalling (12.23), (12.29) to rewrite the resulting $\alpha(0,t)^2$, $\beta(l(t),t)^2$, respectively, we get

$$\dot{V} \leq -\left(\eta_1 - r_7\left|\bar{H}_7\right|^2 - 2r_4\bar{h}_4^2\left|C_0\right|^2 - 4r_8\gamma_{f1} - 4r_9\gamma_{f2} - 4r_{11}\gamma_{f2}\right)|X(t)|^2$$
$$- \left(\eta_2 - \bar{h}_2^2 r_2 - 2r_4\bar{h}_4^2 q^2\right)\beta(0,t)^2$$
$$- \left(\eta_3 - r_5 h_{5\max}^2 L - 4r_8\gamma_{f1} - 4r_9\gamma_{f2} - 4r_{11}\gamma_{f2}\right)\int_0^{l(t)} \beta(x,t)^2 dx$$
$$- \left(\eta_4 - r_6 h_{6\max}^2 L - 5r_8\gamma_{f1} - 4r_9\gamma_{f2} - 4r_{11}\gamma_{f2}\right)\int_0^{l(t)} \alpha(x,t)^2 dx$$
$$- \left(\eta_5 - \bar{h}_3^2 r_3 - r_9\gamma_{f2} - r_{11}\gamma_{f2}\right)\alpha(l(t),t)^2$$
$$- \left(\bar{c}_1 - 1 - \frac{1}{4r_2} - \frac{1}{4r_3} - \frac{1}{4r_4} - \frac{1}{4r_5} - \frac{1}{4r_6} - \frac{1}{4r_7} - \eta_6 \right.$$
$$\left. - \frac{1}{4r_8} - 4r_8\gamma_{f1} - \bar{h}_1 - \left(4 + \frac{2\bar{c}_1^2}{c_3^2}\right)r_9\gamma_{f2} - \left(4 + \frac{2\bar{c}_1^2}{c_3^2}\right)r_{11}\gamma_{f2}\right)y_1(t)^2$$
$$- \left(\bar{c}_2 - \frac{1}{4r_9} - \frac{\bar{c}_1^2}{4c_3^2 r_{10}} - \frac{3}{2} - 2r_{11}\gamma_{f2} - 2r_9\gamma_{f2}\right)y_2(t)^2$$
$$- \left((k_\mathcal{E} - \bar{c}_2) - \frac{1}{2} - \frac{\bar{c}_2^4}{4} - \frac{\bar{c}_2^2}{4r_{11}} - \frac{\bar{c}_2^2\bar{c}_1^2}{4r_{12}c_3^2} - \frac{\bar{c}_2^2 c_3^2}{4} - \frac{c_3^2}{4r_{13}}\right)\mathcal{E}(t)^2$$
$$+ (r_{10} + r_{12} + r_{13})\dot{y}_1(t)^2, \tag{12.173}$$

where $r_1, \ldots, r_{13}$ are positive constants from using Young's inequality,

$$h_{5\max} = \max_{x\in[0,L],l(t)\in[0,L]}\{|h_5(x,l(t))|\}, \tag{12.174}$$

$$h_{6\max} = \max_{x\in[0,L],l(t)\in[0,L]}\{|h_6(x,l(t))|\}, \tag{12.175}$$

and $\bar{h}_1, \bar{h}_2, \bar{h}_3, \bar{h}_4, \bar{H}_7$ are the maximum values of $|h_1(l(t))|$, $|h_2(l(t))|$, $|h_3(l(t))|$, $|h_4(l(t))|$, $|H_7(l(t))|$ for $l(t) \in [0,L]$ in (12.150)–(12.156).

According to (12.29), (12.30), (12.32) and (12.26) with (12.149)–(12.156), we get

$$\dot{y}_1(t)^2 \leq \bar{\xi}_c \left(\bar{c}_1 y_1(t)^2 + y_1(t)^2 + y_2(t)^2 + f_1^2 + \beta(0,t)^2 \right.$$

$$\left. + \alpha(l(t),t)^2 + \alpha(0,t)^2 + \|\beta(\cdot,t)\|^2 + \|\alpha(\cdot,t)\|^2 + X(t)^2 \right) \qquad (12.176)$$

for some positive constants $\bar{\xi}_c$ depending on the kernels $\mathcal{D}, \mathcal{M}, \mathcal{J}$ and gains in (12.150)–(12.156).

Inserting (12.176) into (12.173) to replace $\dot{y}_1(t)^2$ and using (12.171), we arrive at

$$\dot{V} \leq - \left(\eta_1 - r_7 \left| \bar{H}_7 \right|^2 - 2r_4 \bar{h}_4^2 \left| C_0 \right|^2 - 4r_8 \gamma_{f1} - 4r_9 \gamma_{f2} \right.$$

$$- 4r_{11} \gamma_{f2} - (r_{10} + r_{12} + r_{13}) \bar{\xi}_c - 2(r_{10} + r_{12} + r_{13}) \bar{\xi}_c |C_0|^2$$

$$\left. - 4\gamma_{f_1}(r_{10} + r_{12} + r_{13}) \bar{\xi}_c \right) |X(t)|^2 - \left(\eta_2 - \bar{h}_2^2 r_2 - 2r_4 \bar{h}_4^2 q^2 \right.$$

$$\left. - (1 + 2q^2)(r_{10} + r_{12} + r_{13}) \bar{\xi}_c \right) \beta(0,t)^2 - \left(\eta_3 - r_5 h_{5\,\text{max}}^2 L \right.$$

$$- 4r_8 \gamma_{f1} - 4r_9 \gamma_{f2} - 4r_{11} \gamma_{f2} - (r_{10} + r_{12} + r_{13}) \bar{\xi}_c$$

$$\left. - 4\gamma_{f_1}(r_{10} + r_{12} + r_{13}) \bar{\xi}_c \right) \int_0^{l(t)} \beta(x,t)^2 dx - \left(\eta_4 - 5r_8 \gamma_{f1} \right.$$

$$- r_6 h_{6\,\text{max}}^2 L - 4r_9 \gamma_{f2} - 4r_{11} \gamma_{f2} - (r_{10} + r_{12} + r_{13}) \bar{\xi}_c$$

$$\left. - 5\gamma_{f_1}(r_{10} + r_{12} + r_{13}) \bar{\xi}_c \right) \int_0^{l(t)} \alpha(x,t)^2 dx - \left(\eta_5 - \bar{h}_3^2 r_3 \right.$$

$$\left. - r_9 \gamma_{f2} - r_{11} \gamma_{f2} - (r_{10} + r_{12} + r_{13}) \bar{\xi}_c \right) \alpha(l(t),t)^2$$

$$- \left(\bar{c}_1 - 1 - \frac{1}{4r_2} - \frac{1}{4r_3} - \frac{1}{4r_4} - \frac{1}{4r_5} - \frac{1}{4r_6} - \frac{1}{4r_7} \right.$$

$$\left. - \eta_6 - \frac{1}{4r_8} - 4r_8 \gamma_{f1} - \bar{h}_1 - \left(4 + \frac{2\bar{c}_1^2}{c_3^2} \right)(r_9 + r_{11}) \gamma_{f2} \right.$$

$$\left. - (1 + 4\gamma_{f_1} + \bar{c}_1)(r_{10} + r_{12} + r_{13}) \bar{\xi}_c \right) y_1(t)^2$$

$$- \left(\bar{c}_2 - \frac{1}{4r_9} - \frac{\bar{c}_1^2}{4c_3^2 r_{10}} - \frac{3}{2} - 2r_{11} \gamma_{f2} - 2r_9 \gamma_{f2} \right.$$

$$\left. - (r_{10} + r_{12} + r_{13}) \bar{\xi}_c \right) y_2(t)^2 - \left(\bar{a}_0 - c_4 - \bar{c}_2 - \frac{1}{2} \right.$$

$$\left. - \frac{\bar{c}_2^4}{4} - \frac{\bar{c}_2^2}{4r_{11}} - \frac{\bar{c}_1^2 \bar{c}_2^2}{4r_{12} c_3^2} - \frac{\bar{c}_2^2 c_3^2}{4} - \frac{c_3^2}{4r_{13}} \right) \mathcal{E}(t)^2, \qquad (12.177)$$

where

$$k_{\mathcal{E}} = \bar{a}_0 - c_4$$

is recalled from (12.41). Choosing small enough positive $r_2, r_3, r_4, r_5, r_6, r_7, r_8, r_9,$ $r_{10}, r_{11}, r_{12}, r_{13}$ and making the control parameters $\bar{c}_1, \bar{c}_2, \bar{a}_0$ large enough to satisfy

$$\bar{c}_1 > 1 + \frac{1}{4r_2} + \frac{1}{4r_3} + \frac{1}{4r_4} + \frac{1}{4r_5} + \frac{1}{4r_6} + \frac{1}{4r_7} + \eta_6 + \frac{1}{4r_8} + 4r_8 \gamma_{f1} + \bar{h}_1, \quad (12.178)$$

$$\bar{c}_2 > \frac{1}{4r_9} + \frac{\bar{c}_1^2}{4c_3^2 r_{10}} + \frac{3}{2} + 2r_{11}\gamma_{f2} + 2r_9\gamma_{f2} + (r_{10} + r_{12} + r_{13})\bar{\xi}_c, \tag{12.179}$$

$$\bar{a}_0 > c_4 + \bar{c}_2 + \frac{1}{2} + \frac{\bar{c}_2^4}{4} + \frac{\bar{c}_2^2}{4r_{11}} + \frac{\bar{c}_1^2 \bar{c}_2^2}{4r_{12}c_3^2} + \frac{\bar{c}_2^2 c_3^2}{4} + \frac{c_3^2}{4r_{13}}, \tag{12.180}$$

we obtain (12.50) with

$$\hat{\eta}_0 = \eta_2 - \bar{h}_2^2 r_2 - 2r_4\bar{h}_4^2 q^2 - (1 + 2q^2)(r_{10} + r_{12} + r_{13})\bar{\xi}_c > 0, \tag{12.181}$$

$$\hat{\eta}_1 = \eta_5 - \bar{h}_3^2 r_3 - r_9\gamma_{f2} - r_{11}\gamma_{f2} - (r_{10} + r_{12} + r_{13})\bar{\xi}_c > 0. \tag{12.182}$$

D. Proof of lemma 12.1

Differentiating (12.21) and (12.22) with respect to x and differentiating (12.23) with respect to t, we obtain

$$\alpha_{xt}(x,t) = -p_1\alpha_{xx}(x,t) + c_1\beta_x(x,t) - c_1\mathcal{J}'(x)X(t)$$
$$- c_1\mathcal{D}(x,x)\alpha(x,t) - c_1\mathcal{M}(x,x)\beta(x,t)$$
$$- c_1\int_0^x \mathcal{D}_x(x,y)\alpha(y,t)dy - c_1\int_0^x \mathcal{M}_x(x,y)\beta(y,t)dy, \tag{12.183}$$

$$\beta_{xt}(x,t) = p_2\beta_{xx}(x,t), \tag{12.184}$$

$$\alpha_x(0,t) = -q\frac{p_2}{p_1}\beta_x(0,t) - \frac{1}{p_1}\left(C_0(A + B\kappa) - c_1\mathcal{J}(0)\right)X(t)$$
$$- \frac{1}{p_1}(C_0 B - c_1)\beta(0,t). \tag{12.185}$$

Defining

$$\bar{A} = \frac{b_2}{2}\int_0^{l(t)} e^{-\delta_2 x}\alpha_x(x,t)^2 dx + \frac{a_2}{2}\int_0^{l(t)} e^{\delta_2 x}\beta_x(x,t)^2 dx, \tag{12.186}$$

where b_2 is an arbitrary positive constant that can adjust the convergence rate. The positive constants δ_2, a_2 will be chosen later.

Taking the derivative of (12.186) along (12.183), (12.184), we obtain

$$\dot{\bar{A}} = -\frac{p_1}{2}b_2 e^{-\delta_2 l(t)}\alpha_x(l(t),t)^2 + \frac{p_1}{2}b_2\alpha_x(0,t)^2$$
$$+ \frac{b_2\dot{l}(t)}{2}e^{-\delta_2 l(t)}\alpha_x(l(t),t)^2 - \frac{p_1}{2}b_2\delta_2\int_0^{l(t)} e^{-\delta_2 x}\alpha_x(x,t)^2 dx$$
$$+ (p_2 + \dot{l}(t))\frac{a_2}{2}e^{\delta_2 l(t)}\beta_x(l(t),t)^2 - \frac{p_2}{2}a_2\beta_x(0,t)^2$$
$$- \frac{p_2}{2}a_2\delta_2\int_0^{l(t)} e^{\delta_2 x}\beta_x(x,t)^2 dx - \int_0^{l(t)} b_2 e^{-\delta_2 x}\alpha_x(x,t)c_1\mathcal{D}(x,x)\alpha(x,t)dx$$
$$- \int_0^{l(t)} b_2 e^{-\delta_2 x}\alpha_x(x,t)c_1\mathcal{M}(x,x)\beta(x,t)dx$$
$$- \int_0^{l(t)} b_2 e^{-\delta_2 x}\alpha_x(x,t)c_1\int_0^x \mathcal{D}_x(x,y)\alpha(y,t)dydx$$
$$- \int_0^{l(t)} b_2 e^{-\delta_2 x}\alpha_x(x,t)c_1\int_0^x \mathcal{M}_x(x,y)\beta(y,t)dydx$$

$$+ \int_0^{l(t)} b_2 e^{-\delta_2 x} \alpha_x(x,t) c_1 \beta_x(x,t) dx$$

$$- \int_0^{l(t)} b_2 e^{-\delta_2 x} \alpha_x(x,t) c_1 \mathcal{J}'(x) X(t) dx. \tag{12.187}$$

Using Young's inequality and the Cauchy-Schwarz inequality in the last six terms in (12.187) yields the existence of $\xi_2 > 0$ such that

$$\begin{aligned}
\dot{\bar{A}}(t) \leq &- \left(p_1 - \dot{l}(t) \right) \frac{b_2}{2} e^{-\delta_2 l(t)} \alpha_x(l(t),t)^2 + \frac{p_1}{2} b_2 \alpha_x(0,t)^2 \\
&+ \left(p_2 + \dot{l}(t) \right) \frac{a_2}{2} e^{\delta_2 l(t)} \beta_x(l(t),t)^2 - \frac{p_2}{2} a_2 \beta_x(0,t)^2 \\
&- \left(\frac{p_1}{2} b_2 \delta_2 - 4\xi_2 b_2 - \frac{2\xi_2 b_2}{\delta_2} \right) \int_0^{l(t)} e^{-\delta_2 x} \alpha_x(x,t)^2 dx \\
&- \left(\frac{p_2}{2} a_2 \delta_2 - \xi_2 b_2 \right) \int_0^{l(t)} e^{\delta_2 x} \beta_x(x,t)^2 dx \\
&+ \left(\xi_2 b_2 + \frac{\xi_2 b_2}{\delta_2} \right) \int_0^{l(t)} e^{-\delta_2 x} \alpha(x,t)^2 dx + \xi_2 b_2 |X(t)|^2 \\
&+ \left(\xi_2 b_2 + \frac{\xi_2 b_2}{\delta_2} \right) \int_0^{l(t)} e^{\delta_2 x} \beta(x,t)^2 dx. \tag{12.188}
\end{aligned}$$

The function $\alpha_x(0,t)^2$ in (12.188) can be replaced by

$$\begin{aligned}
\alpha_x(0,t)^2 \leq &\, 3 \frac{p_2^2}{p_1^2} q^2 \beta_x(0,t)^2 + \frac{3}{p_1^2} (C_0 B - c_1)^2 \beta(0,t)^2 \\
&+ \frac{3}{p_1^2} |C_0(A + B\kappa) + c_1 \mathcal{J}(0)|^2 |X(t)|^2 \tag{12.189}
\end{aligned}$$

using the Cauchy-Schwarz inequality on (12.185). Recalling (12.22), (12.26) with (12.149)–(12.156), (12.29), (12.30), (12.32), (12.171), using the Cauchy-Schwarz inequality (for the integrals appearing in (12.26), (12.149)), the positive term $\left(p_2 + \dot{l}(t) \right) \frac{a_2}{2} e^{\delta_2 l(t)} \beta_x(l(t),t)^2$ in (12.188) can be replaced as

$$\begin{aligned}
&(p_2 + \dot{l}(t)) \frac{a_2}{2} e^{\delta_2 l(t)} \beta_x(l(t),t)^2 \\
&\leq \bar{\xi}_2 y_2(t)^2 + \bar{\xi}_3 y_1(t)^2 + \bar{\xi}_4 \beta(0,t) + \bar{\xi}_5 \alpha(l(t),t)^2 + \bar{\xi}_6 \alpha(0,t)^2 \\
&\quad + \bar{\xi}_7 \|\alpha(\cdot,t)\|^2 + \bar{\xi}_8 \|\beta(\cdot,t)\|^2 + \bar{\xi}_9 |X(t)|^2 \tag{12.190}
\end{aligned}$$

for some positive $\bar{\xi}_i, i = 2, \ldots, 9$.

We propose a Lyapunov function

$$V_2(t) = \bar{A}(t) + R_1 V(t) \tag{12.191}$$

and define the norm

$$\begin{aligned}
\Omega_2(t) = &\, \|\beta_x(\cdot,t)\|^2 + \|\alpha_x(\cdot,t)\|^2 + \|\beta(\cdot,t)\|^2 \\
&+ \|\alpha(\cdot,t)\|^2 + |X(t)|^2 + y_1(t)^2 + y_2(t)^2 + \mathcal{E}(t)^2 \tag{12.192}
\end{aligned}$$

so that

$$\theta_{2a}\Omega_2(t) \leq V_2(t) \leq \theta_{2b}\Omega_2(t) \tag{12.193}$$

for some positive θ_{2a} and θ_{2b}.

Taking the derivative of (12.191) and recalling (12.188)–(12.190), (12.50), we then get

$$\dot{V}_2(t) = \dot{\bar{A}} + R_1\dot{V}$$

$$\leq -\left(p_1 - l(t)\right)\frac{b_2}{2}e^{-\delta_2 l(t)}\alpha_x(l(t), t)^2$$

$$-\left(\frac{p_2}{2}a_2 - \frac{3p_2^2 q^2}{2b_2 p_1}\right)\beta_x(0, t)^2$$

$$-\left(\frac{p_1}{2}b_2\delta_2 - 4\xi_2 b_2 - \frac{2\xi_2 b_2}{\delta_2}\right)\int_0^{l(t)} e^{-\delta_2 x}\alpha_x(x, t)^2 dx$$

$$-\left(\frac{p_2}{2}a_2\delta_2 - \xi_2 b_2\right)\int_0^{l(t)} e^{\delta_2 x}\beta_x(x, t)^2 dx - \frac{R_1}{2}\lambda V(t)$$

$$-\left(R_1\hat{\eta}_0 - \frac{3b_2}{2p_1}(C_0 B - c_1)^2 - \bar{\xi}_4 - 2\bar{\xi}_6 q^2\right)\beta(0, t)^2$$

$$-\left(\frac{R_1}{2}\theta_{1a}\lambda - \bar{\xi}_2\right)y_2(l)^2 - \left(\frac{R_1}{2}\theta_{1a}\lambda - \bar{\xi}_3\right)y_1(t)^2$$

$$-\left(\frac{R_1}{2}\theta_{1a}\lambda - \xi_2 b_2 - \frac{\xi_2 b_2}{\delta_2} - \bar{\xi}_7\right)\int_0^{l(t)} \alpha(x, t)^2 dx$$

$$-\left(\frac{R_1}{2}\theta_{1a}\lambda - \xi_2 b_2 e^{\delta_2 L} - \frac{\xi_2 b_2}{\delta_2}e^{\delta_2 L} - \bar{\xi}_8\right)\int_0^{l(t)} \beta(x, t)^2 dx$$

$$-\left(\frac{R_1}{2}\theta_{1a}\lambda - \xi_2 b_2 - \frac{3b_2}{2p_1}|c_1\mathcal{J}(0) + C_0(A + B\kappa)|^2\right.$$

$$\left. -\bar{\xi}_9 - 2\bar{\xi}_6|C_0|^2\right)|X(t)|^2 - (R_1\hat{\eta}_1 - \bar{\xi}_5)\alpha(l(t), t)^2$$

$$\leq -\sigma_1 V_2(t) - \left(R_1\hat{\eta}_1 - \bar{\xi}_5\right)\alpha(l(t), t)^2$$

$$-\left(p_1 - l(t)\right)\frac{b_2}{2}e^{-\delta_2 l(t)}\alpha_x(l(t), t)^2$$

$$-\hat{\eta}_2\beta(0, t)^2 - \hat{\eta}_3\beta_x(0, t)^2 \tag{12.194}$$

for some positive σ_1, and

$$\hat{\eta}_2 = R_1\hat{\eta}_0 - \frac{3b_2}{2p_1}(C_0 B - c_1)^2 - \bar{\xi}_4 - 2\bar{\xi}_6 q^2 > 0,$$

$$\hat{\eta}_3 = \frac{p}{2}a_2 - \frac{3p_2^2 q^2}{2b_2 p_1} > 0,$$

$$R_1\hat{\eta}_1 - \bar{\xi}_5 > 0$$

by choosing

$$\delta_2 > \max\left\{1, \frac{12\xi_2}{p_1}\right\}, \quad a_2 > \max\left\{\frac{2\xi_2 b_2}{p_2\delta_2}, \frac{3p_2 q^2}{b_2 p_1}\right\} \tag{12.195}$$

and sufficiently large R_1. According to $p_1 - \dot{l}(t) > 0$ by recalling assumption 12.4, we thus have

$$\dot{V}_2(t) \leq -\sigma_1 V_2(t).$$

It then follows that

$$V_2(t) \leq V_2(0)e^{-\sigma_1 t}.$$

Recalling (12.193), we obtain

$$\Omega_2(t) \leq \frac{\theta_{2b}}{\theta_{2a}}\Omega_2(0)e^{-\sigma_1 t}. \tag{12.196}$$

Differentiating (12.18), (12.19) with respect to x, we get

$$u_x(x,t) = \alpha_x(x,t), \tag{12.197}$$

$$v_x(x,t) = \beta_x(x,t) - \int_0^x \mathcal{D}_x(x,y)\alpha(y,t)dy$$

$$- \int_0^x \mathcal{M}_x(x,y)\beta(y,t)dy - \mathcal{J}'(x)X(t)$$

$$- \mathcal{D}(x,x)\alpha(x,t) - \mathcal{M}(x,x)\beta(x,t). \tag{12.198}$$

Similarly differentiating (12.10), (12.11) with respect to x, together with (12.197), (12.198), using (12.10), (12.11), (12.18), (12.19), (12.29), (12.30), and (12.36), we obtain

$$\bar{\theta}_{2a}\Xi_1(t) \leq \Omega_2(t) \leq \bar{\theta}_{2b}\Xi_1(t)$$

for some positive $\bar{\theta}_{2a}$, $\bar{\theta}_{2b}$, where $\Xi_1(t)$ is defined as

$$\Xi_1(t) = \Xi(t) + \|u_x(\cdot,t)\|^2 + \|v_x(\cdot,t)\|^2.$$

Therefore, we get

$$\|u_x(\cdot,t)\|^2 + \|v_x(\cdot,t)\|^2 \leq \Xi_1(t) \leq \frac{\bar{\theta}_{2b}\theta_{2b}}{\bar{\theta}_{2a}\theta_{2a}}\Xi_1(0)e^{-\sigma_1 t}.$$

Thus, (12.56) is obtained with

$$\Upsilon_{1a} = \frac{\bar{\theta}_{2b}\theta_{2b}}{\bar{\theta}_{2a}\theta_{2a}}, \quad \lambda_{1a} = \sigma_1. \tag{12.199}$$

The proof of lemma 12.1 is complete.

E. Proof of theorem 12.2

Applying the Cauchy-Schwarz inequality in (12.42), we get

$$|U(t)|^2 \leq \xi_d \bigg(|s_1(t)|^2 + |s_2(t)|^2 + |z(t)|^2 + |X(t)|^2$$

$$+ |u(l(t),t)|^2 + \|u(\cdot,t)\|^2 + \|v(\cdot,t)\|^2 \bigg) \tag{12.200}$$

for some positive ξ_d.

Applying the Cauchy-Schwarz inequality and recalling (12.4), (12.5), we obtain

$$
\begin{aligned}
|u(l(t),t)| &\leq |u(0,t)| + \sqrt{L}\|u_x(\cdot,t)\| \\
&\leq |qv(0,t)| + |CX(t)| + \sqrt{L}\|u_x(\cdot,t)\| \\
&\leq |q|\|s_1(t)\| + |q|\sqrt{L}\|v_x(\cdot,t)\| \\
&\quad + |C|\|X(t)\| + \sqrt{L}\|u_x(\cdot,t)\|.
\end{aligned} \tag{12.201}
$$

According to (12.200), (12.201), using theorem 12.1 and lemma 12.1, we know that the control input (12.42) is bounded by

$$
|U(t)| \leq \Upsilon_2 \left(\Xi(0) + \|u_x(\cdot,0)\|^2 + \|v_x(\cdot,0)\|^2 \right)^{\frac{1}{2}} e^{-\lambda_2 t}. \tag{12.202}
$$

Then (12.57) is obtained by recalling (12.52). The proof of theorem 12.2 is complete.

F. Calculation of $\dot{V}_s(t)$

Taking the derivative of (12.120) along (12.110), we obtain

$$
\begin{aligned}
\dot{V}_s(t) &\leq \tilde{S}(t)^T \left((A_s - \mathcal{B}C_2)^T P_0 + P_0(A_s - \mathcal{B}C_2) \right) \tilde{S}(t) \\
&\quad + 2\tilde{S}(t)^T P_0 \tilde{f},
\end{aligned} \tag{12.203}
$$

where

$$
\tilde{f} = \left[\tilde{f}_1, \tilde{f}_2 \right]^T.
$$

Recalling (12.75), (12.76), according to assumption 12.2, we get

$$
\begin{aligned}
\tilde{f}_1^2 &= \left| f_1 \left(s_1, \int_0^{l(t)} u(x,t)dx \right) - f_1 \left(\hat{s}_1, \int_0^{l(t)} \hat{u}(x,t)dx \right) \right|^2 \\
&\leq \gamma_1^2 |s_1 - \hat{s}_1|^2 + \gamma_1^2 \left| \int_0^{l(t)} (u(x,t) - \hat{u}(x,t))dx \right|^2 \\
&\leq \gamma_1^2 \tilde{s}_1^2 + \hat{\gamma}_1^2 \|\tilde{\alpha}(\cdot,t)\|^2, \\
\tilde{f}_2^2 &= |f_2(s_1, s_2, u(l(t),t)) - f_2(\hat{s}_1, \hat{s}_2, \hat{u}(l(t),t))|^2 \\
&\leq \gamma_2^2 |(s_1, s_2, u(l(t),t)) - (\hat{s}_1, \hat{s}_2, \hat{u}(l(t),t))|^2 \\
&\leq \gamma_2^2 \tilde{s}_1^2 + \gamma_2^2 \tilde{s}_2^2 + \gamma_2^2 \tilde{\alpha}(l(t),t)^2,
\end{aligned}
$$

$$
\tag{12.204}
$$
$$
\tag{12.205}
$$

where $\hat{\gamma}_1$ is a positive constant and (12.80) is used. Then,

$$
\begin{aligned}
\tilde{f}^2 &= \tilde{f}_1^2 + \tilde{f}_2^2 \\
&\leq (\gamma_1^2 + \gamma_2^2)\tilde{s}_1(t)^2 + \gamma_2^2 \tilde{s}_2(t)^2 + \hat{\gamma}_1^2 \|\tilde{\alpha}(\cdot,t)\|^2 + \gamma_2^2 \tilde{\alpha}(l(t),t)^2 \\
&\leq (\gamma_1^2 + 2\gamma_2^2) \left| \tilde{S}(t) \right|^2 + \hat{\gamma}_1^2 \|\tilde{\alpha}(\cdot,t)\|^2 + \gamma_2^2 \tilde{\alpha}(l(t),t)^2.
\end{aligned} \tag{12.206}
$$

Thus, we obtain

$$
2\tilde{S}(t)^T P_0 \tilde{f}(t) \leq (\gamma_1^2 + 2\gamma_2^2) \left| P_0 \tilde{S}(t) \right|^2 + \frac{1}{\gamma_1^2 + 2\gamma_2^2} |\tilde{f}(t)|^2
$$

Table 12.1. Comparison of results on the boundary control of sandwich systems

	Types of ODEs of actuation	Types of PDEs	Types of control systems	Delay compensation	Types of control signals
[8]	First-order and scalar	Transport PDE	Output-feedback	$\times$	Continuous-in-time
[49]	First-order and scalar	2×2 coupled transport PDEs	Output-feedback	$\times$	Continuous-in-time
[179]	A chain of integrators	Heat PDE	Output-feedback	$\times$	Continuous-in-time
[42]	$\det(C_1 B) \neq 0$	n coupled parabolic PDEs	Output-feedback	$\times$	Continuous-in-time
[43]	$\det(C_0 B_0) \neq 0$	n coupled transport PDEs	Output-feedback	$\times$	Continuous-in-time
[151]	B_0 being invertible	2×2 coupled transport PDEs	State-feedback	$\times$	Continuous-in-time
[152]	Left invertible	2×2 coupled transport PDEs	State-feedback	$\times$	Continuous-in-time
Chapter 9	A chain of integrators	2×2 coupled transport PDEs	Output-feedback	$\times$	Continuous-in-time
Chapter 10	Left invertible	2×2 coupled transport PDEs	Output-feedback	$\sqrt{}$	Continuous-in-time
Chapter 11	Left invertible	2×2 coupled transport PDEs	Output-feedback	$\times$	Piecewise-constant
Chapter 12	Nonlinear	2×2 coupled transport PDEs	Output-feedback	$\times$	Continuous-in-time

Note: $\sqrt{}$ denotes "included," and $\times$ denotes "not included."

$$\leq (\gamma_1^2 + 2\gamma_2^2) \left| P_0 \tilde{S}(t) \right|^2 + \left| \tilde{S}(t) \right|^2$$

$$+ \frac{\hat{\gamma}_1^2}{\gamma_1^2 + 2\gamma_2^2} \| \tilde{\alpha}(\cdot, t) \|^2 + \frac{\gamma_2^2}{\gamma_1^2 + 2\gamma_2^2} \tilde{\alpha}(l(t), t)^2$$

$$= (\gamma_1^2 + 2\gamma_2^2) \tilde{S}(t)^T P_0^T P_0 \tilde{S}(t) + \tilde{S}(t)^T \tilde{S}(t)$$

$$+ \frac{\hat{\gamma}_1^2}{\gamma_1^2 + 2\gamma_2^2} \| \tilde{\alpha}(\cdot, t) \|^2 + \frac{\gamma_2^2}{\gamma_1^2 + 2\gamma_2^2} \tilde{\alpha}(l(t), t)^2, \qquad (12.207)$$

where Young's inequality is used. Substituting (12.207) into (12.203) yields

$$\dot{V}_s(t) \leq \tilde{S}(t)^T \left((\bar{A} - \mathcal{B}C_2)^T P_0 + P_0(\bar{A} - \mathcal{B}C_2) \right) \tilde{S}(t)$$

$$+ (\gamma_1^2 + 2\gamma_2^2) \tilde{S}^T(t) P_0^T P_0 \tilde{S}(t) + \tilde{S}(t)^T \tilde{S}(t)$$

$$+ \frac{\hat{\gamma}_1^2}{\gamma_1^2 + 2\gamma_2^2} \| \tilde{\alpha}(\cdot, t) \|^2 + \frac{\gamma_2^2}{\gamma_1^2 + 2\gamma_2^2} \tilde{\alpha}(l(t), t)^2$$

$$\leq \tilde{S}(t)^T \left((\bar{A} - \mathcal{B}C_2)^T P_0 + P_0(\bar{A} - \mathcal{B}C_2) \right.$$

$$\left. + (\gamma_1^2 + 2\gamma_2^2) P_0^T P_0 + I^T \right) \tilde{S}(t)$$

$$+ \frac{\hat{\gamma}_1^2}{\gamma_1^2 + 2\gamma_2^2} \| \tilde{\alpha}(\cdot, t) \|^2 + \frac{\gamma_2^2}{\gamma_1^2 + 2\gamma_2^2} \tilde{\alpha}(l(t), t)^2. \qquad (12.208)$$

Recalling (12.121), we arrive at (12.123).

12.7 NOTES

Comparisons of the results on the boundary control of sandwich systems are summarized in table 12.1.

Part III

Adaptive Control of Hyperbolic PDE-ODE Systems

Chapter Thirteen

Adaptive Event-Triggered Control of Hyperbolic PDEs

The adaptive control designs in chapters 5 and 8 employed continuous-in-time approaches. In this part, however, we present triggered-type adaptive control designs for hyperbolic partial differential equation-ordinary differential equation (ODE) systems, with the purpose of alleviating the adaptive learning transient and making the adaptive control laws more user-friendly in string-actuated mechanisms with massive actuators.

In this chapter, we extend the continuous-in-time adaptive controller in chapter 5 to an event-triggered form. Compared with the nonadaptive event-triggered control design in chapter 11, the adaptive update law in this chapter needs an appropriately modified event-triggering mechanism design to guarantee the absence of a Zeno phenomenon.

The content of this chapter is organized as follows. The problem formulation is shown in section 13.1. An observer is designed to estimate the PDE states in section 13.2. Using the approach in chapter 5, the design of a continuous-in-time adaptive observer-based backstepping controller is presented in section 13.3, followed by an observer-based event-triggering mechanism in section 13.4, where the existence of a minimal dwell time is proved. The asymptotic stability of the overall adaptive event-based output-feedback closed-loop system is proved via Lyapunov analysis in section 13.5. In a numerical simulation in section 13.6, the proposed adaptive event-triggered controller is tested in the lateral vibration suppression of a mining cable elevator with a viscoelastic guideway whose stiffness and damping coefficients are unknown.

13.1 PROBLEM FORMULATION

The class of plants considered in this chapter is

$$\dot{X}(t) = AX(t) + Bw(0,t), \tag{13.1}$$

$$z(0,t) = CX(t) + p_1 w(0,t), \tag{13.2}$$

$$z_t(x,t) = -q_1(x)z_x(x,t) + c_1(x)z(x,t) + c_2(x)w(x,t), \tag{13.3}$$

$$w_t(x,t) = q_2(x)w_x(x,t) + c_3(x)z(x,t) + c_4(x)w(x,t), \tag{13.4}$$

$$w(l(t),t) = U(t), \tag{13.5}$$

with $x \in [0, l(t)], t \in [0, \infty)$. The vector $X(t) \in \mathbb{R}^n$ is an ODE state, and the scalars $z(x,t), w(x,t)$ are PDE states. Equation (13.5) is the boundary condition with control. The spatially varying transport speeds $q_1(x), q_2(x) \in C^1$ are positive, and $c_1(x), c_2(x), c_3(x), c_4(x) \in C^0$ are arbitrary. The constant p_1 is nonzero. The matrix

$C \in \mathbb{R}^{1 \times n}$ is arbitrary. The input matrix $B \in \mathbb{R}^{n \times 1}$ and the system matrix $A \in \mathbb{R}^{n \times n}$ with unknown parameters satisfy the following assumptions.

Assumption 13.1. *The matrices A, B are in the form*

$$A = \begin{pmatrix} 0 & 1 & 0 & 0 & \cdots & 0 \\ 0 & 0 & 1 & 0 & \cdots & 0 \\ & & \vdots & & & \\ 0 & 0 & 0 & 0 & \cdots & 1 \\ g_1 & g_2 & g_3 & \cdots & g_{n-1} & g_n \end{pmatrix}, B = \begin{pmatrix} 0 \\ 0 \\ 0 \\ 0 \\ h_n \end{pmatrix}, \tag{13.6}$$

where the constants g_1, g_2, g_3, $\ldots$, g_{n-1}, g_n are unknown and arbitrary, and their lower and upper bounds are known and arbitrary. The constant h_n is nonzero and known.

Assumption 13.1 indicates that the ODE is in the controllable form, which covers many practical models, including the payload dynamics in the string-actuated mechanisms.

The known target matrix

$$A_{\mathrm{m}} = \begin{pmatrix} 0 & 1 & 0 & 0 & \cdots & 0 \\ 0 & 0 & 1 & 0 & \cdots & 0 \\ & & \vdots & & & \\ 0 & 0 & 0 & 0 & \cdots & 1 \\ \bar{g}_1 & \bar{g}_2 & \bar{g}_3 & \cdots & \bar{g}_{n-1} & \bar{g}_n \end{pmatrix} \tag{13.7}$$

indicates the coefficients $\bar{g}_1$, $\bar{g}_2$, $\bar{g}_3$, $\ldots$, $\bar{g}_{n-1}$, $\bar{g}_n$, which are chosen by the user to make A_{m} Hurwitz and to achieve a desired degree of performance for the specific application.

According to assumption 13.1 and (13.7), we know that there exists a unique, though unknown, row vector

$$K_{1 \times n} = [k_1, \ldots, k_n] \tag{13.8}$$

such that

$$A_m = A + BK \tag{13.9}$$

and

$$\bar{g}_i = g_i + h_n k_i, \quad i = 1, 2, \ldots, n. \tag{13.10}$$

By virtue of (13.10), while the k_i's are unknown, the lower and upper bounds on the k_i's—that is, $[\underline{k}_i, \bar{k}_i]$, $i = 1, 2, \ldots, n$—are known because the lower and upper bounds of the system matrix coefficients g_i's are known in assumption 13.1, and the target matrix coefficients $\bar{g}_i$'s are chosen by the user.

The time-varying domain—that is, the moving boundary $l(t)$—is under the following two assumptions.

Assumption 13.2. *The function $l(t)$ is uniformly bounded—that is, $0 < l(t) \leq L$, $\forall t \geq 0$, where L is a positive constant.*

Assumption 13.3. *The function $\dot{l}(t)$ is bounded as*

$$\left|\dot{l}(t)\right| \leq v_m < \min_{0 \leq x \leq L} \{q_1(x), q_2(x)\}, \tag{13.11}$$

where v_m is the maximum velocity of the moving boundary.

The limit of the speed of the moving boundary in assumption 13.3 is to ensure the well-posedness of the plant (13.1)–(13.5) according to [71, 72].

The complete set of the plant parameters is given by

$$\zeta_p = \{p_1, q_1, q_2, c_1, c_2, c_3, c_4, A, B, L, v_m\}, \tag{13.12}$$

where x is omitted for conciseness.

13.2 OBSERVER

To estimate the PDE states $z(x,t)$, $w(x,t)$, which usually cannot be fully measured in practice but are employed in the controller, an observer using the measurements $X(t)$, $z(l(t), t)$ is formulated as

$$\dot{X}(t) = AX(t) + B\hat{w}(0,t) + B\tilde{w}(0,t), \tag{13.13}$$

$$\hat{z}(0,t) = CX(t) + p_1\hat{w}(0,t), \tag{13.14}$$

$$\hat{z}_t(x,t) = -q_1(x)\hat{z}_x(x,t) + c_1(x)\hat{z}(x,t) + c_2(x)\hat{w}(x,t)$$
$$+ \Psi_2(x, l(t))(z(l(t),t) - \hat{z}(l(t),t)), \tag{13.15}$$

$$\hat{w}_t(x,t) = q_2(x)\hat{w}_x(x,t) + c_3(x)\hat{z}(x,t) + c_4(x)\hat{w}(x,t)$$
$$+ \Psi_3(x, l(t))(z(l(t),t) - \hat{z}(l(t),t)), \tag{13.16}$$

$$\hat{w}(l(t),t) = U(t), \tag{13.17}$$

where (13.13) is exactly the ODE (13.1) with

$$w(0,t) = \hat{w}(0,t) + \tilde{w}(0,t),$$

providing the measured signal $X(t)$ into (13.14). It should be noted that (13.13) is not computed online as a part of the observer. The observer gains $\Psi_2(x, l(t))$, $\Psi_3(x, l(t))$ are to be determined.

As we indicate, because the ODE state X is available, the observer (13.14)–(13.17) is only to estimate the PDE states. Let us denote the observer error states as

$$(\tilde{z}(x,t), \tilde{w}(x,t)) = (z(x,t), w(x,t)) - (\hat{z}(x,t), \hat{w}(x,t)). \tag{13.18}$$

According to (13.2)–(13.5) and (13.14)–(13.17), the observer error system is obtained as

$$\tilde{z}(0,t) = p_1\tilde{w}(0,t), \tag{13.19}$$

$$\tilde{z}_t(x,t) = -q_1(x)\tilde{z}_x(x,t) + c_1(x)\tilde{z}(x,t) + c_2(x)\tilde{w}(x,t)$$
$$- \Psi_2(x, l(t))\tilde{z}(l(t),t), \tag{13.20}$$

$$\tilde{w}_t(x,t) = q_2(x)\tilde{w}_x(x,t) + c_3(x)\tilde{z}(x,t) + c_4(x)\tilde{w}(x,t)$$
$$- \Psi_3(x,l(t))\tilde{z}(l(t),t), \tag{13.21}$$
$$\tilde{w}(l(t),t) = 0. \tag{13.22}$$

The observer gains $\Psi_2(x,l(t))$, $\Psi_3(x,l(t))$ are to be designed to ensure convergence to zero of the observer errors (13.18). Next, we postulate the backstepping transformation

$$\tilde{z}(x,t) = \tilde{\alpha}(x,t) - \int_x^{l(t)} \bar{\phi}(x,y)\tilde{\alpha}(y,t)dy$$
$$- \int_x^{l(t)} \check{\phi}(x,y)\tilde{\beta}(y,t)dy, \tag{13.23}$$

$$\tilde{w}(x,t) = \tilde{\beta}(x,t) - \int_x^{l(t)} \bar{\psi}(x,y)\tilde{\alpha}(y,t)dy$$
$$- \int_x^{l(t)} \check{\psi}(x,y)\tilde{\beta}(y,t)dy \tag{13.24}$$

to convert the original observer error system (13.19)–(13.22) to the following target observer error systems:

$$\tilde{\alpha}(0,t) = p_1\tilde{\beta}(0,t), \tag{13.25}$$
$$\tilde{\alpha}_t(x,t) = -q_1(x)\tilde{\alpha}_x(x,t) + c_1(x)\tilde{\alpha}(x,t), \tag{13.26}$$
$$\tilde{\beta}_t(x,t) = q_2(x)\tilde{\beta}_x(x,t) + c_4(x)\tilde{\beta}(x,t), \tag{13.27}$$
$$\tilde{\beta}(l(t),t) = 0. \tag{13.28}$$

Even though the integration interval $[0, l(t)]$ is time-varying, the kernels in (13.23), (13.24) need not include the argument $l(t)$ because the extra terms in which $l(t), \dot{l}(t)$ appear in the course of calculating the kernel conditions will be "absorbed" by the observer gains $\Psi_2(x,l(t))$, $\Psi_3(x,l(t))$. By matching (13.19)–(13.22) and (13.25)–(13.28) via (13.23), (13.24), we obtain PDE conditions on the kernels $\bar{\phi}(x,y)$, $\check{\phi}(x,y)$, $\bar{\psi}(x,y)$, $\check{\psi}(x,y)$. These PDEs are well-posed because they belong to a general class of kernel equations whose well-posedness is proved in theorem 3.2 of [48]. The observer gains are then deduced as

$$\Psi_2(x,l(t)) = \dot{l}(t)\bar{\phi}(x,l(t)) - q_1(l(t))\bar{\phi}(x,l(t)), \tag{13.29}$$
$$\Psi_3(x,l(t)) = \dot{l}(t)\bar{\psi}(x,l(t)) - q_1(l(t))\bar{\psi}(x,l(t)). \tag{13.30}$$

The detailed calculations of matching (13.19)–(13.22) and (13.25)–(13.28) via (13.23), (13.24) and of deriving the conditions on the kernels $\bar{\phi}(x,y)$, $\check{\phi}(x,y)$, $\bar{\psi}(x,y)$, $\check{\psi}(x,y)$ are shown in appendix 5.7A in chapter 5.

Lemma 13.1. *For the observer error system* (13.19)–(13.22), *the state estimation errors* $\tilde{z}(x,t), \tilde{w}(x,t)$ *become zero after*

$$t_f = \frac{L}{\min_{0 \le x \le L}\{q_1(x)\}} + \frac{L}{\min_{0 \le x \le L}\{q_2(x)\}}.$$

Proof. According to the target observer error system (13.25)–(13.28) and the result in [96], we know that $\tilde{\alpha}(x,t), \tilde{\beta}(x,t)$ become zero after a finite time t_f. Applying the Cauchy-Schwarz inequality into (13.23), (13.24), the proof of this lemma is complete.   □

The finite time in which lemma 13.1 establishes that the observer errors vanish depends only on the plant parameters and not on the controller parameters.

In the next section, we first design a continuous-in-time adaptive control law to stabilize the coupled transport PDEs coupled with a highly uncertain ODE at the uncontrolled boundary. Then we design an event-triggering mechanism, which uses the signals from the observer and includes an internal dynamic variable and which produces triggering times based on evaluating the size of the deviation of the control input applied over the interval between the triggers from the continuous-in-time control signal. The combined continuous-in-time adaptive controller and the event-triggering mechanism constitute the adaptive event-triggered boundary controller.

13.3 ADAPTIVE CONTINUOUS-IN-TIME CONTROL DESIGN

In this section, we conduct a state-feedback control backstepping design, with the intent of feeding into this full-state design the observer states from the observer in the previous section. In other words, we will design an observer-based output-feedback controller that we then make adaptive. The output error injections $z(l(t),t)$, $\hat{w}(0,t)$ in the observer are regarded as zero in the state-feedback design, and then the separation principle, which is verified by the fact that the stability of the observer error system is independent of the control design according to lemma 13.1, is applied in the stability analysis of the resulting closed-loop system.

Backstepping

Two transformations are employed to convert (13.13)–(13.17) to a target system, with the purposes of removing the couplings in the PDE domain and of making the ODE system matrix Hurwitz.

a) The first transformation to decouple PDEs

We postulate the backstepping transformation

$$\hat{\alpha}(x,t) = \hat{z}(x,t) - \int_0^x J(x,y)\hat{z}(y,t)dy - \int_0^x G(x,y)\hat{w}(y,t)dy, \qquad (13.31)$$

$$\hat{\beta}(x,t) = \hat{w}(x,t) - \int_0^x F(x,y)\hat{z}(y,t)dy - \int_0^x N(x,y)\hat{w}(y,t)dy \qquad (13.32)$$

to convert (13.13)–(13.17) to the following system:

$$\dot{X}(t) = AX(t) + B\hat{\beta}(0,t), \qquad (13.33)$$

$$\hat{\alpha}(0,t) = CX(t) + p_1\hat{\beta}(0,t), \qquad (13.34)$$

$$\hat{\alpha}_t(x,t) = -q_1(x)\hat{\alpha}_x(x,t) + c_1(x)\hat{\alpha}(x,t) - J(x,0)q_1(0)CX(t), \qquad (13.35)$$

$$\hat{\beta}_t(x,t) = q_2(x)\hat{\beta}_x(x,t) + c_4(x)\hat{\beta}(x,t) - F(x,0)q_1(0)CX(t), \qquad (13.36)$$

$$\hat{\beta}(l(t),t) = U(t) - \int_0^{l(t)} F(l(t),y)\hat{z}(y,t)dy$$

$$\qquad - \int_0^{l(t)} N(l(t),y)\hat{w}(y,t)dy. \qquad (13.37)$$

By matching (13.33)–(13.37) and (13.13)–(13.17) via (13.31), (13.32) through a process similar to that shown in appendix 5.7B in chapter 5, we obtain the conditions of the kernels $J(x,y)$, $G(x,y)$, $F(x,y)$, $N(x,y)$, which is a special case of (6.87)–(6.100) in chapter 6, where (6.88) and (6.95) hold here because the PDE states are scalar in this chapter. The well-posedness proof can be found in lemma 6.1 in chapter 6.

b) The second transformation to form a stable ODE

We postulate the backstepping transformation

$$\hat{\eta}(x,t) = \hat{\beta}(x,t) - \int_0^x \hat{N}(x,y;\hat{K}(t))\hat{\beta}(y,t)dy - D(x;\hat{K}(t))X(t), \qquad (13.38)$$

where $\hat{K}(t) \in R^{1\times n}$ is the estimate of the ideal control gains and will be shown later. The conditions on the kernels $\hat{N}(x,y;\hat{K}(t)), D(x;\hat{K}(t))$ are to be determined next. The inverse transformation is postulated as

$$\hat{\beta}(x,t) = \hat{\eta}(x,t) - \int_0^x \hat{N}_I(x,y;\hat{K}(t))\hat{\eta}(y,t)dy - D_I(x;\hat{K}(t))X(t), \qquad (13.39)$$

where $\hat{N}_I(x,y;\hat{K}(t)), D_I(x;\hat{K}(t))$ are kernels which can be determined after the determination of $\hat{N}(x,y;\hat{K}(t)), D(x;\hat{K}(t))$.

Through the transformation (13.38), we convert (13.33)–(13.37) into the following target system:

$$\dot{X}(t) = A_{\mathrm{m}}X(t) - B\tilde{K}(t)X(t) + B\hat{\eta}(0,t), \qquad (13.40)$$

$$\hat{\alpha}(0,t) = (C + p_1 D(0;\hat{K}(t)))X(t) + p_1\hat{\eta}(0,t), \qquad (13.41)$$

$$\hat{\alpha}_t(x,t) = -q_1(x)\hat{\alpha}_x(x,t) + c_1(x)\hat{\alpha}(x,t) - J(x,0)q_1(0)CX(t), \qquad (13.42)$$

$$\hat{\eta}_t(x,t) = q_2(x)\hat{\eta}_x(x,t) + c_4(x)\hat{\eta}(x,t) - \dot{\hat{K}}(t)R(x,t)$$

$$\qquad + \left(D(x;\hat{K}(t))B\tilde{K}(t) - \dot{\hat{K}}(t)D_{\hat{K}(t)}(x;\hat{K}(t)) \right) X(t), \qquad (13.43)$$

$$\hat{\eta}(l(t),t) = 0, \qquad (13.44)$$

where

$$\tilde{K}(t) = K - \hat{K}(t), \qquad (13.45)$$

and

$$R(x,t) = \int_0^x \hat{N}_{\hat{K}(t)}(x,y;\hat{K}(t))\hat{\beta}(y,t)dy$$

$$\qquad = \int_0^x \hat{N}_{\hat{K}(t)}(x,y;\hat{K}(t))\left[\hat{\eta}(y,t) - \int_0^y \hat{N}_I(y,\sigma;\hat{K}(t))\hat{\eta}(\sigma,t)d\sigma\right.$$

$$\qquad \left. - D_I(y;\hat{K}(t))X(t)\right]dy. \qquad (13.46)$$

The partial derivatives appearing in (13.43)–(13.46), respectively, are

$$D_{\hat{K}(t)}(x; \hat{K}(t)) = \frac{\partial D(x; \hat{K}(t))}{\partial \hat{K}(t)}$$

and

$$\hat{N}_{\hat{K}(t)}(x, y; \hat{K}(t)) = \frac{\partial \hat{N}(x, y; \hat{K}(t))}{\partial \hat{K}(t)}.$$

By matching (13.33)–(13.37) and (13.40)–(13.44) via (13.38) (see part C in appendix 5.7 for the details), the conditions of the kernels $N(x, y; \hat{K}(t))$, $D(x; \hat{K}(t))$ in (13.38) are determined as

$$D(0; \hat{K}(t)) = \hat{K}(t), \tag{13.47}$$

$$-q_2(x)D'(x; \hat{K}(t)) + D(x; \hat{K}(t))(A_\mathrm{m} - c_4(x)I_n - B\hat{K}(t))$$
$$+ F(x, 0)q_1(0)C - \int_0^x \hat{N}(x, y; \hat{K}(t))F(y, 0)q_1(0)C dy = 0, \tag{13.48}$$

$$q_2(y)\hat{N}_y(x, y; \hat{K}(t))$$
$$+ q_2(x)\hat{N}_x(x, y; \hat{K}(t)) + q_2'(y)\hat{N}(x, y; \hat{K}(t)) = 0, \tag{13.49}$$
$$q_2(0)\hat{N}(x, 0; \hat{K}(t)) = D(x; \hat{K}(t))B, \tag{13.50}$$

where I_n is an identity matrix with dimension n. The equation set (13.47)–(13.50) is a transport PDE-ODE coupled system consisting of the transport PDE (13.49) with the boundary condition (13.50) on $\{(x, y)|0 \leq y \leq x \leq l(t)\}$ and the ODE (13.48) with the initial value (13.47) on $\{0 \leq x \leq l(t)\}$. It should be noted that $\hat{K}(t)$ is a parameter rather than a variable in the transport PDE (13.49), (13.50) with respect to the independent variables x, y and in the ODE (13.47), (13.48) with respect to the independent variable x. In the study of the well-posedness of (13.47)–(13.50), the transport PDE state $\hat{N}(x, y; \hat{K}(t))$ can be represented by its boundary value $D(x; \hat{K}(t))B$. Substituting the result into ODE (13.48) to replace $\dot{N}(x, y; \hat{K}(t))$, the unique and continuous solution of the first-order ODE $\hat{D}(x; \hat{K}(t))$ (13.48) can be obtained. Then the unique and continuous solution of the transport PDE $\hat{N}(x, y; \hat{K}(t))$ in (13.49), (13.50) is obtained because of the well-defined and continuous input signal in (13.50). Following section 9.2 in chapter 9, the kernels $\hat{N}_I(x, y; \hat{K}(t))$, $D_I(x; \hat{K}(t))$ in the inverse transformation can then be determined.

Adaptive Update Laws

The objective in this section is to build adaptive update laws to obtain self-tuning of the control gains $\hat{K}(t) = [\hat{k}_1(t), \ldots, \hat{k}_n(t)]$, where normalization and projection operators are used to guarantee boundedness, as is typical in adaptive control designs.

The adaptive update law $\hat{K}(t) = [\hat{k}_1, \ldots, \hat{k}_n]$ is of the form

$$\dot{\hat{k}}_i(t) = \mathrm{Proj}_{[\underline{k}_i, \bar{k}_i]}\left(\tau_i(t), \hat{k}_i(t)\right). \tag{13.51}$$

While projection is applicable for arbitrary convex sets, the set within which the control gain vector K should reside in a hyperrectangle or, as is colloquially said, the estimate $\hat{K}(t)$ should be maintained within box constraints. Given the hyperrectangular set for the feedback gains, for any $m \leq M$ and any r, p, $\mathrm{Proj}_{[m, M]}$ is defined as the operator given by

$$\text{Proj}_{[m,M]}(r,p) = \begin{cases} 0, & \text{if } p=m \text{ and } r<0, \\ 0, & \text{if } p=M \text{ and } r>0, \\ r, & \text{else.} \end{cases}$$

So the projection operator is to keep the scalar components of parameter estimate vector $\hat{K}(t) = [\hat{k}_1, \ldots, \hat{k}_n]$ bounded within the interval $[\underline{k}_i, \bar{k}_i]$. The bounds $\underline{k}_i$ and $\bar{k}_i$ are determined from the bounds on the unknown parameters in A using assumption 13.1, as well as (13.7) and (13.10). We choose the parameter update rate functions τ_i in (13.51) as

$$[\tau_1(t), \ldots, \tau_n(t)]^T = \frac{\Gamma_c}{1 + \Omega(t) - \mu_m m_d(t)} \left[-2X(t)B^T P X(t) \right.$$
$$\left. + r_a \int_0^{l(t)} e^{\delta x} \hat{\eta}(x,t) X(t) B^T D(x; \hat{K}(t))^T dx \right], \qquad (13.52)$$

where $m_d(t) < 0$, a dynamic variable in the event-triggering mechanism, will be defined in the next section, and the adaptation gain matrix is

$$\Gamma_c = \text{diag}\{\gamma_{c1}, \ldots, \gamma_{cn}\}, \qquad (13.53)$$

and where $\Omega(t)$ is defined as

$$\Omega(t) = X(t)^T P X(t) + \frac{1}{2} r_a \int_0^{l(t)} e^{\delta x} \hat{\eta}(x,t)^2 dx$$
$$+ \frac{1}{2} r_b \int_0^{l(t)} e^{-\delta x} \hat{\alpha}(x,t)^2 dx. \qquad (13.54)$$

The determination of the positive constants δ, r_a, and r_b will be shown in the next section. The matrix $P = P^T > 0$ is the unique solution to the Lyapunov equation

$$PA_m + A_m^T P = -Q \qquad (13.55)$$

for some $Q = Q^T > 0$. It should be noted that P is known since A_m is known (chosen by the user). We introduce the normalization $\Omega(t) + 1$ in the denominator in (13.52) in order to keep the rate of change of the parameter estimate $\dot{\hat{K}}(t)$ bounded, which will be used in the following ETM design and stability analysis. The functions $\hat{\eta}(x,t)$ and $\hat{\alpha}(x,t)$ in (13.52)–(13.54) can be represented by the observer states through (13.31), (13.32), (13.38). The complete set of positive design parameters in the parameter update law is defined as

$$\zeta_a = \{\Gamma_c, \delta, r_a, r_b, \mu_m\}. \qquad (13.56)$$

The update law designs in this section will be chosen with the help of a Lyapunov analysis in section 13.5.

Continuous-in-Time Control Law

The continuous-in-time adaptive backstepping control law is derived in this section. For (13.44) to hold, using (13.37), recalling (13.38) and (13.32), we get

$$U(t) = \int_0^{l(t)} \bar{M}(l(t), x; \hat{K}(t))\hat{z}(x,t)dx + D(l(t); \hat{K}(t))X(t)$$

$$+ \int_0^{l(t)} \bar{N}(l(t), x; \hat{K}(t))\hat{w}(x,t)dx, \tag{13.57}$$

where $\bar{M}(l(t), x; \hat{K}(t)), \bar{N}(l(t), x; \hat{K}(t))$ are

$$\bar{M}(l(t), x; \hat{K}(t)) = F(l(t), x) - \int_x^{l(t)} \hat{N}(l(t), y; \hat{K}(t))F(y, x)dy, \tag{13.58}$$

$$\bar{N}(l(t), x; \hat{K}(t)) = N(l(t), x) + \hat{N}(l(t), x; \hat{K}(t))$$

$$- \int_x^{l(t)} \hat{N}(l(t), y; \hat{K}(t))N(y, x)dy. \tag{13.59}$$

In the output-feedback adaptive backstepping control law (13.57), the states $\hat{w}(x,t)$, $\hat{z}(x,t)$ are from the observer (13.14)–(13.17). The kernels $J, F, G, N, \hat{N}, D$ are derived from the backstepping process in this section. The state $X(t)$ is the measurement. The row vector $\hat{K}(t)$ is the adaptive estimate defined in (13.51), (13.52).

13.4 EVENT-TRIGGERING MECHANISM

In this section, we introduce an observer-based event-triggered control scheme for the stabilization of plant (13.1)–(13.5). It relies on both the continuous-in-time adaptive control signal $U(t)$ (13.57) and a dynamic ETM that determines the triggering times t_k (integer $k \geq 0$ and $t_0 = 0$) when the control signal is updated. In other words, the event-triggered control signal $U_d(t)$ is the frozen value of the continuous-in-time $U(t)$ at the time instants t_k, that is,

$$U_d(t) = U(t_k), \qquad t \in [t_k, t_{k+1}). \tag{13.60}$$

Inserting $U_d(t)$ into (13.17), we obtain

$$\hat{w}(l(t), t) = U_d(t). \tag{13.61}$$

A deviation $d(t)$ between the continuous-in-time adaptive control signal and the event-based control signal is given as

$$d(t) = U(t) - U_d(t). \tag{13.62}$$

Then (13.61) can be written as

$$\hat{w}(l(t), t) = U(t) - d(t). \tag{13.63}$$

Recalling the backstepping transformations and designs of $U(t)$ in section 13.3, the target system becomes (13.40)–(13.43) with the right boundary condition

$$\hat{\eta}(l(t), t) = -d(t). \tag{13.64}$$

The ETM to determine the triggering times of U_d is designed, as in [57], using the dynamic triggering condition

$$t_{k+1} = \inf\{t \in R^+ | t > t_k | d(t)^2 \geq \theta\Phi(t) - m_d(t)\}, \qquad (13.65)$$

where the internal dynamic variable $m_d(t)$ satisfies the ODE

$$\dot{m}_d(t) = -\eta m_d(t) + \lambda_d d(t)^2 - \sigma\Phi(t) - \kappa_1\hat{\alpha}(l(t), t)^2$$
$$- \kappa_2\hat{\eta}(0, t)^2 - \kappa_3\hat{\alpha}(0, t)^2, \qquad (13.66)$$

whose initial condition $m_d(0)$ should be chosen to be negative, and which is driven by the norm

$$\Phi(t) = |X(t)|^2 + \|\hat{\eta}(\cdot, t)\|^2 + \|\hat{\alpha}(\cdot, t)\|^2. \qquad (13.67)$$

The signals in (13.67) can be replaced by the observer states via (13.31), (13.32), (13.38). The complete set of ETM parameters is

$$\zeta_e = \{\theta, \eta, \lambda_d, \sigma, \kappa_1, \kappa_2, \kappa_3\}. \qquad (13.68)$$

These positive parameters are to be determined later.

The reason for introducing an internal dynamic variable $m_d(t)$ into the event-triggering condition (13.65) is that the changing rate $\dot{d}(t)$ of the deviation between $U(t)$ and $U_d(t)$, upon which the dwell time relies, includes as the last three terms in (13.66), the boundary states $\hat{\alpha}(l(t), t)$ $\hat{\eta}(0, t)$, $\hat{\alpha}(0, t)$, whose integration should be incorporated into the event-triggering condition (13.65) to avoid the Zeno phenomenon. The internal dynamic variable $m_d(t)$ is kept negative by the choice of θ. The explanation in this paragraph is formalized through the following three lemmas.

Lemma 13.2. *For $d(t)$ defined in (13.62), there exists a positive constant λ_a dependent only on the plant parameters ζ_p and the design parameters $\bar{g}_i$ in A_m in (13.7), such that*

$$\dot{d}(t)^2 \leq \lambda_a(\zeta_p, \bar{g}_i)\left(d(t)^2 + \hat{\alpha}(l(t), t)^2 + \hat{\eta}(0, t)^2 + \hat{\alpha}(0, t)^2 \right.$$
$$+ m_3(\zeta_p, \zeta_a, \bar{g}_i)\|\hat{\alpha}(\cdot, t)\|^2 + m_3(\zeta_p, \zeta_a, \bar{g}_i)\|\hat{\eta}(\cdot, t)\|^2$$
$$\left. + m_3(\zeta_p, \zeta_a, \bar{g}_i)|X(t)|^2 \right) \qquad (13.69)$$

for $t \in (t_k, t_{k+1})$, where m_3 is a positive constant dependent only on the plant parameters ζ_p, the adaptive law parameters ζ_a, and the design parameters $\bar{g}_i$'s in A_m.

Proof. The proof is shown in appendix 13.7A. $\square$

In the proof of lemma 13.2 and the remaining text in this chapter, a constant followed by $(\cdot)$ denotes a constant that depends only on the parameters in the parentheses. For conciseness, after the first appearance of the constant, $(\cdot)$ will be omitted when unnecessary.

Lemma 13.3. *Choosing*

$$\theta \leq \frac{\sigma}{\lambda_d} \qquad (13.70)$$

for the internal dynamic variable $m_d(t)$ defined in (13.66), it holds that $m_d(t) < 0$.

Proof. The proof is shown in appendix 13.7B.   □

Lemma 13.4. *For some κ_1, κ_2, κ_3, there is a minimal dwell time between two triggering times, which is equal to or greater than a positive constant $T_{\min}$, which depends only on the parameters of the plant and the choices of the design parameters.*

Proof. Introduce a function $\psi(t)$,

$$\psi(t) = \frac{d(t)^2 + \frac{1}{2}m_d(t)}{\theta\Phi(t) - \frac{1}{2}m_d(t)}, \tag{13.71}$$

which is proposed in [57]. We have

$$\psi(t_{k+1}) = 1$$

because the event condition in (13.65) is triggered, and

$$\psi(t_k) < 0$$

because $m_d(t) < 0$ (lemma 13.3) and $d(t_k) = 0$. The function $\psi(t)$ is continuous on $[t_k, t_{k+1}]$ due to the continuity and well-posedness of this class of 2×2 hyperbolic PDE-ODE system according to [48]. By the intermediate value theorem, there exists $t^* > t_k$ such that $\psi(t) \in [0,1]$ when $t \in [t^*, t_{k+1}]$. The minimal $T_{\min}$ can be found as the minimal time it takes for $\psi(t)$ to go from 0 to 1—that is, the reciprocal of the absolute value of the maximum changing rate of $\psi(t)$. Taking the derivative of (13.71); recalling lemma 13.2 and (13.66), (13.67); and choosing

$$\kappa_1 \geq \max\{2\lambda_a(\zeta_p, \bar{g}_i), 2\theta\lambda_p(\zeta_p)\}, \tag{13.72}$$

$$\kappa_2 \geq \max\{2\lambda_a(\zeta_p, \bar{g}_i), 2\theta\lambda_p(\zeta_p)\}, \tag{13.73}$$

$$\kappa_3 \geq \max\{2\lambda_a(\zeta_p, \bar{g}_i), 2\theta\lambda_p(\zeta_p)\}, \tag{13.74}$$

for some positive λ_p shown in appendix 13.7C; through a calculation process in appendix 13.7C, we get

$$\dot{\psi}(t) \leq n_1\psi(t)^2 + n_2\psi(t) + n_3 \tag{13.75}$$

with

$$n_1 = \frac{1}{2}\lambda_d + \theta\lambda_p(\zeta_p) > 0, \tag{13.76}$$

$$n_2 = 1 + \lambda_a(\zeta_p, \bar{g}_i) + \lambda_d + \theta\lambda_p(\zeta_p) + \eta + f_1(\sigma, \mu_0(\zeta_p, \zeta_a, \bar{g}_i), \theta) > 0, \tag{13.77}$$

$$n_3 = 1 + \frac{1}{2}\lambda_d + \lambda_a(\zeta_p, \bar{g}_i) + \frac{\lambda_a(\zeta_p, \bar{g}_i)m_3(\zeta_p, \zeta_a, \bar{g}_i)}{\theta} + \eta > 0, \tag{13.78}$$

which are positive constants where

$$f_1 = \begin{cases} \mu_0(\zeta_p, \zeta_a, \bar{g}_i) - \frac{1}{2\theta}\sigma, & \text{if } \sigma < 2\theta\mu_0(\zeta_p, \zeta_a, \bar{g}_i), \\ 0, & \text{if } \sigma \geq 2\theta\mu_0(\zeta_p, \zeta_a, \bar{g}_i). \end{cases}$$

Then it follows that the least time needed by $\psi(t)$ to go from 0 to 1 is

$$T_{\min} = \frac{1}{n_1 + n_2 + n_3} > 0$$

because the maximum changing rate $\dot{\psi}(t)$ is $n_1 + n_2 + n_3$ for $\psi(t) \in [0, 1]$ according to (13.75). The proof of this lemma is complete. $\square$

13.5   STABILITY ANALYSIS OF THE CLOSED-LOOP SYSTEM

The expression of the final adaptive event-triggered control law U_d is

$$U_d(t) = \int_0^{l(t_k)} \bar{M}(l(t_k), x; \hat{K}(t_k))\hat{z}(x, t_k)dx$$

$$+ \int_0^{l(t_k)} \bar{N}(l(t_k), x; \hat{K}(t_k))\hat{w}(x, t_k)dx + D(l(t_k); \hat{K}(t_k))X(t_k) \qquad (13.79)$$

for $t \in [t_k, t_{k+1})$, recalling (13.57) and (13.60). The triggering times t_k (for integer $k \geq 0$) are determined by the ETM in (13.65), (13.66). In (13.79), $\hat{z}, \hat{w}$ are states from the observer (13.14)–(13.17), $\hat{K}$ is the adaptive update law (13.51), (13.52), and X is the ODE measurement. The scalars $l(t_k)$ are the values of the time-varying function $l(t)$, which is known ahead of time, at the times t_k, the functions $\bar{M}, \bar{N}$ are given in (13.58), (13.59), and D is defined in (13.47)–(13.50).

Lemma 13.5. *For all initial values* $(\hat{\alpha}(\cdot, 0), \hat{\eta}(\cdot, 0)) \in L^2(0, L)$, $X(0) \in \mathbb{R}^n$, $m_d(0) < 0$, *the event-based target system* (13.40)–(13.43), (13.64) *is asymptotically stable in the sense of*

$$\lim_{t \to \infty} \left(\|\hat{\alpha}(\cdot, t)\|^2 + \|\hat{\eta}(\cdot, t)\|^2 + |X(t)|^2 + |m_d(t)| \right) = 0.$$

Proof. Define

$$\underline{q_1} = \min_{0 \leq x \leq L} \{q_1(x)\},$$

$$\overline{q_1'} = \max_{0 \leq x \leq L} \{|q_1'(x)|\},$$

$$\underline{q_2} = \min_{0 \leq x \leq L} \{q_2(x)\},$$

$$\overline{q_2'} = \max_{0 \leq x \leq L} \{|q_2'(x)|\},$$

$$\overline{q_1} = \max_{0 \leq x \leq L} \{q_1(x)\},$$

$$\overline{q_2} = \max_{0 \leq x \leq L} \{q_2(x)\},$$

$$\overline{c_1} = \max_{0 \leq x \leq L} \{|c_1(x)|\},$$

$$\overline{c_4} = \max_{0 \leq x \leq L} \{|c_4(x)|\}.$$

Step 1. Choose the Lyapunov function as

$$V(t) = \ln\left(1 + \Omega(t) - \mu_m m_d(t)\right) + \frac{1}{2}\tilde{K}(t)\Gamma_c^{-1}\tilde{K}(t)^T, \tag{13.80}$$

where the terms $\tilde{K}(t)$, $m_d(t)$ are related to the adaptive law and ETM. Because of $m_d(t) < 0$, we have

$$1 + \Omega(t) - \mu_m m_d(t) > 0.$$

According to (13.54) and (13.67), we obtain

$$\mu_1 \Phi(t) \leq \Omega(t) \leq \mu_2 \Phi(t), \tag{13.81}$$

with positive μ_1, μ_2 as

$$\mu_1 = \frac{1}{2}\min\{r_a, r_b e^{-\delta L}, \lambda_{\min}(P)\}, \tag{13.82}$$

$$\mu_2 = \frac{1}{2}\max\{r_a e^{\delta L}, r_b, \lambda_{\max}(P)\}, \tag{13.83}$$

where $\lambda_{\min}$ and $\lambda_{\max}$ denote the minimum and maximum eigenvalues of the corresponding matrix.

Taking the derivative of (13.80) along (13.40)–(13.43) with (13.64) and (13.66) and applying the Young and Cauchy-Schwarz inequalities, through a process in appendix 13.7D, we arrive at

$$\dot{V}(t) \leq \frac{1}{1 + \Omega(t) - \mu_m m_d(t)}\Bigg[-\left(\frac{7}{8}\lambda_{\min}(Q) - q_1(0)r_b\bar{D}^2\right.$$

$$\left. -\frac{1}{2}q_1(0)^2 Lr_b\bar{J}^2\,|C|^2 - \mu_m\sigma - \mu_m\kappa_3\bar{D}^2\right)|X(t)|^2$$

$$-\left(\frac{1}{2}q_2(0)r_a - (q_1(0)r_b + \mu_m\kappa_3)p_1^2\right.$$

$$\left. -\frac{8}{\lambda_{\min}(Q)}|PB|^2 - \mu_m\kappa_2\right)\hat{\eta}(0,t)^2$$

$$-\left[r_a\left(\frac{1}{2}\delta\underline{q_2} - \overline{c_4} - \frac{1}{2}\overline{q_2'}\right)e^{\delta x} - \mu_m\sigma\right]\int_0^{l(t)}\hat{\eta}(x,t)^2 dx$$

$$-\left[\frac{1}{2}(q_1(l(t)) - \dot{l})r_b e^{-\delta l(t)} - \mu_m\kappa_1\right]\hat{\alpha}(l(t),t)^2$$

$$-\left[r_b\left(\frac{1}{2}\underline{q_1}\delta - \overline{c_1} - \frac{1}{2} - \frac{\overline{q_1'}}{2}\right)e^{-\delta x} - \mu_m\sigma\right]\int_0^{l(t)}\hat{\alpha}(x,t)^2 dx$$

$$-r_a\int_0^{l(t)}e^{\delta x}\hat{\eta}(x,t)\left(\dot{\tilde{K}}(t)D_{\hat{K}(t)}(x;\hat{K}(t))X(t) + \dot{\tilde{K}}(t)R(x,t)\right)dx$$

$$-\left(\mu_m\lambda_d - \frac{1}{2}(\overline{q_2} + v_m)r_a e^{\delta L}\right)d(t)^2 + \mu_m\eta m_d(t)\Bigg], \tag{13.84}$$

where

$$\bar{J} = \max_{0 \leq x \leq L}\{|J(x,0)|\},$$

$$\bar{D} = \max_{\underline{k_i} \leq \hat{k}_i(t) \leq \bar{k}_i}\{|C + p_1 D(0;\hat{K}(t))|\},$$

and (13.41), (13.51), (13.52), (13.64) have been used.

Inserting (13.70) into (13.72)–(13.74) to replace θ by $\frac{\sigma}{\lambda_d}$ and adding additional condition

$$\lambda_d \geq 1,$$

we summarize the conditions of the design parameters in adaptive event-triggered backstepping control systems as

$$\min\{\kappa_1, \kappa_2, \kappa_3\} \geq \max\{2\lambda_a, 2\sigma\lambda_p\}, \tag{13.85}$$

$$\delta > \max\left\{ \frac{2\overline{c_4} + \overline{q_2'}}{\underline{q_2}}, \frac{2\overline{c_1} + 1 + \overline{q_1'}}{\underline{q_1}} \right\}, \tag{13.86}$$

$$r_b < \frac{\frac{7}{8}\lambda_{\min}(Q)}{q_1(0)\bar{D}^2 + \frac{1}{2}q_1(0)^2 L\bar{J}^2 |C|^2}, \tag{13.87}$$

$$r_a > \frac{2}{q_2(0)}\left(q_1(0)r_b p_1^2 + \frac{8}{\lambda_{\min}(Q)}|PB|^2 \right), \tag{13.88}$$

$$\mu_m < \min\left\{ \frac{1}{\sigma + \kappa_3 \bar{D}^2}\left[\frac{7}{8}\lambda_{\min}(Q) \right.\right.$$
$$\left. - r_b\left(q_1(0)\bar{D}^2 - \frac{1}{2}q_1(0)^2 L\bar{J}^2 |C|^2 \right) \right],$$
$$\frac{r_a\left(\frac{1}{2}\delta \underline{q_2} - \overline{c_4} - \frac{1}{2}\overline{q_2'} \right)}{\sigma}, \frac{r_b\left(\frac{1}{2}\underline{q_1}\delta - \overline{c_1} - \frac{1}{2} - \frac{1}{2}\overline{q_1'} \right)e^{-\delta L}}{\sigma},$$
$$\frac{\frac{1}{2}q_2(0)r_a - q_1(0)r_b p_1^2 - \frac{8}{\lambda_{\min}(Q)}|PB|^2}{\kappa_2 + \kappa_3 p_1^2},$$
$$\left.\frac{\frac{1}{2}(q_1(l(t)) - v_m)r_b e^{-\delta L}}{\kappa_1} \right\}, \tag{13.89}$$

$$\lambda_d \geq \max\left\{ \frac{(\overline{q_2} + v_m)r_a e^{\delta L}}{2\mu_m}, 1 \right\}, \tag{13.90}$$

$$\theta \leq \frac{\sigma}{\lambda_d}, \tag{13.91}$$

where σ, η are free parameters.

Applying the Young and Cauchy-Schwarz inequalities, recalling (13.46), we get

$$-r_a \int_0^{l(t)} e^{\delta x}\hat{\eta}(x,t)\left(\dot{\hat{K}}D_{\hat{K}(t)}X(t) + \dot{\hat{K}}R(x,t) \right)dx$$
$$\leq \max_{i \in \{1,\ldots,n\}}\{\gamma_{ci}\}\sqrt{n}\lambda_b(\zeta_p, \delta, r_a, r_b, \mu_m, \bar{g}_i)\left(|X(t)|^2 + \|\hat{\eta}\|^2 \right), \tag{13.92}$$

where the positive constant $\lambda_b(\zeta_p, \delta, r_a, r_b, \mu_m, \bar{g}_i) > 0$ depends only on the plant parameters ζ_p and the choices of r_a, r_b, δ, μ_m and the $\bar{g}_i$'s.

By choosing parameters as (13.85)–(13.91) and inserting (13.92), the inequality (13.84) becomes

$$\dot{V}(t) \leq \frac{1}{1 + \Omega(t) - \mu_m m_d(t)}\left[-\lambda_c(\zeta_p, \zeta_e, \delta, r_a, r_b, \mu_m, \bar{g}_i)\left(|X(t)|^2 \right.\right.$$
$$+ \hat{\eta}(0,t)^2 + \|\hat{\eta}(\cdot,t)\|^2 + \hat{\alpha}(l(t),t)^2 + \|\hat{\alpha}(\cdot,t)\|^2 + d(t)^2\right) + \mu_m \eta m_d(t)$$
$$\left. + \max_{i \in \{1,\ldots,n\}}\{\gamma_{ci}\}\sqrt{n}\lambda_b(|X(t)|^2 + \|\hat{\eta}(\cdot,t)\|^2)\right],$$

where the constant $\lambda_c(\zeta_p, \zeta_e, \delta, r_a, r_b, \mu_m, \bar{g}_i) > 0$ depends only on the plant parameters ζ_p, the design parameters $\bar{g}_i$'s in A_m, the event-triggering mechanism parameters ζ_e, and δ, r_a, r_b, μ_m in the adaptive law parameters. The coefficients γ_{ci} are independent of λ_b and λ_c. Choose $\max_{i \in \{1,\dots,n\}} \{\gamma_{ci}\}$ to satisfy

$$\max_{i \in \{1,\dots,n\}} \{\gamma_{ci}\} < \frac{\lambda_c(\zeta_p, \zeta_e, \delta, r_a, r_b, \mu_m, \bar{g}_i)}{\sqrt{n} \lambda_b(\zeta_p, \delta, r_a, r_b, \mu_m, \bar{g}_i)}. \tag{13.93}$$

Finally, we arrive at

$$\dot{V}(t) \leq \frac{-\min\{\bar{\lambda}_c, \eta\}}{1 + \Omega(t)} \Bigg(|X(t)|^2 + \hat{\eta}(0,t)^2 + \|\hat{\eta}(\cdot,t)\|^2$$

$$+ d(t)^2 + \hat{\alpha}(l(t),t)^2 + \|\hat{\alpha}(\cdot,t)\|^2 + \mu_m |m_d(t)| \Bigg) \leq 0, \tag{13.94}$$

where

$$\bar{\lambda}_c = \lambda_c - \max_{i \in \{1,\dots,n\}} \{\gamma_{ci}\} \sqrt{n} \lambda_b > 0$$

and

$$\lambda_{\mathrm{all}} := \min\{\bar{\lambda}_c, \eta\} \tag{13.95}$$

is related to the convergence rate of the closed-loop system.

Step 2. *Boundedness analysis of* $\frac{d}{dt}|X(t)|^2$, $\frac{d}{dt}\|\hat{\eta}(\cdot,t)\|^2$, $\frac{d}{dt}\|\hat{\alpha}(\cdot,t)\|^2$, *and* $\frac{d}{dt}|m_d(t)|$: According to (13.94) obtained in step 1, we thus have

$$V(t) \leq V(0), \ \forall t > 0.$$

One easily gets that $|\tilde{K}(t)|$, $\|\hat{\eta}(\cdot,t)\|$, $\|\hat{\alpha}(\cdot,t)\|$, $|X(t)|$, $|m_d(t)|$ are uniformly bounded and also that $\Phi(t)$ is bounded according to (13.67). Recalling the invertibility of the backstepping transformations (13.31), (13.32), (13.38), the boundedness of the signals $\|\hat{z}(\cdot,t)\|$, $\|\hat{w}(\cdot,t)\|$, $|X(t)|$ is obtained. Therefore, $U(t)$ is bounded according to (13.57). It follows that $d(t)$ is bounded via (13.62). Taking the time derivative of $|X(t)|^2$, $\|\hat{\alpha}(\cdot,t)\|^2$, $\|\hat{\eta}(\cdot,t)\|^2$, and $|m_d(t)|$ along (13.40)–(13.44), (13.66), we obtain

$$\frac{d}{dt}|\hat{X}(t)|^2 = 2X^T(t)(A_m X(t) + B\hat{\eta}(0,t) - B\tilde{K}X(t)), \tag{13.96}$$

$$\frac{d}{dt}\|\hat{\eta}(\cdot,t)\|^2 = (q_2(l(t)) + \dot{l}(t))d(t)^2 - q_2(0)\hat{\eta}(0,t)^2$$

$$- \int_0^{l(t)} (q_2'(x) - 2c_4(x))\hat{\eta}(x,t)^2 dx$$

$$+ 2\int_0^{l(t)} \hat{\eta}(x,t)\Big[D(x;\hat{K}(t))B\tilde{K}(t)$$

$$- \dot{\hat{K}}(t)D_{\hat{K}(t)}(x;\hat{K}(t))X(t) - \dot{\hat{K}}(t)R(x,t)\Big]dx, \tag{13.97}$$

$$\frac{d}{dt}\|\hat{\alpha}(\cdot,t)\|^2 = -(q_1(l(t)) - \dot{l}(t))\hat{\alpha}(l(t),t)^2 + q_1(0)\hat{\alpha}(0,t)^2$$

$$+ \int_0^{l(t)} (q_1'(x) + 2c_1(x))\hat{\alpha}(x,t)^2 dx$$

$$-2 \int_0^{l(t)} \hat{\alpha}(x,t) J(x,0) q_1(0) CX(t) dx, \tag{13.98}$$

$$\frac{d}{dt}|m_d(t)| = -\dot{m}_d(t) = \eta m_d(t) - \lambda_d d(t)^2 + \sigma \Phi(t)$$
$$+ \kappa_1 \hat{\alpha}(l(t),t)^2 + \kappa_2 \hat{\eta}(0,t)^2 + \kappa_3 \hat{\alpha}(0,t)^2. \tag{13.99}$$

Recalling the boundedness results proved above and (13.64), (13.62), (13.60), (13.57), we obtain the boundedness of $\hat{\eta}(l(t),t)$. We then find that $\hat{\eta}(0,t)$ is bounded as a result of the transport PDE (13.43), with recalling the boundedness of $\dot{\hat{K}}$ in (13.115) in appendix 13.7A. The signal $\hat{\alpha}(0,t)$ is bounded as well due to (13.41), and then $\hat{\alpha}(l(t),t)$ is bounded as a result of the transport PDE (13.42). Therefore, by applying the Young and Cauchy-Schwarz inequalities to (13.96)–(13.99), with the boundedness of $\dot{l}(t)$ in assumption 13.3, we get that $\frac{d}{dt}|X(t)|^2$, $\frac{d}{dt}\|\hat{\eta}(\cdot,t)\|^2$, $\frac{d}{dt}\|\hat{\alpha}(\cdot,t)\|^2$, and $\frac{d}{dt}|m_d(t)|$ are uniformly bounded.

Finally, integrating (13.94) obtained in step 1 from 0 to ∞, it follows that $|X(t)|^2$, $\|\hat{\alpha}(\cdot,t)\|^2$, $\|\hat{\eta}(\cdot,t)\|^2$, and $|m_d(t)|$ are integrable. Then using the results obtained in steps 1 and 2, according to Barbalat's lemma, we have that $|X(t)|^2$, $\|\hat{\alpha}(\cdot,t)\|^2$, $\|\hat{\eta}(\cdot,t)\|^2$, and $|m_d(t)|$ tend to zero as $t \to \infty$. $\qquad\square$

The following theorem establishes that, in the closed-loop system, no Zeno phenomenon takes place, namely that

$$\lim_{k \to \infty} t_k = +\infty, \tag{13.100}$$

and the states and the control signal are convergent to zero. The well-posedness of the event-based closed-loop system can be studied in a similar manner as in the proof of property (1) of theorem 11.1.

Theorem 13.1. *For all initial values* $(z(\cdot,0), w(\cdot,0)) \in L^2(0,L)$, $X(0) \in \mathbb{R}^n$, $(\hat{z}(\cdot,0), \hat{w}(\cdot,0)) \in L^2(0,L)$, *and* $m_d(0) < 0$, *with the design parameters satisfying* (13.85)–(13.91), (13.93), *the closed-loop system—that is, the plant* (13.1)–(13.5) *with the proposed observer-based adaptive event-triggered controller* (13.79), *which consists of the observer* (13.14)–(13.17), *the adaptive update law* (13.51), (13.52), *and the ETM* (13.65), (13.66)*—has the following properties:*

1) *There exists a positive constant* $T_{\min}$ *that depends only on the parameters of the plant and the choices of the control parameters such that*

$$\min_{k \geq 0} \{t_{k+1} - t_k\} \geq T_{\min}. \tag{13.101}$$

2) *In the closed-loop system, the states are asymptotically convergent to zero in the sense of*

$$\lim_{t \to \infty} (|X(t)|^2 + \|z(\cdot,t)\|^2 + \|w(\cdot,t)\|^2$$
$$+ \|\hat{z}(\cdot,t)\|^2 + \|\hat{w}(\cdot,t)\|^2 + |m_d(t)|) = 0. \tag{13.102}$$

3) *The adaptive event-triggered control signal is convergent to zero:*

$$\lim_{t \to \infty} U_d(t) = 0. \tag{13.103}$$

Proof. 1) Recalling lemma 13.4, property (1) is obtained.

2) Recalling the asymptotic stability result proved in lemma 13.5 and considering the invertibility and continuity of the backstepping transformations (13.31), (13.32), (13.38), we obtain the asymptotic convergence to zero of $\|\hat{z}(\cdot,t)\|^2 + \|\hat{w}(\cdot,t)\|^2 + |X(t)|^2 + |m_d(t)|$. Recalling lemma 13.1 and (13.18) and applying the separation principle, we obtain property (2).

3) Recalling (13.57) and property (2), we know that the continuous-in-time control signal $U(t)$ is asymptotically convergent to zero. According to the definition (13.60) and property (1), property (3) is then obtained. $\qquad\square$

Conditions on all the control parameters (13.85)–(13.91), (13.93) are cascaded rather than mutually dependent. The optimal choices of these parameters are not studied in this book.

13.6 APPLICATION IN THE FLEXIBLE-GUIDE MINING CABLE ELEVATOR

Simulation Model

According to section 5.1, the wave PDE modeled lateral vibration dynamics of the mining cable elevator are

$$\rho u_{tt} = T(x)u_{xx}(x,t) + T'(x)u_x(x,t) - \bar{c}u_t(x,t), \tag{13.104}$$

$$M_c u_{tt}(0,t) = -k_c u(0,t) - c_d u_t(0,t) + T(0)u_x(0,t), \tag{13.105}$$

$$-T(l(t))u_x(l(t),t) = U(t), \tag{13.106}$$

where $u(x,t)$ denotes the lateral vibration displacements along the cable shown in figure 13.1, and $x \in [0, l(t)]$ are the positions along the cable in a moving coordinate system associated with the motion $l(t)$, whose origin is located at the cage. The function

$$T(x) = M_c g + x\rho g$$

is the static tension along the cable, and ρ is the linear density of the cable. The constant $\bar{c}$ is the material damping coefficient of the cable. The values of the physical parameters of the mining cable elevator tested in the simulation are shown in table 13.1. The constants k_c, c_d are the unknown equivalent stiffness and damping coefficients of the viscoelastic guide.

Through applying the Riemann transformations

$$z(x,t) = u_t(x,t) - \sqrt{\frac{T(x)}{\rho}} u_x(x,t), \tag{13.107}$$

$$w(x,t) = u_t(x,t) + \sqrt{\frac{T(x)}{\rho}} u_x(x,t), \tag{13.108}$$

and defining

$$X(t) = [x_1(t), x_2(t)]^T = [u(0,t), u_t(0,t)]^T,$$

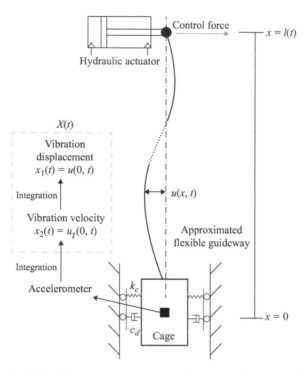

Figure 13.1. Mining cable elevator with viscoelastic guideways.

Table 13.1. Physical parameters of the descending mining cable elevator

Parameters (units)	Values
Initial length L_0 (m)	300
Final length (m)	2460
Cable linear density ρ (kg/m)	8.1
Total hoisted mass M_c (kg)	15000
Gravitational acceleration g (m/s^2)	9.8
Maximum hoisting velocities $\bar{v}_{\max}$ (m/s)	18
Total hoisting time t_f (s)	150
Cable material damping coefficient $\bar{c}$ (N·s/m)	0.4

which physically means the lateral displacement and velocity of the cage, we convert (13.104)–(13.106) into a 2×2 coupled transport PDE-ODE model in the form of (13.1)–(13.5) with the following coefficients:

$$q_1(x) = q_2(x) = \sqrt{\frac{T(x)}{\rho}}, c_1(x) = c_3(x) = \frac{-\bar{c}}{2\rho} - \frac{T'(x)}{4\sqrt{\rho T(x)}}, \tag{13.109}$$

$$c_2(x) = c_4(x) = \frac{-\bar{c}}{2\rho} + \frac{T'(x)}{4\sqrt{\rho T(x)}}, \ p_1 = -1, \tag{13.110}$$

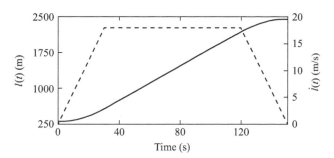

Figure 13.2. Time-varying domain $l(t)$ and the according velocity $\dot{l}(t)$.

$$A = \frac{1}{M_c}\left[\begin{array}{cc} 0 & M_c \\ -k_c & -c_d - \sqrt{M_c\rho g} \end{array}\right], B = \left[\begin{array}{c} 0 \\ \sqrt{\frac{\rho g}{M_c}} \end{array}\right], \quad C = [0,2]. \qquad (13.111)$$

For the lateral vibration model of the mining cable elevator (13.104)–(13.106) with the Riemann transformations (13.107), (13.108), the condition of the controlled boundary in (13.1)–(13.5) should have a simple augmentation, as follows:

$$w(l(t),l) = -\frac{2}{\sqrt{\rho T(l(t))}}U(t) + z(l(t),t). \qquad (13.112)$$

The additional term $z(l(t),t)$ can be canceled at the drum (see figure 2.1), so in the simulation, we consider the controlled boundary as (13.5), that is, $w(l(t),t) = U(t)$, where the designed control input, based on (13.1)–(13.5) with the above coefficients (13.109)–(13.111), should be multiplied by $-\frac{\sqrt{\rho T(l(t))}}{2}$ to convert the input signal computed based on (13.5) into the control force at the head sheave in the mining cable elevator—that is, into the control signal $U(t)$ in (13.112). In the practical mining cable elevator, $l(t)$ is obtained by the product of the radius and the angular displacement of the rotating drum driving the cable, where the angular displacement is measured by the angular displacement sensor at the drum.

   In the simulation, the unknown damping and stiffness coefficients of the flexible guide are set, respectively, as $c_d = 0.4$ and $k_c = 1000$. The target system matrix of the ODE is set as

$$\Lambda_{\mathrm{m}} = \left(\begin{array}{cc} 0 & 1 \\ -2.2 & -5.8 \end{array}\right). \qquad (13.113)$$

The unknown target control parameters k_1, k_2 are sought online by the adaptive mechanism to achieve the target system matrix A_{m}. The bounds of the unknown control parameters k_1, k_2 in the adaptive estimates are defined as $[-50,0], [-100,0]$.

   The time-varying cable length $l(t)$ and its changing rate $\dot{l}(t)$ are shown in figure 13.2. The maximum velocity of the moving boundary—that is, the maximum hoisting velocity $\bar{v}_{\max} = 18$ m/s—satisfies the limit of the changing rate of the time-varying domain proposed in assumption 13.3.

   The initial conditions of the plant (13.1)–(13.5) are defined as

$$w(x,0) = 0.2\sin(2\pi x/L_0),$$
$$z(x,0) = 0.4\sin\left(3\pi x/L_0 + \frac{\pi}{6}\right),$$

Table 13.2. Parameters of the proposed adaptive event-based control system

Parameters	Values
In adaptive law	$\gamma_{c1} = 0.95,\ \gamma_{c2} = 0.46,\ \delta = 3,\ r_b = 0.06,\ r_a = 4,\ \mu_m = 0.00002$
In the ETM	$\theta = 0.118,\ \eta = 41,\ \lambda_d = 1.3,\ \sigma = 0.5,\ \kappa_1 = \kappa_2 = \kappa_3 = 4$

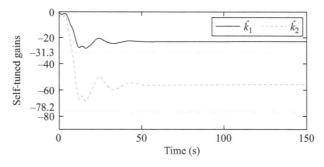

Figure 13.3. Self-tuned control gains $\hat{k}_1, \hat{k}_2$, whose target values are $k_1 = -31.3$, $k_2 = -78.2$.

$$x_2(0) = 0.5w(0,0) + 0.5z(0,0),$$
$$x_1(0) = 0.1.$$

The initial value $m_d(0)$ in the ETM is chosen as -0.03.

Simulation Results

The design parameters in the proposed adaptive event-based control system, including the design parameters in the adaptive law and ETM, are shown in table 13.2.

According to the system matrix, the input matrix in (13.111), and the target system matrix A_{m} (13.113), we know that the ideal control parameters k_1, k_2 are $-31.3, -78.2$, respectively. Figure 13.3 shows that our adaptive design can adjust the control parameters $\hat{k}_1(t), \hat{k}_2(t)$ in an online fashion to approach the ideal values. It often happens in adaptive control that, even though the estimates do not exactly arrive at their actual values, the state convergence is achieved in the closed-loop system, which will be seen shortly.

The proposed adaptive event-based control input and the continuous-in-time adaptive control input are shown in figure 13.4. The internal dynamic variable $m_d(t)$ in the ETM is shown in figure 13.5. The PDE states used in the control law are from the observer (13.14)–(13.17), and we can see from the observer error shown in figure 13.6 that observer errors at the midpoint of the time-varying spatial domain are convergent to zero after $t = 6.4$ s.

The responses of the PDE states and the ODE state are shown in figures 13.7–13.10, where the proposed controller can quickly suppress to zero the oscillations appearing in the open-loop system. The fact that oscillation amplitude decreases in the open-loop system is due to the fact that material damping of the cable is considered in the simulation. Figures 13.7 and 13.8 show that the PDE states at

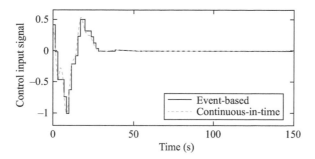

Figure 13.4. Adaptive event-based control input and the continuous-in-time adaptive control input.

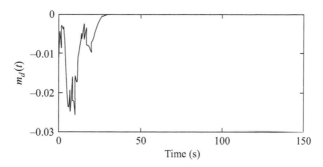

Figure 13.5. Dynamic internal variable $m_d(t)$ in the ETM.

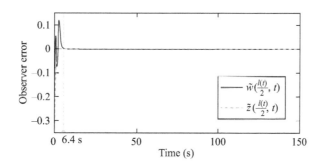

Figure 13.6. Observer errors at the midpoint of the time-varying spatial domain.

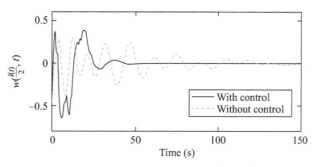

Figure 13.7. Responses of $w\left(\frac{l(t)}{2}, t\right)$.

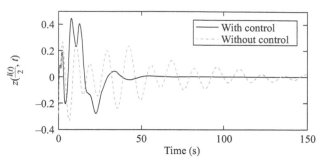

Figure 13.8. Responses of $z\left(\frac{l(t)}{2}, t\right)$.

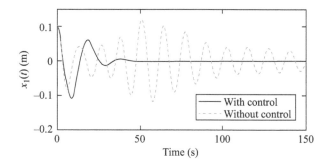

Figure 13.9. Responses of $x_1(t)$.

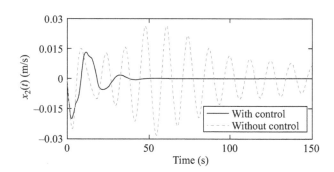

Figure 13.10. Responses of $x_2(t)$.

the midpoint of the time-varying spatial domain are reduced to zero. Figures 13.9 and 13.10 show that the responses of the ODE state $X(t) = [x_1(t), x_2(t)]^T$, which physically represents the displacement and velocity of the lateral vibrations of the cage moving along flexible guideways, are suppressed to zero under the proposed controller.

Represent the responses of $z(x, t), w(x, t)$ in the original cable model (13.104)–(13.106) by using (13.107), (13.108), obtaining the norm $(\|u_t(\cdot, t)\|^2 + \|u_x(\cdot, t)\|^2)^{\frac{1}{2}}$, which physically represents the vibration energy of the cable. A comparison of the performance of the proposed controller with that of a traditional proportional-derivative (PD) controller

$$U_{pd}(t) = 2x_1(t) + 1.2x_2(t),$$

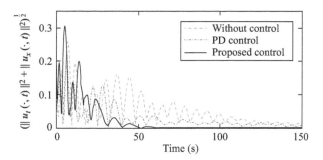

Figure 13.11. Time evolution of the norm $(\|u_t(\cdot,t)\|^2 + \|u_x(\cdot,t)\|^2)^{\frac{1}{2}}$, which physically reflects the vibration energy of the cable modeled by (13.104)–(13.106).

where the PD parameters are chosen by trial and error over many tests, is shown in figure 13.11. From the comparison we observe that both controllers reduce the vibrations compared to the result without control. Even though the vibration energy under the proposed controller is larger at the beginning, which is due to the fact that the self-tuned control gains $\hat{k}_1$, $\hat{k}_2$ start to search for the target values from bad initial values of zero (see figure 13.3), the proposed controller reduces the vibration energy to a much smaller range around zero as time goes on.

13.7 APPENDIX

A. Proof of lemma 13.2

The following notation is used:

$$\bar{M}_{l(t)}(l(t),x;\hat{K}(t)) = \frac{\partial \bar{M}(l(t),x;\hat{K}(t))}{\partial l(t)}.$$

Because $\dot{U}_d = 0$ for $t \in (t_k, t_{k+1})$, recalling (13.62) and taking the time derivative along (13.13)–(13.17), we obtain

$$\dot{d}(t)^2 = \dot{U}(t)^2$$

$$= \Bigg[\Big(\dot{l}(t) - q_1(l(t)) \Big) \bar{M}(l(t),l(t),\hat{K}(t))\hat{z}(l(t),t)$$

$$+ \bar{M}(l(t),0;\hat{K}(t))q_1(0)\hat{z}(0,t)$$

$$+ \Big(q_2(l(t)) + \dot{l}(t) \Big) \bar{N}(l(t),l(t),\hat{K}(t))\hat{w}(l(t),t)$$

$$+ \Big(D(l(t);\hat{K}(t))B - \bar{N}(l(t),0;\hat{K}(t))q_2(0) \Big)\hat{w}(0,t)$$

$$+ \Big(\dot{\hat{K}}(t)D_{\hat{K}(t)}(l(t);\hat{K}(t)) + \dot{l}(t)D'(l(t);\hat{K}(t)) + D(l(t);\hat{K}(t))A \Big)X(t)$$

$$+ \int_0^{l(t)} \Big((\bar{M}(l(t),x;\hat{K}(t))q_1(x))' + \bar{M}(l(t),x;\hat{K}(t))c_1(x)$$

$$+ \bar{N}(l(t),x;\hat{K}(t))c_3(x) + \dot{l}(t)\bar{M}_{l(t)}(l(t),x;\hat{K}(t))$$

$$+ \dot{\hat{K}}(t)\bar{M}_{\hat{K}(t)}(l(t), x; \hat{K}(t)) \bigg) \hat{z}(x,t)dx$$

$$+ \int_0^{l(t)} \bigg(\bar{N}(l(t), x; \hat{K}(t))c_4(x) - (\bar{N}(l(t), x; \hat{K}(t))q_2(x))'$$

$$+ \dot{\hat{K}}(t)\bar{N}_{\hat{K}(t)}(l(t), x; \hat{K}(t)) + \bar{M}(l(t), x; \hat{K}(t))c_2(x)$$

$$+ \dot{l}(t)\bar{N}_{l(t)}(x, l(t), \hat{K}(t)) \bigg) \hat{w}(x,t)dx \bigg]^2$$

$$\leq \lambda_0(\zeta_p, \bar{g}_i) \big[\hat{w}(l(t), t)^2 + \hat{z}(l(t), t)^2 + \hat{w}(0, t)^2 + \hat{z}(0, t)^2$$

$$+ m_3(\zeta_p, \zeta_a)\|\hat{z}(\cdot, t)\|^2 + m_3(\zeta_p, \zeta_a)\|\hat{w}(\cdot, t)\|^2$$

$$+ m_3(\zeta_p, \zeta_a)|X(t)|^2 \big], \tag{13.114}$$

$t \in (t_k, t_{k+1})$ for some positive λ_0, which depends only on the plant parameters ζ_p and the design parameters $\bar{g}_i$'s (the bounds of all kernels depend on the plant parameters ζ_p, and the bounds of $\hat{K}(t)$ depend on the bounds of the unknown parameters g_i's in A and the design parameters $\bar{g}_i$'s in A_m, as mentioned in section 13.1). According to (13.51), (13.52), we know that

$$\left| \dot{\hat{K}}(t) \right|^2 \leq m_3(\zeta_p, \zeta_a, \bar{g}_i), \tag{13.115}$$

where m_3 is a positive constant dependent only on the plant parameters ζ_p, the adaptive law parameters ζ_a, and the design parameters $\bar{g}_i$'s in A_m. (Once again, in this chapter a constant followed by $(\cdot)$ denotes a constant that depends only on the parameters in the parentheses, as in (13.115).)

Recalling the invertibility of the backstepping transformations

$$(\hat{z}, \hat{w}, X(t)) \leftrightarrow (\hat{\alpha}, \hat{\eta}, X(t))$$

and inserting (13.64), we get (13.69). The proof is complete.

B. Proof of lemma 13.3

According to (13.65), events are triggered to guarantee

$$d(t)^2 \leq \theta \Phi(t) - m_d(t). \tag{13.116}$$

Inserting (13.116) into (13.66), one obtains

$$\dot{m}_d(t) \leq -(\eta + \lambda_d)m_d(t) + (\lambda_d\theta - \sigma)\Phi(t) - \kappa_1\hat{\alpha}(l(t), t)^2$$

$$- \kappa_2\hat{\eta}(0, t)^2 - \kappa_3\hat{\alpha}(0, t)^2$$

$$\leq -(\eta + \lambda_d)m_d(t) - \kappa_1\hat{\alpha}(l(t), t)^2$$

$$- \kappa_2\hat{\eta}(0, t)^2 - \kappa_3\hat{\alpha}(0, t)^2, \quad t \geq 0 \tag{13.117}$$

by using (13.70). Hence, by the comparison principle and $m_d(0) < 0$, we conclude that $m_d(t) < 0$. The proof is complete.

C. Calculation of (13.75)

Taking the derivative of (13.71), using lemma 13.2, and inserting the inequality

$$-\dot{\Phi}(t) \leq \mu_0(\zeta_p, \zeta_a, \bar{g}_i)\Phi(t) + \lambda_p(\zeta_p)[\hat{\alpha}(l(t),t)^2 + \hat{\eta}(0,t)^2 + d(t)^2 + \hat{\alpha}(0,t)^2], \quad (13.118)$$

which is obtained by taking the derivative of $\Phi(t)$ along (13.40)–(13.44) for all $t \in [t^*, t_{k+1}]$ and recalling the boundedness of $\tilde{K}(t)$ (specifically, $|\tilde{K}(t)| \leq \sqrt{n} \times \max_{i \in \{1,\dots,n\}}(|\bar{k}_i - \underline{k}_i|))$ and the bound on $\dot{\hat{K}}(t)$ in (13.115), where the constant $\mu_0(\zeta_p, \zeta_a, \bar{g}_i) > 0$ in (13.118) only depends on the plant parameters ζ_p, the adaptive law parameters ζ_a, and the design parameters $\bar{g}_i$'s, and the constant $\lambda_p(\zeta_p) > 0$ in the same inequality depends only on the plant parameters ζ_p, we get

$$\dot{\psi}(t) = \frac{2d(t)\dot{d}(t) + \frac{1}{2}\dot{m}_d(t)}{\theta\Phi(t) - \frac{1}{2}m_d(t)} - \frac{(\theta\dot{\Phi}(t) - \frac{1}{2}\dot{m}_d(t))}{\theta\Phi(t) - \frac{1}{2}m_d(t)}\psi(t)$$

$$\leq \frac{1}{\theta\Phi(t) - \frac{1}{2}m_d(t)}\left[\lambda_a\left(d(t)^2 + \hat{\alpha}(l(t),t)^2 + \hat{\eta}(0,t)^2\right.\right.$$

$$+ \hat{\alpha}(0,t)^2 + m_3(\zeta_p,\zeta_a,\bar{g}_i)\|\hat{\alpha}(\cdot,t)\|^2 + m_3(\zeta_p,\zeta_a,\bar{g}_i)\|\hat{\eta}(\cdot,t)\|^2$$

$$\left.\left. + m_3(\zeta_p,\zeta_a,\bar{g}_i)|X(t)|^2\right) + d(t)^2 + \frac{1}{2}\dot{m}_d(t)\right]$$

$$- \frac{1}{\theta\Phi(t) - \frac{1}{2}m_d(t)}\left[\theta\left(-\mu_0(\zeta_p,\zeta_a,\bar{g}_i)\Phi(t) - \lambda_p\hat{\alpha}(l(t),t)^2\right.\right.$$

$$\left.\left. - \lambda_p\hat{\eta}(0,t)^2 - \lambda_p d(t)^2 - \lambda_p\hat{\alpha}(0,t)^2\right) - \frac{1}{2}\dot{m}_d(t)\right]\psi(t).$$

Inserting (13.66) to rewrite $\dot{m}_d(t)$ and recalling (13.67), we obtain

$$\dot{\psi}(t) \leq \frac{1}{\theta\Phi(t) - \frac{1}{2}m_d(t)}\left[\left(\lambda_a + 1 + \frac{1}{2}\lambda_d\right)d(t)^2 + \left(\lambda_a - \frac{1}{2}\kappa_1\right)\alpha(l(t),t)^2\right.$$

$$+ \left(\lambda_a - \frac{\kappa_3}{2}\right)\alpha(0,t)^2 + \left(\lambda_a - \frac{\kappa_2}{2}\right)\hat{\eta}(0,t)^2 + \lambda_a m_3(\zeta_p,\zeta_a,\bar{g}_i)\Phi(t)$$

$$\left. - \frac{1}{2}\eta m_d(t)\right] - \frac{1}{\theta\Phi(t) - \frac{1}{2}m_d(t)}\left[-\theta\mu_0(\zeta_p,\zeta_a,\bar{g}_i)\Phi(t)\right.$$

$$- \left(\theta\lambda_p - \frac{1}{2}\kappa_1\right)\alpha(l(t),t)^2 - \left(\theta\lambda_p - \frac{1}{2}\kappa_2\right)\hat{\eta}(0,t)^2$$

$$- \left(\theta\lambda_p + \frac{1}{2}\lambda_d\right)d(t)^2 - \left(\theta\lambda_p - \frac{1}{2}\kappa_3\right)\hat{\alpha}(0,t)^2$$

$$\left. + \frac{1}{2}\eta m_d(t) + \frac{1}{2}\sigma\Phi(t)\right]\psi(t). \quad (13.119)$$

Inserting (13.72)–(13.74), (13.119) becomes

$$\dot{\psi}(t) \leq \frac{1}{\theta\Phi(t) - \frac{1}{2}m_d(t)}\left[\left(\lambda_a + 1 + \frac{1}{2}\lambda_d\right)d(t)^2 + \lambda_a(\zeta_p,\bar{g}_i)m_3(\zeta_p,\zeta_a,\bar{g}_i)\Phi(t)\right.$$

$$\left. - \frac{1}{2}\eta m_d(t)\right] - \frac{1}{\theta\Phi(t) - \frac{1}{2}m_d(t)}\left[-\left(\theta\mu_0(\zeta_p,\zeta_a,\bar{g}_i) - \frac{1}{2}\sigma\right)\Phi(t)\right.$$

$$\left. - \left(\theta\lambda_p + \frac{1}{2}\lambda_d\right)d(t)^2 + \frac{1}{2}\eta m_d(t)\right]\psi(t). \quad (13.120)$$

Applying, in (13.120), the inequalities

$$-\frac{\frac{1}{2}\eta m_d(t)}{\theta\Phi(t) - \frac{1}{2}m_d(t)} \leq -\frac{\frac{1}{2}\eta m_d(t)}{-\frac{1}{2}m_d(t)} = \eta,$$

$$\frac{\Phi(t)}{\theta\Phi(t) - \frac{1}{2}m_d(t)} \leq \frac{\Phi(t)}{\theta\Phi(t)} = \frac{1}{\theta},$$

$$\frac{d(t)^2}{\theta\Phi(t) - \frac{1}{2}m_d(t)} = \frac{d(t)^2 + \frac{1}{2}m_d(t) - \frac{1}{2}m_d(t)}{\theta\Phi(t) - \frac{1}{2}m_d(t)} \leq \psi(t) + 1,$$

which hold because $m_d(t) < 0$, we obtain (13.75).

D. Calculation of (13.84)

Taking the derivative of (13.80) along (13.40)–(13.43), employing (13.54), (13.55), (13.64), (13.66), and applying $\dot{\tilde{K}}(t) = -\dot{\hat{K}}(t)$, we obtain

$$
\begin{aligned}
\dot{V}(t) = \frac{1}{1 + \Omega(t) - \mu_m m_d(t)} \Bigg[&-X^T(t)QX(t) \\
&+ 2X^T PB\hat{\eta}(0,t) + \frac{\dot{l}(t)}{2}r_b e^{-\delta l(t)}\hat{\alpha}(l(t),t)^2 \\
&+ \frac{1}{2}(q_2(l(t)) + \dot{l}(t))r_a e^{\delta l(t)}\hat{\eta}(l(t),t)^2 \\
&- \frac{1}{2}q_2(0)r_a\hat{\eta}(0,t)^2 - \frac{1}{2}\delta r_a \int_0^{l(t)} e^{\delta x}q_2(x)\hat{\eta}(x,t)^2 dx \\
&- \frac{1}{2}r_a \int_0^{l(t)} e^{\delta x}q_2{}'(x)\hat{\eta}(x,t)^2 dx + r_a \int_0^{l(t)} c_4(x)e^{\delta x}\hat{\eta}(x,t)^2 dx \\
&- \frac{1}{2}q_1(l(t))r_b e^{-\delta l(t)}\hat{\alpha}(l(t),t)^2 + \frac{1}{2}q_1(0)r_b\hat{\alpha}(0,t)^2 \\
&- \frac{1}{2}\delta r_b \int_0^{l(t)} e^{-\delta x}q_1(x)\hat{\alpha}(x,t)^2 dx \\
&+ \frac{1}{2}r_b \int_0^{l(t)} e^{-\delta x}q_1'(x)\hat{\alpha}(x,t)^2 dx + r_b \int_0^{l(t)} c_1(x)e^{-\delta x}\hat{\alpha}(x,t)^2 dx \\
&- r_a \int_0^{l(t)} e^{\delta x}\hat{\eta}(x,t)\left(\dot{\tilde{K}}(t)D_{\hat{K}(t)}X(t) + \dot{\tilde{K}}R(x,t)\right) dx \\
&- r_b \int_0^{l(t)} e^{-\delta x}\hat{\alpha}(x,t)J(x,0)q_1(0)CX(t) dx \\
&+ \mu_m\eta m_d(t) - \mu_m\lambda_d d(t)^2 + \mu_m\sigma\Phi(t) \\
&+ \mu_m\kappa_1\hat{\alpha}(l(t),t)^2 + \mu_m\kappa_2\hat{\eta}(0,t)^2 + \mu_m\kappa_3\hat{\alpha}(0,t)^2 \Bigg]
\end{aligned}
$$

$$-\tilde{K}(t)\left[\Gamma_c^{-1}\dot{\hat{K}}(t)^T + \frac{1}{1+\Omega(t)-\mu_m m_d(t)}\right.$$

$$\left.\times\left(2X(t)B^T PX(t) - r_a\int_0^{l(t)} e^{\delta x}\hat{\eta}(x,t)X(t)B^T D(x;\hat{K}(t))^T dx\right)\right].$$

(13.121)

Applying (13.64), inserting the adaptive laws (13.51), (13.52) into (13.121), and using Young's inequality, we get

$$\dot{V}(t) \le \frac{1}{1+\Omega(t)-\mu_m m_d(t)}\left[-\frac{7}{8}\lambda_{\min}(Q)|X(t)|^2\right.$$

$$+\frac{8}{\lambda_{\min}(Q)}|PB|^2\hat{\eta}(0,t)^2 - \frac{1}{2}q_2(0)r_a\hat{\eta}(0,t)^2$$

$$-\frac{1}{2}\delta r_a\int_0^{l(t)} e^{\delta x}q_2(x)\hat{\eta}(x,t)^2 dx - \frac{1}{2}r_a\int_0^{l(t)} e^{\delta x}q_2'(x)\hat{\eta}(x,t)^2 dx$$

$$+r_a\int_0^{l(t)} c_4(x)e^{\delta x}\hat{\eta}(x,t)^2 dx$$

$$-\frac{1}{2}(q_1(l(t)) - \dot{l}(t))r_b e^{-\delta l(t)}\hat{\alpha}(l(t),t)^2$$

$$+\frac{1}{2}q_1(0)r_b\hat{\alpha}(0,t)^2 - \frac{1}{2}\delta r_b\int_0^{l(t)} e^{-\delta x}q_1(x)\hat{\alpha}(x,t)^2 dx$$

$$+\frac{r_b}{2}\int_0^{l(t)} e^{-\delta x}q_1'(x)\hat{\alpha}(x,t)^2 dx + r_b\int_0^{l(t)} c_1(x)e^{-\delta x}\hat{\alpha}(x,t)^2 dx$$

$$-r_a\int_0^{l(t)} e^{\delta x}\hat{\eta}(x,t)\left(\dot{\hat{K}}(t)D_{\hat{K}(t)}(x;\hat{K}(t))X(t) + \dot{\hat{K}}(t)R(x,t)\right)dx$$

$$-r_b\int_0^{l(t)} e^{-\delta x}\hat{\alpha}(x,t)J(x,0)q_1(0)CX(t)dx + \mu_m\eta m_d(t)$$

$$-\left(\mu_m\lambda_d - \frac{1}{2}(\overline{q_2} + v_m)r_a e^{\delta L}\right)d(t)^2$$

$$+\mu_m\sigma\Phi(t) + \mu_m\kappa_1\hat{\alpha}(l(t),t)^2$$

$$+\mu_m\kappa_2\hat{\eta}(0,t)^2 + \mu_m\kappa_3\hat{\alpha}(0,t)^2\bigg].$$

(13.122)

Recalling (13.41) and (13.67) applying the Young and Cauchy-Schwarz inequalities, we get (13.84).

13.8 NOTES

The adaptive control of hyperbolic PDEs is a popular research topic, on which recent results are presented in [9], which is limited to continuous inputs and identification. In this chapter, we presented the event-triggered adaptive output-feedback boundary control design of a coupled hyperbolic PDE-ODE system where the control input is piecewise-constant. Another triggered-type adaptive control where the identification employs piecewise-constancy instead of the control input is presented in chapter 14. Developed by nontrivially integrating the adaptive control designs in this chapter (piecewise-constant control inputs) and chapter 14 (piecewise-constant identification), chapter 15 presents the adaptive control of coupled hyperbolic PDEs with piecewise-constancy in both inputs and identification. Recent results on the triggered-type adaptive control of PDEs are summarized in the notes section of chapter 15.

Chapter Fourteen

Adaptive Control with Regulation-Triggered Parameter Estimation of Hyperbolic PDEs

The event-triggered adaptive control design in chapter 13 employed a continuous-in-time parameter estimator, feeding an event-triggered control law, to produce a piecewise-constant input signal. In this chapter, following a recent approach by Karafyllis and coworkers [104], we pursue an event-triggered adaptive design that is a complement of the one in chapter 13. Here, we use a continuous-in-time control law that is fed piecewise-constant parameter estimates from an event-triggered parameter update law that applies a least-squares estimator to data "batches" collected over time intervals between the triggers. A parameter update is triggered by an observed growth in the norm of the partial differential equation (PDE) state. Since the PDE state is being adaptively regulated, this adaptive control approach is called *regulation-triggered*. Since the parameter updates are done on batches of data using least-squares identification, this identifier is referred to as a BaLSI.

As in all conventional adaptive control, in chapters 5, 8, and 13 the adaptive control designs achieve only asymptotic convergence of the plant states and are not guaranteed to identify the true parameters exactly. The regulation-triggered BaLSI adaptive control scheme in this chapter guarantees the exponential regulation of plant states to zero and the finite-time exact identification of the unknown parameters from all but a negligible set of initial conditions, for hyperbolic PDE-ODE (ordinary differential equation) systems with unknown transport speeds.

We start this chapter by formulating the problem in section 14.1 and presenting the nominal control design based on the basic backstepping design in section 14.2. In section 14.3, we propose the regulation-triggered adaptive control scheme, including a certainty-equivalence controller and a least-squares identifier updated in a sequence of times, which are determined by an event trigger designed based on the progress of the regulation of the states. In section 14.4, we prove that the proposed triggering-based adaptive control guarantees: 1) no Zeno phenomenon occurs; 2) parameter estimates are convergent to the true values in finite time (from most initial conditions); 3) the plant states are exponentially regulated to zero. The effectiveness of the proposed design is illustrated with a simulation in section 14.5.

14.1   PROBLEM FORMULATION

In this chapter, we consider the class of plants

$$\dot{\zeta}(t) = (a - q_1 bc)\zeta(t) + b(q_2 + q_1 p)w(0,t), \quad t \geq 0, \tag{14.1}$$

$$z_t(x,t) = -q_1 z_x(x,t), \quad x \in [0,1], \ t \geq 0, \tag{14.2}$$

$$w_t(x,t) = q_2 w_x(x,t), \quad x \in [0,1], \ t \geq 0, \tag{14.3}$$

$$z(0,t) = c\zeta(t) - pw(0,t), \quad t \geq 0, \tag{14.4}$$

$$w(1,t) = \frac{\bar{c}}{q_2} U(t) + \frac{q_1}{q_2} z(1,t), \quad t \geq 0 \tag{14.5}$$

with initial conditions $w(x,0) = w_0(x)$ for $x \in [0,1)$, $z(x,0) = z_0(x)$ for $x \in (0,1]$, $\zeta(0) = \zeta_0$, where $\zeta(t)$ is a scalar ODE state, and scalar $z(x,t), w(x,t)$ are PDE states. The boundary condition (14.5) contains the control input $U(t)$. The class of (14.1)–(14.4) is motivated by a wave PDE converted to Riemann variables. It is through such a transformation process that possibly unmotivated-looking coefficients $a - q_1 bc$ and $b(q_2 + q_1 p)$ in (14.1) arise.

It is the parameters q_1 and q_2, which appear both as transport speeds and in the ODE (14.1) and the boundary condition (14.5), that we consider unknown. The speed q_2 is arbitrary and, of course, positive. The constants a, b, c, p are arbitrary and positive as well. The constant $\bar{c}$ is arbitrary and nonzero.

To make the problem as nontrivial as we can within this class, we only consider the case where the ODE (14.1) is unstable, with $a - q_1 bc > 0$—that is, the case where the unknown propagation speed q_1 satisfies

$$0 < q_1 < \frac{a}{bc}. \tag{14.6}$$

The unknown transport speeds q_1, q_2 are assumed to have known upper bounds $\bar{q}_1 > 0, \bar{q}_2 > 0$ and lower bounds $\underline{q}_1 > 0, \underline{q}_2 > 0$, respectively. The bounds $\underline{q}_2, \bar{q}_2$ are arbitrary, in addition to satisfying the obvious relation $\underline{q}_2 < \bar{q}_2$.

For the bounds $\underline{q}_1, \bar{q}_1$, the following two assumptions are made.

Assumption 14.1. *The upper bound $\bar{q}_1$ satisfies*

$$\bar{q}_1 < \frac{a}{bc}. \tag{14.7}$$

In addition to being consistent with the instability assumption (14.6)—namely, $0 < q_1 \leq \bar{q}_1 < a/(bc)$, assumption 14.1 is also used in the forthcoming design condition (14.12) where the control gain κ is chosen in accordance with the known bounds on the unknown parameters.

Assumption 14.2. *The difference between the upper and lower bounds $\underline{q}_1 \leq \bar{q}_1$ is smaller than the following unknown constant,*

$$\bar{q}_1 - \underline{q}_1 < \sqrt{\frac{q_2 q_1 r_b}{2 r_a}}, \tag{14.8}$$

where r_a, r_b are positive unknown constants, upper and lower bounded, respectively, as

$$r_b < \frac{m}{2 c_0^2 q_1}, \tag{14.9}$$

$$r_a > \frac{2}{q_2} \left(q_1 r_b p^2 + \frac{(q_1 p + q_2)^2 b^2}{2m} \right), \tag{14.10}$$

the unknown constant m in the bounds (14.9) and (14.10) is defined as

$$m = -a + q_1 bc - b\kappa > 0, \tag{14.11}$$

and the constant κ appearing in (14.11), and to be used later in control design, is chosen to satisfy

$$\kappa < \min \left\{ \frac{(a - \underline{q}_1 bc) \left[\bar{q}_2 + (\bar{q}_1 - \underline{q}_1) \left(\frac{cb(\bar{q}_2 + \bar{q}_1 p)}{a - \bar{q}_1 bc} + p \right) \right]}{-\underline{q}_2 b}, \underline{q}_1 c - \frac{a}{b} \right\}. \tag{14.12}$$

The purpose of assumption 14.2—that is, (14.8)—will become evident in section 14.2, with the purpose of (14.8) becoming evident specifically in inequality (14.28), whose role is in estimating the exponential decay rate under nominal feedback. This assumption is not required in the BaLSI design and the exact parameter estimation. It is used in the stability analysis by the Lyapunov method in section 14.4. If there is no unknown parameter staying with the proximal reflection term $z(1,t)$ in (14.5), assumption 14.2—(14.8)—is not required.

Assumption 14.2 is difficult to verify a priori for two reasons: First, because the unknown q_1 and q_2 appear in (14.8), (14.9), (14.10), (14.11). Second, because $\bar{q}_1 - \underline{q}_1$ appears both on the left of (14.8) and on the right of (14.12). But assumption 14.2 can unquestionably be satisfied for sufficiently small $\bar{q}_1 - \underline{q}_1$. Unfortunately, very small $\bar{q}_1 - \underline{q}_1$ essentially means that the transport speed q_1 is known.

Let us now recap that q_2, which appears in the actuated w-PDE in (14.3) and is the transport speed in the direction downstream from the input, is arbitrary (positive), whereas the unactuated z-PDE in (14.2) may have to be nearly perfectly known.

The parameters q_1, q_2 appear in both the PDE as well as the ODE. The structure of the plant and the conditions of the plant parameters come, at least in principle, and as we already indicated above, from writing a wave PDE-ODE coupled model in Riemann coordinates. If the wave PDE's Young modulus were unknown, the transformation into the Riemann variables would contain such an unknown quantity, which would render $z(x,t)$ and $w(x,t)$ unmeasurable. We proceed with an adaptive design for the class of systems (14.2), (14.3) with the expectation that applications do exist in which the transformation step into (14.2), (14.3) is not needed and (z, w) are measurable.

If the original plant were a wave PDE, the ODE (14.1) would be driven by the wave PDE's boundary state of the Neumann type, multiplied by a coefficient associated with the wave PDE's propagation velocity, while the opposite boundary of the wave PDE would be actuated using Neumann actuation with a coefficient associated with the wave PDE's propagation velocity as well. One physical model of this type of system is an oil well-drilling model [155], where

$$a = \frac{c_a}{I_B}, \quad b = \frac{I_d}{2I_B}, \quad q_1 = q_2 = \sqrt{\frac{GJ}{I_d}}, \quad \bar{c} = \frac{2}{I_d}, \quad p = 1, \quad c = 2,$$

with I_d the moment of inertia per unit of length, G the shear modulus, J the geometric moment of inertia of the drill pipe, c_a the anti-damping coefficient at the bit due to the stick-slip instability, and I_B the moment of inertia of the bottom-hole assembly.

As in the previous chapters, we adopt the following notation:

- The symbol Z_+ denotes the set of all nonnegative integers and $\mathbb{R}_+ := [0, +\infty)$.
- Let $U \subseteq \mathbb{R}^n$ be a set with a nonempty interior, and let $\Omega \subseteq \mathbb{R}$ be a set. By $C^0(U; \Omega)$, we denote the class of continuous mappings on U, which take values in Ω.
- We use the notation $\mathbb{N}$ for the set $\{1, 2, \cdots\}$—that is, the natural numbers without 0.
- For an $I \subseteq \mathbb{R}_+$, the space $C^0(I; L^2(0,1))$ is the space of continuous mappings $I \ni t \mapsto u[t] \in L^2(0,1)$.
- Let $u : \mathbb{R}_+ \times [0,1] \to \mathbb{R}$ be given. We use the notation $u[t]$ to denote the profile of u at certain $t \geq 0$—that is, $(u[t])(x) = u(x,t)$, for all $x \in [0,1]$.

14.2   NOMINAL CONTROL DESIGN

We introduce the following backstepping transformation:

$$\alpha(x,t) = z(x,t), \tag{14.13}$$

$$\beta(x,t) = w(x,t) - \int_0^x \phi(x,y)w(y,t)dy - \lambda(x)\zeta(t), \tag{14.14}$$

where

$$\lambda(x; q_1, q_2) = \frac{\kappa}{q_1 p + q_2} e^{\frac{1}{q_2}(a - q_1 bc)x}, \tag{14.15}$$

$$\phi(x, y; q_1, q_2) = \frac{\kappa}{q_2} e^{\frac{1}{q_2}(a - q_1 bc)(x-y)} b, \tag{14.16}$$

and κ is a design parameter, first mentioned in assumption 14.2, and to be chosen according to (14.12).

Writing q_1, q_2 after "; " in $\lambda(x; q_1, q_2)$ and $\phi(x, y; q_1, q_2)$ emphasizes the fact that $\phi(x,y)$, $\lambda(x)$ depend on the unknown parameters q_1, q_2.

By applying the backstepping transformations (14.13), (14.14), we convert the plant (14.1)–(14.5) to the target system

$$\dot{\zeta}(t) = -m\zeta(t) + b(q_1 p + q_2)\beta(0,t), \tag{14.17}$$

$$\alpha(0,t) = c_0\zeta(t) - p\beta(0,t), \tag{14.18}$$

$$\alpha_t(x,t) = -q_1\alpha_x(x,t), \tag{14.19}$$

$$\beta_t(x,t) = q_2\beta_x(x,t), \tag{14.20}$$

$$\beta(1,t) = 0, \tag{14.21}$$

where

$$c_0 = c - p\lambda(0; q_1, q_2). \tag{14.22}$$

The control input $U(t)$ is chosen as

$$U(t) = -\frac{1}{\bar{c}}\left[q_1 z(1,t) - q_2 \int_0^1 \phi(1, y; q_1, q_2)w(y,t)dy - q_2\lambda(1; q_1, q_2)\zeta(t)\right] \tag{14.23}$$

to ensure (14.21).

Define

$$\Omega(t) = \|z[t]\|^2 + \|w[t]\|^2 + \zeta(t)^2 \tag{14.24}$$

and a vector θ containing the two unknown parameters as

$$\theta = [q_1, q_2]^T. \tag{14.25}$$

Through Lyapunov analysis for the target system (14.17)–(14.21) and applying the invertibility of the backstepping transformation, the estimate

$$\Omega(t) \le \Upsilon_\theta \Omega(0) e^{-\lambda_1 t}, \quad t \ge 0, \tag{14.26}$$

is obtained, where the decay rate λ_1 is

$$\lambda_1 = \min\left\{ \frac{1}{2}(q_1 bc - a - b\kappa), \delta q_2, \delta q_1 \right\} \tag{14.27}$$

with the analysis parameter $\delta > 0$ selected as

$$\delta \le \ln\left(\frac{1}{\bar{q}_1 - \underline{q}_1} \sqrt{\frac{q_2 q_1 r_b}{2 r_a}} \right) \tag{14.28}$$

in order to meet the needs of the Lyapunov analysis, which will become evident in section 14.4. For the right-hand side of (14.28) to be positive, we need

$$\frac{1}{(\bar{q}_1 - \underline{q}_1)} \sqrt{\frac{q_2 q_1 r_b}{2 r_a}} > 1. \tag{14.29}$$

This is ensured by assumption 14.2. To recap, δ in (14.28) is only an analysis parameter, which influences the decay rate in (14.26).

The overshoot coefficient Υ_θ obtained in (14.26), through the straightforward and omitted Lyapunov analysis, is

$$\Upsilon_\theta = \frac{\xi_2 \xi_4}{\xi_1 \xi_3}, \tag{14.30}$$

where the positive constants $\xi_1, \xi_2, \xi_3, \xi_4$ are

$$\xi_1 = \min\left\{ \frac{1}{2} r_a, \frac{1}{2} r_b e^{-\delta}, \frac{1}{2} \right\}, \tag{14.31}$$

$$\xi_2 = \max\left\{ \frac{1}{2} r_a e^{\delta}, \frac{1}{2} r_b, \frac{1}{2} \right\}, \tag{14.32}$$

$$\xi_3 = \frac{1}{\max\left\{ 3 + \frac{3\kappa^2 b^2}{q_2^2} \|\bar{m}\|^2, \frac{3\kappa^2}{(q_2 + q_1 p)^2} \|\bar{m}\|^2 + 1 \right\}}, \tag{14.33}$$

$$\xi_4 = \max\left\{ 3 + \frac{3\kappa^2 b^2}{q_2^2} \|\bar{n}\|^2, \frac{3\kappa^2}{(q_2 + q_1 p)^2} \|\bar{n}\|^2 + 1 \right\}, \tag{14.34}$$

with

$$\bar{m}(x) = e^{\frac{a - q_1 bc + b\kappa}{q_2} x}, \quad \bar{n}(x) = e^{\frac{1}{q_2}(a - q_1 bc)x} \tag{14.35}$$

and with the positive constants r_a, r_b required in assumption 14.2 to satisfy (14.10), (14.9).

The relations (14.30)–(14.35), (14.9), (14.10), (14.28) will be used in the proofs of the main results in section 14.4.

We refer to the controller $U(t)$ in (14.23) as the nominal feedback, which requires the knowledge of the values of the parameters q_1, q_2. The adaptive scheme working with the nominal feedback (14.23) and guaranteeing exponential regulation is presented in the next section.

14.3   REGULATION-TRIGGERED ADAPTIVE CONTROL

The regulation-triggered adaptive control includes a certainty-equivalence controller and a least-squares identifier that is updated in a sequence of time instants.

The Certainty-Equivalence Controller

The control action in the interval between two consecutive events is the result of replacing the unknown parameters q_1, q_2 in the nominal control law (14.23) by their estimates $\hat{q}_1, \hat{q}_2$ at the beginning of the interval, with the estimates $\hat{q}_1, \hat{q}_2$ kept constant during the interval. In other words, the adaptive version of (14.23) is given by

$$U(t) = -\frac{1}{\bar{c}}\left[\hat{q}_1(\tau_i)z(1,t) - \hat{q}_2(\tau_i)\int_0^1 \phi(1,y;\hat{\theta}(\tau_i))w(y,t)dy\right.$$

$$\left. - \hat{q}_2(\tau_i)\lambda(1;\hat{\theta}(\tau_i))\zeta(t)\right], \quad t \in [\tau_i, \tau_{i+1}), \quad i \in Z_+, \tag{14.36}$$

$$\hat{\theta}(t) = (\hat{q}_1(t), \hat{q}_2(t))^T = (\hat{q}(\tau_i), \hat{q}_2(\tau_i))^T = \hat{\theta}(\tau_i), \quad t \in [\tau_i, \tau_{i+1}), \quad i \in Z_+, \tag{14.37}$$

where $\{\tau_i \geq 0\}_{i=0}^\infty$ is the sequence of time instants, which, along with the estimates $\hat{\theta}(\tau_i)$, is defined next.

The Event Trigger

The sequence of time instants $\{\tau_i \geq 0\}_{i=0}^\infty$ is chosen to satisfy

$$\tau_{i+1} = \min\{\tau_i + T, r_i\}, \quad i \in Z_+ \tag{14.38}$$

with $\tau_0 = 0$. The constant $T > 0$ is a design parameter with the purpose of avoiding a low update frequency and, more importantly, $r_i > \tau_i$ is a time instant determined by an event trigger which is designed next. The trigger was introduced in [106] and is based on the progress of the regulation of the states.

The event trigger sets $r_i > \tau_i$ to be the smallest value of time $t > \tau_i$ for which

$$\Omega(t) = \Upsilon_{\hat{\theta}(\tau_i)}(1 + \bar{a})\Omega(\tau_i) \tag{14.39}$$

for $\Omega(\tau_i) \neq 0$, where $\Upsilon_{\hat{\theta}(\tau_i)} \geq 1$ is the coefficient defined by (14.30) with q_1, q_2 replaced by $\hat{q}_1, \hat{q}_2$, the design parameter $\bar{a}$ is positive, and Ω is defined by (14.24) with the solutions of (14.1)–(14.5) under (14.36). In simple terms, the parameter estimate update is triggered if the plant norm has grown by a certain factor, specifically, by

$\Upsilon_{\hat{\theta}(\tau_i)}(1+\bar{a})$. Since $\Upsilon_{\hat{\theta}(\tau_i)}$ is the overshoot coefficient already associated with the system transient in accordance with the estimate (14.26), the real net growth factor that triggers the update is $1+\bar{a}$ for any $\bar{a}>0$ chosen by the user.

If a time $t>\tau_i$ satisfying (14.39) does not exist, we set $r_i=+\infty$. For the case that $\Omega(\tau_i)=0$, we set $r_i:=\tau_i+T$. Therefore, the event trigger r_i is built as

$$r_i:=\inf\{t>\tau_i:\Omega(t)=\Upsilon_{\hat{\theta}(\tau_i)}(1+\bar{a})\Omega(\tau_i)\},\ \ \Omega(\tau_i)\neq0, \qquad (14.40)$$

$$r_i:=\tau_i+T,\ \ \Omega(\tau_i)=0. \qquad (14.41)$$

The following lemma shows that the event trigger is well-defined and produces an increasing sequence of events.

Lemma 14.1. *The event trigger (14.38), (14.40), (14.41) is well-defined—that is, $\tau_{i+1}>\tau_i$, for all $i\in Z_+$.*

Proof. If $\Omega(\tau_i)=0$, it follows from (14.38), (14.41) that $\tau_{i+1}=\tau_i+T$. If $\Omega(\tau_i)\neq0$ and r_i defined in (14.40) is less than τ_i+T, the dwell time $\tau_{i+1}-\tau_i$ is greater than zero because $\Omega(\tau_{i+1})=\Upsilon_{\hat{\theta}(\tau_i)}(1+\bar{a})\Omega(\tau_i)>\Omega(\tau_i)$ and $\Omega(t)$ defined in (14.24) is a continuous function on $t\in[\tau_i,\tau_{i+1}]$. If $r_i\geq\tau_i+T$ or r_i is infinite, it follows from (14.38) that $\tau_{i+1}-\tau_i+T$. $\qquad\square$

The above lemma allows us to define the solution on the interval $[0,\lim_{i\to\infty}(\tau_i))$.

Least-Squares Identifier

The least-squares identifier activated by the trigger defined by (14.38)–(14.41) is designed in this subsection. The design idea of the identifier follows from [106]. According to the considered dynamic model, by applying integration, formulating a cost function, and using Fermat's theorem, we construct a matrix equation, with an unknown vector of plant parameters and with the equation's coefficients being the plant states over a time interval. The parameter estimation is then treated as a convex optimization problem with linear equality constraints.

By virtue of (14.1)–(14.5), we get for $\tau>0$ and $n=1,2,\dots$ that

$$\frac{d}{d\tau}\left(\int_0^1\cos(x\pi n)z(x,\tau)dx+\int_0^1\cos(x\pi n)w(x,\tau)dx+\frac{1}{b}\zeta(\tau)\right)$$

$$=-q_1(-1)^nz(1,\tau)+q_1z(0,\tau)-q_1\pi n\int_0^1\sin(x\pi n)z(x,\tau)dx$$

$$+q_2(-1)^nw(1,\tau)-q_2w(0,\tau)+q_2\pi n\int_0^1\sin(x\pi n)w(x,\tau)dx$$

$$+\frac{a}{b}\zeta(\tau)+q_2w(0,\tau)-q_1z(0,\tau)$$

$$=-q_1\pi n\int_0^1\sin(x\pi n)z(x,\tau)dx+q_2\pi n\int_0^1\sin(x\pi n)w(x,\tau)dx$$

$$+(-1)^n\bar{c}U(\tau)+\frac{a}{b}\zeta(\tau), \qquad (14.42)$$

where (14.4) was inserted into (14.1) to replace $q_1bc\zeta(t)$ and to yield

$$\frac{d}{d\tau}\zeta(\tau) = a\zeta(\tau) + b(q_2 w(0,\tau) - q_1 z(0,\tau)). \tag{14.43}$$

Integrating (14.42) from μ_{i+1} to t yields

$$f_n(t, \mu_{i+1}) = q_1 g_{n,1}(t, \mu_{i+1}) + q_2 g_{n,2}(t, \mu_{i+1}), \tag{14.44}$$

where

$$f_n(t, \mu_{i+1}) = \left(\int_0^1 \cos(x\pi n) z(x,t) dx + \int_0^1 \cos(x\pi n) w(x,t) dx + \frac{1}{b}\zeta(t) \right)$$

$$- \left(\int_0^1 \cos(x\pi n) z(x,\mu_{i+1}) dx + \int_0^1 \cos(x\pi n) w(x,\mu_{i+1}) dx + \frac{1}{b}\zeta(\mu_{i+1}) \right)$$

$$- \int_{\mu_{i+1}}^t \left((-1)^n \bar{c} U(\tau) + \frac{a}{b}\zeta(\tau) \right) d\tau,$$

$$g_{n,1}(t, \mu_{i+1}) = - \int_{\mu_{i+1}}^t \pi n \int_0^1 \sin(x\pi n) z(x,\tau) dx d\tau, \tag{14.45}$$

$$g_{n,2}(t, \mu_{i+1}) = \int_{\mu_{i+1}}^t \pi n \int_0^1 \sin(x\pi n) w(x,\tau) dx d\tau \tag{14.46}$$

for $n = 1, 2, \ldots$. The time μ_{i+1} introduced in [104] is

$$\mu_{i+1} := \min\{\tau_f : f \in \{0, \ldots, i\}, \tau_f \geq \tau_{i+1} - \tilde{N}T\}, \tag{14.47}$$

where the positive integer $\tilde{N} \geq 1$ is a design parameter. In practice, a larger $\tilde{N}$ can reduce the effect of measurement noise on the precision of estimation, with a cost of larger computation [104].

Equation (14.44) is written as

$$f_n(t, \mu_{i+1}) = \eta_n(t, \mu_{i+1})\theta, \tag{14.48}$$

where

$$\eta_n(t, \mu_{i+1}) = [g_{n,1}(t, \mu_{i+1}), g_{n,2}(t, \mu_{i+1})], \tag{14.49}$$

and θ is defined in (14.25). Define the function $h_{i,n} : \mathbb{R}^2 \to R_+$ by the formula

$$h_{i,n}(\ell) = \int_{\mu_{i+1}}^{\tau_{i+1}} (f_n(t, \mu_{i+1}) - \eta_n(t, \mu_{i+1})\ell)^2 dt \tag{14.50}$$

for $n = 1, 2, \ldots, \ell = [\ell_1, \ell_2]^T, i \in Z_+$.

According to (14.48), the function $h_{i,n}(\ell)$ (14.50) has a global minimum $h_{i,n}(\theta) = 0$. We get from Fermat's theorem (vanishing gradient at extrema) that the following equations hold for every $i \in Z_+$ and $n = 1, 2, \ldots$:

$$H_{n,1}(\mu_{i+1}, \tau_{i+1}) = q_1 Q_{n,1}(\mu_{i+1}, \tau_{i+1}) + q_2 Q_{n,2}(\mu_{i+1}, \tau_{i+1}), \tag{14.51}$$

$$H_{n,2}(\mu_{i+1}, \tau_{i+1}) = q_1 Q_{n,2}(\mu_{i+1}, \tau_{i+1}) + q_2 Q_{n,3}(\mu_{i+1}, \tau_{i+1}), \tag{14.52}$$

where

$$H_{n,1}(\mu_{i+1}, \tau_{i+1}) = \int_{\mu_{i+1}}^{\tau_{i+1}} g_{n,1}(t, \mu_{i+1}) f_n(t, \mu_{i+1}) dt, \tag{14.53}$$

$$H_{n,2}(\mu_{i+1}, \tau_{i+1}) = \int_{\mu_{i+1}}^{\tau_{i+1}} g_{n,2}(t, \mu_{i+1}) f_n(t, \mu_{i+1}) dt, \tag{14.54}$$

$$Q_{n,1}(\mu_{i+1}, \tau_{i+1}) = \int_{\mu_{i+1}}^{\tau_{i+1}} g_{n,1}(t, \mu_{i+1})^2 dt, \tag{14.55}$$

$$Q_{n,2}(\mu_{i+1}, \tau_{i+1}) = \int_{\mu_{i+1}}^{\tau_{i+1}} g_{n,1}(t, \mu_{i+1}) g_{n,2}(t, \mu_{i+1}) dt, \tag{14.56}$$

$$Q_{n,3}(\mu_{i+1}, \tau_{i+1}) = \int_{\mu_{i+1}}^{\tau_{i+1}} g_{n,2}(t, \mu_{i+1})^2 dt. \tag{14.57}$$

Indeed, (14.51), (14.52) are obtained by differentiating the functions $h_{i,n}(\ell)$ defined by (14.50) with respect to ℓ_1, ℓ_2, respectively, and evaluating the derivatives at the position of the global minimum $(\ell_1, \ell_2) = (q_1, q_2)$. The equations (14.51), (14.52) are organized as

$$Z_n(\mu_{i+1}, \tau_{i+1}) = G_n(\mu_{i+1}, \tau_{i+1})\theta, \tag{14.58}$$

where

$$Z_n(\mu_{i+1}, \tau_{i+1}) = [H_{n,1}(\mu_{i+1}, \tau_{i+1}), H_{n,2}(\mu_{i+1}, \tau_{i+1})]^T, \tag{14.59}$$

$$G_n(\mu_{i+1}, \tau_{i+1}) = \begin{bmatrix} Q_{n,1}(\mu_{i+1}, \tau_{i+1}) & Q_{n,2}(\mu_{i+1}, \tau_{i+1}) \\ Q_{n,2}(\mu_{i+1}, \tau_{i+1}) & Q_{n,3}(\mu_{i+1}, \tau_{i+1}) \end{bmatrix}. \tag{14.60}$$

The parameter update law is defined as

$$\hat{\theta}(\tau_{i+1}) = \operatorname{argmin}\left\{ |\ell - \hat{\theta}(\tau_i)|^2 : \ell \in \Theta, \ Z_n(\mu_{i+1}, \tau_{i+1}) - G_n(\mu_{i|1}, \tau_{i+1})\ell, \ n-1, 2, \ldots \right\}, \tag{14.61}$$

where $\Theta = \{\ell \in \mathbb{R}^2 : \underline{q}_1 \le \ell_1 \le \bar{q}_1, \underline{q}_2 \le \ell_2 \le \bar{q}_2\}$. The estimates are updated at τ_{i+1}— that is, $\hat{\theta}(\tau_{i+1}) = [\hat{q}_1(\tau_{i+1}), \hat{q}_2(\tau_{i+1})]^T$—using the plant states over the time interval $[\mu_{i+1}, \tau_{i+1}]$, where the length of the data acquisition can be adjusted by $\tilde{N}$ in (14.47). The initial values of the estimates $\hat{q}_1(0), \hat{q}_2(0)$ are chosen as $\hat{q}_1(0) = \underline{q}_1, \hat{q}_2(0) = \underline{q}_2$, making $\hat{q}_1(\tau_i) \le q_1$ and $\hat{q}_2(\tau_i) \le q_2$, which will be seen more clearly later. If a more robust identifier with respect to random measurement noise is required, the identifier can be designed in a double integral form as in [104].

For the solution notion, according to definition A.5 in [16], we give the following weak solution definition.

Definition 14.1. *Consider the system*

$$\mathcal{R}_t + \Lambda(x)\mathcal{R}_x + M(x)\mathcal{R} = 0, \quad t \in [0, \infty), \quad x \in [0, L], \tag{14.62}$$

$$\begin{pmatrix} \mathcal{R}^+(t, 0) \\ \mathcal{R}^-(t, L) \end{pmatrix} = K \begin{pmatrix} R^+(t, L) \\ R^-(t, 0) \end{pmatrix} + \begin{pmatrix} N^+ \\ N^- \end{pmatrix} X + \int_0^L \begin{pmatrix} F^+(x) \\ F^-(x) \end{pmatrix} \mathcal{R} dx, \tag{14.63}$$

$$\frac{dX}{dt} = E^+ \mathcal{R}^+(t, L) + E^- \mathcal{R}^+(t, 0) + E_0 X, \quad X \in \mathbb{R}^p, \tag{14.64}$$

$$\mathcal{R}(0,x) = \mathcal{R}_0(x), X(0) = X_0, \tag{14.65}$$

where $\mathcal{R} : [0,+\infty) \times [0,L] \to \mathbb{R}^n$, $M : [0,L] \to \mathcal{M}_{n,n}(\mathbb{R})$, *and the symbol* $\mathcal{M}_{n,n}(\mathbb{R})$, *as usual, denotes the set of* $n \times n$ *real matrices,* $F^+ : [0,L] \to \mathcal{M}_{m,n}(\mathbb{R})$, $F^- : [0,L] \to \mathcal{M}_{n-m,n}(\mathbb{R})$, *and* $\Lambda(x) \triangleq \mathrm{diag}\{\Lambda^+(x), \Lambda^-(x)\}$ *such that*

$$\Lambda^+(x) \triangleq \mathrm{diag}\{\lambda_1(x), \ldots, \lambda_m(x)\}, \tag{14.66}$$

$$\Lambda^-(x) \triangleq -\mathrm{diag}\{\lambda_{m+1}(x), \ldots, \lambda_n(x)\}, \tag{14.67}$$

with $\lambda_i(x) > 0, \forall x \in [0,L]$, *and where*

$$K \triangleq \begin{pmatrix} K_{00} & K_{01} \\ K_{10} & K_{11} \end{pmatrix}, K_{00} \in \mathcal{M}_{m,m}(\mathbb{R}), K_{01} \in \mathcal{M}_{m,n-m}(\mathbb{R}), \tag{14.68}$$

$$K_{10} \in \mathcal{M}_{n-m,m}(\mathbb{R}), \quad K_{11} \in M_{n-m,n-m}(\mathbb{R}), \tag{14.69}$$

$$N^+ \in \mathbb{R}^{m \times p}, \quad N^- \in \mathbb{R}^{(n-m) \times p}, \tag{14.70}$$

$$E^+ \in \mathbb{R}^{p \times m}, \quad E^- \in \mathbb{R}^{p \times (n-m)}, \quad E_0 \in \mathbb{R}^{p \times p}. \tag{14.71}$$

A solution $\mathcal{R} : (0,+\infty) \times (0,L) \to \mathbb{R}^n$, $X : (0,\infty) \to \mathbb{R}^p$ *of the system* (14.62)–(14.65) *is a map* $\mathcal{R} \in C^0([0,+\infty); L^2(0,1); \mathbb{R}^n)$, $X \in C^0([0,+\infty); \mathbb{R}^p)$ *satisfying* (14.65) *such that for every* $T > 0$, *every* $\psi \in C^1([0,T] \times [0,L]; \mathbb{R}^n)$, *and every* $\eta \in C^1([0,T]; \mathbb{R}^p)$ *satisfying*

$$\begin{pmatrix} \psi^+(t,L) \\ \psi^-(t,0) \end{pmatrix} = \begin{pmatrix} \Lambda^+(L)^{-1}K_{00}^T\Lambda^+(0) & \Lambda^+(L)^{-1}K_{10}^T\Lambda^-(L) \\ \Lambda^-(0)^{-1}K_{01}^T\Lambda^+(0) & \Lambda^-(0)^{-1}K_{11}^T\Lambda^-(L) \end{pmatrix}$$

$$\times \begin{pmatrix} \psi^+(t,0) \\ \psi^-(t,L) \end{pmatrix} + \begin{pmatrix} \Lambda^+(L)E^{+T} \\ \Lambda^+(0)E^{-T} \end{pmatrix} \eta \tag{14.72}$$

we have

$$\int_0^L \psi(T,x)^T \mathcal{R}(T,x)dx - \int_0^L \psi(0,x)^T \mathcal{R}_0(x) + \eta^T(T)X(T) - \eta^T(0)X(0)$$

$$= \int_0^T \int_0^L \left[\psi_t^T + \psi_x^T \Lambda + \psi^T(\Lambda_x - M) + \psi^{-T}(t,L)\Lambda^-(L)F^- \right.$$

$$\left. + \psi^{+T}(t,0)\Lambda^+(0)F^+ \right] \mathcal{R} dx dt$$

$$+ \int_0^T \left[\eta_t^T + \eta^T E_0 + \psi^{-T}(t,L)\Lambda^-(L)N^- + \psi^{+T}(t,0)\Lambda^+(0)N^+ \right] X dt. \tag{14.73}$$

Proposition 14.1. *For every* $(z_0, w_0)^T \in L^2((0,1); \mathbb{R}^2)$, $\zeta_0 \in \mathbb{R}$ *and* $\hat{\theta}_0 \in \Theta$, *the initial boundary value problem* (14.1)–(14.5) *with* (14.36), (14.37), (14.38), (14.40), (14.41), (14.47), (14.61) *and initial conditions* $w[0] = w_0$, $z[0] = z_0$, $\zeta(0) = \zeta_0$, $\hat{\theta}(0) = \hat{\theta}_0$, *has a unique (weak) solution* $((z,w)^T, \zeta) \in C^0([0, \lim_{k\to\infty}(\tau_k)); L^2(0,1); \mathbb{R}^2) \times C^0([0, \lim_{k\to\infty}(\tau_k)); \mathbb{R})$.

Proof. The proof is shown in appendix 14.6A. □

The flowchart of the mechanism of the regulation-triggered adaptive controller is shown in figure 14.1, and some system properties are given in the following

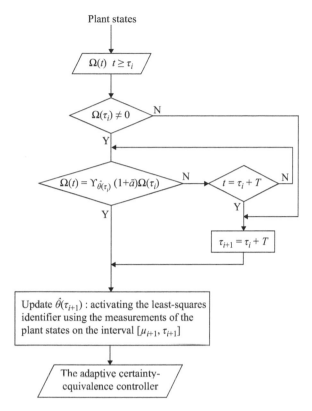

Figure 14.1. The adaptive certainty-equivalence control scheme with regulation-triggered batch least-squares identification.

lemmas. In the rest of this chapter, when we say that $z[t]$, $w[t]$ are equal to zero for $x \in [0, 1], t \in [\mu_{i+1}, \tau_{i+1}]$, or not identically zero on the same domain, we mean "except possibly for finitely many discontinuities of the functions $w[t]$, $z[t]$." These discontinuities are isolated curves in the rectangle $[0, 1] \times [\mu_{i+1}, \tau_{i+1}]$.

Lemma 14.2. *The sufficient and necessary condition of* $Q_{n,1}(\mu_{i+1}, \tau_{i+1}) = 0$ *(or* $Q_{n,3}(\mu_{i+1}, \tau_{i+1}) = 0$*) for* $n = 1, 2, \ldots$ *is* $z[t] = 0$ *(or* $w[t] = 0$*) on* $t \in [\mu_{i+1}, \tau_{i+1}]$.

Proof. Necessity: if $Q_{n,1}(\mu_{i+1}, \tau_{i+1}) = 0$ for $n = 1, 2, \ldots,$ then the definition (14.55) in conjunction with the continuity of $g_{n,1}(t, \mu_{i+1})$ for $t \in [\mu_{i+1}, \tau_{i+1}]$ (a consequence of definition (14.45) and the fact that $z \in C^0([\mu_{i+1}, \tau_{i+1}]; L^2(0, 1)))$ implies

$$g_{n,1}(t, \mu_{i+1}) = 0, \quad t \in [\mu_{i+1}, \tau_{i+1}]. \tag{14.74}$$

According to the definition (14.45) and the continuity of the mapping $\tau \to \int_0^1 \sin(x\pi n)z[\tau]dx$ (a consequence of the fact that $z \in C^0([\mu_{i+1}, \tau_{i+1}]; L^2(0, 1))$), (14.74) implies

$$\int_0^1 \sin(x\pi n)z(x, \tau)dx = 0, \quad \tau \in [\mu_{i+1}, \tau_{i+1}] \tag{14.75}$$

for $n = 1, 2, \ldots$. Since the set $\{\sqrt{2}\sin(n\pi x) : n = 1, 2, \ldots\}$ is an orthonormal basis of $L^2(0, 1)$, we have $z[t] = 0$ for $t \in [\mu_{i+1}, \tau_{i+1}]$.

Similarly, if $Q_{n,3}(\mu_{i+1}, \tau_{i+1}) = 0$ for $n = 1, 2, \ldots$, then $w[t] = 0$ on $t \in [\mu_{i+1}, \tau_{i+1}]$, recalling the definitions (14.57), (14.46) and the fact that $w \in C^0([\mu_{i+1}, \tau_{i+1}]; L^2(0, 1))$, and the set $\{\sqrt{2} \sin(n\pi x) : n = 1, 2, \ldots\}$ being an orthonormal basis of $L^2(0, 1)$.

Sufficiency: if $z[t] = 0$ on $t \in [\mu_{i+1}, \tau_{i+1}]$ (or $w[t] = 0$ on $t \in [\mu_{i+1}, \tau_{i+1}]$), then $Q_{n,1}(\mu_{i+1}, \tau_{i+1}) = 0$ (or $Q_{n,3}(\mu_{i+1}, \tau_{i+1}) = 0$) for $n = 1, 2, \ldots$ is obtained directly, according to (14.45), (14.55) and (14.46), (14.57).

The proof of lemma 14.2 is complete. $\qquad\qquad\qquad\qquad\qquad\qquad\qquad\qquad\qquad$ $\square$

Lemma 14.3. *For the adaptive estimates defined by* (14.61) *based on the data in the interval* $t \in [\mu_{i+1}, \tau_{i+1}]$, *the following statements hold:*

1) *If* $z[t]$ *is not identically zero and* $w[t]$ *is identically zero on* $t \in [\mu_{i+1}, \tau_{i+1}]$, *then* $\hat{q}_1(\tau_{i+1}) = q_1$, $\hat{q}_2(\tau_{i+1}) = \hat{q}_2(\tau_i)$.
2) *If* $w[t]$ *is not identically zero and* $z[t]$ *is identically zero on* $t \in [\mu_{i+1}, \tau_{i+1}]$, *then* $\hat{q}_1(\tau_{i+1}) = \hat{q}_1(\tau_i)$, $\hat{q}_2(\tau_{i+1}) = q_2$.
3) *If* $w[t]$, $z[t]$ *are identically zero on* $t \in [\mu_{i+1}, \tau_{i+1}]$, *then* $\hat{q}_1(\tau_{i+1}) = \hat{q}_1(\tau_i)$, $\hat{q}_2(\tau_{i+1}) = \hat{q}_2(\tau_i)$.
4) *If both* $w[t]$ *and* $z[t]$ *are not identically zero on* $t \in [\mu_{i+1}, \tau_{i+1}]$, *then* $\hat{q}_1(\tau_{i+1}) = q_1$, $\hat{q}_2(\tau_{i+1}) = q_2$.

Moreover, if $\hat{q}_1(\tau_i) = q_1$ *(or* $\hat{q}_2(\tau_i) = q_2$*) for certain* $i \in Z_+$, *then* $\hat{q}_1(t) = q_1$ *(or* $\hat{q}_2(t) = q_2$*) for all* $t \in [\tau_i, \lim_{k \to \infty}(\tau_k))$.

Proof. Define the following set

$$S_i := \left\{ \ell \in \Theta : Z_n(\mu_{i+1}, \tau_{i+1}) = G_n(\mu_{i+1}, \tau_{i+1})\ell, \quad n = 1, 2, \ldots \right\}. \qquad (14.76)$$

If S_i is a singleton, then it is nothing else but the least-squares estimate of the unknown vector of parameters q_1, q_2 on the interval $[\mu_{i+1}, \tau_{i+1}]$.

1) Because $z[t]$ is not identically zero and $w[t]$ is identically zero on $t \in [\mu_{i+1}, \tau_{i+1}]$, there exists $n \in \mathbb{N}$ such that $Q_{n,1}(\mu_{i+1}, \tau_{i+1}) \neq 0$ recalling lemma 14.2. Define the index set I to be the set of all $n \in \mathbb{N}$ with $Q_{n,1}(\mu_{i+1}, \tau_{i+1}) \neq 0$. According to (14.46) and $w[t]$ being identically zero on $t \in [\mu_{i+1}, \tau_{i+1}]$, we know that $g_{n,2}(t, \mu_{i+1}) = 0$ on $t \in [\mu_{i+1}, \tau_{i+1}]$ for $n = 1, 2, \ldots$. It follows that $Q_{n,2}(\mu_{i+1}, \tau_{i+1}) = 0$, $Q_{n,3}(\mu_{i+1}, \tau_{i+1}) = 0$, $H_{n,2}(\mu_{i+1}, \tau_{i+1}) = 0$ for $n = 1, 2, \ldots$, recalling (14.56), (14.57) and (14.54). Then (14.76) implies $S_i = \{(\ell_1, \ell_2) \in \Theta : \ell_1 = \frac{H_{n,1}(\mu_{i+1}, \tau_{i+1})}{Q_{n,1}(\mu_{i+1}, \tau_{i+1})}, n \in I\}$, recalling (14.59), (14.60). Because $(q_1, q_2) \in S_i$ according to (14.58), it follows that $S_i = \{(q_1, \ell_2) \in \Theta : \underline{q}_2 \leq \ell_2 \leq \bar{q}_2\}$. Therefore, (14.61) shows that $\hat{q}_1(\tau_{i+1}) = q_1$ and $\hat{q}_2(\tau_{i+1}) = \hat{q}_2(\tau_i)$.

2) The proof of (2) is very similar to the proof of (1), and thus it is omitted.

3) Because $w[t]$, $z[t]$ are identically zero on $t \in [\mu_{i+1}, \tau_{i+1}]$, then $Q_{n,1}(\mu_{i+1}, \tau_{i+1}) = 0$, $Q_{n,2}(\mu_{i+1}, \tau_{i+1}) = 0$, $Q_{n,3}(\mu_{i+1}, \tau_{i+1}) = 0$, $H_{n,1}(\mu_{i+1}, \tau_{i+1}) = 0$, $H_{n,2}(\mu_{i+1}, \tau_{i+1}) = 0$ for $n = 1, 2, \ldots$ according to (14.45), (14.46), (14.53)–(14.57). It follows that $S_i = \Theta$, and then (14.61) shows that $\hat{q}_1(\tau_{i+1}) = \hat{q}_1(\tau_i)$, $\hat{q}_2(\tau_{i+1}) = \hat{q}_2(\tau_i)$.

4) Because $w[t]$ (or $z[t]$) is not identically zero on $t \in [\mu_{i+1}, \tau_{i+1}]$, there exists $n \in \mathbb{N}$ such that $Q_{n,3}(\mu_{i+1}, \tau_{i+1}) \neq 0$ (or $Q_{n,1}(\mu_{i+1}, \tau_{i+1}) \neq 0$) recalling lemma 14.2. Define the index set I_1 to be the set of all $n \in \mathbb{N}$ with $Q_{n,1}(\mu_{i+1}, \tau_{i+1}) \neq 0$, and define the index set I_2 to be the set of all $n \in \mathbb{N}$ with $Q_{n,3}(\mu_{i+1}, \tau_{i+1}) \neq 0$. Denote the elements in I_1 as $n_1 \in \mathbb{N}$ and those in I_2 as $n_2 \in \mathbb{N}$—that is, $Q_{n_1,1}(\mu_{i+1}, \tau_{i+1}) \neq 0$, $Q_{n_2,3}(\mu_{i+1}, \tau_{i+1}) \neq 0$.

From (14.76), recalling (14.59)–(14.60), we obtain

$$S_i \subseteq \bar{S}_{ai} := \left\{ (\ell_1, \ell_2) \in \Theta : \ell_1 = \frac{H_{n_1,1}(\mu_{i+1}, \tau_{i+1})}{Q_{n_1,1}(\mu_{i+1}, \tau_{i+1})} - \ell_2 \frac{Q_{n_1,2}(\mu_{i+1}, \tau_{i+1})}{Q_{n_1,1}(\mu_{i+1}, \tau_{i+1})}, n_1 \in I_1 \right\},$$
(14.77)

$$S_i \subseteq \bar{S}_{bi} := \left\{ (\ell_1, \ell_2) \in \Theta : \ell_2 = \frac{H_{n_2,2}(\mu_{i+1}, \tau_{i+1})}{Q_{n_2,3}(\mu_{i+1}, \tau_{i+1})} - \ell_1 \frac{Q_{n_2,2}(\mu_{i+1}, \tau_{i+1})}{Q_{n_2,3}(\mu_{i+1}, \tau_{i+1})}, n_2 \in I_2 \right\}.$$
(14.78)

We next prove by contradiction that $S_i = \{(q_1, q_2)\}$. Suppose that on the contrary $S_i \neq \{(q_1, q_2)\}$; that is, S_i defined by (14.76) is not a singleton, which implies the sets $\bar{S}_{ai}, \bar{S}_{bi}$ defined by (14.77), (14.78) are not singletons (because either of $\bar{S}_{ai}, \bar{S}_{bi}$ being a singleton implies that S_i is a singleton). It follows that there exist constants $\bar{\lambda} \in \mathbb{R}$, $\bar{\lambda}_1 \in \mathbb{R}$ such that

$$\frac{Q_{n_1,2}(\mu_{i+1}, \tau_{i+1})}{Q_{n_1,1}(\mu_{i+1}, \tau_{i+1})} = \bar{\lambda}_1, \quad n_1 \in I_1,$$
(14.79)

$$\frac{Q_{n_2,2}(\mu_{i+1}, \tau_{i+1})}{Q_{n_2,3}(\mu_{i+1}, \tau_{i+1})} = \bar{\lambda}, \quad n_2 \in I_2$$
(14.80)

because if there were two different indices $k_1, k_2 \in I_2$ with $\frac{Q_{k_1,2}(\mu_{i+1}, \tau_{i+1})}{Q_{k_1,3}(\mu_{i+1}, \tau_{i+1})} \neq \frac{Q_{k_2,2}(\mu_{i+1}, \tau_{i+1})}{Q_{k_2,3}(\mu_{i+1}, \tau_{i+1})}$, then the set $\bar{S}_{bi}$ defined by (14.78) would be a singleton, and the same would be the case with $\bar{S}_{ai}$ defined by (14.77) if there were two different indices $\bar{k}_1, \bar{k}_2 \in I_1$ with $\frac{Q_{\bar{k}_1,2}(\mu_{i+1}, \tau_{i+1})}{Q_{\bar{k}_1,1}(\mu_{i+1}, \tau_{i+1})} \neq \frac{Q_{\bar{k}_2,2}(\mu_{i+1}, \tau_{i+1})}{Q_{\bar{k}_2,1}(\mu_{i+1}, \tau_{i+1})}$.

Moreover, since S_i is not a singleton, definition (14.76) implies

$$Q_{n,2}(\mu_{i+1}, \tau_{i+1})^2 = Q_{n,1}(\mu_{i+1}, \tau_{i+1}) Q_{n,3}(\mu_{i+1}, \tau_{i+1})$$
(14.81)

for all $n \in I_1 \cup I_2$ ((14.81) naturally holds for $n \notin I_1 \cup I_2$ if $\complement_{\mathbb{N}}\{I_1 \cup I_2\} \neq \emptyset$, because both sides of (14.81) are zero) by recalling (14.60) (because if (14.81) does not hold, it follows from (14.60) that there exists $n \in I_1 \cup I_2$ such that $\det(G_n(\mu_{i+1}, \tau_{i+1})) \neq 0$, which implies S_i defined by (14.76) is a singleton: a contradiction). According to (14.81), (14.55)–(14.57) and the fact that the Cauchy-Schwarz inequality holds as equality only when two functions are linearly dependent, we obtain the existence of constants $\hat{\lambda}_{n_1} \in \mathbb{R}$, $\check{\lambda}_{n_2} \in \mathbb{R}$ such that

$$g_{n_1,2}(t, \mu_{i+1}) = \hat{\lambda}_{n_1} g_{n_1,1}(t, \mu_{i+1}), \quad n_1 \in I_1,$$
(14.82)

$$g_{n_2,1}(t, \mu_{i+1}) = \check{\lambda}_{n_2} g_{n_2,2}(t, \mu_{i+1}), \quad n_2 \in I_2$$
(14.83)

for $t \in [\mu_{i+1}, \tau_{i+1}]$ (notice that $g_{n_1,1}(t, \mu_{i+1})$ and $g_{n_2,2}(t, \mu_{i+1})$ are not identically zero on $t \in [\mu_{i+1}, \tau_{i+1}]$ because $Q_{n_1,1}(\mu_{i+1}, \tau_{i+1}) \neq 0$ and $Q_{n_2,3}(\mu_{i+1}, \tau_{i+1}) \neq 0$). Recalling (14.79), (14.80), we obtain from (14.55)–(14.57) and (14.82), (14.83) that

$$g_{n_1,2}(t, \mu_{i+1}) = \bar{\lambda}_1 g_{n_1,1}(t, \mu_{i+1}), \quad n_1 \in I_1,$$
(14.84)

$$g_{n_2,1}(t, \mu_{i+1}) = \bar{\lambda} g_{n_2,2}(t, \mu_{i+1}), \quad n_2 \in I_2$$
(14.85)

for $t \in [\mu_{i+1}, \tau_{i+1}]$. Equations (14.84), (14.85) holding is a necessary condition of the hypothesis that S_i is not a singleton. The remaining proof of case 4 is divided into the following three claims.

Claim 14.1. *If S_i is not a singleton, then $\bar{\lambda} \neq 0, \bar{\lambda}_1 \neq 0$ and $\bar{\lambda} = \frac{1}{\bar{\lambda}_1}$ in (14.84), (14.85).*

Proof. The proof is shown in appendix 14.6B.   □

Claim 14.2. *Equations (14.84), (14.85) ($\bar{\lambda} \neq 0, \bar{\lambda}_1 \neq 0$ and $\bar{\lambda} = \frac{1}{\bar{\lambda}_1}$) hold if and only if $z[t] + \bar{\lambda} w[t] = 0$ ($\bar{\lambda} \neq 0$) for $t \in [\mu_{i+1}, \tau_{i+1}]$.*

Proof. The proof is shown in appendix 14.6C.   □

Claim 14.3. *The function $z[t] + \bar{\lambda} w[t]$ ($\bar{\lambda} \neq 0$) is not identically zero for $t \in [\mu_{i+1}, \tau_{i+1}]$.*

Proof. The proof is shown in appendix 14.6D.   □

Recalling claims 14.1–14.3, we know that (14.84), (14.85), which is a necessary condition of the hypothesis that S_i not be a singleton, does not hold. Consequently, S_i is a singleton—that is, $S_i = \{(q_1, q_2)\}$. Therefore, (14.61) shows that $\hat{q}_1(\tau_{i+1}) = q_1, \hat{q}_2(\tau_{i+1}) = q_2$. The proof of case 4 is complete.

If $\hat{q}_1(\tau_i) = q_1$ (or $\hat{q}_2(\tau_i) = q_2$) for certain $i \in Z_+$, recalling (14.61) and the analysis in the above four cases, we have $\hat{q}_1(t_{i+1}) = q_1$ (or $\hat{q}_2(t_{i+1}) = q_2$). Repeating the above process, we then have $\hat{q}_1(t) = q_1$ (or $\hat{q}_2(t) = q_2$) for all $t \in [\tau_i, \lim_{k \to \infty}(\tau_k))$.

The proof of lemma 14.3 is complete.   □

14.4   MAIN RESULT

Theorem 14.1. *With arbitrary initial data $(z_0, w_0)^T \in L^2((0,1); \mathbb{R}^2)$, $\zeta_0 \in \mathbb{R}$, and $\hat{\theta}_0 = (\underline{q}_1, \underline{q}_2)^T$, for the plant (14.1)–(14.5) under the adaptive certainty-equivalence boundary controller (14.36) where the regulation-triggered BaLSI is defined by (14.37), (14.61) with (14.38), (14.40), (14.41), (14.47), the closed-loop system satisfies the following properties:*

1) The Zeno phenomenon does not occur—that is,

$$\lim_{i \to \infty} \tau_i = +\infty, \tag{14.86}$$

and the closed-loop system is well-posed.

2) If the finite-time convergence of parameter estimates to the true values does not occur, $\Omega(t)$ reaches zero in finite time $\frac{1}{q_1}$—that is, $\Omega(t) \equiv 0$ on $t \in [\frac{1}{q_1}, \infty)$.

3) If the parameter estimates converge to the true values in finite time, there exist positive constants $M_{\theta, \hat{\theta}(0)}, \lambda_1$ such that

$$\Omega(t) \leq M_{\theta, \hat{\theta}(0)} \Omega(0) e^{-\lambda_1 t}, \quad t \geq 0, \tag{14.87}$$

where $\Omega(t)$ is given in (14.24), and $M_{\theta, \hat{\theta}(0)}$ is a family of constants parameterized by the positive constants $q_1, q_2, \hat{q}_1(0), \hat{q}_2(0)$. The decay rate λ_1 is the same as the nominal control result in (14.26).

Proof. First, we propose the following claim about the sufficient and necessary condition of the finite-time convergence of parameter estimates to the true values.

Claim 14.4. *When $\hat{q}_1(0) \neq q_1$ (or $\hat{q}_2(0) \neq q_2$), the estimate $\hat{q}_1(t)$ (or $\hat{q}_2(t)$) reaches the actual value q_1 (or q_2) in finite time if and only if $z[t]$ (or $w[t]$) is not identically zero on $t = [0, \lim_{i \to \infty}(\tau_i))$.*

Proof. The proof is shown in appendix 14.6E.   □

1) Now we prove the first of the three portions of the theorem. First, if the estimates $\hat{q}_1(t), \hat{q}_2(t)$ reach the true values in finite time τ_ε, we have $\tau_j = \tau_\varepsilon + (j - \varepsilon)T$, $j \in Z_+, j > \varepsilon$. The proof of this is shown next. We prove by induction that $\tau_{i+1} = \tau_i + T$ for $i \geq \varepsilon$. Let $i \geq \varepsilon$ be an integer. Notice that (14.23) holds for all $t \in [\tau_i, \tau_{i+1})$. Assume that $\Omega(\tau_i) \neq 0$. By virtue of (14.26) and since (14.23) holds, we have

$$\Omega(t) \leq \Upsilon_{\hat{\theta}(\tau_i)} \Omega(\tau_i) \tag{14.88}$$

for all $t \in [\tau_i, \tau_{i+1})$. It follows that

$$\Omega(t) \leq \Upsilon_{\hat{\theta}(\tau_i)} \Omega(\tau_i) < \Upsilon_{\hat{\theta}(\tau_i)}(1 + \bar{a})\Omega(\tau_i) \tag{14.89}$$

for $t \in [\tau_i, \tau_{i+1})$, where $\bar{a}$ is positive. Therefore, we get from (14.38), (14.40) that $\tau_{i+1} = \tau_i + T$ for $i \geq \varepsilon$. The same conclusion follows from (14.38), (14.41) if $\Omega(\tau_i) = 0$. Therefore, $\lim_{i \to \infty}(\tau_i) = +\infty$.

If the finite-time convergence of the parameter estimates to the true values is not achieved, the proof is divided into the three cases.

Case 1: We suppose that the estimate $\hat{q}_2(t)$ does not reach q_2 in finite time, but $\hat{q}_1(t)$ does reach q_1 in finite time. The fact that $\hat{q}_2(t)$ does not reach q_2 in finite time implies $w[t] \equiv 0$ on $t \in [0, \lim_{i \to \infty}(\tau_i))$ according to claim 14.4, and $\hat{q}_2(t) = \hat{q}_2(0) \neq q_2$ on $t \in [0, \lim_{i \to \infty}(\tau_i))$ according to lemma 14.3. The fact that $\hat{q}_1(t)$ reaches q_1 in finite time implies $\tilde{q}_1(t) \equiv 0$ after a certain τ_f.

Inserting (14.36) into (14.5), we obtain

$$q_2 w(1, t) = \hat{q}_2(0) \int_0^1 \phi(1, y; \hat{q}_1(t), \hat{q}_2(0))w(y, t)dy$$
$$+ \hat{q}_2(0)\lambda(1; \hat{q}_1(t), \hat{q}_2(0))\zeta(t) + \tilde{q}_1(t)z(1, t). \tag{14.90}$$

Considering $w[t] \equiv 0$ and $\tilde{q}_1(t) \equiv 0$ on $t \in [\tau_f, \lim_{i \to \infty}(\tau_i))$, we obtain from (14.90) that $\zeta(t) \equiv 0$ on $t \in [\tau_f, \lim_{i \to \infty}(\tau_i))$ because $\hat{q}_2(0) \neq 0$ and $\lambda(1; \hat{q}_1(t), \hat{q}_2(0)) \neq 0$. Recalling (14.1), considering $w[t] \equiv 0$ on $t \in [0, \lim_{i \to \infty}(\tau_i))$ and $\zeta(t) = 0$ on $t \in [\tau_f, \lim_{i \to \infty}(\tau_i))$, it further follows that $\zeta(0) = 0$, that is, $\zeta(t) \equiv 0$ on $t \in [0, \lim_{i \to \infty}(\tau_i))$, which means that $z(0, t) \equiv 0$ on $t \in [0, \lim_{i \to \infty}(\tau_i))$. It follows that $\Omega(t)$ defined in (14.24) is nonincreasing on $t \in [0, \frac{1}{q_1}]$. Therefore, we have $\lim_{i \to \infty}(\tau_i) > \frac{1}{q_1}$ according to the definition of triggering times (14.38), (14.40), (14.41). By virtue of (14.2), (14.4) and $z(0, t) \equiv 0$ on $t \in [0, \lim_{i \to \infty}(\tau_i))$, we have $z[t] \equiv 0$ for $t \in [\frac{1}{q_1}, \lim_{i \to \infty}(\tau_i))$. If $z[t]$ is identically zero on $t \in [0, \frac{1}{q_1}]$, then $\hat{q}_1(0) = q_1$ (because $z[t] \equiv 0$ on $t \in [0, \lim_{i \to \infty}(\tau_i))$, and $\hat{q}_1(t)$ reaches q_1 in finite time only when the initial estimate is the true value according to claim 14.4). If $z[t]$ is not identically zero in $t \in [0, \frac{1}{q_1})$, it implies that the initial condition $z(x, 0)$ is not identically zero for $x \in (0, 1]$, moreover, that τ_f must be less than $\frac{1}{q_1}$ and the function $z(x, 0)$ is not identically zero on the interval $(0, 1 - q_1\tau_f]$ for x (otherwise $w[t]$ is not identically zero according to (14.90): a contradiction). The state $z(x, t)$ propagates from its initial condition

$z(x,0)$, which is possibly not identically zero only on $(0, 1 - q_1\tau_f]$, toward the boundary $x = 1$ and finally vanishes not later than $t = \frac{1}{q_1}$ ($z(1,t) = 0$ for $t \in [0, \tau_f)$ and the nonzero values of $z(1,t)$ on $t \in [\tau_f, \frac{1}{q_1}]$ are eliminated by $\tilde{q}_1(t) = 0$ in (14.90)). Together with $w[t] \equiv 0$, $\zeta(t) \equiv 0$ on $t \in [0, \lim_{i \to \infty}(\tau_i))$, we conclude that $\Omega(t)$ is nonincreasing in $t \in [0, \frac{1}{q_1}]$, and $\Omega(t) \equiv 0$ for $t \in [\frac{1}{q_1}, \lim_{i \to \infty}(\tau_i))$. Therefore, $\tau_j = jT$, $j \in Z_+$, according to the definition of triggering times (14.38), (14.40), (14.41).

Case 2: We suppose that the estimate $\hat{q}_1(t)$ does not reach q_1 in finite time, but $\hat{q}_2(t)$ does reach q_2 in finite time. The fact that $\hat{q}_1(t)$ does not reach q_1 in finite time implies that $z(x,t) \equiv 0$ on $t \in [0, \lim_{i \to \infty}(\tau_i))$ according to claim 14.4, and $\hat{q}_1(t) = \hat{q}_1(0) \neq q_1$ on $t \in [0, \lim_{i \to \infty}(\tau_i))$ according to lemma 14.3. Recalling (14.36), then (14.1)–(14.5) become

$$\zeta(t) = e^{[(a - q_1 bc) + b(q_2 + q_1 p)\frac{c}{p}]t}\zeta(0), \tag{14.91}$$

$$w_t(x,t) = q_2 w_x(x,t), \tag{14.92}$$

$$w(0,t) = \frac{c}{p}\zeta(t), \tag{14.93}$$

$$w(1,t) = \frac{\hat{q}_2(t)}{q_2}\left[\int_0^1 \phi(1,y;\hat{q}_1(0),\hat{q}_2(t))w(y,t)dy + \lambda(1;\hat{q}_1(0),\hat{q}_2(t))\zeta(t)\right], \tag{14.94}$$

$t \in [0, \lim_{i \to \infty}(\tau_i))$. If $\zeta(0) = 0$, then $z[t], w[t], \zeta(t)$ are identically zero on $t \in [0, \lim_{i \to \infty}(\tau_i))$ according to (14.91)–(14.93). Next, we discuss the case of $\zeta(0) \neq 0$. Considering $z[t] \equiv 0$ on $t \in [0, \lim_{i \to \infty}(\tau_i))$, the dynamics for $w[t], \zeta(t)$ given as (14.91)–(14.94), and the definition of triggering times (14.38), (14.40), (14.41), we have that $\lim_{i \to \infty}(\tau_i) > \frac{1}{q_2}$. The equation $w(0,t) = \frac{c}{p}\zeta(t)$ (14.93) holding for $t \in [0, \lim_{i \to \infty}(\tau_i))$ requires the initial condition of w to be

$$w(x,0) = \frac{c}{p}e^{[(a - q_1 bc) + b(q_2 + q_1 p)\frac{c}{p}]\frac{1}{q_2}x}\zeta(0)$$

for ensuring that (14.93) holds on $t \in [0, \frac{1}{q_2}]$ and $w(1,t)$ to be

$$w(1,t) = \frac{c}{p}e^{[(a - q_1 bc) + b(q_2 + q_1 p)\frac{c}{p}]\frac{1}{q_2}}\zeta(t), \quad t \in [0, \lim_{i \to \infty}(\tau_i)) \tag{14.95}$$

for ensuring that (14.93) holds on $t \in [\frac{1}{q_2}, \lim_{i \to \infty}(\tau_i))$.

Comparing (14.95), where $w(1,t)$ is a continuous function by virtue of (14.91), with (14.94) which includes possible discontinuities in $\hat{q}_2$, the necessary condition for the equation (14.93) to hold on $t \in [0, \lim_{i \to \infty}(\tau_i))$ is that $w(1,t)$ is a continuous function. In other words, there is no discontinuity in case 2. Considering that the state of the w-PDE in (14.92) propagates from $x = 1$ to $x = 0$ with the propagation speed q_2, by representing the function $w(y,t)$ as the future value of $\zeta(t)$ and using the expression for $\zeta(t)$ given by (14.91), we write the relation (14.94) as

$$w(1,t) = \frac{\hat{q}_2(t)}{q_2}\left[\int_0^1 \phi(1,y;\hat{q}_1(0),\hat{q}_2(t))\frac{c}{p}e^{[(a - q_1 bc) + b(q_2 + q_1 p)\frac{c}{p}]\frac{1}{q_2}y}dy\right.$$

$$\left. + \lambda(1;\hat{q}_1(0),\hat{q}_2(t))\right]\zeta(t), \ t \geq 0. \tag{14.96}$$

Comparing (14.95) and (14.96), applying (14.15), (14.16), the necessary condition of $w(0,t) = \frac{c}{p}\zeta(t)$ (14.93) always holds on $t \in [\frac{1}{q_2}, \lim_{i \to \infty}(\tau_i))$ (when $\zeta(0) \neq 0$), is

$$\frac{1}{q_2}\left[\int_0^1 \frac{c\kappa b}{p}e^{\frac{1}{\hat{q}_2(t)}(a-\hat{q}_1(0)bc)(1-y)}e^{[(a-q_1bc)+b(q_2+q_1p)\frac{c}{p}]\frac{1}{q_2}y}dy\right.$$

$$\left.+\frac{\hat{q}_2(t)\kappa}{\hat{q}_1(0)p+\hat{q}_2(t)}e^{\frac{1}{\hat{q}_2(t)}(a-\hat{q}_1(0)bc)}\right]\equiv\frac{c}{p}e^{[(a-q_1bc)+b(q_2+q_1p)\frac{c}{p}]\frac{1}{q_2}} \qquad (14.97)$$

on $t\in[0,\lim_{i\to\infty}(\tau_i))$. The right-hand side of (14.97) is constant, while the left-hand side of (14.97) includes $\hat{q}_2(t)$, whose potential values are $q_2,\hat{q}_2(0)$ because of lemma 14.3. If the left-hand side of (14.97) is varying with $\hat{q}_2(t)$, then (14.97) does not hold. If the left-hand side of (14.97) is kept constant with $\hat{q}_2(t)=q_2$ and $\hat{q}_2(t)=\hat{q}_2(0)$ (such as $\hat{q}_2(0)=q_2$), since, as we mentioned above, there is no discontinuity in case 2, (14.97) holds only when the design parameter κ is equal to κ^*, where

$$\kappa^*=\frac{cq_2}{p}e^{[(a-q_1bc)+b(q_2+q_1p)\frac{c}{p}]\frac{1}{q_2}}$$

$$\div\left(\int_0^1 \frac{bc}{p}e^{\frac{1}{q_2}(a-\hat{q}_1(0)bc)(1-y)}e^{[(a-q_1bc)+b(q_2+q_1p)\frac{c}{p}]\frac{1}{q_2}y}dy+\frac{q_2e^{\frac{1}{q_2}(a-\hat{q}_1(0)bc)}}{\hat{q}_1(0)p+q_2}\right),$$

$$(14.98)$$

where the symbol $\div$ denotes division. The constant κ^* is positive because $b>0$, $c>0$, $p>0$, $q_1>0$, $q_2>0$, $\hat{q}_1(0)>0$. The positivity of $\kappa=\kappa^*$ contradicts $\kappa<0$. Therefore, case 2 would happen only when $\zeta(0)=0$, where $\Omega(t)\equiv0$ on $t\in[0,\lim_{i\to\infty}(\tau_i))$. Therefore, $\tau_j=jT$, $j\in Z_+$.

Case 3. If neither $\hat{q}_1(t)$ nor $\hat{q}_2(t)$ reaches q_1,q_2, it follows that $z[t],w[t],\zeta(t)$ are identically zero on $t\in[0,\lim_{i\to\infty}(\tau_i))$—that is, $\Omega(t)\equiv0$ on $t\in[0,\lim_{i\to\infty}(\tau_i))$, according to claim 14.4 and (14.4). Therefore, $\tau_j=jT$, $j\in Z_+$ according to (14.38), (14.41).

By virtue of the results in the above discussions, we have $\lim_{i\to\infty}(\tau_i)=+\infty$. The well-posedness of the closed-loop system is then obtained by recalling proposition 14.1 and $\lim_{i\to\infty}(\tau_i)=+\infty$. This completes the proof of portion (1) of the theorem. The fact that $\lim_{i\to\infty}(\tau_i)=+\infty$ allows the solution to be defined on $\mathbb{R}_+$.

2) Now we prove the second of the three portions of the theorem. Recalling the results in the discussions in cases 1–3 in the proof of portion (1) and $\lim_{i\to\infty}(\tau_i)=+\infty$, we conclude that $\Omega(t)$ reaches zero not later than $\frac{1}{q_1}$—that is, $\Omega(t)\equiv0$ on $t\in[\frac{1}{q_1},\infty)$—when the finite-time convergence of the parameter estimates to the true values is not achieved. Thus, portion (2) of the theorem is obtained.

3) Finally, we prove the last of the three portions of the theorem—that is, establish the exponential regulation result when estimates $(\hat{q}_1(t),\hat{q}_2(t))$ reach the true values (q_1,q_2) in finite time τ_ε, that is, when

$$\hat{q}_1(t)=q_1,\quad\hat{q}_2(t)=q_2,\quad t\geq\tau_\varepsilon. \qquad (14.99)$$

Define a Lyapunov function

$$V(t)=\frac{1}{2}r_a\int_0^1 e^{\delta x}\beta(x,t)^2dx+\frac{1}{2}r_b\int_0^1 e^{-\delta x}\alpha(x,t)^2dx+\frac{1}{2}\zeta(t)^2,\quad t\geq0, \qquad (14.100)$$

where the positive constants r_a, r_b, δ are constrained through the inequalities (14.10), (14.9), (14.28). Denoting

$$\Omega_1(t)=\|\alpha[t]\|^2+\|\beta[t]\|^2+\zeta(t)^2,$$

we obtain

$$\xi_1 \Omega_1(t) \leq V(t) \leq \xi_2 \Omega_1(t), \quad t \geq 0, \tag{14.101}$$

where the positive constants ξ_1, ξ_2 are shown in (14.31), (14.32).

Define the errors between the gains in the nominal control law (14.23) and those in the certainty-equivalence controller (14.36), caused by the parameter estimate errors, as

$$\tilde{q}_1(t) = q_1 - \hat{q}_1(t), \tag{14.102}$$

$$\tilde{R}_1(y,t) = q_2 \phi(1, y; q_1, q_2) - \hat{q}_2(t)\phi(1, y; \hat{q}_1(t), \hat{q}_2(t)), \tag{14.103}$$

$$\tilde{R}_2(t) = q_2 \lambda(1; q_1, q_2) - \hat{q}_2(t)\lambda(1; \hat{q}_1(t), \hat{q}_2(t)), \tag{14.104}$$

where $\phi(1, y; \hat{q}_1(t), \hat{q}_2(t)), \lambda(1; \hat{q}_1(t), \hat{q}_2(t))$ are the results of replacing q_1, q_2 with $\hat{q}_1(t), \hat{q}_2(t)$ in $\phi(1, y; q_1, q_2)$ and $\lambda(1; q_1, q_2)$. Because of (14.99), $\tilde{q}_1(t), \tilde{R}_1(\cdot, t), \tilde{R}_2(t)$ are zero for $t \geq \tau_\varepsilon$.

Applying the adaptive control law (14.36), recalling (14.23), the boundary condition (14.21) in the target system (14.17)–(14.21) becomes

$$\beta(1, t) = \frac{1}{q_2}\left[\tilde{q}_1(t)z(1, t) - \int_0^1 \tilde{R}_1(y, t)w(y, t)dy - \tilde{R}_2(t)\zeta(t)\right]. \tag{14.105}$$

Applying the Cauchy-Schwarz inequality into the backstepping transformation (14.13), (14.14) and its inverse

$$z(x, t) = \alpha(x, t), \tag{14.106}$$

$$w(x, t) = \beta(x, t) - \int_0^x \frac{\kappa b}{q_2}e^{\frac{m}{q_2}(x-y)}\beta(y, t)dy - \frac{\kappa}{(q_2 + q_1 p)}e^{\frac{m}{q_2}x}\zeta(t), \tag{14.107}$$

we have that $\Omega_1(t)$ is bounded by

$$\xi_3 \Omega(t) \leq \Omega_1(t) \leq \xi_4 \Omega(t), \quad t \geq 0, \tag{14.108}$$

where the positive constants ξ_3, ξ_4 are defined by (14.33)–(14.35).

Taking the derivative of (14.100) along (14.17)–(14.21), (14.105) and applying Young's inequality and the Cauchy-Schwarz inequality, we get

$$\dot{V}(t) \leq -\left(\frac{1}{2}m - q_1 r_b c_0^2\right)\zeta(t)^2 - \left(\frac{q_2 r_a}{2} - q_1 r_b p^2 - \frac{(q_1 p + q_2)^2 b^2}{2m}\right)\beta(0, t)^2$$

$$- \frac{1}{2}r_a \delta q_2 \int_0^1 e^{\delta x}\beta(x, t)^2 dx - \frac{1}{2}r_b \delta q_1 \int_0^1 e^{-\delta x}\alpha(x, t)^2 dx$$

$$- \frac{1}{2}q_1 r_b e^{-\delta}\alpha(1, t)^2 + \frac{1}{2}q_2 r_a e^{\delta}\beta(1, t)^2 \tag{14.109}$$

for $t \geq 0$. Recalling (14.9) and (14.10), then (14.109) becomes

$$\dot{V}(t) \leq -\lambda_1 V(t) + \frac{1}{2}q_2 r_a e^{\delta}\beta(1, t)^2 \tag{14.110}$$

for $t \geq 0$, where the positive constant λ_1 is shown in (14.27). Multiplying both sides of (14.110) by $e^{\lambda_1 t}$ yields

$$\frac{d(V(t)e^{\lambda_1 t})}{dt} \leq \frac{1}{2}e^{\lambda_1 t}q_2 r_a e^{\delta}\beta(1,t)^2, \quad t \geq 0, \tag{14.111}$$

and then, integrating from τ_ε to t, we obtain

$$V(t)e^{\lambda_1 t} - V(\tau_\varepsilon)e^{\lambda_1 \tau_\varepsilon} \leq \int_{\tau_\varepsilon}^{t} \frac{1}{2}e^{\lambda_1 \varsigma}q_2 r_a e^{\delta}\beta(1,\varsigma)^2 d\varsigma, \quad t \geq \tau_\varepsilon. \tag{14.112}$$

Recalling (14.105) and the fact that $\tilde{q}_1(t)$, $\tilde{R}_1(t)$, $\tilde{R}_2(t)$ are identically zero for $t \geq \tau_\varepsilon$, we get $\beta(1,t) \equiv 0$ for $t \geq \tau_\varepsilon$ according to (14.105). Therefore, the term $\int_{\tau_\varepsilon}^{t} \frac{1}{2}e^{\lambda_1 \varsigma}q_2 r_a e^{\delta}\beta(1,\varsigma)^2 d\varsigma$ in (14.112) is zero. Multiplying both sides of (14.112) by $e^{-\lambda_1 t}$ yields

$$V(t) \leq V(\tau_\varepsilon)e^{-\lambda_1(t-\tau_\varepsilon)}, \quad t \geq \tau_\varepsilon. \tag{14.113}$$

Recalling (14.101), we get

$$\Omega_1(t) \leq \frac{\xi_2}{\xi_1}\Omega_1(\tau_\varepsilon)e^{-\lambda_1(t-\tau_\varepsilon)}, \quad t \geq \tau_\varepsilon.$$

Recalling (14.108), we further have that

$$\Omega(t) \leq \Upsilon_\theta \Omega(\tau_\varepsilon)e^{-\lambda_1(t-\tau_\varepsilon)}, \quad t \geq \tau_\varepsilon, \tag{14.114}$$

where the overshoot coefficient Υ_θ is shown in (14.30).

If $\tau_\varepsilon = 0$, we obtain directly from (14.114) that $\Omega(t) \leq \Upsilon_\theta \Omega(0)e^{-\lambda_1 t}, t \geq 0$. Next, we conduct an analysis for $t \in [0, \tau_\varepsilon]$ when $\tau_\varepsilon \neq 0$. Recalling (14.13), (14.105), (14.109), (14.110), we obtain

$$\dot{V}(t) \leq -\lambda_1 V(t) - \left(\frac{1}{2}q_1 r_b e^{-\delta} - \frac{1}{q_2}r_a e^{\delta}(\bar{q}_1 - \underline{q}_1)^2\right)\alpha(1,t)^2$$
$$+ \frac{9r_a c^{\delta}}{2q_2}\left[\int_0^1 \tilde{R}_1(y,t)^2 w(y,t)^2 dy + \tilde{R}_2(t)^2 \zeta(t)^2\right] \tag{14.115}$$

for $t \in [0, \tau_\varepsilon]$. Recalling (14.28), which makes the coefficient in the parentheses in front of $\alpha(1,t)^2$ positive, and recalling $\tilde{R}_1(y,t)$, $\tilde{R}_2(t)$, defined by (14.103), (14.104), where $\hat{q}_1(t)$ is equal to either $\hat{q}_1(0)$ or q_1 and $\hat{q}_2(t)$ is equal to either $\hat{q}_2(0)$ or q_2 in $t \in [0, \tau_\varepsilon]$ according to lemma 14.3, as well as applying (14.107) and the Cauchy-Schwarz inequality, we obtain from (14.115) that

$$\dot{V}(t) \leq -\lambda_1 V(t) + Q(\hat{q}_1(0), \hat{q}_2(0), q_1, q_2)V(t), \quad t \in [0, \tau_\varepsilon], \tag{14.116}$$

where the positive constant $Q(\hat{q}_1(0), \hat{q}_2(0), q_1, q_2)$, obtained by bounding the last line of (14.115), is a family of constants parameterized by the positive constants $q_1, q_2, \hat{q}_1(0), \hat{q}_2(0)$.

If $\lambda_1 < Q(\hat{q}_1(0), \hat{q}_2(0), q_1, q_2)$, then by defining a positive constant

$$\lambda_2(\hat{q}_1(0), \hat{q}_2(0), q_1, q_2) = Q(\hat{q}_1(0), \hat{q}_2(0), q_1, q_2) - \lambda_1 \tag{14.117}$$

and multiplying both sides of (14.116) by $e^{-\lambda_2(\hat{q}_1(0), \hat{q}_2(0), q_1, q_2)t}$ we obtain

$$\dot{V}(t)e^{-\lambda_2(\hat{q}_1(0),\hat{q}_2(0),q_1,q_2)t} - \lambda_2(\hat{q}_1(0),\hat{q}_2(0),q_1,q_2)V(t)e^{-\lambda_2(\hat{q}_1(0),\hat{q}_2(0),q_1,q_2)t} \leq 0 \tag{14.118}$$

for $t \in [0, \tau_\varepsilon]$, that is,

$$\frac{d(V(t)e^{-\lambda_2(\hat{q}_1(0),\hat{q}_2(0),q_1,q_2)t})}{dt} \leq 0 \tag{14.119}$$

for $t \in [0, \tau_\varepsilon]$. Then integrating from 0 to t yields

$$V(t) \leq V(0)e^{\lambda_2(\hat{q}_1(0),\hat{q}_2(0),q_1,q_2)t}, \quad t \in [0, \tau_\varepsilon]. \tag{14.120}$$

Recalling (14.101), (14.108), we get

$$\Omega(t) \leq \Upsilon_\theta \Omega(0)e^{\lambda_2(\hat{q}_1(0),\hat{q}_2(0),q_1,q_2)t}, \quad t \in [0, \tau_\varepsilon]. \tag{14.121}$$

If $\lambda_1 \geq Q(q_1(0), q_2(0), q_1, q_2)$, then by defining a positive constant

$$\lambda_3(\hat{q}_1(0),\hat{q}_2(0),q_1,q_2) = \lambda_1 - Q(\hat{q}_1(0),\hat{q}_2(0),q_1,q_2) \tag{14.122}$$

we obtain from (14.116) that

$$\Omega(t) \leq \Upsilon_\theta \Omega(0)e^{-\lambda_3(\hat{q}_1(0),\hat{q}_2(0),q_1,q_2)t}, \quad t \in [0, \tau_\varepsilon]. \tag{14.123}$$

Comparing (14.121) and (14.123), we obtain

$$\Omega(\tau_\varepsilon) \leq \Upsilon_\theta e^{\lambda_2(\hat{q}_1(0),\hat{q}_2(0),q_1,q_2)\tau_\varepsilon} \Omega(0), \quad t \in [0, \tau_\varepsilon]. \tag{14.124}$$

Let us now recap that assumption 14.2—that is, (14.8)—is only used to ensure the existence of a positive δ satisfying (14.28), which enables going from (14.115) to (14.116), with the purpose of arriving at (14.124). In plain words, assumption 14.2 is only used in the Lyapunov analysis when the estimate $\hat{q}_1(t)$ has not reached the true value q_1, in order to ensure (14.124).

Recalling (14.114), we conclude that

$$\Omega(t) \leq \Upsilon_\theta^2 e^{\lambda_2(\hat{q}_1(0),\hat{q}_2(0),q_1,q_2)\tau_\varepsilon} e^{\lambda_1 \tau_\varepsilon} \Omega(0)e^{-\lambda_1 t}, \quad t \geq 0. \tag{14.125}$$

Claim 14.5. *If $\hat{\theta}(t)$ reaches θ at τ_ε, then $\tau_\varepsilon \leq \max\{\frac{1}{q_2} + T, 2T\}$.*

Proof. The proof is shown in appendix 14.6F. $\square$

Applying claim 14.5, (14.125) is written as

$$\Omega(t) \leq \Upsilon_\theta^2 e^{\lambda_2(\hat{q}_1(0),\hat{q}_2(0),q_1,q_2)\max\{\frac{1}{q_2}+T,2T\}} e^{\lambda_1 \max\{\frac{1}{q_2}+T,2T\}} \Omega(0)e^{-\lambda_1 t} \tag{14.126}$$

for $t \geq 0$. Denoting

$$M_{\theta,\hat{\theta}(0)} = \Upsilon_\theta^2 e^{\lambda_2(\hat{q}_1(0),\hat{q}_2(0),q_1,q_2)\max\{\frac{1}{q_2}+T,2T\}} e^{\lambda_1 \max\{\frac{1}{q_2}+T,2T\}},$$

we obtain (14.87). This completes the proof of portion (3) of the theorem.

With the proof of theorem 14.1 completed, we thank the reader for sticking with us for the nearly six-page ride and commend the reader's stamina. $\square$

14.5   SIMULATION

The simulation is conducted for the plant (14.1)–(14.5) with the model parameters taken as

$$a = 13, \quad b = 1, \quad c = 2, \quad p = 0.5, \tag{14.127}$$

$$\bar{c} = 1, \quad q_1 = 4, \quad q_2 = 6, \tag{14.128}$$

where q_1, q_2 are treated as unknown, with the known bounds

$$\bar{q}_1 = 6, \quad \underline{q}_1 = 2, \quad \bar{q}_2 = 7, \quad \underline{q}_2 = 3. \tag{14.129}$$

The initial values for the test are chosen as

$$z(x, 0) = \cos\left(\pi x + \frac{\pi}{4}\right) + x^3, \tag{14.130}$$

$$w(x, 0) = \sin\left(1.5\pi x + \frac{\pi}{3}\right) + x^2, \tag{14.131}$$

$$\zeta(0) = \frac{1}{c} z(0, 0) + \frac{p}{c} w(0, 0). \tag{14.132}$$

The finite-difference method is adopted to conduct the simulation with the time and space steps of 0.0001 and 0.01, respectively.

For the regulation-triggered BaLSI defined by (14.37), (14.38), (14.40), (14.41), (14.47), (14.61), we choose $n = 1, 2, \ldots, 7$ and

$$\bar{a} = 0.8, \quad \tilde{N} = 1, \quad T = 8 \tag{14.133}$$

and take the initial values of the estimates as $\hat{q}_1(0) = \underline{q}_1$ $\hat{q}_2(0) = \underline{q}_2$—namely, we start the parameter estimates from their lower bounds. Using the given bounds in (14.129) to determine the gain κ in the controller (14.36) by (14.12), a control gain satisfying $\kappa < -267$ is needed. For $\kappa = -320$, recalling the model parameters in (14.127), (14.128), we obtain $r_b < 0.079$, $r_a > 0.06$ according to (14.9), (14.10), which indicates that $\bar{q}_1 \quad \underline{q}_1$ needs to be smaller than 4.01 according to assumption 14.2 (satisfied by the known bounds of q_1 given in (14.129)). The source of the high gain κ is the first term of (14.12) which is used in claim 14.3 to exclude some rare and extreme situations ($w(x, t) = M$ and $z(x, t) = -\bar{\lambda}M$ for $x \in [0, 1]$, $t \in [\mu_{i+1}, \tau_{i+1}]$ where $M, \bar{\lambda}$ are nonzero constants) affecting the exact parameter estimation. In the simulation, we find that the high gain is actually not needed (the aforementioned extreme situations do not happen), and $\kappa = -6$ derived from the second term in (14.12) is perfectly sufficient to achieve a satisfactory result. According to the initial values of the estimates $\hat{q}_1(0)$ $\hat{q}_2(0)$, we get a large initial value $\Upsilon_{\hat{\theta}(0)}$ by (14.30)–(14.35), (14.9), (14.10), which is a conservative value obtained by the stability analysis in section 14.4. In this simulation, guided by reason rather than by a highly conservative estimate, we adopt a smaller initial value as $\Upsilon_{\hat{\theta}(0)} = 5.5$, which prevents the activation of the identifier from being extremely late (particularly relative to T).

From figure 14.2, we observe that the estimates $\hat{q}_1, \hat{q}_2$ reach the exact values of the unknown parameters $q_1 = 4, q_2 = 6$ at $t = 0.13$ s, in just one trigger. In figure 14.3, the nominal control input applied at the boundary $x = 1$ goes through the PDE domain and reaches the boundary $x = 0$, starting to regulate the ODE state $\zeta(t)$ at $t = \frac{1}{q_2} \approx 0.17$ s. As shown in figure 14.2, the estimates reach the true values

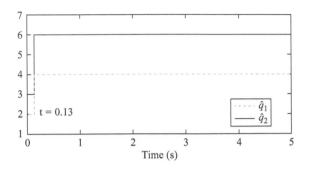

Figure 14.2. Parameter estimates $\hat{q}_1(t), \hat{q}_2(t)$.

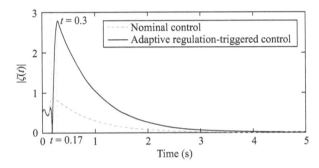

Figure 14.3. The evolution of $|\zeta(t)|$ under the nominal control (14.23) and the proposed adaptive regulation-triggered control (14.36).

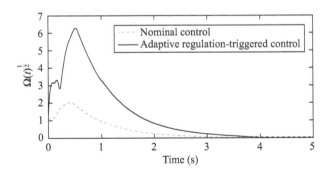

Figure 14.4. The evolution of $\Omega(t)^{\frac{1}{2}}$ under the nominal control (14.23) and the proposed adaptive regulation-triggered control (14.36).

and update the certainty-equivalence controller at $t = 0.13$ s. Then it takes $1/q_2 \approx 0.17$ s for the updated control signal to travel to the ODE; that is, the updated control signal starts to properly regulate the ODE state $\zeta(t)$, as intended by the nominal controller, at $t = 0.3$ s. For the remaining time, as shown in figure 14.3, the performance of the proposed adaptive controller coincides with the nominal feedback, and $|\zeta(t)|$ converges to zero. Similar results are observed in figure 14.4, which shows the evolution of $\Omega(t)^{\frac{1}{2}}$ defined by (14.24) under the nominal control and the proposed adaptive regulation-triggered control. Figures 14.5 and 14.6 show that

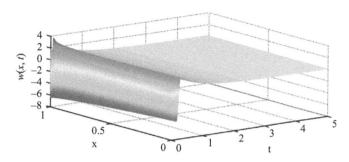

Figure 14.5. The evolution of $w(x,t)$ under the proposed adaptive regulation-triggered control (14.36).

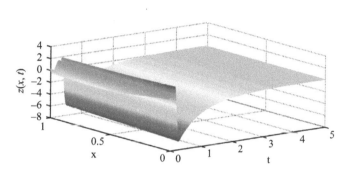

Figure 14.6. The evolution of $z(x,t)$ under the proposed adaptive regulation-triggered control (14.36).

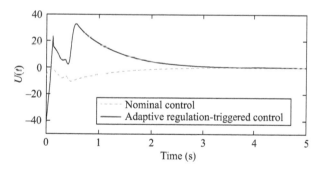

Figure 14.7. The control signals of the nominal control (14.23) and the proposed adaptive regulation-triggered control (14.36).

the PDE states $z(x,t), w(x,t)$ are regulated to zero under the proposed adaptive regulation-triggered controller. The adaptive regulation-triggered control law and the nominal control law are shown in figure 14.7.

At the end of this section, and this chapter, let us reiterate that, as announced at the beginning of this chapter, the BaLSI identifier has ensured the perfect identification of the unknown parameters in finite time and has enabled the regulation-triggered adaptive backstepping controller to achieve exponential regulation, with a decay rate matching the rate corresponding to the case of known parameters.

14.6   APPENDIX

A. Proof of proposition 14.1

Inserting (14.36) into (14.5) yields the closed-loop system

$$\dot{\zeta}(t) = (a - q_1 bc)\zeta(t) + b(q_2 + q_1 p)w(0, t), \tag{14.134}$$

$$z_t(x, t) = -q_1 z_x(x, t), \tag{14.135}$$

$$w_t(x, t) = q_2 w_x(x, t), \tag{14.136}$$

$$z(0, t) = c\zeta(t) - pw(0, t), \tag{14.137}$$

$$w(1, t) = \frac{1}{q_2}\hat{q}_2(\tau_i)\int_0^1 \phi(1, y; \hat{\theta}(\tau_i))w(y, t)dy$$
$$+ \frac{1}{q_2}\hat{q}_2(\tau_i)\lambda(1; \hat{\theta}(\tau_i))\zeta(t) + \frac{q_1 - \hat{q}_1(\tau_i)}{q_2}z(1, t) \tag{14.138}$$

for $t \in [\tau_i, \tau_{i+1})$, $x \in [0, 1]$, $i \in Z_+$, where $\hat{\theta}(\tau_i) = (\hat{q}_1(\tau_i), \hat{q}_2(\tau_i))^T$ is constant and defined by (14.37), (14.38), (14.40), (14.41), (14.47), (14.61). With the purpose of decoupling the ODE and the PDEs, we introduce two transformations. The first is the following Volterra transformation:

$$\bar{z}(x, t) = z(x, t) - \int_0^x \bar{\phi}(x - y)z(y, t)dy - \bar{\varphi}(x)\zeta(t), \tag{14.139}$$

where the functions $\bar{\varphi}$ and $\bar{\phi}$ satisfy

$$q_1\bar{\varphi}'(x) + \left((a - q_1 bc) + \frac{b}{p}(q_2 + q_1 p)c\right)\bar{\varphi}(x) = 0, \tag{14.140}$$

$$\bar{\varphi}(0) = c, \tag{14.141}$$

$$\bar{\phi}(x) = \frac{1}{q_1 p}\bar{\varphi}(x)b(q_2 + q_1 p). \tag{14.142}$$

Through the transformation (14.139), (14.149), the system (14.134)–(14.138) is converted to

$$\dot{\zeta}(t) = (a - q_1 bc)\zeta(t) + b(q_2 + q_1 p)w(0, t), \tag{14.143}$$

$$\bar{z}_t(x, t) = -q_1\bar{z}_x(x, t), \tag{14.144}$$

$$w_t(x, t) = q_2 w_x(x, t), \tag{14.145}$$

$$\bar{z}(0, t) = -pw(0, t), \tag{14.146}$$

$$w(1, t) = \frac{q_1 - \hat{q}_1(\tau_i)}{q_2}\bar{z}(1, t) + \frac{1}{q_2}\hat{q}_2(\tau_i)\int_0^1 \phi(1, y; \hat{\theta}(\tau_i))w(y, t)dy$$
$$- \frac{q_1 - \hat{q}_1(\tau_i)}{q_2}\int_0^1 \bar{\psi}(1 - y)\bar{z}(y, t)dy$$
$$+ \left[\frac{1}{q_2}\hat{q}_2(\tau_i)\lambda(1; \hat{\theta}(\tau_i)) - \frac{q_1 - \hat{q}_1(\tau_i)}{q_2}\bar{\gamma}(1)\right]\zeta(t) \tag{14.147}$$

for $t \in [\tau_i, \tau_{i+1})$, $x \in [0, 1]$. The conditions (14.140)–(14.142) of the functions $\bar{\varphi}$ and $\bar{\phi}$ in the transformation (14.139), (14.149) are obtained through matching (14.134)–(14.138) and (14.143)–(14.147), as follows. Inserting (14.139) into (14.144) and using

(14.134), (14.135), (14.137), we obtain

$$
\bar{z}_t(x,t) + q_1 \bar{z}_x(x,t)
$$
$$
= z_t(x,t) + q_1 \int_0^x \bar{\phi}(x-y) z_x(y,t) dy - \bar{\varphi}(x)(a - q_1 bc)\zeta(t)
$$
$$
- \bar{\varphi}(x) b(q_2 + q_1 p) w(0,t) + q_1 z_x(x,t) - q_1 \bar{\phi}(0) z(x,t)
$$
$$
- \int_0^x q_1 \bar{\phi}'(x-y) z(y,t) dy - q_1 \bar{\varphi}'(x)\zeta(t)
$$
$$
= q_1 \bar{\phi}(0) z(x,t) - q_1 \bar{\phi}(x) z(0,t) - q_1 \bar{\phi}(0) z(x,t) - q_1 \bar{\varphi}'(x)\zeta(t)
$$
$$
- \bar{\varphi}(x)(a - q_1 bc)\zeta(t) - \frac{1}{p}\bar{\varphi}(x) b(q_2 + q_1 p) c\zeta(t) + \frac{1}{p}\bar{\varphi}(x) b(q_2 + q_1 p) z(0,t)
$$
$$
= \left[\frac{1}{p}\bar{\varphi}(x) b(q_2 + q_1 p) - q_1 \bar{\phi}(x) \right] z(0,t)
$$
$$
- \left[\bar{\varphi}(x)(a - q_1 bc) + \frac{1}{p}\bar{\varphi}(x) b(q_2 + q_1 p) c + q_1 \bar{\varphi}'(x) \right] \zeta(t) = 0. \tag{14.148}
$$

For (14.148) to hold, we obtain the conditions (14.140), (14.142). Matching (14.137) and (14.146), we get the condition (14.141). Because $\bar{\phi}$ is a continuous function, we find that the inverse transformation

$$
z(x,t) = \bar{z}(x,t) - \int_0^x \bar{\psi}(x-y) \bar{z}(y,t) dy - \bar{\gamma}(x)\zeta(t) \tag{14.149}
$$

exists (see, e.g., chapter 9.9 in [175]), where the well-posedness of $\bar{\psi}$, $\bar{\gamma}$ is ensured by the well-posedness of (14.140)–(14.142).

Applying the second transformation

$$
\chi(t) = \zeta(t) - \int_0^1 K_{1i}(x) \bar{z}(x,t) dx - \int_0^1 K_{2i}(x) w(x,t) dx \tag{14.150}
$$

for $t \in [\tau_i, \tau_{i+1})$, $i \in Z_+$, where the functions K_{1i}, K_{2i} satisfy the well-posed first-order ODEs

$$
\left[(a - q_1 bc) + K_{2i}(1) \left((q_1 - \hat{q}_1(\tau_i))\bar{\gamma}(1) - \hat{q}_2(\tau_i)\lambda(1;\hat{\theta}(\tau_i)) \right) \right] K_{1i}(x) - q_1 K_{1i}'(x)
$$
$$
= -K_{2i}(1)(q_1 - \hat{q}_1(\tau_i))\bar{\psi}(1-x), \tag{14.151}
$$
$$
\left[(a - q_1 bc) + K_{2i}(1) \left((q_1 - \hat{q}_1(\tau_i))\bar{\gamma}(1) - \hat{q}_2(\tau_i)\lambda(1;\hat{\theta}(\tau_i)) \right) \right] K_{2i}(x) + q_2 K_{2i}'(x)
$$
$$
= K_{2i}(1)\hat{q}_2(\tau_i)\phi(1,x;\hat{\theta}(\tau_i)), \tag{14.152}
$$
$$
K_{1i}(1)q_1 = K_{2i}(1)(q_1 - \hat{q}_1(\tau_i)), \tag{14.153}
$$
$$
K_{2i}(0)q_2 = -b(q_2 + q_1 p) - pK_{1i}(0)q_1, \tag{14.154}
$$

transforms the system (14.143)–(14.147) to

$$
\dot{\chi}(t) = A_i \chi(t), \tag{14.155}
$$
$$
\bar{z}_t(x,t) = -q_1 \bar{z}_x(x,t), \tag{14.156}
$$
$$
w_t(x,t) = q_2 w_x(x,t), \tag{14.157}
$$
$$
\bar{z}(0,t) = -pw(0,t), \tag{14.158}
$$

$$w(1,t) = \frac{q_1 - \hat{q}_1(\tau_i)}{q_2}\bar{z}(1,t) + \int_0^1 D_{1i}(x)w(x,t)dx + \int_0^1 D_{2i}(x)\bar{z}(x,t)dx + D_{3i}\chi(t),$$
$$(14.159)$$

for $t \in [\tau_i, \tau_{i+1})$, $x \in [0,1]$, where

$$A_i = a - q_1 bc + K_{2i}(1)(q_1 - \hat{q}_1(\tau_i))\bar{\gamma}(1) - K_{2i}(1)\hat{q}_2(\tau_i)\lambda(1;\hat{\theta}(\tau_i)),$$

$$D_{1i}(x) = \frac{1}{q_2}\hat{q}_2(\tau_i)\phi(1,x;\hat{\theta}(\tau_i)) + \left(\frac{1}{q_2}\hat{q}_2(\tau_i)\lambda(1;\hat{\theta}(\tau_i)) - \frac{q_1 - \hat{q}_1(\tau_i)}{q_2}\bar{\gamma}(1)\right)K_{1i}(x),$$

$$D_{2i}(x) = -\frac{q_1 - \hat{q}_1(\tau_i)}{q_2}\bar{\psi}(1-x) + \left(\frac{1}{q_2}\hat{q}_2(\tau_i)\lambda(1;\hat{\theta}(\tau_i)) - \frac{q_1 - \hat{q}_1(\tau_i)}{q_2}\bar{\gamma}(1)\right)K_{2i}(x),$$

$$D_{3i} = \frac{1}{q_2}\hat{q}_2(\tau_i)\lambda(1;\hat{\theta}(\tau_i)) - \frac{q_1 - \hat{q}_1(\tau_i)}{q_2}\bar{\gamma}(1).$$

The conditions (14.151)–(14.154) of $K_{1i}(x)$, $K_{2i}(x)$ are defined by matching (14.143)–(14.147) and (14.155)–(14.159), as follows. Inserting (14.150) into (14.155) and using (14.143)–(14.147), we obtain

$$\dot{\chi}(t) - A_i\chi(t)$$
$$= \dot{\chi}(t) - (a - q_1 bc)\chi(t) - \left(K_{2i}(1)(q_1 - \hat{q}_1(\tau_i))\bar{\gamma}(1) - K_{2i}(1)\hat{q}_2(\tau_i)\lambda(1;\hat{\theta}(\tau_i))\right)\chi(t)$$
$$= \dot{\zeta}(t) - \int_0^1 K_{1i}(x)\bar{z}_t(x,t)dx - \int_0^1 K_{2i}(x)w_t(x,t)dx$$
$$\quad - (a - q_1 bc)\zeta(t) + (a - q_1 bc)\int_0^1 K_{1i}(x)\bar{z}(x,t)dx$$
$$\quad + (a - q_1 bc)\int_0^1 K_{2i}(x)w(x,t)dx$$
$$\quad - \left(K_{2i}(1)(q_1 - \hat{q}_1(\tau_i))\bar{\gamma}(1) - K_{2i}(1)\hat{q}_2(\tau_i)\lambda(1;\hat{\theta}(\tau_i))\right)\chi(t)$$
$$= b(q_2 + q_1 p)w(0,t) + q_1\int_0^1 K_{1i}(x)\bar{z}_x(x,t)dx$$
$$\quad - q_2\int_0^1 K_{2i}(x)w_x(x,t)dx + (a - q_1 bc)\int_0^1 K_{1i}(x)\bar{z}(x,t)dx$$
$$\quad + (a - q_1 bc)\int_0^1 K_{2i}(x)w(x,t)dx$$
$$\quad - \left(K_{2i}(1)(q_1 - \hat{q}_1(\tau_i))\bar{\gamma}(1) - K_{2i}(1)\hat{q}_2(\tau_i)\lambda(1;\hat{\theta}(\tau_i))\right)\chi(t)$$
$$= b(q_2 + q_1 p)w(0,t) + K_{1i}(1)q_1\bar{z}(1,t) - K_{1i}(0)q_1\bar{z}(0,t) - \int_0^1 q_1 K_{1i}'(x)\bar{z}(x,t)dx$$
$$\quad - K_{2i}(1)q_2 w(1,t) + K_{2i}(0)q_2 w(0,t) + \int_0^1 q_2 K_{2i}'(x)w(x,t)dx$$
$$\quad + (a - q_1 bc)\int_0^1 K_{1i}(x)\bar{z}(x,t)dx + (a - q_1 bc)\int_0^1 K_{2i}(x)w(x,t)dx$$
$$\quad - \left(K_{2i}(1)(q_1 - \hat{q}_1(\tau_i))\bar{\gamma}(1) - K_{2i}(1)\hat{q}_2(\tau_i)\lambda(1;\hat{\theta}(\tau_i))\right)\chi(t)$$

$$= -K_{2i}(1)q_2 \left[\frac{1}{q_2} \hat{q}_2(\tau_i) \int_0^1 \phi(1,y;\hat{\theta}(\tau_i))w(y,t)dy \right.$$

$$+ \left(\frac{1}{q_2}\hat{q}_2(\tau_i)\lambda(1;\hat{\theta}(\tau_i)) - \frac{q_1 - \hat{q}_1(\tau_i)}{q_2}\bar{\gamma}(1) \right)\zeta(t)$$

$$+ \frac{q_1 - \hat{q}_1(\tau_i)}{q_2}\bar{z}(1,t) - \frac{q_1 - \hat{q}_1(\tau_i)}{q_2} \int_0^1 \bar{\psi}(1-y)\bar{z}(y,t)dy \left.\right] + K_{1i}(1)q_1\bar{z}(1,t)$$

$$+ \left(K_{2i}(0)q_2 + b(q_2 + q_1 p) + pK_{1i}(0)q_1\right)w(0,t)$$

$$+ \int_0^1 \left(q_2 K_{2i}{}'(x) + (a - q_1 bc)K_{2i}(x)\right)w(x,t)dx$$

$$+ \int_0^1 \left((a - q_1 bc)K_{1i}(x) - q_1 K_{1i}{}'(x)\right)\bar{z}(x,t)dx$$

$$- \left(K_{2i}(1)(q_1 - \hat{q}_1(\tau_i))\bar{\gamma}(1) - K_{2i}(1)\hat{q}_2(\tau_i)\lambda(1;\hat{\theta}(\tau_i))\right)\chi(t)$$

$$= \left(K_{1i}(1)q_1 - K_{2i}(1)(q_1 - \hat{q}_1(\tau_i))\right)\bar{z}(1,t) + \left(K_{2i}(0)q_2 + b(q_2\right.$$

$$\left. + q_1 p) + pK_{1i}(0)q_1\right)w(0,t)$$

$$+ \int_0^1 \left[q_2 K_{2i}{}'(x) + (a - q_1 bc)K_{2i}(x) - K_{2i}(1)\hat{q}_2(\tau_i)\phi(1,x;\hat{\theta}(\tau_i)) \right.$$

$$\left. + \left(K_{2i}(1)(q_1 - \hat{q}_1(\tau_i))\bar{\gamma}(1) - K_{2i}(1)\hat{q}_2(\tau_i)\lambda(1;\hat{\theta}(\tau_i))\right)K_{2i}(x) \right]w(x,t)dx$$

$$+ \int_0^1 \left[(a - q_1 bc)K_{1i}(x) - q_1 K_{1i}{}'(x) \right.$$

$$+ K_{2i}(1)(q_1 - \hat{q}_1(\tau_i))\bar{\psi}(1-x) + \left(K_{2i}(1)(q_1 - \hat{q}_1(\tau_i))\bar{\gamma}(1)\right.$$

$$\left.\left. - K_{2i}(1)\hat{q}_2(\tau_i)\lambda(1;\hat{\theta}(\tau_i))\right)K_{1i}(x) \right]\bar{z}(x,t)dx = 0. \tag{14.160}$$

For (14.160) to hold, the conditions (14.151)–(14.154) are obtained.

The equation set (14.156)–(14.159), where $\chi(t)$ is a well-defined external signal generated by (14.155), has an analogous structure with (5), (6) in [35]. According to the result in part 1 of the appendix in [35], the system (14.155)–(14.159) has a unique solution on $t \in [\tau_i, \tau_{i+1})$ for all $(w[\tau_i], \bar{z}[\tau_i])^T \in L^2((0,1); \mathbb{R}^2)$, $\chi(\tau_i) \in \mathbb{R}$. By virtue of the transformations (14.150), (14.149), we see that, for given $(w[\tau_i], z[\tau_i])^T \in L^2((0,1); \mathbb{R}^2)$, $\zeta(\tau_i) \in \mathbb{R}$, the system (14.134)–(14.138) has a unique solution for $t \in [\tau_i, \tau_{i+1})$. Recalling the definition of the weak solution in definition 14.1, we find that for every $(z[\tau_i], w[\tau_i])^T \in L^2((0,1); \mathbb{R}^2)$, $\zeta(\tau_i) \in \mathbb{R}$, and $\hat{\theta}(\tau_i) \in \Theta$, there exists a unique (weak) solution $((z,w)^T, \zeta) \in C^0([\tau_i, \tau_{i+1}]; L^2(0,1); \mathbb{R}^2) \times C^0([\tau_i, \tau_{i+1}]; \mathbb{R})$ to the system (14.1)–(14.5) with (14.36), (14.37), (14.38), (14.40), (14.41), (14.47), (14.61). For every $(z_0, w_0)^T \in L^2((0,1); \mathbb{R}^2)$, $\zeta_0 \in \mathbb{R}$, and $\hat{\theta}_0 \in \Theta$, through iterative constructions between successive triggering times, the proposition is thus obtained.

B. Proof of claim 14.1

If $\bar{\lambda} = 0$, it means that $Q_{n_2,2}(\mu_{i+1}, \tau_{i+1}) = 0$ by recalling (14.85), (14.56), and then $\ell_2 = \frac{H_{n_2,2}(\mu_{i+1}, \tau_{i+1})}{Q_{n_2,3}(\mu_{i+1}, \tau_{i+1})}$ in (14.78). Together with (14.77), we get

$$S_i = \left\{ \left(\frac{H_{n_1,1}(\mu_{i+1}, \tau_{i+1})}{Q_{n_1,1}(\mu_{i+1}, \tau_{i+1})} - \frac{H_{n_2,2}(\mu_{i+1}, \tau_{i+1})}{Q_{n_2,3}(\mu_{i+1}, \tau_{i+1})} \frac{Q_{n_1,2}(\mu_{i+1}, \tau_{i+1})}{Q_{n_1,1}(\mu_{i+1}, \tau_{i+1})}, \frac{H_{n_2,2}(\mu_{i+1}, \tau_{i+1})}{Q_{n_2,3}(\mu_{i+1}, \tau_{i+1})} \right) \right\},$$

$$(14.161)$$

which is a singleton: a contradiction. Similarly, if $\bar{\lambda}_1 = 0$, it means that $Q_{n_1,2}$ $(\mu_{i+1}, \tau_{i+1}) = 0$ by recalling (14.84), (14.56), and then $\ell_1 = \frac{H_{n_1,1}(\mu_{i+1}, \tau_{i+1})}{Q_{n_1,1}(\mu_{i+1}, \tau_{i+1})}$ in (14.77). Together with (14.78), we get

$$S_i = \left\{ \left(\frac{H_{n_1,1}(\mu_{i+1}, \tau_{i+1})}{Q_{n_1,1}(\mu_{i+1}, \tau_{i+1})}, \frac{H_{n_2,2}(\mu_{i+1}, \tau_{i+1})}{Q_{n_2,3}(\mu_{i+1}, \tau_{i+1})} - \frac{H_{n_1,1}(\mu_{i+1}, \tau_{i+1})}{Q_{n_1,1}(\mu_{i+1}, \tau_{i+1})} \frac{Q_{n_2,2}(\mu_{i+1}, \tau_{i+1})}{Q_{n_2,3}(\mu_{i+1}, \tau_{i+1})} \right) \right\},$$

$$(14.162)$$

which is a singleton: a contradiction. Therefore, $\bar{\lambda} \neq 0, \bar{\lambda}_1 \neq 0$. According to (14.45), (14.46) and $\bar{\lambda} \neq 0, \bar{\lambda}_1 \neq 0$, we find from (14.84), (14.85) that

$$\int_{\mu_{i+1}}^{t} \pi n_1 \int_0^1 \sin(x\pi n_1) z(x, \tau) dx d\tau$$

$$= -\frac{1}{\bar{\lambda}_1} \int_{\mu_{i+1}}^{t} \pi n_1 \int_0^1 \sin(x\pi n_1) w(x, \tau) dx d\tau, \quad n_1 \in I_1, \qquad (14.163)$$

$$\int_{\mu_{i+1}}^{t} \pi n_2 \int_0^1 \sin(x\pi n_2) z(x, \tau) dx d\tau$$

$$= -\bar{\lambda} \int_{\mu_{i+1}}^{t} \pi n_2 \int_0^1 \sin(x\pi n_2) w(x, \tau) dx d\tau, \quad n_2 \in I_2 \qquad (14.164)$$

for $t \in [\mu_{i+1}, \tau_{i+1}]$. According to the continuity of the mappings $\tau \to \int_0^1 \sin(x\pi n) z[\tau] dx$ and $\tau \to \int_0^1 \sin(x\pi n) w[\tau] dx$, $n \in \mathbb{N}$ (a consequence of the fact that $z \in C^0$ $([\mu_{i+1}, \tau_{i+1}]; L^2(0,1))$ and $w \in C^0([\mu_{i+1}, \tau_{i+1}]; L^2(0,1))$), (14.163), (14.164) imply

$$\int_0^1 \sin(x\pi n_1) \left(z(x, \tau) + \frac{1}{\bar{\lambda}_1} w(x, \tau) \right) dx = 0, \quad n_1 \in I_1, \qquad (14.165)$$

$$\int_0^1 \sin(x\pi n_2)(z(x, \tau) + \bar{\lambda} w(x, \tau)) dx = 0, \quad n_2 \in I_2 \qquad (14.166)$$

for $\tau \in [\mu_{i+1}, \tau_{i+1}]$. We then prove $I_1 = I_2$ in (14.165), (14.166). If I_2 includes elements not belonging to I_1, there exists $n_2 \in I_2$ with $n_2 \notin I_1$ such that $\int_0^1 \sin(x\pi n_2)$ $z(x, \tau) dx = 0$ on $\tau \in [\mu_{i+1}, \tau_{i+1}]$ due to the fact that $Q_{n,1}(\mu_{i+1}, \tau_{i+1}) = 0$ for $n \notin I_1$ by recalling (14.55), (14.45), and then

$$\int_0^1 \sin(x\pi n_2)(z(x, \tau) + \bar{\lambda} w(x, \tau)) dx = \int_0^1 \sin(x\pi n_2) \bar{\lambda} w(x, \tau) dx, \qquad (14.167)$$

which is not identically zero on $\tau \in [\mu_{i+1}, \tau_{i+1}]$ because of $Q_{n_2,3}(\mu_{i+1}, \tau_{i+1}) \neq 0$ together with (14.57), (14.46) and $\bar{\lambda} \neq 0$. This contradicts (14.166). Similarly, if I_1 includes elements not belonging to I_2, there exists $n_1 \in I_1$ with $n_1 \notin I_2$ such that

$$\int_0^1 \sin(x\pi n_1) \left(z(x, \tau) + \frac{1}{\bar{\lambda}_1} w(x, \tau) \right) dx = \int_0^1 \sin(x\pi n_1) z(x, \tau) dx, \qquad (14.168)$$

which is not identically zero on $\tau \in [\mu_{i+1}, \tau_{i+1}]$ because of $Q_{n_1,1}(\mu_{i+1}, \tau_{i+1}) \neq 0$ together with (14.55), (14.45), where $\int_0^1 \sin(x\pi n_1) \frac{1}{\lambda_1} w(x,\tau) dx = 0$ on $\tau \in [\mu_{i+1}, \tau_{i+1}]$ is due to the fact that $Q_{n,3}(\mu_{i+1}, \tau_{i+1}) = 0$ for $n \notin I_2$ by recalling (14.57), (14.46). This contradicts (14.165). Therefore, we conclude that $I_1 = I_2$ in (14.165), (14.166).

We then prove $\bar{\lambda} = \frac{1}{\lambda_1}$ by contradiction. If $\bar{\lambda} - \frac{1}{\lambda_1} \neq 0$, recalling $I_1 = I_2$ and (14.166), then we obtain

$$
\int_0^1 \sin(x\pi n_1) \left(z(x,\tau) + \frac{1}{\lambda_1} w(x,\tau) \right) dx
$$
$$
= \int_0^1 \sin(x\pi n_1) \left(z(x,\tau) + \left(\frac{1}{\lambda_1} - \bar{\lambda} + \bar{\lambda} \right) w(x,\tau) \right) dx
$$
$$
= \int_0^1 \sin(x\pi n_2)(z(x,\tau) + \bar{\lambda} w(x,\tau)) dx + \int_0^1 \sin(x\pi n_2) \left(\frac{1}{\lambda_1} - \bar{\lambda} \right) w(x,\tau) dx
$$
$$
= \left(\frac{1}{\lambda_1} - \bar{\lambda} \right) \int_0^1 \sin(x\pi n_2) w(x,\tau) dx, \tag{14.169}
$$

which is not identically zero for $\tau \in [\mu_{i+1}, \tau_{i+1}]$ because of $Q_{n_2,3}(\mu_{i+1}, \tau_{i+1}) \neq 0$ with (14.57), (14.46), which contradicts (14.165). Therefore, $\bar{\lambda} - \frac{1}{\lambda_1} = 0$. Claim 14.1 is proven.

C. Proof of claim 14.2

According to (14.45), (14.46), the equations (14.84), (14.85) ($\bar{\lambda} \neq 0, \bar{\lambda}_1 \neq 0$ and $\bar{\lambda} = \frac{1}{\lambda_1}$) are equivalent to

$$
\int_0^1 \sin(x\pi n)(z(x,\tau) + \bar{\lambda} w(x,\tau)) dx = 0, \quad n \in I_2 \cup I_1. \tag{14.170}
$$

If $\mathbb{N} = I_2 \cup I_1$, it means that (14.170) holds for all $n \in \mathbb{N}$. If $I_2 \cup I_1 \subset \mathbb{N}$, recalling the definitions of I_1, I_2, we know that $\int_0^1 \sin(x\pi n) z(x,\tau) dx = \int_0^1 \sin(x\pi n) w(x,\tau) dx = 0$ for $n \in \mathbb{C}_\mathbb{N}\{I_1 \cup I_2\}$ on $\tau \in [\mu_{i+1}, \tau_{i+1}]$, and thus (14.170) is equivalent to

$$
\int_0^1 \sin(x\pi n)(z(x,\tau) + \bar{\lambda} w(x,\tau)) dx = 0, \quad n = 1, 2, \ldots \tag{14.171}
$$

on $\tau \in [\mu_{i+1}, \tau_{i+1}]$. Since the set $\{\sqrt{2} \sin(n\pi x) : n = 1, 2, \ldots\}$ is an orthonormal basis of $L^2(0,1)$, if (14.171) holds, it follows that $z(x,t) + \bar{\lambda} w(x,t) = 0$ for $t \in [\mu_{i+1}, \tau_{i+1}]$.

If $z(x,t) + \bar{\lambda} w(x,t) = 0$ for $t \in [\mu_{i+1}, \tau_{i+1}]$, then (14.171) and (14.84), (14.85) ($\bar{\lambda} \neq 0, \bar{\lambda}_1 \neq 0$ and $\bar{\lambda} = \frac{1}{\lambda_1}$), naturally hold. Claim 14.2 is proven.

D. Proof of claim 14.3

The necessary condition for the equation $z(x,t) + \bar{\lambda} w(x,t) = 0$ ($\bar{\lambda} \neq 0$) to hold on $x \in [0,1], t \in [\mu_{i+1}, \tau_{i+1}]$ is that $z(x,t), w(x,t)$ are kept constant on $x \in [0,1], t \in [\mu_{i+1}, \tau_{i+1}]$ excluding finitely many possible points of discontinuity; that is, $w(x,t) = M$ and $z(x,t) = -\bar{\lambda} M$ on $x \in [0,1], t \in [\mu_{i+1}, \tau_{i+1}]$ excluding finitely many possible points of discontinuity, where M is a nonzero constant (because $z[t], w[t]$ are not identically zero on $t \in [\mu_{i+1}, \tau_{i+1}]$). We prove this by contradiction next.

Taking a spatial interval $[\underline{x}_1, \bar{x}_2] \in [0,1]$ with $\bar{x}_2 - \underline{x}_1 \leq (q_1 + q_2)(\tau_{i+1} - \mu_{i+1})$ (the position of the interval $[\underline{x}_1, \bar{x}_2]$ is arbitrary on $[0,1]$, and $(\underline{x}_1, \mu_{i+1})$, $(\bar{x}_2, \mu_{i+1})$ are not points of discontinuity of the functions $w(x,t), z(x,t))$, suppose that there exist x_a, x_b (without loss of generality we assume $x_a < x_b$) in the interval $[\underline{x}_1, \bar{x}_2]$ with $w(x_a, \mu_{i+1}) \neq w(x_b, \mu_{i+1})$, where (x_a, μ_{i+1}), (x_b, μ_{i+1}) are not points of discontinuity of the functions $w(x,t), z(x,t)$. Also, we know that $z(x_a, \mu_{i+1}) = -\bar{\lambda} w(x_a, \mu_{i+1})$ according to $z(x,t) + \bar{\lambda} w(x,t) = 0$ always holding on $x \in [0,1], t \in [\mu_{i+1}, \tau_{i+1}]$. Because the state of the w-PDE propagates from $x = 1$ to $x = 0$ and the state of the z-PDE propagates from $x = 0$ to $x = 1$, with the respective propagation speeds q_1, q_2, according to the statement in page 60 in [16], which indicates that the system (14.2), (14.3) is equivalent to a pair of scalar delay equations even if the solutions are not differentiable and even not continuous with respect to t and x, we get the following relationships:

$$w(x_b - s_1 q_2, \mu_{i+1} + s_1) = w(x_b, \mu_{i+1}), \qquad (14.172)$$

$$z(x_a + s_1 q_1, \mu_{i+1} + s_1) = z(x_a, \mu_{i+1}) \qquad (14.173)$$

for $s_1 \in [0, \min\{\frac{x_b}{q_2}, \frac{1-x_a}{q_1}\}]$, where $(x_b - s_1 q_2, \mu_{i+1} + s_1)$, $(x_a + s_1 q_1, \mu_{i+1} + s_1)$ are not the points of discontinuity because (x_b, μ_{i+1}), (x_a, μ_{i+1}) are not the points of discontinuity. There exists a $s_1 = \frac{x_b - x_a}{q_1 + q_2}$ such that $x_b - s_1 q_2 = x_a + s_1 q_1 = x_c$, and then we obtain

$$w(x_c, t_c) = w(x_b, \mu_{i+1}), \quad z(x_c, t_c) = z(x_a, \mu_{i+1}), \qquad (14.174)$$

where $x_c = \frac{q_1 x_b + q_2 x_a}{q_1 + q_2} \in (x_a, x_b)$, $t_c = \mu_{i+1} + \frac{x_b - x_a}{q_1 + q_2} \in (\mu_{i+1}, \tau_{i+1}]$, recalling $x_b - x_a \leq \bar{x}_2 - \underline{x}_1 \leq (q_1 + q_2)(\tau_{i+1} - \mu_{i+1})$. Because

$$z(x_a, \mu_{i+1}) = -\bar{\lambda} w(x_a, \mu_{i+1}) \neq -\bar{\lambda} w(x_b, \mu_{i+1}),$$

recalling $\bar{\lambda} \neq 0$ and the hypothesis that $w(x_a, \mu_{i+1}) \neq w(x_b, \mu_{i+1})$ and using (14.174), we have

$$z(x_c, t_c) \neq -\bar{\lambda} w(x_c, t_c)$$

with $x_c \in [0,1], t_c \in (\mu_{i+1}, \tau_{i+1}]$: a contradiction. Therefore, the hypothesis that there exist x_a, x_b in the interval $[\underline{x}_1, \bar{x}_2]$ such that $w(x_a, \mu_{i+1}) \neq w(x_b, \mu_{i+1})$ $((x_a, \mu_{i+1})$, (x_b, μ_{i+1}) are not points of discontinuity) does not hold, and we conclude that $w(x, \mu_{i+1}), z(x, \mu_{i+1})$ are kept constant on $x \in [\underline{x}_1, \bar{x}_2]$ excluding finitely many possible points of discontinuity. Because the position of the interval $[\underline{x}_1, \bar{x}_2]$ is arbitrary on $[0,1]$ (with $(\underline{x}_1, \mu_{i+1})$, $(\bar{x}_2, \mu_{i+1})$ not points of discontinuity of the functions $w(x,t), z(x,t))$, we find that $w(x, \mu_{i+1}), z(x, \mu_{i+1})$ are kept constant for $x \in [0,1]$ excluding finitely many possible points of discontinuity. Taking a time increment s with $0 < s \leq \frac{1}{2 \max\{q_1, q_2\}}$, we have

$$w(x, \mu_{i+1} + s) = w(x + q_2 s, \mu_{i+1}) = w(x, \mu_{i+1})$$

for $x \in [0, \frac{1}{2}]$, excluding the points of discontinuity of the functions $w(x,t), z(x,t)$ along $x \in [0,1]$, $t = \mu_{i+1}$, where $s \in (0, \frac{1}{2 \max\{q_1, q_2\}}]$ ensures $x + q_2 s \in (0,1]$. We also get

$$z(x, \mu_{i+1} + s) = z(x - q_1 s, \mu_{i+1}) = z(x, \mu_{i+1})$$

for $x \in [\frac{1}{2}, 1]$, excluding the points of discontinuity of the functions $w(x,t)$, $z(x,t)$ along $x \in [0,1]$, $t = \mu_{i+1}$, where $s \in (0, \frac{1}{2\max\{q_1, q_2\}}]$ ensures $x - q_1 s \in [0,1)$. Recalling that $z(x,t) + \bar{\lambda} w(x,t) = 0$ always holds on $x \in [0,1]$, $t \in [\mu_{i+1}, \tau_{i+1}]$, we know that z, w are kept constant in $x \in [0,1]$, $t \in [\mu_{i+1}, \mu_{i+1} + \frac{1}{2\max\{q_1, q_2\}}]$ excluding finitely many possible points of discontinuity. If $\mu_{i+1} + \frac{1}{2\max\{q_1, q_2\}} \geq \tau_{i+1}$, we directly obtain the necessary condition for $z(x,t) + \bar{\lambda} w(x,t) = 0$ to hold on $x \in [0,1]$, $t \in [\mu_{i+1}, \tau_{i+1}]$ mentioned at the beginning of the proof of claim 14.3. If $\mu_{i+1} + \frac{1}{2\max\{q_1, q_2\}} < \tau_{i+1}$, then by repeatedly taking the time increments s and conducting the above process k times, based on the fact that w, z are kept constant for $x \in [0,1]$ at the beginning of each time increment, with excluding finitely many possible points of discontinuity until $\mu_{i+1} + \frac{k}{2\max\{q_1, q_2\}} \geq \tau_{i+1}$, we also obtain the necessary condition for $z(x,t) + \bar{\lambda} w(x,t) = 0$ to hold on $x \in [0,1]$, $t \in [\mu_{i+1}, \tau_{i+1}]$ mentioned at the beginning of the proof—namely, that $z(x,t)$, $w(x,t)$ are kept constant on $x \in [0,1]$, $t \in [\mu_{i+1}, \tau_{i+1}]$, excluding finitely many possible points of discontinuity:

$$w(x,t) = M, z(x,t) = -\bar{\lambda} M, \ (x,t) \in ([0,1] \times [\mu_{i+1}, \tau_{i+1}]) \backslash I_d, \qquad (14.175)$$

where I_d denotes a set of finitely many possible points of discontinuity of the functions $w(x,t)$, $z(x,t)$ in $x \in [0,1]$, $t \in [\mu_{i+1}, \tau_{i+1}]$, and where M is a nonzero constant (because $z[t], w[t]$ are not identically zero on $t \in [\mu_{i+1}, \tau_{i+1}]$).

The situation in which $w(x,t) = M$ and $z(x,t) = -\bar{\lambda} M$ for $(x,t) \in ([0,1] \times [\mu_{i+1}, \tau_{i+1}]) \backslash I_d$ means $\zeta(t) = \frac{(-\bar{\lambda}+p)}{c} M$ on $t \in [\mu_{i+1}, \tau_{i+1}]$ according to (14.4), excluding finitely many possible points of discontinuity on $t \in [\mu_{i+1}, \tau_{i+1}]$. Recalling (14.1), it then must be that $(a - q_1 bc)\frac{(-\bar{\lambda}+p)}{c} + b(q_2 + q_1 p) = 0$. It follows that

$$\bar{\lambda} = \frac{cb(q_2 + q_1 p)}{a - q_1 bc} + p > 0 \qquad (14.176)$$

because the constants c, b, q_1, q_2, p, and $a - q_1 bc$ are positive. Inserting the control input (14.36) into the right boundary condition (14.5), recalling (14.15), (14.16) and $\zeta(t) = \frac{(-\bar{\lambda}+p)}{c} M$, we know that a necessary condition of $w(x,t) = M$ and $z(x,t) = -\bar{\lambda} M$ for $(x,t) \in ([0,1] \times [\mu_{i+1}, \tau_{i+1}]) \backslash I_d$ is for the following equation to hold

$$[q_2 + (q_1 - \hat{q}_1(\tau_i))\bar{\lambda}]M = \kappa \Bigg[\hat{q}_2(\tau_i) \int_0^1 \frac{b}{\hat{q}_2(\tau_i)} e^{\frac{1}{\hat{q}_2(\tau_i)}(a - \hat{q}_1(\tau_i)bc)(1-y)} dy$$
$$+ \frac{\hat{q}_2(\tau_i)(-\bar{\lambda}+p)}{c(\hat{q}_1(\tau_i)p + \hat{q}_2(\tau_i))} e^{\frac{1}{\hat{q}_2(\tau_i)}(a - \hat{q}_1(\tau_i)bc)} \Bigg] M, \qquad (14.177)$$

that is,

$$q_2 + (q_1 - \hat{q}_1(\tau_i)) \left(\frac{cb(q_2 + q_1 p)}{a - q_1 bc} + p \right)$$
$$= -\kappa \Bigg[\hat{q}_2(\tau_i) b \Bigg(\frac{1 - e^{\frac{1}{\hat{q}_2(\tau_i)}(a - \hat{q}_1(\tau_i)bc)}}{a - \hat{q}_1(\tau_i)bc} + \frac{e^{\frac{1}{\hat{q}_2(\tau_i)}(a - \hat{q}_1(\tau_i)bc)}}{\hat{q}_1(\tau_i)p + \hat{q}_2(\tau_i)} \frac{(q_2 + q_1 p)}{(a - q_1 bc)} \Bigg) \Bigg] \qquad (14.178)$$

because $M \neq 0$. Recalling $\hat{q}_1(0) = \underline{q}_1$ and $\hat{q}_2(0) = \underline{q}_2$, we have $0 < \underline{q}_1 \leq \hat{q}_1(\tau_i) \leq q_1$, $0 < \underline{q}_2 \leq \hat{q}_2(\tau_i) \leq q_2$ (the consequence of (14.61) and $\hat{\theta} \in S_i$ defined by (14.76)), which implies $\frac{q_2+q_1p}{a-q_1bc} \geq \frac{\hat{q}_2(\tau_i)+\hat{q}_1(\tau_i)p}{a-\hat{q}_1(\tau_i)bc}$. We thus have

$$
\frac{1-e^{\frac{1}{\hat{q}_2(\tau_i)}(a-\hat{q}_1(\tau_i)bc)}}{a-\hat{q}_1(\tau_i)bc} + \frac{e^{\frac{1}{\hat{q}_2(\tau_i)}(a-\hat{q}_1(\tau_i)bc)}}{\hat{q}_1(\tau_i)p+\hat{q}_2(\tau_i)} \frac{(q_2+q_1p)}{(a-q_1bc)}
$$

$$
\geq \frac{1-e^{\frac{1}{\hat{q}_2(\tau_i)}(a-\hat{q}_1(\tau_i)bc)}}{a-\hat{q}_1(\tau_i)bc} + \frac{e^{\frac{1}{\hat{q}_2(\tau_i)}(a-\hat{q}_1(\tau_i)bc)}}{\hat{q}_1(\tau_i)p+\hat{q}_2(\tau_i)} \frac{(\hat{q}_2(\tau_i)+\hat{q}_1(\tau_i)p)}{(a-\hat{q}_1(\tau_i)bc)}
$$

$$
\geq \frac{1}{a-\hat{q}_1(\tau_i)bc} > 0. \tag{14.179}
$$

Therefore, recalling that the constants $c, b, q_1, q_2, p, a - q_1bc$ and $\hat{q}_1(\tau_i), \hat{q}_2(\tau_i)$ are positive, the right-hand side of (14.178) is greater than zero because of $\kappa < 0$ and (14.179), and the left-hand side of (14.178) is also greater than zero because of $\hat{q}_1(\tau_i) \leq q_1$. The necessary condition of $w(x,t) = M$ and $z(x,t) = -\bar{\lambda}M$ for $(x,t) \in ([0,1] \times [\mu_{i+1}, \tau_{i+1}]) \backslash I_d$ with $M \neq 0$ becomes that the design parameter κ is equal to $\bar{\kappa}$ defined as

$$
\bar{\kappa} = - \left(q_2 + (q_1 - \hat{q}_1(\tau_i)) \left(\frac{cb(q_2+q_1p)}{a-q_1bc} + p \right) \right)
$$

$$
\div \left[\hat{q}_2(\tau_i)b \left(\frac{1-e^{\frac{1}{\hat{q}_2(\tau_i)}(a-\hat{q}_1(\tau_i)bc)}}{a-\hat{q}_1(\tau_i)bc} + \frac{e^{\frac{1}{\hat{q}_2(\tau_i)}(a-\hat{q}_1(\tau_i)bc)}}{\hat{q}_1(\tau_i)p+\hat{q}_2(\tau_i)} \frac{(q_2+q_1p)}{(a-q_1bc)} \right) \right] < 0. \tag{14.180}
$$

According to (14.179) and $\hat{q}_1(\tau_i) \geq \underline{q}_1, \hat{q}_2(\tau_i) \geq \underline{q}_2$, we know that $\bar{\kappa}$ defined by (14.180) is in the following range:

$$
\frac{(a-\underline{q}_1bc)[q_2+(q_1-\underline{q}_1)(\frac{cb(q_2+q_1p)}{a-q_1bc}+p)]}{-\underline{q}_2b} \leq \bar{\kappa} \leq 0. \tag{14.181}
$$

Recalling the first term in (14.12), we know that $\kappa \neq \bar{\kappa}$. We thus conclude that $w(x,t) = M$ and $z(x,t) = -\bar{\lambda}M$ with $M \neq 0$ on $(x,t) \in ([0,1] \times [\mu_{i+1}, \tau_{i+1}]) \backslash I_d$ does not hold. Claim 14.3 is proven.

E. Proof of claim 14.4

We first prove sufficiency. If $z[t]$ (or $w[t]$) are not identically zero for $t = [0, \lim_{i\to\infty} (\tau_i))$, there exists an interval $[\mu_{i+1}, \tau_{i+1}]$ on which $z[t]$ (or $w[t]$) are not identically zero. It follows that $\hat{q}_1(\tau_{i+1}) = q_1$ (or $\hat{q}_2(\tau_{i+1}) = q_2$), recalling lemma 14.3.

Next, we prove necessity. When $\hat{q}_1(0) \neq q_1$ (or $\hat{q}_2(0) \neq q_2$), if the estimate reaches the true value at an instant τ_{i+1}—that is, $\hat{q}_1(\tau_{i+1}) = q_1$ (or $\hat{q}_2(\tau_{i+1}) = q_2$)—it follows there exists $n \in \mathbb{N}$ such that $Q_{n,1}(\mu_{i+1}, \tau_{i+1}) \neq 0$ (or $Q_{n,3}(\mu_{i+1}, \tau_{i+1}) \neq 0$). (This is true because if $Q_{n,1}(\mu_{i+1}, \tau_{i+1}) = 0$ (or $Q_{n,3}(\mu_{i+1}, \tau_{i+1}) = 0$) for all $n \in \mathbb{N}$, it would also be true that $g_{n,1}(t, \mu_{i+1}) = 0$ (or $g_{n,2}(t, \mu_{i+1}) = 0$) for all $n \in \mathbb{N}$ on $t \in [\mu_{i+1}, \tau_{i+1}]$, according to (14.55), (14.57). It follows that $Q_{n,2}(\mu_{i+1}, \tau_{i+1}) = 0$, $H_{n,1}(\mu_{i+1}, \tau_{i+1}) = 0$ (or $Q_{n,2}(\mu_{i+1}, \tau_{i+1}) = 0$, $H_{n,2}(\mu_{i+1}, \tau_{i+1}) = 0$) for all $n \in \mathbb{N}$ according to (14.53)–(14.57). Consequently, we have from (14.61) that $\hat{q}_1(\tau_{i+1}) = \hat{q}_1(\tau_i) \neq q_1$ (or $\hat{q}_2(\tau_{i+1}) = \hat{q}_2(\tau_i) \neq q_2$) by recalling (14.59)–(14.60)). We then conclude that $z[t]$ (or $w[t]$) are not identically zero on $t \in [\mu_{i+1}, \tau_{i+1}]$ according to lemma 14.2. That is, $z[t]$ (or $w[t]$) are not identically zero on $t = [0, \lim_{i\to\infty}(\tau_i))$.

The proof of claim 14.4 is complete.

F. Proof of claim 14.5

We prove this claim by estimating the largest convergence time of parameter estimates τ_ε in various situations of the initial conditions $z[0], w[0], \zeta(0)$. Inserting (14.36) into (14.5), we get

$$q_2 w(1,t) = \hat{q}_2(t) \int_0^1 \phi(1,y;\hat{\theta}(t)) w(y,t) dy + \hat{q}_2(t) \lambda(1;\hat{\theta}(t)) \zeta(t) + \tilde{q}_1(t) z(1,t).$$
(14.182)

Case 1: $z[0] \neq 0, w[0] = 0, \zeta(0) = 0$. According to lemma 14.3, we have $\hat{q}_1(\tau_1) = q_1$ and $\tilde{q}_1(t) \equiv 0$ for $t \geq \tau_1$. If $w[t] \equiv 0$ on $t \in [0, \tau_1]$, then $w[t]$ and $\zeta(t)$ are identically zero on $t \geq 0$ according to (14.1), (14.3), and (14.182), with $\tilde{q}_1(t) \equiv 0$ for $t \geq \tau_1$. If $\hat{q}_2(0) \neq q_2$, it follows that $\hat{q}_2(t)$ cannot reach the true value q_2 according to claim 14.4 with property (1): a contradiction with the fact that $\hat{q}_2(t)$ would reach q_2 in finite time. Thus, $w[t]$ is not identically zero on $t \in [0, \tau_1]$ if $\hat{q}_2(0) \neq q_2$. It follows that $\hat{q}_2(t)$ can reach q_2 not later than τ_1 according to lemma 14.3. We know from (14.38) that the dwell time is less than or equal to T. Therefore, $\tau_\varepsilon \leq T$.

Case 2: $w[0] \neq 0, z[0] = 0, \zeta(0) = 0$. The maximum time taken by the nonzero values of $w[0]$ to propagate to $x = 0$ and enter $z(0,t)$ is $\frac{1}{q_2}$. Therefore, the estimate $\hat{q}_1(t)$ would reach the true value q_1 not later than $\tau_f = \min\{\tau_f : f \in Z_+, \tau_f > \frac{1}{q_2}\}$ according to lemma 14.3. Because of $w[0] \neq 0$, we have $\hat{q}_2(\tau_1) = q_2$. It follows that $\tau_\varepsilon \leq \frac{1}{q_2} + T$ because the dwell time is less than or equal to T.

Case 3: $\zeta(0) \neq 0, z[0] = 0, w[0] = 0$. According to (14.182) and (14.4), we know that $w[t], z[t]$ are not identically zero on $t \in [0, \tau_1]$, which implies that the estimates $\hat{\theta}(t)$ reach the true values θ not later than τ_1 according to lemma 14.3. Therefore, $\tau_\varepsilon \leq \tau_1 \leq T$.

Case 4: $\zeta(0) \neq 0, w[0] \neq 0, z[0] = 0$. The necessary condition of the fact that $z[t]$ is identically zero (i.e., $w(0,t) = \frac{c}{p}\zeta(t)$ always holds) for $t \in [0, \tau_f]$ where $\tau_f = \min\{\tau_f : f \in Z_+, \tau_f > \frac{1}{q_2}\}$ is $\kappa > 0$, according to the analysis in case 2 in the proof of portion (1) of the theorem. Recalling $\kappa < 0$ in (14.12), we know that $z[t]$ is not identically zero on $t \in [0, \tau_f]$, which implies that the estimate $\hat{q}_1(t)$ reaches the true value q_1 not later than τ_f according to lemma 14.3. Because $w[0] \neq 0$, we have $\hat{q}_2(\tau_1) = q_2$. Therefore, $\tau_\varepsilon \leq \frac{1}{q_2} + T$.

Case 5: $\zeta(0) \neq 0, z[0] \neq 0, w[0] = 0$. According to lemma 14.3, we have $\hat{q}_1(\tau_1) = q_1$ and $\tilde{q}_1(t) \equiv 0$ for $t \geq \tau_1$. If $w[t] \equiv 0$ on $t \in [0, \tau_2]$, it follows from (14.1) that $\zeta(t) = e^{(a-q_1 bc)t}\zeta(0)$ is not identically zero on $t \in [\tau_1, \tau_2]$. From (14.182) we know that $w(1,t)$ is not identically zero on $t \in [\tau_1, \tau_2]$: a contradiction. Therefore, $w[t]$ are not identically zero on $t \in [0, \tau_2]$, which implies that the estimate $\hat{q}_2(t)$ reaches the true value q_2 not later than τ_2 according to lemma 14.3. Therefore, we find that $\tau_\varepsilon \leq \tau_2 \leq 2T$.

Case 6: $\zeta(0) = 0, z[0] \neq 0, w[0] \neq 0$ and case 7: $\zeta(0) \neq 0, z[0] \neq 0, w[0] \neq 0$. According to lemma 14.3, we see that $\tau_\varepsilon \leq \tau_1 \leq T$.

Case 8: $z[0] = 0, w[0] = 0, \zeta(0) = 0$. According to the plant (14.1)–(14.5) with the control input (14.36), we know that $z[t], w[t], \zeta(t)$ are identically zero for $t \geq 0$. The estimates reach the true values in finite time only when $\hat{q}_1(0) = q_1, \hat{q}_2(0) = q_2$—that is, $\tau_\varepsilon = 0$.

In summary, we have proved for all eight cases that $\tau_\varepsilon \leq \max\{\frac{1}{q_2} + T, 2T\}$. This completes the proof of claim 14.5.

14.7   NOTES

Adaptive control with a regulation-triggered BaLSI was originally introduced in [102, 104]. In this chapter, we designed an adaptive certainty-equivalence controller with regulation-triggered batch least-squares identification for a hyperbolic PDE-ODE system where the unknown parameters are transport speeds. It would be appropriate to regard this chapter as the hyperbolic equivalent of the paper [106] for a parabolic PDE, where the unknown parameters are the reaction coefficient and the high-frequency gain. In chapter 13, triggering was employed for the control law instead of the parameter estimator and only asymptotic convergence to zero of the plant states was achieved. However, this chapter's adaptive control design employed triggering for the parameter update law rather than the control input and achieved finite-time exact identification of the unknown parameters from most initial conditions and, as a result of the finite-time settling of the parameter estimates, exponential regulation of the plant states to zero.

Chapter Fifteen

Adaptive Control of Hyperbolic PDEs with Piecewise-Constant Inputs and Identification

The event-triggered adaptive control design in chapter 13 employs triggering for the control law instead of the parameter estimator, whereas the design in chapter 14 employs triggering for the parameter estimator and not the control law. In this chapter, we pursue the design of an event-triggered adaptive controller where triggering is employed for updating both the parameter estimator and the plant states in the control law simultaneously. As a result, both the parameter estimates and the control input employ piecewise-constant values. The controller consists of a nominal continuous-in-time backstepping feedback law and a triggered batch least-squares identifier (BaLSI). The triggering mechanism is designed based on both evaluating the actuation deviation caused by the difference between the plant states and their sampled values and on evaluating the growth of the plant norm. When either condition is met, recomputing of the parameter estimator and resampling of the plant states in the feedback are done simultaneously.

The problem formulation is presented in section 15.1. The nominal continuous-in-time control design is presented in section 15.2. The design of event-triggered control with piecewise-constant parameter identification is proposed in section 15.3. The results, including the absence of a Zeno phenomenon, parameter convergence, and exponential regulation of the states, are proved in section 15.4. The effectiveness of the proposed design is illustrated with an application in the axial vibration control of a mining cable elevator in section 15.5.

15.1 PROBLEM FORMULATION

We conduct the control design based on the following 2×2 hyperbolic partial differential equation-ordinary differential equation (PDE-ODE) system:

$$\dot{\zeta}(t) = a\zeta(t) + bw(0, t), \tag{15.1}$$

$$z_t(x, t) = -q_1 z_x(x, t) + d_1 z(x, t) + d_2 w(x, t), \tag{15.2}$$

$$w_t(x, t) = q_2 w_x(x, t) + d_3 z(x, t) + d_4 w(x, t), \tag{15.3}$$

$$z(0, t) = c\zeta(t) - pw(0, t), \tag{15.4}$$

$$w(1, t) = c_0 U(t) \tag{15.5}$$

for $x \in [0, 1], t \in [0, \infty)$, where $\zeta(t)$ is a scalar ODE state, and scalar functions $z(x, t)$ and $w(x, t)$ are PDE states. The function $U(t)$ is the control input to be designed. The positive constants q_1, q_2 are transport speeds, and $p \neq 0$, $c_0 \neq 0$ are arbitrary constants. The ODE system parameter a and the coefficients d_2, d_3 of the PDE in-domain couplings are unknown. To make the problem as nontrivial as we can within this class, we focus on the case where in-domain instability exists—that is, $d_3 \neq 0$. Other parameters are arbitrary and satisfy assumptions 15.1, 15.2.

The connection between the general model (15.1)–(15.5) and the mining cable elevator is illustrated in section 15.5. Additionally, the plant (15.1)–(15.5) is inspired by the suppression of vibrations of airplane wings with aeroelastic instability at high Mach numbers [161, 196], which are described by a wave PDE with the in-domain fluid pressure represented by positive stiffness and anti-damping; that is, $u_{tt}(x,t) = c_w^2 u_{xx}(x,t) + \mu_w u_t(x,t) + v_w u_x(x,t)$, where $u(x,t)$ denotes the membrane displacements, and the physical meanings of the positive coefficients c_w, μ_w, v_w are given in [161]. Applying the Riemann transformations $z(x,t) = u_t(x,t) - c_w u_x(x,t)$, $w(x,t) = u_t(x,t) + c_w u_x(x,t)$, the airplane wing vibration dynamics are transformed into a 2×2 hyperbolic PDE system, covered by the considered plant (15.1)–(15.5), where the ODE (15.5) can describe a mass at the wing tip.

Assumption 15.1. *The parameter b is nonzero. For $c \neq 0$, b is not equal to $-\frac{pa}{c}$.*

Assumption 15.2. *For $a \neq 0$, the parameters d_1, d_2, d_3, d_4 satisfy*

$$q_2 \frac{d_3(\frac{bc}{a}+p)^2 - (d_4-d_1)(\frac{bc}{a}+p) - d_2}{(\frac{bc}{a}+p)(q_1+q_2)} - d_3\left(\frac{bc}{a}+p\right) + d_4 \neq 0.$$

For $a = 0$ and $c = 0$, the parameters d_1, d_2, d_3, d_4 satisfy

$$q_2 \frac{d_3 p^2 - (d_4-d_1)p - d_2}{p(q_1+q_2)} - d_3 p + d_4 \neq 0.$$

For the cases not mentioned in assumptions 15.1 and 15.2, no restrictions are imposed for the corresponding parameters. The generically satisfied assumptions 15.1 and 15.2 act as sufficient (but not necessary) conditions for the parameter convergence of the estimates.

Assumption 15.3. *Bounds on the unknown parameters d_3, d_2, a are known though arbitrary; that is, $|d_3| \leq \bar{d}_3$, $|d_2| \leq \bar{d}_2$, $|a| \leq \bar{a}$, where $\bar{d}_3, \bar{d}_2, \bar{a}$ are arbitrary positive constants whose values are known.*

We adopt the following notation.

- The symbol $\mathbb{Z}^+$ denotes the set of all nonnegative integers, and $\mathbb{R}_+ := [0, +\infty)$ and $\mathbb{R}_- := (-\infty, 0]$.
- Let $U \subseteq \mathbb{R}^n$ be a set with a nonempty interior, and let $\Omega \subseteq \mathbb{R}$ be a set. By $C^0(U; \Omega)$, we denote the class of continuous mappings on U, which take values in Ω. By $C^k(U; \Omega)$, where $k \geq 1$, we denote the class of continuous functions on U, which have continuous derivatives of order k on U and take values in Ω.
- We use the notation $L^2(0,1)$ for the standard space of the equivalence class of square-integrable, measurable functions defined on $(0,1)$, and $\|f\| = \left(\int_0^1 f(x)^2 dx\right)^{\frac{1}{2}} < +\infty$ for $f \in L^2(0,1)$.
- We use the notation $\mathbb{N}$ for the set $\{1, 2, \cdots\}$—that is, the natural numbers without 0.
- For an $I \subseteq \mathbb{R}_+$, the space $C^0(I; L^2(0,1))$ is the space of continuous mappings $I \ni t \mapsto u[t] \in L^2(0,1)$.

- Let $u : \mathbb{R}_+ \times [0,1] \to \mathbb{R}$ be given. We use the notation $u[t]$ to denote the profile of u at certain $t \geq 0$—that is, $(u[t])(x) = u(x,t)$, for all $x \in [0,1]$.

15.2   NOMINAL CONTROL DESIGN

We introduce the following backstepping transformation [48]

$$\alpha(x,t) = z(x,t) - \int_0^x \phi(x,y)z(y,t)dy$$

$$- \int_0^x \varphi(x,y)w(y,t)dy - \gamma(x)\zeta(t), \tag{15.6}$$

$$\beta(x,t) = w(x,t) - \int_0^x \Psi(x,y)z(y,t)dy$$

$$- \int_0^x \Phi(x,y)w(y,t)dy - \lambda(x)\zeta(t), \tag{15.7}$$

and its inverse

$$z(x,t) = \alpha(x,t) - \int_0^x \bar{\phi}(x,y)\alpha(y,t)dy$$

$$- \int_0^x \bar{\varphi}(x,y)\beta(y,t)dy - \bar{\gamma}(x)\zeta(t), \tag{15.8}$$

$$w(x,t) = \beta(x,t) - \int_0^x \bar{\Psi}(x,y)\alpha(y,t)dy$$

$$- \int_0^x \bar{\Phi}(x,y)\beta(y,t)dy - \bar{\lambda}(x)\zeta(t) \tag{15.9}$$

to convert the plant (15.1)–(15.5) to the target system

$$\dot{\zeta}(t) = -a_{\mathrm{m}}\zeta(t) + b\beta(0,t), \tag{15.10}$$

$$\alpha(0,t) = -p\beta(0,t), \tag{15.11}$$

$$\alpha_t(x,t) = -q_1\alpha_x(x,t) + d_1\alpha(x,t), \tag{15.12}$$

$$\beta_t(x,t) = q_2\beta_x(x,t) + d_4\beta(x,t), \tag{15.13}$$

$$\beta(1,t) = 0, \tag{15.14}$$

where $a_{\mathrm{m}} = b\kappa - a > 0$ is ensured by a design parameter κ satisfying

$$\kappa > \frac{\bar{a}}{b}, \quad \forall b > 0, \qquad \kappa < \frac{\bar{a}}{b}, \quad \forall b < 0. \tag{15.15}$$

The conditions on the kernels $\phi(x,y)$, $\varphi(x,y)$, $\gamma(x)$, $\Psi(x,y)$, $\Phi(x,y)$, $\lambda(x)$ and $\bar{\phi}(x,y)$, $\bar{\varphi}(x,y)$, $\bar{\gamma}(x)$, $\bar{\Psi}(x,y)$, $\bar{\Phi}(x,y)$, $\bar{\lambda}(x)$ in the backstepping transformations (15.6)–(15.9), which are obtained by matching the original system (15.1)–(15.5) and the target system (15.10)–(15.14), are shown in appendix 15.6A, and the well-posedness has been proved in theorem 4.1 of [48]. The nominal continuous-in-time control is obtained from the right boundary condition (15.14) in the target system as follows:

$$U(t) = \frac{1}{c_0} \int_0^1 \Psi(1, y; \theta) z(y, t) dy$$

$$+ \frac{1}{c_0} \int_0^1 \Phi(1, y; \theta) w(y, t) dy + \frac{1}{c_0} \lambda(1; \theta) \zeta(t). \qquad (15.16)$$

Writing $\theta = [d_3, d_2, a]^T$ after ";" in $\Psi(x, y; \theta)$, $\Phi(x, y; \theta)$, and $\lambda(x; \theta)$ emphasizes the fact that $\Psi(x, y)$, $\Phi(x, y)$, $\lambda(x)$ depend on the unknown parameters d_3, d_2, a, due to the fact that Ψ, Φ, λ is the solution of the kernel PDE set, shown in appendix 15.6A, with the unknown coefficients d_3, d_2, a. Under suitable conditions on (a, b, c), the explicit design from [176] using the generalized Marcum Q-functions of the first order could be employed in the controller (15.16).

15.3   EVENT-TRIGGERED CONTROL DESIGN WITH PIECEWISE-CONSTANT PARAMETER IDENTIFICATION

Based on the nominal continuous-in-time feedback (15.16), we give the form of an adaptive event-triggered control law U_d as follows:

$$U_d(t) = \frac{1}{c_0} \int_0^1 \Psi(1, y; \hat{\theta}(t_i)) z(y, t_i) dy$$

$$+ \frac{1}{c_0} \int_0^1 \Phi(1, y; \hat{\theta}(t_i)) w(y, t_i) dy + \frac{1}{c_0} \lambda(1; \hat{\theta}(t_i)) \zeta(t_i) \qquad (15.17)$$

for $t \in [t_i, t_{i+1})$, where $\hat{\theta} = [\hat{d}_3, \hat{d}_2, \hat{a}]^T$ is an estimate, generated with a triggered BaLSI, of the three unknown parameters d_3, d_2, a. The identifier is presented shortly, and $\{t_i \geq 0\}_{i=0}^{\infty}$, $i \in \mathbb{Z}^+$ is the sequence of time instants, which, along with the parameter estimates and sampled states in (15.17), is defined in the next subsection.

When we mention the continuous-in-state control input U_c, we refer to the control input consisting of triggered parameter estimates and continuous states—that is,

$$U_c(t) = \frac{1}{c_0} \int_0^1 \Psi(1, y; \hat{\theta}(t_i)) z(y, t) dy$$

$$+ \frac{1}{c_0} \int_0^1 \Phi(1, y; \hat{\theta}(t_i)) w(y, t) dy + \frac{1}{c_0} \lambda(1; \hat{\theta}(t_i)) \zeta(t) \qquad (15.18)$$

for $t \in [t_i, t_{i+1})$. Define the difference between the continuous-in-state control input U_c in (15.18) and the event-triggered control input U_d in (15.17) as $d(t)$, given by

$$d(t) = U_c(t) - U_d(t)$$

$$= \frac{1}{c_0} \int_0^1 \Psi(1, y; \hat{\theta}(t_i)) (z(y, t) - z(y, t_i)) dy$$

$$+ \frac{1}{c_0} \int_0^1 \Phi(1, y; \hat{\theta}(t_i)) (w(y, t) - w(y, t_i)) dy$$

$$+ \frac{1}{c_0} \lambda(1; \hat{\theta}(t_i)) (\zeta(t) - \zeta(t_i)), \quad t \in [t_i, t_{i+1}), \qquad (15.19)$$

which reflects the deviation of the plant states from their sampled values.

Define the difference between the continuous-in-state control input $U_c(t)$ in (15.18) and the nominal continuous-in-time control input $U(t)$ in (15.16) as $\xi(t)$, given by

$$\xi(t) = U(t) - U_c(t)$$

$$= \frac{1}{c_0} \int_0^1 (\Psi(1, y; \theta) - \Psi(1, y; \hat{\theta}(t_i))) z(y, t) dy$$

$$+ \frac{1}{c_0} \int_0^1 (\Phi(1, y; \theta) - \Phi(1, y; \hat{\theta}(t_i))) w(y, t) dy$$

$$+ \frac{1}{c_0} (\lambda(1; \theta) - \lambda(1; \hat{\theta}(t_i))) \zeta(t), \quad t \in [t_i, t_{i+1}), \tag{15.20}$$

which reflects the deviation of the estimates from the actual unknown parameters. The deviations $d(t)$ and $\xi(t)$ will be used in the following design and analysis.

Triggering Mechanism

Before showing the triggering mechanism that determines the sequence of time instants $\{t_i \geq 0\}_{i=0}^{\infty}$, $i \in \mathbb{Z}^+$, we introduce the two sets S_i and $\bar{S}_i$, which will be used in the triggering mechanism to judge whether the exact identification of the unknown parameters has been achieved. The set S_i is defined as

$$S_i := \left\{ \bar{\ell} = (\ell_1, \ell_2)^T \in \Theta_1 : \bar{Z}_n(\mu_{i+1}, t_{i+1}) - \bar{G}_n(\mu_{i+1}, t_{i+1}) \bar{\ell}, \right.$$

$$\left. n = 1, 2, \ldots, \right\}, \quad i \in \mathbb{Z}^+, \tag{15.21}$$

where

$$\Theta_1 = \{ \bar{\ell} \in \mathbb{R}^2 : |\ell_1| \leq \bar{d}_3, |\ell_2| \leq \bar{d}_2 \},$$

and $\bar{Z}_n$, $\bar{G}_n$ associated with the plant states over a time interval $[\mu_{i+1}, t_{i+1}]$ (the time instant μ_{i+1} is defined shortly) are

$$\bar{Z}_n(\mu_{i+1}, t_{i+1}) = [H_{n,1}(\mu_{i+1}, t_{i+1}), H_{n,2}(\mu_{i+1}, t_{i+1})]^T, \tag{15.22}$$

$$\bar{G}_n(\mu_{i+1}, t_{i+1}) = \begin{bmatrix} Q_{n,1}(\mu_{i+1}, t_{i+1}) & Q_{n,2}(\mu_{i+1}, t_{i+1}) \\ Q_{n,2}(\mu_{i+1}, t_{i+1}) & Q_{n,3}(\mu_{i+1}, t_{i+1}) \end{bmatrix} \tag{15.23}$$

with

$$H_{n,1}(\mu_{i+1}, t_{i+1}) = \int_{\mu_{i+1}}^{t_{i+1}} g_{n,1}(t, \mu_{i+1}) f_n(t, \mu_{i+1}) dt, \tag{15.24}$$

$$H_{n,2}(\mu_{i+1}, t_{i+1}) = \int_{\mu_{i+1}}^{t_{i+1}} g_{n,2}(t, \mu_{i+1}) f_n(t, \mu_{i+1}) dt, \tag{15.25}$$

$$Q_{n,1}(\mu_{i+1}, t_{i+1}) = \int_{\mu_{i+1}}^{t_{i+1}} g_{n,1}(t, \mu_{i+1})^2 dt, \tag{15.26}$$

$$Q_{n,2}(\mu_{i+1}, t_{i+1}) = \int_{\mu_{i+1}}^{t_{i+1}} g_{n,1}(t, \mu_{i+1}) g_{n,2}(t, \mu_{i+1}) dt, \tag{15.27}$$

$$Q_{n,3}(\mu_{i+1}, t_{i+1}) = \int_{\mu_{i+1}}^{t_{i+1}} g_{n,2}(t, \mu_{i+1})^2 dt, \tag{15.28}$$

where

$$f_n(t, \mu_{i+1}) = \int_0^1 \sin(x\pi n)z(x,t)dx + \int_0^1 \sin(x\pi n)w(x,t)dx$$

$$- \int_0^1 \sin(x\pi n)z(x, \mu_{i+1})dx - \int_0^1 \sin(x\pi n)w(x, \mu_{i+1})dx$$

$$- \pi n \int_{\mu_{i+1}}^t \int_0^1 \cos(x\pi n)(q_1 z(x, \tau) - q_2 w(x, \tau))dxd\tau$$

$$- \int_{\mu_{i+1}}^t \int_0^1 \sin(x\pi n)(d_1 z(x, \tau) + d_4 w(x, \tau))dxd\tau, \qquad (15.29)$$

$$g_{n,1}(t, \mu_{i+1}) = \int_{\mu_{i+1}}^t \int_0^1 \sin(x\pi n)z(x, \tau)dxd\tau, \qquad (15.30)$$

$$g_{n,2}(t, \mu_{i+1}) = \int_{\mu_{i+1}}^t \int_0^1 \sin(x\pi n)w(x, \tau)dxd\tau \qquad (15.31)$$

for $n = 1, 2, \cdots$. The set $\bar{S}_i$ is defined as

$$\bar{S}_i := \left\{ \ell_3 \in [-\bar{a}, \bar{a}] : H_3(\mu_{i+1}, t_{i+1}) = Q_4(\mu_{i+1}, t_{i+1})\ell_3, \right\}, i \in \mathbb{Z}^+, \qquad (15.32)$$

where the function H_3 is defined as

$$H_3(\mu_{i+1}, t_{i+1}) = \int_{\mu_{i+1}}^{t_{i+1}} g_a(t, \mu_{i+1})f_a(t, \mu_{i+1})dt, \qquad (15.33)$$

and the function Q_4 is defined as

$$Q_4(\mu_{i+1}, t_{i+1}) = \int_{\mu_{i+1}}^{t_{i+1}} g_a(t, \mu_{i+1})^2 dt \qquad (15.34)$$

for $i \in \mathbb{Z}^+$, with

$$f_a(t, \mu_{i+1}) = \zeta(t) - \zeta(\mu_{i+1}) - b \int_{\mu_{i+1}}^t w(0, \tau)d\tau, \qquad (15.35)$$

$$g_a(t, \mu_{i+1}) = \int_{\mu_{i+1}}^t \zeta(\tau)d\tau. \qquad (15.36)$$

Based on the above definitions, we introduce the triggering mechanism next. The sequence of time instants $\{t_i \geq 0\}_{i=0}^\infty$ ($t_0 = 0$) is defined as

$$t_{i+1} = \min\{\tau_i, r_i\}, \qquad (15.37)$$

where τ_i is given by

$$\tau_i = \inf\{t > t_i : d(t)^2 > \vartheta V(t) - m(t)\} \qquad (15.38)$$

and where r_i is defined next. If $i \geq 1$ and there exists a singleton S_j for a certain j ($j \in \mathbb{Z}^+$, $j \leq i - 1$) and a singleton $\bar{S}_k$ for a certain k ($k \in \mathbb{Z}^+$, $k \leq i - 1$), then r_i is set as

$$r_i := +\infty, \tag{15.39}$$

or as

$$r_i = \max\left\{t_i + \underline{r}, \min\left\{\inf\{t > t_i : \Omega(t) = (1+\delta)\Omega(t_i)\},\right.\right.$$
$$\left.\left. t_i + T\right\}\right\}, \quad \Omega(t_i) \neq 0, \tag{15.40}$$
$$r_i = t_i + T, \quad \Omega(t_i) = 0, \tag{15.41}$$

where the design parameters δ, T are positive and free, and $\underline{r}$ is a positive constant satisfying

$$0 < \underline{r} < T, \tag{15.42}$$

and where the function $\Omega(t)$ is defined as

$$\Omega(t) = \|z[t]\|^2 + \|w[t]\|^2 + \zeta(t)^2. \tag{15.43}$$

The Lyapunov function $V(t)$ in (15.38) is given as

$$V(t) = \frac{1}{2}\zeta(t)^2 + \frac{1}{2}r_a \int_0^1 e^{\delta_1 x} \beta(x,t)^2 dx$$

$$+ \frac{1}{2}\int_0^1 e^{-\delta_2 x} \alpha(x,t)^2 dx, \tag{15.44}$$

where the positive constants r_a, δ_1, δ_2, which are design parameters, are chosen to satisfy

$$\delta_1 > \frac{2|d_4|}{q_2}, \tag{15.45}$$

$$\delta_2 > \frac{2|d_1|}{q_1}, \tag{15.46}$$

$$r_a > \frac{b^2}{q_2(b\kappa - \bar{a})} + \frac{p^2 q_1}{q_2}, \tag{15.47}$$

which depend only on the design parameter κ in (15.15), the known plant parameters, and the known bounds of the unknown parameters in assumption 15.3.

The internal dynamic variable $m(t)$ in (15.38) satisfies the ordinary differential equation

$$\dot{m}(t) = -\eta m(t) + \lambda_d d(t)^2 - \sigma V(t)$$
$$- \kappa_1 \alpha(1,t)^2 - \kappa_2 \beta(0,t)^2 \tag{15.48}$$

with the initial condition $m(0) < 0$. Choose the positive design parameters κ_1, κ_2 in (15.48) to satisfy

$$\kappa_1 \geq 2\max\{\lambda_a, \lambda_\alpha\}, \tag{15.49}$$
$$\kappa_2 \geq 2\max\{\lambda_a, \lambda_\beta\}, \tag{15.50}$$

where the positive constants λ_α, λ_β, λ_a are

$$\lambda_\alpha = \frac{1}{2}q_1 e^{-\delta_2}, \quad \lambda_\beta = \frac{1}{2}q_2 r_a + \frac{1}{2}|b|, \tag{15.51}$$

$$\lambda_a = \frac{6}{c_0^2} \max_{y \in [0,1], \theta_1 \in \Theta, \theta_2 \in \Theta} \left\{ |M_1(y, \theta_1, \theta_2)|^2, |M_2(y, \theta_1, \theta_2)|^2, \right.$$

$$\left. |M_3(\theta_1, \theta_2)|^2, |M_4(\theta_1, \theta_2)|^2, |M_5(\theta_1, \theta_2)|^2, |M_6(\theta_1, \theta_2)|^2 \right\}. \tag{15.52}$$

The functions $M_1, \ldots, M_6$ are given in appendix 15.6B, and λ_a depends only on the design parameter κ, the known plant parameters, and the known bounds of the unknown parameters in assumption 15.3. Choose the design parameters λ_d, σ, η in (15.48) such that they satisfy

$$\lambda_d \geq \frac{1}{\mu} q_2 r_a e^{\delta_1} c_0^2, \tag{15.53}$$

$$0 < \sigma < \frac{\nu_a}{\mu}, \tag{15.54}$$

$$\eta > \nu_a - \mu\sigma, \tag{15.55}$$

where the positive constants μ, ν_a are

$$\mu \leq \min \left\{ \frac{q_2 r_a}{2\kappa_2} - \frac{b^2}{2\kappa_2(b\kappa - \bar{a})} - \frac{p^2 q_1}{2\kappa_2}, \frac{1}{2\kappa_1} q_1 e^{-\delta_2} \right\}, \tag{15.56}$$

$$\nu_a = \frac{1}{\xi_2} \min \left\{ \frac{1}{2}(b\kappa - \bar{a}), \frac{1}{2}\delta_1 q_2 r_a - r_a |d_4|, \left(\frac{1}{2}\delta_2 q_1 - |d_1| \right) e^{-\delta_2} \right\} \tag{15.57}$$

with

$$\xi_2 = \frac{1}{2} \min\{1, r_a e^{\delta_1}\} > 0. \tag{15.58}$$

The final design parameter ϑ in (15.38) is chosen in accordance with the following condition:

$$0 < \vartheta \leq \max \left\{ 1, \frac{\sigma}{\lambda_d} \right\}. \tag{15.59}$$

The time instant μ_{i+1} used in (15.21), (15.32) is defined as

$$\mu_{i+1} := \min\{t_f : f \in \{0, \ldots, i\}, t_f \geq t_{i+1} - \tilde{N}T\}, \tag{15.60}$$

according to [104], where the positive integer $\tilde{N} \geq 1$ is a design parameter, and the positive constant T is the maximum dwell time according to (15.37)–(15.41). The flowchart of implementing the designed triggering mechanism (15.37)–(15.41) is shown in figure 15.1, and the proof that the triggering mechanism (15.37)–(15.41) is well defined and produces an increasing sequence of events will be shown in lemma 15.3.

Design rationale of the triggering mechanism (15.37)–(15.41): According to (15.19), (15.20)—that is, two deviation signals affecting the stability obtained under the nominal continuous-in-time control—we know the size of $d(t)$ is associated with the deviation of the plant states from their sampled values, and the size of $\xi(t)$ is associated with the deviation of the parameter estimates from the actual unknown parameters. According to the form of the event trigger presented in [57], the triggering condition (15.38) is designed, based on the evolution of the square of

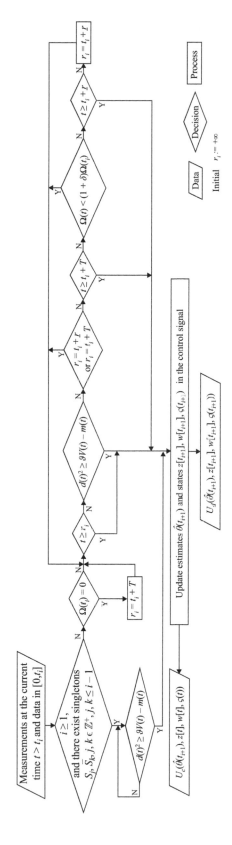

Figure 15.1. Implementing the triggering mechanism (15.37)–(15.41).

$d(t)$, to guarantee that $d(t)^2$ is bounded by the plant norm, and an internal dynamic variable $m(t)$ defined in (15.48) (the introduction of $m(t)$ is to ensure that no Zeno phenomenon occurs, as will be seen in claim 15.2 in lemma 15.3). Even if $d(t)$ is ensured to be small enough in the closed-loop system, a large growth of the plant norm still may appear under a large $\xi(t)$ (as in the analysis of property (3) in theorem 15.1). Therefore, in addition to the triggering condition (15.38) designed to bound $d(t)$, another triggering condition in (15.40) is designed based on evaluating the growth of the plant norm [104], where the updates are triggered if the plant norm has grown by a certain factor, to avoid a large overshoot. Introducing the design parameters $\underline{r}, T$ in (15.40), which set the lower and upper bounds of the interval between r_i, enables the user to adjust the number of r_i. As will be seen later, the condition below (15.38), that there exist the singleton sets $S_j, \bar{S}_k$, implies that the exact identification of θ has been achieved at t_i for $i \geq 1$. Therefore, in this case we set $r_i := +\infty$ in (15.39) in order to avoid unnecessary updates of the control signal after the exact parameter identification has been achieved (when initial estimates are not the true values). Introducing a sufficiently small positive constant $\underline{r}$ in (15.40) is to ensure that no Zeno phenomenon occurs (as will be seen in claim 15.1 in lemma 15.3).

The first triggering condition (15.38) is for resampling the states $z[t], w[t], \zeta(t)$, and the second triggering condition (15.40), (15.41) is for recomputing the estimate $\hat{\theta}(t)$. The synchronous triggering is ensured by the fact that if either condition is met, both the recomputing of $\hat{\theta}(t)$ and the resampling of $z[t], w[t], \zeta(t)$ are done simultaneously, due to (15.37). All design parameters are $\kappa, \underline{r}, \delta_1, \delta_2, r_a, \kappa_1, \kappa_2, \lambda_d, \sigma,$ η, θ, whose conditions are cascaded rather than mutually dependent, and only defined by known parameters. They can be solved in the sequence (15.15), (15.42), (15.45), (15.46), (15.47), (15.49), (15.50), (15.53), (15.54), (15.55), (15.59). The motivation for defining these conditions for the design parameters will become clear in the rest of this chapter.

Least-Squares Identifier

According to (15.2), (15.3), we get the following for $\tau > 0$ and $n = 1, 2, \cdots$:

$$\frac{d}{d\tau}\left(\int_0^1 \sin(x\pi n)z(x,\tau)dx + \int_0^1 \sin(x\pi n)w(x,\tau)dx\right)$$

$$= -q_2\pi n \int_0^1 \cos(x\pi n)w(x,\tau)dx$$

$$+ d_1 \int_0^1 \sin(x\pi n)z(x,\tau)dx + d_2 \int_0^1 \sin(x\pi n)w(x,\tau)dx$$

$$+ q_1\pi n \int_0^1 \cos(x\pi n)z(x,\tau)dx$$

$$+ d_3 \int_0^1 \sin(x\pi n)z(x,\tau)dx + d_4 \int_0^1 \sin(x\pi n)w(x,\tau)dx. \tag{15.61}$$

Integrating (15.61), (15.1) from $\mu_{i\,|\,1}$ to t yields

$$f_n(t, \mu_{i+1}) = d_3 g_{n,1}(t, \mu_{i+1}) + d_2 g_{n,2}(t, \mu_{i+1}), \tag{15.62}$$

$$f_a(t, \mu_{i+1}) = a g_a(t, \mu_{i+1}), \tag{15.63}$$

where f_n, $g_{n,1}$, $g_{n,2}$ are given in (15.29)–(15.31), and f_a, g_a are defined in (15.35), (15.36).

Define the function $h_{i,n} : \mathbb{R}^3 \to \mathbb{R}_+$ by the formula

$$h_{i,n}(\ell) = \int_{\mu_{i+1}}^{t_{i+1}} [(f_n(t, \mu_{i+1}) - \ell_1 g_{n,1}(t, \mu_{i+1}) - \ell_2 g_{n,2}(t, \mu_{i+1}))^2$$

$$+ (f_a(t, \mu_{i+1}) - \ell_3 g_a(t, \mu_{i+1}))^2] dt, \quad i \in \mathbb{Z}^+ \tag{15.64}$$

for $n = 1, 2, \cdots$, $\ell = [\ell_1, \ell_2, \ell_3]^T$.

According to (15.62), (15.63), the function $h_{i,n}(\ell)$ in (15.64) has a global minimum $h_{i,n}(\theta) = 0$. We get from Fermat's theorem (vanishing gradient at extrema) that the following matrix equation holds for every $i \in \mathbb{Z}^+$ and $n = 1, 2, \cdots$:

$$Z_n(\mu_{i+1}, t_{i+1}) = G_n(\mu_{i+1}, t_{i+1})\theta, \tag{15.65}$$

where

$$Z_n = [H_{n,1}, H_{n,2}, H_3]^T, \quad G_n = \begin{bmatrix} Q_{n,1} & Q_{n,2} & 0 \\ Q_{n,2} & Q_{n,3} & 0 \\ 0 & 0 & Q_4 \end{bmatrix}, \tag{15.66}$$

and $H_{n,1}(\mu_{i+1}, t_{i+1})$, $H_{n,2}(\mu_{i+1}, t_{i+1})$, $H_3(\mu_{i+1}, t_{i+1})$, $Q_{n,1}(\mu_{i+1}, t_{i+1})$, $Q_{n,2}$ (μ_{i+1}, t_{i+1}), $Q_{n,3}(\mu_{i+1}, t_{i+1})$, $Q_4(\mu_{i+1}, t_{i+1})$ are given in (15.24)–(15.28), (15.33), (15.34). Indeed, (15.65), (15.66) are obtained by differentiating the functions $h_{i,n}(\ell)$ defined by (15.64) with respect to ℓ_1, ℓ_2, ℓ_3, respectively, and evaluating the derivatives at the position of the global minimum $(\ell_1, \ell_2, \ell_3) = (d_3, d_2, a)$.

The parameter estimator (update law) is defined as

$$\hat{\theta}(t_{i+1}) = \operatorname{argmin}\Big\{ \left|\ell - \hat{\theta}(t_i)\right|^2 : \ell \in \Theta,$$

$$Z_n(\mu_{i+1}, t_{i+1}) = G_n(\mu_{i+1}, t_{i+1})\ell, \quad n = 1, 2, \cdots \Big\}, \tag{15.67}$$

where $\Theta = \{\ell \in \mathbb{R}^3 : |\ell_1| \le \bar{d}_3, |\ell_2| \le \bar{d}_2, |\ell_3| \le \bar{a}\}$.

Proposition 15.1. *For every* $(z[t_i], w[t_i])^T \in L^2((0, 1); \mathbb{R}^2)$, $\zeta(t_i) \in \mathbb{R}$, $m(t_i) \in \mathbb{R}_-$, *and* $\hat{\theta}(t_i) \in \Theta$, *there exists a unique (weak) solution* $((z, w)^T, \zeta) \in C^0([t_i, t_{i+1}];$ $L^2(0, 1); \mathbb{R}^2) \times C^0([t_i, t_{i+1}]; \mathbb{R})$, $m \in C^0([t_i, t_{i+1}]; \mathbb{R}^-)$ *to the system* (15.1)–(15.5), (15.17), (15.48), *and* $\hat{\theta} \in \Theta$ *in* (15.67), *between two time instants* t_i, t_{i+1}.

Proof. The proof is similar to that of proposition 14.1 in chapter 14, and thus it is omitted. $\qquad\square$

15.4   MAIN RESULT

Theorem 15.1. *For all initial data* $(z[0], w[0])^T \in C^1([0, 1])$, $\zeta(0) \in \mathbb{R}$, $\hat{\theta}(0) \in \Theta$, $m(0) < 0$, *the closed-loop system* (15.1)–(15.5) *under the controller* (15.17) *with the triggering mechanism* (15.37)–(15.41) *and the least-squares identifier defined by* (15.67) *has the following properties:*

1) *No Zeno phenomenon occurs—that is, $\lim_{i \to \infty} t_i = +\infty$, and the closed-loop system is well-posed.*

2) *If the finite-time convergence of parameter estimates to the true values is not achieved, it implies $\Omega(t) \equiv 0$ on $t \in [\frac{1}{q_1}, \infty)$, and $m(t)$ is exponentially convergent to zero. If the parameter estimate $\hat{\theta}(t)$ reaches the true value θ in finite time—that is,*

$$\hat{\theta}(t) = \theta, \qquad \forall t \geq t_\varepsilon \tag{15.68}$$

for certain $\varepsilon \in \mathbb{Z}^+$, then

$$t_\varepsilon \leq \frac{1}{q_1} + \frac{1}{q_2} + T. \tag{15.69}$$

3) *If the finite-time convergence of parameter estimates to the true values is achieved, the exponential regulation of the closed-loop system is obtained in the sense that there exist the positive constants M, λ_1 such that*

$$\bar{\Omega}(t) \leq M\bar{\Omega}(0)e^{-\lambda_1 t}, \quad t \geq 0, \tag{15.70}$$

where

$$\bar{\Omega}(t) = \|z[t]\|^2 + \|w[t]\|^2 + \zeta(t)^2 + |m(t)| \tag{15.71}$$

and where M is related to the initial estimate $\hat{\theta}(0)$.

The proof of theorem 15.1 is based on the following technical lemmas. First, we present the two lemmas that will be used in the analysis of the lower bound of the minimal dwell time.

Lemma 15.1. *For $d(t)$ defined in (15.19), there exists a positive constant λ_a such that*

$$\dot{d}(t)^2 \leq \lambda_a \left(d(t)^2 + \alpha(1,t)^2 + \beta(0,t)^2 \right.$$
$$\left. + \|\alpha(\cdot,t)\|^2 + \|\beta(\cdot,t)\|^2 + \zeta(t)^2 \right) \tag{15.72}$$

for $t \in (t_i, t_{i+1})$, where the positive constant λ_a is given in (15.52), which depends only on the design parameter κ in (15.15), the known plant parameters, and the known bounds $\bar{a}, \bar{d}_3, \bar{d}_2$ of the unknown parameters.

Proof. Inserting $U_d(t)$ into (15.5), we have

$$w(1,t) = c_0 U_d(t). \tag{15.73}$$

Applying (15.19), (15.20) allows us to rewrite (15.73) as

$$w(1,t) = c_0(U_c(t) - d(t)) = c_0 U(t) - c_0 \xi(t) - c_0 d(t). \tag{15.74}$$

Inserting (15.8), (15.9) into (15.20), we obtain

$$\xi(t) = \frac{1}{c_0} \left[\int_0^1 \tilde{K}_1(y; \hat{\theta}(t_i), \theta) \alpha(y, t) dy \right.$$

$$\left. + \int_0^1 \tilde{K}_2(y; \hat{\theta}(t_i), \theta) \beta(y, t) dy + \tilde{K}_3(\hat{\theta}(t_i), \theta) \zeta(t) \right] \qquad (15.75)$$

for $t \in [t_i, t_{i+1})$, where the functions $\tilde{K}_1, \tilde{K}_2, \tilde{K}_3$ are shown in appendix 15.6B. Applying the backstepping control design in section 15.2 into (15.1)–(15.4), (15.74), the right boundary condition of the target system (15.10)–(15.14) becomes

$$\beta(1, t) = -c_0 \xi(t) - c_0 d(t). \qquad (15.76)$$

Inserting the inverse backstepping transformations into $U_c(t)$ in (15.18) to replace the original states by the states in the target system, we obtain

$$U_c(t) = \frac{1}{c_0} \int_0^1 K_1(y, \hat{\theta}(t_i), \theta) \alpha(y, t) dy$$

$$+ \frac{1}{c_0} \int_0^1 K_2(y, \hat{\theta}(t_i), \theta) \beta(y, t) dy + \frac{1}{c_0} K_3(\hat{\theta}(t_i), \theta) \zeta(t) \qquad (15.77)$$

for $t \in [t_i, t_{i+1})$, where the functions K_1, K_2, K_3 are shown in appendix 15.6B. The event-triggered control input U_d is constant on $t \in (t_i, t_{i+1})$—that is, $\dot{U}_d(t) = 0$. Taking the time derivative of (15.19) and recalling (15.75), (15.76), (15.77), we obtain

$$\dot{d}(t) = \dot{U}_c(t)$$

$$= \frac{1}{c_0} \left[\int_0^1 M_1(y, \hat{\theta}(t_i), \theta) \alpha(y, t) dy \right.$$

$$- \int_0^1 M_2(y, \hat{\theta}(t_i), \theta) \beta(y, t) dy - M_3(\hat{\theta}(t_i), \theta) \zeta(t)$$

$$\left. + M_4(\hat{\theta}(t_i), \theta) \beta(0, t) - M_5(\hat{\theta}(t_i), \theta) \alpha(1, t) - M_6(\hat{\theta}(t_i), \theta) d(t) \right], \qquad (15.78)$$

where the functions $M_1, \ldots, M_6$ are shown in appendix 15.6B. Applying the Cauchy-Schwarz inequality into (15.78), we obtain (15.72). The proof is complete. $\square$

Lemma 15.2. *For the internal dynamic variable $m(t)$ defined in (15.48), it holds that $m(t) < 0$.*

Proof. According to (15.38), events are triggered to guarantee that $d(t)^2 \leq \vartheta V(t) - m(t)$. Recalling (15.48), we then have

$$\dot{m}(t) \leq -(\eta + \lambda_d) m(t) + (\lambda_d \vartheta - \sigma) V(t) - \kappa_1 \alpha(1, t)^2 - \kappa_2 \beta(0, t)^2$$

$$\leq -(\eta + \lambda_d) m(t), \quad t \geq 0, \qquad (15.79)$$

where $\vartheta \leq \frac{\sigma}{\lambda_d}$ obtained from (15.59) has been used. Together with $m(0) < 0$, we conclude that $m(t) < 0$. $\square$

Relying on lemmas 15.1 and 15.2, we present a lemma that shows that the event trigger is well defined and produces an increasing sequence of events.

Lemma 15.3. *Under the triggering mechanism defined by* (15.37)–(15.41), *the minimal dwell time exists, in the sense of*

$$t_{i+1} - t_i \geq \min\{\underline{\tau}, \underline{r}\} > 0, \quad \forall i \in \mathbb{Z}^+ \tag{15.80}$$

for some positive $\underline{\tau}$.

Proof. At a time instant t_i, the next time instant t_{i+1} is triggered by either r_i in (15.40), (15.41) or τ_i in (15.38). Next we present two claims regarding the lower bound of the minimal dwell time in the two situations in which t_{i+1} is triggered by either r_i or τ_i, respectively.

Claim 15.1. *For an arbitrary time instant t_i, if the next time instant t_{i+1} is triggered by r_i in* (15.40), (15.41), *then $t_{i+1} - t_i \geq \underline{r} > 0$.*

Proof. If $\Omega(t_i) \neq 0$, the interval $t_{i+1} - t_i$ is greater than $\underline{r} > 0$ by virtue of (15.40). If $\Omega(t_i) = 0$, according to (15.41), the interval $t_{i+1} - t_i$ is equal to the positive constant T, which is larger than $\underline{r}$ according to (15.42). $\square$

Claim 15.2. *For an arbitrary time instant t_i, if the next time instant t_{i+1} is triggered by τ_i in* (15.38), *there exists a minimal dwell time $\underline{\tau} > 0$ such that $t_{i+1} - t_i \geq \underline{\tau}$.*

Proof. We know from (15.38) that the events are triggered to guarantee $d(t)^2 \leq \vartheta V(t) - m(t)$ for all $t \geq 0$. We introduce the following function $\psi(t)$, which is similar to (13.71) in chapter 13:

$$\psi(t) = \frac{d(t)^2 + \frac{1}{2}m(t)}{\vartheta V(t) - \frac{1}{2}m(t)}, \tag{15.81}$$

where $\psi(t_{i+1}) = 1$ because the event is triggered, and where $\psi(t_i) < 0$ because of $m(t) < 0$ proved in lemma 15.2 and $d(t_i) = 0$ according to (15.19). The function $\psi(t)$ is a continuous function on $[t_i, t_{i+1}]$ recalling proposition 15.1. By the intermediate value theorem, there exists $t^* > t_i$ such that $\psi(t) \in [0, 1]$ when $t \in [t^*, t_{i+1}]$. The minimal dwell time can be found as the minimal time it takes for $\psi(t)$ from 0 to 1. Defining

$$\Omega_0(t) = \|\alpha(\cdot, t)\|^2 + \|\beta(\cdot, t)\|^2 + \zeta(t)^2 \tag{15.82}$$

and recalling (15.44), we see that the following inequality holds

$$\xi_1 \Omega_0(t) \leq V(t) \leq \xi_2 \Omega_0(t), \tag{15.83}$$

where the positive constant ξ_1 is

$$\xi_1 = \min\left\{\frac{1}{2}, \frac{1}{2}r_a, \frac{1}{2}e^{-\delta_2}\right\}$$

and where the positive constant ξ_2 is defined in (15.58). Taking the time derivative of $V(t)$ in (15.44) and applying (15.10)–(15.13), using integration by parts, we obtain

$$\dot{V}(t) = -a_m\zeta(t)^2 + \zeta(t)b\beta(0,t) + \frac{1}{2}q_2 r_a e^{\delta_1}\beta(1,t)^2$$

$$- \frac{1}{2}\delta_1 q_2 r_a \int_0^1 e^{\delta_1 x}\beta(x,t)^2 dx + r_a d_4 \int_0^1 e^{\delta_1 x}\beta(x,t)^2 dx$$

$$- \frac{1}{2}q_1 e^{-\delta_2}\alpha(1,t)^2 + \frac{1}{2}q_1\alpha(0,t)^2 - \frac{1}{2}q_2 r_a\beta(0,t)^2$$

$$- \frac{1}{2}\delta_2 q_1 \int_0^1 e^{-\delta_2 x}\alpha(x,t)^2 dx + d_1 \int_0^1 e^{-\delta_2 x}\alpha(x,t)^2 dx. \qquad (15.84)$$

We then have

$$\dot{V}(t) \geq -\mu_0 V - \lambda_\alpha \alpha(1,t)^2 - \lambda_\beta \beta(0,t)^2, \qquad (15.85)$$

where the positive constant μ_0 is

$$\mu_0 = \frac{1}{\xi_1} \max\left\{\frac{1}{2}\delta_1 q_2 r_a e^{\delta_1} + r_a|d_4|e^{\delta_1}, \frac{1}{2}\delta_2 q_1 + |d_1|, a_m + \frac{1}{2}|b|\right\} \qquad (15.86)$$

and where $\lambda_\alpha, \lambda_\beta$ are given in (15.51).

Taking the derivative of (15.81) for all $t \in (t^*, t_{i+1})$, applying Young's inequality, using (15.72) in lemma 15.1, inserting (15.48), (15.85), and applying (15.82), (15.83) to rewrite $\|\alpha(\cdot,t)\|^2 + \|\beta(\cdot,t)\|^2 + \zeta(t)^2$, we have

$$\dot{\psi} = \frac{2d(t)\dot{d}(t) + \frac{1}{2}\dot{m}(t)}{\vartheta V(t) - \frac{1}{2}m(t)} - \frac{(\vartheta\dot{V}(t) - \frac{1}{2}\dot{m}(t))}{\vartheta V(t) - \frac{1}{2}m(t)}\psi$$

$$\leq \frac{1}{\vartheta V(t) - \frac{1}{2}m(t)}\left[\left(\lambda_a + 1 + \frac{1}{2}\lambda_d\right)d(t)^2 + \left(\lambda_a - \frac{1}{2}\kappa_1\right)\alpha(1,t)^2\right.$$

$$\left. + \left(\lambda_a - \frac{1}{2}\kappa_2\right)\beta(0,t)^2 + \frac{\lambda_a}{\xi_1}V(t) - \frac{1}{2}\eta m(t)\right]$$

$$- \frac{1}{\vartheta V(t) - \frac{1}{2}m(t)}\left[-\vartheta\mu_0 V(t) - \left(\vartheta\lambda_\alpha - \frac{1}{2}\kappa_1\right)\alpha(1,t)^2\right.$$

$$\left. - \left(\vartheta\lambda_\beta - \frac{1}{2}\kappa_2\right)\beta(0,t)^2 - \frac{1}{2}\lambda_d d(t)^2 + \frac{1}{2}\eta m(t) + \frac{1}{2}\sigma V(t)\right]\psi. \qquad (15.87)$$

Applying

$$\kappa_1 \geq \max\{2\lambda_a, 2\vartheta\lambda_\alpha\}, \quad \kappa_2 \geq \max\{2\lambda_a, 2\vartheta\lambda_\beta\},$$

which are ensured by (15.49), (15.50), (15.59), and applying the following inequalities

$$-\frac{\frac{1}{2}\eta m(t)}{\vartheta V(t) - \frac{1}{2}m(t)} \leq -\frac{\frac{1}{2}\eta m(t)}{-\frac{1}{2}m(t)} = \eta, \qquad (15.88)$$

$$\frac{V(t)}{\vartheta V(t) - \frac{1}{2}m(t)} \leq \frac{V(t)}{\vartheta V(t)} = \frac{1}{\vartheta}, \qquad (15.89)$$

$$\frac{d(t)^2}{\vartheta V(t) - \frac{1}{2}m(t)} = \frac{d(t)^2 + \frac{1}{2}m(t) - \frac{1}{2}m(t)}{\vartheta V(t) - \frac{1}{2}m(t)} \leq \psi(t) + 1, \qquad (15.90)$$

which hold because $m(t) < 0$, then (15.87) becomes

$$\dot{\psi}(t) \leq \frac{1}{2}\lambda_d \psi(t)^2 + (\lambda_a + 1 + \lambda_d + \eta + \mu_0)\psi(t)$$
$$+ 1 + \frac{1}{2}\lambda_d + \lambda_a + \frac{\lambda_a}{\xi_1 \vartheta} + \eta. \tag{15.91}$$

The differential inequality (15.91) has the form

$$\dot{\psi} \leq n_1 \psi^2 + n_2 \psi + n_3,$$

where

$$n_1 = \frac{1}{2}\lambda_d,$$
$$n_2 = \lambda_a + 1 + \lambda_d + \eta + \mu_0,$$
$$n_3 = 1 + \frac{1}{2}\lambda_d + \lambda_a + \frac{\lambda_a}{\xi_1 \vartheta} + \eta$$

are positive constants. It follows that the time needed by ψ to go from 0 to 1 is at least

$$\underline{\tau} = \int_0^1 \frac{1}{n_1 + n_2 s + n_3 s^2} ds > 0. \tag{15.92}$$

The proof of this claim is complete. $\qquad \qquad \square$

According to claims 15.1 and 15.2, the proof of lemma 15.3 is complete. $\qquad \square$

Next, we present the following lemmas regarding parameter convergence. In the rest of this chapter, when we say that $z[t]$, $w[t]$ are equal to zero for $x \in [0, 1], t \in [\mu_{i+1}, t_{i+1}]$, or not identically zero on the same domain, we mean "except possibly for finitely many discontinuities of the functions $w[t], z[t]$." These discontinuities are isolated curves in the rectangle $[0, 1] \times [\mu_{i+1}, t_{i+1}]$.

Lemma 15.4. *The sufficient and necessary conditions of $Q_{n,1}(\mu_{i+1}, t_{i+1}) = 0$, $Q_{n,3}(\mu_{i+1}, t_{i+1}) = 0$ for $n = 1, 2, \ldots$ are $z[t] = 0$, $w[t] = 0$ on $t \in [\mu_{i+1}, t_{i+1}]$, respectively. The sufficient and necessary condition of $Q_4(\mu_{i+1}, t_{i+1}) = 0$ is $\zeta(t) = 0$ on $t \in [\mu_{i+1}, t_{i+1}]$.*

Proof. The proof that the sufficient and necessary conditions of $Q_{n,1}(\mu_{i+1}, t_{i+1}) = 0$, $Q_{n,3}(\mu_{i+1}, t_{i+1}) = 0$ for $n = 1, 2, \ldots$ are $z[t] = 0$, $w[t] = 0$ on $t \in [\mu_{i+1}, t_{i+1}]$, respectively, is the same as the proof of lemma 14.2 in chapter 14, where the fact that $z \in C^0([t_i, t_{i+1}]; L^2(0, 1))$, $w \in C^0([t_i, t_{i+1}]; L^2(0, 1))$, and the set $\{\sqrt{2}\sin(n\pi x) : n = 1, 2, \ldots\}$ is an orthonormal basis of $L^2(0, 1)$ has been used.

By recalling (15.34), (15.36), it is straightforward to see that the sufficient and necessary condition of $Q_4(\mu_{i+1}, t_{i+1}) = 0$ is $\zeta(t) = 0$ on $t \in [\mu_{i+1}, t_{i+1}]$.

The proof of lemma 15.4 is complete. $\qquad \qquad \square$

Lemma 15.5. *For the adaptive estimates defined by (15.67) based on the data in the interval $t \in [\mu_{i+1}, t_{i+1}]$, the following statements hold:*

If $z[t]$ (or $w[t]$ or $\zeta(t)$) is not identically zero for $t \in [\mu_{i+1}, t_{i+1}]$, then $\hat{d}_3(t_{i+1}) = d_3$ (or $\hat{d}_2(t_{i+1}) = d_2$ or $\hat{a}(t_{i+1}) = a$, respectively).

If $z[t]$ (or $w[t]$ or $\zeta(t)$) is identically zero for $t \in [\mu_{i+1}, t_{i+1}]$, then $\hat{d}_3(t_{i+1}) = \hat{d}_3(t_i)$ (or $\hat{d}_2(t_{i+1}) = \hat{d}_2(t_i)$ or $\hat{a}(t_{i+1}) = \hat{a}(t_i)$, respectively).

Proof. We prove the following five results, from which the statements in this lemma are obtained.

1) Result 1: If $z[t]$ is not identically zero and $w[t]$ is identically zero on $t \in [\mu_{i+1}, t_{i+1}]$, then $\hat{d}_3(t_{i+1}) = d_3$, $\hat{d}_2(t_{i+1}) = \hat{d}_2(t_i)$; if $w[t]$ is not identically zero and $z[t]$ is identically zero on $t \in [\mu_{i+1}, t_{i+1}]$, then $\hat{d}_3(t_{i+1}) = \hat{d}_3(t_i)$, $\hat{d}_2(t_{i+1}) = d_2$. The proof of result 1 is very similar to the proof of cases 1 and 2 in lemma 14.3 in chapter 14, and thus they are omitted.

2) Result 2: If $w[t]$, $z[t]$ are identically zero on $t \in [\mu_{i+1}, t_{i+1}]$, then $\hat{d}_3(t_{i+1}) = \hat{d}_3(t_i)$, $\hat{d}_2(t_{i+1}) = \hat{d}_2(t_i)$. The proof of result 2 is shown as follows. In this case, $Q_{n,1}(\mu_{i+1}, t_{i+1}) = 0$, $Q_{n,2}(\mu_{i+1}, t_{i+1}) = 0$, $Q_{n,3}(\mu_{i+1}, t_{i+1}) = 0$, $H_{n,1}(\mu_{i+1}, t_{i+1}) = 0$, $H_{n,2}(\mu_{i+1}, t_{i+1}) = 0$ for $n = 1, 2, \ldots$ according to (15.30), (15.31), (15.24)–(15.28). It follows that $S_i = \Theta$, and then (15.67) shows that $\hat{d}_3(t_{i+1}) = \hat{d}_3(t_i)$, $\hat{d}_2(t_{i+1}) = \hat{d}_2(t_i)$.

3) Result 3: If both $w[t]$ and $z[t]$ are not identically zero on $t \in [\mu_{i+1}, t_{i+1}]$, then $\hat{d}_3(t_{i+1}) = d_3$, $\hat{d}_2(t_{i+1}) = d_2$. The proof of result 3 is shown as follows. According to lemma 15.4, there exists $n \in \mathbb{N}$ such that $Q_{n,3}(\mu_{i+1}, t_{i+1}) \neq 0$ (or $Q_{n,1}(\mu_{i+1}, t_{i+1}) \neq 0$). Define the index set I_1 to be the set of all $n \in \mathbb{N}$ with $Q_{n,1}(\mu_{i+1}, t_{i+1}) \neq 0$, and define the index set I_2 to be the set of all $n \in \mathbb{N}$ with $Q_{n,3}(\mu_{i+1}, t_{i+1}) \neq 0$. Denote the elements in I_1 as $n_1 \in \mathbb{N}$ and those in I_2 as $n_2 \in \mathbb{N}$—that is, $Q_{n_1,1}(\mu_{i+1}, t_{i+1}) \neq 0$, $Q_{n_2,3}(\mu_{i+1}, t_{i+1}) \neq 0$.

For the set S_i defined in (15.21), by virtue of (15.65)–(15.67), if S_i is a singleton then it is nothing but the least-squares estimate of the unknown vector of parameters (d_3, d_2) on the interval $[\mu_{i+1}, t_{i+1}]$, and $S_i = \{(d_3, d_2)\}$ according to (15.65), (15.66). From (15.21)–(15.23), we have

$$S_i \subseteq S_{ai} := \left\{ (\ell_1, \ell_2) \in \Theta_1 : \ell_1 = \frac{H_{n_1,1}(\mu_{i+1}, t_{i+1})}{Q_{n_1,1}(\mu_{i+1}, t_{i+1})} \right.$$
$$\left. - \ell_2 \frac{Q_{n_1,2}(\mu_{i+1}, t_{i+1})}{Q_{n_1,1}(\mu_{i+1}, t_{i+1})}, n_1 \in I_1 \right\}, \tag{15.93}$$

$$S_i \subseteq S_{bi} := \left\{ (\ell_1, \ell_2) \in \Theta_1 : \ell_2 = \frac{H_{n_2,2}(\mu_{i+1}, t_{i+1})}{Q_{n_2,3}(\mu_{i+1}, t_{i+1})} \right.$$
$$\left. - \ell_1 \frac{Q_{n_2,2}(\mu_{i+1}, t_{i+1})}{Q_{n_2,3}(\mu_{i+1}, t_{i+1})}, n_2 \in I_2 \right\}. \tag{15.94}$$

We next prove by contradiction that $S_i = \{(d_3, d_2)\}$. Suppose that on the contrary $S_i \neq \{(d_3, d_2)\}$—that is, S_i defined by (15.21) is not a singleton, which implies that the sets S_{ai}, S_{bi} defined by (15.93), (15.94) are not singletons (because either S_{ai} or S_{bi} being a singleton implies that S_i is a singleton). It follows that there exist the constants $\bar{\lambda} \in \mathbb{R}$, $\bar{\lambda}_1 \in \mathbb{R}$ such that

$$\frac{Q_{n_1,2}(\mu_{i+1}, t_{i+1})}{Q_{n_1,1}(\mu_{i+1}, t_{i+1})} = \bar{\lambda}_1, \ n_1 \in I_1, \quad \frac{Q_{n_2,2}(\mu_{i+1}, t_{i+1})}{Q_{n_2,3}(\mu_{i+1}, t_{i+1})} = \bar{\lambda}, \ n_2 \in I_2 \tag{15.95}$$

because if there were two different indices $k_1, k_2 \in I_2$ with

$$\frac{Q_{k_1,2}(\mu_{i+1}, t_{i+1})}{Q_{k_1,3}(\mu_{i+1}, t_{i+1})} \neq \frac{Q_{k_2,2}(\mu_{i+1}, t_{i+1})}{Q_{k_2,3}(\mu_{i+1}, t_{i+1})},$$

then the set S_{bi} defined by (15.94) would be a singleton, and the same would be the case with S_{ai} defined by (15.93) if there were two different indices $\bar{k}_1, \bar{k}_2 \in I_1$ with

$$\frac{Q_{\bar{k}_1,2}(\mu_{i+1}, t_{i+1})}{Q_{\bar{k}_1,1}(\mu_{i+1}, t_{i+1})} \neq \frac{Q_{\bar{k}_2,2}(\mu_{i+1}, t_{i+1})}{Q_{\bar{k}_2,1}(\mu_{i+1}, t_{i+1})}.$$

Moreover, since S_i is not a singleton, the definition (15.21) implies

$$Q_{n,2}(\mu_{i+1}, t_{i+1})^2 = Q_{n,1}(\mu_{i+1}, t_{i+1}) Q_{n,3}(\mu_{i+1}, t_{i+1}) \tag{15.96}$$

for all $n \in I_1 \cup I_2$ by recalling (15.23) (because if (15.96) does not hold, it follows from (15.23) that there exists $n \in I_1 \cup I_2$ such that $\det(\bar{G}_n(\mu_{i+1}, t_{i+1})) \neq 0$, which implies S_i defined by (15.21) is a singleton: a contradiction). According to (15.96), (15.26)–(15.28) and the fact that the Cauchy-Schwarz inequality holds as an equality only when two functions are linearly dependent, we obtain the existence of the constants $\hat{\lambda}_{n_1} \in \mathbb{R}$, $\check{\lambda}_{n_2} \in \mathbb{R}$ such that

$$g_{n_1,2}(t, \mu_{i+1}) = \hat{\lambda}_{n_1} g_{n_1,1}(t, \mu_{i+1}), \quad n_1 \in I_1, \tag{15.97}$$

$$g_{n_2,1}(t, \mu_{i+1}) = \check{\lambda}_{n_2} g_{n_2,2}(t, \mu_{i+1}), \quad n_2 \in I_2 \tag{15.98}$$

for $t \in [\mu_{i+1}, t_{i+1}]$ (notice that $g_{n_1,1}(t, \mu_{i+1})$ and $g_{n_2,2}(t, \mu_{i+1})$ are not identically zero on $t \in [\mu_{i+1}, t_{i+1}]$ because of $Q_{n_1,1}(\mu_{i+1}, t_{i+1}) \neq 0$ and $Q_{n_2,3}(\mu_{i+1}, t_{i+1}) \neq 0$). Recalling (15.95), we obtain from (15.26)–(15.28) and (15.97), (15.98) that

$$g_{n_1,2}(t, \mu_{i+1}) = \bar{\lambda}_1 g_{n_1,1}(t, \mu_{i+1}), \quad n_1 \in I_1, \tag{15.99}$$

$$g_{n_2,1}(t, \mu_{i+1}) = \bar{\lambda} g_{n_2,2}(t, \mu_{i+1}), \quad n_2 \in I_2 \tag{15.100}$$

for $t \in [\mu_{i+1}, t_{i+1}]$.

Claim 15.3. *If S_i is not a singleton, then $\bar{\lambda} \neq 0, \bar{\lambda}_1 \neq 0$ and $\bar{\lambda} = \frac{1}{\bar{\lambda}_1}$ in (15.99), (15.100).*

Proof. The proof is very similar to that of claim 14.1 in chapter 14, and thus it is omitted. $\square$

Claim 15.4. *Equations (15.99), (15.100) ($\bar{\lambda} \neq 0, \bar{\lambda}_1 \neq 0$ and $\bar{\lambda} = \frac{1}{\bar{\lambda}_1}$) hold if and only if $z(x, t) - \bar{\lambda} w(x, t) = 0$ ($\bar{\lambda} \neq 0$) for $t \in [\mu_{i+1}, t_{i+1}]$, $x \in [0, 1]$.*

Proof. The proof is very similar to the proof of claim 14.2 in chapter 14, and thus it is omitted. $\square$

Claim 15.5. *If $w(x, t)$ is not identically zero on $x \in [0, 1]$, $t \in [\mu_{i+1}, t_{i+1}]$, the function $z(x, t) - \bar{\lambda} w(x, t)$ ($\bar{\lambda} \neq 0$) is not identically zero on $x \in [0, 1]$, $t \in [\mu_{i+1}, t_{i+1}]$.*

Proof. The proof is shown in appendix 15.6C. $\square$

Recalling claims 15.3, 15.4, and 15.5, we know that the equation set (15.99), (15.100) ($\bar{\lambda} \neq 0, \bar{\lambda}_1 \neq 0$ and $\bar{\lambda} = \frac{1}{\bar{\lambda}_1}$), which is a necessary condition of the hypothesis that S_i is not a singleton, does not hold. Consequently, S_i is a singleton—that is, $S_i = \{(d_3, d_2)\}$. Therefore, $\hat{d}_3(t_{i+1}) = d_3, \hat{d}_2(t_{i+1}) = d_2$.

4) Result 4: If $\zeta(t)$ is not identically zero for $t \in [\mu_{i+1}, t_{i+1}]$, then $\hat{a}(t_{i+1}) = a$. The proof of result 4 is shown as follows. By virtue of (15.65)–(15.67), if $\bar{S}_i$ defined in

(15.32) is a singleton, then $\hat{a}(t_{i+1}) = a$. According to lemma 15.4, $Q_4(\mu_{i+1}, t_{i+1}) \neq 0$. It follows that $\bar{S}_i$ is a singleton. Therefore, we obtain $\hat{a}(t_{i+1}) = a$.

5) Result 5: If $\zeta(t)$ is identically zero for $t \in [\mu_{i+1}, t_{i+1}]$, then $\hat{a}(t_{i+1}) = \hat{a}(t_i)$. The proof of result 5 is shown as follows. According to (15.33), (15.34), (15.36), $Q_4(\mu_{i+1}, t_{i+1}) = 0$, $H_3(\mu_{i+1}, t_{i+1}) = 0$. We then find that the set $\bar{S}_i = \{|\ell_3| \leq \bar{a}\}$. Recalling (15.67), we have $\hat{a}(t_{i+1}) = \hat{a}(t_i)$.

From the above five results, we obtain lemma 15.5. □

Lemma 15.6. *If $\hat{d}_3(t_i) = d_3$ (or $\hat{d}_2(t_i) = d_2$ or $\hat{a}(t_i) = a$) for certain $i \in \mathbb{Z}^+$, then $\hat{d}_3(t) = d_3$ (or $\hat{d}_2(t) = d_2$ or $\hat{a}(t) = a$, respectively) for all $t \in [t_i, \lim_{k \to \infty}(t_k))$.*

Proof. According to lemma 15.5, $\hat{d}_3(t_{i+1})$ is equal to either d_3 or $\hat{d}_3(t_i)$. Therefore, if $\hat{d}_3(t_i) = d_3$, then $\hat{d}_3(t) = d_3$ for all $t \in [t_i, \lim_{k \to \infty}(t_k))$. The same is true of $\hat{d}_2$ and $\hat{a}$. The proof is complete. □

We are now ready to provide the proof of theorem 15.1.

Proof. 1) We now prove the first of the three portions of the theorem. Recalling lemma 15.3, we know that

$$t_i \geq \min\{\underline{r}, \underline{\tau}\}(i-1), \quad i \geq 1, \tag{15.101}$$

which yields $\lim_{i \to \infty}(t_i) = +\infty$. Then the well-posedness of the closed-loop system is obtained by recalling proposition 15.1. The property (1) is thus obtained. The fact that $\lim_{i \to \infty}(t_i) = +\infty$ allows the solution to be defined on $\mathbb{R}_+$ in the following analysis.

2) We now prove the second of the three portions of the theorem. First, we propose the following claim about the sufficient and necessary condition of the finite-time convergence of parameter estimates to the true values.

Claim 15.6. *When $\hat{d}_3(0) \neq d_3$ (or $\hat{d}_2(0) \neq d_2$ or $\hat{a}(0) \neq a$), the estimate $\hat{d}_3(t)$ (or $\hat{d}_2(t)$ or $\hat{a}(t)$) reaches the actual value d_3 (or d_2 or a) in finite time if and only if $z[t]$ (or $w[t]$ or $\zeta(t)$, respectively) is not identically zero on $t \in [0, \infty)$.*

Proof. The proof is shown in appendix 15.6D. □

Claim 15.7. *If any one of the three estimates $\hat{d}_3(t), \hat{d}_2(t), \hat{a}(t)$ does not reach the true value in finite time, it implies that $\Omega(t) \equiv 0$ on $t \in [\frac{1}{q_1}, \infty)$ and that $m(t)$ is exponentially convergent to zero.*

Proof. The proof is shown in appendix 15.6E. □

Next, we estimate the maximum convergence time of the parameter estimates when they reach the true values.

Claim 15.8. *If the parameter estimate $\hat{\theta}(t)$ reaches the true value θ in finite time—that is, t_ε—then $t_\varepsilon \leq \frac{1}{q_1} + \frac{1}{q_2} + T$.*

Proof. The proof is shown in appendix 15.6F. □

According to claims 15.7 and 15.8, the proof of property (2) is complete.

3) We now prove the last of the three portions of the theorem. Define a Lyapunov function as

$$V_a(t) = V(t) - \mu m(t), \tag{15.102}$$

where $m(t)$ is defined in (15.48) ($m(t) < 0$ is shown in lemma 15.2), the positive constant μ is defined in (15.56), and $V(t)$ is given in (15.44). Recalling (15.82), (15.83) and applying the Cauchy-Schwarz inequality into the backstepping transformation (15.6), (15.7) and its inverse (15.8), (15.9), we have

$$\xi_3 \bar{\Omega}(t) \leq V_a(t) \leq \xi_4 \bar{\Omega}(t) \tag{15.103}$$

for some positive ξ_3, ξ_4, where the definition of $\bar{\Omega}(t)$ is given in (15.71).

Taking the derivative of (15.102) along (15.10)–(15.13), recalling the equalities (15.48), (15.84), inserting (15.76), and applying Young's inequality and the Cauchy-Schwarz inequality, we have

$$
\begin{aligned}
\dot{V}_a(t) \leq & -\frac{1}{2} a_{\mathrm{m}} \zeta(t)^2 - \left(\frac{1}{2} q_1 e^{-\delta_2} - \mu \kappa_1 \right) \alpha(1,t)^2 \\
& - \left(\frac{1}{2} q_2 r_a - \frac{b^2}{2 a_{\mathrm{m}}} - \frac{p^2}{2} q_1 - \mu \kappa_2 \right) \beta(0,t)^2 \\
& + \left(q_2 r_a e^{\delta_1} c_0^2 - \mu \lambda_d \right) d(t)^2 + q_2 r_a e^{\delta_1} c_0^2 \xi(t)^2 \\
& - \left(\frac{1}{2} \delta_1 q_2 r_a - r_a |d_4| \right) \int_0^1 e^{\delta_1 x} \beta(x,t)^2 dx + \mu \eta m(t) \\
& - \left(\frac{1}{2} \delta_2 q_1 - |d_1| \right) \int_0^1 e^{-\delta_2 x} \alpha(x,t)^2 dx + \mu \sigma V(t)
\end{aligned}
\tag{15.104}
$$

for $t \geq 0$. Because of (15.68), according to (15.20), we have

$$\xi(t) \equiv 0, \quad t \in [t_\varepsilon, \infty). \tag{15.105}$$

Recalling the conditions of δ_1, δ_2, r_a, λ_d, μ in (15.45)–(15.47), (15.53), (15.56), we arrive at

$$
\begin{aligned}
\dot{V}_a(t) \leq & -(\nu_a - \mu \sigma) V + \mu \eta m(t) \\
= & -(\nu_a - \mu \sigma) V_a + [-\mu(\nu_a - \mu \sigma) + \mu \eta] m(t)
\end{aligned}
\tag{15.106}
$$

for $t \geq t_\varepsilon$, where (15.82), (15.83), (15.102), (15.105) and the fact that $a_{\mathrm{m}} \geq b\kappa - \bar{a} > 0$ have been used and where the positive constant ν_a is given in (15.57). Recalling σ, η in (15.54), (15.55), we arrive at

$$\dot{V}_a(t) \leq -\lambda_1 V_a(t), \quad t \geq t_\varepsilon, \tag{15.107}$$

where $\lambda_1 = \nu_a - \mu \sigma > 0$. Multiplying both sides of (15.107) by $e^{\lambda_1 t}$ and integrating both sides of (15.107) from t_ε to t, we obtain

$$V_a(t) \leq V_a(t_\varepsilon) e^{-\lambda_1(t - t_\varepsilon)}, \quad t \geq t_\varepsilon.$$

Recalling (15.103), we have

$$\bar{\Omega}(t) \leq \Upsilon_\theta \bar{\Omega}(t_\varepsilon) e^{-\lambda_1(t - t_\varepsilon)}, \quad t \geq t_\varepsilon, \tag{15.108}$$

where the positive constant Υ_θ is

$$\Upsilon_\theta = \frac{\xi_4}{\xi_3}. \tag{15.109}$$

If $t_\varepsilon = 0$, property (3)—that is, (15.70)—is obtained directly. Next we conduct an analysis for $t \in [0, t_\varepsilon]$ with $t_\varepsilon \neq 0$. Recalling (15.45)–(15.47), (15.53), (15.54), (15.55), (15.56), (15.82), (15.83), (15.102), (15.104), we obtain

$$\dot{V}_a(t) \leq -\lambda_1 V_a(t) + Q(\hat{\theta}(0))V_a(t), \quad t \in [0, t_\varepsilon], \tag{15.110}$$

where the positive constant $Q(\hat{\theta}(0))$ is obtained by bounding $\xi(t)^2$ given by (15.75), and the value of $Q(\hat{\theta}(0))$ depends on the initial estimate $\hat{\theta}(0)$. We thus obtain

$$\bar{\Omega}(t) \leq \Upsilon_\theta \bar{\Omega}(0)e^{\lambda_2(\hat{\theta}(0))t}, \quad t \in [0, t_\varepsilon], \tag{15.111}$$

where $\lambda_2(\hat{\theta}(0)) = |Q(\hat{\theta}(0)) - \lambda_1| > 0$, and the positive constant Υ_θ is given in (15.109). Therefore, we have

$$\bar{\Omega}(t_\varepsilon) \leq \Upsilon_\theta e^{\lambda_2(\hat{\theta}(0))t_\varepsilon}\bar{\Omega}(0). \tag{15.112}$$

Inserting (15.112) into (15.108) and applying claim 15.8 yields

$$\bar{\Omega}(t) \leq \Upsilon_\theta^2 e^{(\lambda_2(\hat{\theta}(0))+\lambda_1)(\frac{1}{q_1}+\frac{1}{q_2}+T)}\bar{\Omega}(0)e^{-\lambda_1 t}, \quad t \geq 0. \tag{15.113}$$

Denoting

$$M = \Upsilon_\theta^2 e^{(\lambda_2(\hat{\theta}(0))+\lambda_1)(\frac{1}{q_1}+\frac{1}{q_2}+T)},$$

we obtain (15.70). This completes the proof of property (3) of the theorem. $\square$

15.5   APPLICATION IN THE MINING CABLE ELEVATOR

In this section, the proposed boundary controller, where both the parameter estimates and the control input employ piecewise-constant values, is applied to the axial vibration control of a mining cable elevator that is 2000 m deep, where the damping coefficients of the cable and the cage are unknown.

Model

The axial vibration model of the mining cable elevator is described by the following wave PDE-ODE model with in-domain damping [185]:

$$\rho u_{tt}(x, t) = \frac{\pi R_d^2}{4} E u_{xx}(x, t) - d_c u_t(x, t), \tag{15.114}$$

$$M_c u_{tt}(0, t) = -c_L u_t(0, t) - \frac{\pi R_d^2}{4} E u_x(0, t), \tag{15.115}$$

$$\frac{\pi R_d^2}{4} E u_x(L, t) = U_d(t), \tag{15.116}$$

where the cable length is considered constant. The PDE state $u(x, t)$ denotes the distributed axial vibration displacements along the cable, and the boundary state

Table 15.1. Physical parameters of the mining cable elevator

Parameters (units)	Values
Cable length L (m)	2000
Cable diameter R_d (m)	0.2
Cable effective Young's modulus E (N/m^2)	1.02×10^9
Cable linear density ρ (kg/m)	8.1
Mass of cage M_c (kg)	15000
Damping coefficient of cage c_L	0.4
Cable material damping coefficient d_c	0.5
Gravitational acceleration g (m/s^2)	9.8

$u(0, t)$ represents the axial vibration displacement of the cage. The physical parameters in (15.114)–(15.116) are shown in table 15.1.

According to [149], we apply the Riemann transformations

$$z(x, t) = u_t(x, t) - \sqrt{\frac{E\pi}{\rho}} \frac{R_d}{2} u_x(x, t), \tag{15.117}$$

$$w(x, t) = u_t(x, t) + \sqrt{\frac{E\pi}{\rho}} \frac{R_d}{2} u_x(x, t) \tag{15.118}$$

and define the new variable $\zeta(t) = u_t(0, t)$, which allows us to rewrite (15.114)–(15.116) as (15.1)–(15.5) with the coefficients

$$q_1 = q_2 = \sqrt{\frac{E\pi}{\rho}} \frac{R_d}{2}, \quad d_1 = d_2 = d_3 = d_4 = \frac{-d_c}{2\rho}, \tag{15.119}$$

$$p = 1, \quad c = 2, \quad c_0 = \frac{4}{R_d \sqrt{E\pi\rho}}, \tag{15.120}$$

$$a = \frac{-c_L}{M_c} + \frac{R_d \sqrt{E\pi\rho}}{2M_c}, b = -\frac{R_d \sqrt{E\pi\rho}}{2M_c}, \tag{15.121}$$

where a reflection term $z(L, t)$ appearing at the controlled boundary (15.5) is considered to be compensated by the control input at the drum [178] independent of the head sheave control input designed in this chapter. The unknown physical parameters are damping coefficients of the cable and the cage, d_c, c_L, which leads to the fact that a is unknown, and $d_1 = d_2 = d_3 = d_4$ are unknown in the 2×2 hyperbolic PDE-ODE system. The designs in this chapter are directly applicable to this problem (with only one slight modification: removing the last term in (15.29) and multiplying (15.30), (15.31) by 2), and the outputs of the identifier are the estimates of d_2, d_3. According to the values in table 15.1, together with (15.119)–(15.121), we know that assumptions 15.1, 15.2 are satisfied. The bounds $\bar{a}, \bar{d}_3, \bar{d}_2$ of the unknown parameters $a = 1.07$, $d_3 = d_2 = -0.025$ are set as, $3, 0.4, 0.5$, respectively. The initial conditions of $z(x, t)$ and $w(x, t)$ are defined as

$$z(x, 0) = 0.5 \sin(2\pi x/L + \pi/6), \quad w(x, 0) = 0.5 \sin(3\pi x/L)$$

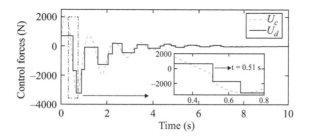

Figure 15.2. The continuous-in-state control signal $U_c(t)$ in (15.18) and the piecewise-constant control signal $U_d(t)$ in (15.17).

and

$$\zeta(0) = \frac{1}{2}(pw(0,0) + z(0,0))$$

according to (15.4). We pick the initial value of $m(t)$ as $m(0) = -10$.

Determination of Design Parameters

The free design parameters δ, T in (15.40) are selected as $\delta = 1.8$, $T = 3$, and the positive integer n in the parameter estimator (15.67) is defined as 7. According to (15.119)–(15.121), the parameter values in table 15.1, and the known bounds $\bar{a}, \bar{d}_3, \bar{d}_2$ defined above, recalling (15.15), (15.45)–(15.47), the design parameters $\kappa, \delta_1, \delta_2, r_a$ are determined to be $\kappa = -10, \delta_1 = 0.1, \delta_2 = 0.1, r_a = 1.5$. Choose the design parameter $\underline{r}$ as $\underline{r} = 0.2$ via (15.42). Next, choose $\kappa_1 = 2100, \kappa_2 = 3000$ according to (15.49), (15.50), where $\lambda_\alpha = 899.6$, $\lambda_\beta = 1490.8$ are obtained from (15.51), and a conservative estimate $\lambda_a = 1000$ comes from (15.52). Then the design parameters λ_d, σ, η are selected as $\lambda_d = 0.005, \sigma = 60, \eta = 120$ to satisfy (15.53)–(15.55), where $\mu = 0.15, \nu_a = 9.66$ are obtained from (15.56), (15.57). Finally, pick the design parameter $\vartheta = 12000$ via (15.59).

Simulation Results

We simulate a mining cable elevator running for a short time period of 10 s where the cable length is regarded as constant. The simulation is conducted based on the finite-difference method with a time step of 0.0015 s and a space step of 0.5 m. By employing the finite-difference method as well with a step length of 0.5 m for y running from 0 to x, the approximate solutions of $\Psi(x, y; \hat{\theta}(t_i))$, $\Phi(x, y; \hat{\theta}(t_i))$, $\lambda(x; \hat{\theta}(t_i))$ in the control law (15.17) are obtained from the conditions (15.128)–(15.133), whose unknown coefficients are replaced by the piecewise-constant estimates. The triggering mechanism (15.37)–(15.41) is implemented as the flowchart in figure 15.1, with the selected design parameters.

The piecewise-constant control input $U_d(t)$ defined in (15.17) is shown in figure 15.2, where the estimate $\hat{\theta}$ is recomputed, and the states z, w, ζ are resampled simultaneously. The first update is triggered at $t = 0.51$ s, the total number of triggering times is 33, the maximum dwell time is 0.595 s, and the minimal dwell time is 0.105 s, which is much larger than the highly conservative minimal dwell time estimate of 5.75×10^{-4} s obtained from (15.80), (15.92) in lemma 15.3, where the analysis parameter μ_0 defined in (15.86) is $\mu_0 = 331$. The continuous-in-state control signal

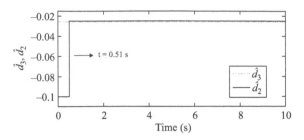

Figure 15.3. Parameter estimates $\hat{d}_3$ and $\hat{d}_2$.

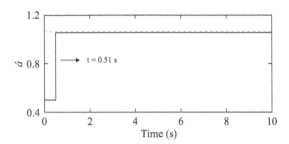

Figure 15.4. Parameter estimate $\hat{a}$.

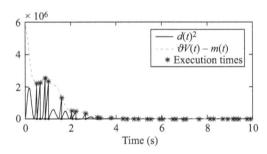

Figure 15.5. The evolution of $d(t)^2$ and $\vartheta V(t) - m(t)$ in (15.38).

$U_c(t)$ (15.18) used in the ETM is also shown in figure 15.2. There is a jump in the continuous-in-state signal $U_c(t)$ at $t = 0.51$ s because $U_c(t)$ defined in (15.18) includes the piecewise-constant parameter estimate $\hat{\theta}$ whose evolution is shown in figures 15.3 and 15.4, where, under the nonzero initial conditions defined at the beginning of this section, the estimates $\hat{d}_3, \hat{d}_2, \hat{a}$ are updated and reach the true values of the unknown parameters $d_3 = -0.025, d_2 = -0.025, a = 1.06$ (figures 15.3 and 15.4, *dashed line*) at the first triggering time, $t = 0.51$ s, and are kept constant in the subsequent recomputations.

Because the estimates have reached the true values at the first triggering time, according to the designed triggering mechanism (15.37)–(15.41), the following execution times are determined by the triggering condition (15.38), which can be seen in figure 15.5 showing the time evolution of the functions in the triggering condition (15.38) and the execution times. Figure 15.5 shows that an event is generated, the control value is updated, and $d(t)$ is reset to zero when the trajectory $d(t)^2$ reaches the trajectory $\vartheta V(t) - m(t)$. Figure 15.6 demonstrates that the ODE state

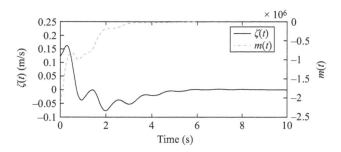

Figure 15.6. The evolution of $\zeta(t)$, $m(t)$ under the control input $U_d(t)$ in (15.17).

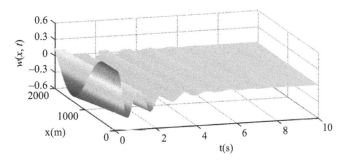

Figure 15.7. The evolution of $w(x,t)$ under the control input $U_d(t)$ in (15.17).

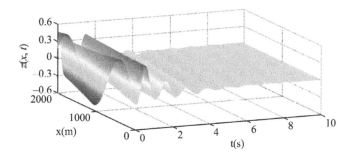

Figure 15.8. The evolution of $z(x,t)$ under the control input $U_d(t)$ in (15.17).

$\zeta(t)$, whose physical meaning is the axial vibration velocity of the cage, is convergent to zero, and the internal variable $m(t)$ in (15.38) is less than zero all the time and convergent to zero as well. Figures 15.7 and 15.8 show that the PDE states $w(x,t), z(x,t)$ are regulated to zero under the control input $U_d(t)$. Therefore, according to the Riemann transformation (15.117), (15.118), we find that the vibration energy of the cable $\frac{1}{2}\rho\|u_t(\cdot,t)\|^2 + \frac{R_d^2\pi}{8}E\|u_x(\cdot,t)\|^2$ also decreases to zero. Finally, we run simulations for eight different initial conditions and compute the inter-execution times between two triggering times. The density of the inter-execution times is shown in figure 15.9, from which we know that the prominent inter-execution times are in the range from 0.1 s to 0.2 s.

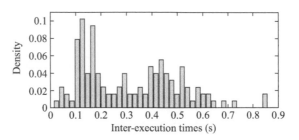

Figure 15.9. Density of the inter-execution times computed for eight different initial conditions given by $z(x,0)=0.5\sin(2\bar{n}\pi x/L+\pi/6)+\bar{k}(x/L)^2$, $w(x,0)=0.5\sin(3\bar{m}\pi x/L)+\bar{k}(x/L)^2$ for $\bar{n}=1,2$, $\bar{m}=1,2$, $\bar{k}=0,1$.

15.6 APPENDIX

A. Conditions of kernels in the backstepping transformation (15.6), (15.7) and its inverse (15.8), (15.9)

The conditions of kernels $\varphi,\phi,\gamma,\Psi,\Phi,\lambda$ in the backstepping transformation (15.6), (15.7) are

$$d_2-(q_1+q_2)\varphi(x,x)=0, \tag{15.122}$$

$$-\gamma(x)b+q_2\varphi(x,0)+q_1p\phi(x,0)=0, \tag{15.123}$$

$$-q_1\phi_x(x,y)-q_1\phi_y(x,y)-d_3\varphi(x,y)=0, \tag{15.124}$$

$$q_2\varphi_y(x,y)-q_1\varphi_x(x,y)-(d_4-d_1)\varphi(x,y)-d_2\phi(x,y)=0, \tag{15.125}$$

$$q_1\gamma'(x)+(a-d_1)\gamma(x)+q_1c\phi(x,0)=0, \tag{15.126}$$

$$\gamma(0)=-p\kappa+c, \tag{15.127}$$

$$d_3+(q_1+q_2)\Psi(x,x)=0, \tag{15.128}$$

$$-\lambda(x)b+q_2\Phi(x,0)+q_1p\Psi(x,0)=0, \tag{15.129}$$

$$-q_1\Psi_y(x,y)+q_2\Psi_x(x,y)+(d_4-d_1)\Psi(x,y)-d_3\Phi(x,y)=0, \tag{15.130}$$

$$q_2\Phi_y(x,y)+q_2\Phi_x(x,y)-d_2\Psi(x,y)=0, \tag{15.131}$$

$$-q_2\lambda'(x)+(a-d_4)\lambda(x)+q_1c\Psi(x,0)=0, \tag{15.132}$$

$$\lambda(0)=-\kappa. \tag{15.133}$$

The conditions of kernels $\bar{\varphi},\bar{\phi},\bar{\gamma},\bar{\Psi},\bar{\Phi},\bar{\lambda}$ in the inverse backstepping transformation (15.8), (15.9) are

$$-d_3+(q_1+q_2)\bar{\Psi}(x,x)=0, \tag{15.134}$$

$$\bar{\lambda}(x)b-q_2\bar{\Phi}(x,0)-q_1p\bar{\Psi}(x,0)=0, \tag{15.135}$$

$$-q_1\bar{\Psi}_y(x,y)+q_2\bar{\Psi}_x(x,y)+(d_4-d_1)\bar{\Psi}(x,y)+d_3\bar{\phi}(x,y)=0, \tag{15.136}$$

$$q_2\bar{\Phi}_y(x,y)+q_2\bar{\Phi}_x(x,y)+d_3\bar{\varphi}(x,y)=0, \tag{15.137}$$

$$-q_2\bar{\lambda}'(x)-(b\kappa-a+d_4)\bar{\lambda}(x)-d_3\bar{\gamma}(x)=0, \tag{15.138}$$

$$\bar{\lambda}(0)=\kappa, \tag{15.139}$$

$$-d_2-(q_1+q_2)\bar{\varphi}(x,x)=0, \tag{15.140}$$

$$\bar{\gamma}(x)b-q_2\bar{\varphi}(x,0)-q_1p\bar{\phi}(x,0)=0, \tag{15.141}$$

$$q_2\bar{\varphi}_y(x,y)-q_1\bar{\varphi}_x(x,y)-(d_4-d_1)\bar{\varphi}(x,y)+d_2\bar{\Phi}(x,y)=0, \tag{15.142}$$

$$-q_1\bar{\phi}_x(x,y) - q_1\bar{\phi}_y(x,y) + d_2\bar{\Psi}(x,y) = 0, \tag{15.143}$$

$$q_1\bar{\gamma}'(x) - (b\kappa - a + d_1)\bar{\gamma}(x) - d_2\bar{\lambda}(x) = 0, \tag{15.144}$$

$$\bar{\gamma}(0) = -p\kappa - c. \tag{15.145}$$

The equation sets (15.122)–(15.133) and (15.134)–(15.145) are well-known coupled linear heterodirectional hyperbolic PDE-ODE systems whose well-posedness has been proved in theorem 4.1 of [48].

B. Expressions of the functions $M_1, \ldots, M_6$, K_1, K_2, K_3, and $\tilde{K}_1, \tilde{K}_2, \tilde{K}_3$

The expressions of the functions $M_1, \ldots, M_6$ are

$$\begin{aligned} M_1(y, \hat{\theta}(t_i), \theta) = {} & q_1 K_{1y}(y, \hat{\theta}(t_i), \theta) + K_1(y, \hat{\theta}(t_i), \theta)d_1 \\ & - q_2 K_2(1, \hat{\theta}(t_i), \theta)\tilde{K}_1(y, \hat{\theta}(t_i), \theta), \end{aligned} \tag{15.146}$$

$$\begin{aligned} M_2(y, \hat{\theta}(t_i), \theta) = {} & q_2 K_{2y}(y, \hat{\theta}(t_i), \theta) - K_2(y, \hat{\theta}(t_i), \theta)d_4 \\ & + q_2 K_2(1, \hat{\theta}(t_i), \theta)\tilde{K}_2(y, \hat{\theta}(t_i), \theta), \end{aligned} \tag{15.147}$$

$$\begin{aligned} M_3(\hat{\theta}(t_i), \theta) = {} & K_3(\hat{\theta}(t_i), \theta)a_m \\ & + q_2 K_2(1, \hat{\theta}(t_i), \theta)\tilde{K}_3(\hat{\theta}(t_i), \theta), \end{aligned} \tag{15.148}$$

$$\begin{aligned} M_4(\hat{\theta}(t_i), \theta) = {} & K_3(\hat{\theta}(t_i), \theta)b - q_2 K_2(0, \hat{\theta}(t_i), \theta) \\ & - pq_1 K_1(0, \hat{\theta}(t_i), \theta), \end{aligned} \tag{15.149}$$

$$M_5(\hat{\theta}(t_i), \theta) = q_1 K_1(1, \hat{\theta}(t_i), \theta), \tag{15.150}$$

$$M_6(\hat{\theta}(t_i), \theta) = q_2 c_0 K_2(1, \hat{\theta}(t_i), \theta), \tag{15.151}$$

where the expressions of the functions K_1, K_2, K_3 are

$$\begin{aligned} K_1(y, \hat{\theta}(t_i), \theta) = {} & \psi(1, y; \hat{\theta}(t_i)) - \int_y^1 \psi(1, \varepsilon; \hat{\theta}(t_i))\bar{\phi}(\varepsilon, y; \theta)d\varepsilon \\ & - \int_y^1 \Phi(1, \varepsilon; \hat{\theta}(t_i))\bar{\psi}(\varepsilon, y; \theta)d\varepsilon, \end{aligned} \tag{15.152}$$

$$\begin{aligned} K_2(y, \hat{\theta}(t_i), \theta) = {} & \Phi(1, y; \hat{\theta}(t_i)) - \int_y^1 \psi(1, \varepsilon; \hat{\theta}(t_i))\bar{\varphi}(\varepsilon, y; \theta)d\varepsilon \\ & - \int_y^1 \Phi(1, \varepsilon; \hat{\theta}(t_i))\bar{\Phi}(\varepsilon, y; \theta)d\varepsilon, \end{aligned} \tag{15.153}$$

$$\begin{aligned} K_3(\hat{\theta}(t_i), \theta) = {} & \lambda(1; \hat{\theta}(t_i)) - \int_0^1 \psi(1, y; \hat{\theta}(t_i))\bar{\gamma}(y; \theta)dy \\ & - \int_0^1 \Phi(1, y; \hat{\theta}(t_i))\bar{\lambda}(y; \theta)dy \end{aligned} \tag{15.154}$$

and where the expressions of the functions $\tilde{K}_1, \tilde{K}_2, \tilde{K}_3$ are

$$\begin{aligned} \tilde{K}_1(y, \hat{\theta}(t_i), \theta) = {} & \psi(1, y; \theta) - \psi(1, y; \hat{\theta}(t_i)) \\ & - \int_y^1 [\psi(1, \varepsilon; \theta) - \psi(1, \varepsilon; \hat{\theta}(t_i))]\bar{\phi}(\varepsilon, y; \theta)d\varepsilon \end{aligned}$$

$$-\int_y^1 [\Phi(1,\varepsilon;\theta)-\Phi(1,\varepsilon;\hat{\theta}(t_i))]\bar{\psi}(\varepsilon,y;\theta)d\varepsilon, \qquad (15.155)$$

$$\tilde{K}_2(y,\hat{\theta}(t_i),\theta)=\Phi(1,y;\theta)-\Phi(1,y;\hat{\theta}(t_i))$$

$$-\int_y^1 [\psi(1,\varepsilon;\theta)-\psi(1,\varepsilon;\hat{\theta}(t_i))]\bar{\varphi}(\varepsilon,y;\theta)d\varepsilon$$

$$-\int_y^1 [\Phi(1,\varepsilon;\theta)-\Phi(1,\varepsilon;\hat{\theta}(t_i))]\bar{\Phi}(\varepsilon,y;\theta)d\varepsilon, \qquad (15.156)$$

$$\tilde{K}_3(\hat{\theta}(t_i),\theta)=\lambda(1;\theta)-\lambda(1;\hat{\theta}(t_i))$$

$$-\int_0^1 [\psi(1,y;\theta)-\psi(1,y;\hat{\theta}(t_i))]\bar{\gamma}(y;\theta)dy$$

$$-\int_0^1 [\Phi(1,y;\theta)-\Phi(1,y;\hat{\theta}(t_i))]\bar{\lambda}(y;\theta)dy. \qquad (15.157)$$

C. Proof of claim 15.5

We prove this by contradiction. Suppose there exists a $\bar{\lambda}\neq 0$ such that

$$z(x,t)-\bar{\lambda}w(x,t)\equiv 0, \quad x\in[0,1],\ t\in[\mu_{i+1},t_{i+1}). \qquad (15.158)$$

Taking the time and spatial derivatives of (15.158) yields

$$z_t(x,t)-\bar{\lambda}w_t(x,t)=0, \qquad (15.159)$$

$$z_x(x,t)-\bar{\lambda}w_x(x,t)=0 \qquad (15.160)$$

for $x\in[0,1], t\in[\mu_{i+1},t_{i+1})$ (except possibly for finitely many discontinuities of the functions $w[t], z[t]$). Recalling (15.2), (15.3), we have

$$w_x(x,t)=\frac{d_3\bar{\lambda}^2+(d_4-d_1)\bar{\lambda}-d_2}{-\bar{\lambda}(q_1+q_2)}w(x,t), \qquad (15.161)$$

$$z_x(x,t)=\frac{d_3\bar{\lambda}^2+(d_4-d_1)\bar{\lambda}-d_2}{-\bar{\lambda}(q_1+q_2)}z(x,t) \qquad (15.162)$$

for $x\in[0,1], t\in[\mu_{i+1},t_{i+1})$. The solutions of (15.161), (15.162) are obtained as

$$w(x,t)=w(0,t)e^{\frac{d_3\bar{\lambda}^2+(d_4-d_1)\bar{\lambda}-d_2}{-\bar{\lambda}(q_1+q_2)}x}, \qquad (15.163)$$

$$z(x,t)=\bar{\lambda}w(0,t)e^{\frac{d_3\bar{\lambda}^2+(d_4-d_1)\bar{\lambda}-d_2}{-\bar{\lambda}(q_1+q_2)}x} \qquad (15.164)$$

for $x\in[0,1], t\in[\mu_{i+1},t_{i+1})$, where $z(0,t)=\bar{\lambda}w(0,t)$, obtained from the hypothesis (15.158), has been used.

By virtue of (15.5), (15.17), we know that $w(1,t)$ is constant on $t\in[\mu_{i+1},t_{i+1})$. It implies that $w(0,t)$ is constant on $t\in[\mu_{i+1},t_{i+1})$ recalling (15.163)—that is,

$$w(0,t)=w(0,\mu_{i+1}), \quad t\in[\mu_{i+1},t_{i+1}). \qquad (15.165)$$

According to the hypothesis (15.158), we also have

$$z(0,t)=\bar{\lambda}w(0,\mu_{i+1}), \quad t\in[\mu_{i+1},t_{i+1}). \qquad (15.166)$$

By virtue of (15.3), (15.158), (15.161)–(15.166), we obtain

$$q_2 \frac{d_3 \bar{\lambda}^2 + (d_4 - d_1)\bar{\lambda} - d_2}{-\bar{\lambda}(q_1 + q_2)} + d_3 \bar{\lambda} + d_4 = 0, \tag{15.167}$$

which is a necessary condition for the hypothesis (15.158) to hold.

For the case that $c \neq 0$, $a \neq 0$, inserting (15.165), (15.166) into (15.4) yields that $\zeta(t) = \frac{(\bar{\lambda}+p)}{c} w(0, \mu_{i+1})$ is also constant on $t \in [\mu_{i+1}, t_{i+1})$. Recalling (15.1), and $w(0, \mu_{i+1}) \neq 0$ which is obtained from the fact that $w(x, t)$ is not identically zero in the interval and (15.163), (15.165), it follows that

$$\bar{\lambda} = -\frac{bc}{a} - p. \tag{15.168}$$

For the cases that $c = 0$, we have

$$\bar{\lambda} = -p \tag{15.169}$$

according to (15.4), (15.165), (15.166), and $w(0, \mu_{i+1}) \neq 0$.

By virtue of assumption 15.2, (15.168), and (15.169), we know that the necessary condition for the hypothesis (15.158) to hold—that is, (15.167)—does not hold for the cases that $a \neq 0$ and the case that $a = 0, c = 0$. For the case that $a = 0$, $c \neq 0$, inserting (15.165), (15.166) into (15.4), recalling (15.1) and $w(0, \mu_{i+1}) \neq 0$, it follows that $\bar{\lambda} = 0$: contradiction. Therefore, the hypothesis (15.158) does not hold. The proof of this claim is complete.

D. Proof of claim 15.6

We first prove sufficiency. If $z[t]$ (or $w[t]$ or $\zeta(t)$) is not identically zero for $t \in [0, \infty)$, there exists an interval $[\mu_{i+1}, t_{i+1}]$ on which $z[t]$ (or $w[t]$ or $\zeta(t)$, respectively) is not identically zero. It follows that $\hat{d}_3(t_{i+1}) = d_3$ (or $\hat{d}_2(t_{i+1}) = d_2$ or $\hat{a}(\tau_{i+1}) = a$, respectively), recalling lemma 15.5.

Next, we prove necessity. When $\hat{d}_3(0) \neq d_3$ (or $\hat{d}_2(0) \neq d_2$ or $\hat{a}(0) \neq a$), if $z[t]$ (or $w[t]$ or $\zeta(t)$) is identically zero for $t \in [0, \infty)$, applying lemma 15.5, we have $\hat{d}_3(t) = \hat{d}_3(0) \neq d_3$ (or $\hat{d}_2(t) = \hat{d}_2(0) \neq d_2$ or $\hat{a}(t) = \hat{a}(0) \neq a$, respectively). Therefore, if the estimate $\hat{d}_3$ (or $\hat{d}_2$ or $\hat{a}$) reaches the true value, it follows that $z[t]$ (or $w[t]$ or $\zeta(t)$, respectively) is not identically zero on $t \in [0, \infty)$.

The proof of claim 15.6 is complete.

E. Proof of claim 15.7

The proof is divided into three cases.

Case 1: We suppose that the estimate $\hat{d}_2(t)$ does not reach d_2 in finite time. This implies that $w[t] \equiv 0$ on $t \in [0, \infty)$ according to claim 15.6. By virtue of (15.3) and $d_3 \neq 0$, we have $z[t] \equiv 0$ on $t \in [0, \infty)$. We then obtain from (15.4) that $\zeta(t) \equiv 0$ on $t \in [0, \infty)$. Therefore, $\Omega(t) \equiv 0$. According to $w[t] \equiv 0$, $z[t] \equiv 0$, and $\zeta(t) \equiv 0$ on $t \in [0, \infty)$, we have that $d(t) \equiv 0$ according to (15.19), and $m(t)$ is exponentially convergent to zero recalling (15.48) and (15.6), (15.7), (15.44).

Case 2: We suppose that the estimate $\hat{d}_3(t)$ does not reach d_3 in finite time. This implies that $z[t] \equiv 0$ on $t \in [0, \infty)$ according to claim 15.6. Then (15.1)–(15.5) become

$$\zeta(t) = e^{(a+b\frac{c}{p})t}\zeta(0), \tag{15.170}$$

$$w_t(x,t) = q_2 w_x(x,t) + d_4 w(x,t), \tag{15.171}$$

$$w(0,t) = \frac{c}{p}\zeta(t), \tag{15.172}$$

$$w(1,t) = c_0 U_d(t), \tag{15.173}$$

$t \in [0, \infty)$, in the closed-loop system.

If $\zeta(0) = 0$, then $z[t], w[t], \zeta(t)$ are identically zero on $t \in [0, \infty)$ according to (15.170)–(15.172)—that is, $\Omega(t) \equiv 0$.

Next we discuss the case when $\zeta(0) \neq 0$. The equation (15.172) holding for $t \in [0, \infty)$ requires the initial condition of w to be

$$w(x,0) = e^{-\frac{d_4}{q_2}x}\frac{c}{p}e^{(a+b\frac{c}{p})\frac{1}{q_2}x}\zeta(0) \tag{15.174}$$

to ensure that (15.172) holds on $t \in [0, \frac{1}{q_2}]$ and $w(1,t)$ to be

$$w(1,t) = e^{-\frac{d_4}{q_2}}\frac{c}{p}e^{(a+b\frac{c}{p})\frac{1}{q_2}}e^{(a+b\frac{c}{p})t}\zeta(0), \quad t \in [0, \infty) \tag{15.175}$$

to ensure that (15.172) holds on $t \in [\frac{1}{q_2}, \infty)$. If $c=0$, we obtain from (15.171), (15.174), (15.175) that $w \equiv 0$ for $t \in [0, \infty)$. Together with $z \equiv 0$ for $t \in [0, \infty)$, recalling (15.17), (15.173), $\zeta(0) \neq 0$ with (15.170), and $\lambda(1) \neq 0$ (according to (15.15), (15.132) with $c=0$, and (15.133)), it follows that $w(1,t)$ is not identically zero on $t \in [0, \infty)$: this contradicts (15.175) under $c=0$. Next, we discuss the cases with $c \neq 0$. According to (15.17) and (15.173), $w(1,t)$ is piecewise-constant, which contradicts (15.175) that is an exponential function ($a + b\frac{c}{p} \neq 0$ ensured by assumption 15.1). Therefore, case 2 only happens when $\zeta(0) = 0$, that is, $z[t], w[t], \zeta(t)$ are identically zero on $t \in [0, \infty)$. This implies that $\Omega(t)$ is identically zero on $t \in [0, \infty)$, and $m(t)$ is exponentially convergent to zero, recalling (15.6), (15.7), (15.19), (15.44), (15.48).

Case 3. We suppose that the estimate $\hat{a}(t)$ does not reach a in finite time. This implies that $\zeta[t] \equiv 0$ on $t \in [0, \infty)$ according to claim 15.6. It means that $\beta(0,t)$ is identically zero on $t \in [0, \infty)$ according to (15.10), and $\alpha(0,t)$ is identically zero on $t \in [0, \infty)$ according to (15.11). It follows that $\beta[t]$ is identically zero on $t \in [0, \infty)$, and $\alpha[t]$ is identically zero on $t \in [\frac{1}{q_1}, \infty)$, recalling (15.12), (15.13). Therefore, for $t \in [\frac{1}{q_1}, \infty)$, $\beta[t], \alpha[t], \zeta(t)$ are identically zero; that is, $w[t], z[t], \zeta(t)$ are identically zero on $t \in [\frac{1}{q_1}, \infty)$, recalling the inverse transformation (15.8), (15.9). Therefore, $\Omega(t)$ is identically zero on $t \in [\frac{1}{q_1}, \infty)$, and $m(t)$ is exponentially convergent to zero, recalling (15.6), (15.7), (15.19), (15.44), (15.48).

The proof of claim 15.7 is complete.

F. Proof of claim 15.8

We estimate the maximum value of t_ε in various situations of initial conditions $z[0], w[0], \zeta(0)$.

Case 1: $z[0] \neq 0, w[0] = 0, \zeta(0) = 0$. According to (15.3), $z[t], w[t]$ are not identically zero on $t \in [0, t_1]$ (if $w[t]$ is identically zero on $t \in [0, t_1]$, by virtue of (15.3) and $d_3 \neq 0$, it implies that $z[t]$ is identically zero on $t \in [0, t_1]$: contradiction). Suppose that $\zeta(t)$ is identically zero on $t \in [0, t_f]$, where $t_f = \min\{t_f : f \in \mathbb{Z}^+, t_f > \frac{1}{q_1} + \frac{1}{q_2}\}$. This means that $\beta(0, t)$ is identically zero on $t \in [0, t_f]$ according to (15.10), and $\alpha(0, t)$ is identically zero on $t \in [0, t_f]$ according to (15.11). It follows that $\beta[t]$ is identically zero on $t \in [0, t_f - \frac{1}{q_2}]$, and $\alpha[t]$ is identically zero on $t \in [\frac{1}{q_1}, t_f]$, recalling (15.12), (15.13). Therefore, for $t \in [\frac{1}{q_1}, t_f - \frac{1}{q_2}]$, $\beta[t]$, $\alpha[t]$, $\zeta(t)$ are identically zero, that is, $w[t]$, $z[t]$, $\zeta(t)$ are identically zero, recalling the inverse transformation (15.8), (15.9). According to (15.5), U_d in (15.17) is kept constant in a time interval. Therefore, U_d is identically zero on $t \in [\frac{1}{q_1}, t_f - \frac{1}{q_2}]$, which implies that U_d is identically zero on $t \in [\frac{1}{q_1}, t_f]$. It follows that $z(x, t), w(x, t), \zeta(t)$ are identically zero for $t \in [\frac{1}{q_1}, t_f]$. By virtue of (15.1)–(15.5), (15.17), through iterative constructions between successive triggering times, we have that $U_d \equiv 0$ and $z(x, t), w(x, t), \zeta(t)$ are identically zero for $t \in [\frac{1}{q_1}, \infty)$. This implies that $\zeta(t) \equiv 0$ on $t \in [0, \infty)$ and that $\hat{a}$ does not reach the true value in finite time: contradiction. Therefore, the nonzero values of $\zeta(t)$ appear not later than t_f. Recalling lemma 15.5, we have $\hat{d}_3(t_1) = d_3$, $\hat{d}_2(t_1) = d_2$, $\hat{a}(t_f) = a$. Therefore, $t_\varepsilon \leq T + \frac{1}{q_1} + \frac{1}{q_2}$, where T is the maximum dwell time between two triggering times.

Case 2: $w[0] \neq 0, z[0] = 0, \zeta(0) = 0$. The maximum time taken by the nonzero values of $w[0]$ propagate to z and $\zeta(t)$ is $\frac{1}{q_2}$. Therefore, the estimate $\hat{d}_3(t), \hat{a}(t)$ would reach the true value d_3, a not later than $t_f = \min\{t_f : f \in \mathbb{Z}^+, t_f > \frac{1}{q_2}\}$ according to lemmas 15.5 and 15.6. Because of $w[0] \neq 0$, we have $\hat{d}_2(t_1) = d_2$. It follows that $t_\varepsilon \leq \frac{1}{q_2} + T$.

Case 3: $\zeta(0) \neq 0, z[0] = 0, w[0] = 0$. If $w[t]$ is identically zero on $t \in [0, t_1]$, this implies that $z[t]$ is identically zero on $t \in [0, t_1]$ according to (15.3) with $d_3 \neq 0$, which means $\zeta(t)$ is identically zero on $t \in [0, t_1]$ by (15.4): contradiction. Therefore, $w[t]$ is not identically zero on $t \in [0, t_1]$. According to the proof in case 2 of claim 15.7, we know that the necessary condition for $z[t]$ to be identically zero on $t \in [0, t_f]$, where $t_f = \min\{t_f : f \in \mathbb{Z}^+, t_f > \frac{1}{q_2}\}$, is that w satisfies (15.174), (15.175), which implies $w[0] \neq 0$ ($c \neq 0$) or $w[t] = 0$ for $t \in [0, t_1]$ ($c = 0$): contradiction. Therefore, $z[t]$ is not identically zero on $t \in [0, t_f]$. Therefore, $\hat{d}_2(t_1) = d_2$, $\hat{a}(t_1) = a$, $\hat{d}_3(t_f) = d_3$. It follows that $\tau_\varepsilon \leq \frac{1}{q_2} + T$.

Case 4: $\zeta(0) \neq 0, w[0] \neq 0, z[0] = 0$. Suppose that $z[t]$ is identically zero on $t \in [0, t_f]$, where $t_f = \min\{t_f : f \in \mathbb{Z}^+, t_f > \frac{1}{q_2}\}$. According to the proof in case 2 of claim 15.7, if $c = 0$, a necessary condition of the above hypothesis is (15.174), which means $w[0] = 0$: contradiction. If $c \neq 0$, a necessary condition of the above hypothesis is (15.175), which does not hold, because the control input U_d applied at (15.173) is piecewise-constant while $w(1, t)$ in (15.175) is an exponential function ($a + b\frac{c}{p} \neq 0$ ensured by assumption 15.1). Therefore, $z[t]$ is not identically zero on $t \in [0, t_f]$. This implies that $\hat{d}_3(\tau_f) = d_3$ according to lemmas 15.5 and 15.6. Because of $w[0] \neq 0$ and $\zeta(0) \neq 0$, we have $\hat{d}_2(t_1) = d_2$, $\hat{a}(t_1) = a$. Therefore, $t_\varepsilon \leq \frac{1}{q_2} + T$.

Case 5: $\zeta(0) \neq 0, z[0] \neq 0, w[0] = 0$. According to (15.3) with $d_3 \neq 0$ and the fact that $z[t]$ is not identically zero on $t \in [0, t_1]$, we know that $w[t]$ is not identically zero on $t \in [0, t_1]$. Recalling lemma 15.5, we have $\hat{\theta}(t_1) = \theta$. Therefore, $t_\varepsilon \leq T$.

Case 6: $\zeta(0) = 0, z[0] \neq 0, w[0] \neq 0$. Following the analysis in case 1, we show that the nonzero values of $\zeta(t)$ appear not later than t_f, where $t_f = \min\{t_f : f \in \mathbb{Z}^+, t_f > \frac{1}{q_1} + \frac{1}{q_2}\}$. Recalling lemmas 15.5 and 15.6, we have $\hat{d}_3(t_1) = d_3$, $\hat{d}_2(t_1) = d_2$, $\hat{a}(t_f) = a$. Therefore, $t_\varepsilon \leq T + \frac{1}{q_1} + \frac{1}{q_2}$.

Table 15.2. Recent results of triggered-type adaptive boundary control of PDEs

	Types of PDEs	Types of control inputs	Types of identification	Types of triggering times
[11]	Hyperbolic PDEs	Continuous except for finite-time instants	Piecewise-constant	Equidistant
[10]	Hyperbolic PDEs	Continuous except for finite-time instants	Piecewise-constant	Equidistant
[106]	Parabolic PDEs	Continuous except for finite-time instants	Piecewise-constant	Nonequidistant
Chapter 13	Coupled hyperbolic PDEs	Piecewise-constant	Continuous	Nonequidistant
Chapter 14	Coupled hyperbolic PDEs	Continuous except for finite-time instants	Piecewise-constant	Nonequidistant
Chapter 15	Coupled hyperbolic PDEs	Piecewise-constant	Piecewise-constant	Nonequidistant

Case 7: $\zeta(0) \neq 0, z[0] \neq 0, w[0] \neq 0$. According to lemma 15.5, $t_\varepsilon \leq t_1 \leq T$.

Case 8: $z[0] = 0, w[0] = 0, \zeta(0) = 0$. According to the plant (15.1)–(15.5) with the control input (15.17), we know that $z[t], w[t], \zeta(t)$ are identically zero for $t \in [0, \infty)$. The estimates reach the true values in finite time only when $\hat{q}_1(0) = q_1, \hat{q}_2(0) = q_2$— that is, $\tau_\varepsilon = 0$.

In summary, we have proved for all eight cases that $t_\varepsilon \leq \frac{1}{q_1} + \frac{1}{q_2} + T$. This completes the proof of claim 15.8.

15.7 NOTES

Recent results of triggered-type adaptive boundary control of PDEs are summarized in table 15.2.

Bibliography

[1]   O. Aamo. Disturbance rejection in 2×2 linear hyperbolic systems. *IEEE Transactions on Automatic Control*, 58(5): 1095–11063, 2013.

[2]   U.J.F. Aarsnes, R. Vazquez, F. Di Meglio, and M. Krstic. Delay robust control design of under-actuated PDE-ODE-PDE systems. In *American Control Conference*, 3200–3205, July 2019.

[3]   T. Ahmed-Ali, I. Karafyllis, and F. Lamnabhi-Lagarrigue. Global exponential sampled-data observers for nonlinear systems with delayed measurements. *Systems & Control Letters*, 62:539–49, 2013.

[4]   H. Anfinsen and O. M. Aamo. Disturbance rejection in the interior domain of linear 2×2 hyperbolic systems. *IEEE Transactions on Automatic Control*, 60(1): 186–91, 2015.

[5]   H. Anfinsen and O. M. Aamo. Adaptive output-feedback stabilization of linear 2×2 hyperbolic systems using anti-collocated sensing and control. *Systems & Control Letters*, 104:86–94, 2017.

[6]   H. Anfinsen and O. M. Aamo. Disturbance rejection in general heterodirectional 1-D linear hyperbolic systems using collocated sensing and control. *Automatica*, 76:230–42, 2017.

[7]   H. Anfinsen and O. M. Aamo. Adaptive control of linear 2×2 hyperbolic systems. *Automatica*, 87:69–82, 2018.

[8]   H. Anfinsen and O. M. Aamo. Stabilization of a linear hyperbolic PDE with actuator and sensor dynamics. *Automatica*, 95:104–11, 2018.

[9]   H. Anfinsen and O. M. Aamo. *Adaptive Control of Hyperbolic PDEs*. Springer, 2019.

[10] H. Anfinsen, H. Holta, and O. M. Aamo. Adaptive control of a linear hyperbolic PDE with uncertain transport speed and a spatially varying coefficient. In *28th Mediterranean Conference on Control and Automation*, 945–51, September 2020.

[11] H. Anfinsen, H. Holta, and O. M. Aamo. Adaptive control of a scalar 1-D linear hyperbolic PDE with uncertain transport speed using boundary sensing. In *American Control Conference*, 1575–81, July 2020.

[12] J. Auriol, U.J.F. Aarsnes, P. Martin, and F. Di Meglio. Delay-robust control design for two heterodirectional linear coupled hyperbolic PDEs. *IEEE Transactions on Automatic Control*, 63(10): 3551–57, 2018.

[13] J. Auriol, F. Bribiesca-Argomedo, D. B. Saba, M. Di Loreto, and F. Di Meglio. Delay-robust stabilization of a hyperbolic PDE-ODE system. *Automatica*, 95:494–502, 2018.

[14] J. Auriol and F. Di Meglio. Minimum time control of heterodirectional linear coupled hyperbolic PDEs. *Automatica*, 71:300–307, 2016.

[15] M. Balas. Feedback control of flexible systems. *IEEE Transactions on Automatic Control*, 23(4): 673–79, 1978.

[16] G. Bastin and J. M. Coron. *Stability and Boundary Stabilization of 1-D Hyperbolic Systems*. Birkhauser, 2016.

[17] H. I. Basturk and M. Krstic. State derivative feedback for adaptive cancellation of unmatched disturbances in unknown strict-feedback LTI systems. *Automatica*, 50:2539–45, 2014.

[18] N. Bekiaris-Liberis and M. Krstic. Compensation of wave actuator dynamics for nonlinear systems. *IEEE Transactions on Automatic Control*, 59(6): 1555–70, 2014.

[19] N. Bekiaris-Liberis and M. Krstic. Compensation of transport actuator dynamics with input-dependent moving controlled boundary. *IEEE Transactions on Automatic Control*, 63(11): 3889–96, 2018.

[20] P. Bernard and M. Krstic. Adaptive output-feedback stabilization of non-local hyperbolic PDEs. *Automatica*, 50: 2692–99, 2014.

[21] M. Bin and F. Di Meglio. Boundary estimation of boundary parameters for linear hyperbolic PDEs. *IEEE Transactions on Automatic Control*, 62(8): 3890–904, 2017.

[22] M. Bohm, M. Krstic, S. Kuchler, and O. Sawodny. Modeling and boundary control of a hanging cable immersed in water. *Journal of Dynamic Systems, Measurement, and Control*, 136:011006, 2014.

[23] S. Bonnabel and X. Claeys. The industrial control of tower cranes: An operator-in-the-loop approach [applications in control]. *IEEE Control Systems Magazine*, 40(5): 27–39, 2020.

[24] D. Bresch-Pietri and M. Krstic. Adaptive output feedback for oil drilling stick-slip instability modeled by wave PDE with anti-damped dynamic boundary. In *American Control Conference*, 386–91, June 2014.

[25] D. Bresch-Pietri and M. Krstic. Adaptive output-feedback for wave PDE with anti-damping application to surface-based control of oil drilling stick-slip instability. In *53rd IEEE Conference on Decision and Control*, 1295–300, December 2014.

[26] D. Bresch-Pietri and M. Krstic. Output-feedback adaptive control of a wave PDE with boundary anti-damping. *Automatica*, 50(6): 1407–15, 2014.

[27] F. Cacace, A. Germani, and C. Manes. An observer for a class of nonlinear systems with time varying observation delay. *Systems & Control Letters*, 59(5): 305–12, 2010.

[28] X. Cai and M. Krstic. Nonlinear control under wave actuator dynamics with time- and state-dependent moving boundary. *International Journal of Robust and Nonlinear Control*, 25(2): 222–51, 2015.

[29] X. Cai and M. Krstic. Nonlinear stabilization through wave PDE dynamics with a moving uncontrolled boundary. *Automatica*, 68:27–38, 2016.

[30] H. Canbolat, D. Dawson, C. Rahn, and S. Nagarkatti. Adaptive boundary control of out-of-plane cable vibration. *Journal of Applied Mechanics*, 65: 963–69, 1998.

[31] N. Challamel. Rock destruction effect on the stability of a drilling structure. *Journal of Sound and Vibration*, 233:235–54, 2000.

[32] J. M. Coron, R. Vazquez, M. Krstic, and G. Bastin. Local exponential H^2 stabilization of a 2×2 quasilinear hyperbolic system using backstepping. *SIAM Journal on Control and Optimization*, 51(3): 2005–35, 2013.

[33] R. Curtain and K. Morris. Transfer functions of distributed parameter systems: A tutorial. *Automatica*, 45:1101–16, 2009.

[34] B. d'Andrea Novel and J. M. Coron. Exponential stabilization of an overhead crane with flexible cable via a back-stepping approach. *Automatica*, 36:587–93, 2000.

[35] M. A. Davo, D. Bresch-Pietri, C. Prieur, and F. Di Meglio. Stability analysis of a 2×2 linear hyperbolic system with a sampled-data controller via backstepping method and looped-functionals. *IEEE Transactions on Automatic Control*, 64(4): 1718–25, 2018.

[36] J. G. De Jalon and E. Bayo. *Kinematic and Dynamic Simulation of Multibody Systems*. Springer, 1994.

[37] J. N. de la Vergne. *Hard Rock Miner's Handbook*. Stantec, 2008.

[38] E. Detournay and P. Defourny. A phenomenological model for the drilling action of drag bits. *International Journal of Rock Mechanics, Mining Science and Geomechanical Abstracts*, 29:13–23, 1992.

[39] J. Deutscher. Finite-time output regulation for linear 2×2 hyperbolic systems using backstepping. *Automatica*, 75:54–62, 2017.

[40] J. Deutscher. Output regulation for general linear heterodirectional hyperbolic systems with spatially-varying coefficients. *Automatica*, 85:34–42, 2017.

[41] J. Deutscher and J. Gabrie. Minimum time output regulation for general linear heterodirectional hyperbolic systems. *International Journal of Control*, 93:1826–38, 2020.

[42] J. Deutscher and N. Gehring. Output feedback control of coupled linear parabolic ODE-PDE-ODE systems. *IEEE Transactions on Automatic Control*, 66(10): 4668–83, 2021.

[43] J. Deutscher, N. Gehring, and R. Kern. Output feedback control of general linear heterodirectional hyperbolic ODE-PDE-ODE systems. *Automatica*, 95:472–80, 2018.

[44]   J. Deutscher, N. Gehring, and R. Kern. Output feedback control of general linear heterodirectional hyperbolic PDE-ODE systems with spatially-varying coefficients. *International Journal of Control*, 92(10): 2274–90, 2019.

[45]   A. Diagne, M. Diagne, S. X. Tang, and M. Krstic. Backstepping stabilization of the linearized Saint-Venant-Exner model. *Automatica*, 76:345–54, 2017.

[46]   M. Diagne, N. Bekiaris-Liberis, and M. Krstic. Time- and state-dependent input delay-compensated Bang-Bang control of a screw extruder for 3D printing. *International Journal of Robust and Nonlinear Control*, 27:3727–57, 2017.

[47]   M. Diagne, S. X. Tang, A. Diagne, and M. Krstic. Control of shallow waves of two unmixed fluids by backstepping. *Annual Reviews in Control*, 44:211–25, 2017.

[48]   F. Di Meglio, F. Bribiesca, L. Hu, and M. Krstic. Stabilization of coupled linear heterodirectional hyperbolic PDE-ODE systems. *Automatica*, 87:281–89, 2018.

[49]   F. Di Meglio, P.-O. Lamare, and U.J.F. Aarsnes. Robust output feedback stabilization of an ODE-PDE-ODE interconnection. *Automatica*, 119:109059, 2020.

[50]   F. Di Meglio, R. Vazquez, and M. Krstic. Stabilization of a system of $n+1$ coupled first-order hyperbolic linear PDEs with a single boundary input. *IEEE Transactions on Automatic Control*, 58:3097–111, 2013.

[51]   V. Dos Santos and C. Prieur. Boundary control of open channels with numerical and experimental validations. *IEEE Transactions on Control Systems Technology*, 16(6): 1252–64, 2008.

[52]   J. Doyle, B. Francis, and A. Tannenbaum. *Feedback Control Theory*. Macmillan, 1992.

[53]   V. Dunayevsky, F. Abbassian, and A. Judzis. Dynamic stability of drill strings under fluctuating weight on bit. *SPE Drilling and Completion*, 8:84–92, 1993.

[54]   N. Espitia. Observer-based event-triggered boundary control of a linear 2×2 hyperbolic systems. *Systems & Control Letters*, 138:104668, 2020.

[55]   N. Espitia, A. Girard, N. Marchand, and C. Prieur. Event-based control of linear hyperbolic systems of conservation laws. *Automatica*, 70:275–87, 2016.

[56]   N. Espitia, A. Girard, N. Marchand, and C. Prieur. Event-based stabilization of linear systems of conservation laws using a dynamic triggering condition. In *10th IFAC Symposium on Nonlinear Control Systems*, 49:362–67, 2016.

[57]   N. Espitia, A. Girard, N. Marchand, and C. Prieur. Event-based boundary control of a linear 2×2 hyperbolic system via backstepping approach. *IEEE Transactions on Automatic Control*, 63(8): 2686–93, 2018.

[58]   N. Espitia, I. Karafyllis, and M. Krstic. Event-triggered boundary control of constant-parameter reaction-diffusion PDEs: A small-gain approach. *Automatica*, 128:109562, 2021.

[59] O. Faltinsen. *Sea Loads on Ships and Offshore Structures*. Cambridge University Press, 1993.

[60] H. Feng and B. Z. Guo. A new active disturbance rejection control to output feedback stabilization for a one-dimensional anti-stable wave equation with disturbance. *IEEE Transactions on Automatic Control*, 62:3774–87, 2017.

[61] T. I. Fossen. *Marine Control Systems: Guidance, Navigation and Control of Ships, Rigs and Underwater Vehicles*. Marine Cybernetics, 2002.

[62] T. I. Fossen. *Marine Craft Hydrodynamics and Motion Control*. John Wiley & Sons, 2011.

[63] B. A. Francis and W. M. Wonham. The internal model principle of control theory. *Automatica*, 12:457–65, 1976.

[64] E. Fridman and A. Blighovsky. Robust sampled-data control of a class of semilinear parabolic systems. *Automatica*, 48(5): 826–36, 2012.

[65] S. S. Ge, W. He, B.V.E. How, and Y. S. Choo. Boundary control of a coupled nonlinear flexible marine riser. *IEEE Transactions on Control Systems Technology*, 18(5): 1080–91, 2010.

[66] M. Geradin and D. Rixen. *Mechanical Vibrations: Theory and Application to Structural Dynamics*. John Wiley & Sons, 1994.

[67] A. Germani, C. Manes, and P. Pepe. A new approach to state observation of nonlinear systems with delayed output. *IEEE Transactions on Automatic Control*, 47(1): 96–101, 2002.

[68] C. Germay, N. Van De Wouw, H. Nijmeijer, and R. Sepulchre. Nonlinear drilling dynamics analysis. *SIAM Journal on Applied Dynamical Systems*, 8:527–53, 2005.

[69] A. Girard. Dynamic triggering mechanisms for event-triggered control. *IEEE Transactions on Automatic Control*, 60(7): 1992–97, 2015.

[70] M. E. Guerrero, D. A. Mercado, R. Lozano, and C. D. Garcia. Passivity based control for a quadrotor UAV transporting a cable-suspended payload with minimum swing. In *54th IEEE Conference on Decision and Control*, 6718–23, December 2015.

[71] M. Gugat. Optimal boundary feedback stabilization of a string with moving boundary. *IMA Journal of Mathematical Control and Information*, 25:111–21, 2007.

[72] M. Gugat. Optimal energy control in finite time by varying the length of the string. *SIAM Journal on Control and Optimization*, 46:1705–25, 2007.

[73] B. Z. Guo and F. F. Jin. The active disturbance rejection and sliding mode control approach to the stabilization of the Euler-Bernoulli beam equation with boundary input disturbance. *Automatica*, 49:2911–18, 2013.

[74] B. Z. Guo and F. F. Jin. Sliding mode and active disturbance rejection control to stabilization of one-dimensional anti-stable wave equations subject

to disturbance in boundary input. *IEEE Transactions on Automatic Control*, 58:1269–74, 2013.

[75] B. Z. Guo and F. F. Jin. Output feedback stabilization for one-dimensional wave equation subject to boundary disturbance. *IEEE Transactions on Automatic Control*, 60:824–30, 2015.

[76] B. Z. Guo and J. J. Liu. Sliding mode control and active disturbance rejection control to the stabilization of one-dimensional Schrödinger equation subject to boundary control matched disturbance. *International Journal of Robust and Nonlinear Control*, 24:2194–212, 2014.

[77] B. Z. Guo and J. M. Wang. *Control of Wave and Beam PDEs*. Springer, 2019.

[78] B. Z. Guo and Z. H. Wu. Output tracking for a class of nonlinear systems with mismatched uncertainties by active disturbance rejection control. *Systems & Control Letters*, 100:21–31, 2017.

[79] B. Z. Guo and C. Z. Xu. Boundary output feedback stabilization of a one-dimensional wave equation system with time delay. In *17th IFAC World Congress*, 41:8755–60, 2008.

[80] B. Z. Guo and H. C. Zhou. The active disturbance rejection control to stabilization for multi-dimensional wave equation with boundary control matched disturbance. *IEEE Transactions on Automatic Control*, 60:143–57, 2015.

[81] W. Guo and B. Z. Guo. Adaptive output feedback stabilization for one-dimensional wave equation with corrupted observation by harmonic disturbance. *SIAM Journal on Control and Optimization*, 51:1679–706, 2013.

[82] W. Guo and B. Z. Guo. Parameter estimation and non-collocated adaptive stabilization for a wave equation subject to general boundary harmonic disturbance. *IEEE Transactions on Automatic Control*, 58:1631–43, 2013.

[83] W. Guo and B. Z. Guo. Stabilization and regulator design for a one-dimensional unstable wave equation with input harmonic disturbance. *International Journal of Robust and Nonlinear Control*, 23:514–33, 2013.

[84] W. Guo and B. Z. Guo. Performance output tracking for a wave equation subject to unmatched general boundary harmonic disturbance. *Automatica*, 68:194–202, 2016.

[85] W. Guo, Z. C. Shao, and M. Krstic. Adaptive rejection of harmonic disturbance anticollocated with control in 1D wave equation. *Automatica*, 79:17–26, 2017.

[86] J. Han. From PID to active disturbance rejection control. *IEEE Transactions on Industrial Electronics*, 56:900–906, 2009.

[87] W. He and S. S. Ge. Robust adaptive boundary control of a vibrating string under unknown time-varying disturbance. *IEEE Transactions on Control Systems Technology*, 20:48–58, 2012.

[88] W. He and S. S. Ge. Vibration control of a flexible string with both boundary input and output constraints. *IEEE Transactions on Control Systems Technology*, 23(4): 1245–54, 2015.

[89] W. He and S. S. Ge. Cooperative control of a nonuniform gantry crane with constrained tension. *Automatica*, 66:146–54, 2016.

[90] W. He, S. S. Ge, and D. Huang. Modeling and vibration control for a nonlinear moving string with output constraint. *IEEE/ASME Transactions on Mechatronics*, 20(4): 1886–97, 2014.

[91] W. He, T. Meng, X. He, and S. S. Ge. Unified iterative learning control for flexible structures with input constraints. *Automatica*, 96:326–36, 2018.

[92] X. Y. He, W. He, J. Shi, and C. Y. Sun. Boundary vibration control of variable length crane systems in two-dimensional space with output constraints. *IEEE/ASME Transactions on Mechatronics*, 22:1952–62, 2017.

[93] W.P.M.H. Heemels, K. H. Johansson, and P. Tabuada. An introduction to event-triggered and self-triggered control. In *51st IEEE Conference on Decision and Control*, 3270–85, December 2012.

[94] B. How, S. S. Ge, and Y. S. Choo. Dynamic load positioning for subsea installation via adaptive neural control. *IEEE Journal of Oceanic Engineering*, 35:366–75, 2010.

[95] B. How, S. S. Ge, and Y. S. Choo. Control of coupled vessel, crane, cable, and payload dynamics for subsea installation operations. *IEEE Transactions on Control Systems Technology*, 19:208–20, 2011.

[96] L. Hu, F. Di Meglio, R. Vazquez, and M. Krstic. Control of homodirectional and general heterodirectional linear coupled hyperbolic PDEs. *IEEE Transactions on Automatic Control*, 61(11): 3301–14, 2016.

[97] L. Hu, R. Vazquez, F. Di Meglio, and M. Krstic. Boundary exponential stabilization of 1-dimensional inhomogeneous quasi-linear hyperbolic systems. *SIAM Journal on Control and Optimization*, 57(2): 963–98, 2019.

[98] J. D. Jansen. *Nonlinear Dynamics of Oilwell Drill Strings*. Delft University Press, 1993.

[99] S. Kaczmarczyk and W. Ostachowicz. Transient vibration phenomena in deep mine hoisting cables. Part 1: Mathematical model. *Journal of Sound and Vibration*, 262(2): 219–44, 2003.

[100] W. Kang and E. Fridman. Sliding mode control of Schrodinger equation-ODE in the presence of unmatched disturbances. *Systems & Control Letters*, 98:65–73, 2016.

[101] I. Karafyllis, N. Bekiaris-Liberis, and M. Papageorgiou. Feedback control of nonlinear hyperbolic PDE systems inspired by traffic flow models. *IEEE Transactions on Automatic Control*, 64(9): 3647–62, 2018.

[102] I. Karafyllis, M. Kontorinaki, and M. Krstic. Adaptive control by regulation-triggered batch least-squares. *IEEE Transactions on Automatic Control*, 65 (7): 2842–55, 2019.

[103] I. Karafyllis and M. Krstic. Sampled-data boundary feedback control of 1-D linear transport PDEs with non-local terms. *Systems & Control Letters*, 107:68–75, 2017.

[104] I. Karafyllis and M. Krstic. Adaptive certainty-equivalence control with regulation-triggered finite-time least-squares identification. *IEEE Transactions on Automatic Control*, 63:3261–75, 2018.

[105] I. Karafyllis and M. Krstic. Sampled-data boundary feedback control of 1-D parabolic PDEs. *Automatica*, 87:226–37, 2018.

[106] I. Karafyllis, M. Krstic, and K. Chrysafi. Adaptive boundary control of constant-parameter reaction-diffusion PDEs using regulation-triggered finite-time identification. *Automatica*, 103:166–79, 2019.

[107] D. Karnopp. Computer simulation of stick-slip friction in mechanical dynamic systems. *ASME Journal of Dynamic Systems, Measurement and Control*, 107:100–103, 1985.

[108] S. Koga, D. Bresch-Pietri, and M. Krstic. Delay compensated control of the Stefan problem and robustness to delay mismatch. *International Journal of Robust and Nonlinear Control*, 30:2304–34, 2020.

[109] S. Koga, M. Diagne, and M. Krstic. Control and state estimation of the one-phase Stefan problem via backstepping design. *IEEE Transactions on Automatic Control*, 64(2): 510–25, 2019.

[110] S. Koga, I. Karafyllis, and M. Krstic. Towards implementation of PDE control for Stefan system: Input-to-state stability and sampled-data design. *Automatica*, 127:109538, 2021.

[111] S. Koga and M. Krstic. *Materials Phase Change PDE Control & Estimation*. Birkhäuser, 2020.

[112] M. Krstic. Systematization of approaches to adaptive boundary stabilization of PDEs. *International Journal of Robust and Nonlinear Control*, 16:801–18, 2006.

[113] M. Krstic. Lyapunov tools for predictor feedbacks for delay systems: Inverse optimality and robustness to delay mismatch. *Automatica*, 44:2930–35, 2008.

[114] M. Krstic. Compensating a string PDE in the actuation or sensing path of an unstable ODE. *IEEE Transactions on Automatic Control*, 54(6): 1362–68, 2009.

[115] M. Krstic. Control of an unstable reaction-diffusion PDE with long input delay. *Systems & Control Letters*, 58:773–82, 2009.

[116] M. Krstic. *Delay Compensation for Nonlinear, Adaptive, and PDE Systems*. Springer, 2009.

[117] M. Krstic. Adaptive control of an anti-stable wave PDE. *Dynamics of Continuous, Discrete and Impulsive Systems Series A: Mathematical Analysis*, 17:853–82, 2010.

[118] M. Krstic. Compensation of infinite-dimensional actuator and sensor dynamics: Nonlinear and delay-adaptive systems. *IEEE Control Systems Magazine*, 30:22–41, 2010.

[119] M. Krstic. Dead-time compensation for wave/string PDEs. *ASME Journal of Dynamic Systems, Measurement and Control*, 133(3): 031004, 2011.

[120] M. Krstic. Adaptive control of anti-stable wave PDE systems: Theory and applications in oil drilling. In *11th IFAC International Workshop on Adaptation and Learning in Control and Signal Processing*, 46:432–39, 2013.

[121] M. Krstic, B. Z. Guo, A. Balogh, and A. Smyshlyaev. Output-feedback stabilization of an anti-stable wave equation. *Automatica*, 44:63–74, 2008.

[122] M. Krstic, I. Kanellakopoulos, and P. Kokotovic. *Nonlinear and Adaptive Control Design*. John Wiley & Sons, 1995.

[123] M. Krstic and A. Smyshlyaev. Adaptive boundary control for unstable parabolic PDEs. Part 1: Lyapunov design. *IEEE Transactions on Automatic Control*, 53:1575–91, 2008.

[124] M. Krstic and A. Smyshlyaev. Adaptive control of PDEs. *Annual Reviews in Control*, 32:149–60, 2008.

[125] M. Krstic and A. Smyshlyaev. Backstepping boundary control for first-order hyperbolic PDEs and application to systems with actuator and sensor delays. *Systems & Control Letters*, 57(9): 750–58, 2008.

[126] M. Krstic and A. Smyshlyaev. *Boundary Control of PDEs: A Course on Backstepping Designs*. SIAM, 2008.

[127] P.-O. Lamare and N. Bekiaris-Liberis. Control of 2×2 linear hyperbolic systems: Backstepping-based trajectory generation and PI-based tracking. *Systems & Control Letters*, 86:24–33, 2015.

[128] I. Landet, A. Pavlov, and O. Aamo. Modeling and control of heave-induced pressure fluctuations in managed pressure drilling. *IEEE Transactions on Control Systems Technology*, 21:1340–51, 2013.

[129] S. Li, J. Yang, W. H. Chen, and X. Chen. Generalized extended state observer based control for systems with mismatched uncertainties. *IEEE Transactions on Industrial Electronics*, 59:4792–802, 2012.

[130] W. J. Liu and M. Krstic. Backstepping boundary control of Burgers' equation with actuator dynamics. *Systems & Control Letters*, 41(4):291–303, 2000.

[131] Z. Liu, J. Liu, and W. He. Modeling and vibration control of a flexible aerial refueling hose with variable lengths and input constraint. *Automatica*, 77:302–10, 2017.

[132] Z. H. Luo, B. Z. Guo, and O. Morgul. *Stability and Stabilization of Infinite Dimensional Systems with Applications.* Springer, 1999.

[133] N. Marchand, S. Durand, and J.F.G. Castellanos. A general formula for event-based stabilization of nonlinear systems. *IEEE Transactions on Automatic Control*, 58(5): 1332–37, 2013.

[134] D. B. McIver. Hamilton's principle for systems of changing mass. *Journal of Engineering Mathematics*, 7(3): 249–61, 1973.

[135] P. J. Moylan. Stable inversion of linear systems. *IEEE Transactions on Automatic Control*, 22(1): 74–78, 1977.

[136] E. Navarro-López and D. Cortés. Sliding-mode of a multi-DOF oilwell drill-string with stick-slip oscillations. In *American Control Conference*, 3837–42, 2007.

[137] Q. C. Nguyen and K.-S. Hong. Asymptotic stabilization of a nonlinear axially moving string by adaptive boundary control. *Journal of Sound and Vibration*, 329(22): 4588–603, 2010.

[138] Q. C. Nguyen and K.-S. Hong. Simultaneous control of longitudinal and transverse vibrations of an axially moving string with velocity tracking. *Journal of Sound and Vibration*, 331(13): 3006–19, 2012.

[139] Q. C. Nguyen and K.-S. Hong. Transverse vibration control of axially moving membranes by regulation of axial velocity. *IEEE Transactions on Control Systems Technology*, 20(4): 1124–31, 2012.

[140] I. Palunko, P. Cruz, and R. Fierro. Agile load transportation: Safe and efficient load manipulation with aerial robots. *IEEE Robotics & Automation Magazine*, 19(3): 69–79, 2012.

[141] C. Prieur, A. Girard, and E. Witrant. Stability of switched linear hyperbolic systems by Lyapunov techniques. *IEEE Transactions on Automatic Control*, 59(8): 2196–202, 2014.

[142] C. Prieur and J. Winkin. Boundary feedback control of linear hyperbolic systems: Application to the Saint-Venant-Exner equations. *Automatica*, 89:44–51, 2018.

[143] C. Prieur, J. Winkin, and G. Bastin. Robust boundary control of systems of conservation laws. *Mathematics of Control, Signals, and Systems*, 20(2): 173–97, 2008.

[144] S. Raghavan and J. K. Hedrick. Observer design for a class of nonlinear systems. *International Journal of Control*, 59:515–28, 1994.

[145] R. Rebarber and G. Weiss. Internal model based tracking and disturbance rejection for stable well-posed systems. *Automatica*, 39:1555–69, 2003.

[146] T. Richarda, C. Germayb, and E. Detournay. A simplified model to explore the root cause of stick-slip vibrations in drilling systems with drag bits. *Journal of Sound and Vibration*, 305:432–56, 2007.

[147] C. Roman, D. Bresch-Pietri, E. Cerpa, C. Prieur, and O. Sename. Back-stepping observer based-control for an anti-damped boundary wave PDE in presence of in-domain viscous damping. In *55th Conference on Decision and Control*, 549–54, December 2016.

[148] C. Roman, D. Bresch-Pietri, C. Prieur, and O. Sename. Robustness of an adaptive output feedback for an anti-damped boundary wave PDE in presence of in-domain viscous damping. In *American Control Conference*, 3455–60, July 2016.

[149] C. Roman, D. Bresch-Pietri, C. Prieur, and O. Sename. Robustness to in-domain viscous damping of a collocated boundary adaptive feedback law for an antidamped boundary wave PDE. *IEEE Transactions on Automatic Control*, 64(8): 3284–99, 2019.

[150] C. Roman, F. Ferrante, and C. Prieur. Parameter identification of a linear wave equation from experimental boundary data. *IEEE Transactions on Control Systems Technology*, 29(5): 2166–79, 2021.

[151] D. B. Saba, F. Bribiesca-Argomedo, M. D. Loreto, and D. Eberard. Backstepping stabilization of 2×2 linear hyperbolic PDEs coupled with potentially unstable actuator and load dynamics. In *56th IEEE Conference on Decision and Control*, 2498–503, December 2017.

[152] D. B. Saba, F. Bribiesca-Argomedo, M. D. Loreto, and D. Eberard. Strictly proper control design for the stabilization of 2×2 linear hyperbolic ODE-PDE-ODE systems. In *58th IEEE Conference on Decision and Control*, 4996–5001, December 2019.

[153] C. Sagert, F. Di Meglio, M. Krstic, and P. Rouchon. Backstepping and flatness approaches for stabilization of the stick-slip phenomenon for drilling. In *5th IFAC Symposium on System Structure and Control*, 46:779–84, 2013.

[154] B. Saldivar, I. Boussaada, H. Mounier, and S. Mondie. An overview on the modeling of oilwell drilling vibrations. In *19th IFAC World Congress*, 47: 5169–74, 2014.

[155] B. Saldivar, S. Mondie, J.-J. Loiseau, and V. Rasvan. Stick-slip oscillations in oilwell drillstrings: Distributed parameter and neutral type retarded model approaches. In *18th IFAC World Congress*, 44(1): 284–89, 2011.

[156] B. Saldivar, S. Mondie, S. I. Niculescu, H. Mounier, and I. Boussaada. A control oriented guided tour in oilwell drilling vibration modeling. *Annual Reviews in Control*, 42:100–113, 2016.

[157] S. H. Sandilo and W. T. van Horssen. On variable length induced vibrations of a vertical string. *Journal of Sound and Vibration*, 333(11): 2432–49, 2014.

[158] A. Selivanov and E. Fridman. Distributed event-triggered control of diffusion semilinear PDEs. *Automatica*, 68:344–51, 2016.

[159] A. Seuret, C. Prieur, and N. Marchand. Stability of non-linear systems by means of event-triggered sampling algorithms. *IMA Journal of Mathematical Control and Information*, 31(3): 415–33, 2014.

[160] A.F.A. Sevrarens, M.J.G. van de Molengraft, J. J. Kok, and L. van den Steen. H∞ control for suppressing stick-slip in oil well drillstring. *IEEE Control Systems*, 18(4): 19–30, 1998.

[161] A. Sezgin and M. Krstic. Boundary backstepping control of flow-induced vibrations of a membrane at high mach numbers. *ASME Journal of Dynamic Systems, Measurement and Control*, 137:081003, 2015.

[162] A. Smyshlyaev, E. Cerpa, and M. Krstic. Boundary stabilization of a 1-D wave equation with in-domain antidamping. *SIAM Journal on Control and Optimization*, 48(6): 4014–31, 2010.

[163] A. Smyshlyaev and M. Krstic. Adaptive boundary control for unstable parabolic PDEs. Part 2: Estimation-based designs. *Automatica*, 43:1543–56, 2007.

[164] A. Smyshlyaev and M. Krstic. Adaptive boundary control for unstable parabolic PDEs. Part 3: Output feedback examples with swapping identifiers. *Automatica*, 43:1557–64, 2007.

[165] A. Smyshlyaev and M. Krstic. Boundary control of an anti-stable wave equation with anti-damping on the uncontrolled boundary. *Systems & Control Letters*, 58:617–23, 2009.

[166] A. Smyshlyaev and M. Krstic. *Adaptive Control of Parabolic PDEs*. Princeton University Press, 2010.

[167] R. G. Standing, B. G. Mackenzie, and R. O. Snell. Enhancing the technology for deepwater installation of subsea hardware. In *Offshore Technology Conference*, 1–6, May 2002.

[168] T. Stensgaard, C. White, and K. Schiffer. Subsea hardware installation from a FDPSO. In *Offshore Technology Conference*, 1–6, May 2010.

[169] L. Su, J. M. Wang, and M. Krstic. Boundary feedback stabilization of a class of coupled hyperbolic equations with non-local terms. *IEEE Transactions on Automatic Control*, 63:2633–40, 2018.

[170] G. A. Susto and M. Krstic. Control of PDE-ODE cascades with Neumann interconnections. *Journal of the Franklin Institute*, 347(1): 284–314, 2010.

[171] P. Tabuada. Event-triggered real-time scheduling of stabilizing control tasks. *IEEE Transactions on Automatic Control*, 52(9): 1680–85, 2007.

[172] S. X. Tang, B. Z. Guo, and M. Krstic. Active disturbance rejection control for a 2×2 hyperbolic system with an input disturbance. In *19th IFAC World Congress*, 47(3): 11385–90, 2014.

[173] S. X. Tang and M. Krstic. Sliding mode control to the stabilization of a linear 2×2 hyperbolic system with boundary input disturbance. In *American Control Conference*, 1027–32, June 2014.

[174] Y. Terumichi, M. Ohtsuka, and M. Yoshizawa. Nonstationary vibrations of a string with time-varying length and a mass-spring system attached at the lower end. *Nonlinear Dynamics*, 12:39–55, 1997.

[175] R. Vazquez. Boundary control laws and observer design for convective, turbulent and magnetohydrodynamic flows. PhD thesis, University of California, San Diego, 2006.

[176] R. Vazquez and M. Krstic. Marcum Q-functions and explicit kernels for stabilization of 2×2 linear hyperbolic systems with constant coefficients. *Systems & Control Letters*, 68:33–42, 2014.

[177] R. Vazquez, M. Krstic, and J. M. Coron. Backstepping boundary stabilization and state estimation of a 2×2 linear hyperbolic system. In *50th IEEE Conference on Decision and Control and European Control Conference*, 4937–42, December 2011.

[178] J. Wang, S. Koga, Y. Pi, and M. Krstic. Axial vibration suppression in a PDE model of ascending mining cable elevator. *ASME Journal of Dynamic Systems, Measurement and Control*, 140:111003, 2018.

[179] J. Wang and M. Krstic. Output-feedback boundary control of a heat PDE sandwiched between two ODEs. *IEEE Transactions on Automatic Control*, 64(11): 4653–60, 2019.

[180] J. Wang and M. Krstic. Delay-compensated control of sandwiched ODE-PDE-ODE hyperbolic systems for oil drilling and disaster relief. *Automatica*, 120:109131, 2020.

[181] J. Wang and M. Krstic. Output-feedback control of an extended class of sandwiched hyperbolic PDE-ODE systems. *IEEE Transactions on Automatic Control*, 66(6): 2588–603, 2021.

[182] J. Wang and M. Krstic. Vibration suppression for coupled wave PDEs in deep-sea construction. *IEEE Transactions on Control Systems Technology*, 29(4): 1733–49, 2021.

[183] J. Wang, M. Krstic, and Y. Pi. Control of a 2×2 coupled linear hyperbolic system sandwiched between two ODEs. *International Journal of Robust and Nonlinear Control*, 28:3987–4016, 2018.

[184] J. Wang, Y. Pi, Y. Hu, and X. Gong. Modeling and dynamic behavior analysis of a coupled multi-cable double drum winding hoister with flexible guides. *Mechanism and Machine Theory*, 108:191–208, 2017.

[185] J. Wang, Y. Pi, and M. Krstic. Balancing and suppression of oscillations of tension and cage in dual-cable mining elevators. *Automatica*, 98:223–38, 2018.

[186] J. Wang, S. X. Tang, and M. Krstic. Adaptive output-feedback control of torsional vibration in off-shore rotary oil drilling systems. *Automatica*, 111:108640, 2020.

[187] J. Wang, S. X. Tang, Y. Pi, and M. Krstic. Disturbance estimation of a wave PDE on a time-varying domain. In *Proceedings of the Conference on Control and Its Applications, SIAM*, 107–11, July 2017.

[188] J. Wang, S. X. Tang, Y. Pi, and M. Krstic. Exponential regulation of the anti-collocatedly disturbed cage in a wave PDE-modeled ascending cable elevator. *Automatica*, 95:122–36, 2018.

[189] J. M. Wang, J. J. Liu, B. Ren, and J. H. Chen. Sliding mode control to stabilization of cascaded heat PDE-ODE systems subject to boundary control matched disturbance. *Automatica*, 52:23–34, 2015.

[190] G. Weiss and X. Zhao. Well-posedness and controllability of a class of coupled linear systems. *SIAM Journal on Control and Optimization*, 48:2719–50, 2009.

[191] J. Willmann, F. Augugliaro, T. Cadalbert, R. D'Andrea, F. Gramazio, and M. Kohler. Aerial robotic construction towards a new field of architectural research. *International Journal of Architectural Computing*, 10:439–59, 2012.

[192] W. Xue and Y. Huang. On performance analysis of ADRC for a class of MIMO lower-triangular nonlinear uncertain systems. *ISA Transactions*, 53:955–62, 2014.

[193] Z. Yao and N. H. El-Farra. Resource-aware model predictive control of spatially distributed processes using event-triggered communication. In *52nd IEEE Conference on Decision and Control*, 3726–31, December 2013.

[194] H. Yu and M. Krstic. Traffic congestion control for Aw-Rascle-Zhang model. *Automatica*, 100:38–51, 2019.

[195] H. Yu and M. Krstic. Bilateral boundary control of moving shockwave in LWR model of congested traffic. *IEEE Transactions on Automatic Control*, 66(3): 1429–36, 2020.

[196] H. Yu, R. Vazquez, and M. Krstic. Adaptive output feedback for hyperbolic PDE pairs with non-local coupling. In *American Control Conference*, 487–92, May 2017.

[197] L. Zhang and C. Prieur. Necessary and sufficient conditions on the exponential stability of positive hyperbolic systems. *IEEE Transactions on Automatic Control*, 62:3610–17, 2017.

[198] H. C. Zhou, B. Z. Guo, and Z. H. Wu. Output feedback stabilisation for a cascaded wave PDE-ODE system subject to boundary control matched disturbance. *International Journal of Control*, 89:2396–405, 2016.

[199] W. Zhu and J. Ni. Energetics and stability of translating media with an arbitrarily varying length. *ASME Journal of Vibration and Acoustics*, 122(3): 295–304, 2000.

[200] W. Zhu and G. Xu. Vibration of elevator cables with small bending stiffness. *Journal of Sound and Vibration*, 263:679–99, 2003.

[201] W. Zhu and N. Zheng. Exact response of a translating string with arbitrarily varying length under general excitation. *ASME Journal of Applied Mechanics*, 75(3): 031003, 2008.

Index